GRAPH THEORY AND
ITS ENGINEERING APPLICATIONS

ADVANCED SERIES IN ELECTRICAL AND COMPUTER ENGINEERING

Editor: W. K. Chen

Published

Volume 1: Broadband Matching: Theory and Implementations (Second edition)
by W. K. Chen

Volume 2: Active Network Analysis
by W. K. Chen

Volume 3: Linear Networks and Systems: Computer-Aided Solutions and Implementations (Second edition)
by W. K. Chen

Volume 4: Introduction to Engineering Electromagnetic Fields
by K. Umashankar

Volume 5: Graph Theory and Its Engineering Applications
by W. K. Chen

Volume 6: Introductory Signal Processing
by R. Priemer

Volume 7: Diagnostic Measurements in LSI/VLSI Integrated Circuits Production
by A. Jakubowski, W. Marciniak, and H. M. Przewlocki

Volume 8: An Introduction to Control Systems (Second edition)
by K. Warwick

Volume 9: Orthogonal Functions in Systems and Controls
by K. B. Datta and B. M. Mohan

Volume 10: Introduction to High Power Pulse Technology
by S. T. Pai and Q. Zhang

Volume 11: Systems and Control: An Introduction to Linear, Sampled and Non-linear Systems
by T. Dougherty

Forthcoming

Protocol Conformance Testing Using Unique Input/Output Sequences
by Y. Shen, C. Feng, X. Sun and F. Lombardi

Infrared Characterization for Microelectronics
by W. S. Lau

Identification, Estimation and Statistical Signal Processing
by Y.-F. Huang

Advanced Series in Electrical and Computer Engineering – Vol. 5

GRAPH THEORY AND ITS ENGINEERING APPLICATIONS

WAI-KAI CHEN
The University of Illinois at Chicago

World Scientific
Singapore • New Jersey • London • Hong Kong

Published by

World Scientific Publishing Co. Pte. Ltd.
P O Box 128, Farrer Road, Singapore 912805
USA office: Suite 1B, 1060 Main Street, River Edge, NJ 07661
UK office: 57 Shelton Street, Covent Garden, London WC2H 9HE

British Library Cataloguing-in-Publication Data
A catalogue record for this book is available from the British Library.

GRAPH THEORY AND ITS ENGINEERING APPLICATIONS

Copyright © 1997 by World Scientific Publishing Co. Pte. Ltd.

All rights reserved. This book, or parts thereof, may not be reproduced in any form or by any means, electronic or mechanical, including photocopying, recording or any information storage and retrieval system now known or to be invented, without written permission from the Publisher.

For photocopying of material in this volume, please pay a copying fee through the Copyright Clearance Center, Inc., 222 Rosewood Drive, Danvers, MA 01923, USA. In this case permission to photocopy is not required from the publisher.

ISBN 981-02-1859-1

Printed in Singapore by Uto-Print

To Shiao-Ling

PREFACE

Because of the intuitive diagrammatic representation, graphs have been found extremely useful in modeling systems in physical science and engineering problems. The fact is that any system involving a binary relation can be represented by a graph. This accelerated pace of development has confirmed the value of the general approach adopted in this text by giving a reasonably deep account of a relatively small part of material that is closely related to engineering applications. As a result, topics on directed-graph solutions of linear algebraic equations, topological analysis of linear systems, trees and their generation, the realizability of directed graphs with prescribed degrees, state equations of networks, squaring rectangles and layouts, and graphic matrices are included.

Depending only on the first chapter, each of the following chapters is virtually self-contained, so that the material may be useful to the persons who are interested in only a single topic. Although the techniques discussed may easily be extended to other disciplines, the dominant theme is nevertheless linear networks and systems. In each of these chapters, the reader is assumed to be familiar with the elementary aspects of the subject and the discussions are devoted to those aspects of the theory that strongly depend on the theory of graphs.

In revising the early editions, I can think of many topics that should be added. Judging from the interest of readers, I have decided to concentrate on areas that have received wide attention in recent years and that depend heavily on the theory of graphs: Squaring Rectangles and VLSI Layout and Graphic Matrices. These subjects will be presented as key areas in chapters 8 and 9, respectively.

One inevitable result in adding key areas is that the book has grown longer. It contains more material than can be adequately presented in a one-semester or a two-quarter three-hours-per-week course in applied graph theory. This added flexibility will allow instructors to select subjects and sections to meet his/her particular needs and environment.

A special feature of the book is that almost all the results are documented in relationship to the known literature, and all the references which have been cited in the text are listed in the bibliography. Thus, the book is especially suitable for those who wish to continue with the study of special topics and to apply graph theory to other fields.

Although basically intended as a reference text for serious researchers, the book may be used equally well as a text for graduate level course on network topology and linear systems and circuits. Some of the later chapters are suitable as topics for advanced seminars. The only prerequisite for this book is really mathematical maturity.

Since the publication of the early editions, many people have been kind enough to give me the benefit of their comments and suggestions, often at the expense of a very considerable amount of their time and energy. Special thanks are due Professor Hui-Yun Wang of Tianjin University, who gave chapters 8 and 9 a careful reading. Finally, I wish to express my appreciation to my wife, Shiao-Ling, for her patience and understanding.

Chicago, Illinois
October, 1996
Wai-Kai Chen

CONTENTS

Preface vii

CHAPTER 1. **Basic theory** 1
    1. Introduction 1
    2. Basic concepts of abstract graphs 3
        2.1. General definitions 3
        2.2. Isomorphism 6
        2.3. Connectedness 8
        2.4. Rank and nullity 11
        2.5. Degrees 12
    3. Operations on graphs 13
    4. Some important classes of graphs 17
        4.1. Planar graphs 17
        4.2. Separable and nonseparable graphs 19
        4.3. Bipartite graphs 22
    5. Directed graphs 23
        5.1. Basic concepts 24
        5.2. Directed-edge sequence 27
        5.3. Outgoing and incoming degrees 29
        5.4. Strongly-connected directed graphs 30
        5.5. Some important classes of directed graphs 31
    6. Mixed graphs 32
    7. Conclusions 32
    Problems 33

CHAPTER 2. **Foundations of electrical network theory** 36
    1. Matrices and directed graphs 37
        1.1. The node-edge incidence matrix 37
        1.2. The circuit-edge incidence matrix 41
        1.3. The cut-edge incidence matrix 46
        1.4. Interrelationships among the matrices A, B_f, and Q_f 53
        1.5. Vector spaces associated with the matrices B_a and Q_a 57
    2. The electrical network problem 58
    3. Solutions of the electrical network problem 62
        3.1. Branch-current and branch-voltage systems of equations 63
        3.2. Loop system of equations 63

	3.3. Cut system of equations	70
	3.4. Additional considerations	76
4.	Invariance and mutual relations of network determinants and the generalized cofactors	77
	4.1. A brief history	77
	4.2. Preliminary considerations	78
	4.3. The loop and cut transformations	83
	4.4. Network matrices	85
	4.5. Generalized cofactors of the elements of the network matrix	95
5.	Invariance and the incidence functions	107
6.	Topological formulas for *RLC* networks	111
	6.1. Network determinants and trees and cotrees	111
	6.2. Generalized cofactors and 2-trees and 2-cotrees	114
	6.3. Topological formulas for *RLC* two-port networks	122
7.	The existence and uniqueness of the network solutions	125
8.	Conclusions	132
	Problems	133

CHAPTER 3. **Directed-graph solutions of linear algebraic equations** 140
1. The associated Coates graph 141
   1.1. Topological evaluation of determinants 142
   1.2. Topological evaluation of cofactors 146
   1:3. Topological solutions of linear algebraic equations 149
   1.4. Equivalence and transformations 155
2. The associated Mason graph 167
   2.1. Topological evaluation of determinants 169
   2.2. Topological evaluation of cofactors 172
   2.3. Topological solutions of linear algebraic equations 174
   2.4. Equivalence and transformations 177
3. The modifications of Coates and Mason graphs 189
   3.1. Modifications of Coates graphs 189
   3.2. Modifications of Mason graphs 197
4. The generation of subgraphs of a directed graph 199
   4.1. The generation of 1-factors and 1-factorial connections 201
   4.2. The generation of semifactors and *k*-semifactors 203
5. The eigenvalue problem 206
6. The matrix inversion 210
7. Conclusions 216
   Problems 216

CHAPTER 4. **Topological analysis of linear systems** 224
1. The equicofactor matrix 225
2. The associated directed graph 230
   2.1. Directed-trees and first-order cofactors 231
   2.2. Directed 2-trees and second-order cofactors 244
3. Equivalence and transformations 251

Contents   xi

  4. The associated directed graph and the Coates graph   262
    4.1. Directed trees, 1-factors, and semifactors   262
    4.2. Directed 2-trees, 1-factorial connections, and 1-semifactors   266
  5. Generation of directed trees and directed 2-trees   269
    5.1. Algebraic formulation   269
    5.2. Iterative procedure   272
    5.3. Partial factoring   279
  6. Direct analysis of electrical networks   281
    6.1. Open-circuit transfer-impedance and voltage-gain functions   281
    6.2. Short-circuit transfer-admittance and current-gain functions   289
    6.3. Open-circuit impedance and short-circuit admittance matrices   294
    6.4. The physical significance of the associated directed graph   297
    6.5. Direct analysis of the associated directed graph   302
  7. Conclusions   311
  Problems   312

CHAPTER 5. **Trees and their generation**   320
  1. The characterizations of a tree   320
  2. The codifying of a tree-structure   325
    2.1. Codification by paths   326
    2.2. Codification by terminal edges   328
  3. Decomposition into paths   330
  4. The Wang-algebra formulation   332
    4.1. The Wang algebra   333
    4.2. Linear dependence   334
    4.3. Trees and cotrees   338
    4.4. Multi-trees and multi-cotrees   340
    4.5. Decomposition   345
  5. Generation of trees by decomposition without duplications   353
    5.1. Essential complementary partitions of a set   353
    5.2. Algorithm   356
    5.3. Decomposition without duplications   359
  6. The matrix formulation   365
    6.1. The enumeration of major submatrices of an arbitrary matrix   365
    6.2. Trees and cotrees   368
    6.3. Directed trees and directed 2-trees   370
  7. Elementary transformations   373
  8. Hamilton circuits in directed-tree graphs   379
  9. Directed trees and directed Euler lines   384
  10. Conclusions   389
  Problems   390

CHAPTER 6. **The realizability of directed graphs with prescribed degrees**   398
  1. Existence and realization as a (p, s)-digraph   398
    1.1. Directed graphs and directed bipartite graphs   400
    1.2. Existence   401

    1.3. A simple algorithm for the realization   413
    1.4. Degree invariant transformations   419
    1.5. Realizability as a connected (p, s)-digraph   422
  2. Realizability as a symmetric (p, s)-digraph   427
    2.1. Existence   428
    2.2. Realization   433
    2.3. Realizability as connected, separable and nonseparable graphs   436
  3. Unique realizability of graphs without self-loops   440
    3.1. Preliminary considerations   441
    3.2. Unique realizability as a connected graph   443
    3.3. Unique realizability as a graph   446
  4. Existence and realization of a (p, s)-matrix   448
  5. Realizability as a weighted directed graph   452
  6. Conclusions   454
  Problems   455

CHAPTER 7. **State equations of networks**   464
  1. State equations in normal form   464
  2. Procedures for writing the state equations   472
  3. The explicit form of the state equation   480
  4. An alternative representation of the state equation   490
  5. Physical interpretations of the parameter matrices   491
  6. Order of complexity   499
    6.1. Relations between det $\mathbf{H}(s)$ and network determinant   504
    6.2. RLC networks   508
    6.3. Active networks   511
  7. Conclusions   514
  Problems   515

CHAPTER 8. **Squaring rectangles and layouts**   518
  1. Introduction   518
  2. The Bouwkamp code   521
  3. The electrical network associated with a dissected rectangle   524
  4. Characterization of the c-nets and c-digraphs   535
  5. Perfect subdivision of the general rectangle   541
    5.1. The perfect rectangles $\mathcal{R}_n$ 542
    5.2. The perfect square S_n 544
    5.3. Sequence of perfect squares   551
  6. Extension to perfect rectangular parallelepiped   553
  7. VLSI layout   554
    7.1. The zero wasted area floorplan with continuous aspect ratios   557
    7.2. Solving the nonlinear nodal equations   574
    7.3. Floorplan area optimization with constrained aspect ratio   581
    7.4. From digraphs to layout   596
    7.5. Graph-theoretic characterization of the minimum area layout   598
  8. Conclusions   605
  Problems   607

CHAPTER 9. **Graphical matrices** 612
    1. Totally unimodular matrix 612
    2. Regular and binary matrices 628
    3. Circuit and cutset matrices 639
    4. 2-isomorphism 653
    5. Graph realization 656
    6. Conclusions 664
    Problems 665

Bibliography 671

Symbol index 684

Subject index 690

CHAPTER 1

BASIC THEORY

The chapter establishes the basic vocabulary for describing graphs and provides a number of basic results that are needed in the subsequent analysis, omitting those aspects of graph theory that are unrelated to the applications discussed in this book. Since the terminology and symbolism currently in use in graph theory are far from standardized, the reader is urged to study the conventions of this chapter carefully before proceeding to the other chapters.

§ 1. Introduction

The term "graph" used in this book denotes something quite different from the graphs that one may be familiar with from analytic geometry or function theory. The graphs that we are about to discuss are simple geometrical figures consisting of points (nodes) and lines (edges) which connect some of these points; they are sometimes called "linear graphs". Because of this diagrammatic representation, graphs have been found extremely useful in modeling systems arising in physical science (BUSACKER and SAATY [1965], and HARARY [1967]), engineering (SESHU and REED [1961], and ROBICHAUD et al. [1962]), social science (HARARY and NORMAN [1953], and FLAMENT [1963]), and economic problems (AVONDO-BODINO [1962], and FORD and FULKERSON [1962]). The fact is that any system involving a binary relation can be represented by a graph.

The first paper on graphs was written by the famous Swiss mathematician Leonhard Euler (1707–1783). He started with a famous unsolved problem of his day called the *Königsberg Bridge Problem*. The city of Königsberg (now Kaliningrad) in East Prussia is located on the banks and on two islands of the river Pregel. The various parts of the city were connected by seven bridges as shown in fig. 1.1. The problem was to cross all seven bridges, passing over each one only once. One can see immediately that there are many ways of trying the problem without solving it. EULER [1736] solved the problem by showing that it was impossible, and laid the foundations of graph theory. We mention here only the formulation, rather than the details.

Replace each part of the city by a point and each bridge by a line joining the points corresponding to these parts. The result is a graph as shown in fig. 1.2. Euler then showed that, no matter at which point one begins, one cannot cover the graph completely and come back to the starting point without retracing one's steps.

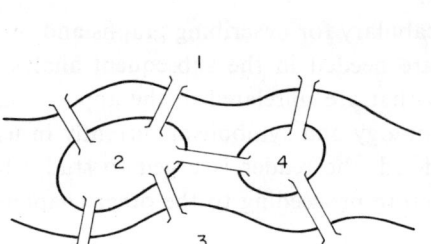

Fig. 1.1. The Königsberg bridge problem.

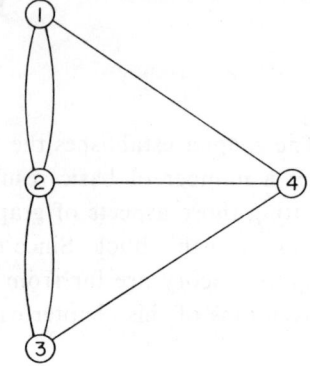

Fig. 1.2. The graph of the Königsberg bridge problem.

The most famous unsolved problem in graph theory is perhaps the celebrated *Four Color Conjecture*. Many centuries ago, makers of maps discovered empirically that in coloring a map of a country, divided into counties, only four distinct colors are required, so that no two adjacent counties should be painted in the same color. At first the problem does not seem to have been taken seriously by the mathematicians until it has withstood every assault by some of the world's most capable mathematicians. HEAWOOD [1890] showed, however, that the conjecture becomes true when "four" is replaced by "five". A counterexample, if ever found, will necessarily be extremely large and complicated, for the conjecture was proved most recently by ORE and STEMPLE [1970] for all maps with fewer than 40 counties.

The problem can easily be transformed into a problem in graph theory because every map yields a graph in which the counties including the exterior region are represented by the points, and two points are joined by a line if and only if their counties have a common boundary.

The most important application of graph theory in the physical science, from our point of view, is its use in the formulation and solution of the electrical network problem by KIRCHHOFF [1847]. His contributions will be treated in great detail in this book; chapters 2 and 4 contain most of his contributions to electrical network theory.

While many of the examples of the graphs arising in applications are geometric, the essential structure in the context of graph theory is combinatorial in nature. In the following sections, we shall introduce the concept of abstract graphs. Aside from stripping the incidental geometric features away from the essential combinatorial characteristics of a graph, the concept enlarges the prospects of applications.

§ 2. Basic concepts of abstract graphs

Like every mathematical theory, we have to begin with a long list of definitions, since we must have a few words to talk about, and in the interest of precision these have to be formally defined. Fortunately, many of these terms that we will define have nearly the same intuitive meaning as in everyday English and so very little conscious effort is required to remember them. In order to relieve the monotony of these necessary preliminaries, we will use diagrams to illustrate our points.

2.1. *General definitions*

DEFINITION 1.1: *Abstract graph.* An *abstract graph* $G(V, E)$, or simply a *graph* G, consists of a set V of elements called *nodes* together with a set E of *unordered* pairs of the form (i, j) or (j, i), i, j in V, called the *edges* of G; the nodes i and j are called the *endpoints* of (i, j).

Other names commonly used for a node are *vertex, point, junction, 0-simplex, 0-cell,* and *element*; and for edges *line, branch, arc, 1-simplex,* and *element*. We say that the edge (i, j) is *connected* between the nodes i and j, and that (i, j) is *incident* with the nodes i and j or conversely that i and j are *incident* with (i, j). In the applications, a graph is usually represented equivalently by a *geometric diagram* in which the nodes are indicated by small circles or dots, while any two of them, i and j, are joined by a continuous curve, or even a straight line, between i and j if and only if (i, j) is in E. This definition of a graph is sufficient for many problems in which graphs make their appearance. However, for our purpose, it is desirable to enlarge the graph concept somewhat.

We extend the graph concept by permitting a pair of nodes to be connected by several distinct edges as indicated by the symbols $(i, j)_1, (i, j)_2, ..., (i, j)_k$; they are called the *parallel edges* of G if $k \geq 2$. If no particular edge is specified, (i, j) denotes any one, but otherwise fixed, of the parallel edges connected between i and j. We also admit edges for which the two endpoints are identical. Such an edge (i, i) shall be called a *self-loop*. If there are two or more self-loops at a node of G, they are also referred to as the parallel edges of G. In the geometric diagram the parallel edges may be represented by continuous lines connected between the same pair of nodes, and a self-loop (i, i) may be introduced as a circular arc returning to the node i and passing through no other nodes.

As an illustration, consider the graph $G(V, E)$ in which

$$V = \{1, 2, 3, 4, 5, 6, 7\},$$
$$E = \{(1, 1), (1, 2), (1, 4), (4, 4)_1, (4, 4)_2, (4, 3), (2, 3)_1,$$
$$(2, 3)_2, (6, 7)_1, (6, 7)_2, (6, 7)_3\}.$$

The corresponding geometric graph is as shown in fig. 1.3 in which we have a self-loop at node 1, two self-loops at node 4, two parallel edges connected between the nodes 2 and 3, and three parallel edges between the nodes 6 and 7. We emphasize that in a graph the order of the nodes i and j in (i, j) is immaterial. In fact we consider $(i, j) = (j, i)$, e.g., $(1, 2) = (2, 1)$ and $(6, 7)_2 = (7, 6)_2$.

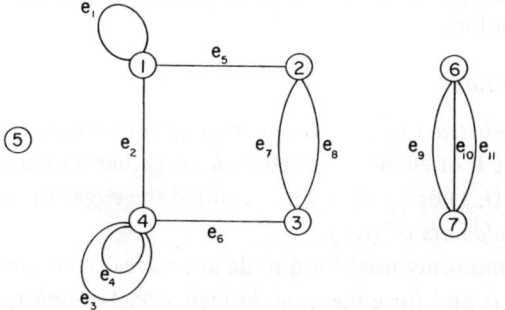

Fig. 1.3. A geometric graph.

A graph $G(V, E)$ is said to be *finite* if both V and E are finite. In this book, we only consider finite graphs. Infinite graphs have some very interesting properties. For interested readers, we refer to KÖNIG [1950] and ORE [1962].

DEFINITION 1.2: *Subgraph.* A *subgraph* of a graph $G(V, E)$ is a graph $G_s(V_s, E_s)$ in which V_s and E_s are subsets of V and E, respectively. If V_s or E_s is a proper subset, the subgraph is called a *proper subgraph* of G. If $V_s = V$, the subgraph is referred to as a *spanning subgraph* of G. If V_s or E_s is empty, the subgraph is called the *null graph*. The null graph is considered as a subgraph of every graph, and is denoted by the symbol $\emptyset$.

DEFINITION 1.3: *Isolated node.* A node not incident with any edge is called an *isolated node*.

In fig. 1.3, for example, the node 5 is an isolated node. Some examples of subgraphs are presented in fig. 1.4. Fig. 1.4(a) is a spanning subgraph since it contains all the nodes of the given graph. Figs. 1.4(b) and (c) are examples of proper subgraphs. A graph itself is also its subgraph.

We say that two subgraphs are *edge-disjoint* if they have no edges in common, and *node-disjoint* if they have no nodes in common. Clearly, two subgraphs are node-disjoint only if they are edge-disjoint, but the converse is not valid in general. For example, in fig. 1.3 the subgraphs (1, 2) and (3, 4) are node-disjoint, and thus they are also edge-disjoint. On the other hand, the subgraphs as shown in figs. 1.4(b) and (c) are edge-disjoint but they are not node-disjoint.

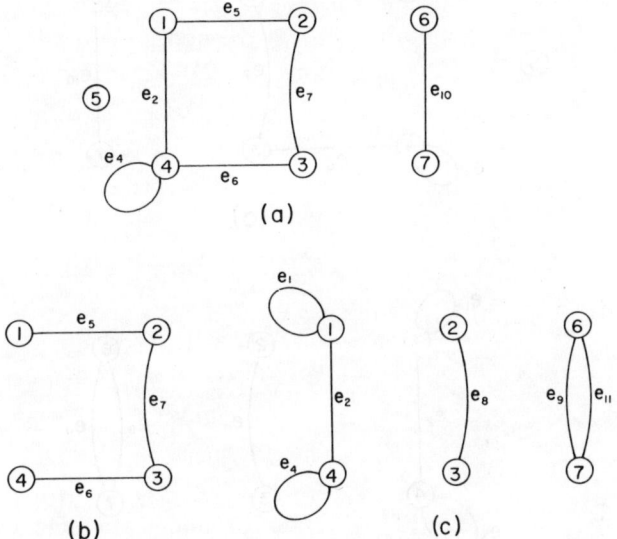

Fig. 1.4. Some examples of the subgraphs of the graph of fig. 1.3.

In a graph G we say that the nodes i and j are *adjacent* if (i, j) is an edge of G. If G_s is a subgraph of G, by the *complement* $\bar{G}_s$ of G_s in G we mean the subgraph of G consisting of all the edges $\bar{E}_s$ that do not belong to G_s and all the nodes of G except those that are in G_s but not in $\bar{E}_s$. Clearly, G_s and $\bar{G}_s$ are edge-disjoint but not necessarily node-disjoint, and their node sets may not be complementary. Thus, the complement of the null graph in G is the graph G itself, and the complement of G in G is the null graph. We also say that G_s and $\bar{G}_s$ are *complementary subgraphs* of G. For example, figs. 1.5(a) and (b) are complementary subgraphs of the graph as shown in fig. 1.3.

In practical applications, it is sometimes convenient to represent the edges of a graph by letters e_i. In this way, a subgraph having no isolated nodes may be expressed by the "product" or by juxtaposition of its edge-designation symbols. For example, in fig. 1.3 the edges of the graph are also represented by the letters e_i: $e_1 = (1, 1)$, $e_2 = (1, 4) = (4, 1), \ldots$, and $e_{11} = (6, 7)_3 = (7, 6)_3$. The subgraphs of figs. 1.4(b) and (c) may be denoted by the products of their edge-designation symbols as $e_5 e_6 e_7$ and $e_1 e_2 e_4 e_8 e_9 e_{11}$, respectively. Of course, we can also use this technique to represent subgraphs having isolated nodes, but then an ambiguity involving the null graph will arise.

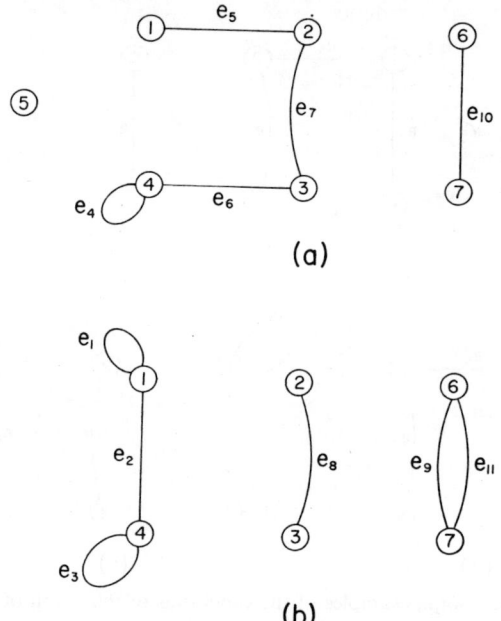

Fig. 1.5. A pair of complementary subgraphs of the graph of fig. 1.3.

2.2. Isomorphism

In the preceding section, we have already pointed out that in drawing the geometric diagram of a graph we have great freedom in the choice of the location of the nodes and in the form of the lines joining them. This may make the diagrams of the same graph look entirely different. In such cases we would like to have a precise way of saying that two graphs are really the same even though they are drawn or labeled differently. The next definition provides the terminology necessary for this purpose.

§2 Basic concepts of abstract graphs

DEFINITION 1.4: *Isomorphism.* Two graphs G_1 and G_2 are said to be *isomorphic*, denoted by $G_1 = G_2$, if there exist a one-to-one correspondence between the elements of their node sets and a one-to-one correspondence between the elements of their edge sets and such that the corresponding edges are incident with the corresponding nodes.

In other words, in two isomorphic graphs the corresponding nodes are connected by the edges in one if, and only if, they are also connected by the same number of edges in the other. Definition 1.4 as stated places two requirements on isomorphism of two graphs. First, they must have the same number of nodes and edges. Second, the incidence relationships must be preserved. The latter is usually difficult to establish.

As an illustration, consider the graphs $G_1(V_1, E_1)$ and $G_2(V_2, E_2)$ as shown in fig. 1.6. These two graphs look quite different, but they are isomorphic. The

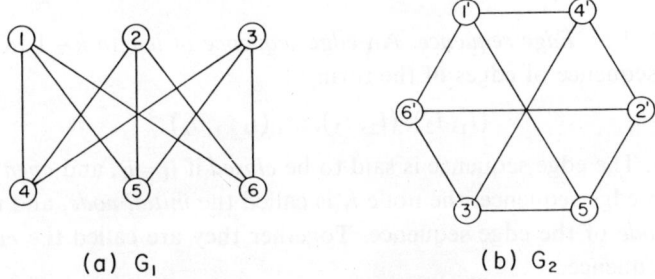

Fig. 1.6. Two isomorphic graphs.

isomorphism of these two graphs can be established by considering the nodes i of V_1 and i' of V_2, $i = 1, 2, 3, 4, 5, 6$, as the corresponding elements of their node sets. It is easy to verify that the corresponding edges are incident with the corresponding nodes. In other words, the incidence relationships are preserved.

As another example, the two graphs given in fig. 1.7 are not isomorphic even though there exists a one-to-one correspondence between their node sets which preserves adjacency. The reason for this is that they do not contain the same

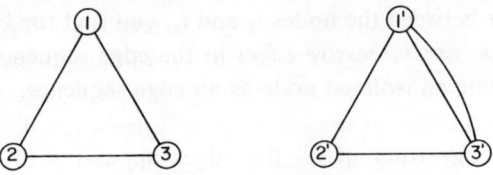

Fig. 1.7. Two non-isomorphic graphs.

number of edges. If we consider the nodes i and i', $i=1, 2, 3$, as the corresponding nodes of the node sets, we have two edges connecting the nodes $1'$ and $3'$ while we have only one connecting the corresponding nodes 1 and 3.

A graph is said to be a *labeled graph* if the nodes or edges of the graph are properly labeled. In this book, the terms *graph* and *labeled graph* are used as synonyms. The graphs that we have witnessed so far are all labeled graphs, the nodes being labeled by the integers $1, 2, \ldots$ or $1', 2', \ldots$, and the edges by $e_1, e_2, \ldots$. A *weighted graph* is a graph in which every edge has been assigned a weight.

2.3. Connectedness

Sequences of edges which form continuous routes play an important role in graph theory. In a geometric graph, a sequence of edges can be visualized as a series of edges connected in a continuous manner. More formally, we define the following.

DEFINITION 1.5: *Edge sequence.* An *edge sequence of length* $k-1$ in a graph G is a finite sequence of edges of the form

$$(i_1, i_2), (i_2, i_3), \ldots, (i_{k-1}, i_k), \tag{1.1}$$

$k \geq 2$, in G. The edge sequence is said to be *closed* if $i_1 = i_k$, and *open* otherwise. In an open edge sequence, the node i_1 is called the *initial node*, and node i_k the *terminal node* of the edge sequence. Together they are called the *endpoints* of the edge sequence.

We mention specifically that not all the nodes in (1.1) are necessarily distinct, and the same edge may appear several times in the edge sequence. For example, in fig. 1.6(a) the sequence of edges

$$(1, 6), (6, 3), (3, 5), (5, 3), (3, 4), (4, 1), (1, 6), (6, 2), (2, 5)$$

is an open edge sequence of length 9. Node 1 is the initial node and node 5 the terminal node of the edge sequence. Similarly, the sequence of edges

$$(4, 2), (2, 6), (6, 3), (3, 5), (5, 2), (2, 6), (6, 3), (3, 5), (5, 2), (2, 4)$$

forms a closed edge sequence of length 10.

We also say that the edge sequence (1.1) is *connected* between its initial and terminal nodes or between the nodes i_1 and i_k, and that for $k > 2$, (i_{x-1}, i_x) and (i_x, i_{x+1}), $1 < x < k$, are *successive edges* in the edge sequence. Sometimes it is convenient to define an isolated node as an edge sequence.

DEFINITION 1.6: *Edge train.* If all the edges appearing in an edge sequence are distinct, the edge sequence is called an *edge train*.

Thus, an edge train can go through a node more than once but cannot retrace parts of itself, as an edge sequence can. An example of an edge train in fig. 1.6(a) is

$$(1, 6), (6, 2), (2, 5), (5, 1), (1, 4), (4, 2).$$

The edge train is open and is of length 6. Clearly, an edge sequence or an edge train is also contained in a subgraph. If in addition we require that all the nodes in an edge train except the initial and the terminal nodes be distinct, we have the usual concepts of a path and a circuit.

DEFINITION 1.7: *Path.* An open edge train, as shown in (1.1), in which all the nodes $i_1, i_2, \ldots, i_k$ are distinct is called a *path of length* $k-1$. An isolated node is considered as a path of zero length.

DEFINITION 1.8: *Circuit.* A closed edge train, as shown in (1.1), in which all the nodes $i_1, i_2, \ldots, i_{k-1}$ are distinct, in this case $i_1 = i_k$, is called a *circuit of length* $k-1$.

Thus, a self-loop is also a circuit of length 1. In the literature, the term *circuit*

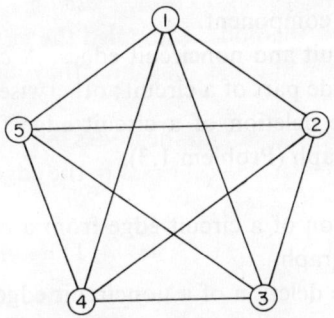

Fig. 1.8. The complete pentagon.

is frequently referred to as a *cycle* or *loop*. The term we adopt here is more commonly used in applications. In fig. 1.8, the open edge train

$$(1, 2), (2, 3), (3, 4), (4, 5)$$

is a path of length 4, and the closed edge train

$$(1, 3), (3, 2), (2, 5), (5, 4), (4, 1)$$

is a circuit of length 5.

DEFINITION 1.9: *Connected graph.* A graph is said to be *connected* if every pair of its nodes are connected by a path.

In other words, a connected graph, intuitively speaking, has only one piece. Fig. 1.5(a) or (b) is an example of a graph which is *not* connected, and fig. 1.6 shows two connected graphs.

DEFINITION 1.10: *Component.* A *component* of a graph is a connected subgraph containing the maximal number of edges. An isolated node is a component.

Thus, if a graph is not connected it must contain a number of components. One or many of these components may each consist of an isolated node. For example, in fig. 1.5(a) the graph has three components; one of them is an isolated node.

Suppose that we define a binary relation R between certain pairs of nodes of a graph G as follows: The relation iRj holds if and only if there is a path connected between the nodes i and j in G where an isolated node is considered as a path of zero length. Then R is an equivalence relation; it uniquely decomposes the node set of G into mutually exclusive equivalence classes of nodes. Each of these classes of nodes, together with the edges of G incident with these nodes, constitutes a component of G. Thus, a graph is connected if and only if it consists of only one component.

We shall speak of circuit and noncircuit edges. A *circuit edge* of a graph is an edge which can be made part of a circuit; otherwise, an edge is called a *noncircuit edge*. Clearly, the deletion of a circuit edge from a connected graph leaves a connected subgraph (Problem 1.3).

THEOREM 1.1: The deletion of a circuit edge from a connected graph leaves a spanning connected subgraph.

It is also clear that the deletion of a noncircuit edge from a graph G results in a graph which has one more component than G. For example, in fig. 1.5(b) if we delete the noncircuit edge e_2 or e_8, the resulting graph has four components, that is one more than the original graph. On the other hand, in fig. 1.7 or 1.8 all the edges are circuit edges; the deletion of any one of these edges results in a spanning connected subgraph.

THEOREM 1.2: Let G be a graph having no parallel edges and self-loops. If G has n nodes and c components, then the maximal number of edges in G is given by

$$\tfrac{1}{2}(n-c)(n-c+1). \tag{1.2}$$

Proof. Let G_i, $i=1, 2, ..., c$, be the components of G, each having n_i nodes. Since the maximal number of edges in each component G_i is $\frac{1}{2}n_i(n_i-1)$, the maximal number of edges in G is given by

$$\frac{1}{2} \sum_{i=1}^{c} n_i(n_i - 1). \tag{1.3}$$

If G has $c-1$ isolated nodes, (1.3) reduces to (1.2) and our proof is complete. Thus, we assume that there are two distinct components G_i and G_j with more than an isolated node. Let $n_i \geq n_j > 1$. If in G we increase the number of nodes in G_i by one and at the same time if we reduce the number of nodes in G_j by one, we obtain a new graph G^* which has the same numbers of nodes and components as G. It is easy to verify that the maximal number of edges in G^* is greater than that in G. Continuing this process, we can show that (1.2) is indeed an upper bound for all such graphs.

COROLLARY 1.1: Let G be an n-node graph having no parallel edges and self-loops. If G has more than $\frac{1}{2}(n-1)(n-2)$ edges, then G must be connected.

2.4. *Rank and nullity*

Rank and nullity are two numbers that are frequently encountered in graph theory. As we shall see in the next chapter, they represent the number of independent "cutsets" and circuits of a graph.

DEFINITION 1.11: *Rank.* The *rank r* of a graph with n nodes and c components is defined as the number $r = n - c$.

DEFINITION 1.12: *Nullity.* The *nullity m* of a graph with b edges, n nodes, and c components is defined as the number $m = b - n + c \, (= b - r)$.

The term *nullity* is also known by the names of *circuit rank, cyclomatic number, cycle rank, connectivity,* and *first Betti number.* The reason that we choose the above names *rank* and *nullity* is that, as we shall see in the next chapter, they are the rank and nullity of the "incidence matrix" associated with the graph. In fig. 1.8 the rank and nullity of the graph are $4(=5-1)$ and $6(=10-5+1)$, respectively. In fig. 1.5(b) we have $r = 6 - 3 = 3$ and $m = 6 - 6 + 3 = 3$. We notice that all these numbers are nonnegative. This is indeed the case as we can see from the following theorem (Problem 1.5).

THEOREM 1.3: For a given graph, its rank and nullity are both nonnegative. The graph is of nullity 0 if and only if it contains no circuit, and is of nullity 1 if and only if it contains a single circuit.

Thus, a connected graph of nullity 1 can also be characterized by the property that the number of its edges is equal to the number of its nodes. For an unconnected graph, its rank and nullity are simply the sums of the ranks and nullities of its components, respectively.

2.5. Degrees

Another term which is commonly used in graph theory is called *degree*. The problem of realizing a graph with prescribed degrees will be treated rather extensively in chapter 6.

DEFINITION 1.13: *Degree*. The *degree* of a node i of a graph, denoted by $d(i)$, is the number of edges incident with the node i.

Thus, an isolated node is a node of zero degree.

Since each edge is incident with two nodes, it contributes 2 to the sum of the degrees of the nodes. Thus, we have

$$\sum_{i=1}^{n} d(i) = 2b. \tag{1.4}$$

THEOREM 1.4: The sum of the degrees of the nodes of a graph is twice the number of its edges.

If V_1 and V_2 are the sets of nodes having odd and even degrees of a graph G, respectively, then the left side of the equation

$$\sum_{i=1}^{n} d(i) - \sum_{j \in V_2} d(j) = \sum_{k \in V_1} d(k) \tag{1.5}$$

is always even. This establishes the following result.

COROLLARY 1.2: The number of nodes of odd degree of a graph is always even.

In fig. 1.9 the degrees of the nodes 1, 2, 3, and 4 are 3, 6, 1, and 4, respectively. There are two nodes of odd degree, i.e., $d(1)=3$ and $d(3)=1$. The sum of all the degrees is twice the number of edges of the graph which is 7.

DEFINITION 1.14: *Regular graph*. A graph is said to be *regular of degree k* if $d(i)=k$ for each node i.

A regular graph of degree 0 is a graph consisting of a set of isolated nodes. A regular graph of degree 1 is a graph, each component containing exactly one edge, and finally a regular graph of degree 2 is a graph, each component being a circuit. These conditions are both necessary and sufficient for their complete characterizations.

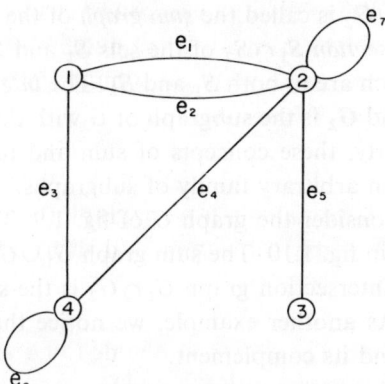

Fig. 1.9. A graph for illustrative purpose.

For example, fig. 1.6 is a regular graph of degree 3 since each node is of degree 3, and fig. 1.8 is a regular graph of degree 4. Other examples of regular graphs are graphs defined by the edges and corners of the five *regular polyhedra*: tetrahedron, cube, octahedron, dodecahedron, and icosahedron.

§ 3. Operations on graphs

Operations defined on graphs are useful in expressing the structure of a given graph in terms of smaller or simpler graphs. It is also convenient to represent graphs which are frequently encountered in analysis. We have already introduced the concept of the complement of a subgraph in a given graph.

By the *removal of a node i* from a graph G we mean the operation that results in the subgraph consisting of all the nodes of G except i and all the edges not incident with i. In other words, it is the subgraph containing all the nodes and edges not incident with i. Similarly, the *removal of an edge* (i,j) from G yields the subgraph containing all the nodes and edges of G except the edge (i,j). The removal of a set of nodes or edges is defined by the removal of the single elements of the set in succession.

Four set-theoretic binary operations *union, intersection, difference,* and *ring sum* represented by $\cup, \cap, -,$ and $\oplus$, respectively, will be defined. They are used in two slightly different contexts. For example, if S_1 and S_2 are two sets, then $S_1 \cup S_2$ denotes the set consisting of all the elements which are either in S_1 or in S_2 or in both. If $G_1(V_1, E_1)$ and $G_2(V_2, E_2)$ are two subgraphs of a graph $G(V, E)$, then $G_1 \cup G_2$ represents the subgraph of G with node set $V_1 \cup V_2$ and edge set $E_1 \cup E_2$. The set $S_1 \cup S_2$ is called the *set union* of the sets S_1 and

S_2, and the graph $G_1 \cup G_2$ is called the *sum graph* of the subgraphs G_1 and G_2. Analogously, the *intersection* $S_1 \cap S_2$ of the sets S_1 and S_2 is the set consisting of all the elements which are in both S_1 and S_2. The *intersection graph* $G_1 \cap G_2$ of the subgraphs G_1 and G_2 is the subgraph of G with the node set $V_1 \cap V_2$ and edge set $E_1 \cap E_2$. Clearly, these concepts of sum and intersection graphs can easily be extended to an arbitrary family of subgraphs.

As an illustration, consider the graph G of fig. 1.9. Two subgraphs G_1 and G_2 of G are presented in fig. 1.10. The sum graph $G_1 \cup G_2$ is the graph G itself, i.e., $G = G_1 \cup G_2$. The intersection graph $G_1 \cap G_2$ is the subgraph consisting of the edges e_1 and e_4. As another example, we notice that a graph is the sum graph of a subgraph and its complement.

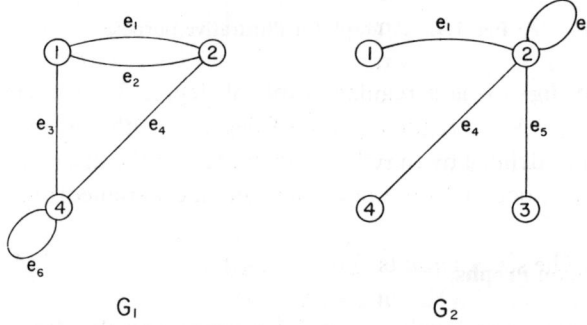

Fig. 1.10. Two subgraphs of the graph of fig. 1.9 to illustrate graph operations.

The set $S_1 - S_2$ consists of all the elements which are in S_1 but not in S_2, and $S_1 \oplus S_2$ denotes the set consisting of all the elements which are either in S_1 or in S_2 but not in both. They are called the *difference* and *ring sum* of the sets S_1 and S_2. The ring sum is also referred to as the *symmetric difference*, since $S_1 \oplus S_2$ is the difference between the union and intersection of the sets S_1 and S_2:

$$S_1 \oplus S_2 = (S_1 \cup S_2) - (S_1 \cap S_2). \tag{1.6}$$

For example, the ring sum of the sets $S_1 = \{a, b, c, d\}$ and $S_2 = \{a, c, e\}$ is given by

$$S_1 \oplus S_2 = \{b, d, e\},$$

and their differences are

$$S_1 - S_2 = \{b, d\},$$
$$S_2 - S_1 = \{e\}.$$

Note that the order in performing the difference operation is important.

§3 Operations on graphs

Similarly, if G_1 and G_2 are subgraphs of G not containing any isolated nodes, then by $G_1 - G_2$ we mean the subgraph consisting of all the edges of G_1 which are not in G_2, and $G_1 \oplus G_2$ is the subgraph consisting of all the edges which are either in G_1 or in G_2 but not in both. In particular, we have $G_1 \oplus \emptyset = \emptyset \oplus G_1 = G_1$. For example, for fig. 1.10 we get

$$G_1 - G_2 = e_1 e_2 e_3 e_4 e_6 - e_1 e_4 e_5 e_7 = e_2 e_3 e_6,$$
$$G_2 - G_1 = e_1 e_4 e_5 e_7 - e_1 e_2 e_3 e_4 e_6 = e_5 e_7,$$
$$G_1 \oplus G_2 = G_2 \oplus G_1 = e_2 e_3 e_5 e_6 e_7.$$

We emphasize that the difference and ring-sum operations of two subgraphs are defined only for those subgraphs containing no isolated nodes. In terms of the difference operation, the removal of an edge e_i from a graph G is equivalent to the operation $G - e_i$, which may involve an ambiguity of an isolated node.

With the aid of ring-sum operation, we can now establish some fundamental properties of the set of circuits of a graph.

THEOREM 1.5: *The ring sum of two circuits is a circuit or an edge-disjoint union of circuits.*

THEOREM 1.6: *The set of circuits and edge-disjoint unions of circuits of a graph is an abelian group under the ring-sum operation.*

The proof of these two theorems is straightforward, and the details are left as exercises (Problems 1.8 and 1.11).

Let (i, j) be an edge of a graph G. By *shorting the edge* (i, j) of G we mean the operation that first removes the edge (i, j) from G and then identifies the

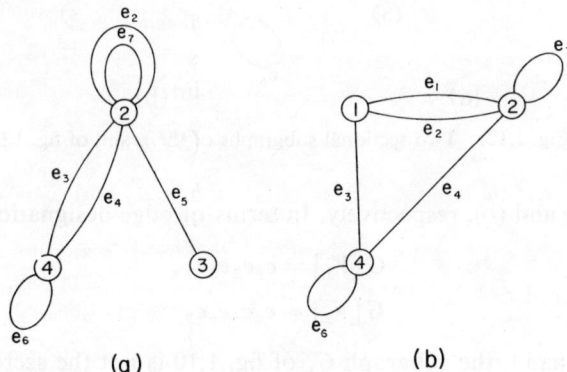

Fig. 1.11. Graphs obtained from the graph of fig. 1.9 by shorting the edge e_1 and e_5, respectively.

nodes i and j in the resulting graph. The operation may produce additional self-loops in the resulting graph. In particular, the removal of a self-loop from G is equivalent to the operation of shorting the self-loop. For example, in fig. 1.9 the shorting of the edge e_1 results in a graph as shown in fig. 1.11(a) and the shorting of the edge e_5 results in a graph of fig. 1.11(b).

A particularly important class of subgraphs that frequently occur in this book are called "sectional subgraphs". These subgraphs are defined below.

DEFINITION 1.15: *Sectional subgraph.* Let V_s be a subset of the node set V of a graph G. The *sectional subgraph*, defined by V_s and denoted by the symbol $G[V_s]$, of G is the subgraph whose node set is V_s and whose edge set consists of all those edges of G connecting two nodes of V_s.

In other words, $G[V_s]$ is the subgraph obtained from G by the removal of all the nodes that are not in V_s. Thus, two nodes of V_s in $G[V_s]$ are adjacent if and only if they are adjacent in G. When $V_s = V$, the sectional subgraph is G itself, i.e., $G[V] = G$. For a single node $V_s = \{i\}$, $G[V_s]$ is either the null graph or the subgraph consisting of all the self-loops at node i.

As an example, consider the graph G as shown in fig. 1.9. The sectional subgraphs defined by the node sets $V_1 = \{1, 2, 3\}$ and $V_2 = \{2, 3, 4\}$ are presented

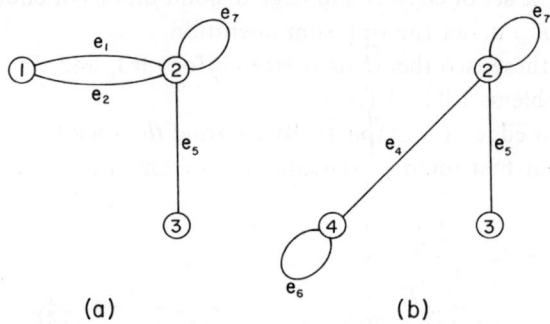

Fig. 1.12. Two sectional subgraphs of the graph of fig. 1.9.

in figs. 1.12(a) and (b), respectively. In terms of edge-designation symbols, we have
$$G[V_1] = e_1 e_2 e_5 e_7,$$
$$G[V_2] = e_4 e_5 e_6 e_7.$$

On the other hand, the subgraph G_1 of fig. 1.10 is not the sectional subgraph defined by the node set $V_s = \{1, 2, 4\}$ since it is not the maximal subgraph defined by these nodes.

§ 4. Some important classes of graphs

Graphs can be classified in many different ways depending upon their structural features which are used as the basis for classification. For example, we have already classified graphs on the basis of their connectivity. In this section, we introduce a number of other useful classifications.

4.1. *Planar graphs*

Our discussions so far have been entirely in terms of the abstract graph; its geometric diagram has served only for illustrative purpose. On the contrary, the planar graphs are defined in terms of their geometric diagrams.

DEFINITION 1.16: *Planar graph.* A graph is said to be *planar* if its geometric diagram can be drawn on a plane such that no two edges have an intersection that is not a node.

Clearly, the geometric diagram of a graph can be drawn on a plane without intersections if and only if it can be drawn on a sphere without intersections. As a matter of fact, using stereographic projection, we can establish a one-to-one correspondence between these two drawings as follows: Let a sphere be kept on a plane such that the point of contact (south pole) is the origin of the coordinate system in the plane. Let z be the intersection (north pole) of the sphere and the line which is perpendicular to the plane and passes through the origin of the plane. Joining z to any point p of the sphere by a straight line and extending the line to meet the plane at p', we establish a one-to-one correspondence between points on the plane and points on the sphere. This procedure is referred to as a *mapping* between the sphere and the plane. Thus, the geometric diagram of a graph on a plane can always be mapped onto a sphere and *vice versa*, provided that z is chosen not on the diagram.

DEFINITION 1.17: *Region* (*window*). The areas into which the geometric diagram of a planar graph divides the plane, when it is drawn on a plane without intersections, are called the *regions* (*windows*) of the planar graph. The unbounded region is called the *outside region*.

Thus, a region is characterized by the edges on its boundary. In electrical network theory, the circuit formed by the boundary edges of a region is also referred to as a *mesh* since it has the appearance of a mesh of a fish net. We remark that the graph corresponding to an electrical network usually does not contain any self-loops, so there is no ambiguity in defining the mesh of a region.

In fig. 1.13 the graph is planar and it has five regions, as indicated, the outside region being number 5. These regions are bounded by the meshes $e_1 e_2 e_3$, $e_2 e_4 e_5$, $e_4 e_6 e_7$, $e_7 e_8 e_9$, and $e_1 e_6 e_8 e_9 e_5 e_3$, respectively. Figs. 1.6 and 1.8 are examples of non-planar graphs.

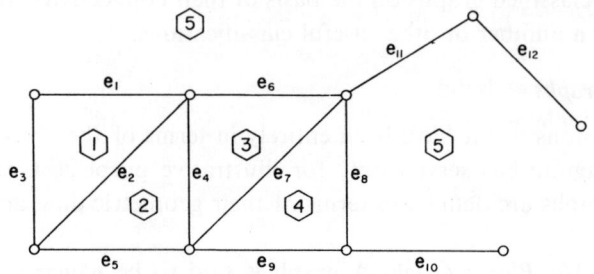

Fig. 1.13. A planar graph.

THEOREM 1.7: Any region of a planar graph can always be the outside region of the planar graph.

Proof. Let the geometric diagram of a planar graph be drawn on a sphere without intersections. Rotate the sphere so that the point z discussed above is inside the given region. Map the geometric diagram back onto the plane, and we obtain the desired result.

The planar graphs are of considerable practical interest. For example, in printed circuits an electrical network can be printed on a single plane surface if and only if it corresponds to a planar graph. Also they are closely related to the concept of duality in electrical network theory. The number of regions of a planar graph is related to the numbers of edges and nodes by the following formula.

THEOREM 1.8: For an n-node connected planar graph with b edges and q regions (including the outside region), we have

$$n - b + q = 2. \tag{1.7}$$

Proof. Let G be the connected planar graph. We shall prove this theorem by induction over the number b of edges of G. For $b = 1$, the theorem is seen to be true. Assume that the assertion is true for any $b-1$, $b \geq 2$. We shall prove that it is also true for any b. If G has no circuit, then G has $n-1$ edges, and (1.7) is satisfied. So let us assume that G has at least one circuit. Let e be a boundary edge of the outside region of G. Let G' be the graph obtained from G by the removal of e. Clearly, G' has n nodes, $b-1$ edges, and $q-1$ regions. By induction hypothesis, we have $n-(b-1)+(q-1)=2$. The theorem follows from here.

The above formula was first given by Euler for polyhedra. As a matter of fact, the subject of planar graphs was discovered by him in his investigation of polyhedra. An easy consequence of Euler's formula is the following which establishes an upper bound for the number of edges in a planar graph not containing parallel edges and self-loops.

COROLLARY 1.3: If G is a planar graph with n nodes and b edges, $n \geq 3$, and if it has no parallel edges and self-loops, then

$$b \leq 3n - 6. \tag{1.8}$$

Proof. The maximum number of edges that can occur in an n-node planar graph G_n is when every edge of G_n is contained in at least one of the circuits formed by the boundary edges of the regions of G_n and when each of these circuits is of length 3. If b' and q' are the numbers of edges and regions of G_n, respectively, then $2b'=3q'$ since each edge of G_n is contained in exactly two such circuits. Using this in conjunction with (1.7) we obtain $b'=3n-6$ which establishes an upper bound for the number of edges in any G_n. So the corollary is proved.

Using this result, we can now confirm an earlier assertion that the graph of fig. 1.8 is non-planar since $b=10>9=3n-6$.

It was first pointed out by KURATOWSKI [1930] that the two non-planar graphs, as shown in figs. 1.6 and 1.8, and their variants are necessary and sufficient to characterize a planar graph.

4.2. *Separable and nonseparable graphs*

Some graphs may be disconnected by the removal of a single node, and others may not. The characterization of this concept will be discussed in this section.

DEFINITION 1.18: *Cutpoint*. In a connected graph G, if there exists a proper non-null subgraph G_s such that G_s and its complement have only one node i in common, then the node i is called a *cutpoint* of G. A node of an unconnected graph is a cutpoint if the node is a cutpoint in one of its components. The term cutpoint is sometimes also called an *articulation point*.

For example, in fig. 1.14 the nodes 4, 6, 7, and 10 are cutpoints, but the nodes 8 and 9 are not. If G has no self-loops, a cutpoint may also be defined as a node whose removal increases the number of components at least by one. If G has self-loops, the statement, however, is not generally true. For example, let $G=(i,j)\cup(i,i)$. The node i is a cutpoint, but its removal does not increase the number of components.

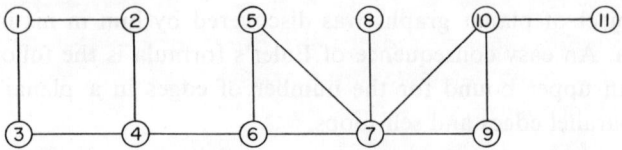

Fig. 1.14. A separable graph.

THEOREM 1.9: A node j of a self-loopless graph is a cutpoint if, and only if, there exist two nodes i and k distinct from j such that every path connecting the nodes i and k contains the node j.

The proof of the theorem is straightforward, and is left as an exercise (Problem 1.15).

DEFINITION 1.19: *Nonseparable graph.* A connected non-null graph is said to be *nonseparable* if it contains no cutpoints. All other non-null graphs are considered as *separable*.

Thus, an unconnected non-null graph is a trivial example of a separable graph. Figs. 1.12–1.14 are other examples of separable graphs. Examples of nonseparable graphs are presented in figs. 1.8 and 1.15. It follows directly from

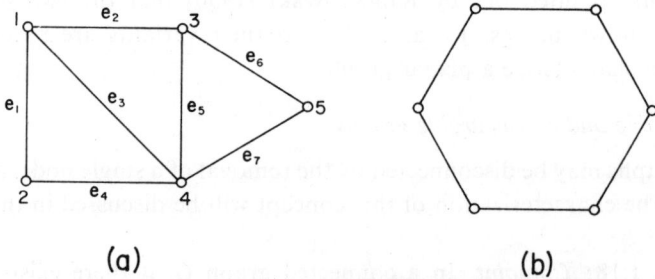

Fig. 1.15. Two nonseparable graphs.

our definition that a connected separable graph must contain a cutpoint, and such a graph must contain at least one non-null subgraph which has only one node in common with its complement. Putting it differently, we obtain the following.

§4 Some important classes of graphs

THEOREM 1.10: A necessary and sufficient condition that a graph containing at least two edges be nonseparable is that every proper non-null subgraph of the graph have at least two nodes in common with its complement.

As an illustration, consider the graph of fig. 1.14. The node 6 is a cutpoint since there exist two nodes, say, 1 and 11 such that every path connecting these two nodes contains the node 6. In fig. 1.15(a), the graph is nonseparable since for each proper non-null subgraph, say, $e_1 e_2 e_3 e_4$ it has at least two nodes in common with its complement $e_5 e_6 e_7$.

A nonseparable graph can also be characterized by the ways in which the nodes are connected.

DEFINITION 1.20: *Cyclically connected.* A graph is said to be *cyclically connected* if any two of its nodes are contained in a circuit.

For example, the graphs of fig. 1.15 are cyclically connected, but the graph of fig. 1.14 is not.

THEOREM 1.11: A self-loopless graph containing at least two edges is nonseparable if, and only if, it is cyclically connected.

Proof. Let G be a nonseparable graph containing at least two edges. Let i and j be any two distinct nodes of G. Clearly, there is a path P connected between i and j of G. We shall prove the necessary part of the theorem by induction over the length of P. If P is of length 1 or 2, then by Theorem 1.9 and the fact that G has no cutpoints there exists a circuit containing i and j. Assume that this is true for any P which is of length $k-1$ or less, $k \geq 3$. We shall show that it is also true for any P of length k. Let (x, j) be the last edge in P. Then by induction hypothesis there exists a circuit L_1 containing the nodes i and x of G. Let (y, x) be an edge of L_1. It follows from Theorem 1.9 that there is a circuit L_2 containing the edges (y, x) and (x, j). In $L_1 \cup L_2$ it is not difficult to see that we can construct another circuit containing the nodes i and j. Thus, any two nodes of G can be placed in a circuit. Conversely, if any two nodes of G can be placed in a circuit, then G must be connected and contains no cutpoints, and so is nonseparable, which completes the proof of the theorem.

We remark that a graph containing at least two edges, one of them being a self-loop, is separable, and that a single edge is nonseparable.

COROLLARY 1.4: A nonseparable graph is a circuit if, and only if, it is of nullity 1.

COROLLARY 1.5: A nonseparable graph containing at least two edges is of nullity greater than 0.

The corollaries follow directly from the fact that each node of a nonseparable graph is at least of degree 2 except the nonseparable graph consisting of one edge and two distinct endpoints (Problems 1.17 and 1.19).

DEFINITION 1.21: *Block*. A *block* of a graph is a nonseparable subgraph containing the maximal number of edges.

If a connected graph G is separable, we can separate G into blocks by splitting off the cutpoints. This process is called the *decomposition* of a separable graph into blocks. For example, the decomposition of the graph of fig. 1.14 is presented in fig. 1.16. Clearly, the decomposition is unique (Problem 1.20).

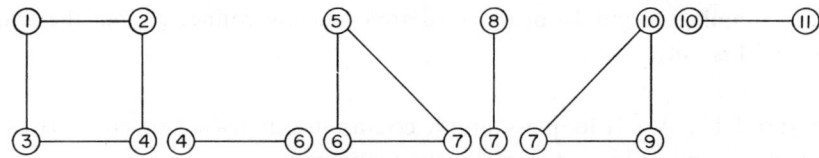

Fig. 1.16. The decomposition of the graph of fig. 1.14 into blocks.

4.3. *Bipartite graphs*

Bipartite graphs frequently occur in matching theorems, and their formal definition is given below.

DEFINITION 1.22: *Bipartite graph*. A graph $G(V, E)$ is said to be *bipartite* if its node set V can be partitioned into two disjoint subsets V_1 and V_2 such that each of its edges has one endpoint in V_1 and the other in V_2.

Thus, a non-null subgraph of a bipartite graph is bipartite. All the paths connected between any two nodes of V_1 or V_2 are of even length, and they are of odd length if they are connected between a node in V_1 and a node in V_2. For example, fig. 1.17 is a bipartite graph with node sets $V_1 = \{1, 2, 3, 4\}$ and $V_2 = \{5, 6, 7, 8, 9\}$. All the paths connecting the nodes, say, 1 and 4 are of even length, and 1 and 8 are of odd length.

A complete characterization of a bipartite graph is given below. However, before we prove this result we need an additional term.

DEFINITION 1.23: *Distance*. For a connected graph G, the *distance* between two nodes i and j of G, denoted by $d(i, j)$, is defined as the length of a shortest path connected between them.

For example, in fig. 1.17 we have $d(1, 4) = 4$ and $d(1, 6) = 1$.

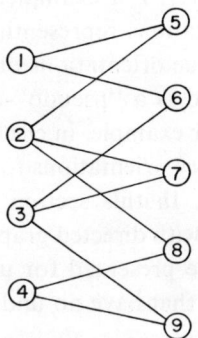

Fig. 1.17. A bipartite graph.

THEOREM 1.12: A non-null graph is bipartite if, and only if, either it has no circuits or each of its circuits is of even length.

Proof. The necessary part is obvious. To prove sufficiency, let us assume that a graph $G(V, E)$ has this property. Without loss of generality, we may further assume that G is connected; for if not we can consider the components of G separately. Also we can assume that G has no self-loops.

Let i be a node of G, and let V_2 be the set consisting of all the nodes of G that are of even distance from i (node i is in V_2), while $V_1 = V - V_2$. Evidently, each node in V_1 is of odd distance from i in G. To show that every edge (u, v) of G is connected between a node u in V_2 and a node v in V_1, we first assume that both u and v are in V_2. Let P_{iu} and P_{iv}^* be the shortest paths of even lengths from i to u and v, respectively. Starting from the nodes u and v, let j be the first common node in P_{iu} and P_{iv}^*. Then the distances $d(j, u)$ and $d(j, v)$ are either both even or both odd. This would imply that the sum graph, formed by the edge (u, v) and the shortest paths from j to u and j to v, is a circuit of odd length, which would contradict to our assumption that every circuit of G is of even length.

In a similar manner, we can show that no two nodes in V_1 are connected by an edge. Thus, G is bipartite, and the proof is complete.

§ 5. Directed graphs

In many applications, it is necessary to associate with each edge of a graph an orientation or direction. In some situations, the orientation of the edges is a "true" orientation in the sense that the system represented by the graph exhibits some unilateral property. For example, the directions of the one-way streets of a city and the orientations representing the unilateral property of a communication network are true orientations of the physical systems. In other situations, the orientation used is a "pseudo"-orientation, used in lieu of an elaborate reference system. For example, in electrical network theory the edges of a graph are assigned arbitrary orientations to represent the references of the branch currents and voltages. In this section, we shall introduce the basic concepts and terms associated with directed graphs. Since many of the concepts are directly analogous to those presented for undirected graphs, they will be mentioned only briefly. Terms that have no undirected counterpart will be discussed in detail.

5.1. Basic concepts

Like the undirected graphs, directed graphs will be formally presented in abstract form.

DEFINITION 1.24: *Abstract directed graph.* An *abstract directed graph* $G_d(V, E)$, or simply a *directed graph* G_d, consists of a set V of elements called *nodes* together with a set E of *ordered* pairs of the form (i, j), i and j in V, called the *directed edges* (or simply *edges*) of G_d; the node i is called the *initial node* and node j the *terminal node*. Together they are called the *endpoints* of (i, j).

Thus, the only difference between a graph and a directed graph is that the edges of a directed graph are ordered pairs of nodes while the edges of a graph are not. Note that we use the term *edge* for both directed and undirected graphs. This should not create any confusion.

We say that the edge (i, j) is *directed* or *oriented* from node i to node j in G_d, and that (i, j) is *incident* with the nodes i and j or alternatively that (i, j) is *directed away* or *outgoing from* i and *directed toward* or *terminating at* j. A directed graph can also be represented equivalently by a geometric diagram in which the nodes are indicated by small circles or dots, while any two of them, i and j, are joined by an arrowheaded continuous curve, or even a straight line, from i to j if and only if (i, j) is in E.

We extend the directed-graph concept by permitting several distinct edges with same initial and terminal nodes; they are called the *parallel edges* of G_d. The parallel edges directed from node i to node j are denoted by the symbols $(i, j)_1, (i, j)_2, \ldots, (i, j)_k, k \geq 2$. If no particular edge is specified, (i, j) denotes any one, but otherwise fixed, of the parallel edges from i to j in G_d. Also we admit edges with the same endpoints; they are called the *self-loops* of G_d.

§ 5 Directed graphs

As an illustration, consider the directed graph $G_d(V, E)$ in which
$$V = \{1, 2, 3, 4, 5, 6, 7\},$$
$$E = \{(1, 1), (1, 2), (1, 4), (4, 4)_1, (4, 4)_2, (4, 3), (2, 3)_1,$$
$$(2, 3)_2, (6, 7)_1, (6, 7)_2, (7, 6)\}.$$

The corresponding geometric graph is as shown in fig. 1.18 in which we have a self-loop at node 1 and two parallel edges directed from node 2 to node 3, and from node 6 to node 7. We emphasize that in a directed graph the order in (i, j) is important. In fact we do not consider (i, j) and (j, i) as the same edge.

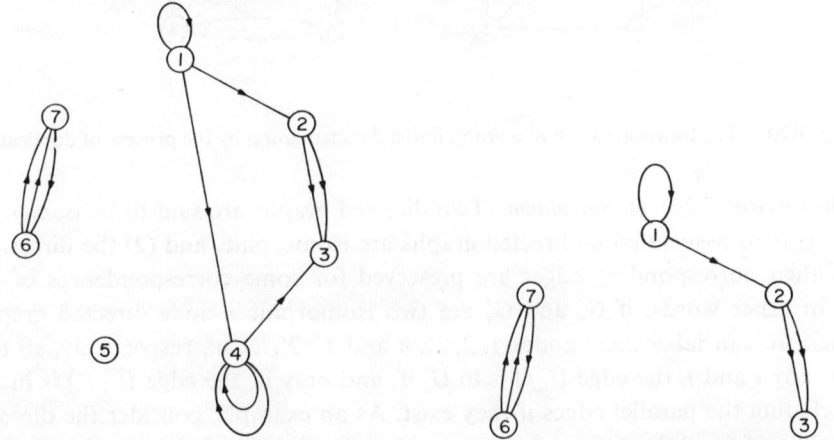

Fig. 1.18. The geometric diagram of a directed graph.

Fig. 1.19. A sectional subgraph of the directed graph of fig. 1.18.

The terms *subgraph, sectional subgraph, rank, nullity*, the *complement* of a subgraph, and other graph operations outlined in § 3 can similarly be defined for G_d. The only difference is that we use G_d instead of G. For example, in fig. 1.18 the sectional subgraph $G_d[V_s]$ defined by the node set $V_s = \{1, 2, 3, 6, 7\}$ is presented in fig. 1.19.

DEFINITION 1.25: *Associated undirected graph*. To every directed graph G_d, there is an *associated undirected graph* G_u whose node and edge sets are the same as those in G_d except that the directions of the edges (orders in the pairs (i, j)) of G_d are removed.

More intuitively, this means that G_u is the graph obtained from G_d by omitting the arrowheads. For example, fig. 1.3 is the associated undirected graph of fig. 1.18. By the same token, sometimes it is desirable to change an undirected graph into a directed one by the process of duplication: to each edge of the graph, we replace it by a pair of edges with the same endpoints but with opposite directions. For example, in fig. 1.20, G_d is the directed graph obtained from G by the procedure just outlined.

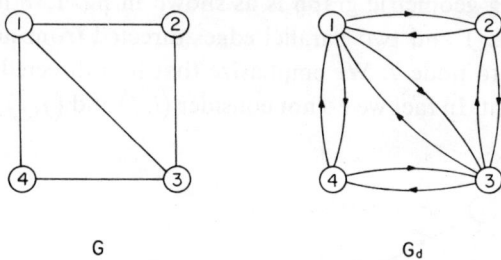

Fig. 1.20. The transformation of a graph into a directed graph by the process of duplication.

DEFINITION 1.26: *Isomorphism*. Two directed graphs are said to be *isomorphic* if (1) their associated undirected graphs are isomorphic, and (2) the directions of their corresponding edges are preserved for some correspondences of (1).

In other words, if G_d and G_d' are two isomorphic n-node directed graphs, then we can label their nodes $1, 2, \ldots, n$ and $1', 2', \ldots, n'$, respectively, so that for any i and j, the edge (i, j) is in G_d if, and only if, the edge (i', j') is in G_d', including the parallel edges if they exist. As an example, consider the directed graphs of fig. 1.21. It is easy to show that G_d and G_d' are isomorphic, but G_d and G_d'' are not even though their associated undirected graphs are isomorphic.

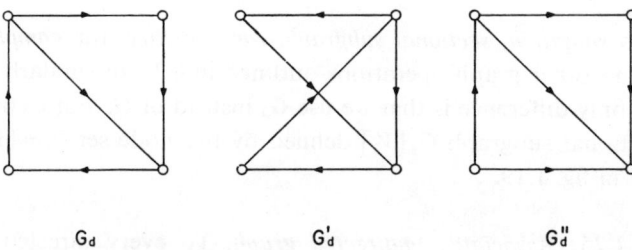

Fig. 1.21. Examples of isomorphic and non-isomorphic directed graphs.

A directed graph G_d is said to be *connected* if its associated undirected graph G_u is connected. This is similarly valid for G_d to be *planar, separable, nonseparable,* or *cyclically connected*. A subgraph G_s of G_d is an *edge sequence* of G_d if the associated undirected graph of G_s is an edge sequence of G_u. Similarly, in

§ 5   Directed graphs   27

G_d we define *edge train, path, circuit,* and *component.* Also, in G_d we speak of *circuit edges, noncircuit edges, cutpoints, blocks, length* of an edge sequence, and the *distance* between two nodes in G_d; they are again defined in terms of G_u.

For example, in fig. 1.22 the edge set $E_s = e_1 e_2 e_3 e_4 e_5 e_6 e_7 e_4 e_8 e_2$ is an edge sequence of length 10. If we delete the last three edges from E_s, we obtain an edge train $E_t = e_1 e_2 e_3 e_4 e_5 e_6 e_7$. The edge train becomes a path P after the removal of the edges e_5, e_6, and e_7 from E_t, i.e., $P = e_1 e_2 e_3 e_4$. G_d has one

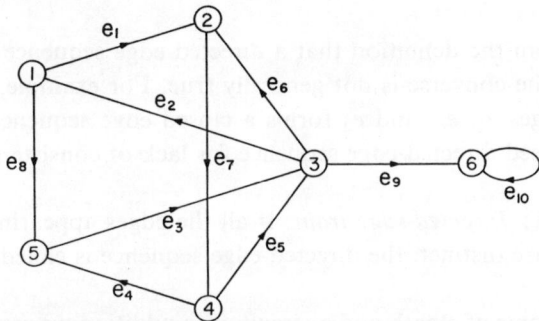

Fig. 1.22. A directed graph for illustrative purpose.

component, and the nodes 3 and 6 are the cutpoints. The blocks of G_d are $e_1 e_2 e_3 e_4 e_5 e_6 e_7 e_8$, e_9, and e_{10}. Nodes 1 and 6 are of distance 2. Finally, the subgraph $e_1 e_8 e_4 e_5 e_6$ is a circuit of length 5, and all the edges of G_d except e_9 are circuit edges, while e_9 is the only noncircuit edge.

5.2. *Directed-edge sequence*

In addition to the terms edge sequence, edge train, path, and circuit defined for a directed graph G_d, we also need special subclasses of these subgraphs known as directed-edge sequence, directed-edge train, directed path, and directed circuit.

DEFINITION 1.27: *Directed-edge sequence.* For a directed graph G_d, a *directed-edge sequence of length* $k-1$ in G_d is an edge sequence in which the edges along the edge sequence are of the form

$$(i_1, i_2), (i_2, i_3), \ldots, (i_{k-1}, i_k), \tag{1.9}$$

$k \geq 2$. The directed-edge sequence is said to be *closed* if $i_1 = i_k$, and *open* otherwise. In an open directed-edge sequence, the node i_1 is called its *initial node*, and node i_k its *terminal node.*

Intuitively speaking, a directed-edge sequence is simply an edge sequence in which all the edges are oriented in the same direction. For example, in fig. 1.22 the sequence of edges

$$(1, 2), (2, 4), (4, 3), (3, 1), (1, 5), (5, 3), (3, 2), (2, 4), (4, 3)$$

constitutes an open directed-edge sequence with initial node 1 and terminal node 3. We also say that the directed-edge sequence (1.9) is directed from i_1 to i_k. In the above example, the directed-edge sequence is directed from node 1 to node 3.

It is clear from the definition that a directed-edge sequence is also an edge sequence, but the converse is not generally true. For example, in fig. 1.22 the sequence of edges e_1, e_6, and e_2 forms a closed edge sequence but it fails to constitute a closed directed-edge sequence for lack of consistent directions.

DEFINITION 1.28: *Directed-edge train.* If all the edges appearing in a directed-edge sequence are distinct, the directed-edge sequence is called a *directed-edge train*.

Like the concepts of a path and a circuit, if in addition we require that all the nodes in a directed-edge train except the initial and terminal nodes be distinct, we have the concepts of a directed path and a directed circuit.

DEFINITION 1.29: *Directed path.* An open directed-edge train as shown in (1.9) in which all the nodes $i_1, i_2, ..., i_k$ are distinct is called a *directed path of length* $k-1$.

DEFINITION 1.30: *Directed circuit.* A closed directed-edge train as shown in (1.9) in which all the nodes $i_1, i_2, ..., i_{k-1}$ are distinct, in this case $i_1 = i_k$, is called a *directed circuit of length* $k-1$.

Thus, a self-loop of a directed graph is also a directed circuit of length 1. As in Definition 1.7, it is convenient to define an isolated node as a directed path of zero length. We shall assume this unless it is stated explicitly otherwise.

As an example, consider the directed graph G_d as shown in fig. 1.22. The sequence of edges

$$(1, 2), (2, 4), (4, 3), (3, 1), (1, 5), (5, 3), (3, 6), (6, 6)$$

is an open directed-edge train of length 8, which is directed from the node 1 to node 6. The open directed-edge train

$$(1, 2), (2, 4), (4, 3), (3, 6)$$

is a directed path of length 4, and the closed directed-edge train

$$(1, 2), (2, 4), (4, 5), (5, 3), (3, 1)$$

is a directed circuit of length 5. Clearly, these edge trains are also subgraphs of the directed graph.

5.3. *Outgoing and incoming degrees*

The local structure of a directed graph is described by the degrees of its nodes.

DEFINITION 1.31: *Outgoing and incoming degrees.* For a directed graph G_d, the number $d^+(i)$ of edges of G_d having node i as their initial node is called the *outgoing degree* of node i in G_d, and the number $d^-(i)$ of edges of G_d having node i as their terminal node is called the *incoming degree* of node i in G_d.

Thus, there are two numbers defined for each node of G_d. These numbers sometimes are also referred to as *positive* and *negative degrees* of a node. If $d(i)$ denotes the number of edges of G_d incident with the node i, then

$$d(i) = d^+(i) + d^-(i). \tag{1.10}$$

Since every edge is outgoing from a node and terminating at another, it is evident that the number b of edges of G_d is related to the degrees of its nodes by the following equation

$$b = \sum_i d^+(i) = \sum_i d^-(i), \tag{1.11}$$

where the summations are over all i of G_d.

As an illustration, consider the directed graph G_d of fig. 1.22, in which we have

$$d^+(3) = 3,\ d^-(3) = 2,\ d^+(6) = 1,\ \text{and}\ d^-(6) = 2.$$

Using (1.11), we have

$$b = 2 + 1 + 3 + 2 + 1 + 1 = 1 + 2 + 2 + 1 + 2 + 2 = 10.$$

DEFINITION 1.32: *Regular directed graph.* A directed graph is said to be *regular of degree* k if $d^+(i) = d^-(i) = k$ for each node i.

Thus, a directed circuit is a connected regular directed graph of degree 1. Clearly, if G_d is regular of degree k, its associated undirected graph G_u must be regular of degree $2k$. An example of a regular directed graph of degree 2 is presented in fig. 1.23.

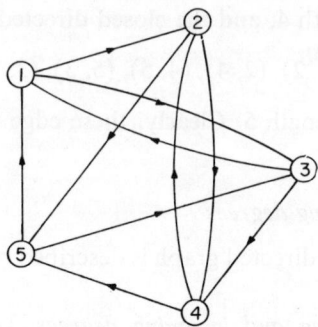

Fig. 1.23. A regular directed graph of degree 2.

5.4. *Strongly-connected directed graphs*

Another term which is useful for describing certain structural features of a directed graph and which has no undirected counterpart is known as strong connectedness. Its formal definition is given below.

DEFINITION 1.33: *Strong connectedness.* A directed graph is said to be *strongly connected* if, for every pair of distinct nodes i and j, there exists a directed path from i to j as well as one from j to i.

It is evident that a strongly-connected directed graph implies the connectedness of the directed graph, but the converse, of course, is not generally true.

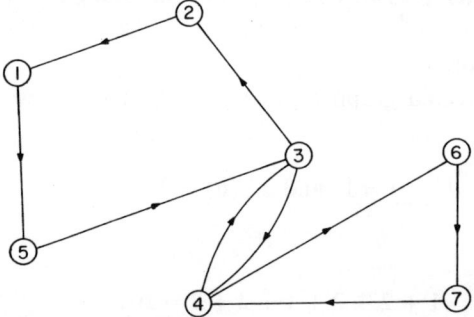
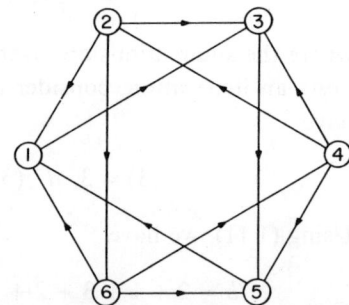

Fig. 1.24. A strongly-connected directed graph.

Fig. 1.25. A connected, but not strongly-connected, directed graph.

For example, fig. 1.24 is a strongly-connected directed graph, but fig. 1.25 which is connected is not strongly connected. From these two examples, we may also conclude that strong connectedness does not depend on the number of edges that a directed graph possesses, but rather their strategic locations.

DEFINITION 1.34: *Strong component.* A *strong component* or a *strongly-connected component* of a directed graph is a strongly-connected subgraph containing the maximal number of edges. An isolated node is a strong component.

In fig. 1.25, the sectional subgraphs $G_d[2, 4, 6]$ and $G_d[1, 3, 5]$ are strong components of G_d.

THEOREM 1.13: A necessary and sufficient condition for a directed graph to be strongly connected is that there exist a closed directed-edge sequence containing all of its nodes.

The proof of the theorem is left as an exercise (Problem 1.25). In the theorem, if we replace the word *nodes* by *edges*, we find that the theorem is still valid; its justification is left as a problem (Problem 1.26).

5.5. *Some important classes of directed graphs*

We have already seen graphs and directed graphs classified on the basis of whether they are planar or nonplanar, and separable or nonseparable. In the following we shall introduce two other useful classifications.

DEFINITION 1.35: *Symmetric directed graph.* A directed graph G_d is said to be *symmetric* if, for every edge (i, j) of G_d, it is matched by an edge (j, i) of G_d, the number of parallel edges in each direction being the same if they exist.

This would imply that, in G_d, if there are k edges directed from i to j then there are k edges directed from j to i. We remark that the number of self-loops at a node has no effect on its symmetry. Clearly, if G_d is symmetric, then the outgoing degree of a node must be equal to its incoming degree. The converse, however, is not generally true. For example, fig. 1.23 is a regular directed graph but it is not symmetric. Fig. 1.26 is an example of a symmetric directed graph.

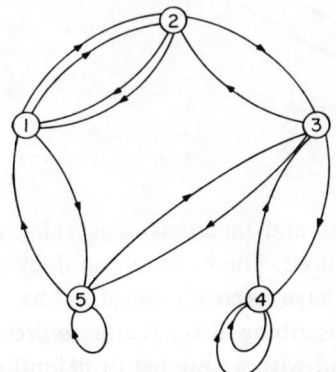

Fig. 1.26. A symmetric directed graph.

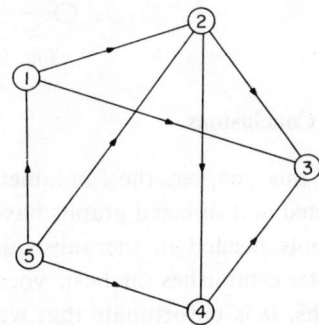

Fig. 1.27. An acyclic directed graph.

DEFINITION 1.36: *Acyclic directed graph.* A directed graph is called *acyclic* if it contains no directed circuits.

Fig. 1.27 is an example of an acyclic directed graph. It is evident that an acyclic directed graph cannot have any self-loops.

THEOREM 1.14: An acyclic directed graph has at least one node of zero outgoing degree and one node of zero incoming degree.

Proof. Let G_d be an acyclic directed graph. Let P be a directed path of maximum length from i to j. Clearly, we cannot have any edges outgoing from j or terminating at i; for otherwise there would be a directed circuit or the directed path would not be of maximum length. It follows that $d^-(i)=0$ and $d^+(j)=0$. So the theorem is proved.

§ 6. Mixed graphs

In many situations, it is natural to consider graphs with both directed and undirected edges; they are referred to as *mixed graphs*. In topological analysis of linear systems to be discussed in chapter 4, we make extensive use of this concept. As an example, a city map with both one-way and two-way streets may be represented by a mixed graph in which the streets are edges and the street intersections the nodes. Fig. 1.28 is an example of a mixed graph.

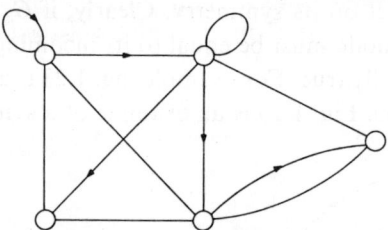

Fig. 1.28. A mixed graph.

§ 7. Conclusions

In this chapter, the fundamental definitions and theorems concerning undirected and directed graphs have been introduced. The basic terminology and symbols needed in the subsequent analysis have been discussed. Thus, the chapter establishes the basic vocabulary for describing directed and undirected graphs. It is unfortunate that we have to begin with a long list of definitions, but it is even more unfortunate that we cannot do anything about it. We have to have a few words to talk about. It is possible that we can reduce the number of definitions, but then each theorem becomes much more complicated and hence nearly incomprehensible.

Problems

1.1. Show that a subgraph G_s of a graph is a circuit if and only if G_s is a connected regular graph of degree 2.
1.2. Let G_s be a subgraph of a connected graph G. Show that the subgraph $\bar{G}_s$ cannot have more components than there are nodes in G_s.
1.3. Prove Theorem 1.1.
1.4. Let P' and P'' be two different paths connected between the nodes i and j. Show that there is a circuit in $P' \cup P''$. Is there a circuit containing the node i or j or both?
1.5. Prove Theorem 1.3.
1.6. Show that an n-node graph which has no parallel edges and self-loops and which has more than $\frac{1}{2}(n-1)(n-2)$ edges is always connected.
1.7. Show that the binary relation R defined in § 2.3 is an *equivalence relation*.
1.8. Prove Theorem 1.5.
1.9. Show that the number of edges b and the number of nodes n of a path are related by $b = n - 1$.
1.10. Show that the nullity of a graph is invariant either by splitting an edge into two edges in series or by merging two edges in series into one.
1.11. Prove Theorem 1.6.
1.12. Let G be a planar graph with n nodes and b edges. If the boundary of each of its regions is a circuit of length k, then

$$b \leq k(n-2)/(k-2). \qquad (1.12)$$

1.13. Show that a graph is planar if and only if each of its blocks is planar.
1.14. Prove that every planar graph with at least four nodes having no parallel edges and self-loops has at least four nodes of degree not exceeding 5.
1.15. Prove Theorem 1.9.
1.16. Show that each node of a nonseparable graph containing at least two edges is of degree at least 2.
1.17. Prove Corollary 1.4.
1.18. Justify the statement that every nonseparable subgraph of a graph G must be contained wholly in one of the components of G.
1.19. Prove Corollary 1.5.

1.20. Show that the decomposition of a connected separable graph into blocks is unique. Can you extend this to any separable graph?

1.21. Let G be a connected graph with at least three nodes. Show that G is a block if, and only if, for every three distinct nodes of G, there is a path connecting any two of them which contains the third.

1.22. Is the statement in Problem 1.21 still true if the last clause is replaced by "which does not contain the third"? Can you formulate other equivalent statements? If possible, give four more.

1.23. Prove that in a nontrivial connected graph having no self-loops there are at least two nodes which are not cutpoint. Is this true for any connected graph?

1.24. Prove that a nontrivial nonseparable graph contains at least one circuit.

1.25. Prove Theorem 1.13.

1.26. Show that a directed graph is strongly connected if, and only if, there exists a closed directed-edge sequence which includes all the edges at least once.

1.27. Let E' and E'' be two directed-edge sequences which have at least one node in common. Show that there exists a directed-edge sequence in $E' \cup E''$ which includes all the edges of $E' \cup E''$. Is $E' \cup E''$ strongly connected?

1.28. Prove that the rank and nullity of a graph G are equal to the sums of the ranks and nullities, respectively, of its components.

1.29. If a graph G is separable, show that its rank and nullity are invariant under the decomposition of G into its components.

1.30. Is it true that if the incoming and outgoing degrees of every node of a directed graph are positive, then every node is contained in at least one directed circuit? If so, prove it.

1.31. For an n-node connected graph, show that it must have at least $n-1$ edges.

1.32. Show that all the edges of a connected graph can be included in an edge sequence. Give an algorithm to construct such an edge sequence.

1.33. Show that in a graph G if there are exactly two nodes i and j of odd degree, then there exists a path connecting the nodes i and j of G.

1.34. Prove that every regular graph of degree 3 has an even number of nodes.

1.35. Show that a graph $G(V, E)$ is connected if, and only if, for any partition of V into two subsets V_1 and V_2, there is an edge connecting a node in V_1 with a node in V_2.

1.36. By inserting nodes in two of the three terminal edges in fig. 1.29, verify that all the longest paths of a connected graph do not have to have a node in common (WALTHER [1969]).

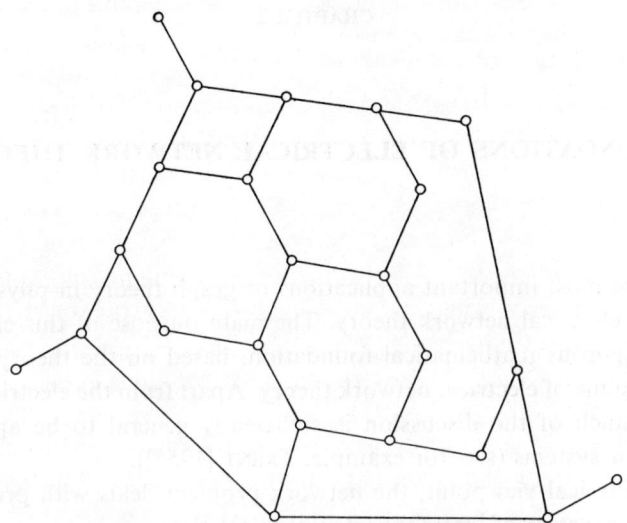

Fig. 1.29. A graph to illustrate the longest paths.

1.37. In contrast to Problem 1.36, show that any two longest paths of a connected graph have a node in common.
1.38. Show that the one-to-one correspondences of a set V onto itself are defined by the regular directed graphs of degree 1 on V.
1.39. Can you extend formula (1.2) to graphs with self-loops?
1.40. In fig. 1.29 which two nodes have the longest distance?
1.41. Using fig. 1.13, construct a planar graph such that the edges e_4, e_6, and e_7 are the boundary edges of the outside region.
1.42. Show that the complement of a path in a circuit is a path.
1.43. For a given graph G, the *block graph* $B(G)$ of G is the graph whose nodes correspond to the blocks of G, and two of these nodes are connected by an edge whenever the corresponding blocks contain a common cutpoint of G. Show that a graph B is the block graph of some graph if, and only if, for every pair of nodes in a block of B there is an edge connecting these two nodes, i.e., every block of B is complete (HARARY [1969]).

CHAPTER 2

FOUNDATIONS OF ELECTRICAL NETWORK THEORY

One of the most important applications of graph theory in physical science is its use in electrical network theory. The main purpose of this chapter is to provide a rigorous mathematical foundation, based on the theory of graphs, for the discipline of electrical network theory. Apart from the electrical network problems, much of the discussion is sufficiently general to be applicable to general linear systems (see, for example, TRENT [1955]).

From a physical viewpoint, the network problem deals with predicting the behavior of a system of interconnected physical elements in terms of the characteristics of the elements and the manner in which these elements are interconnected. The geometrical properties of a network are independent of the constituents of its branches, and so in topological discussions it is usual to replace each branch by a line segment. In other words, from an abstract viewpoint, any lumped electrical network can be represented by a graph with edges denoting, to some extent at least, electrical components and weights representing the constituents of the components. The graph is referred to as a *model* of the physical network. Since the networks that we deal with and analyze are models consisting of an interconnection of *idealized* physical elements such as inductors, capacitors, resistors, and generators, we cannot always assume the existence and uniqueness of their solutions. Thus, it is important to discuss conditions under which a unique solution can be obtained for an electrical network.

KIRCHHOFF [1847] made the first comprehensive investigation of the electrical network problem, and proved the existence of a solution to a resistive network. MAXWELL [1892] pointed out, however, that Kirchhoff's formulation omitted the concept of potential. He remedied this omission and devised two very effective methods for solving the network problem, now termed Maxwell's *mesh* and *node-pair* methods of solution. The latter yields the well-known node-admittance matrix which is the starting point of topological analysis to be discussed in ch. 4. In Maxwell's formulation, however, it is necessary to choose an appropriate set of circuits or node-pairs as independent variables and also to assign an orientation to each branch of the network. It appears, therefore, that the branch voltages and currents induced in the branches of the network may depend upon the choice of circuits or node-pairs and the assignment of branch orientations. Such a dependence would contradict the known electrical network situation. Thus, it is essential to show that they are actually not.

In this chapter we shall justify many of these familiar statements and procedures of network analysis, and provide a rigorous mathematical foundation for the discipline of electrical network theory. It will be shown that many of the invariant characters of the electrical network problems can be derived from purely graph-theoretic considerations. Thus, they are valid not only for the electrical networks, but also for other systems.

§ 1. Matrices and directed graphs

The orientation of the edges of a directed graph, in some applications, is a "true" orientation in the sense that the system represented by the directed graph exhibits some unilateral property such as in the Coates graph, the Mason graph, and in the associated directed graph to be presented in the succeeding two chapters. In electrical network theory, on the other hand, the orientation is a "pseudo-orientation", used in lieu of an elaborate reference system.

Throughout this chapter, the symbols n, b, and c will be used to denote the numbers of nodes, edges, and components of a directed graph G, respectively. Also, we denote the rank and nullity of G by r and m, respectively, where $r = n - c$ and $m = b - n + c$. For convenience, the nodes of G are assumed to be numbered from $1, 2, \ldots, n$, and the edges denoted by $e_1, e_2, \ldots, e_b$.

1.1. The node-edge incidence matrix

A directed graph G without self-loops is completely characterized by its node-edge incidence matrix.

DEFINITION 2.1: *Node-edge incidence matrix* (*incidence matrix*). The *node-edge incidence matrix* or simply *incidence matrix*, denoted by the symbol A_a, of a directed graph G is a matrix of order $n \times b$ such that if $A_a = [a_{ij}]$, then

$a_{ij} = 1$ if edge e_j is incident at node i and is directed away from node i,
$a_{ij} = -1$ if edge e_j is incident at node i and is directed toward node i,
$a_{ij} = 0$ if edge e_j is not incident at node i.

The matrix A_a was first used by KIRCHHOFF [1847], and is the coefficient matrix of Kirchhoff's current equations. Consequently, the properties of this matrix are of considerable interest. Note that in defining A_a, we do not consider directed graphs containing self-loops. These can be considered but we do not find it necessary. In case we find it necessary, and to be consistent with the above rules, we may define a zero column corresponding to a self-loop in A_a since the rules require both a 1 and a -1 entry in the row corresponding to the node at which the self-loop is incident.

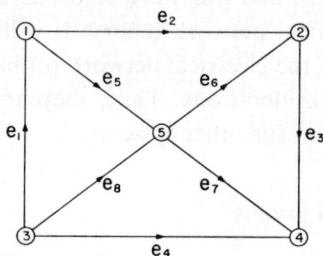

Fig. 2.1. A directed graph to illustrate the incidence matrix.

As an illustration, consider the directed graph G as shown in fig. 2.1. The incidence matrix of G is given by

$$A_a = \begin{array}{c} \\ 1 \\ 2 \\ 3 \\ 4 \\ 5 \end{array} \begin{array}{c} e_1 \quad e_2 \quad e_3 \quad e_4 \quad e_5 \quad e_6 \quad e_7 \quad e_8 \\ \begin{bmatrix} -1 & 1 & 0 & 0 & 1 & 0 & 0 & 0 \\ 0 & -1 & 1 & 0 & 0 & -1 & 0 & 0 \\ 1 & 0 & 0 & 1 & 0 & 0 & 0 & 1 \\ 0 & 0 & -1 & -1 & 0 & 0 & -1 & 0 \\ 0 & 0 & 0 & 0 & -1 & 1 & 1 & -1 \end{bmatrix} \end{array}. \quad (2.1)$$

Obviously, each column of A_a contains exactly two nonzero elements, a 1 and a -1. Hence, the sum of all the rows of A_a is a row of zeros. In other words, not all the rows of A_a are linearly independent.

LEMMA 2.1: For a connected directed graph G, the sum of any k, $k < n$, rows of A_a contains at least one nonzero element.

Proof. In any set of k rows of A_a, there is at least one column containing exactly one nonzero element. For, if not, there is no edge connecting this set of k nodes to the other $(n-k)$ nodes of G, contradicting to our assumption that G is connected. Hence, the sum of any k rows of A_a contains at least one nonzero element.

THEOREM 2.1: The rank of the incidence matrix of a directed graph G is equal to the rank of G.

Proof. We shall only prove the case where G is connected, i.e., $c=1$, and leave the other case as an exercise (Problem 2.1). Let X be an n-vector over the real field. If A'_a is the transpose of the incidence matrix of G, then by Lemma 2.1 the system of linear equations (Problem 2.60)

$$A'_a X = 0 \tag{2.2}$$

has a complete solution βX where β is any real constant and X consists only of 1's. Thus, from matrix algebra any $(n-1)$ columns of A'_a are linearly independent. For, if not, we can construct a nontrivial solution X containing a zero at the row corresponding to the column not included in the $(n-1)$ chosen columns of A'_a, a contradiction. Since not all the columns of A'_a are linearly independent, it follows that the rank of A_a is $(n-1)$. So the theorem is proved.

This property was first established by KIRCHHOFF [1847]. For convenience, we shall use the symbol A to denote a submatrix of A_a of a connected directed graph obtained by deleting an arbitrary row of A_a. The matrix A is called a *basis incidence matrix* of the connected directed graph. The node corresponding to the deleted row of A_a is referred to as the *reference node* of the directed graph because it corresponds to the potential-reference point in the corresponding electrical network.

DEFINITION 2.2: *Tree.* A spanning subgraph of a directed (or undirected) graph is said to be a *tree* if, and only if, it is connected and contains no circuits.

This definition differs slightly from the conventional mathematical definition in that the term "spanning" is usually omitted. The alternative is to use the term *spanning tree* for the concept we need. However, in most practical applications, the terminology is in accordance with Definition 2.2.

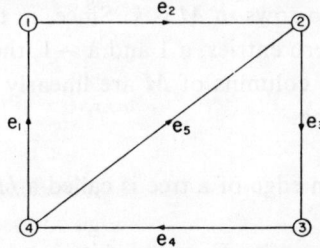

Fig. 2.2. A directed graph to illustrate trees.

As an illustration, consider the directed graph G as shown in fig. 2.2. The set of trees of G is given in fig. 2.3. If G_u is the associated undirected graph of G, the set of trees given in fig. 2.3 with the orientations removed is the set of trees of G_u. Obviously, a tree of a connected graph has $r(=n-1)$ edges.

The concept of a tree is extremely important because of the number of properties of the network that can be related to a tree. For example, the number of independent Kirchhoff's voltage equations, the number of state equations, the methods of choosing the independent equations, and the topological formulas for network functions, may all be stated in terms of trees.

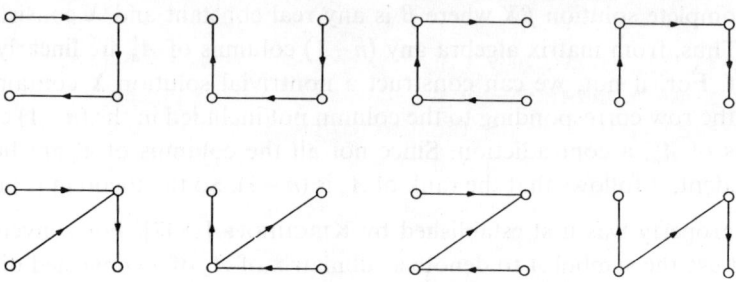

Fig. 2.3. Trees of the directed graph of fig. 2.2.

LEMMA 2.2: The columns of A corresponding to the edges of a circuit in G are linearly dependent.

Proof. Let L be a circuit of length k in G, and let M be the submatrix of A formed by the columns corresponding to the edges of L. Relabel G if necessary so that the columns and the leading rows (corresponding to the nodes in L) of M are arranged in the orders of the edges and nodes appearing in L, respectively. In M it is not difficult to see that if the reference node of G is contained in L, the number of nonzero rows in M is $(k-1)$. Thus, the columns of M are linearly dependent. On the other hand, if the reference node is not contained in L, the number of nonzero rows in M is k. Since, in this case, each of these columns contains two nonzero entries, a 1 and a -1, the rank of M is at most $(k-1)$. In other words, the columns of M are linearly dependent. This completes the proof of the lemma.

DEFINITION 2.3: *Branch.* An edge of a tree is called a *branch*.

DEFINITION 2.4: *Major submatrix.* For an arbitrary matrix F of order $p \times q$ and of rank p, a *major submatrix* of F is a nonsingular submatrix of order p.

THEOREM 2.2: A square submatrix of a basis incidence matrix A of a connected directed graph is a major submatrix if, and only if, the columns of this submatrix correspond to the branches of a tree in the directed graph.

Proof. Let M be the submatrix of A in question. If the columns of M correspond to the branches of a tree t, then, by Theorem 2.1, M is nonsingular since M is a basis incidence matrix of t. Conversely, if M is nonsingular, then by Lemma 2.2 the subgraph corresponding to the columns of M cannot contain any circuits. Since the subgraph has $(n-1)$ edges and contains no circuits, the subgraph must be a tree (Problem 2.2).

COROLLARY 2.1: There exists a one-to-one correspondence between trees of a directed graph G and the major submatrices of a basis incidence matrix of G.

As an illustration, consider the directed graph G as shown in fig. 2.2. The basis incidence matrix A with node 4 used as reference is given by

$$A = \begin{matrix} & e_1 & e_2 & e_3 & e_4 & e_5 \\ 1 \\ 2 \\ 3 \end{matrix} \begin{bmatrix} -1 & 1 & 0 & 0 & 0 \\ 0 & -1 & 1 & 0 & -1 \\ 0 & 0 & -1 & 1 & 0 \end{bmatrix}. \qquad (2.3)$$

It is easy to check that only major submatrices of A are those corresponding to the trees given in fig. 2.3.

1.2. *The circuit-edge incidence matrix*

Since directed graphs are considered, it is natural to orient the circuits.

DEFINITION 2.5: *Oriented circuit.* A circuit of a directed graph with an orientation assigned by a cyclic ordering of nodes along the circuit is called an *oriented circuit*.

For example, in fig. 2.1 the circuit consisting of the edges e_1, e_2, e_3, and e_4 can be oriented as $(1, 2, 4, 3)$ or as $(1, 3, 4, 2)$. Also, we can represent the orientation pictorially by an arrowhead. We shall say that the orientations of an edge of a circuit and the circuit *coincide* if the nodes of the edge appear in the same order both in the ordered-pair representation of the edge and in the ordered-node representation of the circuit. Otherwise, they are *opposite*. Pictorially, the meaning is obvious. For example, the edge $(1, 2)$ coincides with the circuit $(1, 2, 4, 3)$, and $(3, 4)$ is opposite to the circuit. Note that a circuit of a directed graph need not be a directed circuit.

DEFINITION 2.6: *Circuit-edge incidence matrix (circuit matrix).* The *circuit-edge incidence matrix* or simply *circuit matrix*, denoted by the symbol B_a, of a directed graph G is a matrix of order $p \times b$, where p is the number of circuits in G, such that if $B_a = [b_{ij}]$, then

$b_{ij} = 1$ if edge e_j is in circuit i and the orientations of the circuit and the edge coincide,

$b_{ij} = -1$ if edge e_j is in circuit i and the orientations of the circuit and the edge are opposite,

$b_{ij} = 0$ if the edge e_j is not in circuit i.

As an example, consider the set of all possible circuits in the directed graph G as shown in fig. 2.4. The matrix $\boldsymbol{B}_a$ is given by

$$\boldsymbol{B}_a = \begin{array}{c} \\ 1 \\ 2 \\ 3 \end{array} \begin{array}{ccccc} e_1 & e_2 & e_3 & e_4 & e_5 \\ \left[\begin{array}{ccccc} 1 & 1 & 0 & 0 & -1 \\ 0 & 0 & 1 & 1 & 1 \\ -1 & -1 & -1 & -1 & 0 \end{array}\right] \end{array}. \qquad (2.4)$$

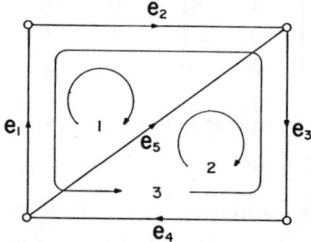

Fig. 2.4. A directed graph to illustrate the circuit matrix.

Note that in $\boldsymbol{B}_a$ we do not include circuits that are obtained merely by reversing the orientations, the reason being that by reversing an orientation it only changes the sign of a row of $\boldsymbol{B}_a$. Also, $\boldsymbol{B}_a$ is the coefficient matrix of Kirchhoff's voltage equations.

DEFINITION 2.7: *Cotree.* The complement of a tree in a directed or undirected graph is called a *cotree.*

DEFINITION 2.8: *Chord (link).* An edge in a cotree is called a *chord (link).*

Thus, a cotree of a connected graph contains $m(=b-n+1)$ chords.

Since there is a unique path connected between any two nodes in a tree, the addition of a chord to the tree produces a unique circuit in the resulting graph. Thus, each of the chords in a cotree defines a circuit (with respect to the chosen tree) in the directed graph in a unique way.

DEFINITION 2.9: *Fundamental circuits (f-circuits).* The *fundamental circuits* or simply *f-circuits* of a connected directed graph of nullity m with respect to a tree t are the m circuits, each being formed by a chord and the unique tree path connecting the two endpoints of the chord in t. The f-circuit orientation is chosen to agree with that of the defining chord.

The submatrix B_f of B_a corresponding to a set of f-circuits has a very special form. For example, if the f-circuits are numbered in some arbitrary manner as $1, 2, \ldots, m$, and if the chord that appears in circuit i is numbered as edge e_i for $i = 1, 2, \ldots, m$, then

$$B_f = [U_m \quad B_{12}], \qquad (2.5)$$

where U_m is the identity matrix of order m. As an example, consider the directed graph as shown in fig. 2.1. If we choose the tree consisting of the edges e_5, e_6, e_7, and e_8, the f-circuits are presented in fig. 2.5. The circuit matrix of these f-circuits is given by

$$B_f = \begin{array}{c} \\ 1 \\ 2 \\ 3 \\ 4 \end{array} \begin{array}{cccccccc} e_1 & e_2 & e_3 & e_4 & e_5 & e_6 & e_7 & e_8 \\ \left[\begin{array}{cccc|cccc} 1 & 0 & 0 & 0 & 1 & 0 & 0 & -1 \\ 0 & 1 & 0 & 0 & -1 & -1 & 0 & 0 \\ 0 & 0 & 1 & 0 & 0 & 1 & -1 & 0 \\ 0 & 0 & 0 & 1 & 0 & 0 & -1 & -1 \end{array}\right] \end{array}. \qquad (2.6)$$

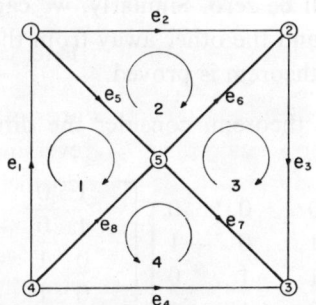

Fig. 2.5. An example for the f-circuits.

DEFINITION 2.10: *Fundamental circuit matrix (f-circuit matrix)*. For a connected directed graph G of nullity m, the submatrix of the circuit matrix of G corresponding to a set of m f-circuits in G is called a *fundamental circuit matrix* or simply *f-circuit matrix* of G. The f-circuit matrix is denoted by the symbol B_f.

The matrix B_f obviously has rank m. Thus, the circuit matrix B_a has rank at least m. To show that m is also an upper bound for the rank of B_a, we need the following theorem. Apart from establishing the rank of B_a, the result is extremely important in network theory, as is evident from the rest of the chapter. The result was first established by VEBLEN [1931].

THEOREM 2.3: If the columns of the matrices A_a and B_a of a directed graph are arranged in the same edge order, then

$$A_a B'_a = 0 \text{ and } B_a A'_a = 0, \qquad (2.7)$$

where the prime indicates the transpose of a matrix.

Proof. We shall only prove the first identity; the second one follows immediately after we take transpose on both sides of the first one.

Consider the ith row of A_a and the jth column of B'_a, i.e., the jth row of B_a. The entries in the corresponding positions in the ith row of A_a and the jth row of B_a are nonzero if and only if the corresponding edge is incident at node i and is also contained in circuit j. If node i is not contained in circuit j, then there is no such edge and the product is zero, giving the required result. If circuit j does include node i, there will be precisely two edges of the circuit j incident at node i, i.e., there are only two nonzero entries in the ith row of A_a which have the property that the corresponding entries in the jth row of B_a are also nonzero. If both edges are incident toward or away from the node i, then the corresponding entries in the ith row of A_a will be of the same sign, and those in the jth row of B_a of opposite signs, so the product of the ith row of A_a and jth row of B_a will be zero. Similarly, we can show that if one of the edges is incident toward and the other away from the node i, the product will again be zero. Thus, the theorem is proved.

As an example of the theorem, consider the directed graph as shown in fig. 2.2. We have

$$A_a B'_a = \begin{bmatrix} -1 & 1 & 0 & 0 & 0 \\ 0 & -1 & 1 & 0 & -1 \\ 0 & 0 & -1 & 1 & 0 \\ 1 & 0 & 0 & -1 & 1 \end{bmatrix} \begin{bmatrix} 1 & 0 & -1 \\ 1 & 0 & -1 \\ 0 & 1 & -1 \\ 0 & 1 & -1 \\ -1 & 1 & 0 \end{bmatrix} = \begin{bmatrix} 0 & 0 & 0 \\ 0 & 0 & 0 \\ 0 & 0 & 0 \\ 0 & 0 & 0 \end{bmatrix}. \quad (2.8)$$

Throughout the chapter, we shall assume that the columns of A_a and B_a are always arranged in the same edge order. Using Theorem 2.3, we immediately have the rank of B_a.

THEOREM 2.4: The rank of the circuit matrix of a directed graph G is equal to the nullity of G.

Proof. We have already established that the rank of B_a is at least m for $c=1$. It is not difficult to see that the rank of B_a in general is at least $m=b-n+c$ (Problem 2.3). It remains to be shown that m is also an upper bound for its rank.

Since $A_a B_a' = 0$, the columns of B_a' must be in the null-space of A_a. Since the rank of A_a is $(n-c)$, the nullity of A_a, i.e., the dimension of the null-space of A_a, is m. Thus, the maximum number of linearly independent rows of B_a is m. The theorem follows directly from here.

Since the rank of B_a is m, there is really no need to write down all the rows of B_a. For our purpose, we shall reserve the symbol B to represent a submatrix of B_a, of order $m \times b$ and of rank m. The matrix B is termed as a *basis circuit matrix* of G because it contains all the information that is contained in B_a. Obviously, the f-circuit matrix is a basis circuit matrix of G.

LEMMA 2.3: There is a nonsingular $m \times m$ matrix C of 1's, -1's and 0's such that $B = CB_f$.

THEOREM 2.5: A square submatrix of a basis circuit matrix B of a connected directed graph is a major submatrix if, and only if, the columns of this submatrix correspond to the edges of a cotree in the directed graph.

Proof. Necessity. By permuting the columns of B if necessary, we may assume that the first m columns of B form a major submatrix. Let B be partitioned as

$$B = [B_{11} \quad B_{12}], \qquad (2.9)$$

where B_{11} is nonsingular and of order m. Since there are r columns in B_{12}, it is sufficient to show that the subgraph corresponding to these columns does not contain any circuit. Assume otherwise, and let B_i be the row in B_a corresponding to this circuit. Consider the submatrix of B_a consisting of the rows of B and B_i. Then the submatrix can be partitioned as

$$\begin{bmatrix} B \\ B_i \end{bmatrix} = \begin{bmatrix} B_{11} & B_{12} \\ 0 & B_{22} \end{bmatrix}. \qquad (2.10)$$

It is obvious that there is at least one nonzero element in B_{22}. Since B_{11} is nonsingular, the rank of the matrix of (2.10) is $(m+1)$. But the matrix is also a submatrix of B_a which by Theorem 2.4 is of rank m. Hence this is impossible.

Sufficiency. Assume that the columns of a submatrix of B correspond to a cotree. Let B_f be the f-circuit matrix formed by the chords of this cotree in the directed graph G. By permuting the columns of B and B_f if necessary, the matrices B and B_f can be partitioned as

$$B = [B_{11} \quad B_{12}], \qquad (2.11\text{a})$$

$$B_f = [U_m \quad B_{f12}]. \qquad (2.11\text{b})$$

Since the rows of B and B_f are the two bases for the vector space spanned by the rows of B_a, it follows that there exists a nonsingular matrix C of order m such that (Lemma 2.3)

$$B = CB_f. \tag{2.12}$$

Thus, $B_{11} = CU_m = C$. So B_{11} is nonsingular.

COROLLARY 2.2: There exists a one-to-one correspondence between cotrees of a directed graph G and the major submatrices of a basis circuit matrix B of G.

As an example, consider the directed graph G as shown in fig. 2.4. We have $m = 5 - 4 + 1 = 2$. Thus, a basis circuit matrix B of G must be of order 2×5. It is easy to check that the submatrix of B_a formed by any two of the three circuits of G is a basis circuit matrix of G. The basis circuit matrix B formed by the circuits 1 and 2 is given by

$$B = \begin{matrix} & e_1 & e_2 & e_3 & e_4 & e_5 \\ & \begin{bmatrix} 1 & 1 & 0 & 0 & -1 \\ 0 & 0 & 1 & 1 & 1 \end{bmatrix} \end{matrix}. \tag{2.13}$$

The set $\bar{T}$ of cotrees of G can easily be obtained, and is given by

$$\bar{T} = \{e_1 e_5, e_2 e_5, e_3 e_5, e_4 e_5, e_1 e_4, e_2 e_3, e_1 e_3, e_2 e_4\}. \tag{2.14}$$

Each of the terms in $\bar{T}$ is the complement of a tree given in fig. 2.3. The submatrix formed by the columns corresponding to each of the cotrees given in $\bar{T}$ is a major submatrix of B, and these are the only major submatrices of B.

1.3. *The cut-edge incidence matrix*

In the preceding discussions, we have certainly indicated that the subgraph formed by a set of edges incident at a node is important because it is closely related to the Kirchhoff's current law. A second class of subgraphs called the *cutset* also finds important use in electrical network theory because it generalizes the Kirchhoff's current law. The concept of cutsets was originally introduced by WHITNEY [1933] and systematically developed by SESHU and REED [1956]. Our discussion here is to show that cutsets bear the same relationship to circuits that circuits bear to incidence relationships. Thus we will find the duals of a number of theorems proved earlier for the circuits. It is also the purpose of this section to show that the cutsets contain the same essential information as the directed graph itself.

DEFINITION 2.11: *Cutset*. For a directed (undirected) graph G, a *cutset* of G

is a subgraph consisting of a minimal collection of edges whose removal reduces the rank of G by one.

Since the rank of G is $(n-c)$, the removal of a cutset from G increases the number of components by one. Thus, if G has more than one component, a cutset can only be formed from the edges of one of its components. For, if not, the collection will not be minimal. Intuitively, if we "cut" the edges of a cutset, one of the components of G will be cut into two pieces. The name *cutset* has its origin in this interpretation. Note that a component of G may consist of an isolated node. For examples of cutsets, consider the directed graph G as shown in fig. 2.6. The subgraphs $e_1 e_5 e_4$, $e_2 e_5 e_4$, e_6, and $e_1 e_5 e_3$ are examples of cutsets. The broken lines of fig. 2.6 show how these cutsets "cut" the graph. However, the subgraph $e_2 e_5 e_3 e_6$ is not a cutset because if we remove these edges from G, the rank of G is reduced from 4 to 2 ($=5-3$), a reduction of two instead of one as required. As another example, consider the subgraph $e_3 e_6$. Although the removal of e_3 and e_6 from G will reduce the rank of G by one, the set is not a cutset because it is not minimal; the removal of e_6 from G will also reduce the rank of G by one. Thus, the set of edges incident at a node, in general, does not constitute a cutset.

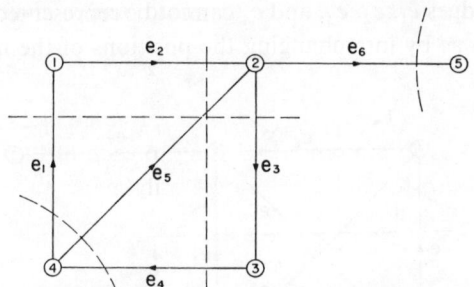

Fig. 2.6. Some examples of cutsets.

DEFINITION 2.12: *Incidence cut.* For a directed (undirected) graph G, the subgraph formed by the edges incident at a node of G is termed an *incidence cut* of G.

Thus, an incidence cut may either be a cutset or an edge-disjoint union of cutsets. It is also obvious that the nonzero entries in each row of the incidence matrix A_a correspond to an incidence cut in the graph. In general, we define

DEFINITION 2.13: *Cut.* A cutset or an edge-disjoint union of cutsets of a directed (undirected) graph is called a *cut.*

The term *cut* is also referred to as a *seg (segregate)* by REED [1961]. Since the circuit and the cutset are dual concepts, the dual of a cut may also be considered, i.e., an edge-disjoint union of circuits, but we do not find it necessary. A cut can also be interpreted in another useful fashion. Let V_1 be a nonempty proper subset of the node set V of G, and let $V_2 = V - V_1$. Then the set of edges of G each of which is incident with one of its two endpoints in V_1 and the other in V_2 is a cut of G. In particular, if the removal of these edges from G increases the number of components of G by one, then the cut is also a cutset. As an example, consider the directed graph G of fig. 2.6. Let $V_1 = \{3, 4, 5\}$ and $V_2 = \{1, 2\}$. Then the set of edges e_1, e_5, e_3, and e_6 forms a cut of G. On the other hand, if we let $V_1 = \{3, 4\}$ and $V_2 = \{1, 2, 5\}$, the set of edges e_1, e_5, and e_3 forms a cutset. Of course, the cutsets and the incidence cuts are special types of cuts. Like the circuit, it is more natural to orient the cuts in a directed graph.

DEFINITION 2.14: *Cut orientation.* For a directed graph G, let V_1 and V_2 be the sets of nodes partitioned by a cut C of G. The cut C is said to be *oriented* if the sets V_1 and V_2 are ordered either as (V_1, V_2) or as (V_2, V_1).

In most cases, the orientation of a cut may be represented by an arrow. For example, we can place an arrow near the broken line defining the cut. In fig. 2.7, the orientations of the cuts of G are as indicated. However, the cutset consisting of the edges e_1, e_2, e_3, and e_4 cannot be represented in this way unless we redraw G such as by interchanging the positions of the nodes 3 and 4.

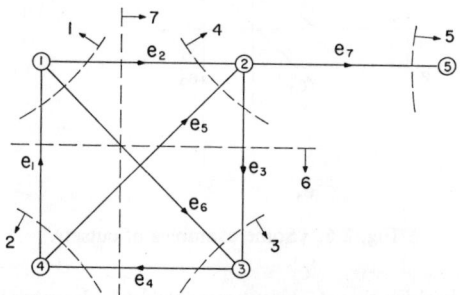

Fig. 2.7.   Some examples of oriented cuts.

Let a cut C of G be ordered as (V_1, V_2). We shall say that the orientations of the edge (i, j) and the cut C *coincide* if i is in V_1 and j in V_2. Otherwise, they are *opposite*. Pictorially, the meaning is obvious. For example, in fig. 2.7, the orientations of the edge $e_2 = (1, 2)$ and the cutsets 4 and 7 coincide, and the orientations of edge e_2 and the cutset 1 are opposite.

LEMMA 2.4: There are $(2^r - 1)$ nonempty cuts in G.

§1 Matrices and directed graphs

LEMMA 2.5: *A cut and a circuit of G have an even number of edges in common.*

The above two lemmas follow directly from our interpretation of a cut. The details are left as exercises (Problems 2.4 and 2.5).

Once again, we can discuss cuts most conveniently by means of a cut matrix.

DEFINITION 2.15: *Cut-edge incidence matrix (cut matrix).* The *cut-edge incidence matrix* or simply the *cut matrix*, denoted by the symbol Q_a, of a directed graph G is a matrix of order $q \times b$, where q is the number of nonempty cuts in G, such that if $Q_a = [q_{ij}]$, then

$q_{ij} = 1$ if edge e_j is in cut i and the orientations of the cut and the edge coincide,

$q_{ij} = -1$ if edge e_j is in cut i and the orientations of the cut and the edge are opposite,

$q_{ij} = 0$ if the edge e_j is not in cut i.

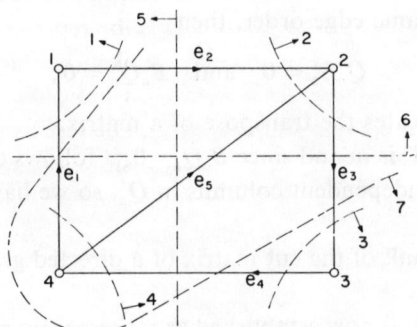

Fig. 2.8. A directed graph to illustrate the cut matrix.

It should be noted that in Q_a we do not include cuts that are obtained merely by reversing the orientations. For the directed graph of fig. 2.8, there are seven ($=2^{4-1}-1$) cuts with the orientations of the cuts as indicated. The cut matrix appears as

$$Q_a = \begin{matrix} & \begin{matrix} e_1 & e_2 & e_3 & e_4 & e_5 \end{matrix} \\ \begin{matrix} 1 \\ 2 \\ 3 \\ 4 \\ 5 \\ 6 \\ 7 \end{matrix} & \begin{bmatrix} 1 & -1 & 0 & 0 & 0 \\ 0 & 1 & -1 & 0 & 1 \\ 0 & 0 & 1 & -1 & 0 \\ 1 & 0 & 0 & -1 & 1 \\ 0 & -1 & 0 & 1 & -1 \\ 1 & 0 & -1 & 0 & 1 \\ 1 & -1 & 1 & -1 & 0 \end{bmatrix} \end{matrix}. \qquad (2.15)$$

Obviously, not all the rows of Q_a are linearly independent. A question naturally asked at this point is then "what is the rank of Q_a?". Since Q_a contains the incidence matrix A_a as a submatrix, we know right away that the rank of Q_a is at least r. For example, in (2.15) the submatrix formed by the first four rows of Q_a is the incidence matrix of G except the signs of some rows.

To show that r is also an upper bound for the rank of Q_a, again we need to establish a relationship between B_a and Q_a, similar to that between B_a and A_a given in (2.7).

From Lemma 2.5, we know that the number of edges common to a cut and a circuit is always even. If a cut has $2k$ edges in common with a circuit, then it is not difficult to see that k of these edges have the same relative orientation in the cut and in the circuit, and the other k have one orientation in the cut and the opposite orientation in the circuit (Problems 2.6 and 2.7). Thus, we have

THEOREM 2.6: If the columns of the matrices B_a and Q_a of a directed graph are arranged in the same edge order, then

$$Q_a B_a' = 0 \quad \text{and} \quad B_a Q_a' = 0, \tag{2.16}$$

where the prime indicates the transpose of a matrix.

Since the rank of B_a is m and since $B_a Q_a' = 0$, it follows that there are at most $r\ (=b-m)$ linearly independent columns in Q_a', so we have

THEOREM 2.7: The rank of the cut matrix of a directed graph G is equal to the rank of G.

Since the rank of Q_a is now established as r, there is no need to write down all the rows of Q_a. We shall reserve the symbol Q to represent a submatrix of Q_a, of order $r \times b$ and of rank r. The matrix Q is called a *basis cut matrix* of G. Like the basis circuit matrix, a basis cut matrix contains all the information that is contained in Q_a.

Earlier, we have indicated how a cut of G can be used to partition its node set. In the following, we shall show how this partitioning of nodes of G can be done by means of a tree. Here we have implicitly assumed that G is connected. For unconnected G, the extension is straightforward, i.e., we apply the technique to each of the components of G (Problem 2.13).

Let t be a tree of G, and let e be a branch of t. Since t is connected and contains no circuits, the removal of e from t results in a subgraph consisting of two components. If V_1 and V_2 are the node sets of these two components, then V_1 and V_2 are mutually exclusive and together include all the nodes of G. Therefore, the branch e of t defines a partition of the nodes of G in a unique way.

The subgraph consisting of the edges each of which is incident with one of its two endpoints in V_1 and the other in V_2 is a cutset of G. This cutset contains only one branch of t, namely e, and some chords in the cotree (with respect to t). Obviously, there are r such cutsets.

DEFINITION 2.16: *Fundamental cutsets* (*f-cutsets*). The *fundamental cutsets* or simply *f-cutsets* of a connected directed graph with respect to a tree t are the r cutsets in which each cutset includes only one branch of t. The *f*-cutset orientation is chosen to agree with that of the defining branch.

DEFINITION 2.17: *Fundamental cutset matrix* (*f-cutset matrix*). For a connected directed graph G, the submatrix of the cut matrix of G corresponding to a set of r *f*-cutsets in G is called a *fundamental cutset matrix* or simply *f-cutset matrix* of G. The *f*-cutset matrix is denoted by the symbol $\boldsymbol{Q}_f$.

Like the *f*-circuit matrix, if the edges of G are numbered in such a way that the last r columns of $\boldsymbol{Q}_f$ correspond to the branches of t and if the *f*-cutsets are numbered correspondingly (or equivalently, rearranging the rows and columns of $\boldsymbol{Q}_f$), the *f*-cutset matrix $\boldsymbol{Q}_f$ can be partitioned as

$$\boldsymbol{Q}_f = [\boldsymbol{Q}_{f11} \quad \boldsymbol{U}_r], \tag{2.17}$$

where $\boldsymbol{U}_r$ is the identity matrix of order r. Thus, the *f*-cutset matrix is also a basis cut matrix of G.

As an illustration, consider the directed graph G given in fig. 2.9. Let us choose a tree t consisting of the edges e_4, e_5, e_6, and e_7. Then the *f*-cutsets are the cutsets 1, 2, 3, and 4 as shown in fig. 2.9. The *f*-cutset matrix is given by

$$\boldsymbol{Q}_f = \begin{array}{c} \\ 1 \\ 2 \\ 3 \\ 4 \end{array} \begin{array}{c} e_1 \quad e_2 \quad e_3 \quad e_4 \quad e_5 \quad e_6 \quad e_7 \\ \left[\begin{array}{ccc|cccc} -1 & -1 & 0 & 1 & 0 & 0 & 0 \\ -1 & -1 & -1 & 0 & 1 & 0 & 0 \\ -1 & 0 & -1 & 0 & 0 & 1 & 0 \\ 0 & 0 & 0 & 0 & 0 & 0 & 1 \end{array}\right] \end{array}. \tag{2.18}$$

The *f*-circuit matrix with respect to t in G is given by

$$\boldsymbol{B}_f = \begin{array}{c} \\ 1 \\ 2 \\ 3 \end{array} \begin{array}{c} e_1 \quad e_2 \quad e_3 \quad e_4 \quad e_5 \quad e_6 \quad e_7 \\ \left[\begin{array}{ccccccc} 1 & 0 & 0 & 1 & 1 & 1 & 0 \\ 0 & 1 & 0 & 1 & 1 & 0 & 0 \\ 0 & 0 & 1 & 0 & 1 & 1 & 0 \end{array}\right] \end{array}. \tag{2.19}$$

Thus, we have $\boldsymbol{Q}_f \boldsymbol{B}_f' = 0$.

Unless otherwise specified, throughout the chapter we shall assume that the columns of the cut matrix and the circuit matrix of a directed graph are always arranged in the same edge order. In particular, we assume B_f and Q_f are given in the forms as shown in (2.5) and (2.17), respectively.

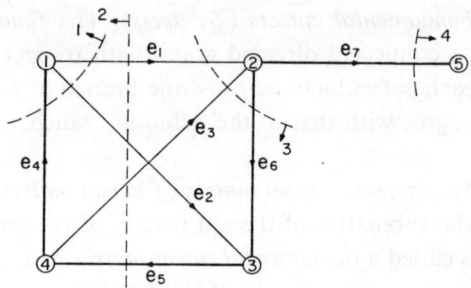

Fig. 2.9. A directed graph to illustrate the f-cutset matrix.

Since the rows of Q and Q_f are the two bases for the vector space spanned by the rows of Q_a, it follows that one can be obtained from the other by a nonsingular transformation.

LEMMA 2.6: There is a nonsingular matrix D of order r whose elements are 1, -1, or 0 such that
$$Q = DQ_f. \tag{2.20}$$

THEOREM 2.8: A square submatrix of a basis cut matrix Q of a connected directed graph is a major submatrix if, and only if, the columns of this submatrix correspond to the branches of a tree of the directed graph.

Proof. From Lemma 2.6, it follows that there exists a nonsingular matrix D of order r such that $Q = DA$. Hence, by Theorem 2.2, we have the desired result.

COROLLARY 2.3: There exists a one-to-one correspondence between trees of a connected directed graph G and the major submatrices of a basis cut matrix Q of G.

As an example, the set T of trees of the directed graph G of fig. 2.9 is given by

$$T = \{e_2e_5e_6e_7, e_1e_4e_5e_7, e_3e_4e_5e_7, e_1e_2e_3e_7, e_2e_3e_6e_7, e_2e_3e_4e_7,$$
$$e_4e_5e_6e_7, e_1e_5e_6e_7, e_2e_4e_6e_7, e_3e_4e_6e_7, e_1e_2e_4e_7, e_1e_2e_5e_7,$$
$$e_1e_3e_5e_7, e_1e_3e_6e_7, e_2e_3e_5e_7, e_1e_4e_6e_7\}. \tag{2.21}$$

The f-cutset matrix Q_f of G is given in (2.18). It is easy to check that each of the major submatrices of Q_f consists of columns corresponding to a term in (2.21).

1.4. Interrelationships among the matrices A, B_f, and Q_f

As mentioned in § 1.1, a basis incidence matrix A of a directed graph completely characterizes the directed graph. In other words, if A is given the directed graph can be drawn in a straightforward manner. Thus, it is logical to expect that we should be able to derive formulas relating the matrices A, B_f, and Q_f.

THEOREM 2.9: If the columns of A, B_f, and Q_f of a directed graph G are arranged in the order of chords and branches for the tree t defining the f-circuit matrix B_f and f-cutset matrix Q_f such that

$$A = [A_{11} \quad A_{12}], \tag{2.22a}$$

$$B_f = [U_m \quad B_{f12}], \tag{2.22b}$$

$$Q_f = [Q_{f11} \quad U_r], \tag{2.22c}$$

then we have

$$B_f = [U_m \quad -A'_{11}A'^{-1}_{12}], \tag{2.23a}$$

$$Q_f = A_{12}^{-1}A = [-B'_{f12} \quad U_r], \tag{2.23b}$$

where the prime denotes the transpose of a matrix, and U_m and U_r are identity matrices of orders m and r, respectively.

The above theorem follows directly from the identities given in (2.7) and (2.16). The details are left as an exercise (Problem 2.8). Since $Q_{f11} = -B'_{f12}$ we have the following simple interpretation which relates the f-cutsets and the f-circuits of G (Problem 2.11).

COROLLARY 2.4: Let t be a tree of a connected directed graph G, and also let e be a branch of t. Then the f-cutset determined by e contains exactly those chords of G for which e is in each of the f-circuits determined by these chords.

As an example, in fig. 2.9 let $t = e_4 e_5 e_6 e_7$. Also let $e = e_4$. The f-circuits defined by the chords e_1, e_2, and e_3 and containing the branch e_4 are $e_2 e_5 e_4$ and $e_1 e_6 e_5 e_4$. Thus, the f-cutset determined by the branch e_4 contains exactly those chords in $e_2 e_5 e_4$ and $e_1 e_6 e_5 e_4$, i.e., e_2 and e_1, so the f-cutset consists of the edges e_4, e_1, and e_2.

Next, let us verify the formulas given in (2.23) for the directed graph G as shown in fig. 2.9. A basis incidence matrix A of G is given by

$$A = \begin{matrix} 1 \\ 2 \\ 3 \\ 4 \end{matrix} \begin{bmatrix} \overset{e_1}{1} & \overset{e_2}{1} & \overset{e_3}{0} & | & \overset{e_4}{-1} & \overset{e_5}{0} & \overset{e_6}{0} & \overset{e_7}{0} \\ -1 & 0 & -1 & | & 0 & 0 & 1 & 1 \\ 0 & -1 & 0 & | & 0 & 1 & -1 & 0 \\ 0 & 0 & 1 & | & 1 & -1 & 0 & 0 \end{bmatrix}. \quad (2.24)$$

Now to verify Theorem 2.9 we have

$$A_{12}^{-1}A_{11} = \begin{bmatrix} -1 & 0 & 0 & 0 \\ 0 & 0 & 1 & 1 \\ 0 & 1 & -1 & 0 \\ 1 & -1 & 0 & 0 \end{bmatrix}^{-1} \begin{bmatrix} 1 & 1 & 0 \\ -1 & 0 & -1 \\ 0 & -1 & 0 \\ 0 & 0 & 1 \end{bmatrix}$$

$$= \begin{bmatrix} -1 & 0 & 0 & 0 \\ -1 & 0 & 0 & -1 \\ -1 & 0 & -1 & -1 \\ 1 & 1 & 1 & 1 \end{bmatrix} \begin{bmatrix} 1 & 1 & 0 \\ -1 & 0 & -1 \\ 0 & -1 & 0 \\ 0 & 0 & 1 \end{bmatrix} = \begin{bmatrix} -1 & -1 & 0 \\ -1 & -1 & -1 \\ -1 & 0 & -1 \\ 0 & 0 & 0 \end{bmatrix}.$$

(2.25)

Hence,

$$B_f = [U_3 \quad -A'_{11}A'^{-1}_{12}] = \begin{bmatrix} 1 & 0 & 0 & | & 1 & 1 & 1 & 0 \\ 0 & 1 & 0 & | & 1 & 1 & 0 & 0 \\ 0 & 0 & 1 & | & 0 & 1 & 1 & 0 \end{bmatrix} \quad (2.26)$$

and

$$Q_f = [A_{12}^{-1}A_{11} \quad U_4] = \begin{bmatrix} -1 & -1 & 0 & | & 1 & 0 & 0 & 0 \\ -1 & -1 & -1 & | & 0 & 1 & 0 & 0 \\ -1 & 0 & -1 & | & 0 & 0 & 1 & 0 \\ 0 & 0 & 0 & | & 0 & 0 & 0 & 1 \end{bmatrix}, \quad (2.27)$$

which are indeed the same matrices as given in (2.18) and (2.19).

From the example we have just worked out, one soon realizes that the inversion of a major submatrix A_{12} of A is not a trivial matter. In order to overcome this difficulty, BRANIN [1962] presented an interesting interpretation of the inverse of A_{12}.

DEFINITION 2.18: *Node-to-datum path matrix.* For an n-node tree t, the *node-to-datum path matrix*, denoted by the symbol P, of t with reference node n is a matrix of order $(n-1)$ such that if $P = [p_{ij}]$, then

$p_{ij} = 1$ if branch e_i of t is contained in the unique path connected between the nodes j and n, and is directed toward the node n in the path,

$p_{ij} = -1$ if branch e_i of t is contained in the unique path connected between the nodes j and n, and is directed away from the node n,
$p_{ij} = 0$ if branch e_i of t is not contained in the unique path connected between the nodes j and n.

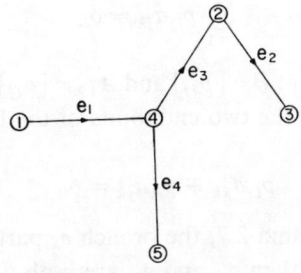

Fig. 2.10. A tree to illustrate the node-to-datum path matrix.

As an illustration, consider the tree t as shown in fig. 2.10. The node-to-datum path matrix P of t with reference node 5 is given by

$$P = \begin{array}{c} \\ e_1 \\ e_2 \\ e_3 \\ e_4 \end{array} \begin{array}{cccc} 1 & 2 & 3 & 4 \\ \left[\begin{array}{cccc} 1 & 0 & 0 & 0 \\ 0 & 0 & -1 & 0 \\ 0 & -1 & -1 & 0 \\ 1 & 1 & 1 & 1 \end{array}\right] \end{array}. \qquad (2.28)$$

LEMMA 2.7: *The nonzero entries of any row of a node-to-datum path matrix must all have the same sign.*

Proof. Let P be the node-to-datum path matrix of a tree t with reference node n. Also let e_k be a branch of t. Since e_k is a cutset of t, the removal of e_k partitions the nodes of t into two sets V_1 and V_2. For convenience, let V_2 contain the node n and $P = [p_{ij}]$. Then p_{kj} is nonzero if and only if node j is in V_1. Since any path originating from a node in V_1 must go through e_k in the same relative orientation, it follows that all the nonzero entries in the kth row of P must have the same sign. So the lemma is proved.

THEOREM 2.10: *The inverse of a major submatrix of the basis incidence matrix A of a connected directed graph with reference node n is equal to the node-to-datum path matrix of the tree corresponding to the columns of the major submatrix with n used as the reference node of the tree.*

Proof. Let A_{12} be the submatrix of A corresponding to the tree t. Also let P

be the node-to-datum path matrix of t with reference node n. Then we shall show that

$$P = A_{12}^{-1}, \qquad (2.29)$$

i.e.,

$$\sum_{j=1}^{n-1} p_{ij}a_{jk} = \delta_{ik} \qquad (2.30)$$

for $i, k = 1, 2, \ldots, n-1$, where $P = [p_{ij}]$ and $A_{12} = [a_{ij}]$, and δ_{ik} is the Kronecker delta. If nodes x and y are the two endpoints of the branch e_k, then (2.30) can be reduced to

$$p_{ix}a_{xk} + p_{iy}a_{yk} = \delta_{ik}. \qquad (2.31)$$

Like in the proof of Lemma 2.7, the branch e_k partitions the nodes of t into two sets V_1 and V_2. If $i \neq k$, then p_{ix} and p_{iy} are both nonzero or both zero since the two endpoints of the branch e_i are both in V_1 or in V_2. Since from Lemma 2.7 p_{ix} and p_{iy} must have the same sign, and since a_{xk} and a_{yk} are of opposite signs, it follows that

$$p_{ix}a_{xk} + p_{iy}a_{yk} = 0 \qquad (2.32)$$

for $i \neq k$. If $i = k$, two cases are considered: $p_{kx} = 0$ and $p_{kx} = \pm 1$. For $p_{kx} = 0$, we have $p_{ky} = \pm 1$ and $a_{yk} = \pm 1$. For $p_{kx} = \pm 1$, we have $p_{ky} = 0$ and $a_{xk} = \pm 1$. Thus,

$$p_{kx}a_{xk} + p_{ky}a_{yk} = 1. \qquad (2.33)$$

So the theorem is proved.

As an example, in fig. 2.10 the basis incidence matrix A with node 5 used as the reference node is given by

$$A = \begin{matrix} & \begin{matrix} e_1 & e_2 & e_3 & e_4 \end{matrix} \\ \begin{matrix} 1 \\ 2 \\ 3 \\ 4 \end{matrix} & \begin{bmatrix} 1 & 0 & 0 & 0 \\ 0 & 1 & -1 & 0 \\ 0 & -1 & 0 & 0 \\ -1 & 0 & 1 & 1 \end{bmatrix} \end{matrix}. \qquad (2.34)$$

It is easy to check that the matrix P given in (2.28) is the inverse of A. As another example, the matrix A_{12}^{-1} given in (2.25) is the node-to-datum path matrix of the tree $e_4 e_5 e_6 e_7$ in the directed graph given in fig. 2.9 with reference node 5.

So far we have presented formulas which express B_f and Q_f in terms of the submatrices of A. The reverse process of expressing A in terms of the submatrices of B_f or Q_f is difficult; it amounts to realizing a directed graph with a prescribed f-circuit or f-cutset matrix. The problem is very important because it is closely related to the synthesis of conventional electrical networks and combinational switching circuits by algebraic methods. Its complete characterizations were first established by TUTTE [1958, 1959].

1.5. *Vector spaces associated with the matrices* B_a *and* Q_a

The last major algebraic concept to be introduced in this section is that of a *linear vector space*. An abstract algebraic discussion of linear vector spaces may by found in any text on modern algebra. (See, for example, MACLANE and BIRKHOFF [1967].) The application of linear vector spaces to linear graphs has been considered by WHITNEY [1935], DOYLE [1955], GOULD [1958], and CHEN [1970b]. In this section, we shall only discuss those aspects of the linear vector spaces which are closely related to the electrical network problem.

DEFINITION 2.19: *B-space.* For a directed graph G, the linear vector space, denoted by the symbol $\mathscr{V}_B$, spanned by the columns of the transpose of its circuit matrix B_a over the real field is called the *B-space* of G.

DEFINITION 2.20: *Q-space.* For a directed graph G, the linear vector space, denoted by the symbol $\mathscr{V}_Q$, spanned by the columns of the transpose of its cut matrix Q_a over the real field is called the *Q-space* of G.

Obviously, $\mathscr{V}_B$ is an m-dimensional space, and the columns of the transpose of any B of G will serve as a basis for $\mathscr{V}_B$. Similarly, $\mathscr{V}_Q$ is an r-dimensional space, and the columns of the transpose of any Q of G will serve as a basis for $\mathscr{V}_Q$.

The following theorem is a direct consequence of (2.16).

THEOREM 2.11: *The B-space and the Q-space of a directed graph G form orthogonal subspaces of a b-dimensional space.*

Thus, if $\mathscr{V}_G$ is the b-dimensional space discussed above, then any vector in $\mathscr{V}_G$ is a linear combination of the basis vectors of $\mathscr{V}_B$ and $\mathscr{V}_Q$, since $\mathscr{V}_B$ is of dimension m, $\mathscr{V}_Q$ is of dimension r, and $b=m+r$. It follows that the basis vectors of $\mathscr{V}_B$ together with the basis vectors of $\mathscr{V}_Q$ constitute a basis for $\mathscr{V}_G$. In other words, $\mathscr{V}_G$ is the direct sum of the subspaces $\mathscr{V}_B$ and $\mathscr{V}_Q$.

COROLLARY 2.5: *The subspaces $\mathscr{V}_B$ and $\mathscr{V}_Q$ form orthogonal complements of the space $\mathscr{V}_G$.*

THEOREM 2.12: *Any vector X in $\mathscr{V}_G$ can be decomposed uniquely as the sum of two vectors X_B and X_Q with X_B in $\mathscr{V}_B$ and X_Q in $\mathscr{V}_Q$.*

Proof. From Corollary 2.5, it is obvious that X can be expressed as

$$X = B'C_1 + Q'C_2, \qquad (2.35)$$

where C_1 is an m-vector and C_2 is an r-vector, and the prime denotes the transpose of a matrix. Now if we let

$$X_B = B'C_1, \qquad (2.36a)$$
$$X_Q = Q'C_2, \qquad (2.36b)$$

we have the decomposition. To show this is unique, let

$$X = B'C_3 + Q'C_4, \qquad (2.37)$$

where C_3 is an m-vector and C_4 an r-vector. Then from (2.35) and (2.37) we have

$$B'(C_1 - C_3) + Q'(C_2 - C_4) = 0. \qquad (2.38)$$

Since the columns of B' and Q' are linearly independent, it follows that

$$C_1 = C_3 \quad \text{and} \quad C_2 = C_4. \qquad (2.39)$$

Hence the theorem is proved.

COROLLARY 2.6: There are $(2^m - 1)$ nonzero vectors in $\mathscr{V}_B$ with components 1, -1, and 0, each of which corresponds to a circuit or an edge-disjoint union of circuits of G; there are $(2^r - 1)$ nonzero vectors in $\mathscr{V}_Q$ with components 1, -1, and 0, each of which corresponds to a cut of G.

COROLLARY 2.7: Any vector X in $\mathscr{V}_G$ which satisfies $QX = 0$ is contained in $\mathscr{V}_B$, and any vector Y in $\mathscr{V}_G$ which satisfies $BY = 0$ is contained in $\mathscr{V}_Q$.

We emphasize that in Corollary 2.6 only those vectors in $\mathscr{V}_B$ ($\mathscr{V}_Q$) with components 1, -1, and 0 correspond to circuits or edge-disjoint unions of circuits (cuts) of G; others do not. As a matter of fact, both $\mathscr{V}_B$ and $\mathscr{V}_Q$ contain an infinite number of vectors.

§ 2. The electrical network problem

Electrical network theory is formulated in terms of two variables, called *current* $i(t)$ and *voltage* $v(t)$, associated with each network element called a *(network) branch*. The current and voltage variables are real functions of time t. The voltages $v(t)$ may be thought of as "across-variables", in the sense that they exist across the two terminals of the branch, while currents may be thought of as "through-variables", in the sense that they flow in at one terminal through the branch and out the other terminal. Thus, the voltage $v(t)$ and the current $i(t)$ are both oriented variables, and we use the edge-orientation arrows of a directed graph to define the positive orientations of the voltages and currents. These orientations are usually called the voltage and current *references*. The convention adopted here is as follows: If edge $e_k = (i, j)$, then the voltage $v_k(t)$ and current $i_k(t)$ associated with the edge e_k are taken to be positive if they are going from node i to node j, and negative if they are going from j to i. In other words, the voltage $+$-references are assumed to be at the tails of the current-reference arrows.

An *(ideal) voltage generator* is a branch in which the voltage is specified in the form of an imposed function of time, and an *(ideal) current generator* is a branch in which the current is specified in the form of an imposed function of time. Together, they are referred to as *sources*. For other branches, each electrical element imposes relationships between its current and voltage variables, the precise nature of which depends upon the type of electrical element.

In this chapter, we shall only consider time-invariant and linear electrical elements. For convenience, we take Laplace transforms of all our time-dependent variables, so that only systems of linear algebraic equations are involved. With these definitions, a formal statement of the electrical network problem may now be given.

DEFINITION 2.21: *Electrical network*. An *electrical network* is a directed graph G with two functions $v(s)$ and $i(s)$ of a complex variable s associated with each edge of G, satisfying the following three postulates:
1. Kirchhoff's current law (KIRCHHOFF [1847]):

$$A_a I(s) = 0, \qquad (2.40)$$

where $I(s)$ is a b-vector denoting the currents of the edges, and is called the *branch-current vector* of G.

2. Kirchhoff's voltage law (KIRCHHOFF [1847]):

$$B_a V(s) = 0, \qquad (2.41)$$

where $V(s)$ is a b-vector denoting the voltages of the edges, and is called the *branch-voltage vector* of G.

3. Element defining equations or generalized Ohm's law (OHM [1827]):

$$V(s) = E(s) + Z(s) I(s) \qquad (2.42a)$$

or

$$I(s) = J(s) + Y(s) V(s), \qquad (2.42b)$$

where $Z(s)$ and $Y(s)$ are given matrices of order b, and are called the *branch-impedance* and the *branch-admittance matrices*, respectively; $E(s)$ and $J(s)$ are given b-vectors denoting the sources and the initial conditions, and are called the *branch voltage-source* and the *branch current-source vectors* of G, respectively.

Here we have implicitly assumed that if (2.42a) is used, then G has no current generators, and if (2.42b) is used, then G has no voltage generators. This is not a serious restriction since a voltage generator can easily be transformed into an equivalent current generator, and vice versa (see, for example, SESHU and BALABANIAN [1959]). Also we assume that the columns of A_a and B_a and the rows of $I(s)$ and $V(s)$ are arranged in the same edge order.

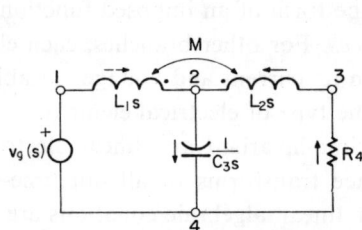

Fig. 2.11. A network diagram.

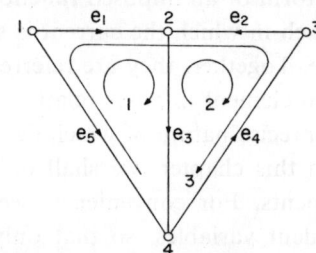

Fig. 2.12. The corresponding directed graph.

As an illustration, consider the electrical network diagram N as shown in fig. 2.11. The associated directed graph G of N is given in fig. 2.12. Then the branch-current vector $I(s)$ is required to satisfy the Kirchhoff's current law:

$$A_a I(s) = \begin{array}{c} \\ 1 \\ 2 \\ 3 \\ 4 \end{array} \begin{array}{c} e_1 \quad e_2 \quad e_3 \quad e_4 \quad e_5 \\ \begin{bmatrix} 1 & 0 & 0 & 0 & 1 \\ -1 & -1 & 1 & 0 & 0 \\ 0 & 1 & 0 & -1 & 0 \\ 0 & 0 & -1 & 1 & -1 \end{bmatrix} \end{array} \begin{bmatrix} i_1(s) \\ i_2(s) \\ i_3(s) \\ i_4(s) \\ i_5(s) \end{bmatrix} = \begin{bmatrix} 0 \\ 0 \\ 0 \\ 0 \end{bmatrix}. \quad (2.43)$$

The branch-voltage vector $V(s)$ is required to satisfy the Kirchhoff's voltage law:

$$B_a V(s) = \begin{array}{c} \\ 1 \\ 2 \\ 3 \end{array} \begin{array}{c} e_1 \quad e_2 \quad e_3 \quad e_4 \quad e_5 \\ \begin{bmatrix} 1 & 0 & 1 & 0 & -1 \\ 0 & -1 & -1 & -1 & 0 \\ 1 & -1 & 0 & -1 & -1 \end{bmatrix} \end{array} \begin{bmatrix} v_1(s) \\ v_2(s) \\ v_3(s) \\ v_4(s) \\ v_5(s) \end{bmatrix} = \begin{bmatrix} 0 \\ 0 \\ 0 \\ 0 \\ 0 \end{bmatrix}. \quad (2.44)$$

§2 The electrical network problem

The branch-voltage and the branch-current vectors $V(s)$ and $I(s)$ are related by the branch-impedance matrix $Z(s)$ and the branch voltage-source vector $E(s)$:

$$\begin{bmatrix} v_1(s) \\ v_2(s) \\ v_3(s) \\ v_4(s) \\ v_5(s) \end{bmatrix} = \begin{bmatrix} L_1 s & Ms & 0 & 0 & 0 \\ Ms & L_2 s & 0 & 0 & 0 \\ 0 & 0 & \dfrac{1}{C_3 s} & 0 & 0 \\ 0 & 0 & 0 & R_4 & 0 \\ 0 & 0 & 0 & 0 & 0 \end{bmatrix} \begin{bmatrix} i_1(s) \\ i_2(s) \\ i_3(s) \\ i_4(s) \\ i_5(s) \end{bmatrix} + \begin{bmatrix} 0 \\ 0 \\ 0 \\ 0 \\ v_g(s) \end{bmatrix}. \qquad (2.45)$$

We notice that the rows and columns in $Z(s)$ corresponding to the edges of voltage generators consist only of zeros. Hence, $Z(s)$ in general may be singular. This is similarly valid for $Y(s)$.

For a given electrical network, the problem is then to find the vectors $V(s)$ and $I(s)$ such that (2.40)–(2.42) hold true. Theoretically, the process is straightforward; it amounts to solving a system of algebraic equations. In practice, we seldom go through the normal process in getting the solutions because the number of equations involved is rather large. For example, the simple network in fig. 2.11, we would need to solve twelve simultaneous equations. From our earlier experience, we know that not all the equations given in (2.40) and (2.41) are linearly independent. Thus, our first step is to try to remove as many redundant equations as possible, and then to organize procedures for further reduction of the number of equations.

Since the properties of the incidence and circuit matrices are known, it follows immediately from our previous discussions (Theorems 2.1 and 2.4) that there are only r $(=n-c)$ linearly independent Kirchhoff's current equations, and m $(=b-n+c)$ linearly independent Kirchhoff's voltage equations.

Since a basis incidence matrix A and a basis cut matrix Q differ only by a nonsingular transformation, a generalization of Kirchhoff's current law may now be stated in terms of the rows of the cut matrix (Problem 2.14).

COROLLARY 2.8: $A_a I(s) = 0$ if and only if

$$Q_a I(s) = 0. \qquad (2.46)$$

Using the interpretation of a cut given in § 1.3, the physical meaning of (2.46) is not difficult to find. Since a cut of a directed graph G corresponds to a set of wires in the physical network, the cutting of these wires will separate the network into at least two connected parts. It is a familiar fact that the sum of the currents in these wires with references taken into account is zero.

COROLLARY 2.9: If the columns of Q and B of a connected electrical network are arranged in the order of chords and branches for a tree t such that

$$Q = [Q_{11} \quad Q_{12}], \qquad (2.47)$$

$$B = [B_{11} \quad B_{12}], \qquad (2.48)$$

then

$$I_2(s) = - Q_{12}^{-1} Q_{11} I_1(s), \qquad (2.49)$$

$$V_1(s) = - B_{11}^{-1} B_{12} V_2(s), \qquad (2.50)$$

where $I_1(s)$ and $I_2(s)$ are the subvectors of the branch-current vector $I(s)$ corresponding to the chords and branches defined by t, respectively; and $V_1(s)$ and $V_2(s)$ are the subvectors of the branch-voltage vector $V(s)$ corresponding to the chords and branches defined by t, respectively.

The corollary follows directly from Theorems 2.5 and 2.8; its proof is left as an exercise (Problem 2.15).

The implication of Corollary 2.9 is that if the current functions of the chords are arbitrarily chosen and if the current functions of the branches are obtained by (2.49), then the set of current functions will always satisfy Kirchhoff's current law. Similarly, if the voltage functions of the branches are arbitrarily chosen and if the voltage functions of the chords are obtained by (2.50), then the set of voltage functions will always satisfy Kirchhoff's voltage law. Thus, we have m degrees of freedom in choosing the currents, and r degrees of freedom in choosing the voltages if they are only required to satisfy Kirchhoff's two laws. However, we have b $(=m+r)$ more constraints specified by (2.42). In other words, we have $2b$ unknowns in $I(s)$ and $V(s)$, and also we have $2b$ equations (after deleting the redundant ones in (2.40) and (2.41)). If these equations are consistent, we are assured of a solution; if, furthermore, they are linearly independent, then the solution will be unique. However, for most practical networks the number is still too large to handle. In the following section, we shall show how these variables can be organized, so that the number of equations involved will be greatly reduced.

For convenience, from here on we shall drop the complex variable s in writing the variables and vectors with the understanding that they stand for the Laplace transforms of the real variables and vectors.

§ 3. Solutions of the electrical network problem

To summarize what we have discussed in the preceding section, the three fundamental systems of equations given in (2.40)–(2.42) are equivalent to the following three systems of equations:

$$QI = 0, \tag{2.51}$$

$$BV = 0, \tag{2.52}$$

$$V = \begin{bmatrix} V_s \\ V_p \end{bmatrix} = \begin{bmatrix} E_s \\ E_p \end{bmatrix} + \begin{bmatrix} 0 & 0 \\ 0 & Z_{pp} \end{bmatrix} \begin{bmatrix} I_s \\ I_p \end{bmatrix} = E + ZI \tag{2.53a}$$

or

$$I = \begin{bmatrix} I_s \\ I_p \end{bmatrix} = \begin{bmatrix} J_s \\ J_p \end{bmatrix} + \begin{bmatrix} 0 & 0 \\ 0 & Y_{pp} \end{bmatrix} \begin{bmatrix} V_s \\ V_p \end{bmatrix} = J + YV, \tag{2.53b}$$

where the subscript s is used to denote the voltages or currents of the edges corresponding to the sources, and the subscript p for all others.

3.1. *Branch-current and branch-voltage systems of equations*

If we substitute (2.53a) into (2.52), and then combine with those in (2.51), we get the *branch-current system of equations*,

$$\begin{bmatrix} BZ \\ A \end{bmatrix} I = \begin{bmatrix} -BE \\ 0 \end{bmatrix}. \tag{2.54}$$

The system is presented here because of its historical importance. Early workers, including Kirchhoff himself, used this system in their work. The system is a set of b equations in b unknown branch currents, which can be solved in the usual manner.

Similarly, if we substitute (2.53b) into (2.51), and then combine with those in (2.52), we get the *branch-voltage system of equations*,

$$\begin{bmatrix} AY \\ B \end{bmatrix} V = \begin{bmatrix} -AJ \\ 0 \end{bmatrix}. \tag{2.55}$$

The system is a set of b equations in b unknown branch voltages. These systems are not of much practical use, because the number of equations involved is still large and there are other simpler techniques available.

3.2. *Loop system of equations*

The loop and nodal systems of equations were originally devised by MAXWELL [1892] for solving the electrical network problem effectively. They amount to a systematic organization of the variables, so that the number of equations needed can be reduced. In this section, we shall discuss the formulation of the loop system. The formulation of the nodal system will be presented in the following section.

Since any I that satisfies (2.51) must be contained in the B-space, and since the columns of B' are a basis of the B-space, it follows that there exists an m-vector I_m such that
$$I = B'I_m. \tag{2.56}$$

The vector I_m is called the *loop-current vector*, and the equations are referred to as the *loop transformation*. The reason for the name *loop current* is that the elements of I_m can be interpreted physically as a set of circulating currents around the set of circuits defined by the rows of B. The term *mesh current* is also frequently used in the literature for the loop current. In this book, we shall reserve the term *mesh current* for planar networks which have the appearance of the meshes in a fish net.

In mathematics, the columns of B' are also called a *reference system* in the B-space, and the elements in I_m are the *coordinates* of I with respect to this reference system. Thus, our problem is to obtain the coordinates of I so that it will satisfy (2.52) and (2.53a) simultaneously. However, before we do this, let us summarize the above result as

COROLLARY 2.10: $QI = 0$ if, and only if, there exists an m-vector I_m such that
$$I = B'I_m. \tag{2.57}$$

If we use (2.57) in conjunction with (2.52) and (2.53a), we get the *loop system of equations*:
$$Z_m I_m = E_m, \tag{2.58}$$
where
$$Z_m = BZB', \tag{2.59a}$$
$$E_m = -BE. \tag{2.59b}$$

The coefficient matrix Z_m is referred to as the *loop-impedance matrix* of the network, and the vector E_m is called the *loop voltage-source vector*. (2.58) represents a system of m equations in m loop currents. Knowing the loop currents we can find the branch currents from (2.57) and the branch voltages from (2.53a). The time functions are found by inverting the Laplace transforms.

Apart from reducing the number of equations from $2b$ to m in the present formulation, it is significant to point out that the matrices Z_m and E_m, in most cases, can be written down directly from the network by inspection. The rules are usually simple (Problem 2.17), and can be found in many texts on modern network theory (see, for example, CARLIN and GIORDANO [1964]). Any electrical engineer would hardly go through the matrix multiplication of (2.59) to get the loop system of equations.

We shall illustrate the above by the following example.

Example 2.1: Consider the ladder network as shown in fig. 2.13. Its corresponding directed graph G is given in fig. 2.14. Thus, we have $m = 3$. The choice of the three circuits as indicated in fig. 2.14 satisfies the condition that B is a basis circuit matrix.

$$B = \begin{array}{c} \phantom{\begin{bmatrix}}\begin{matrix} e_1 & e_2 & e_3 & e_4 & e_5 & e_6 & e_7 & e_8 \end{matrix} \\ \begin{bmatrix} -1 & 1 & 0 & 1 & 0 & 0 & 0 & 0 \\ 0 & 0 & 1 & -1 & -1 & 0 & 0 & 0 \\ 0 & 0 & 0 & 0 & 1 & -1 & 1 & -1 \end{bmatrix} \end{array}, \quad (2.60)$$

$$Z = \begin{bmatrix} 0 & 0 & 0 & 0 & 0 & 0 & 0 & 0 \\ 0 & L_2 s & Ms & 0 & 0 & 0 & 0 & 0 \\ 0 & Ms & L_3 s & 0 & 0 & 0 & 0 & 0 \\ 0 & 0 & 0 & \dfrac{1}{C_4 s} & 0 & 0 & 0 & 0 \\ 0 & 0 & 0 & 0 & R_5 & 0 & 0 & 0 \\ 0 & 0 & 0 & g_m & 0 & 0 & 0 & 0 \\ 0 & 0 & 0 & 0 & 0 & 0 & R_7 & 0 \\ 0 & 0 & 0 & 0 & 0 & 0 & 0 & z_8 \end{bmatrix}, \quad (2.61)$$

and the transpose of E is given by

$$E' = [v_g \ \ 0 \ \ 0 \ \ 0 \ \ 0 \ \ 0 \ \ 0 \ \ 0]. \quad (2.62)$$

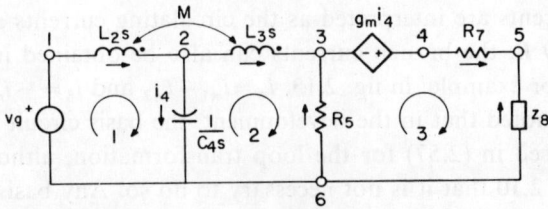

Fig. 2.13. A ladder network diagram.

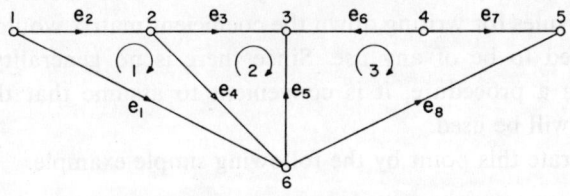

Fig. 2.14. The corresponding directed graph of the network of fig. 2.13.

Performing the indicated matrix multiplications of (2.59), we obtain the desired loop system of equations:

$$\begin{bmatrix} L_2 s + \dfrac{1}{C_4 s} & Ms - \dfrac{1}{C_4 s} & 0 \\ Ms - \dfrac{1}{C_4 s} & L_3 s + R_5 + \dfrac{1}{C_4 s} & -R_5 \\ -g_m & g_m - R_5 & R_5 + R_7 + z_8 \end{bmatrix} \begin{bmatrix} i_{m1} \\ i_{m2} \\ i_{m3} \end{bmatrix} = \begin{bmatrix} v_g \\ 0 \\ 0 \end{bmatrix}. \quad (2.63)$$

The loop-impedance matrix Z_m can also be obtained from the network by inspection, as a moment of study of the structure of the matrix Z_m and the three circuits in fig. 2.13 will show. Now if Z_m is nonsingular, the loop-current vector I_m can be computed by inverting the matrix Z_m. The branch currents and voltages are given by

$$\begin{bmatrix} i_1 \\ i_2 \\ i_3 \\ i_4 \\ i_5 \\ i_6 \\ i_7 \\ i_8 \end{bmatrix} = \begin{bmatrix} -i_{m1} \\ i_{m1} \\ i_{m2} \\ i_{m1} - i_{m2} \\ i_{m3} - i_{m2} \\ -i_{m3} \\ i_{m3} \\ -i_{m3} \end{bmatrix} \text{ and } \begin{bmatrix} v_1 \\ v_2 \\ v_3 \\ v_4 \\ v_5 \\ v_6 \\ v_7 \\ v_8 \end{bmatrix} = \begin{bmatrix} v_g \\ L_2 s i_{m1} + M s i_{m2} \\ M s i_{m1} + L_3 s i_{m2} \\ (i_{m1} - i_{m2})/C_4 s \\ (i_{m3} - i_{m2}) R_5 \\ (i_{m1} - i_{m2}) g_m \\ i_{m3} R_7 \\ -i_{m3} z_8 \end{bmatrix}. \quad (2.64)$$

If the loop currents are interpreted as the circulating currents around the circuits defined by B, the branch currents can also be obtained intuitively from the network. For example, in fig. 2.13, $i_4 = i_{m1} - i_{m2}$ and $i_6 = -i_{m3}$.

It should be noted that in the development, the basis circuit matrix used in (2.52) is also used in (2.57) for the loop transformation, although it is clear from Corollary 2.10 that it is not necessary to do so. Any basis circuit matrix can be used for the loop transformation. However, if two different basis circuit matrices are used, one for (2.52) and one for (2.57), the coefficient matrix of the loop system of equations will be asymmetric even for a reciprocal network, and the rules for writing down the coefficient matrix would be in general too complicated to be of any use. Since there is no generality achieved by following such a procedure, it is convenient to assume that the same basis circuit matrix will be used.

Let us illustrate this point by the following simple example.

Example 2.2: Consider the network and its corresponding directed graph G

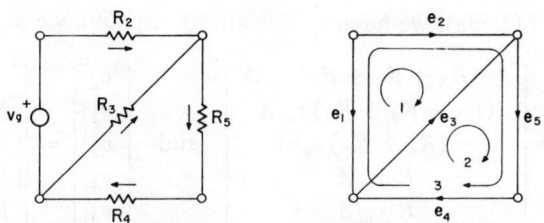

Fig. 2.15. A network and its corresponding directed graph.

as shown in fig. 2.15. Two basis circuit matrices B and B^* of G can easily be found, and they are given by

$$B = \begin{matrix} \\ 1 \\ 2 \end{matrix} \begin{matrix} e_1 & e_2 & e_3 & e_4 & e_5 \\ \begin{bmatrix} -1 & 1 & -1 & 0 & 0 \\ 0 & 0 & 1 & 1 & 1 \end{bmatrix} \end{matrix}, \qquad (2.65)$$

$$B^* = \begin{matrix} 1 \\ 3 \end{matrix} \begin{bmatrix} -1 & 1 & -1 & 0 & 0 \\ -1 & 1 & 0 & 1 & 1 \end{bmatrix}. \qquad (2.66)$$

The transpose E' of the branch voltage-source vector E is given by

$$E' = [v_g \ 0 \ 0 \ 0 \ 0]. \qquad (2.67)$$

The branch-impedance matrix Z of the network is obtained as follows:

$$Z = \begin{bmatrix} 0 & 0 & 0 & 0 & 0 \\ 0 & R_2 & 0 & 0 & 0 \\ 0 & 0 & R_3 & 0 & 0 \\ 0 & 0 & 0 & R_4 & 0 \\ 0 & 0 & 0 & 0 & R_5 \end{bmatrix}. \qquad (2.68)$$

If the basis circuit matrix B given in (2.65) is used both for (2.52) and (2.57), the loop system of equations is given by

$$\begin{bmatrix} R_2 + R_3 & -R_3 \\ -R_3 & R_3 + R_4 + R_5 \end{bmatrix} \begin{bmatrix} i_{m1} \\ i_{m2} \end{bmatrix} = \begin{bmatrix} v_g \\ 0 \end{bmatrix}. \qquad (2.69)$$

The loop currents can now be obtained by inverting the coefficient matrix.

$$i_{m1} = (R_3 + R_4 + R_5) v_g / \Delta, \qquad (2.70a)$$

$$i_{m2} = R_3 v_g / \Delta, \qquad (2.70b)$$

where

$$\Delta = R_2 (R_3 + R_4 + R_5) + R_3 (R_4 + R_5). \qquad (2.70c)$$

From (2.57) and (2.53a), we have

$$\begin{bmatrix} i_1 \\ i_2 \\ i_3 \\ i_4 \\ i_5 \end{bmatrix} = \begin{bmatrix} -(R_3 + R_4 + R_5)\,v_g/\Delta \\ (R_3 + R_4 + R_5)\,v_g/\Delta \\ -(R_4 + R_5)\,v_g/\Delta \\ R_3 v_g/\Delta \\ R_3 v_g/\Delta \end{bmatrix} \quad \text{and} \quad \begin{bmatrix} v_1 \\ v_2 \\ v_3 \\ v_4 \\ v_5 \end{bmatrix} = \begin{bmatrix} v_g \\ R_2 i_2 \\ R_3 i_3 \\ R_4 i_4 \\ R_5 i_5 \end{bmatrix}. \quad (2.71)$$

Now if B is used in (2.52) and if the transpose $B^{*\prime}$ of B^* is used in (2.57), the loop system of equations becomes

$$BZB^{*\prime}I_m = -BE. \quad (2.72)$$

Performing the needed matrix multiplications, we have

$$\begin{bmatrix} R_2 + R_3 & R_2 \\ -R_3 & R_4 + R_5 \end{bmatrix} \begin{bmatrix} i_{m1} \\ i_{m3} \end{bmatrix} = \begin{bmatrix} v_g \\ 0 \end{bmatrix}. \quad (2.73)$$

The loop currents are given by

$$i_{m1} = (R_4 + R_5)\,v_g/\Delta, \quad (2.74\text{a})$$

$$i_{m3} = R_3 v_g/\Delta, \quad (2.74\text{b})$$

where

$$\Delta = (R_2 + R_3)(R_4 + R_5) + R_2 R_3. \quad (2.74\text{c})$$

The branch currents can now be computed as follows:

$$\begin{bmatrix} i_1 \\ i_2 \\ i_3 \\ i_4 \\ i_5 \end{bmatrix} = \begin{bmatrix} -1 & -1 \\ 1 & 1 \\ -1 & 0 \\ 0 & 1 \\ 0 & 1 \end{bmatrix} \begin{bmatrix} i_{m1} \\ i_{m3} \end{bmatrix} = \begin{bmatrix} -(R_3 + R_4 + R_5)\,v_g/\Delta \\ (R_3 + R_4 + R_5)\,v_g/\Delta \\ -(R_4 + R_5)\,v_g/\Delta \\ R_3 v_g/\Delta \\ R_3 v_g/\Delta \end{bmatrix}, \quad (2.75)$$

which is the same as that given in (2.71). The branch voltages can be obtained from (2.53a), and are the same as those given in (2.71). Notice that the coefficient matrix given in (2.73) is asymmetric, while the one given in (2.69) is symmetric. Since symmetry is much more convenient, we seldom choose two different basis circuit matrices for the loop system of equations.

Another interesting point worth mentioning is that the coefficient matrices of (2.63), (2.69), and (2.73) are all independent of the sources. This is to be expected since all the sources have been moved to the right-hand side of the equations. In other words, we should be able to find the loop-impedance matrix in the simplified network in which all the voltage generators have been shorted.

§ 3   Solutions

Since shorting an edge in a directed or undirected graph does not change its nullity, we have

COROLLARY 2.11: Let G^* be the directed graph obtained from an electrical network G by shorting all the edges corresponding to the independent voltage generators, while keeping the numberings of their corresponding circuits unaltered. If B and B^* are the basis circuit matrices of G and G^* formed by the corresponding sets of circuits or edge-disjoint unions of circuits, respectively, and if there are no circuits consisting only of voltage generators in G, then

$$BZB' = B^*Z_{pp}B^{*'}, \qquad (2.76)$$

where Z_{pp} is defined in (2.53a).

The corollary follows directly from our partitioning of the matrices given in (2.53). The proof is straightforward, and is left as an exercise (Problem 2.18). However, as mentioned in § 2, we have implicitly assumed that G has no current generators, i.e., E in (2.53a) is a given source vector.

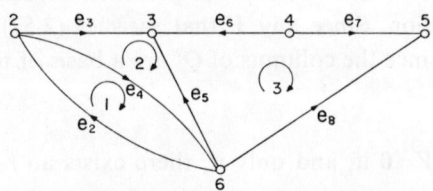

Fig. 2.16. The simplified directed graph G^* of G of fig. 2.14.

As an illustration, consider the problem given in Example 2.1. The simplified directed graph G^* of G is presented in fig. 2.16. Using the corresponding set of circuits as indicated, we have

$$B^* = \begin{matrix} & e_2 & e_3 & e_4 & e_5 & e_6 & e_7 & e_8 \\ 1 \\ 2 \\ 3 \end{matrix} \begin{bmatrix} 1 & 0 & 1 & 0 & 0 & 0 & 0 \\ 0 & 1 & -1 & -1 & 0 & 0 & 0 \\ 0 & 0 & 0 & 1 & -1 & 1 & -1 \end{bmatrix}. \qquad (2.77)$$

Z_{pp} is the matrix obtained from (2.61) by deleting the first row and the first column from Z. It is easy to check that after performing the indicated multiplication the matrix product $B^*Z_{pp}B^{*'}$ gives the coefficient matrix of (2.63).

Before we leave this section, let us summarize the result by

THEOREM 2.13: For an electrical network G, if the loop-impedance matrix is nonsingular, then the solution of G is given by

$$I = - B'(BZB')^{-1} BE \qquad (2.78)$$

and

$$V = - ZB'(BZB')^{-1} BE + E. \qquad (2.79)$$

Observe that in the solution for I and V (excluding the voltage-source branches), the branch-voltage sources of E are included only as the product $-BE\,(=E_m)$. This means that the actual branch distribution of the voltage generators is not important; only their circuit distribution E_m matters.

3.3. *Cut system of equations*

The concept of cutsets of a graph was originally introduced by WHITNEY [1933]. A study of their properties and applications to network theory has been made by SESHU and REED [1955, 1956]. A summary of their work was given by BRYANT [1961]. The present section is mainly based on their work.

The cut system of equations is exactly dual to the loop system of equations, and, therefore, we can afford to keep our present discussion shorter than that of the preceding section. Since any V that satisfies (2.52) must be contained in the Q-space, and since the columns of Q' are a basis of the Q-space, we have (Problem 2.23)

COROLLARY 2.12: $BV=0$ if, and only if, there exists an r-vector V_c such that

$$V = Q'V_c. \qquad (2.80)$$

The vector V_c is called the *cut-voltage vector*, and equations (2.80) are referred to as the *cut transformation*. The reason for the name *cut voltage* is that the elements in V_c can be identified or interpreted physically as a set of voltages associated with the cuts defined by the rows of Q. For some choices of the cuts, the cut voltages become the voltages between some pairs of nodes in the network, and are known as the *node-pair voltages* (KRON [1939]). If Q becomes Q_f, the variables in V_c are the voltages of the branches of the tree defining Q_f. If Q becomes A, the variables are known as the *node-to-datum voltages*. In the cases where the interpretation as node-pair voltages is possible, the cut transformation is also referred to as the *node-pair transformation*.

Thus, like the formulation of the loop system of equations, our problem is then to obtain the coordinates of V with respect to the reference system defined by the columns of Q', so that (2.51) and (2.53b) are also satisfied. This can easily be achieved by simple substitutions, and the result is given by

$$Y_c V_c = J_c, \qquad (2.81)$$

where
$$Y_c = QYQ', \qquad (2.82a)$$
$$J_c = -QJ. \qquad (2.82b)$$

The system of equations given in (2.81) is known as the *cut system of equations*. The coefficient matrix Y_c is called the *cut-admittance matrix* of the network, and J_c the *cut current-source vector*. (2.81) represents a system of r equations in r cut voltages. Knowing the cut voltages we can find the branch voltages from (2.80) and the branch currents from (2.53b). The time functions are then found by inverting the Laplace transforms.

Like the loop case, it is not necessary to use the same basis cut matrix Q in both (2.51) and (2.80). However, if two different basis cut matrices are used, one for (2.51) and one for (2.80), the coefficient matrix of the cut system of equations will be asymmetric even for a reciprocal network. Since there is no generality achieved by such a procedure, it is convenient to assume that the same basis cut matrix will be used.

If Q becomes A, the cut system of equations is also known as the *nodal system of equations*, and was first devised by MAXWELL [1892]. The coefficient matrix of the nodal system of equations, denoted by the symbol $Y_n = AYA'$, is the well-known *node-admittance matrix*, and the vector $J_n = -AJ$ is called the *nodal current-source vector*. Because of the special form of the matrix Y_n (see ch. 4, § 2) and because of the facts that the elements of the matrices Y_n and J_n can easily be obtained by inspection (Problem 2.21), and that the elements in V_c can be identified as the node-to-datum voltages in the network, the nodal technique has been widely used in practical analysis of electrical networks.

THEOREM 2.14: For an electrical network G, if the cut-admittance matrix is nonsingular, then the solution of G is given by
$$V = -Q'(QYQ')^{-1}QJ \qquad (2.83)$$
and
$$I = -YQ'(QYQ')^{-1}QJ + J. \qquad (2.84)$$

It is of interest to note that, in the solution for V and I (excluding the current-source branches), the branch-current sources in J are included only as the product $-QJ$ ($=J_c$). This means that the actual branch distribution of the current sources is not important; only the cut distribution J_c matters.

We shall illustrate the above procedures by the following examples.

Example 2.3: Consider the network and its corresponding directed graph G

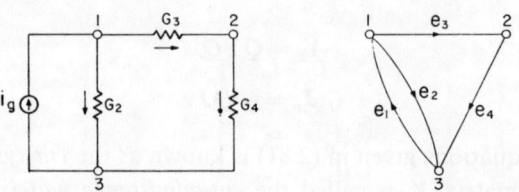

Fig. 2.17. A network and its corresponding directed graph.

as shown in fig. 2.17. The two basis incidence matrices A and A^* of G are given by

$$A = \begin{matrix} & e_1 & e_2 & e_3 & e_4 \\ \begin{matrix}1\\2\end{matrix} & \begin{bmatrix} -1 & 1 & 1 & 0 \\ 0 & 0 & -1 & 1 \end{bmatrix} \end{matrix}, \quad (2.85)$$

$$A^* = \begin{matrix}2\\3\end{matrix}\begin{bmatrix} 0 & 0 & -1 & 1 \\ 1 & -1 & 0 & -1 \end{bmatrix}. \quad (2.86)$$

The transpose J' of the branch current-source vector J is given by

$$J' = [i_g \ \ 0 \ \ 0 \ \ 0]. \quad (2.87)$$

The branch-admittance matrix Y of the network is obtained as follows:

$$Y = \begin{bmatrix} 0 & 0 & 0 & 0 \\ 0 & G_2 & 0 & 0 \\ 0 & 0 & G_3 & 0 \\ 0 & 0 & 0 & G_4 \end{bmatrix}. \quad (2.88)$$

If the basis incidence matrix A of (2.85) is used for both (2.51) and (2.80), the nodal system of equations can now be obtained by performing the needed matrix multiplications, and is given by

$$\begin{bmatrix} G_2 + G_3 & -G_3 \\ -G_3 & G_3 + G_4 \end{bmatrix} \begin{bmatrix} v_{n1} \\ v_{n2} \end{bmatrix} = \begin{bmatrix} i_g \\ 0 \end{bmatrix}. \quad (2.89)$$

The node-to-datum voltages v_{n1} and v_{n2} can be obtained by inverting the node-admittance matrix, and are given by

$$v_{n1} = (G_3 + G_4) i_g / \Delta, \quad (2.90a)$$

$$v_{n2} = G_3 i_g / \Delta, \quad (2.90b)$$

where

$$\Delta = G_2(G_3 + G_4) + G_3 G_4. \quad (2.90c)$$

From (2.80) and (2.53b), we have

$$\begin{bmatrix} v_1 \\ v_2 \\ v_3 \\ v_4 \end{bmatrix} = \begin{bmatrix} -(G_3+G_4)\,i_g/\Delta \\ (G_3+G_4)\,i_g/\Delta \\ G_4 i_g/\Delta \\ G_3 i_g/\Delta \end{bmatrix} \quad \text{and} \quad \begin{bmatrix} i_1 \\ i_2 \\ i_3 \\ i_4 \end{bmatrix} = \begin{bmatrix} i_g \\ G_2 v_2 \\ G_3 v_3 \\ G_4 v_4 \end{bmatrix}. \tag{2.91}$$

If the variables v_{n1} and v_{n2} are interpreted as the potential differences from nodes 1 and 2 to the reference node 3, respectively, i.e., the node-to-datum voltages, the branch voltages can also be obtained intuitively from the network. For example, in fig. 2.17, the branch voltage v_3 is equal to the potential difference from node 1 to node 2, i.e., $v_3 = v_{n1} - v_{n2}$. Similarly, the branch voltage v_2 is the same as the node-to-datum voltage v_{n1}.

Now, if A is used in (2.51) and if the transpose $A^{*\prime}$ of A^* is used in (2.80), the nodal system of equations becomes

$$AYA^{*\prime}V_n = -AJ. \tag{2.92}$$

Performing the indicated matrix multiplications, we have

$$\begin{bmatrix} -G_3 & -G_2 \\ G_3+G_4 & -G_4 \end{bmatrix} \begin{bmatrix} v_{n2} \\ v_{n3} \end{bmatrix} = \begin{bmatrix} i_g \\ 0 \end{bmatrix}. \tag{2.93}$$

The node-to-datum voltages v_{n2} and v_{n3} are given by

$$v_{n2} = -G_4 i_g/\Delta, \tag{2.94a}$$

$$v_{n3} = -(G_3+G_4)\,i_g/\Delta, \tag{2.94b}$$

where

$$\Delta = G_3 G_4 + G_2(G_3+G_4). \tag{2.94c}$$

The branch voltages can be computed from (2.80) via $V = A^{*\prime}V_n$, and the branch currents from (2.53b), and they are the same as those given in (2.91). The branch voltages can also be obtained intuitively from the network given in fig. 2.17 if the variables v_{n2} and v_{n3} are interpreted as the potential differences from the nodes 2 and 3 to the reference (datum) node 1, respectively. For example, the branch voltage v_3 is simply the negative of the potential difference from node 2 to node 1, i.e., $v_3 = -v_{n2}$. Similarly, the branch voltage v_4 is equal to the potential difference from node 2 to node 3, i.e., $v_4 = v_{n2} - v_{n3}$. Obviously, the physical interpretation of the variables in V_n as the node-to-datum voltages is valid only for the incidence matrix. Earlier, we have mentioned that if A is the matrix obtained from A_a by deleting the row i, the node corresponding to the row i is called the reference node. The reason for this name is now apparent, because it corresponds to the reference point for the node-to-datum voltages.

Again, it is significant to point out that the coefficient matrix in (2.93) is asymmetric, while the one given in (2.89) is symmetric. Since symmetry is much more convenient, we seldom choose two different basis cut matrices for the cut system of equations.

For illustrative purposes, let us formulate the cut system of equations for a more complicated network which will involve a unilateral network element.

Example 2.4: Consider the network as shown in fig. 2.18 which is equivalent to the network considered in Example 2.1. The corresponding directed graph

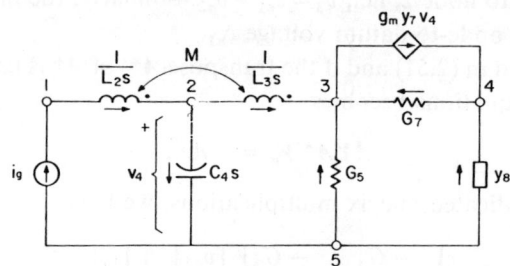

Fig. 2.18. A network diagram.

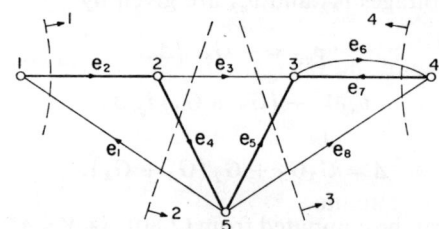

Fig. 2.19. The corresponding directed graph of the network of fig. 2.18.

G of the network is given in fig. 2.19. Thus, we have $r=4$. Consider the f-cutset matrix Q_f of G with respect to the tree $t=e_2e_4e_5e_7$ as indicated by the heavy lines in fig. 2.19. Then

$$Q_f = \begin{matrix} 1 \\ 2 \\ 3 \\ 4 \end{matrix} \begin{matrix} e_2 & e_4 & e_5 & e_7 & e_1 & e_3 & e_6 & e_8 \\ \begin{bmatrix} 1 & 0 & 0 & 0 & -1 & 0 & 0 & 0 \\ 0 & 1 & 0 & 0 & -1 & 1 & 0 & 0 \\ 0 & 0 & 1 & 0 & 0 & 1 & 0 & 1 \\ 0 & 0 & 0 & 1 & 0 & 0 & -1 & -1 \end{bmatrix} \end{matrix}. \qquad (2.95)$$

The branch-admittance matrix Y of the network is obtained as follows:

$$Y = \begin{bmatrix} L_3/ys & 0 & 0 & 0 & 0 & -M/ys & 0 & 0 \\ 0 & C_4 s & 0 & 0 & 0 & 0 & 0 & 0 \\ 0 & 0 & G_5 & 0 & 0 & 0 & 0 & 0 \\ 0 & 0 & 0 & G_7 & 0 & 0 & 0 & 0 \\ 0 & 0 & 0 & 0 & 0 & 0 & 0 & 0 \\ -M/ys & 0 & 0 & 0 & 0 & L_2/ys & 0 & 0 \\ 0 & g_m y_7 & 0 & 0 & 0 & 0 & 0 & 0 \\ 0 & 0 & 0 & 0 & 0 & 0 & 0 & y_8 \end{bmatrix}, \quad (2.96)$$

where $y = L_2 L_3 - M^2$ ($y \neq 0$), and $y_7 = G_7 C_4 s$. The transpose J' of the branch current-source vector is given by

$$J' = [0 \ 0 \ 0 \ 0 \ i_g \ 0 \ 0 \ 0]. \quad (2.97)$$

Performing the indicated matrix multiplications of (2.82), we obtain the desired cut system of equations:

$$\begin{bmatrix} L_3/ys & -M/ys & -M/ys & 0 \\ -M/ys & C_4 s + L_2/ys & L_2/ys & 0 \\ -M/ys & L_2/ys & G_5 + L_2/ys + y_8 & -y_8 \\ 0 & -g_m y_7 & -y_8 & G_7 + y_8 \end{bmatrix} \begin{bmatrix} v_{c2} \\ v_{c4} \\ v_{c5} \\ v_{c7} \end{bmatrix} = \begin{bmatrix} i_g \\ i_g \\ 0 \\ 0 \end{bmatrix}.$$

(2.98)

Note that the subscripts of the variables in V_c correspond to the tree branches of t. The reason for this is that they may be identified as the branch voltages of the tree. For example, the branch voltage v_x is the same as the cut voltage v_{cx}, i.e., $v_x = v_{cx}$ for $x = 2, 4, 5, 7$. Now if the coefficient matrix Y_c is nonsingular, the cut voltages can be computed. However, the inversion of the matrix Y_c is rather tedious; a short-cut method for this known as the topological analysis will be presented in ch. 4.

Like the loop case discussed in the preceding section, the cut-admittance matrix can also be obtained from the simplified network with all the independent current generators removed. Recall that we assume there are no voltage generators in the original network.

COROLLARY 2.13: Let G^* be the directed graph obtained from an electrical network G by the removal of all the edges corresponding to the independent current generators, while keeping the numberings of their corresponding cuts unaltered. If Q and Q^* are the basis cut matrices of G and G^* formed by the corresponding sets of cuts, respectively, and if there are no cuts consisting only of current generators in G, then

$$QYQ' = Q^*Y_{pp}Q^{*'},\qquad(2.99)$$

where Y_{pp} is defined in (2.53b).

The proof of the corollary is straightforward, and is left as an exercise (Problem 2.24).

3.4. *Additional considerations*

In the preceding two sections, we have presented two powerful techniques for solving the network problems. However, the formal procedures are not always the simplest methods for all problems. In many common problems, inspection and short cuts may provide answers much more easily than solving the loop and cut systems of equations. The real value of these formulations lies in their generality, and thus they can be programmed for a digital computer. Although any one of these methods may be used to solve a general network problem, the number of equations involved is different. Thus, the choice of the systems depends upon the types of networks at hand.

The discussion up to this point has been in terms of the branch currents and branch voltages associated with the network elements. Since these quantities are more basic in the sense of being directly measurable, they are usually referred to as the *primary variables*. The loop currents and the cut voltages introduced in the foregoing are examples of the *secondary variables* used to compute the primary variables. Another type of secondary variables is defined as the product of a current and a voltage. For the reference convention that we have adopted here, the product

$$p_k(t) = i_k(t)\, v_k(t) \qquad(2.100)$$

is defined as the *power function*, which corresponds to the *power absorbed in branch* k. If we integrate (2.100) between any two limits t_0 and t,

$$w_k(t) = \int_{t_0}^{t} p_k(x)\,\mathrm{d}x, \qquad(2.101)$$

we create yet another secondary variable called the *energy function*. In the following, we shall show that if the energy function is well behaved, the Kirchhoff's two laws imply the conservation of energy. In other words, the conservation of energy need not be added as a postulate for the discipline of electrical network theory (Problems 2.61 and 2.62).

THEOREM 2.15 (TELLEGEN's Theorem [1952]): If the currents and voltages in a network G satisfy Kirchhoff's current law and voltage law, respectively, then

$$\sum_{k=1}^{b} p_k(t) = 0. \qquad (2.102)$$

Proof. Since

$$\sum_{k=1}^{b} p_k(t) = I'(t) V(t), \qquad (2.103)$$

the theorem follows from the fact that $I(t)$ and $V(t)$ are orthogonal (Corollaries 2.5 and 2.7).

THEOREM 2.16: If the energy function is absolutely continuous so that (2.101) and its derivative exist, then Kirchhoff's laws imply conservation of energy:

$$\sum_{k=1}^{b} w_k(t) = \text{constant}. \qquad (2.104)$$

§ 4. Invariance and mutual relations of network determinants and the generalized cofactors

The zeros of the determinant of the loop-impedance matrix or the cut-admittance matrix of a network are called the *natural frequencies* of the network. It is well known that these frequencies characterize its natural response. It appears, therefore, that its natural response may also depend upon the choice of circuits or cuts. Such a dependence would contradict the known electrical network situation. Thus, it is essential to show that this is not the case.

The main objective of this section is to show that the network functions are invariant with respect to the general transformations of the reference frame. We shall also show that many of the invariant characters of a network are consequences of its topology rather than the characteristics of its network elements. Thus, they are valid not only for the electrical networks, but also for other systems.

4.1. A brief history

The invariant character of the determinant of the node-admittance matrix with regard to the choice of the reference node was first considered by JEANS [1925] p. 328, and rediscovered by PERCIVAL [1954] and SHEKEL [1954]. The same statement is also made very often about the determinant of the loop-impedance matrix (see, for example, TSANG [1954]). It was pointed out by SESHU [1955] that the determinant of the loop-impedance matrix in general is not invariant under the transformation of circuits. However, they are related by a real constant depending only on the two choices of the sets of circuits. OKADA [1955] showed that the determinant of the loop-impedance matrix formed by the f-circuits is invariant, but the determinant of the cut-admittance matrix does not remain invariant when general cut variables are used. CEDERBAUM [1956] considered the problem in a more general nature and showed that these determinants, although they are not invariant for a general loop system or for a general cutset representation, are invariant for the most common and important types of network representations such as f-circuits, f-cutsets, and circuits formed by the regions of a planar network. He also showed that the ratio of the determinants of the loop-impedance and the cutset-admittance matrices under the invariant condition is equal to the determinant of the branch-impedance matrix of the network. This is a generalization of an earlier work of TSANG [1954]. The extension of the invariant character of the network determinants to cofactors was recently given by CHEN [1970a], who also showed that many of these invariant characters could be derived from a purely graph-theoretic viewpoint. Thus, they are consequences of the network topology rather than the characteristics of the network elements. The present section is based mainly on the work of Cederbaum and Chen.

4.2. *Preliminary considerations*

Let $S_x = \{1, 2, ..., x\}$. For a given matrix F of order $p \times b$, let $F(I_u, J_v)$ be the submatrix of F consisting of the rows and columns corresponding to the integers in the sets I_u and J_v, respectively, where $I_u = \{i_1, i_2, ..., i_u\}$ and $J_v = \{j_1, j_2, ..., j_v\}$ are subsets of S_p and S_b, respectively. In particular, if $p \leq b$ we denote $F(I_p, J_p)$ by $F(J_p)$. $\det F(J_p)$ is referred to as a *major determinant* of F. Clearly, if $F(J_p)$ is nonsingular, it is a major submatrix of F (see Definition 2.4). Unless specified to the contrary, throughout the chapter by $\bar{J}_v$ we mean the complement of the subset J_v in S_b.

It is well known that the determinant of the product of square matrices is equal to the product of their determinants, but for the product of nonsquare matrices, its determinant may not be so familiar. For our purpose, we shall state a theorem of matrix theory known as the *Binet-Cauchy theorem*, which will be used rather extensively for many of our derivations.

THEOREM 2.17 (Binet-Cauchy Theorem): If C, D, and F are matrices of orders $p \times q$, $q \times s$, and $s \times p$, respectively, over a field ($p \leq q, s$), then

$$\det CDF = \sum_{(I_p)} \sum_{(J_p)} \det C(I_p) \det D(I_p, J_p) \det F(J_p, I_p), \qquad (2.105)$$

§4 Invariance and mutual relations 79

where I_p and J_p are subsets of the sets S_q and S_s, respectively, and the summations are taken over all possible such subsets I_p and J_p.

The theorem is of fundamental importance in network theory; it serves as a bridge between topology and network matrices. Many of the topological formulas in network theory are derived by applying this theorem.

Another theorem which we shall need is known as the *Jacobi's theorem*, which may be stated as follows:

THEOREM 2.18 (Jacobi's Theorem): For a nonsingular matrix D of order b if C is the inverse of D, then

$$\det D \det C(I_u, J_u) = (-1)^{\Sigma i + \Sigma j} \det D(\bar{J}_u, \bar{I}_u), \qquad (2.106)$$

where $\sum i$ and $\sum j$ denote the sums of integers contained in I_u and J_u, respectively, and I_u and J_u are subsets of S_b.

The proofs of the above two theorems may be found in any text on matrix theory (see, for example, AITKEN [1962]), so it will not be given here.

As an example of Jacobi's theorem, let us consider the matrix

$$D = \begin{bmatrix} 8 & 2 & 1 \\ 4 & 8 & 2 \\ 2 & 4 & 8 \end{bmatrix}. \qquad (2.107)$$

The inverse of D is given by

$$C = D^{-1} = \tfrac{1}{392}\begin{bmatrix} 56 & -12 & -4 \\ -28 & 62 & -12 \\ 0 & -28 & 56 \end{bmatrix}, \qquad (2.108)$$

where $\det D = 392$. Thus, we have

$$\det D \det C(\{1,2\}, \{2,3\}) = \tfrac{1}{392}\begin{vmatrix} -12 & -4 \\ 62 & -12 \end{vmatrix} = 1$$

and

$$(-1)^{1+2+2+3} \det D(\{1\}, \{3\}) = 1,$$

which confirms Jacobi's theorem.

After this digression into a discussion of two theorems in matrix theory, let us return to the circuit and cut matrices of a directed graph. Since the columns of B' and B'_f, and Q' and Q'_f, are the two bases, respectively, of the B-space and the Q-space, it is evident that there exist nonsingular matrices C and D of orders m and r, respectively, such that (Lemmas 2.3 and 2.6)

$$B = CB_f \text{ and } Q = DQ_f. \qquad (2.109)$$

Since the major submatrices of Q and B are in one-to-one correspondence with the trees and cotrees of the directed graph G (Theorems 2.5 and 2.8), it follows that

$$B(I_m) = CB_f(I_m), \qquad (2.110)$$

$$Q(J_r) = DQ_f(J_r). \qquad (2.111)$$

In particular, if M and N are the major submatrices of B and Q corresponding to the cotree $\bar{t}$ and tree t defining the identity matrices U_m and U_r of B_f and Q_f, respectively, then

$$C = M \quad \text{and} \quad D = N. \qquad (2.112)$$

DEFINITION 2.22: *Totally unimodular matrix.* A matrix with the property that the determinant of each of its square submatrices is 1, -1, or 0 is called a *totally unimodular matrix.*

A direct consequence of this definition and Jacobi's theorem is the following.

COROLLARY 2.14: If a square nonsingular matrix is totally unimodular, so is its inverse.

THEOREM 2.19: The incidence matrix A_a, the f-circuit matrix B_f, and the f-cutset matrix Q_f of a connected directed graph G are totally unimodular.

Proof. Part I. A_a *is totally unimodular.* Consider any nonsingular submatrix of A_a. Since each column of this submatrix has at most two nonzero entries, a 1 and a -1, and since not every column can have both a 1 and -1, it follows that there must exist at least one column with just one nonzero entry, for otherwise the submatrix would be singular. Expanding the determinant of this submatrix by this column, we obtain the desired result by induction.

Part II. B_f *is totally unimodular.* Let A be partitioned according to (2.22a), where A_{12} is the major submatrix corresponding to the branches of the tree t defining B_f. Consider the nonsingular matrix

$$F = \begin{bmatrix} U_m & 0 \\ A_{11} & A_{12} \end{bmatrix}. \qquad (2.113)$$

Since in Part I we have shown that A is totally unimodular, by applying the Laplace expansion to F it is easy to show that F is also totally unimodular. Now invert F to give

$$F^{-1} = \begin{bmatrix} U_m & 0 \\ -A_{12}^{-1}A_{11} & A_{12}^{-1} \end{bmatrix}, \qquad (2.114)$$

which, by Corollary 2.14, is still totally unimodular. Since a submatrix of a totally unimodular matrix is totally unimodular, we see that the submatrix

$$\begin{bmatrix} U_m \\ -A_{12}^{-1}A_{11} \end{bmatrix}, \quad (2.115)$$

which by (2.23a) is the transpose of B_f, is totally unimodular.

Part III. Q_f is totally unimodular. From (2.23b), we have

$$Q_f = [A_{12}^{-1}A_{11} \quad U_r]. \quad (2.116)$$

Since by (2.114) $A_{12}^{-1}A_{11}$ is totally unimodular, it is easy to show that Q_f is also totally unimodular. This completes the proof of the theorem.

The first part of the theorem was first established by VEBLEN and FRANKLIN [1921]. The assumption that G is connected is necessary for B_f and Q_f since they are defined in terms of a tree of G. However, it is not necessary for A_a. Clearly, these results can easily be extended to unconnected directed graphs by considering the f-circuits and f-cutsets of each component.

Using Theorem 2.19 in conjunction with (2.110) and (2.111), we have

COROLLARY 2.15: *The determinants of all the major submatrices of either B or Q of a directed graph have the same magnitude.*

Thus, if $B(I_m)$ and $Q(J_r)$ are major submatrices of B and Q, respectively, let

$$k(B) = [\det B(I_m)]^2 \quad (2.117)$$

and

$$k(Q) = [\det Q(J_r)]^2. \quad (2.118)$$

Since the elements of $B(I_m)$ and $Q(J_r)$ are 1, -1, and 0, the values of $k(B)$ and $k(Q)$ must be squared integers.

As an example, the basis circuit matrix B as defined by the circuits formed by the subgraphs $e_1e_5e_6e_3e_4$, $e_1e_2e_6e_7e_4$, $e_1e_2e_3e_7e_8$, and $e_5e_2e_3e_4e_8$ in the directed graph G of fig. 2.1 is

$$B = \begin{matrix} & \begin{matrix} e_1 & e_2 & e_3 & e_4 & e_5 & e_6 & e_7 & e_8 \end{matrix} \\ \begin{matrix}(e_1e_5e_6e_3e_4)\\(e_1e_2e_6e_7e_4)\\(e_1e_2e_3e_7e_8)\\(e_5e_2e_3e_4e_8)\end{matrix} & \begin{bmatrix} 1 & 0 & 1 & -1 & 1 & 1 & 0 & 0 \\ 1 & 1 & 0 & -1 & 0 & -1 & 1 & 0 \\ 1 & 1 & 1 & 0 & 0 & 0 & -1 & -1 \\ 0 & 1 & 1 & -1 & -1 & 0 & 0 & 1 \end{bmatrix} \end{matrix}. \quad (2.119)$$

It is easy to check that the determinant of all the major submatrices of B has the magnitude of 3. Thus, $k(B)=9$.

OKADA [1955] has shown that if the circuits corresponding to the rows of B together with m surfaces, obtained by assuming each circuit to be the boundary of a surface, form what topologists call a *non-orientable manifold*, then the determinant of any major submatrix of B is given by $\pm 2^i$ where i is a nonnegative integer fixed for a given B. In other words, for this type of circuits, we have $k(B) = 2^{2i}$. Obviously, not all the basis circuit matrices have this property; a counterexample was just given above.

From the above discussion, it is evident that $k(A) = 1$, $k(B_f) = 1$, and $k(Q_f) = 1$, and this is the minimum value that can be obtained for any B and Q of G. Nevertheless, they are not the only ones for which the determinant of the major submatrices has this minimum value. As a matter of fact, $k(B)$ and $k(Q)$ have this minimum value if and only if there exist major submatrices $B(I_m)$ and $Q(J_r)$ such that $\det B(I_m) = \pm 1$ and $\det Q(J_r) = \pm 1$, respectively. For another characterization, see Problem 2.26.

Since $k(B)$ and $k(Q)$ are invariant with respect to the choice of the major submatrices $B(I_m)$ and $Q(J_r)$, a proper selection of these submatrices may greatly simplify their evaluation. One way to accomplish this is the following (Problem 2.25).

COROLLARY 2.16: If B and Q contain a column which has only one nonzero entry, and if B_1 and Q_1 are the matrices obtained from B and Q, respectively, by deleting this column and the row containing the nonzero entry of this column, then

$$k(B) = k(B_1) \text{ and } k(Q) = k(Q_1). \qquad (2.120)$$

Obviously, the process may be repeated in B_1 and Q_1 until there are no such columns. It may also be interpreted as follows: If e_1 is an edge of G contained in only one of the circuits, say L_1, then B contains a column consisting of a 1 or a -1 in the row corresponding to the circuit L_1 and all other entries are zero. Thus, we may remove e_1 from G and ignore L_1 in setting up the circuit matrix B_1. Evidently, B_1 can be obtained from B by deleting the row corresponding to L_1 and the column corresponding to e_1. In other words, B_1 is a basis circuit matrix of the directed graph G_1 which is obtained from G by removing the edge e_1 and by ignoring the circuit L_1 in G. If after this process, another circuit, say L_2, appears alone in some edge of G_1, or equivalently the matrix B_1 has a column consisting only of one nonzero entry and 0's, we may remove the edge from G_1 or delete the corresponding row and column from B_1 as well, and this process may be continued until all such edges and circuits have been removed. Let the final directed graph and its corresponding basis circuit matrix be denoted by G_α and B_α. If G_α or B_α is empty, then $k(B)=1$, and if not then $k(B)=k(B_\alpha)$. This process is similarly valid for the cuts of G or the cut matrix formed by these cuts. The only difference is that each time we short an edge in G instead of removing an edge from G. If we apply this process to A, B_f, and Q_f, we find G_α's are empty. Thus, $k(A)=k(B_f)=k(Q_f)=1$.

The discussion up to now is based on any directed graph G. If, in particular, G is planar, then an important class of circuits known as *meshes* may be used. Recall the definition of a region defined for a planar graph (Definition 1.17). The *mesh* of a region is the circuit formed by the boundary edges of the region. For convenience, we shall use the symbol B_p to represent the submatrix of the circuit matrix B_a of a planar graph G formed by the rows corresponding to the set of meshes of G.

For example, in fig. 2.16 the circuits formed by the subgraphs e_2e_4, $e_3e_5e_4$, and $e_5e_6e_7e_8$ are meshes of the directed graph. The matrix B_p corresponding to these meshes are given by

$$B_p = \begin{matrix} 1 \\ 2 \\ 3 \end{matrix} \begin{bmatrix} 1 & 0 & 1 & 0 & 0 & 0 & 0 \\ 0 & 1 & -1 & -1 & 0 & 0 & 0 \\ 0 & 0 & 0 & 1 & -1 & 1 & -1 \end{bmatrix}. \qquad (2.121)$$

Now, if we apply the above mentioned simplification procedure to B_p of (2.121), we find $k(B_p)=1$. Because of the ways with which the meshes are defined, it is not difficult to see that this is valid in general (Problem 2.25). Thus, we may state

COROLLARY 2.17: If a directed graph G is planar, then B_p is a basis circuit matrix of G, and furthermore $k(B_p)=1$.

4.3. *The loop and cut transformations*

In many instances, a linear transformation of network variables is useful in simplifying network calculations or visualizing network properties. The simplest of such a transformation is the change from a given set of m loop currents or r cut voltages to an alternate group of m loop currents or r cut voltages.

Let

$$Z_m I_m = E_m, \qquad (2.122)$$

where $Z_m = BZB'$ and $E_m = -BE$, and

$$Z_m^* I_m^* = E_m^*, \qquad (2.123)$$

where $Z_m^* = B^*ZB^{*\prime}$ and $E_m^* = -B^*E$, be two loop systems of equations of a given network G corresponding to the two different choices of the basis circuit

matrices B and B^*, respectively. Since B and B^* are related by a nonsingular transformation C,
$$B = CB^*, \tag{2.124}$$
it follows after a simple substitution that
$$Z_m = CZ_m^* C', \tag{2.125a}$$
$$E_m = CE_m^*, \tag{2.125b}$$
$$I_m = C'^{-1} I_m^*. \tag{2.125c}$$
The last equation is obtained by the following argument. Since
$$I = B' I_m = B^{*'} C' I_m = B^{*'} I_m^*, \tag{2.126}$$
and since B^* is a basis circuit matrix, we have $C' I_m = I_m^*$. Another possibility is that, as mentioned in § 3.2, we can interpret the elements of I_m^* and I_m as two different sets of coordinates of the branch-current vector I in the B-space with respect to the two different reference systems formed by the columns of $B^{*'}$ and B', respectively. Since $B' = B^{*'} C'$, it follows that $I_m^* = C' I_m$. This is known as the transformations of coordinates in matrix theory.

As an illustration, consider the two basis circuit matrices given in (2.65) and (2.66) for the directed graph G as shown in fig. 2.15. Since B is also the f-circuit matrix with respect to the tree $t = e_1 e_3 e_5$, from (2.112) we have
$$C = \begin{bmatrix} 1 & 0 \\ 1 & 1 \end{bmatrix}^{-1} = \begin{bmatrix} 1 & 0 \\ -1 & 1 \end{bmatrix} \tag{2.127}$$
and $B = CB^*$. Thus, from (2.125) we obtain
$$Z_m = CZ_m^* C' = \begin{bmatrix} 1 & 0 \\ -1 & 1 \end{bmatrix} \begin{bmatrix} R_2 + R_3 & R_2 \\ R_2 & R_2 + R_4 + R_5 \end{bmatrix} \begin{bmatrix} 1 & -1 \\ 0 & 1 \end{bmatrix}$$
$$= \begin{bmatrix} R_2 + R_3 & -R_3 \\ -R_3 & R_3 + R_4 + R_5 \end{bmatrix}, \tag{2.128a}$$
$$E_m = CE_m^* = \begin{bmatrix} 1 & 0 \\ -1 & 1 \end{bmatrix} \begin{bmatrix} v_g \\ v_g \end{bmatrix} = \begin{bmatrix} v_g \\ 0 \end{bmatrix}, \tag{2.128b}$$
which are the same as those given in (2.69), and
$$I_m = C'^{-1} I_m^* = \begin{bmatrix} 1 & 1 \\ 0 & 1 \end{bmatrix} \begin{bmatrix} (R_4 + R_5) v_g / \Delta \\ R_3 v_g / \Delta \end{bmatrix} = \begin{bmatrix} (R_3 + R_4 + R_5) v_g / \Delta \\ R_3 v_g / \Delta \end{bmatrix}, \tag{2.128c}$$

where Δ is defined in (2.70c). The result is the same as that given in (2.70). Similarly, if
$$Y_c V_c = J_c, \qquad (2.129)$$
where $Y_c = QYQ'$ and $J_c = -QJ$, and
$$Y_c^* V_c^* = J_c^*, \qquad (2.130)$$
where $Y_c^* = Q^* Y Q^{*\prime}$ and $J_c^* = -Q^* J$, are two systems of cut equations of the network G corresponding to the two different choices of the basis cut matrices Q and Q^*, respectively, then there exists a nonsingular matrix D of order r such that
$$Q = DQ^*. \qquad (2.131)$$
Thus, we have
$$Y_c = D Y_c^* D', \qquad (2.132a)$$
$$J_c = D J_c^*, \qquad (2.132b)$$
$$V_c = D'^{-1} V_c^*. \qquad (2.132c)$$

4.4. Network matrices

In the preceding section, we have discussed the transformation from one loop or cut system of equations to another. Since the zeros of the determinant of the loop-impedance or the cut-admittance matrix of a network G are also the natural frequencies of G, one would anticipate that they should be invariant with respect to the choice of circuits or cuts. In this section, we shall discuss the conditions under which the value of these determinants is invariant with respect to the transformation from one such system to another, and their interrelationships.

DEFINITION 2.23: *Network matrix.* For a directed graph G, the matrix triple products BZB' and QYQ' are called the *network matrices* of G, where Z and Y are given matrices of order b.

THEOREM 2.20: Let B_1 and B_2 be two basis circuit matrices of G, and also let Q_1 and Q_2 be two basis cut matrices of G. Then
$$k(B_2) \det B_1 Z B_1' = k(B_1) \det B_2 Z B_2' \qquad (2.133)$$
and
$$k(Q_2) \det Q_1 Y Q_1' = k(Q_1) \det Q_2 Y Q_2'. \qquad (2.134)$$

Proof. We shall only prove the circuit part since the cut part can be proved in an entirely similar manner. Let M_1 and M_2 be the corresponding major submatrices of B_1 and B_2, respectively. Then from (2.112) we have

$$B_f = M_1^{-1} B_1 = M_2^{-1} B_2, \qquad (2.135)$$

where B_f is defined with respect to the tree corresponding to the columns of M_1 or M_2. Thus, we get

$$\begin{aligned}
\det B_1 Z B_1' &= \det\left[(M_1 M_2^{-1} B_2) Z (M_1 M_2^{-1} B_2)'\right] \\
&= \det\left[(M_1 M_2^{-1})(B_2 Z B_2')(M_1 M_2^{-1})'\right] \\
&= (\det M_1/\det M_2)^2 \det B_2 Z B_2' \\
&= [k(B_1)/k(B_2)] \det B_2 Z B_2', \qquad (2.136)
\end{aligned}$$

which completes the proof of the theorem.

COROLLARY 2.18: $\det B_1 Z B_1' = \det B_2 Z B_2' (\neq 0)$ and $\det Q_1 Y Q_1' = \det Q_2 Y Q_2' (\neq 0)$ if, and only if, $k(B_1) = k(B_2)$ and $k(Q_1) = k(Q_2)$, respectively.

COROLLARY 2.19: The determinant of the network matrix associated with a system of f-circuits or f-cutsets is invariant with respect to the transformation from one such system to another.

Obviously, for a given Z or Y, the value of the determinant of the network matrix depends upon the choice of circuits or cuts in the given G. For convenience, we shall use the value associated with a system of f-circuits or f-cutsets as our standard for comparison since it is invariant with respect to these transformations. By *network determinant* we shall mean the determinant of the network matrix.

DEFINITION 2.24: *Basic value.* The value of the determinant of the network matrix associated with a set of f-circuits or f-cutsets of a network is called the *basic value* of the network matrix.

Note that there are two basic values for a network, one for the circuit formulation and the other for the cut formulation.

COROLLARY 2.20: For given matrices Z and Y, the determinants of the network matrices BZB' and QYQ', if not identically zero, achieve their minimum absolute values if, and only if, $k(B) = 1$ and $k(Q) = 1$, respectively.

COROLLARY 2.21: The network determinant is equal to its basic value multiplied by the square of an integer. The absolute value of the basic value is its minimum absolute value.

The term "absolute" is necessary because the elements in Z and Y may be complex. The squared integer is actually $k(B)$ or $k(Q)$.

COROLLARY 2.22: The determinant of the network matrix AYA' or $B_p ZB'_p$ has its basic value.

Thus, we may conclude that the determinants of any two loop-impedance or cut-admittance matrices are related by a real constant. In particular, the determinant of the loop-impedance or the cut-admittance matrix associated with a system of f-circuits or f-cutsets is invariant with respect to the transformations from one such system to another.

COROLLARY 2.23: The determinant of the node-admittance matrix is invariant with respect to the choice of the reference node in the network.

Proof. Let A be the basis incidence matrix of the network with reference node i. Since by Corollary 2.22 $\det AYA'$ is invariant with respect to any i, the corollary follows from the fact that the node-admittance matrix AYA' uses node i as the reference node.

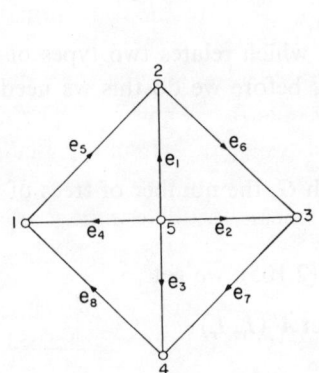

Fig. 2.20. A directed graph.

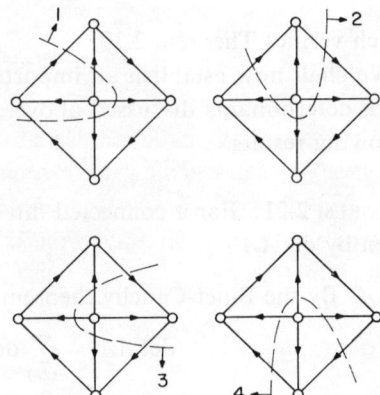

Fig. 2.21. The four chosen cutsets.

Example 2.5: Consider the directed graph G as shown in fig. 2.20. The basis cutset matrix Q corresponding to the four cutsets as indicated in fig. 2.21 is given by

$$Q = \begin{bmatrix} 1 & 1 & 1 & 0 & 1 & 0 & 0 & -1 \\ 0 & 1 & 1 & 1 & -1 & 1 & 0 & 0 \\ 1 & 0 & 1 & 1 & 0 & -1 & 1 & 0 \\ 1 & 1 & 0 & 1 & 0 & 0 & -1 & 1 \end{bmatrix}. \quad (2.137a)$$

For simplicity, let Y be the identity matrix of order 8. The cutset-admittance matrix Y_c is given by

$$Y_c = QYQ' = \begin{bmatrix} 5 & 1 & 2 & 1 \\ 1 & 5 & 1 & 2 \\ 2 & 1 & 5 & 1 \\ 1 & 2 & 1 & 5 \end{bmatrix}. \qquad (2.137b)$$

Thus, we have $k(Q) = 3^2$ and $\det Y_c = 405$. If A is the basis incidence matrix of G with reference node 5, then the node-admittance matrix Y_n of G is given by

$$Y_n = AYA' = \begin{bmatrix} 3 & -1 & 0 & -1 \\ -1 & 3 & -1 & 0 \\ 0 & -1 & 3 & -1 \\ -1 & 0 & -1 & 3 \end{bmatrix}, \qquad (2.138)$$

and $\det Y_n = 45$, which is also the basic value of the cut-admittance matrix. So

$$\det Y_c = k(Q) \det Y_n, \qquad (2.139)$$

which verifies Theorem 2.20.

We shall now establish an important result which relates two types of network determinants discussed above. However, before we do this we need the following results.

THEOREM 2.21: For a connected directed graph G, the number of trees of G is given by $\det AA'$.

Proof. By the Binet-Cauchy theorem given in (2.105), we get

$$\begin{aligned} \det AA' &= \sum_{(I_r)} \det A(I_r) \det A'(I_r, I_r) \\ &= \sum_{(I_r)} \det A(I_r) \det A(I_r) \\ &= \sum_{(I_r)} 1. \end{aligned} \qquad (2.140)$$

The third line follows from Theorem 2.19. Since $A(I_r)$ is nonsingular if, and only if, the columns of $A(I_r)$ correspond to the branches of a tree in G (Theorem 2.2), the theorem follows from here.

COROLLARY 2.24: The number of trees in G is given by $(\det BB')/k(B)$ or $(\det QQ')/k(Q)$.

Proof. Let N be the major submatrix of Q corresponding to a tree t of G. Let Q_f be the f-cutset matrix defined with respect to the tree t. Then from (2.112) we have $Q = NQ_f$. It follows that

$$\det QQ' = \det NQ_f Q'_f N' = (\det NN') \det Q_f Q'_f$$
$$= k(Q) \det AA'. \qquad (2.141)$$

The second line follows from Corollary 2.18. Thus, from Theorem 2.21 we have the desired result for the cutset case. Similarly, we can prove the circuit case (Problem 2.29). Notice that (2.141) also follows directly from (2.134).

LEMMA 2.8:

$$(-1)^{\Sigma i} \det Q(I_r) \det B(\bar{I}_r) = (-1)^{\Sigma j} \det Q(J_r) \det B(\bar{J}_r), \qquad (2.142)$$

where $\sum i$ and $\sum j$ denote the sums of integers contained in the sets I_r and J_r, respectively, and $Q(I_r)$ and $Q(J_r)$ are major submatrices of Q.

Proof. Let

$$H' = [Q' \quad B']. \qquad (2.143)$$

Then $\det HH' = (\det QQ')(\det BB')$. If q denotes the number of trees in G, then by Corollary 2.24 we have

$$\det HH' = (\det H)^2 = q^2 k(Q) k(B) \qquad (2.144a)$$

or

$$\det H = \pm q [k(Q) k(B)]^{\frac{1}{2}}. \qquad (2.144b)$$

Expanding the $\det H$ according to the minor determinants of the first r rows by Laplace's expansion, we get

$$\sum_{(I_r)} (-1)^{\Sigma i} \det Q(I_r) \det B(\bar{I}_r) = \sum_{(I_r)} (-1)^{\Sigma i} (\pm 1) [k(Q) k(B)]^{\frac{1}{2}}$$
$$= [k(Q) k(B)]^{\frac{1}{2}} \sum_{(I_r)} (-1)^{\Sigma i} (\pm 1)$$
$$= \pm q [k(Q) k(B)]^{\frac{1}{2}} \quad \text{(from (2.144b))}. \qquad (2.145)$$

Since there are exactly q terms in the summation, for the last equality to hold it is necessary that all the terms inside the summation sign agree in algebraic sign. The lemma follows immediately.

COROLLARY 2.25:

$$(-1)^{\Sigma i + \Sigma j} \det Q(I_r) \det Q(J_r) = [k(Q)/k(B)] \det B(\bar{I}_r) \det B(\bar{J}_r). \qquad (2.146)$$

Using these results, we are now in a position to state and prove an identity which relates the two types of network determinants.

THEOREM 2.22: If Z is a nonsingular matrix of order b, then

$$k(Q) \det BZB' = k(B)(\det Z) \det QZ^{-1}Q'. \qquad (2.147)$$

Proof. Let $Y = Z^{-1}$. Applying the Binet-Cauchy theorem to the expansions of $\det BZB'$ and $\det QYQ'$, we obtain

$$\det BZB' = \sum_{(I_r)} \sum_{(J_r)} \det B(I_r) \det Z(I_r, J_r) \det B(J_r)'$$

$$= \sum_{(I_r)} \sum_{(J_r)} \det B(I_r) \det B(J_r) \det Z(I_r, J_r) \qquad (2.148)$$

and

$$\det QYQ' = \sum_{(I_r)} \sum_{(J_r)} \det Q(I_r) \det Q(J_r) \det Y(I_r, J_r)$$

$$= [k(Q)/k(B)] \sum_{(I_r)} \sum_{(J_r)} (-1)^{\Sigma i + \Sigma j} \det B(I_r) \det B(J_r) \det Y(I_r, J_r)$$

$$= [k(Q)/k(B)] \sum_{(I_r)} \sum_{(J_r)} \det B(I_r) \det B(J_r) \det Z(J_r, I_r)/(\det Z)$$

$$= \frac{k(Q)}{k(B) \det Z} \sum_{(I_r)} \sum_{(J_r)} \det B(I_r) \det B(J_r) \det Z(I_r, J_r)$$

$$= k(Q)(\det BZB')/[k(B) \det Z]. \qquad (2.149)$$

The third line in (2.149) is obtained by Jacobi's theorem given in (2.106). Thus, the theorem is proved.

COROLLARY 2.26: If Z is nonsingular, then

$$\det B_f ZB'_f = (\det Z)(\det Q_f Z^{-1} Q'_f) = (\det Z)(\det AZ^{-1} A'). \qquad (2.150)$$

Note that Z and Y are arbitrarily given matrices, and they need not be those given in (2.42).

In the following, we shall apply the above results to the loop-impedance and the cut-admittance matrices, and show how their determinants are related.

As mentioned earlier, the loop-impedance and the cut-admittance matrices of a network G depend only upon the network parameters, not on the driving sources. Thus, without loss of generality, we may assume that there are no independent sources in G. If Z is the branch-impedance matrix of G and if it is nonsingular, then from (2.42) the inverse of Z must be the branch-admittance matrix Y of G, i.e., $Y = Z^{-1}$. From (2.150) we have

COROLLARY 2.27: The ratio of the basic values of the determinants of the loop-

§4   Invariance and mutual relations   91

impedance and the cut-admittance matrices of a network, if it exists, is equal to the determinant of the branch-impedance matrix.

In the case where there are unilateral *transimmittances*[†] which do not form any closed loop, we can obviously renumber the branches of the network such that there is a coupling from the branch with a lower index to a branch with a higher index but not vice versa. This results in a triangular branch-impedance matrix all of whose entries above the main diagonal are zeros (Problem 2.30). Hence, we get

COROLLARY 2.28: If the transimmittances of a network do not form any closed loop, then the ratio of the basic values of the determinants of the loop-impedance and the cut-admittance matrices of the network, if it exists, is equal to the product of all the branch driving-point impedances. The ratio is independent of the transimmittances present.

In terms of the poles and zeros of the two types of network determinants that are not identically zero, the result is given by

COROLLARY 2.29: The poles and zeros of the determinant of the loop-impedance matrix and those in the determinant of the cut-admittance matrix differ only by those contained in the determinant of the branch-impedance matrix.

DEFINITION 2.25: *RLC network*. A network consisting only of resistors, capacitors, and self-inductors is called an *RLC network*.

For *RLC* networks, Z and Y become diagonal and nonsingular. Thus we have

COROLLARY 2.30: For an *RLC* network, the determinants of the loop-impedance and the cut-admittance matrices have the same zeros, excluding those at the origin and at infinity.

We shall illustrate the above results by the following example.

Example 2.6: Consider the directed graph G as shown in fig. 2.22. Let

$$Z = \begin{bmatrix} z_1 & 0 & 0 & 0 & 0 \\ 0 & z_2 & z_0 & 0 & 0 \\ 0 & z_0 & z_3 & 0 & 0 \\ 0 & 0 & 0 & z_4 & 0 \\ 0 & 0 & 0 & 0 & z_5 \end{bmatrix}. \qquad (2.151)$$

[†] A general term for transconductance. It includes transimpedance, transadmittance, etc.

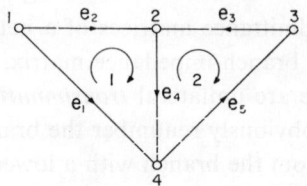

Fig. 2.22. A directed graph for Example 2.6.

Then the inverse Y of Z is given by

$$Y = Z^{-1} = \begin{bmatrix} y_1 & 0 & 0 & 0 & 0 \\ 0 & z_3/z & -z_0/z & 0 & 0 \\ 0 & -z_0/z & z_2/z & 0 & 0 \\ 0 & 0 & 0 & y_4 & 0 \\ 0 & 0 & 0 & 0 & y_5 \end{bmatrix}, \qquad (2.152)$$

where $z = z_2 z_3 - z_0^2$ and $y_i = 1/z_i$ for $i = 1, 4,$ and 5. The determinant of the loop-impedance matrix associated with the set of meshes of G as shown in fig. 2.22 is given by

$$\det Z_m = \det \begin{bmatrix} z_1 + z_2 + z_4 & -z_0 - z_4 \\ -z_0 - z_4 & z_3 + z_4 + z_5 \end{bmatrix}$$

$$= (z_1 + z_2)(z_3 + z_4 + z_5) + z_4(z_3 + z_5) - 2z_0 z_4 - z_0^2. \qquad (2.153)$$

The determinant of the node-admittance matrix of G with reference node 4 is

$$\det Y_n = \det AYA' = \det \left(\frac{1}{z} \begin{bmatrix} y_1 z + z_3 & z_0 - z_3 & -z_0 \\ z_0 - z_3 & z_2 + z_3 + y_4 z - 2z_0 & z_0 - z_2 \\ -z_0 & z_0 - z_2 & z_2 + y_5 z \end{bmatrix} \right)$$

$$= [y_4 + z_3 y_5 y_4 + y_1 z_2 y_4 + y_5 + y_1 + y_1 y_5 (z_3 + z_2 - 2z_0) + y_1 y_5 y_4 z]/z. \qquad (2.154)$$

Now it is easy to verify that

$$\det Z_m = (\det Z)(\det Y_n), \qquad (2.155)$$

where $\det Z = z_1 z_4 z_5 z$.

It should be emphasized that the relation between the determinants of the loop-impedance and the cut-admittance matrices of a network is specified in that part of the network when all the independent sources have been removed. To show that these determinants are indeed related by (2.147), it is necessary to define precisely the corresponding networks used in both formulations. As mentioned earlier, in the loop analysis we insist that all the sources be voltage generators, and in the cut analysis all the sources be current generators. This is not a serious restriction since a voltage generator can easily be transformed into an equivalent current generator, and vice versa.

Let G be a given network without independent sources. In the loop formulation, let each of the voltage generators be inserted in series with a passive element (resistor, capacitor, or self-inductor) of G. For convenience, let the source e_x be connected in series with the passive element e_{s+x} for $x=1, 2, ..., s$, where s is the number of independent voltage generators inserted, and let the resulting network thus obtained be denoted by G_v. In the cut formulation, let G_i be the network obtained from G by connecting each of the s independent current generators in parallel with a passive element e_{s+x} of G. If the strength and polarity of the corresponding voltage and current generators are properly chosen, it can easily be shown that the networks G_v and G_i are completely equivalent, since one can be obtained from the other by applying the well-known Thévenin's and Norton's theorems in electrical network theory (see, for example, SESHU and BALABANIAN [1959]).

In G_v or G_i, let the branch-impedance and the branch-admittance matrices Z and Y be partitioned as follows:

$$Z = \begin{bmatrix} 0 & 0 \\ 0 & Z_{pp} \end{bmatrix} \quad \text{and} \quad Y = \begin{bmatrix} 0 & 0 \\ 0 & Y_{pp} \end{bmatrix}, \quad (2.156)$$

where Z_{pp} and Y_{pp} are the branch-impedance and the branch-admittance matrices of G, respectively. If B and Q are the basis circuit and cut matrices of G_v and G_i, respectively, and if they are arranged in the same edge order as Z or Y and partitioned accordingly,

$$B = [B_{11} \quad B_{12}] \quad \text{and} \quad Q = [Q_{11} \quad Q_{12}], \quad (2.157)$$

then we have

COROLLARY 2.31: If Y_{pp} is the inverse of Z_{pp}, then

$$k(Q) \det BZB' = k(B) (\det Z_{pp}) (\det QYQ') \quad (2.158a)$$

and

$$BZB' = B_{12}Z_{pp}B'_{12} \quad \text{and} \quad QYQ' = Q_{12}Y_{pp}Q'_{12}. \quad (2.158b)$$

The corollary is obvious since B_{12} and Q_{12} are actually the basis circuit and cut matrices of G. Putting it differently, B_{12} and Q_{12} are the basis circuit and cut matrices B^* and Q^* of the simplified networks G^* defined in Corollaries

2.11 and 2.13, respectively. The simplified network G^* can be obtained from either G_v or G_i by removing all the independent sources. Note that by removing all the independent sources we mean by shorting all the voltage generators and by opening (removing) all the current generators.

In fig. 2.22, let the edge e_1 denote a voltage generator rather than a passive element, i.e., $z_1 = 0$ in (2.151). The corresponding G_i is presented in fig. 2.23.

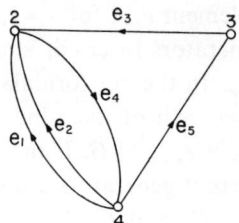 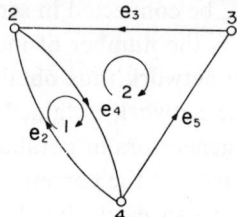

Fig. 2.23. The corresponding G_i.   Fig. 2.24. The simplified network G^*.

The branch-admittance matrix Y of G_i is given in (2.152) with $y_1 = 0$. Note that Y is no longer the inverse of Z since Z is now singular. The determinant of the loop-impedance matrix Z_m corresponding to the set of meshes as indicated in fig. 2.22 and the determinant of the node-admittance matrix Y_n of the network as shown in fig. 2.23 with reference node 4 are given by

$$\det Z_m = \det \begin{bmatrix} z_2 + z_4 & -z_0 - z_4 \\ -z_0 - z_4 & z_3 + z_4 + z_5 \end{bmatrix}$$
$$= z_2(z_3 + z_4 + z_5) + z_4(z_3 + z_5) - z_0^2 - 2z_0 z_4, \quad (2.159)$$

$$\det Y_n = \det \left(\frac{1}{z} \begin{bmatrix} z_2 + z_3 + y_4 z - 2z_0 & z_0 - z_2 \\ z_0 - z_2 & z_2 + y_5 z \end{bmatrix} \right)$$
$$= (1 + z_2 y_4 + y_5 z_3 + y_5 z_2 + y_5 y_4 z - 2z_0 y_5)/z. \quad (2.160)$$

It is easy to check that

$$\det Z_m = (\det Z_{pp})(\det Y_n), \quad (2.161)$$

where $\det Z_{pp} = z_4 z_5 z$, $z = z_2 z_3 - z_0^2$, and $z_4 = 1/y_4$ and $z_5 = 1/y_5$. Equivalently, Z_m and Y_n are the loop-impedance and the node-admittance matrices of the simplified network G^*, as shown in fig. 2.24, obtained either from G_v by shorting the edge e_1 or from G_i by removing the edge e_1.

4.5. Generalized cofactors of the elements of the network matrix

In this section, we shall look into the conditions under which some precisely defined quantities called the "generalized cofactors" of the elements of the network matrix are invariant, so that they may be used to show that the network functions are invariant with respect to the choice of circuits or cuts.

For a given matrix F, let the symbol F_{-i} be used to denote the submatrix obtained from F by deleting the row i from F. Since the determinants of the major submatrices of either B_{-i} or Q_{-i}, in general, do not have the same magnitude, the cofactors of the elements of a network matrix are not invariant with respect to the choice of circuits or cuts. For instance, consider the circuit matrix B given in (2.119). If we delete the fourth row from B, we get

$$B_{-4} = \begin{bmatrix} 1 & 0 & 1 & -1 & 1 & 1 & 0 & 0 \\ 1 & 1 & 0 & -1 & 0 & -1 & 1 & 0 \\ 1 & 1 & 1 & 0 & 0 & 0 & -1 & -1 \end{bmatrix}. \quad (2.162)$$

It is easy to check that the determinant of the major submatrix consisting of the columns 1, 2, and 3 of B_{-4} has value 1, while the one corresponding to the columns 2, 3, and 4 has value -2. Thus, they do not have the same magnitude. As an example for the cofactors, the loop-impedance matrices Z_m and Z_m^* of the directed graph G of fig. 2.15, corresponding to the basis circuit matrices B and B^* of (2.65) and (2.66), respectively, are obtained as follows:

$$Z_m = \begin{bmatrix} R_2 + R_3 & -R_3 \\ -R_3 & R_3 + R_4 + R_5 \end{bmatrix}, \quad (2.163)$$

$$Z_m^* = \begin{bmatrix} R_2 + R_3 & R_2 \\ R_2 & R_2 + R_4 + R_5 \end{bmatrix}. \quad (2.164)$$

The cofactors of the (1, 2)-element[†] of Z_m and Z_m^* are R_3 and $-R_2$, respectively. Since Z_m and Z_m^* are also the loop-impedance matrices of G with respect to the sets of f-circuits formed by the trees $e_1 e_3 e_4$ and $e_1 e_2 e_4$ (except the orientation of one f-circuit which will only effect the sign of the off-diagonal elements of Z_m^*), respectively, it may be concluded that the cofactors are not even invariant with respect to the transformation from one system of f-circuits to another. This is similarly valid for the cofactors of the elements of the cut-admittance matrix.

[†] We mean the element in the first row and the second column of Z_m or $Z_m{}^*$.

Let
$$B = [b_{ij}] \quad \text{and} \quad Q = [q_{ij}]. \tag{2.165}$$
Define
$$M_j(B; I_{m-1}) = \sum_{i=1}^{m} (-1)^i b_{ij} \det B_{-i}(I_{m-1}), \tag{2.166a}$$

$$M_j(Q; I_{r-1}) = \sum_{i=1}^{r} (-1)^i q_{ij} \det Q_{-i}(I_{r-1}). \tag{2.166b}$$

LEMMA 2.9: Let B and B^* be two basis circuit matrices of a directed graph G, and also let Q and Q^* be two basis cut matrices of G. Then

$$k(B^*)^{\frac{1}{2}} |M_j(B; I_{m-1})| = k(B)^{\frac{1}{2}} |M_j(B^*; I_{m-1})|, \tag{2.167a}$$

$$k(Q^*)^{\frac{1}{2}} |M_j(Q; I_{r-1})| = k(Q)^{\frac{1}{2}} |M_j(Q^*; I_{r-1})|, \tag{2.167b}$$

if j is not contained in I_{m-1} and I_{r-1}; and they all vanish if j is contained in I_{m-1} and I_{r-1}.

Proof. We shall only prove the circuit part since the cut case can be proved in an entirely similar manner. If j is not in I_{m-1}, then

$$M_j(B; I_{m-1}) = \pm \det B(I_{m-1} \cup \{j\}), \tag{2.168a}$$

$$M_j(B^*; I_{m-1}) = \pm \det B^*(I_{m-1} \cup \{j\}). \tag{2.168b}$$

The lemma follows from here. If j is in I_{m-1}, let us consider the matrix R which is obtained from $B(I_{m-1})$ by inserting the jth column of B next to the left of the column corresponding to j in $B(I_{m-1})$. Now if we expand $\det R$ along the inserted column of R, we find $\det R = \pm M_j(B; I_{m-1})$. Since R has two identical columns, it follows that $\det R = 0$. This completes the proof of the lemma.

Thus, if $M_j(B; I_{m-1})$ and $M_j(B; J_{m-1})$ are nonzero, from (2.168a) and Corollary 2.15 they must have the same magnitude. This is similarly valid for the cut case.

COROLLARY 2.32: The magnitude of $M_j(B_f; I_{m-1})$ and $M_j(Q_f; I_{r-1})$, if not identically zero, is equal to unity, thus being invariant with respect to the transformation from one such system to another.

COROLLARY 2.33:

$$k(B^*) M_i(B; I_{m-1}) M_j(B; J_{m-1}) = k(B) M_i(B^*; I_{m-1}) M_j(B^*; J_{m-1}), \tag{2.169a}$$

$$k(Q^*) M_i(Q; I_{r-1}) M_j(Q; J_{r-1}) = k(Q) M_i(Q^*; I_{r-1}) M_j(Q^*; J_{r-1}).$$
(2.169b)

COROLLARY 2.34: (1) $M_j(Q; I_{r-1})$ is nonzero if, and only if, the subgraph corresponding to the integers in I_{r-1} is a tree in the directed graph G_α obtained from G by shorting the edge e_j corresponding to the integer j.

(2) $M_j(B; I_{m-1})$ is nonzero if, and only if, the subgraph corresponding to the integers in I_{m-1} is a cotree in $G - e_j$.

Proof. Since $M_j(Q; I_{r-1})$ is nonzero if, and only if, $Q(I_{r-1} \cup \{j\})$ is a major submatrix of Q where j is not in I_{r-1}, and since from Corollary 2.3 det $Q(I_{r-1} \cup \{j\}) \neq 0$ if, and only if, the subgraph corresponding to the integers in $I_{r-1} \cup \{j\}$ is a tree containing the edge e_j in G, the first part of the corollary follows from the fact that trees containing the edge e_j in G are in one-to-one correspondence with the trees in G_α (another relation will be given in Lemma 2.11). Similarly, we can prove (2).

DEFINITION 2.26: *Generalized cofactor.* For arbitrarily given $b \times b$ matrices Z and Y, and for a given directed graph G, let $(B_{-j})' = B'_{-j}$, $(Q_{-j})' = Q'_{-j}$, and

$$M_{uv}(B) = \sum_{i=1}^{m} \sum_{j=1}^{m} (-1)^{i+j} b_{iu} b_{jv} \det B_{-i} Z B'_{-j},$$
(2.170a)

$$M_{uv}(Q) = \sum_{i=1}^{r} \sum_{j=1}^{r} (-1)^{i+j} q_{iu} q_{jv} \det Q_{-i} Y Q'_{-j}$$
(2.170b)

for $u, v = 1, 2, \ldots, b$. The scalar quantities $M_{uv}(B)$ and $M_{uv}(Q)$ are called the *generalized cofactors* of the elements of the matrices BZB' and QYQ', respectively.

The reason for the name "generalized cofactor" is that the generalized cofactors reduce to the usual cofactors under additional constraints. For example, if edges e_u and e_v of G corresponding to the columns u and v in B are contained only in the circuits L_k and L_h corresponding to the rows k and h in B, and oriented in the same directions as the circuits L_k and L_h, respectively, then we have

$$M_{uv}(B) = (-1)^{k+h} \det B_{-k} Z B'_{-h},$$
(2.171)

which in fact is the cofactor of the (k, h)-element of BZB'. If in addition the edges and circuits of G are labeled in such a way that $k = u$ and $h = v$, then $M_{uv}(B)$ is simply the cofactor of the (u, v)-element of BZB'. This is similarly valid for the generalized cofactors $M_{uv}(Q)$ of the elements of QYQ'.

As an illustration, consider the directed graph G as shown in fig. 2.15. The two basis circuit matrices of G are given below:

$$B = \begin{bmatrix} -1 & 1 & -1 & 0 & 0 \\ 0 & 0 & 1 & 1 & 1 \end{bmatrix}, \quad (2.172a)$$

$$B^* = \begin{bmatrix} -1 & 1 & -1 & 0 & 0 \\ -1 & 1 & 0 & 1 & 1 \end{bmatrix}. \quad (2.172b)$$

If Z is the same as that given in (2.68), the generalized cofactors $M_{12}(B)$ and $M_{12}(B^*)$ of the elements of BZB' and $B^*ZB^{*\prime}$, respectively, are given by

$$M_{12}(B) = \sum_{i=1}^{2} \sum_{j=1}^{2} (-1)^{i+j} b_{i1} b_{j2} \det B_{-i} ZB'_{-j}$$
$$= -\det B_{-1} ZB'_{-1} = -(R_3 + R_4 + R_5), \quad (2.173a)$$

$$M_{12}(B^*) = -\det B^*_{-1} ZB^{*\prime}_{-1} + \det B^*_{-1} ZB^{*\prime}_{-2} + \det B^*_{-2} ZB^{*\prime}_{-1} - \det B^*_{-2} ZB^{*\prime}_{-2}$$
$$= -(R_2 + R_4 + R_5) + R_2 + R_2 - (R_2 + R_3)$$
$$= -(R_3 + R_4 + R_5). \quad (2.173b)$$

In the following, we shall show that under certain conditions the generalized cofactors are invariant with respect to the choice of the circuits and cuts.

THEOREM 2.23: Let B and B^* be two basis circuit matrices of a directed graph G, and also let Q and Q^* be two basis cut matrices of G. Then for arbitrarily given square matrices Z and Y of order b, we have

$$k(B^*) M_{uv}(B) = k(B) M_{uv}(B^*), \quad (2.174a)$$

$$k(Q^*) M_{uv}(Q) = k(Q) M_{uv}(Q^*) \quad (2.174b)$$

for $u, v = 1, 2, ..., b$.

Proof. Let

$$B^* = [b^*_{ij}] \text{ and } Q^* = [q^*_{ij}]. \quad (2.175)$$

Then we have

$$M_{uv}(B) = \sum_{i=1}^{m} \sum_{j=1}^{m} (-1)^{i+j} b_{iu} b_{jv} \det B_{-i} ZB'_{-j}$$
$$= \sum_{i=1}^{m} \sum_{j=1}^{m} \left[(-1)^{i+j} b_{iu} b_{jv} \left(\sum_{(I_{m-1})} \sum_{(J_{m-1})} \det B_{-i}(I_{m-1}) \right. \right.$$
$$\left. \left. \times \det Z(I_{m-1}, J_{m-1}) \det B_{-j}(J_{m-1}) \right) \right]$$

§ 4   Invariance and mutual relations   99

$$\begin{aligned}
&= \sum_{(I_{m-1})} \sum_{(J_{m-1})} \left[\left(\sum_{i=1}^{m} (-1)^i b_{iu} \det \boldsymbol{B}_{-i}(I_{m-1}) \right) \det \boldsymbol{Z}(I_{m-1}, J_{m-1}) \right. \\
&\qquad\qquad\qquad \left. \times \left(\sum_{j=1}^{m} (-1)^j b_{jv} \det \boldsymbol{B}_{-j}(J_{m-1}) \right) \right] \\
&= \sum_{(I_{m-1})} \sum_{(J_{m-1})} M_u(\boldsymbol{B}; I_{m-1}) \det \boldsymbol{Z}(I_{m-1}, J_{m-1}) M_v(\boldsymbol{B}; J_{m-1}) \\
&= \sum_{(I_{m-1})} \sum_{(J_{m-1})} k(\boldsymbol{B}) M_u(\boldsymbol{B}^*; I_{m-1}) \\
&\qquad\qquad\qquad \times \det \boldsymbol{Z}(I_{m-1}, J_{m-1}) M_v(\boldsymbol{B}^*; J_{m-1})/k(\boldsymbol{B}^*) \\
&= [k(\boldsymbol{B})/k(\boldsymbol{B}^*)] \sum_{(I_{m-1})} \sum_{(J_{m-1})} M_u(\boldsymbol{B}^*; I_{m-1}) \\
&\qquad\qquad\qquad \times \det \boldsymbol{Z}(I_{m-1}, J_{m-1}) M_v(\boldsymbol{B}^*; J_{m-1}) \\
&= k(\boldsymbol{B}) M_{uv}(\boldsymbol{B}^*)/k(\boldsymbol{B}^*). \qquad\qquad (2.176)
\end{aligned}$$

Similarly, we can prove the other case. This completes the proof of the theorem.

COROLLARY 2.35: *The generalized cofactors of the elements of the network matrix associated with a system of f-circuits or f-cutsets are invariant with respect to the transformation from one such system to another.*

Like the network determinants, we shall use the values of the generalized cofactors associated with a system of f-circuits or f-cutsets as our standard for comparison since they are invariant with respect to these transformations.

DEFINITION 2.27: *Basic value of a generalized cofactor.* The value of a generalized cofactor of the elements of a network matrix associated with a system of f-circuits or f-cutsets of the network is called the *basic value* of the generalized cofactor.

Obviously, for each choice of the integers u and v, there is a unique basic value associated with it. In general, we have b^2 generalized cofactors.

COROLLARY 2.36: *For arbitrarily given square matrices Z and Y, the generalized cofactors $M_{uv}(\boldsymbol{B})$ and $M_{uv}(\boldsymbol{Q})$ of the elements of $\boldsymbol{BZB}'$ and $\boldsymbol{QYQ}'$, if not identically zero, achieve their minimum absolute values if, and only if, $k(\boldsymbol{B})=1$ and $k(\boldsymbol{Q})=1$, respectively.*

COROLLARY 2.37: *The value of a generalized cofactor of the elements of a network matrix is equal to the basic value of the generalized cofactor multiplied by the square of an integer. The absolute value of the basic value is the minimum absolute value possible for the generalized cofactor.*

COROLLARY 2.38: The generalized cofactors $M_{uv}(B_p)$ and $M_{uv}(A)$ have the basic values.

A direct consequence of Theorems 2.20 and 2.23 is the following.

COROLLARY 2.39: The ratio of the determinant of a network matrix and one of its generalized cofactors or the ratio of two of its generalized cofactors is invariant with respect to the transformation from one system of circuits or cuts to another (corresponding to a basis circuit or cut matrix).

We shall illustrate the above results by the following example.

Example 2.7: Consider the directed graph G as shown in fig. 2.20. The basis cutset matrix Q corresponding to the four cutsets as indicated in fig. 2.21 is given in (2.137a). For simplicity, let Y be the identity matrix of order 8. The generalized cofactors $M_{57}(Q)$ and $M_{57}(A)$ associated with Q and A (with node 5 used as the reference node) are given by

$$M_{57}(Q) = \sum_{i=1}^{4} \sum_{j=1}^{4} (-1)^{i+j} q_{i5} q_{j7} \det Q_{-i} Q'_{-j}$$
$$= \det Q_{-1} Q'_{-3} + \det Q_{-1} Q'_{-4} + \det Q_{-2} Q'_{-3} + \det Q_{-2} Q'_{-4}$$
$$= -36 + 9 + 9 - 36 = 3^2 \cdot (-6), \tag{2.177a}$$

$$M_{57}(A) = \det A_{-1} A'_{-3} + \det A_{-1} A'_{-4} + \det A_{-2} A'_{-3} + \det A_{-2} A'_{-4}$$
$$= 6 - 9 - 9 + 6 = -6. \tag{2.177b}$$

Thus, the basic value of the generalized cofactor $M_{57}(Q)$ is -6, and

$$M_{57}(Q) = k(Q) M_{57}(A), \tag{2.178}$$

where $k(Q) = 9$, which verifies the identity (2.174b).

In the following, we shall establish an important identity which relates the generalized cofactors associated with two types of network matrices.

Let $S = \{1, 2, ..., b\}$. For u and v in S, let $S^{uv} = S - \{u, v\}$. For convenience, let the symbols I_k^{uv} and J_k^{uv} denote the subsets of the set S^{uv} each of which contains k elements. By $\overline{I_k^{uv}}$ and $\overline{J_k^{uv}}$ we mean the complements of I_k^{uv} and J_k^{uv} in S^{uv}, respectively. With these definitions, we have

LEMMA 2.10: If u and v are adjacent columns in B or Q, then

$$k(Q) M_u(B; I_{m-1}^{uv}) M_v(B; J_{m-1}^{uv})$$
$$= (-1)^{\alpha+\beta+1} k(B) M_u(Q; \overline{J_{m-1}^{uv}}) M_v(Q; \overline{I_{m-1}^{uv}}), \tag{2.179}$$

where α and β are the sums of the integers contained in $\overline{I_{m-1}^{uv}}$ and $\overline{J_{m-1}^{uv}}$, respectively.

§ 4   Invariance and mutual relations

Proof. Without loss of generality, let us assume that $v > u$. Since

$$M_u(B; I^{uv}_{m-1}) = (-1)^{u'} \det B(I^{uv}_{m-1} \cup \{u\}), \tag{2.180a}$$

$$M_v(B; J^{uv}_{m-1}) = (-1)^{v'} \det B(J^{uv}_{m-1} \cup \{v\}), \tag{2.180b}$$

where u' and v' are the column indices of the columns corresponding to the integers u and v in $B(I^{uv}_{m-1} \cup \{u\})$ and $B(J^{uv}_{m-1} \cup \{v\})$, respectively, it follows from (2.146) that

$$k(Q) M_u(B; I^{uv}_{m-1}) M_v(B; J^{uv}_{m-1})$$
$$= (-1)^{u'+v'+\alpha+\beta+u+v} k(B) \det Q(\overline{I^{uv}_{m-1}} \cup \{v\}) \det Q(\overline{J^{uv}_{m-1}} \cup \{u\}). \tag{2.181}$$

Since

$$M_v(Q; \overline{I^{uv}_{m-1}}) = (-1)^{v''} \det Q(\overline{I^{uv}_{m-1}} \cup \{v\}), \tag{2.182a}$$

$$M_u(Q; \overline{J^{uv}_{m-1}}) = (-1)^{u''} \det Q(\overline{J^{uv}_{m-1}} \cup \{u\}), \tag{2.182b}$$

where v'' and u'' denote the column indices of the columns corresponding to the integers v and u in $Q(\overline{I^{uv}_{m-1}} \cup \{v\})$ and $Q(\overline{J^{uv}_{m-1}} \cup \{u\})$, respectively, and since $u = u' + v'' - 1$ and $v = v' + u''$, it follows that

$$k(Q) M_u(B; I^{uv}_{m-1}) M_v(B; J^{uv}_{m-1})$$
$$= (-1)^{\alpha+\beta+1} k(B) M_u(Q; \overline{J^{uv}_{m-1}}) M_v(Q; \overline{I^{uv}_{m-1}}). \tag{2.183}$$

So the lemma is proved.

DEFINITION 2.28: *Uncoupled edges.* For a given square matrix Z of order b and for a given directed graph G, the edges e_u and e_v of G corresponding to the rows (or colums) u and v of Z are said to be *uncoupled* with respect to Z if $z_{ui} = z_{iu} = 0$ for $i \neq u$ and $i = 1, 2, \ldots, b$; and $z_{vj} = z_{jv} = 0$ for $j \neq v$ and $j = 1, 2, \ldots, b$, where $Z = [z_{ij}]$.

In other words, if $z_{uu} \neq 0$ and $z_{vv} \neq 0$ they are the only nonzero entries in the rows and columns u and v of Z. For convenience, by $Z_{uu,vv}$ we mean the submatrix obtained from Z by deleting the rows and the columns u and v.

THEOREM 2.24: If Z is nonsingular such that $Y = Z^{-1}$, then

$$k(Q) M_{uv}(B) = -k(B) (\det Z_{uu,vv}) M_{uv}(Q) \tag{2.184}$$

for all uncoupled edges e_u and e_v, $u \neq v$, of a directed graph with respect to Z.

Proof. Let $H = Z_{uu,vv}$. From Problem 2.35, we may assume, without loss of generality, that $u=b-1$ and $v=b$. From the fourth line of (2.176) we have

$$\begin{aligned}
-M_{uv}(B) &= -\sum_{(I_{m-1})}\sum_{(J_{m-1})} M_u(B; I_{m-1}) \det Z(I_{m-1}, J_{m-1}) M_v(B; J_{m-1}) \\
&= -\sum_{(I^{uv}_{m-1})}\sum_{(J^{uv}_{m-1})} M_u(B; I^{uv}_{m-1}) M_v(B; J^{uv}_{m-1}) \det Z(I^{uv}_{m-1}, J^{uv}_{m-1}) \\
&= \sum_{(I^{uv}_{m-1})}\sum_{(J^{uv}_{m-1})} [k(B)/k(Q)] M_v(Q; \overline{I^{uv}_{m-1}}) \\
&\qquad \times M_u(Q; \overline{J^{uv}_{m-1}})(-1)^{\alpha+\beta} \det Z(I^{uv}_{m-1}, J^{uv}_{m-1}) \\
&= [k(B)/k(Q)]\sum_{(I^{uv}_{m-1})}\sum_{(J^{uv}_{m-1})} M_v(Q; \overline{I^{uv}_{m-1}}) \\
&\qquad \times M_u(Q; \overline{J^{uv}_{m-1}})(-1)^{\alpha+\beta} \det H(I^{uv}_{m-1}, J^{uv}_{m-1}) \\
&= [k(B)/k(Q)]\sum_{(I^{uv}_{m-1})}\sum_{(J^{uv}_{m-1})} M_v(Q; \overline{I^{uv}_{m-1}}) \\
&\qquad \times M_u(Q; \overline{J^{uv}_{m-1}}) \det H^{-1}(\overline{J^{uv}_{m-1}}, \overline{I^{uv}_{m-1}}) \det H \\
&= [k(B)(\det H)/k(Q)]\sum_{(I^{uv}_{m-1})}\sum_{(J^{uv}_{m-1})} M_v(Q; \overline{I^{uv}_{m-1}}) \\
&\qquad \times M_u(Q; \overline{J^{uv}_{m-1}}) \det Y(\overline{J^{uv}_{m-1}}, \overline{I^{uv}_{m-1}}) \\
&= [k(B)(\det H)/k(Q)]\sum_{(J^{uv}_{m-1})}\sum_{(I^{uv}_{m-1})} M_u(Q; \overline{J^{uv}_{m-1}}) \\
&\qquad \times \det Y(\overline{J^{uv}_{m-1}}, \overline{I^{uv}_{m-1}}) M_v(Q; \overline{I^{uv}_{m-1}}) \\
&= k(B)(\det H) M_{uv}(Q)/k(Q). \qquad (2.185)
\end{aligned}$$

The second line follows from Lemma 2.9 and the fact that the triple product is zero if v is in I_{m-1} or u is in J_{m-1} since e_u and e_v are uncoupled edges of G. The third line follows from (2.179) and the fifth line is obtained by means of Jacobi's theorem of (2.106). All the symbols are defined the same as those in Lemma 2.10. This completes the proof of the theorem.

COROLLARY 2.40: If Z is nonsingular such that $Y = Z^{-1}$, then

$$-M_{uv}(B_f) = (\det Z_{uu,vv}) M_{uv}(Q_f) = (\det Z_{uu,vv}) M_{uv}(A) \qquad (2.186)$$

for all uncoupled edges e_u and e_v, $u \neq v$, of G with respect to Z.

COROLLARY 2.41: If $e_s, e_t, e_u,$ and e_v are uncoupled edges of G with respect to Z, and if Z is nonsingular such that $Y = Z^{-1}$, then for $s \neq t$ and $u \neq v$ we have

$$M_{uv}(B)/M_{st}(B) = K_1 M_{uv}(Q)/M_{st}(Q), \qquad (2.187a)$$

$$M_{uv}(B)/(\det BZB') = K_2 M_{uv}(Q)/(\det QYQ'), \qquad (2.187b)$$

where $K_1 = z_{ss}z_{tt}/z_{uu}z_{vv}$ and $K_2 = -1/z_{uu}z_{vv}$.

Like the case for the network determinants discussed in the preceding section, let G be a network having no independent sources, whose branch-impedance and branch-admittance matrices are Z_{pp} and Y_{pp}. If Z, B, Y, and Q are the matrices of the associated directed graphs G_v and G_i, respectively, of G defined in the preceding section, and if they are partitioned according to (2.156) and (2.157), then we have

COROLLARY 2.42: Let Y_{pp} be the inverse of Z_{pp}. If e_s and e_t are the edges of G corresponding to the columns or rows s and t of Y_{pp}, then for $s \neq t$ we have

$$k(Q) M_{st}(B) = -k(B)(\det Z_{pp,ss,tt}) M_{st}(Q) \qquad (2.188a)$$

and

$$M_{st}(B) = M_{st}(B_{12}) \quad \text{and} \quad M_{st}(Q) = M_{st}(Q_{12}) \qquad (2.188b)$$

for all uncoupled edges e_s and e_t of G with respect to Z_{pp}, where $Z_{pp,ss,tt}$ is the submatrix obtained from Z_{pp} by deleting the rows and the columns s and t.

We shall now use the above results to show that the network functions are invariant with respect to the transformations of the reference frame, i.e., the choice of loop or cut technique of analysis and the choice of circuits or cuts.

Without loss of generality, we assume that we have removed all the independent sources from G_v or G_i of G except the one corresponding to the edge e_u. Suppose that we wish to compute the transfer admittance function $y_{uu,vv}$ between the edges e_u and e_v of G_v using a loop system of equations. From (2.78) the current i_v in e_v with the direction of e_v taken as the positive orientation of the current, due to a voltage generator v_{gu} at the edge e_u with reference + at the tail of the current-reference arrow, is given by (Problem 2.32)

$$i_v = -v_{gu} \sum_{i=1}^{m} \sum_{j=1}^{m} b_{iu} b_{jv} (-1)^{i+j} (\det B_{-i} Z B'_{-j})/(\det BZB')$$
$$= -v_{gu} M_{uv}(B)/(\det BZB'). \qquad (2.189)$$

According to the convention we have adopted in this book for the current and voltage references, the transfer admittance function $y_{uu,vv}$ is defined as the ratio of i_v to v_{gu},

$$y_{uu,vv} = i_v/v_{gu} = -M_{uv}(B)/(\det BZB'). \qquad (2.190)$$

Similarly, if a cut system of equations is used, the transfer impedance function $z_{uu,vv}$ between the edges e_u and e_v, defined as the ratio of the voltage at e_v to the current at e_u, is given by (Problem 2.33)

$$z_{uu,vv} = -M_{uv}(Q)/(\det QYQ'). \qquad (2.191)$$

Since the driving-point functions are special cases of the transfer functions, and since other types of network functions can also be obtained in a similar manner, it may be concluded that a network function can always be expressed as the ratio of the determinant of a network matrix and one of its generalized cofactors or the reciprocal of this ratio or the ratio of two of its generalized cofactors. For example, the transfer voltage-ratio function and the transfer admittance function between the edges e_u and e_v are related by the impedance associated with the edge e_v of G.

Again, it should be emphasized that the relationship given in (2.184) between the generalized cofactors of the elements of the loop-impedance and the cut-admittance matrices of a network is specified in that part of the network when all the independent sources have been removed. If the independent sources are included, then the associated directed graphs G_v and G_i of G, defined in the preceding section, must be used for the loop and the cut formulations, respectively.

Before we illustrate the above results, we may conclude from the discussions given in this and the preceding sections that we have

THEOREM 2.25: The network functions are invariant with respect to the ways in which the loop currents or the cut voltages are chosen as the variables. In other words, they are invariant with respect to the transformations of the reference frame.

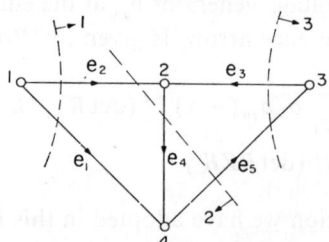

Fig. 2.25. A network considered in Example 2.8.

Example 2.8: Consider the network G as shown in fig. 2.25. The basis cut matrix Q corresponding to the three cutsets as indicated in fig. 2.25 is given by

$$Q = \begin{matrix} & e_1 & e_2 & e_3 & e_4 & e_5 \\ 1 \\ 2 \\ 3 \end{matrix} \begin{bmatrix} 1 & 1 & 0 & 0 & 0 \\ 0 & -1 & 0 & 1 & -1 \\ 0 & 0 & -1 & 0 & 1 \end{bmatrix}. \qquad (2.192)$$

§4 Invariance and mutual relations

The generalized cofactor $M_{15}(Q)$ is obtained as follows:

$$\begin{aligned}M_{15}(Q) &= \det Q_{-1}YQ'_{-2} + \det Q_{-1}YQ'_{-3} \\&= -(z_2/z + y_5)z_3/z + (z_0/z + y_5)z_0/z + (z_0/z + y_5)z_3/z \\&\quad - (z_3/z + y_5 + y_4)z_0/z \\&= -(1 + z_0 y_4)/z,\end{aligned} \qquad (2.193)$$

where Y is given in (2.152). Similarly, the basis circuit matrix B of G corresponding to the two circuits as indicated in fig. 2.22 is given by

$$B = \frac{1}{2}\begin{array}{c}\begin{array}{ccccc}e_1 & e_2 & e_3 & e_4 & e_5\end{array}\\\left[\begin{array}{ccccc}-1 & 1 & 0 & 1 & 0 \\ 0 & 0 & -1 & -1 & -1\end{array}\right].\end{array} \qquad (2.194)$$

The generalized cofactor $M_{15}(B)$ is given by

$$M_{15}(B) = -\det B_{-1}ZB'_{-2} = z_0 + z_4, \qquad (2.195)$$

where Z is given in (2.151). Now it is easy to verify that

$$M_{15}(B) = -(\det Z_{11,55})M_{15}(Q), \qquad (2.196)$$

where $\det Z_{11,55} = z_4 z$, and $k(B) = k(Q) = 1$.

As another example, the generalized cofactors $M_{45}(Q)$ and $M_{45}(B)$ of the elements of QYQ' and BZB', respectively, are computed as follows:

$$\begin{aligned}M_{45}(Q) &= -\det Q_{-2}YQ'_{-2} - \det Q_{-2}YQ'_{-3} \\&= -(y_1 + z_3/z)(y_5 + z_2/z) + z_0^2/z^2 - z_3 z_0/z^2 \\&\quad + (y_1 + z_3/z)(y_5 + z_0/z) \\&= (y_1 z_0 - y_1 z_2 - 1)/z,\end{aligned} \qquad (2.197)$$

$$\begin{aligned}M_{45}(B) &= \det B_{-1}ZB'_{-2} + \det B_{-2}ZB'_{-2} \\&= -z_0 - z_4 + z_1 + z_2 + z_4 \\&= z_1 + z_2 - z_0.\end{aligned} \qquad (2.198)$$

Again, they satisfy the identity

$$M_{45}(B) = -(\det Z_{44,55})M_{45}(Q), \qquad (2.199)$$

where $\det Z_{44,55} = z_1 z$.

Example 2.9: In fig. 2.25, let the edge e_1 denote a voltage generator v_g rather than a passive element, i.e., $z_1 = 0$ in (2.151). Then the corresponding G_i of $G(=G_v)$ is as shown in fig. 2.23 where the edge e_1 corresponds to a current generator i_g. For G_v and G_i to be equivalent, we must have $i_g = v_g y_2$ with references for v_g and i_g as indicated (Thévenin and Norton theorems). Using G_v, the transfer admittance function $y_{11,55}$ of the network can now be obtained from (2.190), and is given by

$$y_{11,55} = -M_{15}(B)/(\det BZB')$$
$$= (\det B_{-1}ZB'_{-2})/(\det BZB')$$
$$= -(z_0 + z_4)/[z_2(z_3 + z_4 + z_5) + z_4(z_3 + z_5) - z_0^2 - 2z_0 z_4], \quad (2.200)$$

where B is given in (2.194), and Z in (2.151) with $z_1 = 0$. Similarly, if we use G_i the transfer impedance function $z_{11,55}$ of the network can be obtained from (2.191), and is given by

$$z_{11,55} = -M_{15}(A)/(\det AYA')$$
$$= (\det A_{-1}YA'_{-2})/(\det AYA')$$
$$= (z_0 - z_2)/(1 + z_2 y_4 + y_5 z_3 + y_5 z_2 + y_5 y_4 z - 2z_0 y_5), \quad (2.201)$$

where A is the basis incidence matrix of G_i with reference node 4, and Y is given in (2.152) with $y_1 = 0$.

* The driving-point admittance function $y_{11,22}$ faced by the voltage generator in G_v is given by

$$y_{11,22} = i_2/v_g = -M_{12}(B)/(\det BZB')$$
$$= (\det B_{-1}ZB'_{-1})/(\det BZB')$$
$$= (z_3 + z_4 + z_5)/[z_2(z_3 + z_4 + z_5) + z_4(z_3 + z_5) - z_0^2 - 2z_0 z_4].$$
$$(2.202)$$

The driving-point impedance function $z_{11,22}$ faced by the current generator in G_i is given by

$$z_{11,22} = -v_2/i_g = M_{12}(A)/(\det AYA')$$
$$= (\det A_{-1}YA'_{-1})/(\det AYA')$$
$$= (z_2 + y_5 z)/(1 + z_2 y_4 + y_5 z_3 + y_5 z_2 + y_5 y_4 z - 2z_0 y_5). \quad (2.203)$$

Note that $z_{11,22}$ is not the reciprocal of $y_{11,22}$ since the driving-point functions are different in G_v and G_i. However, if we assume that there is no mutual coupling between the edges e_2 and e_3, i.e., $z_0 = 0$, then

$$1/y_{11,22} = z_2 + z_4(z_3 + z_5)/(z_3 + z_4 + z_5), \quad (2.204a)$$
$$1/z_{11,22} = y_2 + (z_3 + z_4 + z_5)/z_4(z_3 + z_5). \quad (2.204b)$$

Thus, the second terms in (2.204a) and (2.204b) are reciprocal to each other.

Evidently, if e_2 has no coupling with the rest of the network, the statement is true in general since they are the driving-point impedance and admittance functions faced by e_2 in the network with all the sources removed.

Finally, let us compute $M_{45}(A)$ and $M_{45}(B)$ in G_i and G_v, respectively,

$$M_{45}(A) = \det A_{-1} Y A'_{-2} = (z_0 - z_2)/z, \tag{2.205a}$$

$$\begin{aligned} M_{45}(B) &= \det B_{-1} Z B'_{-2} + \det B_{-2} Z B'_{-2} \\ &= -(z_0 + z_4) + (z_2 + z_4) = -z_0 + z_2. \end{aligned} \tag{2.205b}$$

It is easy to verify that

$$k(A) M_{45}(B) = -k(B) (\det Z_{pp, 44, 55}) M_{45}(A), \tag{2.205c}$$

where $\det Z_{pp, 44, 55} = z$, and $k(A) = k(B) = 1$.

§ 5. Invariance and the incidence functions

In the foregoing, we have considered the problem of invariance of the network determinants and the generalized cofactors with respect to the choice of circuits or cuts in a given network G. In this section, we shall consider the corresponding problem of invariance of these determinants and generalized cofactors with respect to the choice of the "incidence functions" of G.

As mentioned in § 2, in electrical network theory, the orientations or directions of the edges of G are "pseudo-orientations", used in lieu of an elaborate reference system. In other words, they are used only for the purpose of giving references for the branch voltages and currents. The choice of these references can best be described by a function.

DEFINITION 2.29: *Incidence function.* For a given directed graph G, a function which specifies the orientations of the edges of G is called the *incidence function* of G.

By altering the incidence function of G we mean by reversing the orientations for some of the edges of G. In the following we shall show that if the incidence function of G is altered, the determinants of the network matrices BZB' and QYQ' and their generalized cofactors, in general, *do not* remain invariant. However, if matrices Z and Y are the branch-impedance and the branch-admittance matrices of an electrical network, respectively, then they are invariant with respect to the choice of the incidence functions.

Consider the directed graph G as shown in fig. 2.26(a). Let B be the basis circuit matrix corresponding to the two meshes of G. Also let $z_{ii} = z_i$ and $z_{ij} = 0$,

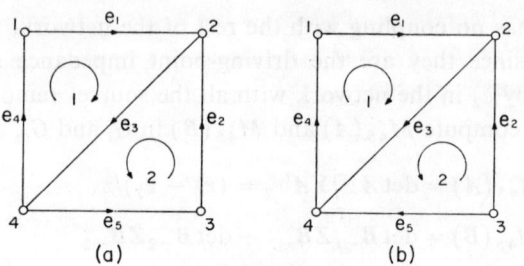

Fig. 2.26. Two choices of the incidence functions of a network.

$i \neq j$, except z_{35} which is z_6, where $Z = [z_{ij}]$ and $i, j = 1, 2, 3, 4, 5$. Then we have

$$BZB' = \begin{bmatrix} 1 & 0 & 1 & 1 & 0 \\ 0 & 1 & 1 & 0 & 1 \end{bmatrix} \begin{bmatrix} z_1 & 0 & 0 & 0 & 0 \\ 0 & z_2 & 0 & 0 & 0 \\ 0 & 0 & z_3 & 0 & z_6 \\ 0 & 0 & 0 & z_4 & 0 \\ 0 & 0 & 0 & 0 & z_5 \end{bmatrix} \begin{bmatrix} 1 & 0 \\ 0 & 1 \\ 1 & 1 \\ 1 & 0 \\ 0 & 1 \end{bmatrix}$$

$$= \begin{bmatrix} z_1 + z_3 + z_4 & z_3 + z_6 \\ z_3 & z_2 + z_3 + z_5 + z_6 \end{bmatrix}. \quad (2.206)$$

Thus, we have

$$\det BZB' = (z_1 + z_4)(z_2 + z_3 + z_5 + z_6) + z_3(z_2 + z_5). \quad (2.207)$$

Now if we reverse the orientation of the edge e_5 in G, we obtain a new directed graph G^* as shown in fig. 2.26(b). If B^* is the corresponding basis circuit matrix of G in G^*, then

$$\det B^*ZB^{*'} = (z_1 + z_4)(z_2 + z_3 + z_5 - z_6) + z_3(z_2 + z_5). \quad (2.208)$$

The generalized cofactors $M_{21}(B)$ and $M_{21}(B^*)$ of the elements of BZB' and $B^*ZB^{*'}$, respectively, are given by

$$M_{21}(B) = -\det B_{-2}ZB'_{-1} = -(z_3 + z_6), \quad (2.209a)$$

$$M_{21}(B^*) = -\det B^*_{-2}ZB^{*'}_{-1} = -(z_3 - z_6). \quad (2.209b)$$

Thus, we conclude that the determinant of the network matrix and its generalized cofactors do not in general remain invariant with respect to the altering of the incidence function of G. However, we see that they remain invariant under certain conditions, and for the most common and important type of directed graphs which represent the electrical networks, they are invariant.

For a given directed graph G, let G^* be the directed graph obtained from G

§ 5 Invariance and the incidence functions

by reversing the orientations of some of the edges of G. If B and B^* are the basis circuit matrices of G and G^* corresponding to the same set of circuits in G and G^*, respectively, then

$$B = B^*D, \qquad (2.210)$$

where D is a $b \times b$ diagonal matrix whose iith entry d_{ii} corresponds to the edge e_i of G or G^*; and $d_{ii} = 1$ if the orientation of the edge e_i in G is the same as that of e_i in G^*, and $d_{ii} = -1$ otherwise. A slight generalization of this is that if B and B^* correspond to two different sets of circuits in G and G^*, then

$$B = CB^*D, \qquad (2.211)$$

where C is an $m \times m$ nonsingular matrix.

We shall now consider some sufficient conditions under which the determinant of the network matrix and its generalized cofactors will remain invariant with respect to the choice of the incidence functions.

THEOREM 2.26: If the matrices Z and Y are diagonal, then the determinants of the network matrices BZB' and QYQ' and their generalized cofactors $M_{uv}(B)$ and $M_{uv}(Q)$ associated with a directed graph G are invariant with respect to the altering of the incidence function of G.

Proof. We shall only consider the determinant of the matrix BZB', since the others can be proved in a similar manner. For our purpose, it is sufficient to show that if B and B^* are the basis circuit matrices of G and G^* corresponding to the same set of circuits in G and G^*, respectively, then $\det BZB' = \det B^*ZB^{*\prime}$. This is evident since

$$\det BZB' = \det B^*DZD'B^{*\prime}$$
$$= \det B^*DD'ZB^{*\prime}. \qquad (2.212)$$

The second line of (2.212) follows from the fact that Z is diagonal. Since DD' is the identity matrix of order b, the theorem follows from here.

Let G be an electrical network, and also let $I(I^*)$ and $V(V^*)$ be the branch-current and the branch-voltage vectors of $G(G^*)$. Without loss of generality, we assume that the same set of circuits is chosen for both G and G^*. If E and E^* are the branch voltage-source vectors of G and G^*, respectively, then

$$B = B^*D, \qquad (2.213a)$$
$$V = ZI + E, \qquad (2.213b)$$
$$V^* = Z^*I^* + E^*, \qquad (2.213c)$$

where Z and Z^* are the branch-impedance matrices of G and G^*, respectively. Since

$$E = DE^*, \quad (2.214a)$$

$$V = DV^*, \quad (2.214b)$$

$$I = DI^*, \quad (2.214c)$$

it follows that, after a simple substitution and the fact that DD is the identity matrix, we have

$$Z^* = DZD. \quad (2.215)$$

Thus, the invariant character of the determinant of the loop-impedance matrix of a network with respect to the choice of the incidence functions can easily be established as follows:

$$\det BZB' = \det B^*DZD'B^{*\prime}$$
$$= \det B^*Z^*B^{*\prime}. \quad (2.216)$$

This is of course similarly valid for the determinant of the cut-admittance matrix of the network, and the generalized cofactors of the elements of these matrices (Problem 2.36).

THEOREM 2.27: The determinant of the loop-impedance or the cut-admittance matrix of a network and its generalized cofactors are invariant with respect to the choice of the incidence functions of the network.

As an illustration, in fig. 2.26(a) let G be an electrical network, and also let Z in (2.206) be the branch-impedance matrix of G. If G^* as shown in fig. 2.26(b) is the directed graph obtained from G by reversing the orientation of the edge e_5 in G, then we have

$$D = \begin{bmatrix} 1 & 0 & 0 & 0 & 0 \\ 0 & 1 & 0 & 0 & 0 \\ 0 & 0 & 1 & 0 & 0 \\ 0 & 0 & 0 & 1 & 0 \\ 0 & 0 & 0 & 0 & -1 \end{bmatrix}. \quad (2.217)$$

From (2.215), the branch-impedance matrix Z^* of G^* is given by

$$Z^* = \begin{bmatrix} z_1 & 0 & 0 & 0 & 0 \\ 0 & z_2 & 0 & 0 & 0 \\ 0 & 0 & z_3 & 0 & -z_6 \\ 0 & 0 & 0 & z_4 & 0 \\ 0 & 0 & 0 & 0 & z_5 \end{bmatrix}. \quad (2.218)$$

Thus, we have

$$\det BZB' = \det B^*Z^*B^{*\prime}$$
$$= (z_1 + z_4)(z_2 + z_3 + z_5 + z_6) + z_3(z_2 + z_5), \quad (2.219\text{a})$$
$$M_{21}(B) = M_{21}(B^*) = -(z_3 + z_6). \quad (2.219\text{b})$$

§ 6. Topological formulas for *RLC* networks

The name *topological formulas* is applied to the formulas for writing down the driving-point and transfer functions of a network by inspection of the network diagram without actually expanding various determinants and cofactors. As such, these formulas have applications to both network analysis and network synthesis. In analysis, they provide a short-cut method of evaluating network determinants and cofactors because the usual cancellations inherent in evaluating these determinants are avoided. In synthesis, they present the possibilities of discovering radically new synthesis procedures for lumped networks which can be extended to cover all lumped systems.

Like many other topics discussed in this book, the basic concepts of topological formulas are not new. More than a century ago, KIRCHHOFF [1847] stated the basic formulas for the loop system of equations, and MAXWELL [1892] gave the basic formulas for the nodal system of equations. Although the rules are fundamentally correct, they are far from complete. Using some of the methods of modern algebra, FRANKLIN [1925] gave a proof of Kirchhoff's impedance rules. Following Franklin, KU [1932] contributed to the theory by giving a useful interpretation of Maxwell's admittance rules. Recently, the formulas have also been proved by NERODE and SHANK [1961] and SLEPIAN [1968] using algebraic topology as a tool. However, all these formulas were stated only for the resistive networks. Obviously, they can easily be extended to ordinary *RLC* networks as well. Although it is possible to obtain the general formulas for the general networks directly and treat the *RLC* networks as a special case, the special case will be considered first in the following discussion. The reason for this is that the formulas for the *RLC* networks are much the simplest, and that they are sufficiently important to be considered separately. We postpone the treatment of the general case to ch. 4.

6.1. *Network determinants and trees and cotrees*

In this and following sections, we consider only *RLC* networks G with $r \neq 0$ and $m \neq 0$. From the discussion given in § 4, it is evident that the results can also be applied to *RLC* networks with independent sources. Since the branch-impedance and the branch-admittance matrices Z and Y of G are diagonal and nonsingular, it is convenient to associate the iith entry z_{ii} (y_{ii}) of Z (Y) with the edge e_i of G for $i = 1, 2, \ldots, b$. The weight z_{ii} (y_{ii}) associated with the edge e_i is is called its *impedance* (*admittance*).

DEFINITION 2.30: *Tree product*. For a connected *RLC* network G, the product of the weights associated with the edges of a tree of G is called a *tree product* of G. If the weights are all admittances, then the product is referred to as a *tree-admittance product* of G; and if they are all impedances, then a *tree-impedance product* of G.

For convenience, let

$$V(Y) = \sum \text{tree-admittance products}, \qquad (2.220\text{a})$$

$$V(Z) = \sum \text{tree-impedance products}, \qquad (2.220\text{b})$$

where the summations are taken over all possible tree-admittance and tree-impedance products of G, respectively.

THEOREM 2.28: The determinant of the node-admittance matrix Y_n of an *RLC* network is given by

$$\det Y_n = V(Y). \qquad (2.221)$$

Proof. By the Binet-Cauchy theorem of (2.105), we have

$$\begin{aligned}
\det Y_n = \det AYA' &= \sum_{(I_r)} \sum_{(J_r)} \det A(I_r) \det Y(I_r, J_r) \det A(J_r)' \\
&= \sum_{(I_r)} \det A(I_r) \det Y(I_r, I_r) \det A(I_r)' \\
&= \sum_{(I_r)} \det Y(I_r, I_r) = V(Y). \qquad (2.222)
\end{aligned}$$

The second line follows from the fact that Y is diagonal, the third line from Theorem 2.19, and the last equality from Theorem 2.2, where the summations are over all the major submatrices $A(I_r)$ of A. So the theorem is proved.

The above result was first used by MAXWELL [1892] as a short-cut method for the evaluation of the nodal determinant. Its first formal proof was given by Brooks et al. [1940]. A direct consequence of this and (2.134) is the following.

COROLLARY 2.43: The determinant of the cut-admittance matrix Y_c of an *RLC*

network is given by
$$\det Y_c = k(Q) V(Y), \qquad (2.223)$$
where $Y_c = QYQ'$.

Let us illustrate the theorem by means of an example. For the network G of fig. 2.26(a), the set T of trees of G is given by

$$T = \{e_1e_2e_4, e_2e_4e_5, e_1e_4e_5, e_1e_2e_5, e_3e_4e_5, e_1e_2e_3, e_2e_3e_4, e_1e_3e_5\}. \qquad (2.224)$$

By Theorem 2.28, we have

$$\det Y_n = y_1y_2y_4 + y_2y_4y_5 + y_1y_4y_5 + y_1y_2y_5 + y_3y_4y_5 \\ + y_1y_2y_3 + y_2y_3y_4 + y_1y_3y_5, \qquad (2.225)$$

where y_i is the admittance of the edge e_i for $i = 1, 2, 3, 4, 5$.

Three important facts should be noted. Firstly, the node-admittance matrix need not be written; its determinant can be found directly. Secondly, there was no cancellation, and so no unnecessary work was done. One need only write down the node-admittance matrix for fig. 2.26(a) and compute its determinant to appreciate this fact. Finally, the determinant of the node-admittance matrix is independent of the choice of the reference node, a fact that was proved in Corollary 2.23.

Let us now investigate the dual rules given by KIRCHHOFF [1847] for the computation of network response. We interpret his rules in terms of impedances and the loop system of equations, even though Kirchhoff's rules were stated in terms of the branch-current system of equations discussed in § 3.1.

DEFINITION 2.31: *Cotree product.* For a connected *RLC* network G, the product of the weights associated with the edges of a cotree of G is called a *cotree product* of G. If the weights are all admittances, then the product is referred to as a *cotree-admittance product* of G; and if they are all impedances, then a *cotree-impedance product* of G.

Let

$$C[V(Z)] = \sum \text{cotree-impedance products}, \qquad (2.226)$$

where the summation is taken over all cotree-impedance products of G.

THEOREM 2.29: The determinant of the loop-impedance matrix Z_m of an *RLC* network is given by

$$\det Z_m = k(B) C[V(Z)], \qquad (2.227)$$

where $Z_m = BZB'$.

Proof. Since Y is the inverse of Z, from Theorems 2.22 and 2.28 we have

$$\det Z_m = k(B)(\det Z)(\det AYA')$$
$$= k(B)(z_1 z_2 \ldots z_b) V(Y)$$
$$= k(B) C[V(Z)], \qquad (2.228)$$

where z_i is the impedance associated with the edge e_i for $i = 1, 2, \ldots, b$. So the theorem is proved.

The expressions given in (2.221) and (2.227) are known as the *node* and *mesh discriminants*, respectively.

COROLLARY 2.44:

$$\det B_f Z B'_f = \det B_p Z B'_p = C[V(Z)] = z_1 z_2 \ldots z_b V(Y), \qquad (2.229a)$$
$$\det Q_f Y Z'_f = \det AYA' = V(Y) = y_1 y_2 \ldots y_b C[V(Z)], \qquad (2.229b)$$

where $y_i = 1/z_i$ for $i = 1, 2, \ldots, b$.

For example, in fig. 2.26(a) the set $\bar{T}$ of cotrees of G is given by

$$\bar{T} = \{e_3 e_5, e_1 e_3, e_2 e_3, e_3 e_4, e_1 e_2, e_4 e_5, e_1 e_5, e_2 e_4\}. \qquad (2.230)$$

By Theorem 2.29, we have

$$\det Z_m = z_3 z_5 + z_1 z_3 + z_2 z_3 + z_3 z_4 + z_1 z_2 + z_4 z_5 + z_1 z_5 + z_2 z_4. \qquad (2.231)$$

6.2. Generalized cofactors and 2-trees and 2-cotrees

In this section, we present formulas that express the generalized cofactors in terms of the weight products of certain subgraphs of the network.

DEFINITION 2.32: *2-tree.* A spanning subgraph of a directed (undirected) graph G is said to be a *2-tree* of G, if, and only if, it has two components and contains no circuits. One (or, in trivial cases, both) of the components may consist of an isolated node.

Very often, 2-trees in which certain designated nodes are required to be in different components, are needed. For convenience, let the subscripts be used for this purpose. For example, $t_{ab,cde}$ is the symbol for a 2-tree in which the nodes a and b are in one component, and the nodes c, d, and e in the other component. As an example, the sets of 2-trees $t_{12,4}$ and $t_{2,4}$ of the graph of fig. 2.26(a) are presented in figs. 2.27 and 2.28, respectively.

§6 Topological formulas for RLC networks

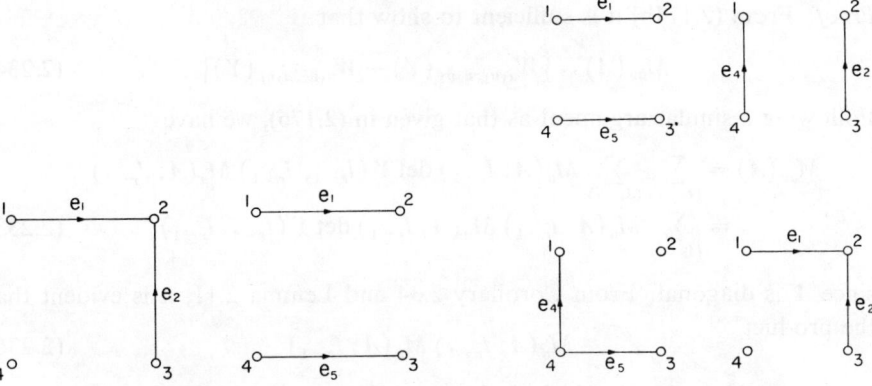

Fig. 2.27. The 2-trees $t_{12,4}$. Fig. 2.28. The 2-trees $t_{2,4}$.

DEFINITION 2.33: *2-tree product*. For an RLC network G, the product of the weights associated with the edges of a 2-tree of G is called a *2-tree product* of G. If the weights are all admittances, then the product is referred to as a *2-tree-admittance product* of G; and if they are all impedances, then a *2-tree-impedance product* of G. A 2-tree such as $t_{a,a}$ in which the same node is required to be in different components has by definition a zero product.

In trivial case, a 2-tree may consist of two isolated nodes. The product of such a 2-tree can be defined, but we do not find it necessary.

Let
$$W_{ab,cd}(Y) = \sum \text{2-tree-admittance products}, \tag{2.232a}$$
$$W_{ab,cd}(Z) = \sum \text{2-tree-impedance products}, \tag{2.232b}$$

where the summations are taken over all possible 2-tree-admittance and 2-tree-impedance products of all the 2-trees $t_{ab,cd}$ of G, respectively.

LEMMA 2.11: For a directed graph G, let G_α be the directed graph obtained from G by shorting the edge e_α. Then there exists a one-to-one correspondence between 2-trees separating the two endpoints of e_α in G and trees in G_α. Furthermore, the corresponding subgraphs in G and G_α have the same product.

The proof is straightforward, and is left as an exercise (Problem 2.37).

THEOREM 2.30: For a connected RLC network G, let $e_u = (u_1, u_2)$ and $e_v = (v_1, v_2)$ be two edges of G. Then the generalized cofactor $M_{uv}(Q)$ of the elements of the cut-admittance matrix Y_c of G is given by

$$M_{uv}(Q) = k(Q)[W_{u_1v_1, u_2v_2}(Y) - W_{u_1v_2, u_2v_1}(Y)], \tag{2.233}$$

where $Y_c = QYQ'$.

Proof. From (2.174b) it is sufficient to show that

$$M_{uv}(A) = [W_{u_1v_1, u_2v_2}(Y) - W_{u_1v_2, u_2v_1}(Y)]. \tag{2.234}$$

Following a similar argument as that given in (2.176), we have

$$\begin{aligned} M_{uv}(A) &= \sum_{(I_{r-1})} \sum_{(J_{r-1})} M_u(A; I_{r-1}) \det Y(I_{r-1}, J_{r-1}) M_v(A; J_{r-1}) \\ &= \sum_{(I_{r-1})} M_u(A; I_{r-1}) M_v(A; I_{r-1}) \det Y(I_{r-1}, I_{r-1}) \end{aligned} \tag{2.235}$$

since Y is diagonal. From Corollary 2.34 and Lemma 2.11, it is evident that the product

$$M_u(A; I_{r-1}) M_v(A; I_{r-1}) \tag{2.236}$$

is nonzero if, and only if, the subgraph of G corresponding to the integers in I_{r-1} is a 2-tree t_{u_1, u_2} as well as a 2-tree t_{v_1, v_2}. Thus, the nonzero product corresponds to a 2-tree $t_{u_1v_1, u_2v_2}$ or $t_{u_1v_2, u_2v_1}$. Since Y is diagonal, it introduces no complications in (2.235). In other words, the triple product in the summation of (2.235) is nonzero if, and only if, it corresponds to a 2-tree-admittance product for the 2-tree $t_{u_1v_1, u_2v_2}$ or $t_{u_1v_2, u_2v_1}$ in G.

It remains to establish the signs to be prefixed to the 2-tree-admittance products. If e_u and e_v are parallel edges or if $e_u = e_v$, the theorem is seen to be true. So let us assume that $e_u \neq e_v$ and that they are not connected in parallel. From Problems 2.35 and 2.59, without loss of generality, let us assume that $u_1 = 1$, $u_2 = 4$, $v_1 = 2$, $v_2 = 3$, $u = 1$, and $v = 2$. Also let the reference node of G be node 4. Then

$$A(I_{r-1} \cup \{1, 2\}) = \begin{matrix} & e_1 & e_2 & (I_{r-1}) \\ & \begin{bmatrix} 1 & 0 & A_1 \\ 0 & 1 & A_2 \\ 0 & -1 & A_3 \\ 0 & 0 & F \end{bmatrix} \end{matrix}, \tag{2.237}$$

where A_1, A_2, and A_3 are row vectors, and F is a matrix of order $(r-3) \times (r-1)$. Note that the integers 1 and 2 cannot be contained in I_{r-1} for (2.236) to be nonzero. Thus, we have

$$M_u(A; I_{r-1}) = -\det A(I_{r-1} \cup \{1\}) = -\det F_1, \tag{2.238a}$$

$$M_v(A; I_{r-1}) = -\det A(I_{r-1} \cup \{2\}) = \det F_2, \tag{2.238b}$$

where

$$F_1 = \begin{bmatrix} A_2 \\ A_3 \\ F \end{bmatrix} \quad \text{and} \quad F_2 = \begin{bmatrix} A_1 \\ A_2 + A_3 \\ F \end{bmatrix}. \tag{2.239}$$

Two cases are considered; they correspond to two types of 2-trees discussed above.

Case 1: *2-trees are of the type* $t_{12,34}$. Now the sum of all the rows of $A(I_{r-1})$ which correspond to the nodes in the same component of the 2-tree $t_{12,34}$ as the nodes 1 and 2 is zero, since these rows form the incidence (not basis incidence) matrix of the component. Denote this sum by the row matrix $\sum A_i$. Then $A_1 = -(\sum A_i - A_1)$. Hence, since adding rows of a matrix does not affect the value of its determinant, we may substitute the first row of F_1 by $-A_1$ and the second row by $A_2 + A_3$. In other words, we have

$$\det F_1 = -\det F_2. \tag{2.240}$$

It follows from (2.238) that the product in (2.236) has the value 1.

Case 2: *2-trees are of the type* $t_{13,24}$. Now the sum of all the rows of $A(I_{r-1})$ which correspond to the nodes in the same component of the 2-tree $t_{13,24}$ as the nodes 1 and 3 is zero. Again denote this sum by the row matrix $\sum A_i$. Then $A_1 = -(\sum A_i - A_1)$. Hence, we may substitute the first row of F_1 by $A_2 + A_3$, and the second row by $-A_1$. This leads to

$$\det F_1 = \det F_2. \tag{2.241}$$

From (2.238) and (2.241), it follows that the product in (2.236) has the value -1. Thus, we have

$$M_{uv}(A) = [W_{12,34}(Y) - W_{13,24}(Y)]. \tag{2.242}$$

Similarly, we can prove the case where $u_1 = v_1$ or v_2, or $u_2 = v_1$ or v_2 (Problem 2.59). This completes the proof of the theorem.

Note that all the terms in (2.233) are distinct. This means that one does not calculate any superfluous terms in following the formula, as one does in evaluating network determinants. Only those terms which do not cancel are included. This is one of the typical characters of all the topological formulas.

The term 2-tree was first used by PERCIVAL [1953]. Using his geometrical symbol for a 2-tree, the formula (2.233) may be expressed in the intuitive fashion as shown in fig. 2.29.

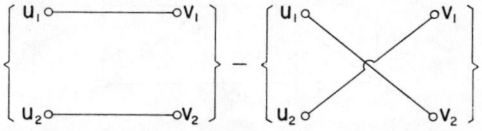

Fig. 2.29. A geometric representation of the formula (2.233).

COROLLARY 2.45:

$$M_{uv}(Q_f) = M_{uv}(A) = [W_{u_1v_1, u_2v_2}(Y) - W_{u_1v_2, u_2v_1}(Y)]. \quad (2.243)$$

COROLLARY 2.46: Let k be the reference node of G. Then the cofactor Δ_{ij}^n of the (i,j)-element of the node-admittance matrix of G is given by

$$\Delta_{ij}^n = W_{ij,k}(Y). \quad (2.244)$$

Proof. In Theorem 2.30, let $u_1 = i$, $v_1 = j$, and $u_2 = v_2 = k$. Then

$$M_{uv}(A) = (-1)^{i+j} \det A_{-i} Y A'_{-j} = \Delta_{ij}^n. \quad (2.245)$$

The corollary follows from (2.233) and the fact that $W_{ik,kj}(Y) = 0$.

COROLLARY 2.47: If the edges $e_u = (u_1, u_2)$ and $e_v = (v_1, v_2)$ are contained only in the cuts i and j, respectively, and if they are oriented in the same directions as the cuts, then the cofactor Δ_{ij}^c of the (i,j)-element of the cut-admittance matrix of G formed by the basis cut matrix Q (including the cuts i and j) is given by

$$\Delta_{ij}^c = k(Q) [W_{u_1v_1, u_2v_2}(Y) - W_{u_1v_2, u_2v_1}(Y)]. \quad (2.246)$$

Proof. From (2.170b) we have

$$M_{uv}(Q) = (-1)^{i+j} \det Q_{-i} Y Q'_{-j} = \Delta_{ij}^c. \quad (2.247)$$

The corollary follows from here.

Obviously, the set of f-cutsets with respect to a tree containing the edges e_u and e_v of G satisfies the requirements of Corollary 2.47.

Example 2.10: Consider the *RLC* network G as shown in fig. 2.30. The generalized cofactor $M_{13}(Q)$ of the elements of the cutset-admittance matrix Y_c

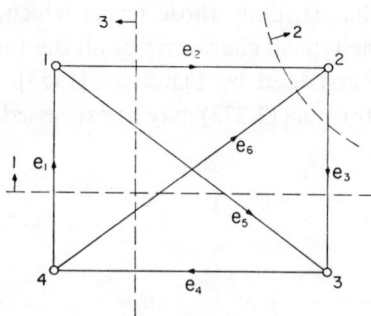

Fig. 2.30. An *RLC* network considered in Example 2.10.

formed by the three cutsets, as indicated in fig. 2.30, can now be obtained directly from G by means of (2.233), where $e_1 = (4, 1)$ and $e_3 = (2, 3)$. The result is given by

$$M_{13}(Q) = [W_{42,13}(Y) - W_{43,12}(Y)] = y_5 y_6 - y_2 y_4, \qquad (2.248)$$

where y_i is the admittance of the edge e_i for $i = 1, 2, \ldots, 6$. For illustrative purpose, let us compute $M_{13}(Q)$ in the usual way:

$$\begin{aligned}M_{13}(Q) &= -\det Q_{-1} Y Q'_{-1} + \det Q_{-1} Y Q'_{-2} \\ &= -(y_2 + y_3 + y_6)(y_2 + y_5 + y_4 + y_6) + (y_2 + y_6)^2 \\ &\quad + (y_3 + y_6)(y_2 + y_5 + y_4 + y_6) + (y_2 + y_6)(y_5 - y_6) \\ &= y_5 y_6 - y_2 y_4. \end{aligned} \qquad (2.249)$$

The cofactor Δ^c_{13} of the $(1, 3)$-element of Y_c can be obtained directly from (2.246), where $e_u = (4, 1)$ and $e_v = (3, 4)$:

$$\begin{aligned}\Delta^c_{13} &= [W_{43,14}(Y) - W_{44,13}(Y)] = -W_{4,13}(Y) \\ &= -y_2 y_3 - y_2 y_5 - y_3 y_5 - y_5 y_6.\end{aligned} \qquad (2.250)$$

If the generalized cofactor $M_{24}(A)$ of the elements of the node-admittance matrix Y_n formed by the basis incidence matrix A of G with reference node 4 is needed, it can easily be obtained from (2.243), and is given by

$$M_{24}(A) = [W_{13,24}(Y) - W_{14,23}(Y)] = y_5 y_6 - y_1 y_3. \qquad (2.251)$$

Again for comparison, $M_{24}(A)$ is also computed by the usual expansion:

$$\begin{aligned}M_{24}(A) &= \det A_{-1} Y A'_{-3} + \det A_{-2} Y A'_{-3} \\ &= y_2 y_3 + y_5(y_2 + y_3 + y_6) - y_3(y_1 + y_2 + y_5) - y_2 y_5 \\ &= y_5 y_6 - y_1 y_3.\end{aligned} \qquad (2.252)$$

Finally, the cofactor Δ^n_{13} of the $(1, 3)$-element of Y_n can be computed directly from (2.244), and is given by

$$\begin{aligned}\Delta^n_{13} &= W_{13,4}(Y) \\ &= y_2 y_3 + y_2 y_5 + y_3 y_5 + y_5 y_6.\end{aligned} \qquad (2.253)$$

Let us now consider the dual of the above, i.e., the formulas for the generalized cofactors $M_{uv}(B)$ of the elements of the loop-impedance matrix of an RLC network.

DEFINITION 2.34: *2-cotree*. For a directed (undirected) graph G, the complement of a 2-tree in G is called a *2-cotree* of G.

Like cotrees, if $t_{ab,cd}$ is a 2-tree of G, the symbol $\bar{t}_{ab,cd}$ denotes a 2-cotree with respect to the 2-tree $t_{ab,cd}$.

DEFINITION 2.35: *2-cotree product*. For an *RLC* network G, the product of the weights associated with the edges of a 2-cotree of G is called a 2-*cotree product* of G. If the weights are all admittances, then the product is referred to as a 2-*cotree-admittance product* of G; and if they are all impedances, then a 2-*cotree-impedance product* of G.

In trivial case, if G itself is a 2-tree, the corresponding 2-cotree is null. Like the 2-tree case, the product of such a 2-cotree can be defined, but we do not find it necessary.

Let
$$C[W_{ab,cd}(Y)] = \sum 2\text{-cotree-admittance products}, \qquad (2.254a)$$
$$C[W_{ab,cd}(Z)] = \sum 2\text{-cotree-impedance products}, \qquad (2.254b)$$

where the summations are taken over all possible 2-cotree-admittance and 2-cotree-impedance products of all the 2-cotrees $\bar{t}_{ab,cd}$ of G, respectively.

THEOREM 2.31: For a connected *RLC* network G, let $e_u = (u_1, u_2)$ and $e_v = (v_1, v_2)$ be two edges of G. Then the generalized cofactor $M_{uv}(B)$ of the elements of the loop-impedance matrix Z_m of G is given by

$$M_{uv}(B) = \frac{k(B)}{z_u z_v} (C[W_{u_1v_2,u_2v_1}(Z)] - C[W_{u_1v_1,u_2v_2}(Z)]), \qquad (2.255)$$

where $Z_m = BZB'$, and z_i is the impedance of the edge e_i for $i = 1, 2, \ldots, b$.

Proof. From Theorem 2.24 and Corollary 2.45, we have

$$\begin{aligned}
M_{uv}(B) &= -k(B)(\det Z_{uu,vv}) M_{uv}(A) \\
&= k(B)(z_1 z_2 \cdots z_b / z_u z_v)[W_{u_1v_2,u_2v_1}(Y) - W_{u_1v_1,u_2v_2}(Y)] \\
&= k(B)(1/z_u z_v)(C[W_{u_1v_2,u_2v_1}(Z)] - C[W_{u_1v_1,u_2v_2}(Z)]), \qquad (2.256)
\end{aligned}$$

which completes the proof of the theorem.

Again, the formula inside the parentheses of (2.255) can be written in an intuitive fashion following Percival's symbol as shown in fig. 2.31.

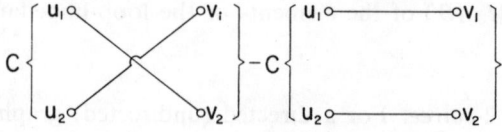

Fig. 2.31. A geometric representation of the formula (2.255).

COROLLARY 2.48:

$$M_{uv}(B_f) = M_{uv}(B_p) = (1/z_u z_v)(C[W_{u_1 v_2, u_2 v_1}(Z)] - C[W_{u_1 v_1, u_2 v_2}(Z)]). \quad (2.257)$$

COROLLARY 2.49: Let G^* be the directed graph obtained from G by removing the edges e_u and e_v. If $C[W^*_{ab,cd}(Z)]$ denotes the sum of all possible 2-cotree-impedance products of all the 2-cotrees $\hat{t}_{ab,cd}$ in G^*, then the generalized cofactor $M_{uv}(B)$ of the elements of the loop-impedance matrix Z_m is given by

$$M_{uv}(B) = k(B)(C[W^*_{u_1 v_2, u_2 v_1}(Z)] - C[W^*_{u_1 v_1, u_2 v_2}(Z)]), \quad (2.258)$$

where $Z_m = BZB'$.

COROLLARY 2.50: If the edges $e_u = (u_1, u_2)$ and $e_v = (v_1, v_2)$ are contained only in the circuits i and j, respectively, and if they are oriented in the same directions as the circuits, then the cofactor Δ^m_{ij} of the (i, j)-element of the loop-impedance matrix of G formed by the basis circuit matrix B (including the circuits i and j) is given by

$$\Delta^m_{ij} = \frac{k(B)}{z_u z_v}(C[W_{u_1 v_2, u_2 v_1}(Z)] - C[W_{u_1 v_1, u_2 v_2}(Z)]) \quad (2.259a)$$

or

$$\Delta^m_{ij} = k(B)(C[W^*_{u_1 v_2, u_2 v_1}(Z)] - C[W^*_{u_1 v_1, u_2 v_2}(Z)]), \quad (2.259b)$$

where $C[W^*_{ab,cd}(Z)]$ denotes the sum of all possible 2-cotree-impedance products of all 2-cotrees $\hat{t}_{ab,cd}$ of G^*, and G^* is obtained from G by removing the edges e_u and e_v.

The proof of the corollary is similar to that of Corollary 2.47, and is left as an exercise (Problem 2.38). Obviously, the set of f-circuits with respect to a tree not containing the edges e_u and e_v of G satisfies the conditions of the above corollary.

Example 2.11: Consider the *RLC* network G as shown in fig. 2.32. The gener-

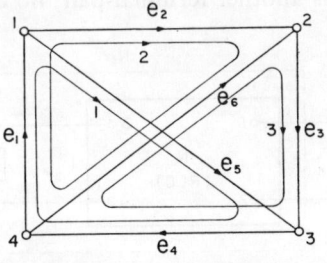

Fig. 2.32. An *RLC* network considered in Example 2.11.

alized cofactor $M_{13}(B)$ of the elements of the loop-impedance matrix Z_m formed by the three circuits, as indicated in fig. 2.32, can now be obtained directly from G by means of (2.255), and is given by

$$M_{13}(B) = (1/z_1 z_3)(C[W_{43,12}(Z)] - C[W_{42,13}(Z)])$$
$$= (1/z_1 z_3)(z_1 z_3 z_5 z_6 - z_1 z_2 z_3 z_4) = z_5 z_6 - z_2 z_4, \quad (2.260)$$

where z_k is the impedance associated with the edge e_k for $k = 1, 2, \ldots, 6$. The cofactor Δ_{23}^m of the $(2, 3)$-element of Z_m can be found directly from G by (2.259), where $e_u = (1, 2)$ and $e_v = (2, 3)$:

$$\Delta_{23}^m = (1/z_2 z_3)(C[W_{13,22}(Z)] - C[W_{12,23}(Z)])$$
$$= (1/z_2 z_3)(z_2 z_3 z_4 z_6 + z_1 z_2 z_3 z_6 + z_1 z_2 z_3 z_4 + z_2 z_3 z_5 z_6)$$
$$= z_4 z_6 + z_1 z_6 + z_1 z_4 + z_5 z_6. \quad (2.261)$$

Finally, the generalized cofactor $M_{16}(B_p)$ of the elements of the loop-impedance matrix formed by the three meshes of G, oriented in clockwise, is given by

$$M_{16}(B_p) = (1/z_1 z_6)(C[W_{42,14}(Z)] - C[W_{44,12}(Z)])$$
$$= -(1/z_1 z_6) C[W_{4,12}(Z)]$$
$$= -(z_3 z_4 + z_4 z_5 + z_2 z_4 + z_3 z_5), \quad (2.262)$$

which is equal to the negative of the cofactor Δ_{13}^m of the $(1, 3)$-element of $B_p Z B_p'$, whose three corresponding meshes are $e_1 e_4 e_5$, $e_2 e_3 e_5$, and $e_3 e_4 e_6$.

6.3. *Topological formulas for RLC two-port networks*

A *two-port network* is defined as a network with two accessible pairs of terminals, the network being electrically and magnetically isolated except for these two pairs of terminals. By calling these "terminal-pairs" or "ports", we imply that external connections are not to be made between one terminal of a terminal-pair and another terminal of a different terminal-pair. For example, if the black box of fig. 2.33 denotes a two-port network N, where $(1, 2)$ is one terminal-pair and $(3, 4)$ is another terminal-pair, no connection is to be made,

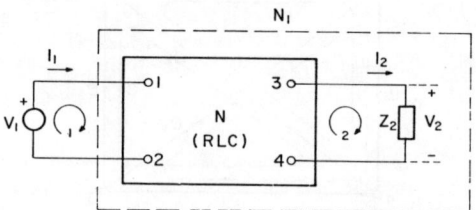

Fig. 2.33. A two-port network excited by a voltage generator.

externally, between terminals 1 and 3 or 2 and 4 or 1 and 4 or 2 and 3. An *RLC two-port network* is a two-port network consisting only of resistors, capacitors, and self-inductors.

The application of the results of the two preceding sections to the analysis of *RLC* two-port networks is straightforward. Therefore, we simply state the results and leave the details as exercises (Problem 2.39).

For simplicity, we shall label the nodes and branches of a two-port network *N* as shown in fig. 2.33 or 2.34. Formulas using the loop system of equations can be derived by assuming a voltage generator at the input as shown in fig. 2.33. Alternative formulas can also be derived using the nodal system of equations and exciting the network with a current generator as shown in fig. 2.34. The choice of circuits and the reference node is indicated in figs. 2.33 and 2.34. The simplest way to apply the foregoing results to the derivation of the

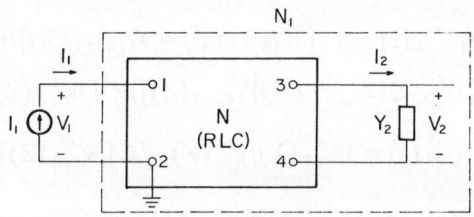

Fig. 2.34. A two-port network excited by a current generator.

formulas is that we consider the voltage and current generators as edges with zero impedance and admittance, respectively. Note that the reference convention for the voltage generator in fig. 2.33 is different from our convention defined earlier. Thus, in applying the preceding results, it will introduce a negative sign. Using the references defined in fig. 2.33, the four transfer functions for the terminated two-port network *N*, under zero initial conditions, are defined as follows[†]:

$$z_{13,24}(s) = \text{transfer impedance function} = v_2(s)/i_1(s), \quad (2.263\text{a})$$

$$\alpha_{13,24}(s) = \text{transfer current-ratio function} = i_2(s)/i_1(s), \quad (2.263\text{b})$$

$$y_{13,24}(s) = \text{transfer admittance function} = i_2(s)/v_1(s), \quad (2.263\text{c})$$

$$g_{13,24}(s) = \text{transfer voltage-ratio function} = v_2(s)/v_1(s). \quad (2.263\text{d})$$

[†] Since the endpoints of the edges are specified, the symbols defined here are more convenient than those used in § 4.5. This should not create any difficulty. The symbols used here, however, are consistent with those used in ch. 4, § 6.

The two driving-point functions defined for the network N_1 of fig. 2.33, again under zero initial conditions, are

$z_{11,22}(s)$ = driving-point impedance function at the terminal-pair (1, 2)
$= v_1(s)/i_1(s),$ (2.264a)

$y_{11,22}(s)$ = driving-point admittance function at the terminal-pair (1, 2)
$= i_1(s)/v_1(s).$ (2.264b)

THEOREM 2.32: For an RLC two-port network N as shown in fig. 2.33 or 2.34, we have

$$z_{13,24}(s) = z_2\alpha_{13,24} = [W_{13,24}(Y) - W_{14,23}(Y)]/V^*(Y) \quad (2.265a)$$

$$= z_2(C[W_{13,24}(Z)] - C[W_{14,23}(Z)])/C[V^*(Z)]; \quad (2.265b)$$

$$y_{13,24}(s) = y_2 g_{13,24}(s) = y_2[W_{13,24}(Y) - W_{14,23}(Y)]/W_{1,2}^*(Y) \quad (2.266a)$$

$$= (C[W_{13,24}(Z)] - C[W_{14,23}(Z)])/C[W_{1,2}^*(Z)]; \quad (2.266b)$$

$$z_{11,22}(s) = 1/y_{11,22}(s) = W_{1,2}^*(Y)/V^*(Y) = C[W_{1,2}^*(Z)]/C[V^*(Z)]; \quad (2.267)$$

where $z_2 = 1/y_2$, and the star indicates that the tree and 2-tree products and their complements are computed for the network N_1, and all other terms are computed for the two-port network N alone.

Example 2.12: Consider the two-port network N as shown in fig. 2.35. The needed terms are computed as follows:

$$W_{13,24}(Y) - W_{14,23}(Y) = y_1 y_3 y_5 y_7 - 0 = y_1 y_3 y_5 y_7, \quad (2.268a)$$

$$C[W_{13,24}(Z)] - C[W_{14,23}(Z)] = z_4 z_6, \quad (2.268b)$$

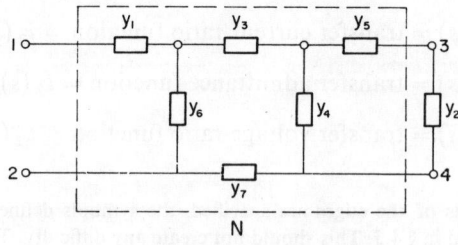

Fig. 2.35. A ladder network.

$$W^*_{1,2}(Y) = y_1y_3y_5y_7 + y_1y_3y_2y_7 + y_1y_3y_4y_5 + y_1y_2y_3y_4 + y_1y_2y_3y_5 +$$
$$y_1y_4y_5y_7 + y_1y_2y_4y_7 + y_1y_2y_5y_7 + y_3y_5y_6y_7 + y_2y_3y_6y_7 +$$
$$y_3y_4y_5y_6 + y_2y_3y_4y_6 + y_2y_3y_5y_6 + y_4y_5y_6y_7 + y_2y_4y_6y_7 +$$
$$y_2y_5y_6y_7 + y_3y_4y_5y_7 + y_2y_3y_4y_7 + y_2y_3y_5y_7, \qquad (2.269a)$$

$$C[W^*_{1,2}(Z)] = z_2z_4z_6 + z_4z_5z_6 + z_2z_6z_7 + z_5z_6z_7 + z_4z_6z_7 + z_2z_3z_6 +$$
$$z_3z_5z_6 + z_3z_4z_6 + z_1z_2z_4 + z_1z_4z_5 + z_1z_2z_7 + z_1z_5z_7 +$$
$$z_1z_4z_7 + z_1z_2z_3 + z_1z_3z_5 + z_1z_3z_4 + z_1z_2z_6 + z_1z_5z_6 +$$
$$z_1z_4z_6, \qquad (2.269b)$$

$$V^*(Y) = y_1y_3y_5y_6y_7 + y_1y_2y_3y_4y_6 + y_1y_3y_4y_5y_6 + y_1y_2y_3y_6y_7 +$$
$$y_1y_2y_3y_5y_6 + y_1y_4y_5y_6y_7 + y_1y_2y_4y_6y_7 + y_1y_2y_5y_6y_7 +$$
$$y_1y_3y_4y_5y_7 + y_1y_2y_3y_4y_7 + y_1y_2y_3y_5y_7, \qquad (2.270a)$$

$$C[V^*(Z)] = z_2z_4 + z_5z_7 + z_2z_7 + z_4z_5 + z_4z_7 + z_2z_3 + z_3z_5 + z_3z_4 +$$
$$z_2z_6 + z_5z_6 + z_4z_6. \qquad (2.270b)$$

The required network functions can now be obtained from (2.265)–(2.267).

§ 7. The existence and uniqueness of the network solutions

In the foregoing discussions we have tacitly assumed that the inverse of the network matrix exists. It is of great interest, therefore, to determine the conditions under which unique solutions can be obtained for the electrical networks.

KIRCHHOFF [1847] was the first to prove the existence of a solution for a network consisting only of resistors. This proof was elaborated by WEYL [1923] and ECKMANN [1944], and they also gave a deeper insight into the problem. SESHU and REED [1961] have presented the necessary and sufficient conditions for the existence of a unique solution for an *RLC* network with mutual inductances and independent sources. In an attempt to cover a more general case, ROTH [1955, 1959] and MALIK and HALE [1967] have also examined this question and proved some necessary and sufficient conditions for the unique solvability of a large class of networks.

The three fundamental systems of equations (2.51)–(2.53) constitute the starting point of the development. First, we partition the columns of Q and B according to those of Z or Y in (2.53). Next we write the three systems of equations together so that the known quantities can be separated from the unknowns:

$$\begin{bmatrix} Q_s & Q_p & 0 & 0 \\ 0 & 0 & B_s & B_p \\ 0 & 0 & U_s & 0 \\ 0 & -Z_{pp} & 0 & U_p \end{bmatrix} \begin{bmatrix} I_s \\ I_p \\ V_s \\ V_p \end{bmatrix} = \begin{bmatrix} 0 \\ 0 \\ E_s \\ E_p \end{bmatrix}, \quad (2.271)$$

where U_x denotes the identity matrix of order x. Since V_s is merely an alternative symbol for E_s, the third-row equations become trivial. Deleting these trivial equations and transposing the known V_s to the right, we find

$$\begin{bmatrix} Q_s & Q_p & 0 \\ 0 & 0 & B_p \\ 0 & -Z_{pp} & U_p \end{bmatrix} \begin{bmatrix} I_s \\ I_p \\ V_p \end{bmatrix} = \begin{bmatrix} 0 \\ -B_s E_s \\ E_p \end{bmatrix}. \quad (2.272)$$

For the system of equations (2.272) to be consistent, it is necessary and sufficient that the coefficient matrix and the augmented matrix have the same rank. Thus, in a given network G if there exists a circuit consisting only of independent and dependent voltage generators, then for G to have a solution it is necessary that the algebraic sum of these voltages be zero, since the row corresponding to this circuit in the coefficient matrix consists only of zeros. Similarly, if (2.53b) is used in (2.271) instead of (2.53a), then for G to have a solution it is necessary that the algebraic sum of the currents associated with a cut consisting only of independent and dependent current generators must be zero. Clearly, if G contains such a circuit or cut, its solutions, if they exist, cannot be unique, since the rank of the coefficient matrix is less than the number of unknowns.

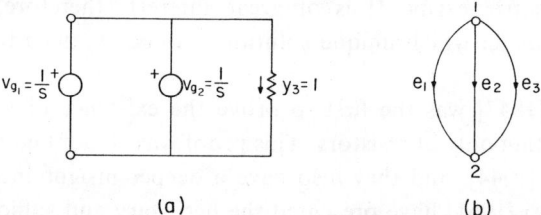

(a)   (b)

Fig. 2.36. A network having infinitely many solutions.

As an illustration, consider the network and its corresponding directed graph G as shown in fig. 2.36. The matrix equations corresponding to (2.272) for the network are given by

$$\begin{bmatrix} 1 & 1 & | & 1 & | & 0 \\ \hline 0 & 0 & | & 0 & | & 0 \\ 0 & 0 & | & 0 & | & 1 \\ 0 & 0 & | & -1 & | & 1 \end{bmatrix} \begin{bmatrix} i_1 \\ i_2 \\ i_3 \\ v_3 \end{bmatrix} = \begin{bmatrix} 0 \\ 0 \\ 1/s \\ 0 \end{bmatrix}. \quad (2.273)$$

§ 7 Existence and uniqueness of solutions 127

The second equation in (2.273) corresponds to the circuit consisting of the voltage generators v_{g1} and v_{g2}. Then a complete solution for the network is given by

$$V = \begin{bmatrix} v_1 \\ v_2 \\ v_3 \end{bmatrix} = \begin{bmatrix} 1/s \\ 1/s \\ 1/s \end{bmatrix} \quad \text{and} \quad I = \begin{bmatrix} i_1 \\ i_2 \\ i_3 \end{bmatrix} = \begin{bmatrix} x \\ -x - 1/s \\ 1/s \end{bmatrix}, \qquad (2.274)$$

where x is an arbitrary constant over the complex field. Thus, the network has infinitely many solutions.

LEMMA 2.12: An arbitrary square matrix Z can always be decomposed uniquely into two parts

$$Z = H + jF, \qquad (2.275)$$

where H and F are hermitian.

Proof. Let

$$H = \tfrac{1}{2}(Z + \bar{Z}'), \qquad (2.276a)$$

$$jF = \tfrac{1}{2}(Z - \bar{Z}'), \qquad (2.276b)$$

where the bar denotes the complex conjugate. The lemma follows immediately from here.

For convenience, here we shall call H the *hermitian part* of Z. Note that matrices H and F are, in general, complex. For a symmetric Z, however, both H and F are real.

LEMMA 2.13: If the branch-voltage and the branch-current vectors V and I of an initially relaxed network G are partitioned according to (2.53), and if the voltage or current generators are sinusoidal having the same frequency, then the quadratic form

$$\bar{I}'_p(j\omega) H_1(j\omega) I_p(j\omega) \quad \text{or} \quad \bar{V}'_p(j\omega) H_2(j\omega) V_p(j\omega) \qquad (2.277)$$

is simply twice the average power delivered by the independent voltage or current generators to the remaining network of G, where H_1 and H_2 are the hermitian parts of Z_{pp} and Y_{pp}, respectively, and the bar denotes the complex conjugate.

Proof. We shall only prove the case for Z; the other case can be proved in a similar manner. Since G is initially relaxed, from (2.53a) we have $E_p = 0$ and

$$\bar{I}'(j\omega) V(j\omega) = \bar{I}'_s(j\omega) V_s(j\omega) + \bar{I}'_p(j\omega) V_p(j\omega)$$
$$= \bar{I}'_s(j\omega) V_s(j\omega) + \bar{I}'_p(j\omega) Z_{pp}(j\omega) I_p(j\omega). \qquad (2.278)$$

$$\mathrm{Re}\,\bar{I}'(j\omega)\,V(j\omega) = \mathrm{Re}\,\bar{I}'_s(j\omega)\,V_s(j\omega) + \bar{I}'_p(j\omega)\,H_1(j\omega)\,I_p(j\omega). \quad (2.279)$$

Note that the quadratic forms in (2.277) are always real. It is well known in network theory that Re $\bar{I}'(j\omega)\,V(j\omega)$ and $-\mathrm{Re}\,\bar{I}'_s(j\omega)\,V_s(j\omega)$ are simply twice the average power absorbed in G and delivered by the independent voltage generators of G, respectively. Thus, from Theorem 2.15, we conclude that Re $\bar{I}'(j\omega)\,V(j\omega) = 0$. The lemma follows from here.

Since Z_{pp} and Y_{pp} are actually the branch-impedance and the branch-admittance matrices of the network G^* obtained from G by removing all the independent sources, respectively, it follows that if either one of the two quadratic forms in (2.277) is positive definite or semidefinite at a certain frequency, the network G^* cannot be used to amplify the power of an input signal having that frequency. On the other hand, if it is not positive definite or semidefinite, the network G^* can always be connected to operate either as an amplifier or as an oscillator.

In the following we shall show that if the quadratic forms in (2.277) are not identically zero, the network has a unique solution.

DEFINITION 2.36: *Ohmicness.* A square matrix Z over the complex field is said to be *ohmic* if the quadratic form $\bar{X}'ZX \neq 0$ for each nonzero vector X over the complex field.

THEOREM 2.33: If Z is ohmic, then Z is nonsingular. The inverse of Z is ohmic if, and only if, Z is ohmic.

Proof. Suppose that Z is singular. Then there exists a nontrivial solution X_1 such that $ZX_1 = 0$. Thus, $\bar{X}'_1 Z X_1 = 0$ for $X_1 \neq 0$, a contradiction. So Z must be nonsingular. If Z^{-1} is not ohmic, then there exists a nonzero vector X_1 such that $\bar{X}'_1 Z^{-1} X_1 = 0$. Let $X_2 = Z^{-1} X_1$. Since $X_1 \neq 0$, $X_2 \neq 0$. Thus, we have

$$\bar{X}'_1 Z^{-1} X_1 = \bar{X}'_2 \bar{Z}' Z^{-1} Z X_2 = \bar{X}'_2 \bar{Z}' X_2 = 0. \quad (2.280)$$

Now, if we take the complex conjugate and the transpose of the last relation, we see that $\bar{X}'_2 Z X_2 = 0$ for $X_2 \neq 0$. This contradicts to the assumption that Z is ohmic. Therefore, Z^{-1} is ohmic. Conversely, if Z^{-1} is ohmic then $(Z^{-1})^{-1} = Z$ must be ohmic which completes the proof of the theorem.

LEMMA 2.14: Let F be a real matrix of order $p \times q$ and of rank p. If Z is an ohmic matrix of order q, then FZF' is also ohmic.

Proof. Let X be a nonzero p-vector. We need to show that $\bar{X}'FZF'X \neq 0$. To

this end, define
$$X_1 = F'X. \tag{2.281}$$

Then X_1 is a q-vector. Since F' is of rank p and since $X \neq 0$, it follows that $X_1 \neq 0$. Since Z is ohmic, we have $\bar{X}_1' Z X_1 \neq 0$, i.e.,
$$\bar{X}_1' Z X_1 = \bar{X}' F Z F' X \neq 0. \tag{2.282}$$

So the lemma is proved.

THEOREM 2.34: Let G be a network which does not contain any circuit (cut) consisting only of independent and dependent voltage (current) generators. Let G^* be the network obtained from G by removing all the independent sources. If the branch-impedance (branch-admittance) matrix Z_{pp} (Y_{pp}) of G^* is ohmic, then the loop-impedance (cut-admittance) matrix BZB' (QYQ') of G is nonsingular, and the solution of the network G is unique.

Proof. Let B^* and Q^* be the basis circuit and cut matrices of G^*, respectively. Since Z_{pp} and Y_{pp} are ohmic, it follows from Theorem 2.33 and Lemma 2.14 that the network matrices $B^* Z_{pp} B^{*'}$ and $Q^* Y_{pp} Q^{*'}$ are both ohmic and thus nonsingular. The first part of the theorem follows from Corollaries 2.11 and 2.13.

To show that the solution is unique, let us assume that the pairs (V_1, I_1) and (V_2, I_2) are two different sets of the branch-voltage and the branch-current vectors of G which satisfy (2.40)–(2.42). Also let
$$V = V_1 - V_2 \quad \text{and} \quad I = I_1 - I_2. \tag{2.283}$$

Since $BV_1 = 0$ and $BV_2 = 0$, we find $BV = 0$. From Corollary 2.12, there exists an r-vector V_c such that
$$V = Q'V_c. \tag{2.284}$$
Next, since
$$I = I_1 - I_2 = YV_1 - YV_2 = YV, \tag{2.285}$$
and since $QI = 0$, it follows from (2.284) and (2.285) that
$$QYQ'V_c = 0. \tag{2.286}$$

But, from the first part of the proof, QYQ' is nonsingular. Hence, (2.286) has only the trivial solution $V_c = 0$. From (2.284) and (2.285), we have $V = 0$ and $I = 0$, showing that the pairs (V_1, I_1) and (V_2, I_2) are identical and that the solution is indeed unique.

COROLLARY 2.51: If the hermitian part of Z_{pp} or Y_{pp} of G^* is positive definite or negative definite, then G has a unique solution.

We shall now apply the above results to "*RLCM* networks".

DEFINITION 2.37: *RLCM network.* A network consisting of resistors, capacitors, and self and mutual inductors is called an *RLCM* network.

For an *RLCM* network G^*, its branch-impedance matrix Z_{pp} can be partitioned, according to the resistors, inductors and capacitors of G^*, into the form

$$Z_{pp} = \begin{bmatrix} R & 0 & 0 \\ 0 & sL & 0 \\ 0 & 0 & D/s \end{bmatrix} \quad (2.287)$$

by simultaneously permuting the rows and columns if necessary, where R is diagonal with positive diagonal entries, and so is D, and L is positive definite or semidefinite. The restriction that the matrix L be positive definite or semidefinite is equivalent to requiring that a passive system be stable. In other words, it can be shown that if a set of inductors can be found such that the matrix L of these inductors is neither positive definite nor semidefinite, then we can build a passive network consisting of these inductors and some passive resistors which is unstable. Certainly, one does not expect a physical network such as this to go up in smoke if it contains no generators. Thus, the postulate on L is justified.

Let us now consider the semidefiniteness of L. If L is positive semidefinite, it contains some principal submatrix which is singular. If any inductors are so coupled that their matrix is singular, while no principal submatrix, corresponding to some subset of the inductors, is singular, then the set of inductors is called *perfectly coupled*. If an *RLCM* network contains no perfectly coupled inductors, the matrix L will be nonsingular and so positive definite. We shall assume this unless it is stated to the contrary.

THEOREM 2.35: Let G be an imperfectly coupled *RLCM* network with independent generators. Then G has a unique solution if, and only if, there is no circuit consisting only of voltage generators and there is no cut consisting only of current generators.

Proof. It is sufficient to show that the conditions are necessary and sufficient for the network matrices BZB' and QYQ' to be nonsingular.

From our earlier discussions for (2.272), it is evident that the conditions are necessary for the network matrices to be nonsingular. It remains to be shown that the conditions are also sufficient. To this end, let the branch-impedance and the branch-admittance matrices Z and Y of G be partitioned as in (2.53). Also let the columns of the basis circuit and cut matrices B and Q be arranged in the same edge order as Z or Y and partitioned accordingly:

$$B = [B_{11} \quad B_{12}] \quad \text{and} \quad Q = [Q_{11} \quad Q_{12}]. \tag{2.288}$$

It is easy to check that

$$BZB' = B_{12}Z_{pp}B'_{12} \quad \text{and} \quad QYQ' = Q_{12}Y_{pp}Q'_{12}. \tag{2.289}$$

Without loss of generality, we assume that the network G is connected. If the loop system of equations is used, then by a condition of the theorem there exists a tree t containing all the voltage generators of G (Problem 2.43). Thus, the columns corresponding to the chords of the cotree $\bar{t}$ (complement of t in G) must be in B_{12}. From Theorem 2.5, B_{12} contains a major submatrix of order m. Since Z_{pp} is in fact the branch-impedance matrix of the network obtained from G by removing all the voltage generators, from (2.287) Z_{pp} is positive definite for positive real s. Therefore, Z_{pp} is ohmic. From (2.289) and Lemma 2.14, it follows that BZB' is nonsingular. Similarly, using Problem 2.44 we can show that QYQ' is also nonsingular. This completes the proof of the theorem.

It is worthwhile to justify our earlier assumption that for loop analysis there are no current generators, while for cut analysis there are no voltage generators. Although each method can be used when both current and voltage generators are present, the assumption would simplify the solutions obtained for each case. If both current and voltage generators are present, then, as mentioned earlier, to analyze the network by either the loop or cut method requires some simple network transformations (Thévenin and Norton equivalent networks). A more elegant method of solution, however, makes use of the linearity of the network and applies the principle of superposition to the solutions already obtained in § 3.2 and § 3.3 (Problem 2.42). Obviously, if the solution for each case is unique, the solution of the network due to two types of generators must also be unique.

Finally, we mention two things that might otherwise go unnoticed. First, we have no exceptional networks to which loop or cut system of equations cannot be applied. Secondly, a network branch may be taken to be either a simple resistor, inductor, capacitor, or independent or dependent generator, or a series and/or parallel connection of these elements, as we choose. If we choose the series and/or parallel connections of elements as the network branches, the branch-impedance or the branch-admittance matrix of the network can usually be made nonsingular, which in turn would help us to determine its unique solvability. To illustrate the latter remark let us consider the network elements

V_g and L_2s in fig. 2.13 as a single branch. Suppose that we do the same for $g_m i_4$ and R_7. Then the branch-impedance matrix of the network becomes

$$Z = \begin{bmatrix} L_2s & Ms & 0 & 0 & 0 & 0 \\ Ms & L_3s & 0 & 0 & 0 & 0 \\ 0 & 0 & \dfrac{1}{C_4s} & 0 & 0 & 0 \\ 0 & 0 & 0 & R_5 & 0 & 0 \\ 0 & 0 & -g_m & 0 & R_7 & 0 \\ 0 & 0 & 0 & 0 & 0 & z_8 \end{bmatrix},$$

where the directions of the current references of the elements L_2s and R_7 are used as the references of the combined branches. If the two inductors are not perfectly coupled, Z is nonsingular, but the corresponding branch-impedance matrix of (2.61) is singular.

§ 8. Conclusions

In this chapter, we began our discussion on the fundamental properties of the matrices associated with a directed graph. Using these results, the electrical network problem, based on Kirchhoff's and Ohm's laws, has been formulated in a precise postulational manner. Much of the discussion is sufficiently general to be applicable to general linear systems. Also we have introduced two powerful techniques, the loop and the cut methods, for solving the network problem.

We have shown that many of the invariant characters of the network determinants can be derived from purely graph-theoretic considerations rather than from physical arguments. Thus, they are valid not only for the electrical networks but also for other systems. Since the cofactors of the elements of the network matrix are in general not invariant with respect to the transformations of either circuits or cuts, the invariance of their values can only be stated in terms of the values of some precisely defined quantities called the generalized cofactors. It has been shown that although the network determinants or the generalized cofactors are not invariant for general systems of circuits or cuts, they are related by a real constant depending only upon the two choices of circuits or cuts; and for the most common and important types of representations such as the f-circuits and f-cutsets, meshes of a planar graph, and the set of incidence cuts, they all are invariant. It has also been shown that the ratio of the determinant of a network matrix and one of its generalized cofactors or the ratio of two of its generalized cofactors is invariant with respect to the choice of circuits or cuts. Thus, in particular, the network functions are invariant with respect to general transformations of the reference frame. In other words, we can state that many of the invariant characters of a network are consequences of its topology rather than the characteristics of its branches.

We have also shown that the determinant of the network matrix and its generalized cofactors, although they are not in general invariant with respect to the choice of the incidence functions, are invariant for general electrical networks.

Using the established identities, topological formulas for *RLC* networks have been derived. They have applications to both network analysis and network synthesis. In analysis, they provide a short-cut method of evaluating network determinants and cofactors because the usual cancellations inherent in evaluating these determinants are avoided. In synthesis, they present the possibilities of discovering radically new procedures for lumped networks which can be extended to cover all lumped systems.

Finally, we have presented a sufficient condition for the existence of a unique solution to the general network problem as well as necessary and sufficient conditions for the unique solvability of an imperfectly coupled *RLCM* network.

Problems

2.1. Prove Theorem 2.1 for an unconnected directed graph G.
2.2. Let g be a subgraph of a connected n-node graph G. Show that if g has $(n-1)$ edges and contains no circuit, then g is a tree of G.
2.3. Show that the rank of the circuit matrix B_a of an unconnected directed graph G is at least equal to the nullity of G.
2.4. Prove Lemma 2.4.
2.5. Prove Lemma 2.5.
2.6. Show that the number of edges common to a cut and a circuit of a directed graph G is always even. Use this result to show that if a cut has $2k$ edges in common with a circuit, then k of these edges have the same relative orientation in the cut and in the circuit, and the other k edges have one orientation in the cut and the opposite orientation in the circuit.
2.7. Prove Theorem 2.6.
2.8. Write out the details of the proof of Theorem 2.9.
2.9. Show that if a directed graph G has several components, then we can renumber the nodes and edges of G in such a way that A_a is the direct sum of the incidence matrices of the components.
2.10. Show that if G is an n-node directed graph of c components, then a submatrix of order $(n-c)$ of Q is nonsingular if, and only if, the columns of

this submatrix correspond to a node-disjoint union of c trees of the components, i.e., a *forest* of G.

2.11. Prove Corollary 2.4.

2.12. Show that a cut of a connected graph defined by a segregation of the nodes into V_1 and V_2, say, is a cutset if, and only if, the two sectional subgraphs defined by the sets V_1 and V_2 are each connected subgraphs.

2.13. Define a set of f-cutsets and a set of f-circuits for an unconnected directed graph.

2.14. Prove Corollary 2.8.

2.15. Derive the identities given in (2.49) and (2.50).

2.16. Let A be the basis incidence matrix of a directed graph G with reference node k. Let G^* be the directed graph obtained from G by identifying the nodes i and k. Show that if A_{-i} is the submatrix obtained from A by deleting the ith row, then A_{-i} is a basis incidence matrix of G^*.

2.17. Let $Z_m = [z_{ij}]$ be the loop-impedance matrix of an RLC network G. Show that

$$z_{ii} = \sum \text{(all branch impedances in circuit } i\text{)}, \qquad (2.290\text{a})$$

$z_{ij} = \sum$(all branch impedances common to circuits i and j and traversed by these circuits in the same direction) − (all branch impedances common to circuits i and j and traversed by these circuits in the opposite directions) \hfill (2.290b)

for $i \neq j$. What are the rules for the elements of E_m?

2.18. Write out the details of the proof of Corollary 2.11.

2.19. Let $Y_c = [y_{ij}]$ be the cut-admittance matrix of an RLC network G. Show that

$$y_{ii} = \sum \text{(all branch admittances contained in cut } i\text{)}, \qquad (2.291\text{a})$$

$y_{ij} = \sum$ (all branch admittances common to cut i and cut j, which have these cuts pointing in the same direction through the edges) − (all branch admittances common to cut i and cut j, which have these cuts pointing in opposite directions through the edges) \hfill (2.291b)

for $i \neq j$. What are the rules for the elements of J_c?

2.20. Applying the rules given in Problem 2.17, write down the loop system of equations corresponding to the circuits of the network as shown in fig. 2.15. Compare your result with (2.69).

2.21. Show that the rules given in Problem 2.19 for the elements of Y_c can be simplified as follows if incidence cuts are used:

$$y_{ii} = \sum \text{(all branch admittances incident at node } i\text{)}, \quad (2.292a)$$

$$y_{ij} = -\sum \text{(all branch admittances connected between the nodes } i \text{ and } j\text{)} \quad (2.292b)$$

for $i \neq j$. In other words, the elements of the node-admittance matrix of an RLC network can be obtained as above.

2.22. Applying the rules given in Problem 2.21, write down the nodal system of equations for the network as shown in fig. 2.17.
2.23. Prove Corollary 2.12.
2.24. Prove Corollary 2.13.
2.25. Give a detailed justification for Corollaries 2.16 and 2.17.
2.26. Show that $k(B)$ and $k(Q)$ are unity if, and only if, the entries of the inverses of the major submatrices $B(I_m)$ and $Q(J_r)$ are all integers, respectively.
2.27. Show that the transformation from one system of f-circuits or f-cutsets into another involves a matrix with determinant equal to 1 or -1.
2.28. Show that the determinant of a node-admittance matrix of an RLC network G can be regarded as the total number of trees in G if the admittance associated with each of the edges of G is unity.
2.29. Show that the number of trees of a connected directed graph is given by $(\det BB')/k(B)$.
2.30. Show that if the unilateral transimmittances of a network do not form any closed loop, then we can always renumber the branches of the network in such a way that its corresponding branch-impedance matrix is triangular.
2.31. Prove the identities (2.169a) and (2.169b).
2.32. Using (2.78), derive the formula (2.189).
2.33. Using (2.83), derive the formula (2.191).
2.34. Show that the network determinant of a network G is invariant with respect to the labeling of the nodes and edges of G. In other words, by permuting the rows or columns of B (Q) and the corresponding rows and columns of Z (Y), the network determinant is invariant.
2.35. Show that the statement given in Problem 2.34 is also valid for the generalized cofactors of the elements of a network matrix.
2.36. Show that the generalized cofactors of the elements of the loop-impedance or the cut-admittance matrix of a network are invariant with respect to the choice of the incidence functions of the network.

2.37. Prove Lemma 2.11.
2.38. Prove Corollary 2.50.
2.39. Derive the identities given in Theorem 2.32.
2.40. Derive the quadratic forms similar to those given in (2.277) for imaginary power delivered by the independent voltage or current generators of the network. (*Hint*: Use the imaginary part of (2.275).)
2.41. Show that each circuit edge of a connected planar graph G is contained in two and only two meshes of G, including the mesh formed by the boundary edges of the outside region.
2.42. Making use of the linearity of the network and applying the principle of superposition to the solutions already obtained in § 3.2 and § 3.3, derive formulas expressing the branch-voltage and branch-current vectors in terms of the branch voltage-source and branch current-source vectors for networks containing both voltage and current generators.
2.43. Show that a subgraph of a connected graph G can be made part of a tree of G if, and only if, the subgraph contains no circuit.
2.44. Show that a subgraph of a connected graph G can be made part of a cotree of G, if, and only if, the subgraph contains no cut of G.
2.45. Let B be a basis circuit matrix of a directed graph G. Show that if F is any matrix of order $r \times b$ and of rank r, having entries 1, -1, and 0, and satisfying $BF' = 0$, then each row of F corresponds to a cut of G.
2.46. Let Q be a basis cut matrix of a directed graph G. Show that if F is any matrix of order $m \times b$ and rank m, having as entries only 1, -1, and 0, and satisfying $QF' = 0$, then each row of F corresponds to a circuit or an edge-disjoint union of circuits (SESHU and REED [1961]).
2.47. Applying Theorem 2.10, find the inverse of a major submatrix of a basis incidence matrix of the directed graph as shown in fig. 2.2.
2.48. Let Z be a diagonal matrix with nonnegative real entries. Show that the network matrix $B_f Z B_f'$ and any matrix of the form $P B_f Z B_f' P'$, where P is a permutation matrix with entries 1, -1, and 0, cannot contain a principal submatrix of order greater than 3 with all negative off-diagonal entries (BROWN [1968]).
2.49. In Problem 2.48, if all the entries in $B_f Z B_f'$ are nonzero, then there exists a permutation matrix P with entries 1, -1, and 0 such that

$$P B_f Z B_f' P' = \begin{bmatrix} R_1 & R_2 \\ R_2' & R_3 \end{bmatrix}, \qquad (2.293)$$

where R_1 and R_3 contain all positive entries and all the entries in any row of R_2 have the same sign (BROWN [1968]).

2.50. Fig. 2.37 shows a simple feedback amplifier and its equivalent network. Use this network to verify Corollary 2.28.

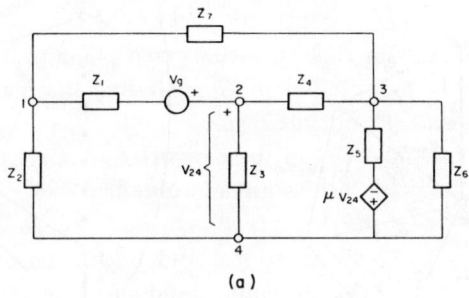

(a)

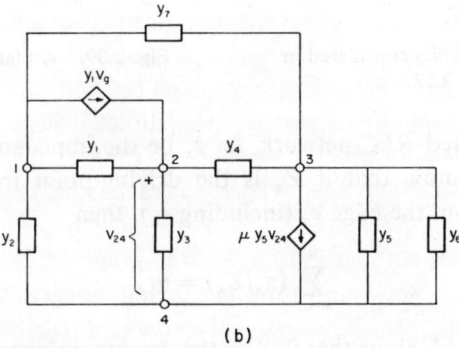

(b)

Fig. 2.37. A feedback amplifier and its equivalent network.

2.51. Show that if M is any square matrix with entries 1, −1, and 0, containing an even number of nonzero entries in each column or row, then $\det M$ is either zero or an even integer.

2.52. Let B be the basis circuit matrix of the planar directed graph, as shown in fig. 2.38, corresponding to the m circuits as indicated. Show that $k(B) = (m-1)^2$. (BROWN and BUDNER [1965].)

2.53. In fig. 2.39, let the node set of the directed graph G be partitioned into the disjoint subsets $V_k = \{k, n\}$ and $\bar{V}_k$ where $\bar{V}_k$ is the complement of V_k in $\{1, 2, ..., n\}$. Also let C_k be the cutset of G separating the nodes of G into two subsets V_k and $\bar{V}_k$ and oriented from V_k to $\bar{V}_k$. Show that if Q is the basis cutset matrix corresponding to the $(n-1)$ cutsets C_k of G, then $k(Q) = (n-2)^2$, where $k = 1, 2, ..., n-1$.

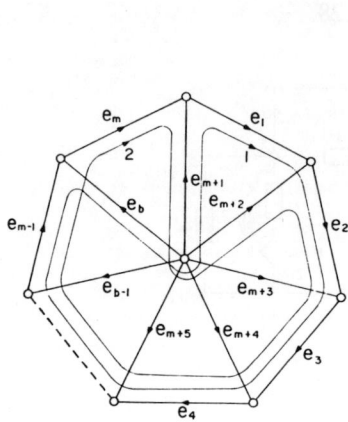
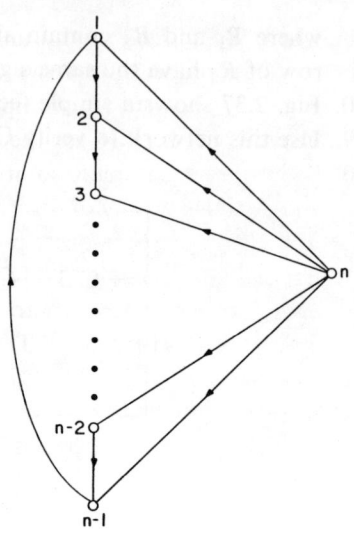

Fig. 2.38. The m circuits considered in Problem 2.52.

Fig. 2.39. A planar directed graph.

2.54. For a connected RLC network, let z_x be the impedance associated with its edge e_x. Show that if Z_x is the driving-point impedance function looking in from the edge e_x (including e_x), then

$$\sum_{x=1}^{b}(z_x/Z_x) = m. \qquad (2.294)$$

2.55. In Problem 2.54, show that if Y_x is the driving-point admittance function looking in from the two endpoints of e_x, then

$$\sum_{x=1}^{b}(y_x/Y_x) = r, \qquad (2.295)$$

where $y_x = 1/z_x$.

2.56. Show that if a network G has a unique solution for its branch-current and branch-voltage vectors, then there exists a tree t of G such that the independent and dependent current generators are chords of the cotree $\bar{t}$ and the independent and dependent voltage generators are branches of t (SESHU and REED [1961]). (*Hint*: See the discussion following (2.272) and use Theorem 5.4.)

2.57. Show that there are $(2^m - 1)$ nonempty circuits and edge-disjoint unions of circuits in a directed or undirected graph G. Is it necessary that G be connected?

Problems

2.58. Show that if F is totally unimodular, then so are F', matrices obtained by a permutation of rows or columns of F, all submatrices of F, and matrices obtained by multiplying rows or columns of F by -1.

2.59. Prove Theorem 2.30 for the case where $u_1 = v_1$ or v_2, or $u_2 = v_1$ or v_2.

2.60. Show that a complete solution of (2.2) is βX where β is a real constant and X consists only of 1's. (*Hint*: Apply Lemma 2.1 and the result stated in the first sentence of its proof.)

2.61. For networks such as two identical capacitors charged to different voltages and connected in parallel, we know that the energy is not conservative. What happens to Theorem 2.16 for these cases?

2.62. In view of Problem 2.61, give another network whose energy is not conservative. If possible, give a simple characterization of such networks.

2.63. Prove that of the three postulates

$$QI(t) = 0, \quad BV(t) = 0, \quad I'(t)V(t) = 0, \qquad (2.296)$$

any two would imply the third.

CHAPTER 3

DIRECTED-GRAPH SOLUTIONS OF LINEAR ALGEBRAIC EQUATIONS

The analysis of a linear system reduces ultimately to the solution of a system of simultaneous linear algebraic equations. As an alternative to conventional algebraic methods of solving the system, it is possible to obtain a solution by considering the properties of certain directed graphs associated with the system. The unknowns of the equations correspond to the nodes of the graph, while the linear relations between them appear in the form of directed edges connecting the nodes. The directed-graph approach is of particular interest in the analysis of engineering problems, linear problems especially, since the associated directed graphs in many cases can be set up directly by inspection of the physical system without the necessity of first formulating the associated equations; it offers a visual structure upon which causal relationships among several variables may be laid out and compared.

The basic idea of associating a directed graph with a system of linear equations was first introduced by MASON [1953], and the graph is called a *signal-flow graph*. Since then, an alternative representation of the equations as a directed graph called a *flow graph* has been described by COATES [1959]. The two formulations are very closely related both in the directed graphs that result and in the manipulations or topological formulas, as the case may be. Since a signal-flow graph has also been called a flow graph in the literature, in order to avoid possible confusion, a signal-flow graph is here referred to as a *Mason graph*, as distinguished from a flow graph which is here referred to as a *Coates graph*.

The present chapter is mainly concerned with the fundamental aspects of the problem, and not with specific applications of these graphs to physical systems. Many examples of the applications of the Mason graph are to be found in CHOW and CASSIGNOL [1962], ROBICHAUD et al. [1962], and TRUXAL [1955]. The emphasis is rather on the justification of these procedures.

As mentioned earlier, a weighted directed graph G is a directed graph in which each edge has been assigned a weight. We denote by $f(i,j)$ the weight associated with the edge (i,j) of G. If G_s is a subgraph of G, by $f(G_s)$ we mean

$$f(G_s) = \prod f(i,j) \tag{3.1a}$$

if $G_s \neq \emptyset$, the null graph, where the product is taken over all possible edges (i,j) of G_s; and
$$f(\emptyset) = 0. \qquad (3.1b)$$

As an alternative, if the weights of G are from a field, the function f can be considered as a mapping function with domain in the set of all subgraphs of G and range in the field such that for each subgraph G_s of G, (3.1) holds. In other words, $f(G_s)$ denotes the product of the weights associated with the edges of G_s.

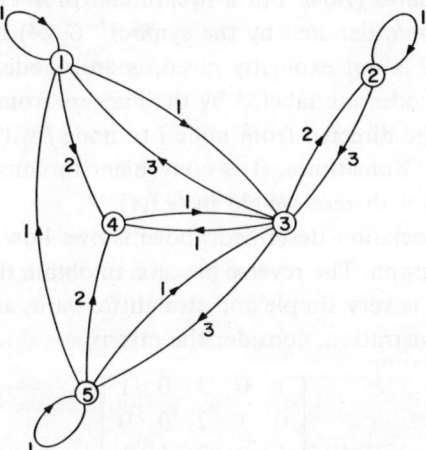

Fig. 3.1. A weighted directed graph.

For example, in fig. 3.1, let G_s be the subgraph consisting of the edges (3,1), (2,2), (5,4), and (3,5). Then

$$f(G_s) = f(3,1)f(2,2)f(5,4)f(3,5)$$
$$= 3 \cdot 1 \cdot 2 \cdot 3 = 18.$$

For a special class of subgraphs G_{uv} of G, the symbol

$$\sum_{G_{uv}} f(G_{uv}) \quad \text{or} \quad \sum_{u,v} f(G_{uv})$$

denotes that the summation or summations are taken over all possible such subgraphs G_{uv} of the class.

§ 1. The associated Coates graph

The basic idea of the correspondence between the terms in the expansion of a determinant and the corresponding subgraphs in an associated directed graph is not new. KÖNIG [1916] was the first to apply a graph-theoretic approach to the evaluation of a determinant. Since then, it has been utilized for various purposes by many authors. (See, for instances, LUCE and PERRY [1949], KAC and WARD [1952], HOHN and SCHISSLER [1955], and COPI [1958].) Our purpose is to present a rigorous and systematic development of these correspondences based on the results of CHEN [1967a].

DEFINITION 3.1: *Coates graph*. For a square matrix $A = [a_{ij}]$ of order n, the associated *Coates graph*, denoted by the symbol[†] $G_c(A)$ or simply G_c if A is clearly understood or is not explicitly given, is an n-node, weighted, labeled, directed graph. The nodes are labeled by the integers from 1 to n such that if $a_{ji} \neq 0$, there is an edge directed from node i to node j with associated weight a_{ji} for $i, j = 1, 2, ..., n$. Sometimes, it is convenient to consider edges that are not in $G_c(A)$ as edges with zero weight in $G_c(A)$.

The process of association described above shows how one goes from the matrix to the Coates graph. The reverse process, to obtain the associated matrix from a Coates graph, is very simple and straightforward, and needs no further elaboration. As an illustration, consider the matrix

$$A = \begin{bmatrix} 1 & 0 & 3 & 0 & 1 \\ 0 & 1 & 2 & 0 & 0 \\ 1 & 3 & 0 & 1 & 1 \\ 2 & 0 & 1 & 0 & 2 \\ 0 & 0 & 3 & 0 & 1 \end{bmatrix}. \tag{3.2}$$

The associated Coates graph $G_c(A)$ of the matrix A is as shown in fig. 3.1. Conversely, if fig. 3.1 is given, the associated matrix A of $G_c(A)$ can easily be obtained. Thus, there is a one-to-one correspondence between the matrix and the Coates graph. Actually, in the language of variable adjacency matrix used by HARARY [1962a], the associated matrix of a Coates graph G_c is simply the transpose of its *variable adjacency matrix*, or of its *primitive connection matrix* in the sense of HOHN et al. [1957].

1.1. *Topological evaluation of determinants*

In this section, we shall demonstrate the close relationships between the terms in the expansion of the determinant of a matrix A and certain types of

---

[†] In contrast to the symbols used in ch. 2, A does not represent a basis incidence matrix of a directed graph. This should not create any confusion since it will not appear in this chapter.

subgraphs of $G_c(A)$. The following results were first exploited by KÖNIG [1916] and were rigorously, systematically developed by COATES [1959] by using the incidence matrix of a directed graph as a tool. A simple derivation of Coates' formulas was given by DESOER [1960].

DEFINITION 3.2: 1-*factor*. A 1-*factor* of a directed graph G is a spanning subgraph of G which is regular of degree 1.

In general topological analysis of linear systems, 1-factors play a considerable role in various applications. More intuitively, a 1-factor is a set of node-disjoint directed circuits which include all the nodes of G. Sometimes, a 1-factor is also referred to as a *connection* or a *P-set of cycles* in engineering literature (DESOER [1960], and SESHU and REED [1961] § 10.2). However, we prefer the above usage since the term 1-factor is more descriptive and was used long before the term connection or P-set of cycles was coined. (See, for example, TUTTE [1953].) Consider the Coates graph as shown in fig. 3.1. For illustrative purposes, the four 1-factors of fig. 3.1 are presented in fig. 3.2.

The first link between the determinant of A and the Coates graph $G_c(A)$ is obtained by referring to the definition of a determinant:

$$\det A = \sum_{(j)} \varepsilon_{j_1 j_2 \ldots j_n} a_{1j_1} a_{2j_2} \cdots a_{nj_n}, \tag{3.3}$$

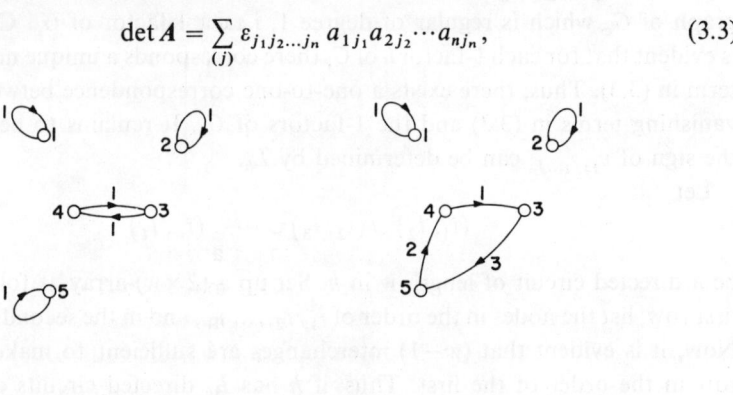

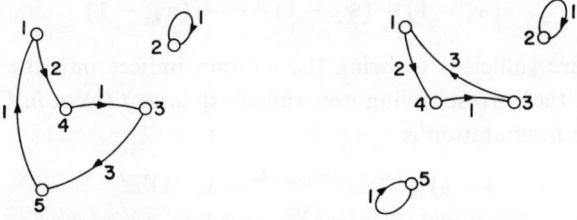

Fig. 3.2. The set of 1-factors of the Coates graph of fig. 3.1.

where $j_1 j_2 ... j_n$ is a permutation of $1, 2, ..., n$; the sum $\sum_{(j)}$ is taken over all such permutations; and $\varepsilon_{j_1 j_2 ... j_n}$ is 1 if $j_1 j_2 ... j_n$ is an even permutation and -1 if it is an odd permutation.

THEOREM 3.1: Let $G_c(A)$ be the associated Coates graph of a matrix A of order n. Then

$$\det A = (-1)^n \sum_h (-1)^{L_h} f(h), \qquad (3.4)$$

where h is a 1-factor in $G_c(A)$, and L_h denotes the number of directed circuits in h.

Proof. Let

$$a_{1j_1} a_{2j_2} \cdots a_{nj_n} \neq 0 \qquad (3.5)$$

for some permutation. It is clear that the subgraph h of G_c corresponding to this non-vanishing term consists of the edges $(j_1, 1), (j_2, 2), ..., $ and (j_n, n). Since each k appears exactly twice in the sequence, once as a first element and once as a second element in some ordered pairs, it follows that the incoming and the outgoing degrees of each node of h are 1. Thus, h is a spanning subgraph of G_c which is regular of degree 1, i.e., a 1-factor of G_c. Conversely, it is evident that for each 1-factor h of G_c there corresponds a unique non-vanishing term in (3.3). Thus, there exists a one-to-one correspondence between the non-vanishing terms in (3.3) and the 1-factors of G_c. It remains to be shown that the sign of $\varepsilon_{j_1 j_2 ... j_n}$ can be determined by L_h.

Let

$$(i_1, i_2) \cup (i_2, i_3) \cup \cdots \cup (i_w, i_1) \qquad (3.6a)$$

be a directed circuit of length w in h. Set up a $(2 \times w)$-array as follows: In the first row, list the nodes in the order of $i_1, i_2, ..., i_w$; and in the second, $i_2, i_3, ..., i_1$. Now, it is evident that $(w-1)$ interchanges are sufficient to make the second row in the order of the first. Thus, if h has L_h directed circuits consisting of $w_1, w_2, ..., w_{L_h}$ edges, a total of

$$(w_1 - 1) + (w_2 - 1) + \cdots + (w_{L_h} - 1) \qquad (3.6b)$$

interchanges are sufficient to bring the column indices into the order of the row indices in the corresponding non-vanishing term $f(h)$ of h. Consequently, the sign of the permutation is

$$(-1)^{w_1 + w_2 + \cdots + w_{L_h} - L_h} = (-1)^{n - L_h}, \qquad (3.6c)$$

which completes the proof of the theorem.

The above result is really a variant of a well-known result in the theory of symmetric groups of degree n. It states that every permutation can be resolved into a family of disjoint "cycles", and this resolution is unique. (See, for example, MACLANE and BIRKHOFF [1967].)

As an alternative, the sign of a term in (3.4) may be determined by the number of even components of the corresponding 1-factor.

DEFINITION 3.3: *Even (odd) component.* A component of a directed graph G is said to be *even (odd)* if it contains an even (odd) number of edges. An isolated node is considered as an even component.

COROLLARY 3.1:
$$(-1)^{n+L_h} f(h) = (-1)^{q_h} f(h), \qquad (3.7)$$

where q_h is the number of even components in the 1-factor h.

This result is useful in that we can associate each directed circuit of a 1-factor with either a plus or a minus sign, according to whether the directed circuit is of odd or even length.

In linear system analysis, it is usually necessary to compute a symbolic system function, i.e., a function of the elements in *literal* rather than in numerical form. For example, in the frequency domain analysis of an electrical network, the elements of a network matrix are usually rational functions in a complex variable s, being the ratios of polynomials in s. Although its determinant can be evaluated by the usual Gaussian-elimination or pivotal technique, the process is slow and complicated. The directed-graph approach provides an effective solution to the problem. However, its effectiveness depends upon the efficiency with which the 1-factors of the directed graph are generated. Since, for a complicated Coates graph G_c, the number of its 1-factors is usually large, it is desirable to have a simple formula to determine this number in advance, so that no terms will be left out in the expansion of (3.4). One way to achieve this is the following (CHEN [1965a]).

For a given square matrix, there exists, in addition to the determinant of the matrix, also another less used quantity, the permanent of the matrix. The *permanent* of a square matrix A, denoted by per A, is defined exactly the same as the determinant of the matrix except that all the terms are taken with positive signs:

$$\text{per } A = \sum_{(j)} a_{1j_1} a_{2j_2} \cdots a_{nj_n}. \qquad (3.8)$$

Some properties of permanents are presented in Problem 3.62. A computer program for its evaluation was recently written by SHRIVER et al. [1969].

For an n-node directed graph G, let $C(G) = [c_{ij}]$ be the $n \times n$ matrix denoting the connection of G such that $c_{ij} = k$ if and only if there are k edges directed from node i to node j in G, and $c_{ij} = 0$ otherwise. Clearly, $C(G)$ is a matrix of nonnegative integers. Now we show that the permanent of $C(G_c)$ is equal to the number of 1-factors in G_c.

COROLLARY 3.2: The number of 1-factors in a Coates graph G_c is equal to the value of the permanent of the matrix $C(G_c)$.

Proof. Let A be the associated matrix of G_c. It is easy to check that $C(G_c)$ is the transpose of the matrix obtained from A by replacing each nonzero element of A by 1. Since the permanent of a matrix is equal to the permanent of its transpose, it follows from Theorem 3.1 that each 1-factor of G_c contributes a 1 in the summation of (3.8). Thus, the permanent of $C(G_c)$ denotes the total number of 1-factors of G_c. So the corollary is proved.

Since $C(G_c)$ is a matrix of 1's and 0's, its permanent is not computationally difficult to evaluate. As an illustration, consider the Coates graph of fig. 3.1. The number of 1-factors of G_c is given by

$$\operatorname{per} C(G_c) = \operatorname{per} \begin{bmatrix} 1 & 0 & 1 & 1 & 0 \\ 0 & 1 & 1 & 0 & 0 \\ 1 & 1 & 0 & 1 & 1 \\ 0 & 0 & 1 & 0 & 0 \\ 1 & 0 & 1 & 1 & 1 \end{bmatrix} = 4. \tag{3.9}$$

From the four 1-factors of G_c as shown in fig. 3.2, we have

$$\det A = -a_{11}a_{22}a_{34}a_{43}a_{55} + a_{11}a_{22}a_{34}a_{53}a_{45} - a_{22}a_{41}a_{34}a_{53}a_{15} + \\ + a_{22}a_{41}a_{34}a_{13}a_{55} = -1 + 6 - 6 + 6 = 5. \tag{3.10}$$

1.2. *Topological evaluation of cofactors*

The cofactor of an element of a matrix may be evaluated in a similar manner as a determinant. However, we shall need a definition for this.

DEFINITION 3.4: 1-*factorial connection*. A 1-*factorial connection* from node i to node j of a directed graph G is a spanning subgraph of G which contains (1) a directed path from node i to node j; (2) a set of node-disjoint directed circuits which include all the nodes of G except those contained in (1).

An example of a set of 1-factorial connections from node 1 to node 4 in G_c

§1 The associated Coates graph

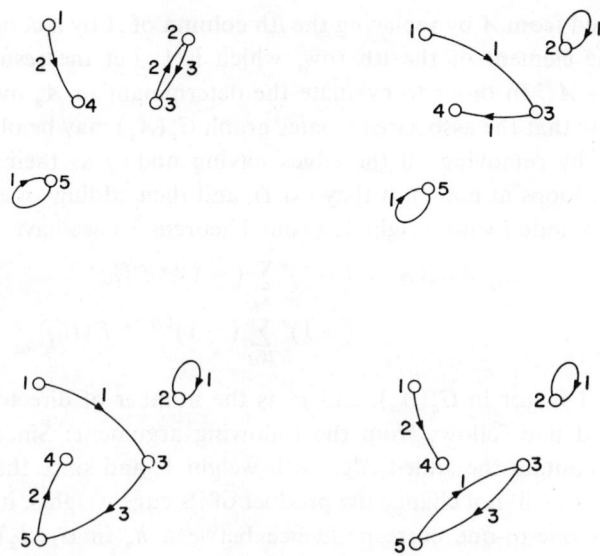

Fig. 3.3. The set of 1-factorial connections from node 1 to node 4 in the Coates graph of fig. 3.1.

of fig. 3.1 is presented in fig. 3.3. A 1-factorial connection is also known as a *one-connection* (COATES [1959]).

THEOREM 3.2: Let $G_c(A)$ be the associated Coates graph of a matrix A of order n. Then

$$\Delta_{ii} = (-1)^{n-1} \sum_{h'} (-1)^{L'_h} f(h') = \sum_{h'} (-1)^{q'_h} f(h'), \quad (3.11)$$

$$\Delta_{ij} = (-1)^{n-1} \sum_{H_{ij}} (-1)^{L_H} f(H_{ij}) = \sum_{H_{ij}} (-1)^{q_H - 1} f(H_{ij}) \quad (3.12)$$

for $i \neq j$, where Δ_{ij} is the cofactor of the (i,j)-element[†] of A; h' is a 1-factor in $G_c(A_{ii})$; H_{ij} is a 1-factorial connection from node i to node j in $G_c(A)$; L'_h and L_H are the numbers of directed circuits in h' and H_{ij}, respectively; A_{ij} is the matrix obtained from A by striking out the ith row and the jth column; and q'_h and q_H are the numbers of even components in h' and H_{ij}, respectively.

Proof. We shall only prove the first part of (3.12); the others follow directly from Theorem 3.1 and Corollary 3.1.

To prove the first part of (3.12), we note that Δ_{ij} is the determinant of the

---

[†] We mean the element in the ith row and jth column of A.

matrix obtained from A by replacing the jth column of A by a column of zeros except for the element of the ith row, which is 1. Let the resulting matrix be denoted by A_α. In order to evaluate the determinant of A_α by topological means, we note that the associated Coates graph $G_c(A_\alpha)$ may be obtained from $G_c(A)$ simply by removing all the edges having node j as their initial node (including self-loops at node j, if they exist), and then adding an edge directed from node j to node i with weight 1. From Theorem 3.1 we have

$$\Delta_{ij} = \det A_\alpha = (-1)^n \sum_{h_\alpha} (-1)^{L_\alpha} f(h_\alpha)$$
$$= (-1)^n \sum_{H_{ij}} (-1)^{L_H-1} f(H_{ij}), \qquad (3.13)$$

where h_α is a 1-factor in $G_c(A_\alpha)$, and L_α is the number of directed circuits in h_α. The second line follows from the following argument: Since each h_α in $G_c(A_\alpha)$ must contain the added edge with weight 1, and since the removal of this edge from h_α will not change the product of its edge weights, it follows that there exists a one-to-one correspondence between h_α in $G_c(A_\alpha)$ and H_{ij} in $G_c(A)$ such that $f(h_\alpha) = f(H_{ij})$. Since the number of directed circuits in H_{ij} is one less than that in h_α, we have $L_H = L_\alpha - 1$, which completes the proof of the theorem.

As an alternative to the sign problem, we may associate each component of H_{ij} (in the present case, either a directed circuit or a directed path) with either a plus or a minus sign, according to whether the component is odd or even. Then $(-1)^{q_H} f(H_{ij})$ is simply the signed product of the weights of the edges of H_{ij}.

COROLLARY 3.3: *The number of 1-factorial connections from node i to node j in a Coates graph G_c is equal to the value of the permanent of the matrix $C(G_c)_{ji}$ obtained from the matrix $C(G_c)$ by striking out the jth row and ith column.*

Since the determinant of a matrix is equal to the determinant of the transpose of the matrix, it is obvious that Theorems 3.1 and 3.2 still hold if we replace $G_c(A)$ by $G_c(A')$ where the prime indicates the transpose of a matrix. The only exception is that we have to interpret H_{ij} as a 1-factorial connection from the node j to node i instead of from i to j. This is why in Definition 3.1 we define that there is an edge directed from i to j in G_c if $a_{ji} \neq 0$ rather than $a_{ij} \neq 0$. Thus, we can also use Theorems 3.1 and 3.2 to evaluate the determinant of the variable adjacency matrix (HARARY [1962a]) or the connection matrix (HOHN et al. [1957]) of a directed graph and the cofactors of its elements.

Again, consider the Coates graph $G_c(A)$ of fig. 3.1. The number of 1-factorial connections from node 1 to node 4 is given by

$$\operatorname{per} C(G_c)_{41} = \operatorname{per} \begin{bmatrix} 0 & 1 & 1 & 0 \\ 1 & 1 & 0 & 0 \\ 1 & 0 & 1 & 1 \\ 0 & 1 & 1 & 1 \end{bmatrix} = 4. \qquad (3.14)$$

From the four 1-factorial connections of $G_c(A)$ as shown in fig. 3.3, we have

$$\begin{aligned} \Delta_{14} &= a_{41}a_{32}a_{23}a_{55} + a_{31}a_{43}a_{22}a_{55} - a_{31}a_{53}a_{45}a_{22} + a_{41}a_{53}a_{35}a_{22} \\ &= 12 + 1 - 6 + 6 = 13. \end{aligned} \qquad (3.15)$$

1.3. Topological solutions of linear algebraic equations

The above results will now be applied to the solutions of a system of simultaneous (consistent) linear algebraic equations. It is convenient to write the system in matrix form:

$$AX = B, \qquad (3.16)$$

where $A = [a_{ij}]$ is the coefficient matrix of order n and X and B are column vectors. The transposes of X and B are given by $[x_1, x_2, ..., x_n]$ and $[b_1, b_2, ..., b_n]$, respectively.

THEOREM 3.3: If the coefficient matrix A is nonsingular, then the solution of (3.16) is given by

$$x_k = \frac{\sum\limits_{H_{(n+1)k}} (-1)^{L_H} f(H_{(n+1)k})}{\sum\limits_{h'} (-1)^{L'_h} f(h')} \qquad (3.17)$$

for $k = 1, 2, ..., n$, where $H_{(n+1)k}$ is a 1-factorial connection from node $n+1$ to node k in $G_c(A_u)$, where A_u is the augmented matrix obtained from A by attaching $-B$ to the right of A, and then attaching a row of zeros at the bottom of the resultant matrix; h' is a 1-factor in $G_c(A)$; and L_H and L'_h are the numbers of directed circuits in $H_{(n+1)k}$ and h', respectively.

Proof. Since the coefficient matrix A is nonsingular, the solution of the system (3.16) is given by

$$x_k = \frac{\sum\limits_{i=1}^{n} b_i \Delta_{ik}}{\det A}, \qquad (3.18)$$

where Δ_{ik} is the cofactor of a_{ik} in det A. By interchanging the columns of A_u

if necessary, it is not difficult to see that the numerator of (3.18) can be expressed in terms of the minor determinant of A_u.

$$\sum_{i=1}^{n} b_i \Delta_{ik} = (-1)^{(n+1)-k-1}(-1)\det(A_u)_{(n+1)k}, \qquad (3.19)$$

where $(A_u)_{ij}$ is the matrix obtained from A_u by striking out the ith row and the jth column. Using Theorem 3.2 in conjunction with (3.19) we obtain

$$\sum_{i=1}^{n} b_i \Delta_{ik} = (-1)^{(n+1)-1} \sum_{H_{(n+1)k}} (-1)^{L_H} f(H_{(n+1)k}). \qquad (3.20)$$

From Theorem 3.1 we have

$$\det A = (-1)^n \sum_{h'} (-1)^{L'_h} f(h'). \qquad (3.21)$$

Substituting (3.20) and (3.21) into (3.18), we obtain the desired result. This completes the proof of the theorem.

A few comments, at this point, might be in order. Instead of considering the system (3.16), we may consider the equivalent system

$$A_u X'' = 0, \qquad (3.22)$$

where the transpose of X'' is $[x_1, x_2, \ldots, x_n, x_{n+1}]$ with $x_{n+1}=1$. The solution x_k of this system may again be written as the ratio of (3.19) to (3.21).

Since the nodes of $G_c(A_u)$ are associated with the columns or rows of A_u, it is convenient to assign the variables $x_1, x_2, \ldots, x_n, x_0$ to the nodes $1, 2, \ldots, n+1$ of $G_c(A_u)$, respectively. These variables may be considered as *weights* or *signals* associated with the nodes of the Coates graph. For convenience, we shall speak loosely of a node x_k rather than of the node k associated with the node weight or signal x_k. The last node $x_0 = x_{n+1}$ is associated with the input variable x_0 which is taken to be unity, and is referred to as the *source node* because in applications it corresponds to a source and there are no edges terminating at it. The edge weights are also called *transmittances*[†]. In this chapter, edge *weight* and *transmittance* are used as synonyms. Physically, we may think of the nodes as high-gain operational amplifiers whose feedback loop is open. Because, in that case, if the output voltage is in the linear range of the amplifier, the sum of the currents into the input node must be very nearly zero. Thus, the system

---

[†] Sometimes, they are also called *edge gains* by various authors.

of equations may be obtained by equating to zero the sum of the products of the transmittances of the incoming edges of each node (excluding the source node) and the variables these edges outgoing from. Perhaps this is one of the reasons that the name *flow graph* was coined. Finally, it should be noted that if $G_c(A)$ is given, $G_c(A_u)$ may be obtained from $G_c(A)$ simply by attaching a source node x_0 to $G_c(A)$ such that if $b_k \neq 0$, there is an edge directed from the source node to node k with edge transmittance $-b_k$.

Example 3.1: Consider the following system of linear equations,

$$\begin{bmatrix} a_{11} & 0 & a_{13} & 0 & a_{15} \\ 0 & a_{22} & a_{23} & 0 & 0 \\ a_{31} & a_{32} & 0 & a_{34} & a_{35} \\ a_{41} & 0 & a_{43} & 0 & a_{45} \\ 0 & 0 & a_{53} & 0 & a_{55} \end{bmatrix} \begin{bmatrix} x_1 \\ x_2 \\ x_3 \\ x_4 \\ x_5 \end{bmatrix} = \begin{bmatrix} b_1 \\ 0 \\ 0 \\ 0 \\ b_5 \end{bmatrix}. \quad (3.23)$$

Let us solve for x_4. The associated Coates graph $G_c(A_u)$ is as shown in fig. 3.4. All the 1-factors of $G_c(A)$ except weights are listed in fig. 3.2. The set of all 1-factorial connections from x_0 to x_4 in $G_c(A_u)$ is presented in fig. 3.5. By Theorem 3.3, we have

$$\begin{aligned} x_4 = (&- b_1 a_{41} a_{32} a_{23} a_{55} - b_1 a_{31} a_{43} a_{22} a_{55} + b_1 a_{31} a_{53} a_{45} a_{22} \\ &- b_1 a_{41} a_{35} a_{53} a_{22} - b_5 a_{45} a_{23} a_{32} a_{11} - b_5 a_{45} a_{13} a_{31} a_{22} \\ &- b_5 a_{35} a_{43} a_{11} a_{22} + b_5 a_{35} a_{13} a_{41} a_{22} + b_5 a_{15} a_{41} a_{32} a_{23} \\ &+ b_5 a_{15} a_{31} a_{43} a_{22})/(a_{11} a_{22} a_{34} a_{43} a_{55} - a_{11} a_{22} a_{34} a_{45} a_{53} \\ &+ a_{22} a_{41} a_{34} a_{53} a_{15} - a_{22} a_{41} a_{34} a_{13} a_{55}). \end{aligned}$$

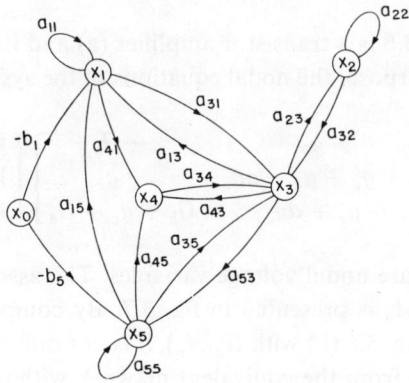

Fig. 3.4. The associated Coates graph of the system of equations (3.23).

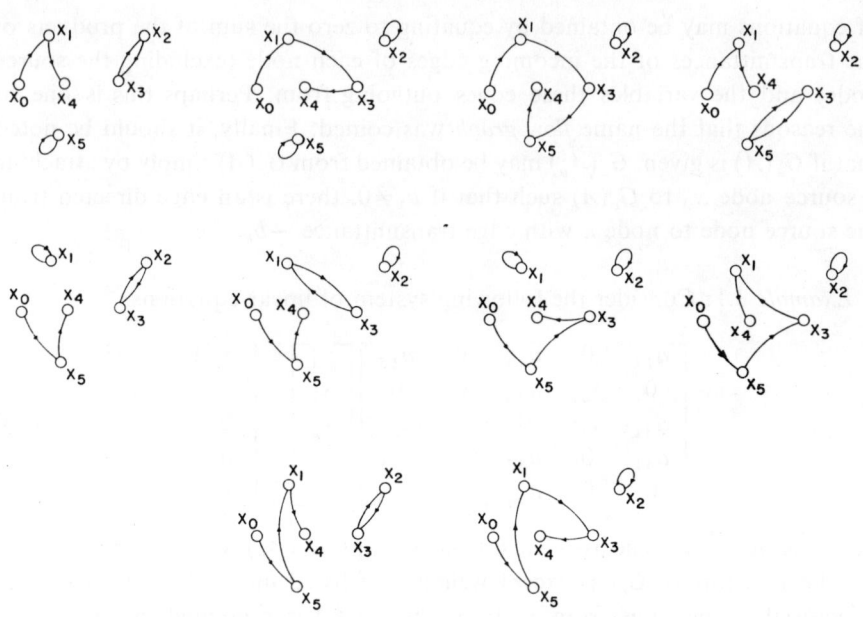

Fig. 3.5. The set of 1-factorial connections from x_0 to x_4 in the Coates graph of fig. 3.4.

In the analysis of a linear system, electrical networks in particular, the associated Coates graph, in many cases, can be drawn directly from the system diagram without the necessity of first setting up the equations in matrix form (CHEN [1967b]). The following example illustrates this for a common type of transistor feedback amplifiers.

Example 3.2: Fig. 3.6 is a transistor amplifier (a) and its equivalent network (b). For illustrative purpose, the nodal equations of the system are given below:

$$\begin{bmatrix} G_b + g_b & -g_b & -G_b \\ -g_b & g_b + g_e + g_c - ag_e & -g_c \\ -G_b & -g_c + ag_e & G_b + g_c + G_d \end{bmatrix} \begin{bmatrix} V_1 \\ V_2 \\ V_3 \end{bmatrix} = \begin{bmatrix} I_1 \\ 0 \\ 0 \end{bmatrix}, \quad (3.24)$$

where V_1, V_2, and V_3 are nodal voltage variables. The associated Coates graph $G_c(A_u)$ of the matrix A_u is presented in fig. 3.7. By comparing the equivalent network as shown in fig. 3.6 (b) with $G_c(A_u)$, it is not difficult to see that $G_c(A_u)$ can be drawn directly from the equivalent network without first writing down the nodal equations. Suppose that the current-gain function of the amplifier

§ 1   The associated Coates graph

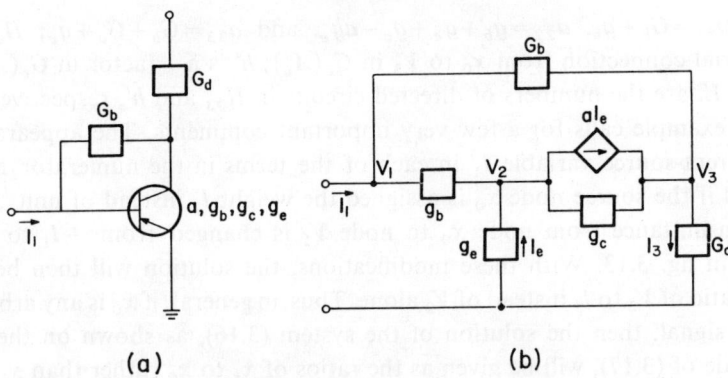

Fig. 3.6. A transistor feedback amplifier and its equivalent network.

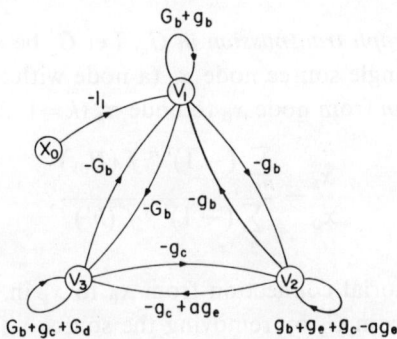

Fig. 3.7. The associated Coates graph of the amplifier of fig. 3.6.

is needed which is defined as the ratio of I_3 to I_1. Then

$$\begin{aligned}
\frac{I_3}{I_1} &= \frac{G_d V_3}{I_1} = \frac{G_d \sum_{H_{03}} (-1)^{L_H} f(H_{03})}{I_1 \sum_{h'} (-1)^{L'_h} f(h')} \\
&= \frac{G_d [-(-I_1)(-G_b) a_{22} + (-I_1)(-g_b)(ag_e - g_c)]}{I_1 [a_{22} G_b G_b + a_{33} g_b g_b + a_{11} g_c (g_c - ag_e) \\
\quad - G_b g_b (ag_e - g_c) + g_c g_b G_b - a_{11} a_{22} a_{33}]} \\
&= \frac{G_d [g_b (g_c - ag_e) + G_b (g_b + g_e + g_c - ag_e)]}{G_b [g_b + g_c + (1-a) g_e] G_d + g_b g_e (G_b + g_c + G_d)} \\
&\qquad\qquad\qquad\qquad\qquad\qquad\qquad + G_b g_e g_c + g_b G_d (g_c - ag_e)
\end{aligned}$$

(3.25)

where $a_{11}=G_b+g_b$, $a_{22}=g_b+g_e+g_c-ag_e$, and $a_{33}=G_b+G_d+g_c$; H_{03} is a 1-factorial connection from x_0 to V_3 in $G_c(A_u)$; h' is a 1-factor in $G_c(A)$; and L_H and L'_h are the numbers of directed circuits in H_{03} and h', respectively.

This example calls for a few very important comments. The appearance of the current-source variable I_1 in each of the terms in the numerator may be avoided if the source node x_0 is assigned the weight I_1 instead of unity, and if the transmittance from node x_0 to node V_1 is changed from $-I_1$ to -1 as shown in fig. 3.13. With these modifications, the solution will then be given as the ratio of V_3 to I_1 instead of V_3 alone. Thus, in general, if x_0 is any arbitrarily chosen signal, then the solution of the system (3.16), as shown on the right-hand side of (3.17), will be given as the ratios of x_k to x_0 rather than x_k alone. The ratio x_k/x_0 is referred to as the graph transmission from node x_0 to node x_k in $G_c(A_u)$. Formally, it is defined as

DEFINITION 3.5: *Graph transmission in G_c*. Let G_c be an $(n+1)$-node Coates graph containing a single source node x_0 (a node with zero incoming degree). The *graph transmission* from node x_0 to node x_k ($k=1, 2, ..., n$) in G_c is defined as the ratio

$$\frac{x_k}{x_0} = \frac{\sum_{H_{0k}} (-1)^{L_H} f(H_{0k})}{\sum_{h'} (-1)^{L'_h} f(h')}, \qquad (3.26)$$

where H_{0k} is a 1-factorial connection from x_0 to x_k in G_c; h' is a 1-factor in the graph obtained from G_c by removing the source node x_0; and L_H and L'_h are the numbers of directed circuits in H_{0k} and h', respectively.

This seemingly simple new concept is extremely useful in that, in most engineering problems, it is necessary to compute the symbolic system functions, which are defined as the ratios of two functions with the parameters of the system given in *literal* rather than in numerical form, instead of solving a system of linear algebraic equations in the usual sense. Thus, the advantages of this approach are clear, and we shall use it whenever appropriate. Mathematically, the evaluation of the graph transmissions defined in (3.26) is equivalent to solving the system of equations

$$AX = x_0 B \qquad (3.27)$$

for X in terms of x_0. In the definition, we require that G_c contain only a single source. This is not a serious restriction since for G_c with more than one source the graph transmission due to each of these sources can be computed separately with all other sources being removed.

The second point is that if we consider all the elements in the coefficient matrix to be distinct, in general, there can be no cancellation of terms in (3.17). However, in many engineering applications, this is not always true as we can see from the example given above. Thus, in this sense, the formula (3.17) is not very efficient, but it can be made more efficient by simple modifications of the Coates graph. We shall discuss this further in § 3. Finally, it should be pointed out that, to every system of linear equations such as (3.16), there corresponds a unique associated Coates graph $G_c(A_u)$, and vice versa. If, however, the order in which the equations appear in (3.16) is changed, the associated Coates graph changes in a nontrivial manner since the matrix A_u is also changed. Thus, the unique association of a system of equations with a Coates graph is meaningful only when the system of equations has been ordered in a fixed way.

1.4. *Equivalence and transformations*

Once a Coates graph has been constructed, it is possible to write down a solution for the graph transmission from the source node to any other node immediately. However, it is usually advantageous to carry out some transformations and simplifications of the graph before applying the general formulas derived earlier. There are a number of techniques available for this, each of which has its analogue in the conventional algebraic operations on systems of linear equations.

(A) *Elimination of nodes*

A standard algebraic procedure for solving a system of linear equations for any set of variables as a function of some other set of variables is to reduce systematically the number of equations and the number of variables until a set of equations involving only the desired variables remains. To such a process in the algebraic procedure there corresponds a topological reduction process on the directed graphs associated with the equations such that the set of equations associated with the resultant graph is that which results from applying the algebraic reduction process. To accomplish this, COATES [1959] showed how to obtain equivalent Coates graphs using one less node than the original one. The operation is repeated until the necessary number of nodes is removed. In the following, we shall present a generalization of Coates' procedure. The generalization may be regarded as the multiple-node removal algorithm for a Coates graph, and was first given by CHEN [1964a].

For simplicity, by the *graph determinant* of G_c, denoted by $\det G_c$, we mean the determinant of the matrix associated with the Coates graph G_c. Recall the definition and symbol (Definition 1.15 and ch. 1, § 5.1) defined for a sectional subgraph of a directed graph in chapter 1. With these preliminaries, we now proceed to discuss the reduction process for a Coates graph.

Let V_c be a nonempty proper subset of the node set V of a Coates graph G_c. Let G_{cr} be the reduced Coates graph obtained from G_c by the following process:

(i) Remove the sectional subgraph $G_c[V_c]$ from G_c, i.e., remove all nodes and edges incident with any node in V_c.

(ii) The weight or transmittance c_{ij} associated with the edge (i,j) in G_{cr} is given by[†]

$$c_{ij} = (-1)^\alpha (1/K_c) \sum_{H_{ij}''} (-1)^{L_H''} f(H_{ij}'')$$

$$= (1/K_c) \sum_{H''_{ij}} (-1)^{q''_H} f(H_{ij}'') \qquad (3.28)$$

for all i and j in $V - V_c$; where H_{ij}'' is a 1-factorial connection from node i to node j in the sectional subgraph $G_c[V_c^{ij}]$, where V_c^{ij} is the set union of the sets V_c and $\{i,j\}$; α, L_H'', and q_H'' are the numbers of nodes, directed circuits, and even components in H_{ij}'', respectively; and K_c is the determinant of the sectional subgraph $G_c[V_c]$, which can be evaluated by the formula

$$K_c = \det G_c[V_c] = (-1)^\beta \sum_{h''} (-1)^{L_h''} f(h''), \qquad (3.29)$$

where h'' is a 1-factor in $G_c[V_c]$, and β and L_h'' are the numbers of nodes and directed circuits in V_c and h'', respectively. When $i=j$, $G_c[V_c^{ij}]$ reduces to $G_c[V_c^i]$ and the 1-factorial connection H_{ij}'' from node i to node j becomes a 1-factor in $G_c[V_c^i]$ by definition.

Fortunately, the process outlined above is easier to visualize than it is to describe in words, as a moment's study of its topological structure will verify. In the following, we shall show that the transmittances c_{ij} given in (3.28) can be interpreted as the graph transmissions in some modified sectional subgraphs of G_c (CHEN [1969a]).

For $i \neq j$, let $G_c''[V_c^{ij}]$ be the Coates graph obtained from the sectional subgraph $G_c[V_c^{ij}]$ first by removing all the edges terminating at node i and all the edges outgoing from node j, and then by adding a self-loop of transmittance -1 at the node j. Then c_{ij} is simply the graph transmission from node i to node j in $G_c''[V_c^{ij}]$.

For $i=j$, there is a similar physical interpretation. In the sectional subgraph $G_c[V_c^i]$, we first replace the node i by two distinct nodes i' and i'' such that all the edges originally outgoing from node i in $G_c[V_c^i]$ now outgo from node i', and all the edges originally terminating at node i now terminate at node i''.

[†] Edges not in G_c are considered as edges with zero weight. Thus, we may have edges in G_{cr} but not in G_c.

Then we add a self-loop of transmittance -1 at the node i''. Let the resulting graph be denoted by $G_c''[V_c^i]$. The transmittance c_{ii} is simply the graph transmission from node i' to i'' in $G_c''[V_c^i]$. The operation of replacing node i by two distinct nodes i' and i'' used above is known as the *node-splitting* technique, and is extremely useful in feedback systems (see, for example, HOSKINS [1960]).

The justification for the above physical interpretations is not hard to find, and is left as an exercise (Problem 3.5).

THEOREM 3.4: If K_c is nonzero, then

$$\det G_c = K_c \det G_{cr}. \tag{3.30}$$

Proof. Let A be the associated matrix of G_c. Since a relabeling of the nodes of G_c corresponds to simultaneous interchanges of the rows and the corresponding columns of A, without loss of generality we may assume that the matrix A can be partitioned in such a way that A_{11} corresponds to the node set V_c:

$$A = \begin{bmatrix} A_{11} & A_{12} \\ A_{21} & A_{22} \end{bmatrix}, \tag{3.31}$$

where A_{11} and A_{22} are square submatrices of orders β and $n-\beta$, respectively; n is the number of nodes in G_c; and A_{11} is nonsingular since by hypothesis K_c is nonzero. (Note that A_{ij} no longer represents the submatrix obtained from A by striking out the ith row and jth column. This should not create any confusion.)

Also, it is easy to show that the identities

$$\begin{bmatrix} U_\beta & 0 \\ -A_{21}A_{11}^{-1} & U_{n-\beta} \end{bmatrix} \begin{bmatrix} A_{11} & A_{12} \\ A_{21} & A_{22} \end{bmatrix} = \begin{bmatrix} A_{11} & A_{12} \\ 0 & A_{22} - A_{21}A_{11}^{-1}A_{12} \end{bmatrix} \tag{3.32a}$$

and

$$\det A = (\det A_{11}) \det (A_{22} - A_{21}A_{11}^{-1}A_{12}) \tag{3.32b}$$

hold, where U_x is the identity matrix of order x and 0 is the zero matrix of appropriate order, and A_{11}^{-1} denotes the inverse of A_{11}.

Let

$$A_{kt} = [a_{ij}^{kt}] \tag{3.33}$$

for $k, t = 1, 2$, where the subscripts i and j are the row and column indices; and

$$A_{22} - A_{21}A_{11}^{-1}A_{12} = (1/K_c)[c_{uv}''], \tag{3.34}$$

where $K_c = \det A_{11}$. It follows that

$$c_{uv}'' = K_c a_{uv}^{22} - \sum_{k_1, k_2} a_{uk_1}^{21} \Delta_{k_2 k_1}^{11} a_{k_2 v}^{12} \tag{3.35}$$

for $u, v = 1, 2, \ldots, n-\beta$, where $\Delta^{11}_{k_2 k_1}$ is the cofactor of the (k_2, k_1)-element of A_{11}, and k_1 and k_2 range from 1 to β. By Theorem 3.2, it is evident that

$$\Delta^{11}_{k_2 k_1} = \sum_{h^{11}} (-1)^{q_{11}} f(h^{11}) \tag{3.36}$$

for $k_1 = k_2$, and

$$\Delta^{11}_{k_2 k_1} = \sum_{H^{11}_{k_2 k_1}} (-1)^{q'_{11}-1} f(H^{11}_{k_2 k_1}) \tag{3.37}$$

for $k_1 \neq k_2$, where h^{11} and $H^{11}_{k_2 k_1}$ are a 1-factor and a 1-factorial connection from k_2 to k_1 in $G_c[V_c - \{k_1\}]$ and $G_c[V_c]$, respectively; and q_{11} and q'_{11} are the numbers of the even components in h^{11} and $H^{11}_{k_2 k_1}$, respectively.

Let $i = u + \beta$ and $j = v + \beta$. From (3.35)–(3.37), we have

$$c''_{uv} = \sum_{h''} (-1)^{q''} f(h'') f(j,i) + \sum_{\substack{k_1, k_2 \\ k_1 \neq k_2}} \sum_{H^{11}_{k_2 k_1}} (-1)^{q'_{11}} f(j, k_2) f(H^{11}_{k_2 k_1}) f(k_1, i)$$

$$+ \sum_{k_1} \sum_{h^{11}} (-1)^{q_{11}+1} f(j, k_1) f(h^{11}) f(k_1, i), \tag{3.38}$$

where q'' denotes the number of even components in h''. Note that we have used the following identity:

$$(-1)^{\beta + L''_h} f(h'') = (-1)^{q''} f(h''). \tag{3.39}$$

Next, observe the one-to-one correspondence between 1-factors in $G_c[V_c]$ and 1-factorial connections H''_{ji} in $G_c[V^{ij}_c]$ whose directed path P''_{ji} consists of a single edge (j, i); and also note that $q''_H = q''$ in this case.

Observe also, when $k_1 \neq k_2$, the one-to-one correspondence between 1-factorial connections $H^{11}_{k_2 k_1}$ in $G_c[V_c]$ and 1-factorial connections H''_{ji} in $G_c[V^{ij}_c]$ whose directed path P''_{ji} consists of the edge (j, k_2), the directed path $P^{11}_{k_2 k_1}$ of $H^{11}_{k_2 k_1}$, and the edge (k_1, i); and that the relation between numbers of even components is again $q''_H = q'_{11}$.

Finally, when $k_1 = k_2$, observe the one-to-one correspondence between 1-factors of $G_c[V_c - \{k_1\}]$ and 1-factorial connections H''_{ji} of $G_c[V^{ij}_c]$ whose directed path P''_{ji} consists of the two edges (j, k_1) and (k_1, i); and that in this case $q''_H = q_{11} + 1$.

Now, since every 1-factorial connection H''_{ji} of $G_c[V^{ij}_c]$ is one of the above three types, it follows that

$$c''_{uv} = \sum_{H''_{ji}} (-1)^{q''_H} f(H''_{ji}) \tag{3.40a}$$

$$= (-1)^{\alpha} \sum_{H''_{ji}} (-1)^{L''_H} f(H''_{ji}). \tag{3.40b}$$

The second line follows directly from Corollary 3.1.

At this stage, it is clear that the reduced Coates graph G_{cr} is actually the associated Coates graph of the matrix $(1/K_c) [c''_{uv}]$ defined in (3.34). Thus, from (3.32b) and (3.34) we have

$$\det G_c = \det A = K_c \det \{(1/K_c) [c''_{uv}]\} = K_c \det G_{cr},$$

which completes the proof of the theorem.

In particular, if V_c contains only a single node, the process outlined above can be simplified, and is given by

COROLLARY 3.4: If w_{ij} is the weight associated with the edge (i,j) of G_c, and if $w_{kk} \neq 0$, then

$$\det G_c = w_{kk} \det G_{cr}, \qquad (3.41)$$

where G_{cr} is obtained from G_c by the following process:
(i) Remove the node k from G_c.
(ii) The weight c_{ij} associated with the edge (i,j) in G_{cr} is given by

$$c_{ij} = w_{ij} - w_{ik} w_{kj}/w_{kk}. \qquad (3.42)$$

The usefulness of Theorem 3.4 really lies in the following. (Also, see Problem 3.55.)

THEOREM 3.5: If G_{cr} is the reduced Coates graph of G_c, and if $K_c \neq 0$ and $i \neq j$, then

$$\sum_{H_{ij}} (-1)^{L_H} f(H_{ij}) = K_c \sum_{H_{ij}^r} (-1)^{L_H^r + \beta} f(H_{ij}^r) \qquad (3.43)$$

or

$$\sum_{H_{ij}} (-1)^{q_H} f(H_{ij}) = K_c \sum_{H_{ij}^r} (-1)^{q_H^r} f(H_{ij}^r) \qquad (3.44)$$

for all i and j not in V_c, where H_{ij} and H_{ij}^r are the 1-factorial connections from i to j in G_c and G_{cr}, respectively; L_H and L_H^r are the numbers of directed circuits in H_{ij} and H_{ij}^r, respectively; q_H and q_H^r are the numbers of even components in H_{ij} and H_{ij}^r, respectively; and β and K_c are defined in (3.29).

Proof. Let G_c^* be the Coates graph obtained from G_c by deleting all the edges having node j as their initial node, and all edges having node i as their terminal node, and then adding an edge from node j to node i with weight 1. Also let G_{cr}^* be the reduced Coates graph of G_c^* obtained from G_c^* by the removal of $G_c^* [V_c]$ using the procedure outlined earlier. Then

$$\det G_c^* = K_c \det G_{cr}^*. \qquad (3.45)$$

Since the determinants of the matrices associated with G_c^* and G_{cr}^* are equal to the cofactors of the (i,j)-element of the matrices associated with the graphs G_c and G_{cr}, respectively, it follows from Theorem 3.2 that $\det G_c^*$ and $\det G_{cr}^*$ can be evaluated by the subgraphs of G_c and G_{cr}, respectively. The theorem follows after a simple substitution.

Let us now apply the above results to the solution of the system (3.16). Let $G_c = G_c(A_u)$. Assume that the source node x_0 is not in V_c. Then

COROLLARY 3.5: *The solution of the system (3.16) may be evaluated either in G_c or in G_{cr} for all the variables not associated with the nodes in V_c.*

COROLLARY 3.6: *The graph transmission from the source node x_0 to any other node remains the same in both G_c and G_{cr} for all the nodes not in V_c.*

We shall illustrate the reduction procedure by the following example.

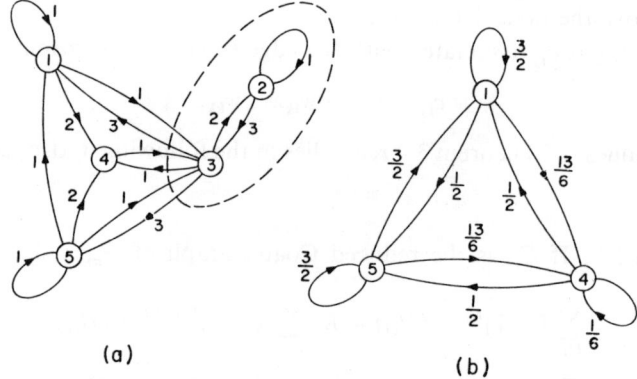

Fig. 3.8. An example to illustrate the reduction procedure for a Coates graph.

Example 3.3: Consider the Coates graph G_c as shown in fig. 3.8(a). The dotted part of the figure is the sectional subgraph $G_c[2, 3]$ to be removed. Fig. 3.8(b) is the reduced Coates graph G_{cr} of G_c. For illustrative purpose, we shall compute the weights associated with two typical edges $(5,1)$ and $(4,4)$ of G_{cr}. The sectional subgraph $G_c[1, 2, 3, 5]$ together with all the 1-factorial connections H_{51}'' from node 5 to node 1 in $G_c[1, 2, 3, 5]$ is presented in fig. 3.9(a). Thus, the weight c_{51} associated with the edge $(5,1)$ of G_{cr} with w_{ij} being the weight of the edge (i,j) in G_c is given by

$$c_{51} = (-1)^4 (1/K_c) [(-1)^1 w_{53}w_{31}w_{22} + (-1)^1 w_{51}w_{23}w_{32}]$$
$$= \tfrac{1}{6}[1 \cdot 3 \cdot 1 + 1 \cdot 3 \cdot 2] = \tfrac{3}{2},$$

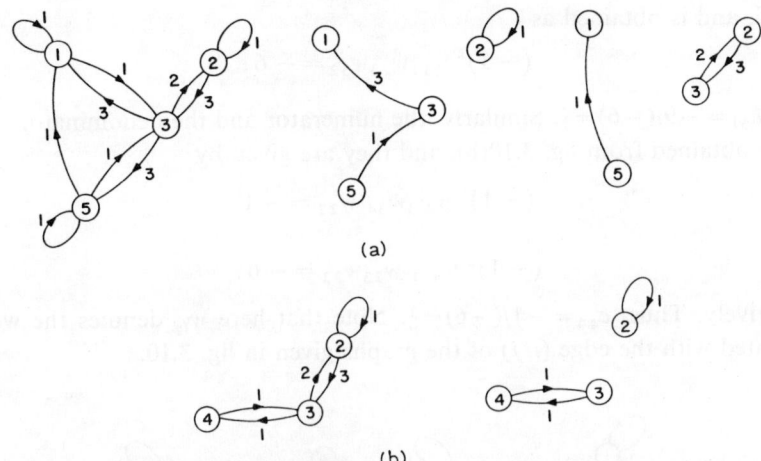

Fig. 3.9. The sectional subgraphs and their 1-factorial connections for the evaluation of the transmittances c_{51} and c_{44}.

where $K_c = -6$. Similarly, the sectional subgraph $G_c[2, 3, 4]$ together with all the 1-factors in $G_c[2, 3, 4]$ is given in fig. 3.9(b). The weight c_{44} associated with the self-loop $(4, 4)$ in G_{cr} is then obtained as follows:

$$c_{44} = (-1)^3 (1/K_c) [(-1)^2 w_{34}w_{43}w_{22}] = \tfrac{1}{6}(1 \cdot 1 \cdot 1) = \tfrac{1}{6}.$$

The graph determinant of G_{cr} may be computed by the formula given in Theorem 3.1, and its value is equal to $-\tfrac{5}{6}$. The graph determinant of G_c was computed in (3.10) and is equal to 5. Thus, we have

$$\det G_c = (-6) \det G_{cr}. \tag{3.46}$$

As an alternative, the transmittances c_{ij} of G_{cr} can also be computed as the graph transmissions from node i to node j in $G_c''[V_c^{ij}]$. For example, c_{51} and c_{44} are the graph transmissions from node 5 to node 1, and from node 4' to node 4'', in the Coates graphs $G_c''[1, 2, 3, 5]$ and $G_c''[2, 3, 4]$, as shown in fig. 3.10, respectively. For illustrative purposes, we shall compute these two graph transmissions as follows:

From (3.26), the numerator of c_{51} is simply the algebraic sum of all the products of the weights associated with the 1-factorial connections from node 5 to node 1 in fig. 3.10(a), and is given by

$$(-1)^1 w_{53}w_{31}w_{22} + (-1)^1 w_{51}w_{23}w_{32} = -9.$$

The denominator of c_{51} is the algebraic sum of the weight products of all the 1-factors in the Coates graph obtained from fig. 3.10(a) by removing the source node 5, and is obtained as

$$(-1)^2 w_{11} w_{23} w_{32} = -6.$$

Thus, $c_{51} = -9/(-6) = \frac{3}{2}$. Similarly, the numerator and the denominator of c_{44} can be obtained from fig. 3.10(b), and they are given by

$$(-1)^1 w_{4'3} w_{34''} w_{22} = -1$$

and

$$(-1)^2 w_{4''4''} w_{23} w_{32} = -6,$$

respectively. Thus, $c_{44} = -1/(-6) = \frac{1}{6}$. Note that here w_{ij} denotes the weight associated with the edge (i, j) of the graphs given in fig. 3.10.

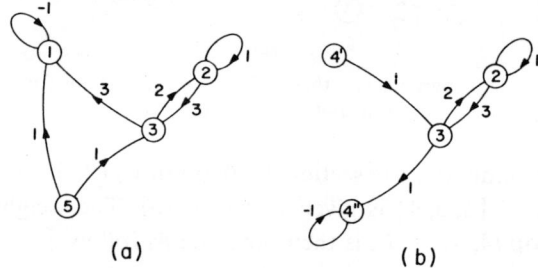

Fig. 3.10. The graphs for the evaluation of the transmittances c_{51} and c_{44} by means of graph transmissions.

From the above discussion, we conclude that the formula used to evaluate the graph transmissions in a Coates graph can also be used to evaluate the transmittances associated with the edges of the reduced Coates graph. The only difference is that it is now applied to certain modified sectional subgraphs of the original Coates graph.

(B) *Decomposition into subgraphs*

In many cases, a Coates graph G_c may be decomposed into unconnected subgraphs such that the graph determinant of G_c is equal to the product of the graph determinants of the subgraphs, where $f(\emptyset) = 0$ by definition.

DEFINITION 3.6: *Decomposable Coates graph*. A Coates graph G_c is said to be *decomposable* if the associated matrix A of G_c can be partitioned into the form

$$A = \begin{bmatrix} A_{11} & 0 \\ A_{21} & A_{22} \end{bmatrix} \quad \text{or} \quad \begin{bmatrix} A_{11} & A_{12} \\ 0 & A_{22} \end{bmatrix}, \tag{3.47}$$

where A_{11} and A_{22} are square submatrices, by permuting the rows and columns of A if necessary; these rows and columns are not required to permute simultaneously.

DEFINITION 3.7: *Deficiency and converse deficiency.* The *deficiency* (*converse deficiency*) of a directed graph G defined by a subset V_c of its node set is $\alpha - w(V_c)$ ($\alpha - w^*(V_c)$), where α is the number of elements in V_c, and $w(V_c)$ ($w^*(V_c)$) denotes the number of terminal (initial) nodes of the edges of G having their initial (terminal) nodes in V_c.

As an illustration, consider the Coates graph G_c as shown in fig. 3.11. Let $V_c = \{3, 4\}$. The set of terminal nodes of the edges having their initial nodes in V_c is $\{2, 3\}$, and the set of initial nodes of the edges having their terminal nodes in V_c is the node set of G_c. Thus, $\alpha = 2$, $w(V_c) = 2$, and $w^*(V_c) = 5$. The deficiency and the converse deficiency of G_c defined by the nodes 3 and 4 are 0 and -3, respectively.

THEOREM 3.6: A Coates graph G_c is decomposable if and only if there exists a nonempty proper subset of the node set such that its deficiency or converse deficiency is nonnegative. Furthermore, if either its deficiency or converse deficiency is positive, then the graph determinant of G_c vanishes.

Proof. Suppose there exists a nonempty proper subset V_c such that

$$\alpha - w(V_c) \geq 0,$$

where α and $w(V_c)$ are defined the same as in Definition 3.7. In terms of the rows and columns of the associated matrix A of G_c, which is of order n, $w(V_c)$ is the number of rows in A having at least one nonzero element in common with one of the columns corresponding to the elements in V_c. The element a_{ij} of A is zero if i is in $(V - R(V_c))$ and j in V_c, where $R(V_c)$ denotes the set of terminal nodes of the edges having their initial nodes in V_c and V is the node set of G_c. But, by hypothesis,

$$(n - w(V_c)) + \alpha \geq n,$$

thus, there is always a zero submatrix of order $(n - w(V_c)) \times \alpha$ in A since n, $w(V_c)$, and α are the numbers of elements in V, $R(V_c)$, and V_c, respectively. Therefore, G_c is decomposable. The converse is also true. Similarly, we can prove the case for nonnegative converse deficiency.

If the deficiency is positive, then $(n - w(V_c)) + \alpha$ is greater than n. This means that if we expand the determinant of A along the rows or columns corresponding to the zero submatrix by Laplace expansion, each minor contains at least one zero row or zero column. Thus, $\det A = 0$. Similarly, this is valid for positive converse deficiency. So the theorem holds.

So far, we have only characterized the Coates graphs that are decomposable. The usefulness of the theorem depends largely upon the efficiency with which a Coates graph can be decomposed into subgraphs. In the sequel, we shall discuss one of such decomposition procedures.

The interchange of any two rows i and j of A corresponds to moving the set of edges of $G_c(A)$ having their terminal node j to node i, and to moving the set of edges having their terminal node i to node j, while keeping all their initial nodes and edge weights unaltered. We also include the case where no edges terminate at the node i or j. For simplicity, this type of operations will be termed as the *interchange of the incoming edges of the nodes i and j in $G_c(A)$*. Similarly, the interchange of any two columns i and j of A corresponds to the *interchange of the outgoing edges of the nodes i and j in $G_c(A)$*. Thus, if V_c is a nonempty proper subset of G_c with zero deficiency, then a series of interchanges of the incoming edges of the nodes i and j, i in $V_c - (V_c \cap R(V_c))$ and j in $R(V_c) - (R(V_c) \cap V_c)$, will transform G_c into a Coates graph with no edges directed from a node in V_c to any node in $V - V_c$ (justify this). Consequently, all the edges of the form (u, v), u in $(V - V_c)$ and v in V_c, may be removed from $G_c(A)$ without changing its graph determinant. This process decomposes G_c into at least two components G_{c1} and G_{c2} such that the union of the node sets of the components is V, and furthermore,

$$\det G_c = (-1)^\beta (\det G_{c1})(\det G_{c2}), \tag{3.48}$$

where β is the number of such interchanges of the incoming edges. A similar procedure can be obtained if V_c has zero converse deficiency (Problem 3.54).

Example 3.4: Consider the Coates graph G_c given in fig. 3.11. Since the set $V_c = \{3, 4\}$ is of zero deficiency, it follows from the theorem that G_c can be decomposed into components. An interchange of the incoming edges of the nodes 2 and 4 in G_c results in a Coates graph G_c^* which has no edges directed from any node in V_c to any node in $\{1, 2, 5\}$. Thus, the edges $(5, 3)$, $(2, 4)$, and $(1, 4)$ may be removed from G_c^* without changing its determinant. This process decomposes G_c into two components G_{c1} and G_{c2}, as shown in fig. 3.12, such that

$$\det G_c = (-1)^1 (\det G_{c1})(\det G_{c2}).$$

A subsidiary result of some interest has to do with edges that are not con-

§1 The associated Coates graph

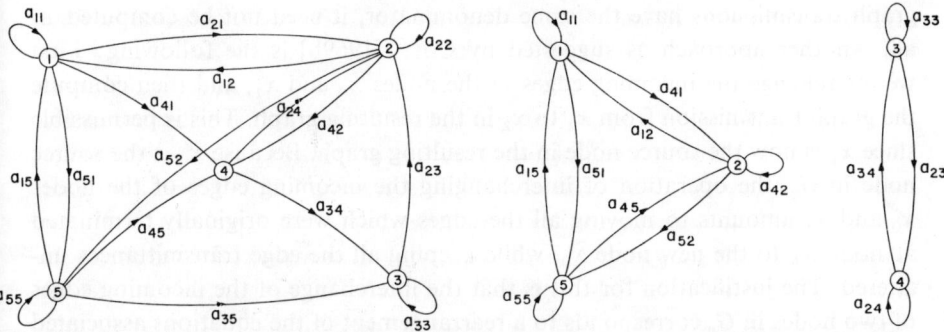

Fig. 3.11. A decomposable Coates graph.   Fig. 3.12. The decomposed components.

tained in any directed circuit of G_c. Clearly, these edges of G_c can be removed without effecting its graph determinant because they cannot be contained in any 1-factors of G_c. On the other hand, for edges contained in directed circuits of G_c the above procedure shows how some of these edges can be removed again without effecting its graph determinant. More specifically, we state the following corollary; its justification is left as an exercise (Problem 3.47).

COROLLARY 3.7: Let V_c be a nonempty proper subset of the node set V of G_c. If V_c is of zero deficiency, then the set of edges (i,j), i in $(V-V_c)$ and j in $R(V_c)$, can be removed from G_c without effecting its graph determinant. Similarly, if V_c is of zero converse deficiency, then the set of edges (i,j), i in $R^*(V_c)$ and j in $(V-V_c)$, can be removed from G_c without effecting its graph determinant, where $R(V_c)$ and $R^*(V_c)$ are the sets of terminal and initial nodes of the edges having their initial and terminal nodes in V_c, respectively.

(C) *Inversion*

In the foregoing, we have used the operation of interchanging the incoming edges of two nodes to decompose a Coates graph into components. In this section, we shall show how this type of operations can be used to invert a directed path or circuit, and to compute the graph transmission between any two of its nodes.

Let G_c be a Coates graph having a single source node x_0. The graph transmission from x_0 to any other node x_k in G_c is defined as the ratio of x_k to x_0. Suppose that we wish to compute the ratio of, say, x_2 to x_1. One way to accomplish this is simply that we first compute the graph transmissions from the source node x_0 to the nodes x_2 and x_1, and then take the ratio. Since these graph transmissions have the same denominator, it need not be computed at all. Another approach as suggested by CHEN [1969b] is the following: First we interchange the incoming edges of the nodes x_0 and x_1, and then compute the graph transmission from x_1 to x_2 in the resulting graph. This is permissible since x_1 is now the source node in the resulting graph. Because x_0 is the source node in G_c, the operation of interchanging the incoming edges of the nodes x_0 and x_1 amounts to moving all the edges which were originally terminated at node x_1 to the new node x_0, while keeping all the edge transmittances unaltered. The justification for this is that the interchange of the incoming edges of two nodes in G_c corresponds to a rearrangement of the equations associated with G_c, and thus will not change the solutions of the system.

As an illustration, consider the feedback amplifier as shown in fig. 3.6. Suppose that we are interested in finding the voltage ratio of V_3 to V_2. The Coates graph $G_c(A_u)$ of (3.24) is presented in fig. 3.13. Since we wish to compute the ratio V_3/V_2, let us move all the edges terminated at node V_2 to node I_1.

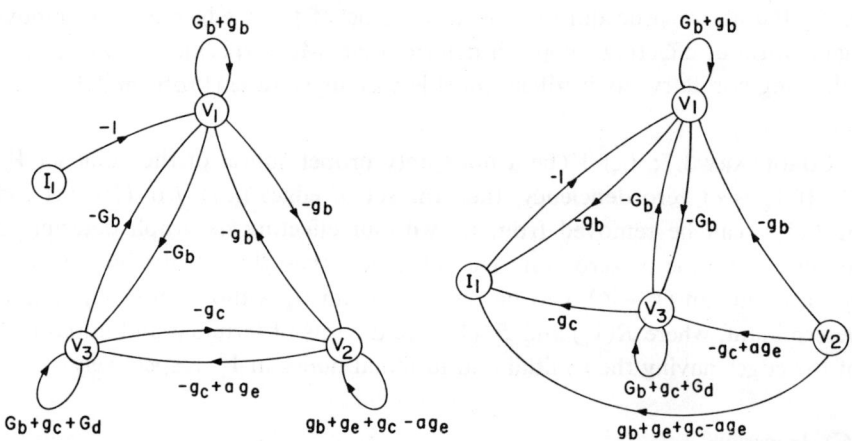

Fig. 3.13. The Coates graph of (3.24).

Fig. 3.14. The conversion of the node V_2 into a source node.

The process results in a new Coates graph as shown in fig. 3.14. The graph transmission from V_2 to V_3 in the new Coates graph is then obtained by (3.26) as follows:

$$\frac{V_3}{V_2} = \frac{(g_c - ag_e) g_b + (g_b + g_e + g_c - ag_e) G_b}{(G_b + G_d + g_c) g_b + G_b g_c}. \tag{3.49}$$

The terms in the numerator correspond to the 1-factorial connections from V_2 to V_3 in the new Coates graph, and the terms in the denominator correspond to the 1-factors in the new Coates graph after the node V_2 has been removed.

Finally, if the operation of interchanging the incoming edges of two nodes is applied successively to a directed path which originates from a source node or to a directed circuit, the directed path or circuit will, in many but not all cases, appear in inverted form. For example, if the operation is applied successively to the nodes i_1 and V_1, and then V_1 and V_3 in fig. 3.13, the original path is then inverted in the resulting Coates graph as shown in fig. 3.15. The

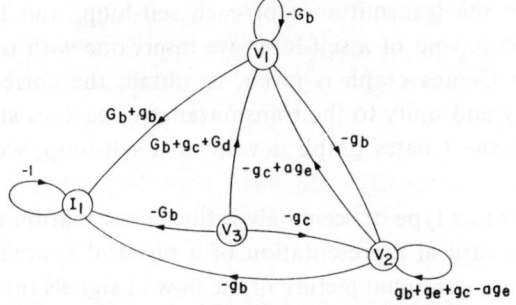

Fig. 3.15. The Coates graph obtained by inverting a directed path in the Coates graph of fig. 3.13.

graph transmission from V_3 to I_1 is, of course, the reciprocal of that given in (3.25) (verify this). On the other hand, had the self-loop (V_1, V_1) been removed from the original graph, the corresponding directed path in the resulting Coates graph would have been destroyed.

§ 2. The associated Mason graph

There is another way to associate a directed graph with a given matrix, known as a *signal-flow graph*. The concept of a signal-flow graph was originally worked out by SHANNON [1942] in dealing with analog computers. The greatest credit for the formulation and organization of signal-flow graphs is normally extended to MASON [1953, 1956]. He showed how to use the signal-flow graph technique to solve some difficult electronic problems in a relatively simple manner. The term *signal-flow graph* was used because of its original application to electronic problems and the association with electronic signals and flowcharts of the systems under study. Since 1956, there has been considerable progress in formulating signal-flow graphs and broadening their applications to statistical, mechanical, heat transfer, pneumatic, microwave, and multiple-loop feedback systems (LORENS [1964] and RIEGLE and LIN [1972]). For our purposes, a signal-flow graph will be called a *Mason graph*, as distinguished from a flow graph which is referred to as a Coates graph.

DEFINITION 3.8: *Mason graph*. For a given square matrix A of order n, the associated *Mason graph* of A, denoted by the symbol $G_m(A)$ or simply G_m if A is clearly understood or is not explicitly given, is the associated Coates graph of $A + U_n$, where U_n is the identity matrix of order n.

Thus, to obtain a Coates graph from a given Mason graph, we simply subtract unity from the transmittance of each self-loop, and for each node of the Mason graph devoid of a self-loop, we insert one with transmittance -1. Conversely, if a Coates graph is given, to obtain the corresponding Mason graph, we simply add unity to the transmittance of each existing self-loop and to each node of the Coates graph devoid of a self-loop, we insert one with transmittance 1.

The reason for this type of seemingly artificial association is that the Mason graph is a more natural representation of a physical system than the Coates graph. It presents a continual picture of the flow of signals through the physical system, and permits a physical evaluation and a heuristic proof of some basic theorems. Like the Coates graph, the associated Mason graph of a physical system, in many cases, can be drawn directly from the system diagram without the necessity of first setting up the equations in matrix form (CHEN [1967b]).

Consider the system of linear equations of (3.16), which may be written in a slightly different form as follows:

$$(A_u + U_{n+1}) X'' = X'', \qquad (3.50)$$

where A_u is defined the same as in Theorem 3.3, and the transpose of X'' is $[x_1, x_2, ..., x_n, x_{n+1}]$ with $x_{n+1} = 1$. Now, we can interpret the equations of the above system

$$(a_{kk} + 1) x_k - b_k x_{n+1} + \sum_{\substack{j=1 \\ j \neq k}}^{n} a_{kj} x_j = x_k \qquad (3.51)$$

for $k = 1, 2, ..., n$ as the flow of signals in $G_m(A_u)$. Since the nodes of $G_m(A_u)$ are associated with the columns or rows of A_u, like the Coates graph it is convenient to assign the variables $x_1, x_2, ..., x_{n+1}$ to the nodes $1, 2, ..., n+1$ of $G_m(A_u)$, respectively. The associated weights of the nodes may be considered as *nodal signals* x_k. Then each equation gives the nodal signal as the algebraic sum of the signals entering via the incoming edges. Each entering signal is the product of the edge transmittance and the signal at the node from which that edge originates. Since the signal at a given node is the sum of the incoming signals, the presence of outgoing edges does not directly affect the signal at that node. Specifically, the flow of signals in $G_m(A_u)$ is to be interpreted as that shown in fig. 3.16. This interpretation follows directly from the equations (3.51). The source-node signal $x_0 = x_{n+1}$ is considered as an independent variable in the algebraic equations and all other variables can be expressed in terms of the independent variable. Thus, the formulation is very convenient and intuitive for a large class of physical problems. This is why Mason's formulation is more popular than that of Coates.

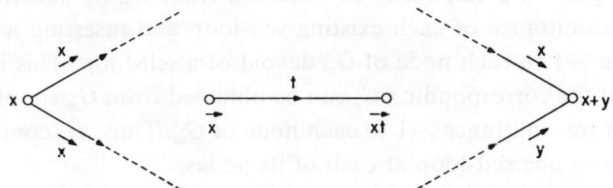

Fig. 3.16. A physical interpretation of the Mason graph.

By *graph determinant* of a Mason graph $G_m(A)$, denoted by det $G_m(A)$, we mean the determinant of the matrix A. In other words, the graph determinant of $G_m(A)$ is the same as that of $G_c(A)$. However, the formulas for their evaluation are completely different.

2.1. *Topological evaluation of determinants*

The rules for the evaluation of the graph determinant of a Mason graph were first given and proved by SHANNON [1942] using mathematical induction. His work remained essentially unknown even after Mason published his classical work in 1953. Three years later, MASON [1956] rediscovered the rules and proved them by considering the value of a determinant and how it changes as variables are added to the graph. Subsequently, a third proof was found by LORENS [1964] who showed the relationships between the rules and the Cramer's rule for the solution of linear algebraic equations. However, the arguments that are followed in these derivations are heuristic, and sometimes quite devious. A formal proof was given by ASH [1959] and NATHAN [1961] based on some simple properties of determinants which can be correlated with a directed graph. In the following, we shall present yet another simple topological derivation of det G_m based on the Coates graph.

The Coates graphs that we have considered so far are directed graphs having no parallel edges. It is easy to show (Problem 3.1) that if some of the edges (i, j) of G_c are replaced by the parallel edges $(i, j)_x$, $x = 1, 2, ..., k_{ij}$, (including the case where $i=j$) and such that

$$f(i, j) = \sum_{x=1}^{k_{ij}} f((i, j)_x), \qquad (3.52)$$

then the resulting Coates graph has the same graph determinant and graph transmissions as those of G_c. This extension is found to be very useful in proving many of the subsequent theorems, and we shall use it whenever convenient.

For example, we have stated earlier that the corresponding Coates graph G_c of a Mason graph G_m can easily be obtained from G_m by substracting unity from the transmittance of each existing self-loop and inserting a self-loop of transmittance -1 to each node of G_m devoid of a self-loop. This is equivalent to saying that the corresponding G_c can be obtained from G_m simply by adding a self-loop of transmittance -1 to each node of G_m. Thus, G_c contains at most two and at least one self-loop at each of its nodes.

With these preliminaries, we now proceed to prove the main theorem of this section.

THEOREM 3.7: Let $G_m(A)$ be the associated Mason graph of a square matrix A of order n. Then

$$\det A = (-1)^n \left[1 + \sum_{\substack{u,v \\ v>0}} (-1)^v f(C_{uv})\right], \qquad (3.53)$$

where C_{uv} is the uth subgraph of v node-disjoint directed circuits in $G_m(A)$.

Proof. Let G_c be the corresponding Coates graph of G_m obtained from G_m by adding a self-loop of transmittance -1 to each node of G_m. If h_{ij} is the ith 1-factor in G_c which contains j added self-loops of transmittance -1, and if Q_{ij} is the subgraph obtained from h_{ij} with the j added self-loops removed, then

$$f(h_{ij}) = (-1)^j f(Q_{ij}) \qquad (3.54\text{a})$$

for $j=0, 1, \ldots, n-1$, and

$$f(h_{ij}) = (-1)^j \qquad (3.54\text{b})$$

for $j=n$. By Theorem 3.1 and Problem 3.1, the totality of h_{ij} with appropriate signs is precisely the determinant of A,

$$\det A = (-1)^n \sum_{i,j} (-1)^{L_Q+j} f(h_{ij})$$

$$= (-1)^n \left[1 + \sum_i \sum_{j=0}^{n-1} (-1)^{L_Q} f(Q_{ij})\right], \qquad (3.55)$$

where L_Q is the number of directed circuits in Q_{ij}. Since for each choice of Q_{ij} in G_c, there corresponds a unique subgraph C_{uv} in G_m with $v = L_Q$, and vice versa, it follows that the determinant of A can be evaluated by the subgraphs C_{uv} in G_m rather than Q_{ij} in G_c. Merely by rearranging the terms of (3.55), we obtain the alternative and better known expansion,

$$\det A = (-1)^n \left[1 + \sum_{\substack{u,v \\ v>0}} (-1)^v f(C_{uv}) \right]. \tag{3.56}$$

This completes the proof of the theorem.

Like the Coates graph, it is desirable to know the number of terms in (3.53) in advance so that we may avoid the possibility of missing a term in the expansion (Problem 3.53).

COROLLARY 3.8: The number of terms in (3.53) is equal to the value of the permanent of the matrix $C(G_m) + U_n$, where $C(G_m)$ is defined the same as in Corollary 3.2.

As an example, consider the matrix (3.2). The associated Mason graph $G_m(A)$ is given in fig. 3.17. If the formula (3.53) is used to evaluate the determinant of A, then the number of terms expected in (3.53) is simply the value of the permanent of the matrix $C(G_m) + U_5$, and is given by

$$\text{per}\,[C(G_m) + U_5] = \text{per} \begin{bmatrix} 2 & 0 & 1 & 1 & 0 \\ 0 & 2 & 1 & 0 & 0 \\ 1 & 1 & 2 & 1 & 1 \\ 0 & 0 & 1 & 2 & 0 \\ 1 & 0 & 1 & 1 & 2 \end{bmatrix} = 78, \tag{3.57}$$

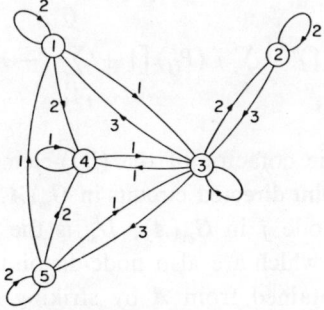

Fig. 3.17. The associated Mason graph of the matrix (3.2).

which is about twenty times the number of terms in the corresponding Coates graph. This brings out a very important point in that the Mason graphs suffer from unnecessary redundancies in the calculation of determinants from the graphs, and that these are obviated by the use of Coates graphs. However, as we shall see, it is possible to reduce the number of redundancies considerably by merely making appropriate modifications to the self-loop transmittances. Hence, the issue between the Coates and the Mason graphs cannot be simply decided on this ground alone. In engineering analysis, there is some important ground for preferring the Mason graph to the Coates graph, especially when we are considering systems in which physical feedback of some kind takes place. In such circumstances, as mentioned earlier, the appropriate Mason graph can often provide a direct representation of the feedback process. (See, for example, TRUXAL [1955].) In practical applications, the number of feedback loops are usually small, and also the elements of the matrix are given in *literal* rather than in numerical form. Thus, the advantages of the Mason-graph formulation are clear. The numerical example given above is served only for illustrative purposes. The elements of the matrix are usually given as rational functions in a complex variable s.

2.2. *Topological evaluation of cofactors*

The formula for the evaluation of the cofactors of the elements of a matrix by means of certain subgraphs in the associated Mason graph can be derived in a manner similar to that of a determinant.

THEOREM 3.8: Let $G_m(A)$ be the associated Mason graph of a square matrix A of order n, $n \geq 2$. Then

$$\Delta_{ii} = (-1)^{n-1} \left[1 + \sum_{\substack{u,v \\ v>0}} (-1)^v f(C'_{uv}) \right], \qquad (3.58a)$$

$$\Delta_{ij} = (-1)^{n-1} \sum_k f(P^k_{ij}) \left[1 + \sum_{\substack{s,t \\ t>0}} (-1)^t f(D^k_{st}) \right] \qquad (3.58b)$$

for $i \neq j$, where Δ_{ij} is the cofactor of the (i,j)-element of A; C'_{uv} is the uth subgraph of v node-disjoint directed circuits in $G_m(A_{ii})$; P^k_{ij} is the kth directed path from node i to node j in $G_m(A)$; D^k_{st} is the sth subgraph of t node-disjoint directed circuits which are also node-disjoint with P^k_{ij} in $G_m(A)$; and A_{ij} is the submatrix obtained from A by striking out the ith row and jth column.

§2 The associated Mason graph

Proof. The first part of the theorem follows directly from Theorem 3.7. The second part may be proved in a manner similar to that of Theorem 3.7.

Let $G_c(A)$ be the corresponding Coates graph of $G_m(A)$ obtained from $G_m(A)$ by inserting a self-loop of transmittance -1 to each node of $G_m(A)$. If H_{ij}^{sx} is the sth 1-factorial connection from i to j in $G_c(A)$ which contains x added self-loops of transmittance -1, and if W_{ij}^{sx} is the subgraph obtained from H_{ij}^{sx} by removing the x added self-loops, then

$$f(H_{ij}^{sx}) = (-1)^x f(W_{ij}^{sx}) \tag{3.59}$$

for $x = 0, 1, \ldots, n-2$. By Theorem 3.2 and Problem 3.1, the totality of H_{ij}^{sx} with appropriate signs is precisely the cofactor Δ_{ij},

$$\Delta_{ij} = (-1)^{n-1} \sum_{s,x} (-1)^{L_W+x} (-1)^x f(W_{ij}^{sx}), \tag{3.60}$$

where L_W is the number of directed circuits in W_{ij}^{sx}. If P_{ij}^{sx} is the directed path from i to j in W_{ij}^{sx}, and if D_{ij}^{sx} is the subgraph corresponding to the set of node-disjoint directed circuits of W_{ij}^{sx}, then (3.60) can be rearranged as

$$\Delta_{ij} = (-1)^{n-1} \sum_{s} f(P_{ij}^{sx}) \left[1 + \sum_{\substack{x \\ L_W \neq 0}} (-1)^{L_W} f(D_{ij}^{sx})\right]. \tag{3.61}$$

Since for each choice of P_{ij}^{sx} and D_{ij}^{sx} in $G_c(A)$ there exist a unique corresponding directed path P_{ij}^k of P_{ij}^{sx} in $G_m(A)$ and a unique corresponding set of node-disjoint directed circuits D_{ut}^k of D_{ij}^{sx} in $G_m(A)$ with $t = L_W$, and vice versa, it follows that the cofactor Δ_{ij} can now be evaluated by the subgraphs P_{ij}^k and D_{ut}^k of $G_m(A)$ rather than the subgraphs P_{ij}^{sx} and D_{ij}^{sx} of $G_c(A)$. Merely by substituting the proper terms in (3.61), we obtain the desired expansion (3.58b). This completes the proof of the theorem.

COROLLARY 3.9: The number of terms in the expansions (3.58) is equal to the value of the permanent of the matrix obtained from the matrix $C(G_m) + U_n$ by striking out the jth row and ith column.

Thus, in fig. 3.17, if the formula (3.58) is used to evaluate Δ_{14}, then the number of terms expected in the expansion is simply the value of the permanent of the matrix

$$\begin{bmatrix} 0 & 1 & 1 & 0 \\ 2 & 1 & 0 & 0 \\ 1 & 2 & 1 & 1 \\ 0 & 1 & 1 & 2 \end{bmatrix}, \tag{3.62}$$

which is equal to 18. By Theorem 3.8, we have

$$\Delta_{14} = w_{14}(1 - w_{22} - w_{23}w_{32} - w_{35}w_{53} - w_{55} - w_{33} + w_{32}w_{23}w_{55} +$$
$$+ w_{35}w_{53}w_{22} + w_{22}w_{55} + w_{33}w_{55} + w_{22}w_{33} - w_{22}w_{33}w_{55}) +$$
$$+ w_{13}w_{34}(1 - w_{22} - w_{55} + w_{22}w_{55}) + w_{13}w_{35}w_{54}(1 - w_{22}) = 13,$$
(3.63)

where w_{ij} denotes the transmittance of the edge (i, j) in $G_m(A)$.

2.3. *Topological solutions of linear algebraic equations*

The application of the Mason-graph technique to the solutions of a system of linear equations is straightforward. The development parallels that of § 1.3. We shall only state the result, and leave the proof as an exercise (Problem 3.46).

THEOREM 3.9: If the coefficient matrix A is nonsingular, then the solution of (3.16) is given by

$$x_j = \frac{\sum_k f(P^k_{(n+1)j})[1 + \sum_{s,t}(-1)^t f(D^k_{st})]}{1 + \sum_{w,v}(-1)^v f(C_{wv})}$$
(3.64)

for $j = 1, 2, ..., n$; where $P^k_{(n+1)j}$ is the kth directed path from node $n+1$ to node j in $G_m(A_u)$; D^k_{st} is the sth subgraph of t ($t > 0$) node-disjoint directed circuits which are also node-disjoint with $P^k_{(n+1)j}$ in $G_m(A_u)$; A_u is the augmented matrix defined in Theorem 3.3; and C_{wv} is the wth subgraph of v ($v > 0$) node-disjoint directed circuits in $G_m(A_u)$.

As mentioned in the physical interpretation of (3.51), if each node k of $G_m(A_u)$ is associated with a nodal signal x_k, then each equation of (3.51) gives the nodal signal as the algebraic sum of the signals entering via the incoming edges. Each entering signal is the product of the edge transmittance and the signal at the node from which that edge originates. Thus, the solution (3.64) represents the signal at node j when the signal x_{n+1} at the last node $n+1$ is taken to be unity. Like the Coates graph, on the other hand, if x_{n+1} is an arbitrarily chosen signal, then the right-hand side of (3.64) denotes the ratio of x_j to x_{n+1} in $G_m(A_u)$. The ratio is referred to as the graph transmission from node x_{n+1} to node x_j in $G_m(A_u)$. Formally, it is defined as

DEFINITION 3.9: *Graph transmission in G_m*. Let G_m be an $(n+1)$-node Mason graph containing at least one source node x_0 (a node with zero incoming degree). The *graph transmission* from node x_0 to node x_j ($j = 1, 2, ..., n$) in G_m is defined

§ 2 The associated Mason graph

as the ratio
$$\frac{x_j}{x_0} = \frac{\sum_k f(P_{0j}^k)\left[1 + \sum_{s,t}(-1)^t f(D_{st}^k)\right]}{1 + \sum_{u,v}(-1)^v f(C_{uv})}, \tag{3.65}$$

where P_{0j}^k is the kth directed path from node x_0 to node x_j in G_m; D_{st}^k is the sth subgraph of t ($t>0$) node-disjoint directed circuits which are also node-disjoint with P_{0j}^k in G_m; and C_{uv} is the uth subgraph of v ($v>0$) node-disjoint directed circuits in G_m.

In Theorem 3.9, if $G_m(A)$ is given, $G_m(A_u)$ can easily be obtained from $G_m(A)$ simply by attaching a source node $x_0 = x_{n+1}$ to $G_m(A)$ such that if $b_k \neq 0$, there is an edge directed from the source node to node k with edge transmittance $-b_k$. Strictly speaking, we should add a self-loop of transmittance 1 at the node x_{n+1}, so that, according to Definition 3.8, $G_m(A_u) = G_c(A_u + U_{n+1})$. In practice, this is seldom done because the addition of a self-loop of transmittance 1 at the node x_{n+1} is equivalent to adding a trivial equation $x_{n+1} = x_{n+1}$ to the system (3.27). Thus, it may be omitted from the graph without effecting its graph transmissions. Also, the omission would make the node x_{n+1} a truly source node in the sense that it has zero incoming degree. Throughout the remaining part of this chapter, we shall assume this for all $G_m(A_u)$.

We emphasize that (3.65) is applicable only to the computation of the graph transmission from a source node to a non-source node. However, unlike the Coates graph in which only one source node is permitted at a time for the computation of its graph transmissions, the Mason graph may possess several source nodes in evaluating its graph transmissions.

Finally, we mention that in (3.64) the term $P_{(n+1)j}^k$ is also called a *forward path*, and $f(P_{(n+1)j}^k)$ a *path transmission*. A directed circuit is also called a *feedback loop*, and its weight product a *loop transmission*. Since the term

$$1 + \sum_{s,t}(-1)^t f(D_{st}^k) \tag{3.66}$$

is actually the graph determinant of the Mason graph obtained from $G_m(A_u)$ by the removal of all the nodes in $P_{(n+1)j}^k$ and all the edges incident with this path, it is usually referred to as the *cofactor* of $P_{(n+1)j}^k$ in $G_m(A_u)$.

Example 3.5: Consider the transistor feedback amplifier given in fig. 3.6. The associated Mason graph can simply be obtained from the corresponding Coates graph (fig. 3.13) by adding unity to each of the self-loops, as shown in fig. 3.18. A simple computation shows that there will be sixteen terms in the denominator of (3.64). In order to reduce the number of such terms, let us rewrite the system (3.24) in a slightly different form:

$$\begin{bmatrix} -1 & g_b/a_{11} & G_b/a_{11} & 1/a_{11} \\ g_b/a_{22} & -1 & g_c/a_{22} & 0 \\ G_b/a_{33} & (g_c - ag_e)/a_{33} & -1 & 0 \end{bmatrix} \begin{bmatrix} V_1 \\ V_2 \\ V_3 \\ I_1 \end{bmatrix} = \begin{bmatrix} 0 \\ 0 \\ 0 \end{bmatrix}, \quad (3.67)$$

where a_{11}, a_{22}, and a_{33} are defined the same as in (3.25). The associated Mason graph G_m of (3.67) is presented in fig. 3.19. By Corollary 3.8, it is easy to show that there will be six terms in the denominator of (3.64), a net reduction of ten.

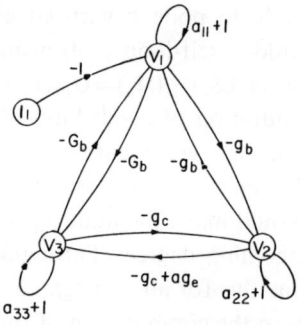
Fig. 3.18. The Mason graph of (3.24).

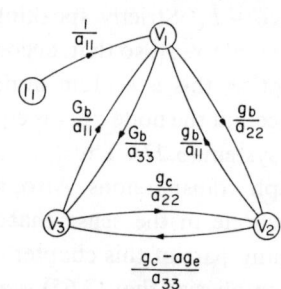
Fig. 3.19. The Mason graph of (3.67).

Suppose that we wish to compute the current-gain function of the amplifier which is defined as I_3/I_1 or $G_d V_3/I_1$. By Theorem 3.9, the graph transmission from I_1 to V_3 is given by

$$\frac{V_3}{I_1} = \frac{(1/a_{11})(G_b/a_{33}) + (1/a_{11})(g_b/a_{22})(g_c/a_{33} - ag_e/a_{33})}{1 - \dfrac{G_b}{a_{11}} \dfrac{G_b}{a_{33}} - \dfrac{g_b}{a_{11}} \dfrac{g_b}{a_{22}} - \dfrac{g_c}{a_{22}} \left(\dfrac{g_c}{a_{33}} - \dfrac{ag_e}{a_{33}} \right) - \dfrac{G_b}{a_{11}} \dfrac{g_b}{a_{22}} \left(\dfrac{g_c}{a_{33}} - \dfrac{ag_e}{a_{33}} \right) - \dfrac{g_c}{a_{22}} \dfrac{g_b}{a_{11}} \dfrac{G_b}{a_{33}}}$$

$$= \frac{a_{22}G_b + g_b(g_c - ag_e)}{a_{11}a_{22}a_{33} - a_{22}G_b^2 - a_{33}g_b^2 - a_{11}g_c(g_c - ag_e) - G_b g_b(g_c - ag_e) - g_c g_b G_b},$$
(3.68)

which is equivalent to (3.25). Again, we notice that when the corresponding quantities of a_{11}, a_{22}, and a_{33} are substituted in (3.68), many terms in the final expansion will cancel one another. Thus, in this sense, the formula is not very efficient for electrical network problems at least. However, it can be made more efficient by simple modifications of the Mason graph. We shall discuss this further in § 3.

Another point worth mentioning is that the associated Mason graph obtained by rewriting the original system of equations in the form of (3.67), in which all the iith elements of the coefficient matrix are -1, is termed as the *normalized Mason graph* of the original system by HOSKINS [1961] since no self-loops will appear in it. Clearly, every Mason graph can be put in normal form.

2.4. *Equivalence and transformations*

Like the Coates graph, it is usually advantageous to carry out some transformations and simplifications on the Mason graph before applying the general formulas derived earlier. The development in this section parallels that of § 1.4.

(A) *Elimination of nodes*

The node-elimination procedure for a Mason graph is similar to that for a Coates graph, and may be derived directly from Theorem 3.4 by a strategy similar to that used in the proof of Theorem 3.7. The following procedure is a generalization of Mason's technique (CHEN [1964b]).

Let V_m be a nonempty proper subset of the node set V of a Mason graph G_m. Let G_{mr} be the reduced Mason graph obtained from G_m by the following process:

(i) Remove the sectional subgraph $G_m[V_m]$ from G_m, i.e., remove all nodes and edges incident with any node in V_m.

(ii) The weight or transmittance m_{ij} associated with the edge (i, j) in G_{mr} is given by

$$m_{ij} = (-1)^{\alpha+\delta_{ij}} (1/K_m) \sum_k f(P_{ij}^{\prime\prime k}) \left[1 + \sum_{s,t} (-1)^t f(D_{st}^{\prime\prime k})\right] \quad (3.69)$$

for all i and j in $V - V_m$; where $P_{ij}^{\prime\prime k}$ is the kth directed path from node i to node j in the sectional subgraph $G_m[V_m^{ij}]$, where V_m^{ij} is the set union of the sets V_m and $\{i, j\}$; $D_{st}^{\prime\prime k}$ is the sth subgraph of t ($t > 0$) node-disjoint directed circuits which are also node-disjoint with $P_{ij}^{\prime\prime k}$ in $G_m[V_m^{ij}]$; α is the number of nodes in V_m^{ij}; δ_{ij} is the Kronecker's delta; and K_m is the determinant of the sectional subgraph $G_m[V_m]$, which can be evaluated by the formula

$$K_m = \det G_m[V_m] = (-1)^\beta \left[1 + \sum_{u,v} (-1)^v f(C_{uv}^{\prime\prime})\right], \quad (3.70)$$

where C''_{uv} is the uth subgraph of v ($v>0$) node-disjoint directed circuits in $G_m[V_m]$, and β is the number of nodes in V_m. When $i=j$, $G_m[V_m^{ij}]$ reduces to $G_m[V_m^i]$ and the kth directed path $P_{ij}^{''k}$ becomes a directed circuit containing the node i in $G_m[V_m^i]$ by definition.

Like the Coates-graph reduction, the process outlined above is easier to visualize than it is to describe in words. A simple physical interpretation of this process will now be presented below (CHEN [1969a]).

For $i \neq j$, m_{ij} is the graph transmission from node i to node j in the graph $G''_m[V_m^{ij}]$ obtained from the sectional subgraph $G_m[V_m^{ij}]$ by removing all the edges terminating at node i and outgoing from node j.

For $i=j$, it is necessary that we first split the node i in $G_m[V_m^i]$ into two distinct nodes i' and i'' such that all the edges outgoing from i in $G_m[V_m^i]$ now outgo from i', and all the edges originally terminating at i now terminate at i''. Then m_{ii} is simply the graph transmission from node i' to node i'' in the resulting graph $G''_m[V_m^i]$. In other words, the formula used to evaluate the transmittances m_{ij} in the reduced Mason graph G_{mr} is the same as the original graph-transmission formula (3.65) for G_m except that it is now applied to the modified sectional subgraphs $G''_m[V_m^{ij}]$.

THEOREM 3.10: If K_m is nonzero, then

$$\det G_m = K_m \det G_{mr}. \tag{3.71}$$

Proof. Let G_c be the corresponding Coates graph of G_m obtained from G_m by inserting a self-loop of transmittance -1 to each node of G_m. Since the graph transmission evaluated in G_c is the same as that (between the same pair of nodes) evaluated in G_m, and since $G_c[V_c^{ij}]$ is the corresponding Coates graph of $G_m[V_m^{ij}]$, where $V_c = V_m$, it follows that $K_c = K_m$, and for $i \neq j$, $c_{ij} = m_{ij}$, where c_{ij} is defined in (3.28). For $i=j$ we have

$$\begin{aligned}
c_{ii} &= (-1)^\alpha (1/K_c) \sum_{h^*} (-1)^{L^*h} f(h^*) \\
&= (-1)^\alpha (1/K_c) \left[\sum_{h^*_1} (-1)^{L^*h_1} f(h^*_1) + \sum_{h^*_2} (-1)^{L^*h_2} f(h^*_2) \right] \\
&= (1/K_c) \left[-K_c + (-1)^\alpha \sum_{h^*_2} (-1)^{L^*h_2} f(h^*_2) \right] \\
&= -1 - (-1)^\alpha (1/K_m) \left[\sum_k f(P_{ii}^{''k}) \{1 + \sum_{s,t} (-1)^t f(D_{st}^{''k})\} \right] \\
&= -1 + m_{ii}, \tag{3.72}
\end{aligned}$$

where h^* is a 1-factor in $G_c[V_c^i]$; h_1^* is a 1-factor in $G_c[V_c^i]$ that contains the added self-loop of transmittance -1 at node i; h_2^* is a 1-factor in $G_c[V_c^i]$ that does not contain the added self-loop of transmittance -1 at the node i; and $L_h^*, L_{h_1}^*$, and $L_{h_2}^*$ are the numbers of directed circuits in h^*, h_1^* and h_2^*, respectively. The fourth line follows from a reasoning similar to that given in the proof of Theorem 3.7.

Thus, G_{cr} as defined in (3.28) is the corresponding Coates graph of G_{mr}, and by Theorem 3.4 we have

$$\det G_m = \det G_c = K_c \det G_{cr} = K_m \det G_{mr}, \qquad (3.73)$$

which completes the proof of the theorem.

In particular, when V_m contains only a single node, the theorem reduces to that of MASON [1953].

COROLLARY 3.10: If w_{ij} is the weight associated with the edge (i, j) in G_m, and if $w_{kk} \neq 1$, then

$$\det G_m = (w_{kk} - 1) \det G_{mr}, \qquad (3.74)$$

where G_{mr} is obtained from G_m by the following process:
(i) Remove node k from G_m.
(ii) The weight m_{ij} associated with the edge (i, j) in G_{mr} is given by

$$m_{ij} = w_{ij} + w_{ik}w_{kj}/(1 - w_{kk}). \qquad (3.75)$$

The usefulness of the above theorem lies in the following.

THEOREM 3.11: If G_{mr} is the reduced Mason graph of G_m, and if $K_m \neq 0$ and $i \neq j$, then

$$(-1)^\beta \sum_k f(P_{ij}^k) \left[1 + \sum_{s,t} (-1)^t f(D_{st}^k)\right]$$
$$= K_m \sum_w f(P_{ij}^{rw}) \left[1 + \sum_{u,v} (-1)^v f(D_{uv}^{rw})\right] \qquad (3.76)$$

for all i and j not in V_m; where P_{ij}^k and P_{ij}^{rw} are the kth and wth directed paths from i to j in G_m and G_{mr}, respectively; D_{st}^k and D_{uv}^{rw} are the sth and uth subgraphs of t $(t>0)$ and v $(v>0)$ node-disjoint directed circuits which are also node-disjoint with P_{ij}^k in G_m and P_{ij}^{rw} in G_{mr}, respectively; and β and K_m are defined in (3.70).

The theorem follows directly from Theorem 3.5, and may be proved in a manner similar to that of Theorem 3.10. The details are left as an exercise (Problem 3.6).

COROLLARY 3.11: The solution of the system (3.16) may be evaluated either in G_m or G_{mr} for all the variables not associated with the nodes in V_m.

COROLLARY 3.12: The graph transmission from a source node to a non-source node remains the same in both G_m and G_{mr} for all the nodes not in V_m. We shall illustrate the node-elimination procedure by the following example.

Example 3.6: Consider the Mason graph G_m as shown in fig. 3.17. Suppose that we wish to remove the sectional subgraph $G_m[2, 3]$ from G_m. Fig. 3.20 is the reduced Mason graph G_{mr} of G_m. For illustrative purpose, we shall compute the transmittances associated with two typical edges (5, 1) and (4, 4)

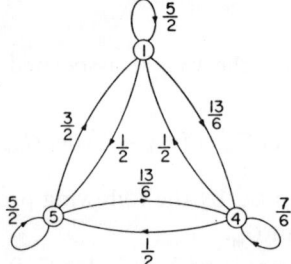

Fig. 3.20. A reduced Mason graph of the Mason graph of fig. 3.17.

of G_{mr}. The sectional subgraph $G_m[1, 2, 3, 5]$ together with the needed subgraphs is presented in fig. 3.21. Thus, from (3.69) the transmittance m_{51} of the edge (5, 1) in G_{mr} with w_{ij} being the weight of (i, j) in G_m is

$$m_{51} = (-1)^4 (1/K_m) \{w_{53}w_{31}[1 + (-1)^1 w_{22}] \\
+ w_{51}[1 + (-1)^1 w_{23}w_{32} + (-1)^1 w_{33} \\
+ (-1)^1 w_{22} + (-1)^2 w_{22}w_{33}]\} \\
= -\tfrac{1}{6}\{(1 \cdot 3) \cdot [1 + (-1)^1 \cdot 2] + 1 \cdot [1 + (-1)^1 (3 \cdot 2) \\
+ (-1)^1 \cdot 1 + (-1)^1 \cdot 2 + (-1)^2 (2 \cdot 1)]\} \\
= \tfrac{3}{2},$$

where K_m is the determinant of the sectional subgraph $G_m[2, 3]$, which can be evaluated by the formula (3.70):

$$K_m = (-1)^2 [1 + (-1)^1 w_{22} + (-1)^1 w_{33} \\
+ (-1)^1 w_{23}w_{32} + (-1)^2 w_{22}w_{33}] \\
= 1 - 2 - 1 - 6 + 2 = -6.$$

§ 2   The associated Mason graph   181

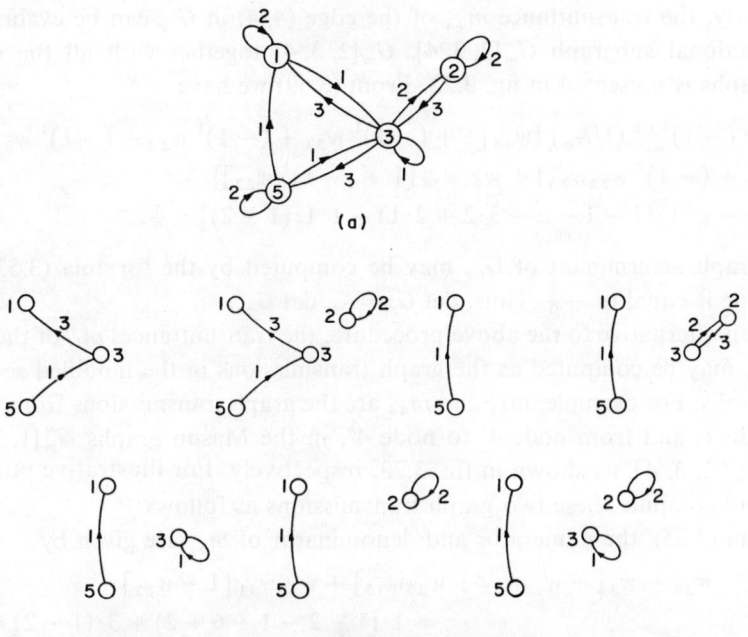

Fig. 3.21. The sectional subgraph $G_m[1, 2, 3, 5]$ and its subgraphs for the evaluation of the transmittance m_{51}.

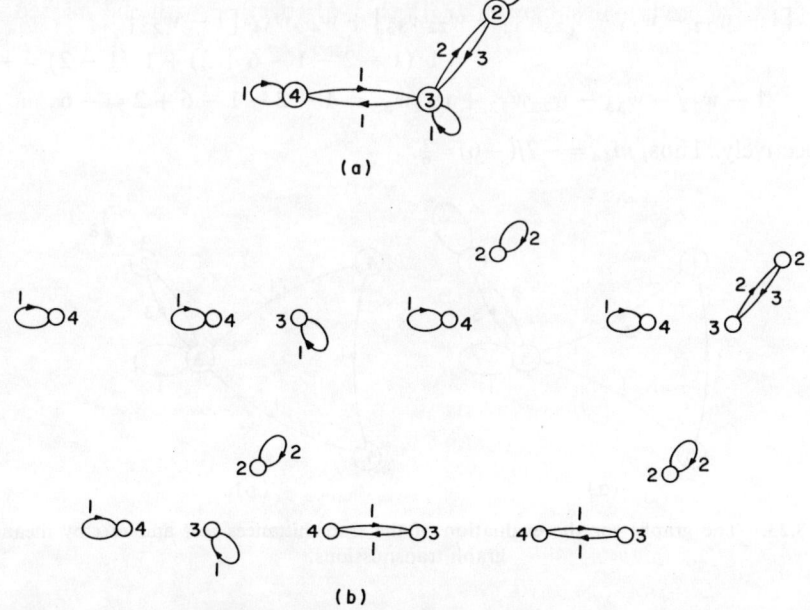

Fig. 3.22. The sectional subgraph $G_m[2, 3, 4]$ and its subgraphs for the evaluation of the transmittance m_{44}.

Similarly, the transmittance m_{44} of the edge (4, 4) in G_{mr} can be evaluated in the sectional subgraph $G_m[2, 3, 4]$. $G_m[2, 3, 4]$ together with all the needed subgraphs is presented in fig. 3.22. From (3.69) we have

$$m_{44} = (-1)^{3+1}(1/K_m)\{w_{44}[1 + (-1)^1 w_{33} + (-1)^1 w_{22} + (-1)^1 w_{23}w_{32}$$
$$+ (-1)^2 w_{22}w_{33}] + w_{43}w_{34}[1 + (-1)^1 w_{22}]\}$$
$$= -\tfrac{1}{6}[1\cdot(1 - 1 - 2 - 3\cdot 2 + 2\cdot 1) + 1\cdot 1\cdot(1 - 2)] = \tfrac{7}{6}.$$

The graph determinant of G_{mr} may be computed by the formula (3.53), and its value is equal to $-\tfrac{5}{6}$. Thus, det $G_m = K_m$ det G_{mr}.

As an alternative to the above procedure, the transmittances m_{ij} of the edges of G_{mr} may be computed as the graph transmissions in the modified sectional subgraphs. For example, m_{51} and m_{44} are the graph transmissions from node 5 to node 1, and from node 4' to node 4", in the Mason graphs $G''_m[1, 2, 3, 5]$ and $G''_m[2, 3, 4]$, as shown in fig. 3.23, respectively. For illustrative purposes, we shall compute these two graph transmissions as follows:

From (3.65), the numerator and denominator of m_{51} are given by

$$w_{51}[1 - w_{22} - w_{33} - w_{23}w_{32} + w_{22}w_{33}] + w_{53}w_{31}[1 - w_{22}]$$
$$= 1\cdot(1 - 2 - 1 - 6 + 2) + 3\cdot(1 - 2) = -9,$$
$$1 - w_{22} - w_{33} - w_{23}w_{32} + w_{22}w_{33} = 1 - 2 - 1 - 6 + 2 = -6,$$

respectively. Thus, $m_{51} = -9/(-6) = \tfrac{3}{2}$. Similarly, the numerator and denominator of m_{44} are given by

$$w_{4'4''}[1 - w_{22} - w_{33} - w_{23}w_{32} + w_{22}w_{33}] + w_{4'3}w_{34''}[1 - w_{22}]$$
$$= 1\cdot(1 - 2 - 1 - 6 + 2) + 1\cdot(1 - 2) = -7,$$
$$1 - w_{22} - w_{33} - w_{23}w_{32} + w_{22}w_{33} = 1 - 2 - 1 - 6 + 2 = -6,$$

respectively. Thus, $m_{44} = -7/(-6) = \tfrac{7}{6}$.

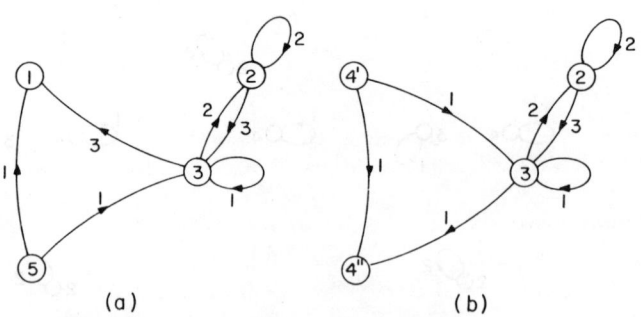

Fig. 3.23. The graphs for the evaluation of the transmittances m_{51} and m_{44} by means of graph transmissions.

§2 The associated Mason graph

Example 3.7: Consider the Mason graph G_m presented in fig. 3.24(a), which is associated with a transistor amplifier. Suppose that we wish to remove the nodes αi_1, v_2, and i_5 from G_m. To avoid multiple subscripts, let v_g, i_1, i_3, and v_4 be also denoted by g, 1, 3, and 4, respectively. The reduced Mason graph G_{mr} of G_m is given in fig. 3.24(b). For illustrative purposes, we shall compute the transmittances m_{14} and m_{11} associated with the edges (i_1, v_4) and (i_1, i_1) in G_{mr}, respectively.

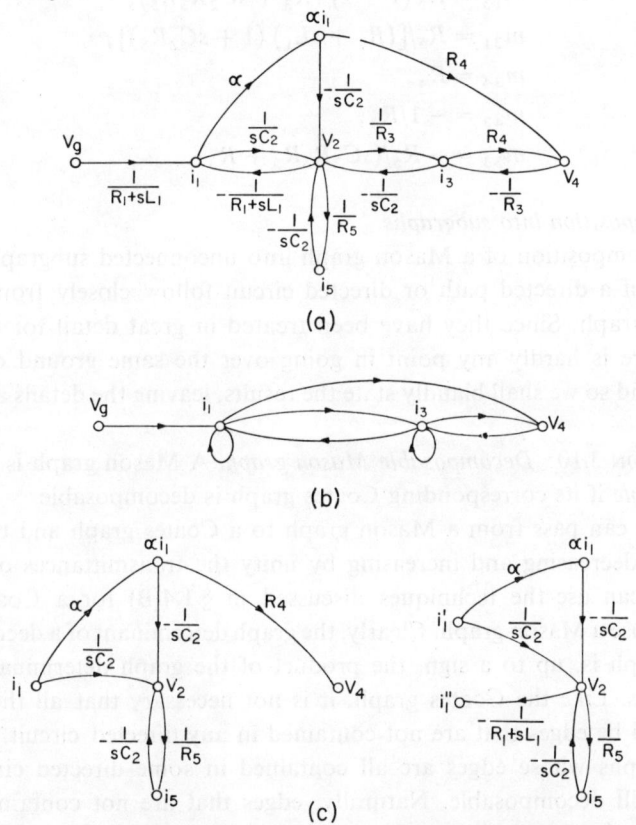

Fig. 3.24. (a) A Mason graph, (b) its reduced Mason graph, and (c) the graphs for the evaluation of m_{14} and m_{11}.

The transmittances m_{14} and m_{11} are the graph transmissions from node i_1 to node v_4, and from node i_1' to node i_1'', in the modified sectional subgraphs, as shown in fig. 3.24(c), respectively. The results are

$$m_{14} = \alpha R_4 [1 - (1/R_5)(-1/sC_2)]/[1 - (1/R_5)(-1/sC_2)]$$
$$= \alpha R_4,$$

$$m_{11} = \{\alpha(-1/sC_2)[-1/(R_1 + sL_1)]$$
$$+ (1/sC_2)[-1/(R_1 + sL_1)]\}/[1 - (1/R_5)(-1/sC_2)]$$
$$= (\alpha - 1) R_5/[(R_1 + sL_1)(1 + sC_2R_5)].$$

Similarly, we have

$$m_{g1} = 1/(R_1 + sL_1),$$
$$m_{13} = R_5(1 - \alpha)/(R_3 + sC_2R_3R_5),$$
$$m_{31} = R_5/[(R_1 + sL_1)(1 + sC_2R_5)],$$
$$m_{34} = R_4,$$
$$m_{43} = -1/R_3,$$
$$m_{33} = -R_5/(sC_2R_3R_5 + R_3).$$

(B) *Decomposition into subgraphs*

The decomposition of a Mason graph into unconnected subgraphs and the inversion of a directed path or directed circuit follow closely from those of a Coates graph. Since they have been treated in great detail for the Coates graph, there is hardly any point in going over the same ground once again in detail, and so we shall blandly state the results, leaving the details as obvious.

DEFINITION 3.10: *Decomposable Mason graph.* A Mason graph is said to be *decomposable* if its corresponding Coates graph is decomposable.

Since we can pass from a Mason graph to a Coates graph and back again by merely decreasing and increasing by unity the transmittances of the self-loops, we can use the techniques discussed in § 1.4(B) for a Coates graph to decompose a Mason graph. Clearly, the graph determinant of a decomposable Mason graph is, up to a sign, the product of the graph determinants of the components. Like the Coates graph, it is not necessary that all the edges to be removed be edges that are not contained in any directed circuit. There are Mason graphs whose edges are all contained in some directed circuits, but they are still decomposable. Naturally, edges that are not contained in any directed circuits can always be removed from the Mason graph without affecting its graph determinant since they will not appear in any term of (3.53). Remember that if one of the components of a decomposed Mason graph is an isolated node, then its graph determinant is zero since $f(\emptyset) = 0$ by definition.

As an illustration, assume that the directed graph of fig. 3.25(a) is a given Mason graph G_m. The corresponding Coates graph G_c of G_m is obtained by inserting a self-loop of transmittance -1 to each node of G_m. Since by Theorem 3.6 there exists a subset $\{3, 4\}$ of the node set of G_c which is of zero deficiency,

G_c is decomposable. Using a procedure outlined in § 1.4(B), we can decompose G_c into components. These component Coates graphs are then converted back to the desired component Mason graphs, as shown in fig. 3.25(b). Since one interchange of the incoming edges of two nodes is needed in decomposing G_c, the graph determinant of G_m is simply the negative of the product of the graph determinants of the component Mason graphs. We remark that each of the edges of G_m is contained in some directed circuits.

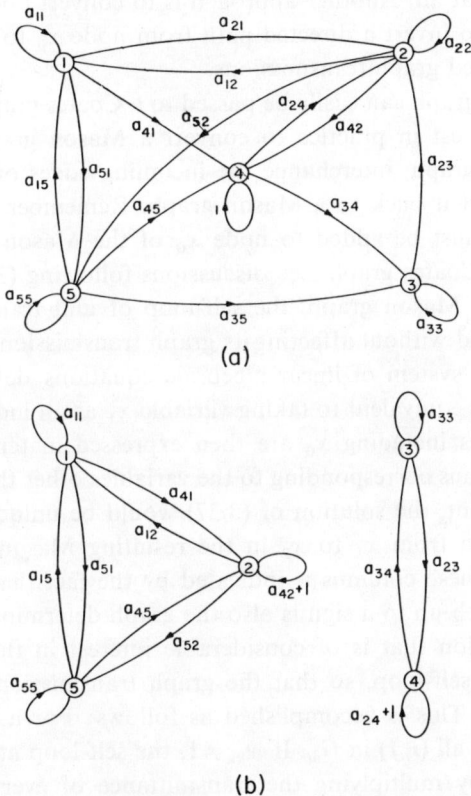

Fig. 3.25. (a) A Mason graph and (b) its decomposed components.

Finally, we mention that if in a Mason graph G_m there exists a cut consisting only of edges that are not in any directed circuits, the removal of this cut decomposes G_m into components. This type of decomposition is called a *factorization* by MASON and ZIMMERMANN [1960].

(C) *Inversion*

As mentioned earlier, the formula for graph transmissions defined in (3.65) is applicable only to the computation of the graph transmission from a source node x_0 to any other non-source node. However, in many applications, it is necessary to compute the graph transmission from a non-source node to another, say x_1 to x_2. One way to accomplish this is simply first to compute the graph transmissions from the source node x_0 to nodes x_1 and x_2, and then take the ratio. Since these two graph transmissions have the same denominator, it need not be computed at all. Another approach is to convert node x_1 into a source node directly or to invert a directed path from node x_0 to node x_1, and then compute the needed graph transmissions.

Since a Mason graph can easily be passed to a Coates graph and back again, it is probably easiest in practice to convert a Mason graph into the corresponding Coates graph, interchange the incoming edges of nodes x_0 and x_1, and finally convert it back to a Mason graph. Remember that a self-loop of transmittance 1 must be added to node x_0 of the Mason graph before it is converted into a Coates graph (see discussions following (3.65)). However, in the final resulting Mason graph, the self-loop of unit transmittance at node x_1 may be removed without affecting its graph transmissions.

In terms of the system of linear algebraic equations defined in (3.27), the above operation is equivalent to taking variable x_1 as an independent variable. All other variables including x_0 are then expressed in terms of x_1 through (3.27). If the columns corresponding to the variables other than x_1 in (3.27) are linearly independent, the solution of (3.27) would be unique. In terms of the graph transmission from x_1 to x_2 in the resulting Mason graph, the linear independence of these columns is indicated by the fact that it has a nonzero denominator, which up to a sign is also the graph determinant.

Another operation that is of considerable interest in the Mason graph is the removal of a self-loop, so that the graph transmissions are not affected by this operation. This is accomplished as follows: For a Mason graph G_m, let $w_{ij} = f(i, j)$ for all (i, j) in G_m. If $w_{kk} \neq 1$, the self-loop at the node k can be removed simply by multiplying the transmittance of every edge which terminates at node k by the factor

$$\frac{1}{1 - w_{kk}}$$

(Problem 3.7). Thus, in G_m if there are no self-loops of transmittance 1, the operation of removing each of its self-loops is equivalent to that of transforming the given Mason graph into its normal form (see § 2.3).

Finally, if the operation of interchanging the incoming edges of two nodes is applied successively to a directed path or circuit in the corresponding Coates graph of a given Mason graph, the directed path or circuit will, in most cases, appear in inverted form. Because of the special feature of the Mason graph, it is a simple matter to identify the general inversion rules which correspond to the successive interchanges of the incoming edges of two nodes along a directed path or circuit in the corresponding Coates graph. The general rules can be expressed as follows:

To invert an edge (i, j) outgoing from a source node i in G_m, we first move the set of edges having their terminal node j to node i and move the initial node of the self-loop thus created at node i to node j. Then we replace the transmittance w_{ji} associated with the edge (j, i) in the resulting Mason graph by $1/w_{ji}$, and multiply by $(-1/w_{ji})$ the transmittances of all other edges whose terminal nodes have been moved. To invert a directed path outgoing from a source node or a directed circuit, we simply invert all the edges in that directed path or circuit successively.

Again, the inversion is easier to visualize than it is to describe in words. The justification for these rules is not difficult to find, and is left as an exercise (Problem 3.9).

Example 3.8: Consider the Mason graph G_m given in fig. 3.26(a). Suppose that we wish to compute the graph transmission from node x_3 to node x_6. In order to apply formula (3.65), it is necessary that we convert node x_3 into a source node. This can easily be accomplished as follows:

The corresponding Coates graph G_c of G_m is presented in fig. 3.26(b). In G_c, if we interchange the incoming edges of the nodes x_1 and x_3, we have G_c^* as shown in fig. 3.26(c). Finally, the corresponding Mason graph G_m^* of G_c^* is given in fig. 3.26(d). Note that, to obtain G_m^*, we have omitted the self-loop of transmittance 1 at node x_3. Thus, the graph transmission from node x_3 to node x_6 is given by

$$x_6/x_3 = [-jgh(1 + dm - 1 - dm1) + cde(1 - 1 + i - ab - i1 - iab)$$
$$- 1agh(1 + dm) - ckagh - cdpagh]/\Delta$$
$$= [-abcde(1 + i) - agh(1 + dm + ck + cpd)]/\Delta,$$

where

$$\Delta = 1 - (-dm - en - i + ab + 1 - kaghnm + npagh)$$
$$+ (-dm1 - en1 - i1 - iab - abdm - aben + idm + ien)$$
$$- (i1dm + i1en + abidm + abien)$$
$$= -ab(1 + i + dm + en + idm + ien) + agh(nmk - np).$$

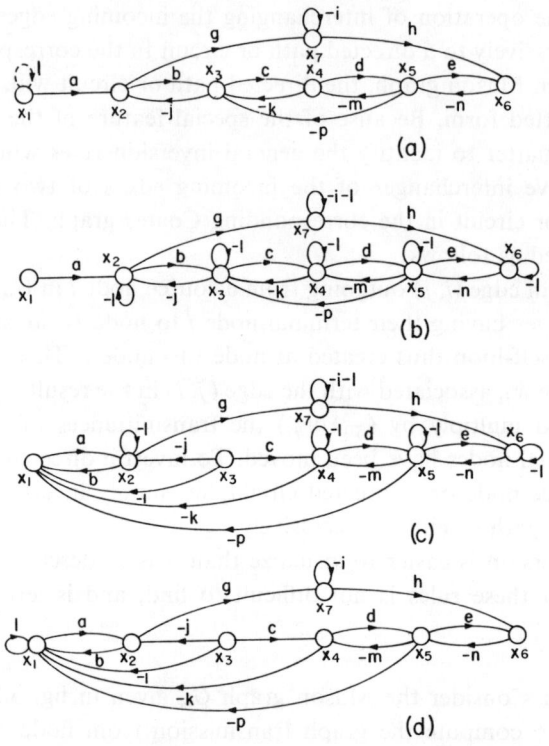

Fig. 3.26. The conversion of the node x_3 into a source node.

Similarly, if the graph transmission from node x_3 to node x_4 is needed, we have

$$x_4/x_3 = [-jghnm(1-1) + c(1 + en + i - 1 - ab - npagh - i1 - en1 \\ - iab - aben + ien - i1en - iaben) - 1aghnm]/\Delta \\ = [-abc(1 + en + i + ien) - agh(nm + npc)]/\Delta.$$

Example 3.9: Consider the Mason graph G_m of fig. 3.19 associated with the transistor feedback amplifier of fig. 3.6. Suppose that we wish to compute the voltage-gain function V_3/V_2 of the amplifier. In order to apply the graph transmission formula (3.65), it is necessary that we convert node V_3 into a source node. To accomplish this, let us invert the directed path consisting of the edges (I_1, V_1) and (V_1, V_2) in G_m by the general rules outlined above. The resulting Mason graph is presented in fig. 3.27. The graph transmission from V_2 to V_3

in the new Mason graph is then obtained by (3.65), and is given by

$$\frac{V_3}{V_2} = \frac{(g_c - ag_e)g_b + (g_b + g_e + g_c - ag_e)G_b}{(G_b + G_d + g_c)g_b + G_b g_c}. \qquad (3.77)$$

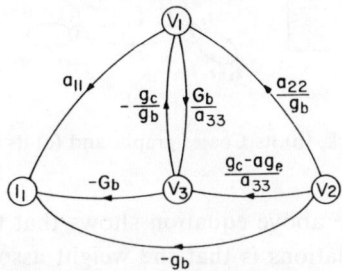

Fig. 3.27. The Mason graph obtained by inverting a directed path in the Mason graph of fig. 3.19.

§ 3. The modifications of Coates and Mason graphs

The topological methods presented in the preceding two sections are of particular interest in linear system analysis, since the associated directed graphs can usually be set up directly by inspecting the system without the necessity of first formulating the associated equations. However, as pointed out earlier, these formulas are not very efficient in the calculation of electrical network functions, the main reason for their inefficiency being that there exists a large number of cancellations. In this section, we shall show that if the weights associated with the self-loops of a Coates graph or a Mason graph are slightly modified, then more efficient formulas can be obtained, and they are especially efficient when the nodal system of equations is used in the analysis (see ch. 2, § 3.3 and Problems 3.10 and 3.11). The development follows closely that of CHEN [1965b].

3.1. *Modifications of Coates graphs*

Consider the network diagram of fig. 3.28(a). The corresponding Coates graph G_c associated with the loop-impedance matrix of the network is presented in fig. 3.28(b). By Theorem 3.1, the determinant Δ of the loop-impedance matrix is given by

$$\begin{aligned}
\Delta &= (Z_1 + Z_3)(Z_3 + Z_2 + Z_4)(Z_4 + Z_5 + Z_6) \\
&\quad - (-Z_3)(-Z_3 - Z_2)(Z_4 + Z_5 + Z_6) - (-Z_4)(-Z_4)(Z_1 + Z_3) \\
&= Z_1(Z_2 + Z_3)(Z_4 + Z_5 + Z_6) + (Z_1 + Z_3)Z_4(Z_5 + Z_6). \qquad (3.78)
\end{aligned}$$

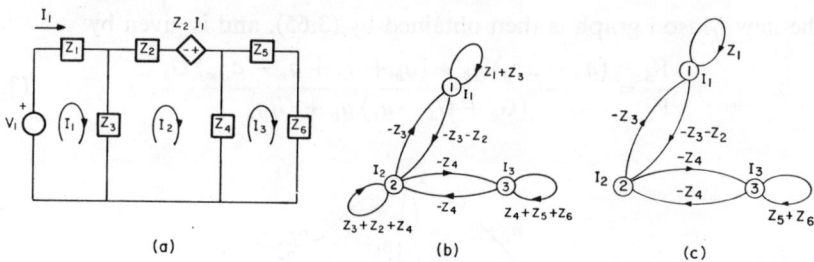

Fig. 3.28. (a) A network, (b) its Coates graph, and (c) its modified Coates graph.

A careful check of the above equation shows that the main reason for the existence of such cancellations is that the weight associated with each of the self-loops contains some of the weights of the edges having that particular node as their terminal node. This is similarly valid for the Coates graph associated with the node-admittance matrix of a network such as the one given in Example 3.2. In order to obtain more efficient formulas, we shall modify a Coates graph in the following way.

DEFINITION 3.11: *Modified Coates graph.* For a Coates graph G_c, the *modified Coates graph* of G_c, denoted by the symbol G'_c, is the directed graph obtained from G_c by changing the weight associated with each of its self-loops (i, i) to the sum of the weights of the edges having node i as their terminal node, while keeping all other edge weights unaltered.

Mathematically, it is equivalent to saying that if w_{ij} and w'_{ij} are the weights associated with the edges (i, j) in G_c and G'_c, respectively, then

$$w'_{ij} = w_{ij} \tag{3.79}$$

for $i \neq j$, and

$$w'_{ij} = \sum_{k=1}^{n} w_{kj} \tag{3.80}$$

for $i = j$, where n is the number of nodes in G_c. For example, fig. 3.28(c) is the modified Coates graph of fig. 3.28(b). In fig. 3.28(c), we have $w'_{11} = Z_1$, $w'_{22} = 0$, and $w'_{33} = Z_5 + Z_6$.

DEFINITION 3.12: *Semifactor.* A *semifactor* of a directed graph is a spanning subgraph which does not contain any directed circuits of length greater than one, and such that each of its nodes is of incoming degree 1.

As an illustration, the set of all possible semifactors of the modified Coates graph G'_c of fig. 3.28(c) is presented in fig. 3.29. A direct consequence of the above definition is the following.

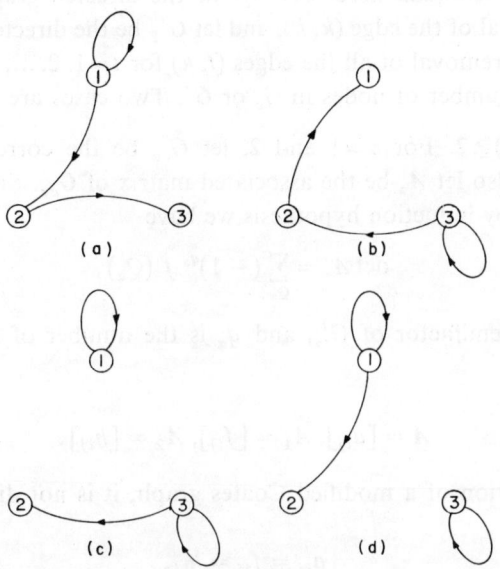

Fig. 3.29. The set of semifactors in the directed graph of fig. 3.28(c).

LEMMA 3.1: Each component of a semifactor of a directed graph contains exactly one self-loop, and the number of edges in each of its components is the same as the number of nodes in that component.

Now, we shall proceed to prove the following theorem.

THEOREM 3.12: Suppose that G'_c is the modified Coates graph of a given Coates graph G_c. If A is the matrix associated with G_c, then

$$\det A = \sum_R (-1)^{q_R} f(R), \qquad (3.81)$$

where R is a semifactor of G'_c, and q_R is the number of even components in R.

Proof. We shall prove (3.81) by induction over the number of edges of G'_c. Without loss of generality, assume that A does not contain a zero row.

If G'_c has only one edge, the theorem is seen to be true. Assume that the theorem is true for any G'_c which has $b-1$ edges or less, $b \geq 2$. To complete the induction, we need to show that it is also true for b edges.

Assume in G' there exists an edge (k, k) such that $d^-(k) \geq d^-(x) \geq 1$ for all the nodes x of G'_c, where $d^-(x)$ denotes the incoming degree of the node x in G'_c. For, if not, the theorem holds for this case since the sum of the elements in every row of A equals zero. Let G'_{c1} be the directed graph obtained from G'_c by the removal of the edge (k, k), and let G'_{c2} be the directed graph obtained from G'_c by the removal of all the edges (t, k) for $t=1, 2, \ldots, k-1, k+1, \ldots, n$, where n is the number of nodes in G_c or G'_c. Two cases are being considered.

Case 1. $d^-(k) \geq 2$. For $\alpha = 1$ and 2, let $G_{c\alpha}$ be the corresponding Coates graph of $G'_{c\alpha}$. Also let A_α be the associated matrix of $G_{c\alpha}$. Since $G'_{c\alpha}$ has fewer edges than G'_c, by induction hypothesis we have

$$\det A_\alpha = \sum_{Q_\alpha} (-1)^{q_\alpha} f(Q_\alpha), \tag{3.82}$$

where Q_α is a semifactor of $G'_{c\alpha}$, and q_α is the number of even components in Q_α.

Let

$$A = [a_{ij}], A_1 = [f_{ij}], A_2 = [h_{ij}].$$

From the definition of a modified Coates graph, it is not difficult to see that for all $i \neq k$

$$a_{ij} = f_{ij} = h_{ij},$$

and for $i = k$

$$f_{kj} = a_{kj}, j = 1, 2, \ldots, k-1, k+1, \ldots, n,$$

$$f_{kk} = a_{kk} - \sum_{x=1}^{n} a_{kx}$$

and

$$h_{kj} = 0, j = 1, 2, \ldots, k-1, k+1, \ldots, n,$$

$$h_{kk} = \sum_{x=1}^{n} a_{kx}.$$

Since the kth row of A is equal to the sum of the kth rows of A_1 and A_2, and since all other corresponding rows of A, A_1, and A_2 are the same, it follows that

$$\det A = \det A_1 + \det A_2 = \sum_{Q_1} (-1)^{q_1} f(Q_1) + \sum_{Q_2} (-1)^{q_2} f(Q_2)$$

$$= \sum_{R} (-1)^{q_R} f(R). \tag{3.83}$$

The second line follows from the fact that there exists a one-to-one correspondence between the semifactors in G'_{c2} (G'_{c1}) and those semifactors in G'_c, each of which contains (does not contain) the self-loop (k, k) at node k.

§3 Modifications of Coates and Mason graphs

Case 2. $d^-(k)=1$. This implies that if there is a self-loop (x, x) at node x in G'_c, then (x, x) is the only edge having node x as its terminal node. For otherwise $d^-(k) \geq d^-(x) > 1$, contradicting to our assumption.

Let A_{kk} be the matrix obtained from A by striking out the kth row and kth column. Let G'_{c3} be the modified Coates graph of the associated Coates graph G_{c3} of A_{kk}. From the above discussion, it is not difficult to see that G'_{c3} can also be obtained from G'_c by first removing the node k, and then adding a self-loop of weight $-a_{jk}$ to each of the nodes j of G'_c devoid of a self-loop. Note that the weight associated with the self-loop (j, j) of G_c is a_{jj}. Since G'_{c3} has fewer edges than G'_c, by induction hypothesis we have

$$\det A_{kk} = \sum_{Q_3} (-1)^{q_3} f(Q_3), \tag{3.84}$$

where Q_3 is a semifactor of G'_{c3}, and q_3 is the number of even components in Q_3.

Next, observe the one-to-one correspondence between the semifactors Q_3 in G'_{c3} and the semifactors R in G'_c such that

$$f(R) = (-1)^{\beta_3} a_{kk} f(Q_3), \tag{3.85a}$$

where β_3 is the number of added self-loops of weights $-a_{jk}$ in Q_3. Since from Lemma 3.1 each component of Q_3 contains exactly one self-loop, let G_{3u}, $u=1, 2, \ldots, \beta$, be the components of Q_3. Without loss of generality, let us assume that $G_{31}, G_{32}, \ldots, G_{3y}$, $y=\beta_3$, are the components, each of which contains an added self-loop of weight $-a_{jk}$. (Note that $k \neq j$.) Clearly, $G_{3(y+1)}$, $G_{3(y+2)}, \ldots, G_{3\beta}$ are also the components of R. If n_u is the number of nodes in G_{3u}, then again from Lemma 3.1 we obtain

$$\begin{aligned}(-1)^{q_3} &= (-1)^{(n_1-1)+(n_2-1)+\cdots+(n_\beta-1)} \\ &= (-1)^{[(n_1+\cdots+n_y+1)-1]+(n_{y+1}-1)+\cdots+(n_\beta-1)-y} \\ &= (-1)^{q_R-\beta_3}. \end{aligned} \tag{3.85b}$$

The third line follows from the facts that R has $(\beta-\beta_3+1)$ components, and that $(n_1+\cdots+n_y+1)$ is the number of nodes in that component of R containing the self-loop (k, k) of G'_c. Substituting (3.85a) and (3.85b) in (3.84), we get

$$\det A_{kk} = \sum_R (-1)^{q_R} (1/a_{kk}) f(R). \tag{3.85c}$$

Since a_{kk} is the only nonzero element in the kth row of A, it follows that $\det A = a_{kk} \det A_{kk}$. After a simple substitution, we obtain the desired result. This completes the proof of the theorem.

Like the Coates graph, if we associate with each component of R either a plus or a minus sign, according to whether the component is odd or even, the resulting product will give the same sign as $(-1)^{q_R}$.

Consider the modified Coates graph G'_c of fig. 3.28(c). The set of its semifactors is given in fig. 3.29. Thus, from the above theorem we have

$$\det A = Z_1(Z_3 + Z_2)Z_4 + Z_3Z_4(Z_5 + Z_6) + Z_1(Z_5 + Z_6)Z_4$$
$$+ Z_1(Z_5 + Z_6)(Z_3 + Z_2)$$
$$= Z_1(Z_2 + Z_3)(Z_4 + Z_5 + Z_6) + (Z_1 + Z_3)Z_4(Z_5 + Z_6). \qquad (3.86)$$

It is significant to observe that cancellations due to the passive elements do not exist. In general, there may be some cancellations due to the active elements, but in this particular example such terms do not appear. However, it should be emphasized that the effectiveness of the formulas depends largely upon the choice of the variables involved. One way to accomplish this is to draw the corresponding Coates graph from the node-admittance matrix of a network. (See ch. 2, § 3.3 and Problems 3.10 and 3.11.) In particular, for passive networks without mutual couplings, this method does not calculate any superfluous terms as do Coates and Mason graph techniques.

DEFINITION 3.13: *k-semifactor.* A spanning subgraph, denoted by the symbol $R(j_1;j_2;\ldots;j_k)$, of a directed graph G is said to be a *k-semifactor* of G if, when the self-loops (j_u, j_u), $u=1, 2, \ldots, k$, are added to $R(j_1;j_2;\ldots;j_k)$, the resulting graph becomes a semifactor in the directed graph obtained from G by adding a self-loop to each of the nodes j_u of G devoid of a self-loop.

Thus, each component of a k-semifactor contains either a self-loop or no self-loop at all; the number of components which do not contain any self-loop is precisely k. In many cases, k-semifactors in which two designated nodes j_1 and i_1 are required to appear in the same component are used. Then the symbol $R(j_1 i_1; j_2; \ldots; j_k)$ denotes a k-semifactor $R(j_1; j_2; \ldots; j_k)$ in which the nodes j_1

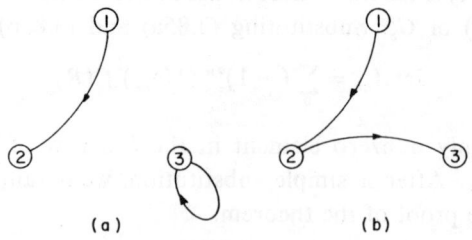

Fig. 3.30. The set of 1-semifactors $R(12)$ in the directed graph of fig. 3.28(c).

and i_1 appear in the same component. Note that the order of j_1 and i_1 in $R(j_1i_1;j_2;\ldots;j_k)$ is important, and also that in a k-semifactor one or more of its components may each consist of an isolated node.

Examples of 1-semifactors $R(1)$ of fig. 3.28(c) are presented in fig. 3.30 or 3.31. All of these 1-semifactors except that of fig. 3.31(a) are also 1-semifactors $R(12)$, since the nodes 1 and 2 are contained in the same component.

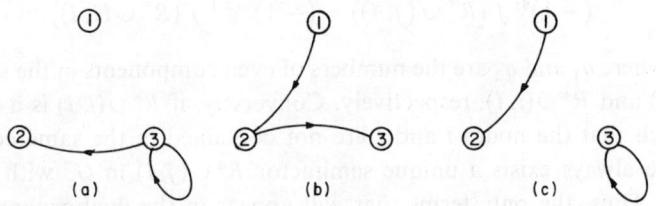

Fig. 3.31. The set of 1-semifactors $R(1)$ in the directed graph of fig. 3.28(c).

THEOREM 3.13: Let G'_c be the modified Coates graph of a Coates graph G_c. If Δ_{ij} is the cofactor of the (i, j)-element of the associated matrix of G_c, then

$$\Delta_{ij} = \sum_{R(ij)} (-1)^{q_{ij}-1} f(R(ij)), \qquad (3.87)$$

where $R(ij)$ is a 1-semifactor of G'_c, and q_{ij} is the number of even components in $R(ij)$.

Proof. Let G''_c be the directed graph obtained from G'_c by first removing all the edges having node i as their terminal node, and then adding two edges (i, i) and (j, i) of weight 1 to the resulting graph. Also, let G^α_c be the Coates graph obtained from G_c by first removing all the edges having node i as their terminal node, and then adding an edge (j, i) of weight 1 to the resulting graph. It is not difficult to see that G''_c is the modified Coates graph of G^α_c. Since Δ_{ij} is the same as the graph determinant of G^α_c, it follows from Theorem 3.12 that

$$\Delta_{ij} = \sum_{R''} (-1)^{q''} f(R''), \qquad (3.88)$$

where R'' is a semifactor in G''_c, and q'' is the number of the even components of R''. Since every semifactor R'' of G''_c must contain the edge (j, i) or (i, i), it can always be expressed as

$$R'' = R^* \cup (k, i), \qquad (3.89)$$

where $k = j$ or i, and R^* and (k, i) denote the complementary subgraphs of the semifactor R''. Clearly, the corresponding subgraph of R^* in G'_c forms a 1-semifactor $R(i)$ of G'_c and such that

$$f(R'') = f(R(i)). \tag{3.90}$$

Next, consider a semifactor of the form $R^* \cup (j, i)$ in G''_c. There always exists a unique semifactor $R^* \cup (i, i)$ in G''_c such that

$$(-1)^{q_1} f(R^* \cup (j, i)) = (-1)^{q_2 \pm 1} f(R^* \cup (i, i)) \tag{3.91}$$

for $i \neq j$, where q_1 and q_2 are the numbers of even components in the semifactors $R^* \cup (j, i)$ and $R^* \cup (i, i)$, respectively. Conversely, if $R^* \cup (i, i)$ is a semifactor of G''_c such that the nodes i and j are not contained in the same component, then there always exists a unique semifactor $R^* \cup (j, i)$ in G''_c with the above property. Thus, the only terms that will appear in the final expansion of Δ_{ij} are those semifactors $R^* \cup (i, i)$ with nodes i and j belonging to the same component.

Finally, observe the one-to-one correspondence between the uncancelled semifactors $R^* \cup (i, i)$ in G''_c and 1-semifactors $R(ij)$ in G'_c. Observe also that the removal of the edge (i, i) from $R^* \cup (i, i)$ increases or decreases the number of the even components of $R^* \cup (i, i)$ by one, so it follows that

$$(-1)^{q''} f(R^* \cup (i, i)) = (-1)^{q_{ij}-1} f(R(ij)), \tag{3.92}$$

which completes the proof of the theorem.

As an illustration, let us again consider the modified Coates graph G'_c given in fig. 3.28(c). The sets of 1-semifactors $R(12)$ and $R(1)$ are presented in figs. 3.30 and 3.31, respectively. Thus, we have

$$\begin{aligned}
\Delta_{12} &= \sum_{R(12)} (-1)^{q_{12}-1} f(R(12)) \\
&= (Z_3 + Z_2)(Z_5 + Z_6) + (Z_3 + Z_2) Z_4 \\
&= (Z_3 + Z_2)(Z_4 + Z_5 + Z_6),
\end{aligned} \tag{3.93}$$

$$\begin{aligned}
\Delta_{11} &= \sum_{R(1)} (-1)^{q_{11}-1} f(R(1)) \\
&= (Z_5 + Z_6) Z_4 + (Z_3 + Z_2) Z_4 + (Z_3 + Z_2)(Z_5 + Z_6) \\
&= (Z_5 + Z_6)(Z_2 + Z_3 + Z_4) + Z_4 (Z_3 + Z_2).
\end{aligned} \tag{3.94}$$

We shall now apply these results to the solutions of a system of linear equations. The theorem given below may be proved in a manner similar to that of Theorem 3.3. Thus, only an outline of a proof will be presented.

§ 3 Modifications of Coates and Mason graphs

THEOREM 3.14: If the coefficient matrix A is nonsingular, then the solution of (3.16) is given by

$$x_k = \frac{\sum_{R(sk)} (-1)^{q_{sk}-1} f(R(sk))}{\sum_R (-1)^{q_R} f(R)} \qquad (3.95)$$

for $k=1, 2, \ldots, n$, where $s=n+1$; $R(sk)$ is a 1-semifactor of $G'_c(A_u)$; R is a semifactor in $G'_c(A)$; $G'_c(A)$ and $G'_c(A_u)$ are the modified Coates graphs of the Coates graphs $G_c(A)$ and $G_c(A_u)$, respectively; q_{sk} and q_R are the numbers of the even components of $R(sk)$ and R, respectively; and A_u is the augmented matrix obtained from A by attaching $-B$ to the right of A, and then attaching a row of zeros at the bottom of the resultant matrix.

Proof. Consider the equivalent system

$$A_u X'' = 0, \qquad (3.96)$$

where the transpose of X'' is $[x_1, x_2, \ldots, x_n, x_{n+1}]$ with $x_{n+1}=1$. Following a similar argument as that given in the proof of Theorem 3.3, the solution of the system can be written as

$$\begin{aligned} x_k &= (-1)^{(n+1)-k-1}(-1)\det(A_u)_{(n+1)k}/(\det A) \\ &= (-1)^{n+1+k}\det(A_u)_{(n+1)k}/(\det A), \end{aligned} \qquad (3.97)$$

where $(A_u)_{ij}$ is the matrix obtained from A_u by striking out the ith row and jth column. The theorem follows immediately after substituting the results (3.81) and (3.87) in (3.97). This completes the proof of the theorem.

3.2. Modifications of Mason graphs

The extension of the modifications of Coates graphs to Mason graphs is straightforward. The procedures are identical and need no further elaboration.

DEFINITION 3.14: *Modified Mason graph.* For a Mason graph G_m, the *modified Mason graph* of G_m, denoted by the symbol G'_m, is the directed graph obtained from G_m by changing the weight associated with each of its self-loops (i, i) to the sum of the weights of the edges having node i as their terminal node, while keeping all other edge weights unaltered.

THEOREM 3.15: Let G'_m be the modified Mason graph of an n-node Mason graph G_m. Then

$$\det G_m = (-1)^n + \sum_{v=0}^{n-1} \sum_u (-1)^{q_m} f(R^{uv}), \qquad (3.98)$$

where R^{uv} is the uth v-semifactor of G'_m; q_m is the number of even components of R^{uv}; and R^{u0} is defined to be the uth semifactor of G'_m.

Proof. Let G_c be the corresponding Coates graph of G_m. Also, let G'_c be the modified Coates graph of G_c. It is not difficult to see that G'_c can be obtained from G'_m simply by adding a self-loop of weight -1 to each node of G'_m. If W^{ik} is the ith semifactor of G'_c which contains k added self-loops of weight -1, then from Lemma 3.1 we have $f(W^{in}) = (-1)^n$ and

$$(-1)^q f(W^{ik}) = (-1)^{q_t + k}(-1)^k f(Q^{ik}), \tag{3.99}$$

where k ranges from 0 to $n-1$; q and q_t are the numbers of even components in W^{ik} and Q^{ik}, respectively; and Q^{ik} is obtained from W^{ik} with the k added self-loops removed. By Theorem 3.12, the totality of W^{ik} with appropriate signs together with the term $(-1)^n$ is precisely the determinant of the matrix associated with G_c or det G_m:

$$\det G_m = (-1)^n + \sum_{\substack{i,k \\ k<n}} (-1)^{q_t} f(Q^{ik}). \tag{3.100}$$

It is readily recognized that the corresponding subgraph R^{ik} of Q^{ik} in G'_m is actually a k-semifactor of G'_m. Since there exists a one-to-one correspondence between the subgraphs Q^{ik} in G'_c and k-semifactors R^{ik} in G'_m, it follows that we may use R^{ik} to evaluate the graph determinant of G_m instead of Q^{ik}. Thus, from (3.100) we obtain the desired expansion

$$\det G_m = (-1)^n + \sum_{v=0}^{n-1} \sum_{u} (-1)^{q_m} f(R^{uv}), \tag{3.101}$$

which completes the proof of the theorem.

The following theorem follows directly from Theorem 3.13, and can be proved in a manner similar to that in the proof of Theorem 3.15. We shall only state the result, and leave the details as an exercise (Problem 3.42).

THEOREM 3.16: Let G'_m be the modified Mason graph of an n-node Mason graph G_m. If Δ_{ij} is the cofactor of the (i,j)-element of the matrix associated with G_m, then for $i \neq j$

$$\Delta_{ij} = \sum_{v=1}^{n-1} \sum_{(i_1 i_2 \ldots i_v)} \sum_u (-1)^{q_{ij}-1} f(R_{ij}^{uv}), \tag{3.102}$$

where R_{ij}^{uv} is the uth v-semifactor $R(i_1 j; i_2; \ldots; i_v)$ of G'_m with $i = i_1$, and q_{ij} is the number of even components in R_{ij}^{uv}, and $\{i_1, i_2, \ldots, i_v\}$ is a subset of the node set of G'_m containing the node $i = i_1$.

The application of the above results to the solution of a system of linear algebraic equations is straightforward, and the formula can be derived directly from (3.97).

THEOREM 3.17: If the coefficient matrix A of system (3.16) is nonsingular, then the solution of the system is given by

$$x_k = \frac{\sum_{v=1}^{n} \sum_{(i_1 i_2 \ldots i_v)} \sum_u (-1)^{q_{sk}-1} f(R_{sk}^{uv})}{(-1)^n + \sum_{\substack{x,y \\ 0 \le y < n}} (-1)^{q_m} f(R^{xy})} \qquad (3.103)$$

for $k=1, 2, \ldots, n$; where $s=n+1$; R_{sk}^{uv} is the uth v-semifactor $R(i_1 k; i_2; \ldots; i_v)$ of $G'_m(A_u)$ with $i_1 = s$; R^{xy} is the xth y-semifactor of $G'_m(A)$; $G'_m(A_u)$ and $G'_m(A)$ are the modified Mason graphs of the Mason graphs $G_m(A_u)$ and $G_m(A)$, respectively; q_{sk} and q_m are the numbers of even components in R_{sk}^{uv} and R^{xy}, respectively; and $\{i_1, i_2, \ldots, i_v\}$ is a subset of the node set of $G'_m(A_u)$ containing the node $s = i_1$. (A_u is defined in Theorem 3.14.)

§ 4. The generation of subgraphs of a directed graph

The usefulness of the results presented above depends upon the efficiency with which certain subgraphs such as 1-factors, 1-factorial connections, semifactors, and k-semifactors of a directed graph are generated. In this section, we shall present systematic ways of generating these subgraphs. The techniques to be described below are readily adaptable for digital computation. For simplicity, we assume that all the directed graphs considered in this section have no parallel edges. However, all the results obtained here can be extended easily, and the details of these extensions are left as exercises (Problems 3.50 and 3.51).

Let G be an n-node directed graph in which every edge has been assigned an edge-designation symbol e_k or e_{ij}. By *variable adjacency matrix* of G we mean the square matrix of order n such that the ijth entry is e_k if and only if there is an edge in G directed from i to j with edge-designation symbol e_k, and the entry is 0 otherwise. As pointed out earlier in § 1, if we interpret G as a given Coates graph, then the associated matrix of G is identical to the transpose of the variable adjacency matrix, where the edge-designation symbols are considered as the edge weights of G.

THEOREM 3.18: In an n-node acyclic directed graph G all its directed paths P_{ij} from node i to node j are represented by the (i,j)-element of $(U_n - D)^{-1}$

for all $i \neq j$, where U_n is the identity matrix of order n, and D is the variable adjacency matrix of G.

Proof. Consider G as a given Mason graph. Let G_c be the corresponding Coates graph of G. If A is the associated matrix of G_c, and if Δ_{ij} is the cofactor of the (i,j)-element of A, then by Theorems 3.7 and 3.8 we have

$$\det A = (-1)^n \tag{3.104}$$

and

$$\Delta_{ij} = (-1)^{n-1} \sum_k f(P_{ij}^k), \tag{3.105}$$

where P_{ij}^k is the kth directed path from node i to node j in G. Since the transpose A' of A is

$$A' = D - U_n, \tag{3.106}$$

it follows that

$$-\frac{(-1)^{i+j}\det(U_n - D)_{ji}}{\det(U_n - D)} = \frac{(-1)^{i+j}\det(D - U_n)_{ji}}{\det(D - U_n)}$$

$$= \frac{\Delta_{ij}}{\det A'} = \frac{\Delta_{ij}}{\det A} = -\sum_k f(P_{ij}^k), \tag{3.107}$$

where $(D - U_n)_{ji}$ is the matrix obtained from $(D - U_n)$ by striking out the jth row and ith column. Since the left side of (3.107) is the negative of the ijth entry of $(U_n - D)^{-1}$, the theorem follows.

As an illustration, consider the directed graph G of fig. 3.32. The matrix $U_5 - D$ is given by

$$U_5 - D = \begin{bmatrix} 1 & 0 & 0 & 0 & 0 \\ -e_4 & 1 & 0 & 0 & 0 \\ 0 & -e_6 & 1 & -e_7 & 0 \\ -e_3 & -e_5 & 0 & 1 & -e_2 \\ -e_1 & 0 & 0 & 0 & 1 \end{bmatrix}. \tag{3.108}$$

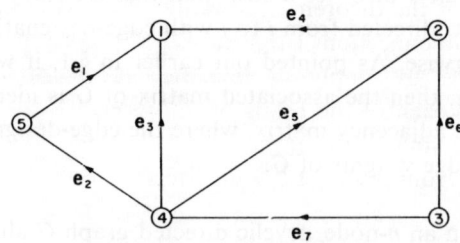

Fig. 3.32. An acyclic directed graph.

§4   Subgraphs of a directed graph

The set of all directed paths P_{ij}, $i \neq j$, from i to j in G is represented by the ijth off-diagonal element of the inverse of $(U_5 - D)$ which is given by

$$\begin{bmatrix} 1 & 0 & 0 & 0 & 0 \\ e_4 & 1 & 0 & 0 & 0 \\ e_3 e_7 + e_1 e_2 e_7 + e_4 e_6 + e_4 e_5 e_7 & e_6 + e_5 e_7 & 1 & e_7 & e_2 e_7 \\ e_4 e_5 + e_3 + e_1 e_2 & e_5 & 0 & 1 & e_2 \\ e_1 & 0 & 0 & 0 & 1 \end{bmatrix}. \quad (3.109)$$

4.1. *The generation of 1-factors and 1-factorial connections*

For a given directed graph G, let the symbols S_i^+ and S_i^- be used to denote the sets of edge-designation symbols of the edges of G having node i as their initial and terminal nodes, respectively.

THEOREM 3.19: For an n-node directed graph G, the set X of 1-factors and the set X_{ij} of 1-factorial connections from node i to node j are represented by

$$X = (\prod_{k=1}^{n} S_k^+) \cap (\prod_{t=1}^{n} S_t^-) \quad (3.110)$$

and

$$X_{ij} = (\prod_{\substack{k=1 \\ k \neq j}}^{n} S_k^+) \cap (\prod_{\substack{t=1 \\ t \neq i}}^{n} S_t^-), \quad (3.111)$$

respectively, where the products denote the Cartesian products of the sets.

Proof. Without loss of generality, we assume that, for each node i of G, $d^+(i) \neq 0$ and $d^-(i) \neq 0$. Since each element of $\prod_{k=1}^{n} S_k^+$ corresponds to a spanning subgraph of G such that the outgoing degree of each of its nodes is 1, and each element of $\prod_{t=1}^{n} S_t^-$ corresponds to a spanning subgraph of G such that the incoming degree of each of its nodes is 1, it follows that only 1-factors of G can appear in X. This completes the proof of the first part of the theorem.

The second part of the theorem follows immediately from the fact that the set X_{ij} of 1-factorial connections from i to j in G can be obtained from the set X' of 1-factors of G' by simply deleting the edge-designation symbol e_{ji} of the edge (j, i) from each of the terms of X', where G' is the graph obtained from G by first removing all the edges (j, u) and (u, i), $u = 1, 2, \ldots, n$, and then adding an edge e_{ji} from node j to node i in the resulting graph. Since in G'

$$X' = (\prod_{k=1}^{n} S_k'^+) \cap (\prod_{t=1}^{n} S_t'^-), \quad (3.112)$$

where

$$S_k'^+ = S_k^+ - \{e_{ki}\} \tag{3.113}$$

for $k = 1, 2, \ldots, j-1, j+1, \ldots, n$, and

$$S_t'^- = S_t^- - \{e_{jt}\} \tag{3.114}$$

for $t = 1, 2, \ldots, i-1, i+1, \ldots, n$, and $S_j'^+ = S_i'^- = \{e_{ji}\}$ where e_{ij} is the edge-designation symbol of the edge (i, j) in G. Then (3.111) follows directly from (3.112) and the fact that

$$\{e_{ji}\} \times X_{ij} = X', \tag{3.115}$$

which completes the proof of the theorem.

The sets of subgraphs of the types needed in Theorems 3.7 and 3.8 for the Mason graphs can also be obtained by using a procedure similar to the above. The only difference is that we have to insert an identity element, denoted by e_0, to each set such that $e_0 e_{ij} = e_{ij}$ for all i and j.

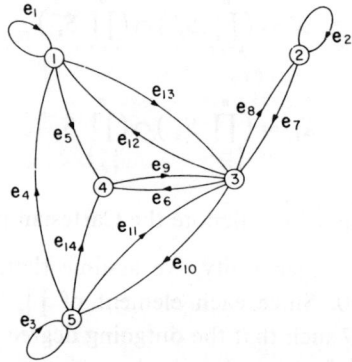

Fig. 3.33. A directed graph used to illustrate the generation of its 1-factors and 1-factorial connections.

Consider the directed graph G of fig. 3.33, which is the same as that of fig. 3.1. Since

$$\prod_{k=1}^{5} S_k^+ = \{e_5, e_{13}, e_1\} \times \{e_7, e_2\} \times \{e_6, e_8, e_{10}, e_{12}\} \times \{e_9\} \times \{e_3, e_4, e_{11}, e_{14}\}$$

and

$$\prod_{t=1}^{5} S_t^- = \{e_1, e_4, e_{12}\} \times \{e_2, e_8\} \times \{e_7, e_9, e_{11}, e_{13}\} \times \{e_5, e_6, e_{14}\} \times \{e_3, e_{10}\},$$

the set X of 1-factors of G is obtained as follows:

$$\begin{aligned}X &= \{e_9\} \times \{e_2\} \times [(\{e_1, e_5\} \times \{e_6, e_{10}, e_{12}\} \times \{e_3, e_4, e_{14}\}) \\
&\quad \cap (\{e_1, e_4, e_{12}\} \times \{e_5, e_6, e_{14}\} \times \{e_3, e_{10}\})] \\
&= \{e_2 e_9\} \times (\{e_3\} \times [(\{e_1, e_5\} \times \{e_6, e_{12}\}) \cap (\{e_1, e_{12}\} \times \{e_5, e_6\})] \\
&\quad \cup \{e_{10}\} \times [(\{e_1, e_5\} \times \{e_4, e_{14}\}) \cap (\{e_1, e_4\} \times \{e_5, e_{14}\})]) \\
&= \{e_2 e_9\} \times \{e_1 e_3 e_6, e_3 e_5 e_{12}, e_1 e_{10} e_{14}, e_5 e_{10} e_4\}.\end{aligned}$$ (3.116)

It should be noted that it is advantageous not to expand the Cartesian products completely before taking the intersection operations, and also in performing the intersection operation of two sets expressed in Cartesian-product form, much of the labor work could be saved if all the elements which are in one of the sets but not in the other are deleted in advance. Other types of deletions may also be performed, as we did above. Similarly, the set X_{41} of 1-factorial connections from node 4 to node 1 in G is given by

$$\begin{aligned}X_{41} &= (\{e_7, e_2\} \times \{e_6, e_8, e_{10}, e_{12}\} \times \{e_9\} \times \{e_3, e_4, e_{11}, e_{14}\}) \\
&\quad \cap (\{e_1, e_4, e_{12}\} \times \{e_2, e_8\} \times \{e_7, e_9, e_{11}, e_{13}\} \times \{e_3, e_{10}\}) \\
&= \{e_9\} \times [(\{e_2\} \times \{e_8, e_{10}, e_{12}\} \times \{e_3, e_4\}) \cap (\{e_4, e_{12}\} \times \{e_2, e_8\} \\
&\quad \times \{e_3, e_{10}\})] \\
&= \{e_2 e_9\} \times [(\{e_{10}, e_{12}\} \times \{e_3, e_4\}) \cap (\{e_4, e_{12}\} \times \{e_3, e_{10}\})] \\
&= \{e_2 e_9\} \times \{e_3 e_{12}, e_4 e_{10}\}.\end{aligned}$$ (3.117)

4.2. The generation of semifactors and k-semifactors

The technique to be presented in this section is a variant of the Wang-algebra formulation. Since the Wang algebra will be formally introduced in ch. 5, § 4.1, for the present purpose we shall only define the Wang-product operation on sets of subgraphs.

For a directed graph G, a subgraph of G as defined in ch. 1, § 5.1 and Definition 1.2 is a sub-collection of nodes and edges of G. Thus, a subgraph of G may contain isolated nodes. The Wang product of two sets of subgraphs, on the other hand, is defined only for those sets of subgraphs that do not contain any isolated nodes. In the following, we shall assume this.

Let S_1 and S_2 be two sets of subgraphs of G with each of their elements being represented by the product of its edge-designation symbols. The *Wang product* of the sets S_1 and S_2, denoted by the symbol $S_1 @ S_2$, is defined as

$$S_1 @ S_2 = \{w_1\} \oplus \{w_2\} \oplus \cdots \oplus \{w_\alpha\},$$ (3.118)

where α denotes the number of elements in the set $W = \{w_i\}$, and

$$W = \{g_1 \cup g_2; g_1 \text{ is in } S_1, g_2 \text{ in } S_2, \text{ and } g_1 \cap g_2 = \emptyset\}, \quad (3.119)$$

where $\emptyset$ is the null graph of G. In other words, $S_1 @ S_2$ is the set of subgraphs of G obtained by taking the ring-sum operations [see (1.6)] of all the sets, each of which contains only one element obtained by taking the union of two edge-disjoint subgraphs, one from S_1 and one from S_2.

It can easily be shown that the Wang-product operation of the sets is commutative and associative. Other properties of the operation will be presented in ch. 5, § 4.1, and thus will not be elaborated here.

With the above definition, we can now proceed to generate the sets of semifactors and k-semifactors $R(j_1; j_2; \ldots; j_k)$ of G. The method to be described below is a modification of a technique for generating the sets of "directed trees" and "directed k-trees" (CHEN [1966a]). Since the directed tree and k-tree cases will be treated in great detail in ch. 4, § 5.1, for the present case we shall only state the results, and leave the derivations as obvious after ch. 4, § 5.1.

As before, let e_{ij} be the edge-designation symbol for the edge (i, j) of G. For simplicity, we shall speak loosely of an edge e_{ij} rather than of the edge-designation symbol e_{ij} of the edge (i, j). Again, let S_i^- denote the set of edges of G having node i as their terminal node.

THEOREM 3.20: In an n-node symmetric directed graph G, let

$$Y = S_1^- @ S_2^- @ \cdots @ S_n^-, \quad (3.120a)$$

$$Y(i_1; i_2; \ldots; i_{n-k}) = S_{i_1}^- @ S_{i_2}^- @ \cdots @ S_{i_{n-k}}^-, \quad (3.120b)$$

where $i_1 i_2 \ldots i_{n-k}$ and $j_1 j_2 \ldots j_k$ are the complementary sets of indices of the set of integers $1, 2, \ldots, n$. Then the set of semifactors and the set of k-semifactors $R(j_1; j_2; \ldots; j_k)$, up to a set of isolated nodes, of G are represented by the sets of elements in the final expansions of Y and $Y(i_1; i_2; \ldots; i_{n-k})$, respectively, provided that the edges e_{ij} and e_{ji} are considered as identical elements in all the operations.

If G is not symmetric, it is always possible to make it symmetric. Let G_s be the symmetric directed graph obtained from G by adding a minimum number of edges to G. In order that we may distinguish the added edges from those already in G, let the primes be used for this purpose. For example, if (i, j) is in G and if (j, i) is in G_s but not in G, then in G_s let (i, j) be labeled by e_{ij} and (j, i) by e'_{ji}.

For an arbitrary directed graph G, suppose that the edges of its corresponding symmetric directed graph G_s have been labeled according to the rules just adopted. The sets of semifactors and k-semifactors of G_s are then obtained by the procedures outlined in Theorem 3.20 provided that e_{ij} and e_{ji} or e_{uv} and e'_{vu} are considered as identical elements in all the operations. Finally, if we eliminate all those elements in the final expansions of (3.120) which contain at least one primed letter, then the remaining elements in the expansions are the desired sets of semifactors and k-semifactors of G.

We shall illustrate the above procedures by the following example.

Example 3.10: Consider the coefficient matrix A of (3.24), which is associated with the feedback amplifier as shown in fig. 3.6. The modified Coates graph G'_c of $G_c(A)$ is presented in fig. 3.34. Suppose that we wish to compute the current-gain function I_3/I_1, which can also be expressed as

$$I_3/I_1 = G_d \Delta_{13}/(\det A), \tag{3.121}$$

where Δ_{13} is the cofactor of the (1, 3)-element of A. Thus, the set of semifactors and the set of 1-semifactors $R(1)$ of G'_c must be generated in advance. By Theorem 3.20, they are represented by the sets

$$Y = \{e_{21}, e_{31}\} @ \{e_{12}, e_{22}, e_{32}\} @ \{e_{13}, e_{23}, e_{33}\}$$
$$= \{e_{21}e_{32}e_{33}, e_{21}e_{22}e_{33}, e_{21}e_{22}e_{23}, e_{21}e_{22}e_{13}, e_{31}e_{32}e_{33}, e_{31}e_{22}e_{33},$$
$$e_{31}e_{22}e_{23}, e_{31}e_{12}e_{33}\}, \tag{3.122}$$

$$Y(2;3) = \{e_{12}, e_{22}, e_{32}\} @ \{e_{13}, e_{23}, e_{33}\}$$
$$= \{e_{12}e_{13}, e_{12}e_{23}, e_{12}e_{33}, e_{22}e_{13}, e_{22}e_{23}, e_{22}e_{33}, e_{32}e_{13}, e_{32}e_{33}\}, \tag{3.123}$$

respectively. The desired set of 1-semifactors $R(13)$ of G'_c is a subset of $Y(2;3)$, and is given by

$$\{R(13)\} = \{e_{12}e_{13}, e_{12}e_{23}, e_{22}e_{13}, e_{32}e_{13}\}. \tag{3.124}$$

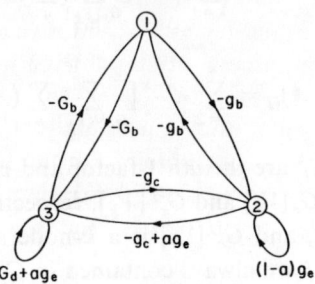

Fig. 3.34. A directed graph used to illustrate the generation of its semifactors and 1-semifactors.

The gain function of the amplifier is then obtained as follows:

$$\frac{I_3}{I_1} = \frac{-G_d \sum_{R(13)} (-1)^{q_{13}} f(R(13))}{\sum_R (-1)^{q_R} f(R)}$$

$$= G_d [g_b G_b + g_b(g_c - ag_e) + (1-a) g_e G_b + g_c G_b]/\Delta, \quad (3.125)$$

where

$$\Delta = (1-a) g_e [(G_d + g_c)(g_b + G_b) + g_b G_b]$$
$$+ (G_d + ag_e)[g_c(g_b + G_b) + g_b G_b], \quad (3.126)$$

where $R(13)$ and R are a 1-semifactor and a semifactor of G'_c, respectively, and q_{13} and q_R are the numbers of even components in $R(13)$ and R, respectively. The gain function of (3.125) is of course identical to that obtained by the Coates graph technique given in (3.25). However, the number of cancellation terms in the present case is reduced.

§ 5. The eigenvalue problem

The relationship between the coefficients of the eigenvalue equation of a matrix A and certain types of subgraphs of $G_c(A)$ was investigated by CHEN [1967a] and PONSTEIN [1966]. The use of graph-theoretic method for the complete reduction of a matrix with a view toward finding its eigenvalues was considered by HARARY [1959]. In this section, we shall summarize these results, and put them in a more convenient form.

THEOREM 3.21: Let G_c be the associated Coates graph of a given matrix A of order n. Then

$$\det(\lambda U_n - A) = \lambda^n + \sum_{k=1}^{n} \lambda^{n-k} [\sum_{G_c[V_k]} \sum_u (-1)^{L_h} f(h_{uk})] \quad (3.127)$$

and

$$(-1)^{i+j} \det(\lambda U_n - A)_{ij} = \sum_{k=2}^{n} \lambda^{n-k} [\sum_{G_c^{ij}[V_k]} \sum_v (-1)^{L_H} f(H_{ij}^{vk})] \quad (3.128)$$

for $i \neq j$, where h_{uk} and H_{ij}^{vk} are the uth 1-factor and vth 1-factorial connection from node i to node j in $G_c[V_k]$ and $G_c^{ij}[V_k]$, respectively; $G_c[V_k]$ is a k-node sectional subgraph of G_c, and $G_c^{ij}[V_k]$ is a k-node sectional subgraph of G_c in which the nodes i and j are always contained in $G_c^{ij}[V_k]$; L_h and L_H are the numbers of directed circuits in h_{uk} and H_{ij}^{vk}, respectively; and $(\lambda U_n - A)_{ij}$ is the matrix obtained from $(\lambda U_n - A)$ by striking out the ith row and jth column.

§5 The eigenvalue problem

Proof. Since the identities can be established in a manner similar to that given in the proofs of Theorems 3.7 and 3.8, we shall only prove (3.127), and leave the other as an exercise (Problem 3.17).

Let G_c^α be the graph obtained from G_c by adding a self-loop of weight $-\lambda$ to each of the nodes of G_c. Clearly, G_c^α is the associated Coates graph of $(A - \lambda U_n)$. If $h_{u(n-x)}^\alpha$ is the uth 1-factor of G_c^α which contains $(n-x)$ added self-loops of weight $-\lambda$, and if $Q_{u(n-x)}^\alpha$ is the graph obtained from $h_{u(n-x)}^\alpha$ with the $(n-x)$ added self-loops removed, then

$$f(h_{un}^\alpha) = (-1)^n \lambda^n \tag{3.129a}$$

and

$$f(h_{u(n-k)}^\alpha) = (-1)^{n-k} f(Q_{u(n-k)}^\alpha) \lambda^{n-k} \tag{3.129b}$$

for $k = 1, 2, \ldots, n$. By Theorem 3.1 and Problem 3.1, the totality of $h_{u(n-x)}^\alpha$, $x = 0, 1, \ldots, n$, with appropriate signs is precisely the determinant of $(A - \lambda U_n)$. Thus, we have

$$\det(A - \lambda U_n) = (-1)^n \left[\lambda^n + \sum_{k=1}^{n} \sum_u (-1)^{L_\alpha + n - k}(-\lambda)^{n-k} f(Q_{u(n-k)}^\alpha)\right] \tag{3.130a}$$

or

$$\det(\lambda U_n - A) = \lambda^n + \sum_{k=1}^{n} \sum_u (-1)^{L_\alpha} f(Q_{u(n-k)}^\alpha) \lambda^{n-k}, \tag{3.130b}$$

where L_α is the number of directed circuits in $Q_{u(n-k)}^\alpha$. Since for each choice of $Q_{u(n-x)}^\alpha$ in G_c^α, there corresponds a unique 1-factor h_{ux} in a sectional subgraph $G_c[V_x]$ of G_c, and vice versa, it follows that the determinant of $(\lambda U_n - A)$ can be evaluated by the 1-factors h_{ux} of $G_c[V_x]$ rather than by 1-factors $h_{u(n-x)}^\alpha$ of G_c^α. Merely by rearranging the terms in (3.130b) in a slightly different form we obtain the desired result of (3.127). This completes the proof of the theorem.

In particular, if we let $\lambda = 0$, (3.127) and (3.128) reduce to (3.4) and (3.12), respectively. In this sense, the present case may be considered as a generalization of the Coates-graph formulation. Obviously, Theorem 3.21 is still valid if A is the associated variable adjacency matrix of any n-node directed graph. The only difference is that the term $\det(\lambda U_n - A)_{ij}$ should be replaced by $\det(\lambda U_n - A)_{ji}$.

Next, we shall consider the eigenvalues of a sparse matrix A. The problem is to reduce A, if possible, to matrices with lower orders so that the set of eigenvalues of A is simply the union of the sets of eigenvalues of these lower-order matrices, including multiplicity.

THEOREM 3.22: Let G_c be the associated Coates graph of a matrix A or its transpose. If G_c is not strongly connected, then the set of eigenvalues of A is the union (including multiplicity) of the sets of eigenvalues of the submatrices corresponding to the strongly-connected components of G_c.

Proof. Let A be of order n. Let G_c^* be the directed graph obtained from G_c by adding a self-loop of weight $-\lambda$ to each of its nodes. Clearly, G_c^* is the associated Coates graph of $(A - \lambda U_n)$. Also, let $G_{cx}, x = 1, 2, \ldots, k$, be the strongly-connected components of G_c. It is not difficult to see that the corresponding sectional subgraph G_{cx}^* of G_{cx} in G_c^* is a strongly-connected component of G_c^* if, and only if, G_{cx} is a strongly-connected component of G_c. Thus, if A_{xx} and A_{xx}^* are the associated matrices of G_{cx} and G_{cx}^*, respectively, then from Problem 3.36 we have

$$\det G_c^* = \det(A - \lambda U_n) = (\det A_{11}^*)(\det A_{22}^*) \cdots (\det A_{kk}^*), \quad (3.131\text{a})$$

$$\det G_c = \det A = (\det A_{11})(\det A_{22}) \cdots (\det A_{kk}), \quad (3.131\text{b})$$

since A_{xx} and A_{xx}^* are the submatrices associated with the strongly-connected components of G_c and G_c^*, respectively. But A_{xx} and A_{xx}^* are related by

$$A_{xx}^* = A_{xx} - \lambda U_p, \quad (3.132)$$

where U_p is the identity matrix of appropriate order. Substituting (3.132) into (3.131a), the theorem follows immediately. So the theorem is proved.

Example 3.11: Suppose that we wish to compute the eigenvalue equation and eigenvalues of the following sparse matrix:

$$A = \begin{bmatrix} 0 & 0 & 0 & 2 & 1 & 0 & 0 & 0 & 0 & 3 \\ 5 & 0 & 1 & 3 & 0 & 0 & 0 & 0 & 0 & 0 \\ 0 & 0 & 0 & -1 & 0 & 0 & -1 & 1 & 0 & 0 \\ 1 & 0 & 0 & 1 & 1 & 0 & -2 & 0 & 0 & 0 \\ 1 & 0 & 0 & 1 & -1 & 4 & 1 & 0 & 0 & 0 \\ 0 & 0 & 0 & 0 & 0 & 0 & -1 & 0 & 0 & 0 \\ 0 & 0 & 0 & 0 & 0 & 0 & 1 & 0 & 0 & 1 \\ 0 & 0 & 0 & -1 & 0 & 0 & 0 & 0 & 1 & 1 \\ 0 & 1 & 0 & 0 & 0 & 0 & 0 & 0 & 0 & 2 \\ 0 & 0 & 0 & 0 & 0 & 2 & 0 & 0 & 0 & 0 \end{bmatrix}. \quad (3.133)$$

The associated Coates graph of the transpose of the matrix A is presented

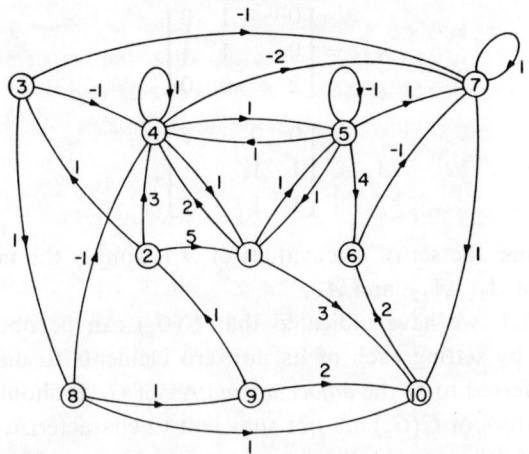

Fig. 3.35. The associated Coates graph of the transpose of the matrix (3.133).

in fig. 3.35. The eigenvalue equation is obtained as follows:

$$\begin{aligned}\det(\lambda U_{10} - A) &= \lambda^{10} + (b_2 - a_{77})\lambda^9 + (b_3 - b_2 a_{77})\lambda^8 \\ &+ (b_4 + b_0 - b_3 a_{77})\lambda^7 + (b_1 + b_0 b_2 - b_4 a_{77})\lambda^6 \\ &+ (b_0 b_3 + b_1 b_2 - b_1 a_{77})\lambda^5 + (b_0 b_4 + b_1 b_3 - b_1 b_2 a_{77})\lambda^4 \\ &+ (b_0 b_1 + b_1 b_4 - b_1 b_3 a_{77})\lambda^3 + (b_0 b_1 b_2 - b_1 b_4 a_{77})\lambda^2 \\ &+ b_0 b_1 b_3 \lambda + b_0 b_1 b_4 \\ &= \lambda^{10} - \lambda^9 - 5\lambda^8 - 3\lambda^7 + 3\lambda^6 - 9\lambda^5 - 3\lambda^4 - 3\lambda^3 \\ &- 4\lambda^2 + 10\lambda + 8, \end{aligned} \qquad (3.134)$$

where a_{ij} is the (i,j)-element of A; and

$$\begin{aligned} b_0 &= -a_{67} a_{7,10} a_{10,6} = 2, \\ b_1 &= -a_{23} a_{38} a_{89} a_{92} = -1, \\ b_2 &= -(a_{44} + a_{55}) = 0, \\ b_3 &= a_{44} a_{55} - a_{15} a_{51} - a_{45} a_{54} - a_{14} a_{41} = -5, \\ b_4 &= -a_{41} a_{15} a_{54} - a_{14} a_{45} a_{51} + a_{15} a_{51} a_{44} + a_{14} a_{41} a_{55} = -4. \end{aligned}$$

In G_c, there are three strongly-connected components corresponding to the sectional subgraphs $G_c[2, 3, 8, 9]$, $G_c[6, 7, 10]$, and $G_c[1, 4, 5]$. The matrices A_{11}, A_{22}, and A_{33} associated with these components are given by

$$A_{11} = \begin{bmatrix} 0 & 1 & 0 & 0 \\ 0 & 0 & 1 & 0 \\ 0 & 0 & 0 & 1 \\ 1 & 0 & 0 & 0 \end{bmatrix}, \qquad (3.135a)$$

$$A_{22} = \begin{bmatrix} 0 & -1 & 0 \\ 0 & 1 & 1 \\ 2 & 0 & 0 \end{bmatrix}, \qquad (3.135b)$$

$$A_{33} = \begin{bmatrix} 0 & 2 & 1 \\ 1 & 1 & 1 \\ 1 & 1 & -1 \end{bmatrix}, \qquad (3.135c)$$

respectively. Thus, the set of eigenvalues of A is simply the union of the sets of eigenvalues of A_{11}, A_{22}, and A_{33}.

Earlier, in § 1.1, we have indicated that $C(G_c)$ can be obtained from the transpose of A by setting each of its nonzero elements to unity. The matrix $C(G_c)$ is also referred to as the *adjacency matrix* of G_c. It should be mentioned that the eigenvalues of $C(G_c)$ are not sufficient to characterize G_c to within an isomorphism. In other words, there exist two non-isomorphic directed graphs whose adjacency matrices have the same eigenvalues (Problem 3.19). However, it can be verified by exhaustive methods that the counterexamples do not exist for symmetric directed graphs up to six nodes. (See, for example, HARARY [1962a]). For a related treatment, see MOWSHOWITZ [1972].

§ 6. The matrix inversion

Since each element in the inverse of a matrix is the ratio of the cofactor of an element of the matrix and its determinant, the application of the directed-graph techniques to the matrix inversion is straightforward, and they are especially useful for large sparse matrices with elements given in literal rather than in numerical form such as frequently occur in engineering problems.

Earlier, in § 1.3 and § 2.3, we have defined the graph transmissions for both the Coates and Mason graphs associated with a system of linear algebraic equations; they are defined in terms of the ratios of a dependent and an independent variable called the source node. In this section, we shall extend this idea to Coates and Mason graphs associated with any matrix and then show how it can be used to invert a matrix.

Let A be a square matrix. Let $G_c^i(A)$ and $G_m^i(A)$ be the Coates and Mason graphs obtained from $G_c(A)$ and $G_m(A)$, respectively, by attaching a source node x_0 to $G_c(A)$ and $G_m(A)$, and then by adding an edge with transmittance -1 from node x_0 to node i in the resulting graphs. With this simple alteration of $G_c(A)$ and $G_m(A)$, the following corollaries are obvious.

COROLLARY 3.13: Let A be a nonsingular matrix. Then the (i,j)-element of the inverse of A is simply the graph transmission from node x_0 to node j in $G_c^i(A')$ or in $G_m^i(A')$, where the prime again indicates the matrix transpose.

COROLLARY 3.14: Let G_{cr} and G_{mr} be the reduced Coates and Mason graphs obtained from $G_c^i(A')$ and $G_m^i(A')$ by the removal of the node sets V_c and V_m by the processes outlined in (3.28) and (3.69), respectively. If the source node x_0 is not in V_c or V_m, then the (i, j)-element of the inverse of A is simply the graph transmission from node x_0 to node j in G_{cr} or G_{mr} for all j ($\neq x_0$) not in V_c or V_m.

A combination of the above two approaches is also possible. Let A be partitioned into the form

$$A = \begin{bmatrix} A_{11} & A_{12} \\ A_{21} & A_{22} \end{bmatrix} = [a_{ij}], \tag{3.136}$$

where A_{11} is nonsingular and of order β. The inverse of A is given by

$$A^{-1} = \begin{bmatrix} B_{11} & B_{12} \\ B_{21} & B_{22} \end{bmatrix} = [b_{ij}], \tag{3.137}$$

where

$$B_{11} = A_{11}^{-1} + A_{11}^{-1} A_{12} B_{22} A_{21} A_{11}^{-1}, \tag{3.138a}$$

$$B_{12} = -A_{11}^{-1} A_{12} B_{22}, \tag{3.138b}$$

$$B_{21} = -B_{22} A_{21} A_{11}^{-1}, \tag{3.138c}$$

$$B_{22} = (A_{22} - A_{21} A_{11}^{-1} A_{12})^{-1}. \tag{3.138d}$$

Let G_{cr} be the reduced Coates graph obtained from $G_c(A')$ by the removal of the node set V_c corresponding to the submatrix A_{11}. Also let c_{ij} and w_{ij} be the transmittances of the edge (i, j) in G_{cr} and $G_c(A')$, respectively. If G_c^* is the Coates graph obtained from $G_c(A')$ such that $w_{ij}^* = c_{ij}$ for both i and j not in V_c, and $w_{ij}^* = w_{ij}$ otherwise, where w_{ij}^* is the transmittance of the edge (i, j) in G_c^*, then the elements in B_{12} or B_{21} may be evaluated in G_c^*.

THEOREM 3.23: For b_{ij} in B_{12} or B_{21} of (3.137), we have

$$b_{ij} = (1/K) \sum_{H^*_{ij}} (-1)^{q^*_H - 1} f(H^*_{ij}), \tag{3.139}$$

where $K = \det A$; H_{ij}^* is a 1-factorial connection from i to j in G_c^*; and q_H^* is the number of even components in H_{ij}^*.

Proof. There are two cases to be considered: b_{ij} is in B_{21} and b_{ij} in B_{12}

Since the proofs are similar, the case where b_{ij} is in B_{21} will be shown here. Let A be of order n, and let

$$A_{uv} = [a_{ij}^{uv}], \quad B_{uv} = [b_{ij}^{uv}] \tag{3.140}$$

for $u, v = 1, 2$, where the subscripts i and j are the row and column indices of the elements. Also, let Δ_{ij}^{11} and Δ^{11} be the cofactor of the (i, j)-element of A_{11} and its determinant, respectively. From (3.138c) we obtain

$$b_{ij}^{21} = -\sum_{k_1, k_2} b_{ik_1}^{22} a_{k_1 k_2}^{21} \Delta_{jk_2}^{11} / \Delta^{11}, \tag{3.141}$$

where $k_1 = 1, 2, \ldots, n - \beta$ and $k_2 = 1, 2, \ldots, \beta$. Since from (3.34) and (3.138d) G_{cr} is the associated Coates graph of the transpose of $(B_{22})^{-1}$, it follows from Theorems 3.1 and 3.2 that

$$b_{ik_1}^{22} = \frac{\sum_{H'_{xy}} (-1)^{q'_H - 1} f(H'_{xy})}{\det G_{cr}}, \tag{3.142}$$

where $x = i + \beta$ and $y = k_1 + \beta$, and H'_{xy} is a 1-factorial connection from x to y in G_{cr}, and q'_H is the number of even components in H'_{xy}. Remember that G_{cr} and G_c^* are directed graphs obtained from the associated Coates graph of A' rather than A. Similarly, since the sectional subgraph $G_c[V_c]$ is the associated Coates graph of $(A_{11})'$, it follows again from Theorems 3.1 and 3.2 that

$$\frac{\Delta_{jk_2}^{11}}{\Delta^{11}} = \frac{\sum_{H''_{k_2 j}} (-1)^{q''_H - 1} f(H''_{k_2 j})}{\det A_{11}}, \tag{3.143}$$

where $H''_{k_2 j}$ is a 1-factorial connection from k_2 to j in $G_c[V_c]$, and q''_H is the number of even components in $H''_{k_2 j}$. Also, in G_c^* the weight associated with the edge (y, k_2) is a_{yk_2}, which is also the same as $a_{k_1 k_2}^{21}$,

$$w_{yk_2}^* = a_{yk_2} = a_{k_1 k_2}^{21}. \tag{3.144}$$

Substituting (3.142)–(3.144) into (3.141) and using the fact that

$$K = \det A = \det A' = \det G_c(A') = (\det A_{11})(\det G_{cr}), \tag{3.145}$$

we obtain the desired result (3.139). We leave the details of the last part of the proof as an exercise (Problem 3.38). This completes the proof of the theorem.

§6   The matrix inversion   213

Example 3.12: Let us compute the inverse of the matrix

$$A = \begin{bmatrix} 0 & 0 & 0 & 0 & b & k & 0 \\ 0 & 0 & 0 & 0 & 0 & f & 0 \\ 0 & 0 & 0 & j & 0 & 0 & d \\ 0 & e & 0 & 0 & 0 & 0 & 0 \\ 0 & 0 & c & i & 0 & 0 & 0 \\ 0 & 0 & 0 & g & 0 & h & 0 \\ a & 0 & 0 & 0 & 0 & 0 & 0 \end{bmatrix}. \tag{3.146}$$

The associated Coates graph $G_c(A')$ is presented in fig. 3.36. The elements b_{ij} in the inverse of A are obtained as follows:

$$b_{62} = (ge)(abcd)/(abcdefg) = 1/f,$$
$$b_{42} = -(e)(h)(abcd)/(abcdefg) = -h/fg,$$
$$b_{46} = (ef)(abcd)/(abcdefg) = 1/g,$$
$$b_{24} = (fg)(abcd)/(abcdefg) = 1/e,$$
$$b_{52} = -(cdakge)/(abcdefg) = -k/bf,$$
$$b_{32} = (dabie)(h)/(abcdefg) = hi/cfg,$$
$$b_{72} = (abcje)(h)/(abcdefg) = jh/dfg,$$
$$b_{36} = -(dabief)/(abcdefg) = -i/cg,$$
$$b_{76} = -(abcjef)/(abcdefg) = -j/dg,$$
$$b_{17} = (bcd)(efg)/(abcdefg) = 1/a,$$
$$b_{51} = (cda)(efg)/(abcdefg) = 1/b,$$
$$b_{35} = (dab)(efg)/(abcdefg) = 1/c,$$
$$b_{73} = (abc)(efg)/(abcdefg) = 1/d,$$

and all other elements are zero.

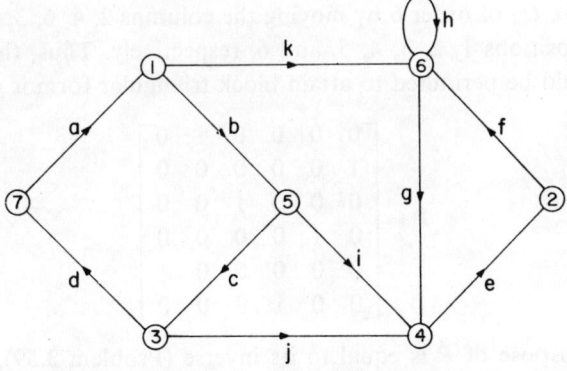

Fig. 3.36. A Coates graph illustrating the inversion of its associated matrix.

In many cases, a matrix A may be put in the form (3.47) by simultaneous interchange of the rows and columns of A. If this can be done, then the inversion formulas given in (3.138) can be greatly simplified. In other words, the problem is equivalent to that of finding a permutation matrix P such that PAP^{-1} is partitioned in block triangular form. In the following, we shall describe one of such procedures which was also considered by HARARY [1962b].

It can be shown that A has the property just described if, and only if, the associated Coates graph $G_c(A')$ or $G_c(A)$ is not strongly connected. (See Problem 3.35.) Assume that $G_c(A')$ is not strongly connected. Let us relabel the nodes of $G_c(A')$ in such a way that the nodes in any strongly-connected component are labeled in a consecutive order. Then the permutation matrix P can easily be obtained from the identity matrix simply by interchanging the columns corresponding to the new and old labelings of the nodes. As an illustration, consider the matrix

$$A = \begin{bmatrix} 11 & 0 & -3 & -4 & 11 & 0 \\ 0 & -1 & 1 & 0 & 0 & 0 \\ 0 & 0 & -1 & 0 & 0 & 1 \\ 3 & 5 & -1 & -1 & 2 & 2 \\ 3 & 2 & 0 & -1 & 3 & 1 \\ 0 & 0 & -2 & 0 & 0 & 3 \end{bmatrix}. \qquad (3.147)$$

The associated Coates graph $G_c(A')$ is as shown in fig. 3.37. The following table shows one possible way to relabel the nodes of $G_c(A')$ according to the above rule:

Old labelings   1   2   3   4   5   6
New labelings   2   4   6   3   1   5.

According to this table, the permutation matrix P can be obtained from the identity matrix U_6 of order 6 by moving the columns 2, 4, 6, 3, 1, and 5 of U_6 to the new positions 1, 2, 3, 4, 5, and 6, respectively. Thus, the matrix P by which A should be permuted to attain block triangular form is given by

$$P = \begin{bmatrix} 0 & 0 & 0 & 0 & 1 & 0 \\ 1 & 0 & 0 & 0 & 0 & 0 \\ 0 & 0 & 0 & 1 & 0 & 0 \\ 0 & 1 & 0 & 0 & 0 & 0 \\ 0 & 0 & 0 & 0 & 0 & 1 \\ 0 & 0 & 1 & 0 & 0 & 0 \end{bmatrix}. \qquad (3.148)$$

Since the transpose of P is equal to its inverse (Problem 3.59), the partition of PAP^{-1} using the strongly-connected components of $G_c(A')$ can easily be

§ 6   The matrix inversion   215

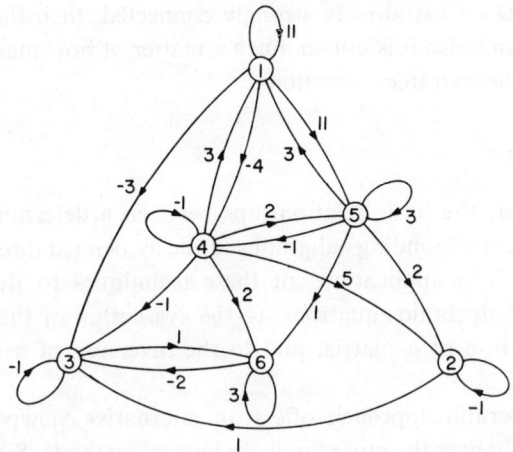

Fig. 3.37. The associated Coates graph of the transpose of the matrix (3.147).

accomplished, and is given by

$$PAP^{-1} = \begin{bmatrix} 3 & 3 & -1 & 2 & 1 & 0 \\ 11 & 11 & -4 & 0 & 0 & -3 \\ 2 & 3 & -1 & 5 & 2 & -1 \\ \hdashline 0 & 0 & 0 & -1 & 0 & 1 \\ 0 & 0 & 0 & 0 & 3 & -2 \\ 0 & 0 & 0 & 0 & 1 & -1 \end{bmatrix}. \quad (3.149)$$

Using the formulas given in (3.138), the inverse of PAP^{-1} is given by

$$(PAP^{-1})^{-1} = \begin{bmatrix} 1 & 0 & -1 & -3 & 3 & -8 \\ 3 & -1 & 1 & 11 & -18 & 49 \\ 11 & -3 & 0 & 22 & -42 & 115 \\ 0 & 0 & 0 & -1 & 1 & -3 \\ 0 & 0 & 0 & 0 & 1 & -2 \\ 0 & 0 & 0 & 0 & 1 & -3 \end{bmatrix}. \quad (3.150)$$

Hence, the inverse of A is

$$A^{-1} = P^{-1}(PAP^{-1})^{-1}P = \begin{bmatrix} -1 & 11 & 49 & 1 & 3 & -18 \\ 0 & -1 & -3 & 0 & 0 & 1 \\ 0 & 0 & -3 & 0 & 0 & 1 \\ -3 & 22 & 115 & 0 & 11 & -42 \\ 0 & -3 & -8 & -1 & 1 & 3 \\ 0 & 0 & -2 & 0 & 0 & 1 \end{bmatrix}. \quad (3.151)$$

Note that if $G_c(A')$ is already strongly connected, then the technique does not help at all, and also it is not so much a matter of how many zeros a matrix has but rather their strategic locations.

§ 7. Conclusions

In this chapter, the basic relationships between a determinant and the cofactors and the corresponding subgraphs in the associated directed graphs have been discussed. The applications of these techniques to the solutions of a system of linear algebraic equations, to the evaluation of the eigenvalues and eigenvalue equation of a matrix, and to the inversion of a matrix have also been presented.

The directed-graph approach offers an alternative viewpoint which complements and enhances the more familiar classical methods. Since this approach not only displays in a very intuitive manner the causal relationships among the elements of the matrices under study, but also provides a great insight into the properties of the systems being described by such matrices, it has been widely used as a tool for analysis of engineering problems. The methods are especially useful for large sparse matrices with elements given in literal rather than in numerical form such as frequently occur in engineering problems.

Methods for generating 1-factors, 1-factorial connections, semifactors, and k-semifactors have been introduced. The main advantage of these methods lies in the fact that once a desired set of sets, which can be obtained easily by inspection, is derived from a given directed graph all the remaining manipulations are algebraic. It is not necessary for one to go back to the original graph to see whether a generated term forms a desired subgraph. However, duplications necessarily occur, and thus they are not very efficient.

Problems

3.1. In a Coates graph G_c, let some of its edges (i, j) be replaced by the parallel edges $(i, j)_x$, $x = 1, 2, \ldots, k_{ij}$, such that (3.52) holds. Show that Theorems 3.1, 3.2, and 3.3 are still valid in the resulting Coates graph. Extend Corollaries 3.2 and 3.3 to include this case.
3.2. Repeat Problem 3.1 for the Mason graphs.
3.3. Prove Corollary 3.1.
3.4. For a Mason graph having p different directed circuits L_i, $i = 1, 2, \ldots, p$, show that the formula (3.53) may be written as

$$(-1)^n \det A = [1 - f(L_1)][1 - f(L_2)] \cdots [1 - f(L_p)] \qquad (3.152)$$

with the understanding that we shall drop terms containing weight products formed by non-node-disjoint directed circuits.

3.5. Show that the term c_{ij}, $i \neq j$, in (3.28) is actually the graph transmission from node i to node j in the sectional subgraph $G_c''[V_c^{ij}]$, and that c_{ii} is the graph transmission from node i' to node i'' in $G_c''[V_c^i]$ where $G_c''[V_c^{ij}]$ and $G_c''[V_c^i]$ are Coates graphs obtained from $G_c[V_c^{ij}]$ and $G_c[V_c^i]$, respectively, and are defined in § 1.4(A).

3.6. Prove Theorem 3.11. (*Hint*: Follow the strategy used to prove Theorem 3.5.)

3.7. If G_m is the Mason graph associated with a system of linear equations, show that we can remove the self-loop of transmittance $w_{kk} \neq 1$ at node x_k simply by multiplying the transmittance of every edge which terminates at node x_k by the factor $1/(1-w_{kk})$.

3.8. If G_c is the Coates graph associated with a system of linear equations, show that we can reduce the transmittance w_{kk} of the self-loop at node x_k to unity simply by dividing the transmittance of every edge terminating at node x_k by w_{kk}.

3.9. Justify the inversion rules stated in § 2.4(C) for a directed path and a directed circuit in a Mason graph.

3.10. Give simple rules for constructing both the Coates and Mason graphs associated with the nodal system of equations of a linear network directly by inspecting the network without the necessity of first writing out the nodal equations.

3.11. Repeat Problem 3.10 for the loop system of equations of a network.

3.12. Like Problem 3.1, if some of the edges of a modified Coates graph are replaced by the parallel edges such that (3.52) holds, show that Theorems 3.12, 3.13, and 3.14 are still valid in the resulting graph.

3.13. Repeat Problem 3.12 for Theorems 3.15, 3.16, and 3.17 for modified Mason graphs.

3.14. Repeat Problem 3.10 for a general ladder network without mutual couplings.

3.15. A *k-factorial connection* $H_{\alpha,\beta}$ of a Coates graph G_c, where α and β are disjoint subsets of the node set of G_c containing the same number of elements, is a subgraph of G_c which contains: (1) a set of k node-disjoint directed paths p_{ij} from i in α to j in β, where k is the number of elements in α or β, and (2) a set of node-disjoint directed circuits which include all the nodes of G_c except those contained in (1). If P is a set of k such directed paths $p_{i_x j_x}$, $x=1, 2, \ldots, k$ and $i_y < i_{y+1}$ for $y=1, 2, \ldots, k-1$, in $H_{\alpha,\beta}$, the array $j_1 j_2 \ldots j_k$ is said to be the *accessible permutation* of the integers of j's by P. Show that if $A_{\alpha,\beta}$ is the matrix obtained from the associated matrix A of G_c by striking out rows and columns contained in α and β, respectively, then

$$(-1)^s \det A_{\alpha,\beta} = \sum_{H_{\alpha,\beta}} (-1)^{q+\theta} f(H_{\alpha,\beta}), \tag{3.153}$$

where s is the sum of indices of the rows and columns contained in the sets α and β; q is the number of even components in $H_{\alpha,\beta}$; and θ is the number of inversions in the accessible permutation β' of the integers in β by P.

3.16. Extend the definition of a k-factorial connection also to include the case where $\alpha \cap \beta$ is not empty. Modify (3.153) accordingly (CHEN [1965a]).

3.17. Derive the formula (3.128).

3.18. For $i=j$, derive a formula similar to (3.128).

3.19. Give an example to show that two non-isomorphic directed graphs have the same eigenvalues for their adjacency matrices.

3.20. By replacing each undirected edge by a pair of oppositely directed edges, any undirected graph may be transformed into a directed one by this process. With this show that the values of the determinants of the adjacency matrices of tetrahedron, hexahedron, and octahedron are -3, 9, and 0, respectively (PÓLYA and SZEGÖ [1945]).

3.21. Simplify the formulas (3.127) and (3.128) for symmetric Coates graphs in which $f(i,j)=f(j,i)$ for all i and j.

3.22. Let A be the associated matrix of a symmetric Coates graph G_c in which $f(i,j)=f(j,i)=y_{ij}$ for all i and j. Let G_{cu} be the undirected graph obtained from G_c such that there is an edge between i and j in G_{cu} if and only if (i,j) and (j,i) are in G_c. Also, let the weight associated with the edge (i,j) of G_{cu} be y_{ij}. By a *linear subgraph* of G_{cu} we mean a spanning subgraph whose components are either edges or circuits of G_{cu}. Show that Theorem 3.1 may be reduced to the following:

$$\det A = \sum_i (-1)^{q_1+q_2} 2^{q_3} f(E_i) f(L_i), \tag{3.154}$$

where G_i is a linear subgraph of G_{cu}; E_i and L_i are the subgraphs of G_i consisting of all the single edges and circuits, respectively; q_1, q_2, and q_3 are the numbers of edges in E_i, the even components in L_i, and the circuits of length greater than 2 in L_i, respectively; and the summation is taken over all possible G_i in G_{cu} (HARARY [1962a]).

3.23. Repeat Problem 3.22 for the cofactors of the elements of A.

3.24. Show that the determinant of the adjacency matrix of a symmetric

Coates graph defined in Problem 3.22 is equal to

$$\sum_{i=1}^{k} (-1)^{q_1+q_2} 2^{q_3}, \qquad (3.155)$$

where k is the number of linear subgraphs of G_{cu}.

3.25. Repeat Problem 3.19 for symmetric directed graphs.
3.26. Let G_{cr} be the reduced Coates graph of the Coates graph G_c discussed in Problem 3.15 by the removal of the node set V_c. If α and β are in $V-V_c$, where V is the node set of G_c, show that the minor det $A_{\alpha,\beta}$ evaluated in G_c is equal to K_c times the corresponding minor evaluated in G_{cr}, where $K_c = \det G_c[V_c]$.
3.27. Applying the results of Problems 3.15 and 3.26, show that the ratio of the determinant of A and one of its minors or the ratio of two of its minors remains the same in both G_c and G_{cr} for all α and β in $V-V_c$.
3.28. For a Coates graph G_c, there is an equivalent representation as a bipartite graph B_c: to the node set V of G_c, we construct a replica V' which is in a one-to-one correspondence with V. The (undirected) edge (i, j') or (j', i), connected between the nodes i and j' where i is in V and j' in V', is in B_c if, and only if, there exists an edge (i,j) in G_c such that $f(i,j')$ or $f(j', i) = f(i,j)$. Show that a spanning subgraph h_n of G_c is a regular subgraph of degree n (called an *n-factor*) if, and only if, the corresponding subgraph h'_n of h_n in B_c is a regular subgraph of degree n (CHEN [1965c]).
3.29. In Problem 3.28, show that a spanning subgraph h_2 of G_c is a 2-factor of G_c if, and only if, the corresponding subgraph h'_2 of h_2 in B_c is a set of node-disjoint circuits which include all the nodes of B_c.
3.30. A *directed bipartite graph* B_d is a bipartite graph in which the edges are oriented. Let

$$A = [a_{ij}]_{n \times p} \quad \text{and} \quad C = [c_{ij}]_{p \times n}, \, p \geq n, \qquad (3.156)$$

be two given matrices, and let B_d be their associated directed bipartite graph in which if V_1 and V_2 are the node sets of B_d, then
1. V_1 has n nodes and V_2 has p nodes;
2. there is a directed edge (i,j), i in V_1 and j in V_2, in B_d with $f(i,j) = a_{ij}$ if and only if $a_{ij} \neq 0$;
3. there is a directed edge (i,j), i in V_2 and j in V_1, in B_d with $f(i,j) = c_{ij}$ if and only if $c_{ij} \neq 0$.

Show that

$$\det AC = (-1)^n \sum_{h_d} (-1)^L f(h_d), \qquad (3.157)$$

where h_d is a subgraph consisting of a set of node-disjoint directed circuits of B_d which contains all the nodes of V_1, and L is the number of directed circuits in h_d (CHEN [1965c]).

3.31. In Problem 3.30, show that if $p=n$, then

$$\det AC = (-1)^n \sum_h (-1)^{L_h} f(h), \tag{3.158}$$

where h is a 1-factor of B_d, and L_h is the number of directed circuits in h.

3.32. Evaluate the determinant and the cofactor of the (3, 1)-element of the matrix given in (3.133) by both the Coates-graph and the Mason-graph techniques.

3.33. Repeat Problem 3.32, but remove the nodes 6, 7, and 10 first.

3.34. Solve the system of linear equations

$$\begin{bmatrix} a_{11} & a_{12} & a_{13} & 0 \\ 0 & a_{22} & 0 & a_{24} \\ a_{31} & 0 & a_{33} & 0 \\ 0 & a_{42} & a_{43} & a_{44} \end{bmatrix} \begin{bmatrix} x_1 \\ x_2 \\ x_3 \\ x_4 \end{bmatrix} = \begin{bmatrix} b_1 \\ 0 \\ b_3 \\ 0 \end{bmatrix} \tag{3.159}$$

for x_3 and x_4 (a) by Cramer's rule, (b) by Coates-graph technique, (c) by Mason-graph technique, and (d) by Coates and Mason graph techniques, using reduction and transformation procedures.

3.35. In matrix theory we say a matrix A is *reducible* if it can be transformed into the form (3.47) by simultaneous interchange of the rows and columns; otherwise A is *irreducible*. If in addition A_{12} and A_{21} are both zero, then A is said to be *decomposable*; otherwise A is *indecomposable*. Show that the matrix A is irreducible if and only if its associated Coates graph $G_c(A)$ is strongly connected, and A is indecomposable if and only if $G_c(A)$ is connected (HARARY [1962b]).

3.36. Show that the product of the determinants of the matrices associated with the strongly-connected components of a Coates graph $G_c(A)$ is equal to $\det A$.

3.37. Prove Corollaries 3.13 and 3.14.

3.38. Supply the detailed proof of Theorem 3.23.

3.39. Partition the matrix A (3.146) into block triangular form by the associated Coates graph $G_c(A)$.

3.40. Let G_c be the associated Coates graph of a system of linear equations of (3.16), i.e., $G_c = G_c(A_u)$. Also let G_c^* be the Coates graph obtained from G_c such that for each edge (i, j) of G_c^*,

$$f(i, j) = 0 \tag{3.160a}$$

for $i \leq j$, and

$$f(i, j) = \frac{\sum_{H^*_{ij}} (-1)^{L^*_H - 1} f(H^*_{ij})}{\sum_{h^*} (-1)^{L^*_h} f(h^*)} \qquad (3.160b)$$

for $i > j$, where H^*_{ij} is a 1-factorial connection from i to j in the sectional subgraph $G_c[V_{ji}]$; V_{ji} is the set union of the nodes $1, 2, \ldots, j$ and i; h^* is a 1-factor in $G_c[V_{jj}]$; and L^*_H and L^*_h are the numbers of directed circuits in H^*_{ij} and h^*, respectively. Modify $G_c[V_{ji}]$ so that (3.160b) can be stated as a graph transmission in the modified graph. If the determinants of the matrices associated with the sectional subgraphs $G_c[V_{jj}]$ are nonzero, show that the solution of (3.16) is given by

$$x_k = \sum_{P_{sk}} (-1)^{b_P} f(P_{sk}) \qquad (3.161)$$

for $k = 1, 2, \ldots, n$, where $s = n+1$; P_{sk} is a directed path from node s to node k in G_c^*; and b_P is the number of edges in P_{sk}. (*Hint*: Use (3.22) and the Gaussian-elimination technique.)

3.41. Using (3.161), solve the following system of linear equations:

$$\begin{bmatrix} 1 & 1 & 1 & 1 \\ 2 & 1 & 2 & 1 \\ 2 & 2 & -1 & 2 \\ 1 & 2 & 3 & 3 \end{bmatrix} \begin{bmatrix} x_1 \\ x_2 \\ x_3 \\ x_4 \end{bmatrix} = \begin{bmatrix} 1 \\ 1 \\ 5 \\ 3 \end{bmatrix}. \qquad (3.162)$$

3.42. Prove Theorem 3.16. (*Hint*: Follow the proof of Theorem 3.15.)

3.43. Give a simple physical interpretation on the effect of a self-loop at some node upon the path transmission through that node in a Mason graph. If the absolute value of the transmittance of a self-loop (i, i) is less than unity, show that the effect of (i, i) upon the outgoing signals of the node i is equivalent to changing the signal at each of the outgoing edges to the infinite sum of signals which circulate zero, once, twice, ... around the self-loop before proceeding onward. (*Hint*: Expand the reciprocal of $(1 - w_{ii})$ in infinite geometric series.)

3.44. By attaching an additional node x'_k to a Mason graph G_m and then by adding an edge of transmittance 1 from node x_k to node x'_k, show that the graph transmission from the source node x_0 to x_k in G_m is the same as that from x_0 to x'_k in the new Mason graph G^*_m. If the directions of all the edges in G^*_m are reversed, show that the graph transmission from x'_k to x_0 in the resulting graph is the same as that from x_0 to x_k in G^*_m.

3.45. Modify the topological formulas for both the Coates graph and the Mason graph when all the self-loops are deleted and when each of their nodes is assigned the weight of the self-loop at that node.
3.46. Prove Theorem 3.9. (*Hint*: Follow the proof of Theorem 3.3.)
3.47. Prove Corollary 3.7.
3.48. Justify the following procedure for removing a node in a Coates graph. Let k be the node with a self-loop of transmittance w_{kk}. Divide all the transmittances of the edges having their terminal node k by $-w_{kk}$. Delete the self-loop (k, k). Then remove the node k as in the Mason graph. The operation leaves the graph transmissions in the resulting Coates graph unaltered.
3.49. Let G_c be the Coates graph associated with a system of equations of (3.16). If G_c has a self-loop of transmittance w_{kk} at each of its nodes k, show that if all the transmittances of the edges terminated at each nonsource node k are divided by $-w_{kk}$, and then the self-loops are removed, the resulting graph is the Mason graph associated with an equivalent system of (3.16).
3.50. Extend Theorems 3.18 and 3.19 to directed graphs containing parallel edges.
3.51. Extend Theorem 3.20 to symmetric directed graphs containing parallel edges, and also to any directed graphs.
3.52. Prove Corollary 3.3.
3.53. Prove Corollaries 3.8 and 3.9.
3.54. Give a procedure similar to that given in (3.48) for decomposing a Coates graph G_c into components if there is a nonempty proper subset V_c of the node set which has zero converse deficiency.
3.55. Derive a formula similar to that of (3.43) or (3.44) for $i=j$.
3.56. Find the eigenvalues of the matrix A of (3.133).
3.57. Prove Lemma 3.1.
3.58. Derive an identity similar to that of (3.76) for $i=j$.
3.59. Show that the inverse of a permutation matrix is equal to its transpose.
3.60. Consider the matrix $C(G_c)$ of Corollary 3.2. Show that the permanent of $C(G_c)$ is equal to its determinant if, and only if, the number of even components in every 1-factor of G_c is even. Give a similar statement for the cofactors of the elements of $C(G_c)$.
3.61. In Problem 3.60, show that the permanent of $C(G_c)$ is equal to its determinant if, and only if, (1) within each strongly-connected component of G_c the numbers of even components of the 1-factors must all be even or odd, and (2) the number of strongly-connected components, each of

which contains at least one 1-factor that has an odd number of even components, is even. Also show that the statement is similarly valid for any real matrix with nonnegative entries. Is this statement valid for any real matrix? If not, give a counterexample.

3.62. Show that the permanents of matrices possess the following properties:
(a) the permanent is a multilinear function of the rows and columns;
(b) the permanent of a matrix is equal to the permanent of its transpose;
(c) if P_1 and P_2 are permutation matrices of appropriate order, then

$$\text{per } P_1 A P_2 = \text{per } A; \qquad (3.163a)$$

(d) if D_1 and D_2 are diagonal matrices of appropriate order, then

$$\text{per } D_1 A D_2 = (\text{per } D_1)(\text{per } A)(\text{per } D_2). \qquad (3.163b)$$

3.63. Assume that all the directed circuits of an n-node Coates graph $G_c(A)$ are of even length. Show that if n is odd, then the matrix A is singular, and for all n if λ is an eigenvalue of A, so is $-\lambda$.

3.64. Let G_c be a symmetric Coates graph which is regular of degree k and which has no parallel edges and self-loops. Assume that all the transmittances of G_c are unity. Show that for $k=2, 3$, and 7, there exist unique Coates graphs G_c (within an isomorphism) having (k^2+1) nodes.

3.65. In Problem 3.64, let A be the associated matrix of G_c. Show that A satisfies the equation

$$A^2 + A - (k-1) U_n = E \qquad (3.164)$$

for $k=2, 3$, and 7 where E is the matrix of order n whose entries are all unity, and n is the number of nodes of G_c.

CHAPTER 4

TOPOLOGICAL ANALYSIS OF LINEAR SYSTEMS

In the preceding chapter, we have shown how to obtain the solutions of a system of simultaneous linear algebraic equations by considering the properties of the associated Coates graph or the Mason graph. However, as mentioned in ch. 3, § 3, the derived formulas are not very efficient for the evaluation of system determinants and cofactors because of the existence of a large number of cancellation terms. Thus, they suffer the same drawback as the conventional methods of analysis such as the pivotal or Gaussian elimination technique. On the other hand, in dealing with large sparse matrices with elements given in *literal* rather than in numerical form such as frequently occur in engineering problems, the directed-graph technique provides an effective solution to the problem, since it is usually necessary to compute a symbolic system function.

In the present chapter, we are mainly concerned with a special class of system matrices known as the *node-admittance matrices*. It arises when the nodal system of equations is used in analyzing an electrical network as discussed in ch. 2. The approach to be presented in this chapter is similar to that of the modified Coates graph presented in the preceding chapter, but in a much more convenient form. Although these matrices are closely related to the theory of electrical networks, their applications can be extended to other systems as well. For example, there is a complete analogy insofar as system equations are concerned between a linear lumped mechanical system and its electrical counterpart. For detailed discussions of these analogies, the reader is referred to GARDNER and BARNES [1945]. Because of the special feature of these matrices, their determinants are not computationally difficult to evaluate. The primary objective of this chapter is to present formulas for writing down certain classes of functions (driving-point and transfer functions) by inspection of the system diagram or its associated directed graphs without actually expanding various determinants and cofactors. They are referred to as the *topological formulas* of the system. In ch. 2, § 6.3, we have presented topological formulas for *RLC* two-port networks. The reason for the popularity and usefulness of the topological analysis of linear systems is that it not only displays in a very intuitive manner the causal relationships among the several variables of the system under study, but it also provides a short-cut in evaluating system functions, because the usual cancellations inherent in the evaluation of system determinants and cofactors are avoided. For example, the formulas have been used rather effectively in the study of the order of complexity of electrical networks by BRYANT [1959a, b and c]. A recent resurgence of interest in topological concepts, on the part of engineers in general and circuit theorists in particular, is due mainly to this fact.

Like many other topics discussed in this book, the basic concepts of topological formulas are not new. More than a century ago, the basic formulas for the loop system of equations were stated by KIRCHHOFF [1847]. Some forty years later, MAXWELL [1892] reviewed Kirchhoff's rules, and presented the basic formulas for the nodal system of equations. Although the rules are fundamentally correct, they were far from being complete. Using some of the methods of modern algebra, FRANKLIN [1925] gave a proof of Kirchhoff's impedance law. Following Franklin, KU [1952] contributed to the theory by giving a useful interpretation of Maxwell's rules. All these formulas were stated only for the resistive networks. Obviously, they can be applied to ordinary RLC networks as well. However, the extension of the formulas to active networks is relatively recent (PERCIVAL [1955]).

In this chapter, we shall present the fundamental relationships between the terms in the expansions of the determinant of a node-admittance matrix and the cofactors of its elements and certain types of subgraphs in the associated directed graphs of the system, and then apply these results to the evaluation of system functions. Our entire development is primarily based on the work of MASON [1957] and CHEN [1965d, 1967b].

§ 1. The equicofactor matrix

In the analysis of linear systems, there arises a class of matrices known as the "equicofactor matrices" because all the (first-order) cofactors are equal. In this section, we shall first examine some of the fundamental properties of an equicofactor matrix, and then show how its "second-order cofactors" are related.

For a square matrix A, denote by A_{ij} the submatrix obtained from A by deleting the ith row and the jth column, and by A_{ij} the cofactor of the (i, j)-element of A. In other words, we have $A_{ij}=(-1)^{i+j} \det A_{ij}$. These symbols will be used throughout this chapter.

DEFINITION 4.1: *Equicofactor matrix*. A square matrix is said to be an *equi-*

cofactor matrix if it has the properties that the sum of the elements of every row and every column equals zero.

For example, the indefinite-admittance matrix of a linear electrical network is an equicofactor matrix. This property is a direct consequence of Kirchhoff's current law coupled with the fact that these currents are invariant to a change of all nodal potentials by the same amount.

LEMMA 4.1: If A is a square matrix with the property that the sum of the elements of every row equals zero, then the cofactors of the elements of every row of A are equal.

Proof. Let $A = [a_{ij}]$ be a matrix of order n. We shall show that

$$A_{ix} = A_{iy} \qquad (4.1)$$

for all i, x, and y. Without loss of generality, let us assume that $x > y$. Since $\sum_{k=1}^{n} a_{ik} = 0$ for $i = 1, 2, \ldots, n$, we may replace the elements a_{jy} ($j = 1, 2, \ldots, n$) in A_{ix} by

$$-\sum_{\substack{k=1 \\ k \neq y}}^{n} a_{jk} \qquad (4.2)$$

without changing the value of A_{ix}. Let the submatrix thus obtained be denoted by A''_{ix}. Now we add all the columns of A''_{ix} to the column y of A''_{ix}, and then shift column y to the right of column $x-1$ if $y \neq x-1$. It follows that $\det A''_{ix}$ is equal to $(-1)^{x-y-1}(-1) M_{iy}$, where M_{iy} is the minor of A obtained by striking out the ith row and the yth column from A. Thus,

$$A_{ix} = (-1)^{i+x}(-1)^{x-y-1}(-1) M_{iy} = A_{iy}, \qquad (4.3)$$

which completes the proof of the lemma.

THEOREM 4.1: If A is an equicofactor matrix, then all the cofactors of the elements of A are equal. (Also see Problem 4.69.)

Proof. Since A is an equicofactor matrix, it follows from Lemma 4.1 that $A_{ix} = A_{iy}$ for all i, x, and y. Next, let us consider the transpose A' of A. Since A' is an equicofactor matrix, it follows that $A_{xi} = A_{yi}$ for all i, x, and y. Thus, $A_{ij} = A_{xy}$ for all i, j, x, and y. This completes the proof of the theorem.

As an illustration, consider the following matrix which is the indefinite-admittance matrix of an amplifier sometimes called a Miller integrator:

$$A = \begin{bmatrix} G_1 & -G_1 & 0 & 0 \\ -G_1 & G_1 + y_c & -y_c & 0 \\ 0 & g_m - y_c & G_2 + y_c & -(g_m + G_2) \\ 0 & -g_m & -G_2 & g_m + G_2 \end{bmatrix}. \qquad (4.4)$$

It is easy to check that the matrix A is an equicofactor matrix, and thus that all its cofactors are equal. For illustrative purposes, let us compute the cofactors of the elements a_{44} and a_{41} of A:

$$A_{44} = (-1)^{4+4} [G_1 (G_1 + y_c)(G_2 + y_c) + y_c(g_m - y_c) G_1 - G_1 G_1 (G_2 + y_c)]$$
$$= G_1 y_c (G_2 + g_m), \qquad (4.5)$$

$$A_{41} = (-1)^{4+1} [-G_1 y_c (g_m + G_2)] = G_1 y_c (g_m + G_2). \qquad (4.6)$$

This important property of an equicofactor matrix is known for the indefinite-admittance matrix of a linear network. JEANS [1925] stated this result for cofactors of the elements on the main diagonal. BARABASCHI and GATTI [1954] and PERCIVAL [1954] proved the general case.

COROLLARY 4.1: If A is a square matrix with the property that the sum of the elements of every row equals zero, then the matrix obtained by multiplying the ith row of A by A_{ij} for all i is an equicofactor matrix.

The proof of this corollary is left as an exercise (Problem 4.1).

Now we shall show that not all the "second-order cofactors" obtained from an equicofactor matrix are independent.

DEFINITION 4.2: *Second-order cofactor.* For a square matrix $A = [a_{ij}]$, the *second-order cofactor*, denoted by $A_{pq,rs}$, of the elements a_{pq} and a_{rs} of A is the determinant of the submatrix $A_{pq,rs}$ obtained from A by striking out the rows p and r and the columns q and s, and then prefixing the sign

$$\operatorname{sgn}(p-r) \operatorname{sgn}(q-s) (-1)^{p+q+r+s}, \qquad (4.7)$$

where $p \neq r$ and $q \neq s$; and $\operatorname{sgn} w = +1$ if $w > 0$, and $\operatorname{sgn} w = -1$ if $w < 0$.

Consider the matrix A given in (4.4). The second-order cofactor $A_{12,34}$ of the elements a_{12} and a_{34} of A is given by

$$A_{12,34} = \operatorname{sgn}(1-3) \operatorname{sgn}(2-4) (-1)^{1+2+3+4} \begin{vmatrix} -G_1 & -y_c \\ 0 & -G_2 \end{vmatrix}$$
$$= G_1 G_2. \qquad (4.8)$$

LEMMA 4.2: If A is a square matrix with the property that the sum of the elements of every row equals zero, then

$$A_{pq,rs} = A_{pq,rv} - A_{ps,rv} \qquad (4.9)$$

for all p, q, r, s, and v, where $p \neq r$ and $q \neq s \neq v$.

Proof. Let $A_{pq,rs}$ be the submatrix of A obtained by deleting the rows p

and r and the columns q and s. Also let $M_{pq,rs} = \det A_{pq,rs}$. In $A_{pq,rs}$, replace the column corresponding to the vth column of A by the sum of all the columns of $A_{pq,rs}$. In virtue of $\sum_{k=1}^{n} a_{ik} = 0$, the element of this column in the ith row is now $-(a_{iq} + a_{is})$. If we remove the -1 from this column and let the resulting matrix be denoted by $A''_{pq,rs}$, then $M_{pq,rs} = -\det A''_{pq,rs}$. In $A''_{pq,rs}$, let A''_1 and A''_2 be the matrices obtained from $A''_{pq,rs}$ by setting $a_{is} = 0$ and $a_{iq} = 0$ ($i = 1, 2, ..., n$), respectively. It is obvious that we have $\det A''_{pq,rs} = \det A''_1 + \det A''_2$.

At this stage the numerical order in which the columns q, s, and v occur is very relevant. Apart from a sign, $\det A''_1$ and $\det A''_2$ are the second-order cofactors $A_{ps,rv}$ and $A_{pq,rv}$, respectively.

In A''_1, let us shift the column containing the elements a_{iq} to the position, so that the columns in the resulting matrix will appear in the same relative order as those in A. In doing so, it requires $(v-q-1)$ adjacent transpositions for the cases $s < q < v$ and $q < v < s$; $(q-v-1)$ for $s < v < q$ and $v < q < s$; $(v-q-2)$ for $q < s < v$; and $(q-v-2)$ for $v < s < q$. It is not difficult to check that all the cases are clearly contained in the single equation,

$$\det A''_1 = \operatorname{sgn}(q-s) \operatorname{sgn}(v-s) (-1)^{v-q-1} M_{ps,rv}. \tag{4.10}$$

Thus, we have

$$\begin{aligned} A_{ps,rv} &= \operatorname{sgn}(p-r) \operatorname{sgn}(s-v) (-1)^{p+s+r+v} M_{ps,rv} \\ &= \operatorname{sgn}(p-r) \operatorname{sgn}(q-s) (-1)^{p+q+r+s} \det A''_1. \end{aligned} \tag{4.11}$$

In the same way, we obtain an identity similar to that of (4.11) for A''_2:

$$A_{pq,rv} = \operatorname{sgn}(p-r) \operatorname{sgn}(s-q) (-1)^{p+s+r+q} \det A''_2. \tag{4.12}$$

Combining these results, we have

$$\begin{aligned} A_{pq,rs} &= \operatorname{sgn}(p-r) \operatorname{sgn}(q-s) (-1)^{p+q+r+s} M_{pq,rs} \\ &= \operatorname{sgn}(p-r) \operatorname{sgn}(q-s) (-1)^{p+q+r+s-1} (\det A''_1 + \det A''_2) \\ &= A_{pq,rv} - A_{ps,rv}. \end{aligned} \tag{4.13}$$

So the lemma is proved.

THEOREM 4.2: If A is a square matrix with the property that the sum of the elements of every row equals zero, then

$$A_{ux} A_{pq,rs} = A_{ry} A_{pq,uv} + A_{pz} A_{rs,uv} - A_{ry} A_{ps,uv} - A_{pz} A_{rq,uv} \tag{4.14}$$

for all positive integers p, q, r, s, u, v, x, y and z, where $p \neq u \neq r$ and $q \neq v \neq s$.

Proof. Let A'' be the matrix obtained from A by multiplying the ith row of

A by A_{ij} ($i=1, 2, ..., n$) for some $j \leq n$, where n is the order of the matrix A. From Corollary 4.1, A'' is an equicofactor matrix.

Consider the transpose of A''. By Lemma 4.2, the second-order cofactors $A''_{pq,rs}$ of the elements of A'' are related by

$$A''_{pq,rs} = A''_{pq,vs} - A''_{rq,vs}. \tag{4.15}$$

Since

$$A''_{pq,rs} = \left(\prod_{\substack{k=1 \\ k \neq p, r}}^{n} A_{kt_k} \right) A_{pq,rs}, \tag{4.16}$$

where t_k are positive integers less than $n+1$, after a simple substitution we have

$$A_{vt_v} A_{pq,rs} = A_{rt_r} A_{pq,vs} - A_{pt_p} A_{rq,vs}$$

or

$$A_{ux} A_{pq,rs} = A_{ry} A_{pq,us} - A_{pz} A_{rq,us}. \tag{4.17}$$

Using (4.17) in conjunction with (4.9), we obtain the desired result. This completes the proof of the theorem.

COROLLARY 4.2: If A is an equicofactor matrix, then

$$A_{pq,rs} = A_{pq,uv} + A_{rs,uv} - A_{ps,uv} - A_{rq,uv}. \tag{4.18}$$

The identity given in (4.18) is also referred to as Jeans' theorem by SHARPE and SPAIN [1960] because the "genesis" of the formula is found in JEANS [1925]. It is significant to note that, in virtue of (4.18), all the second-order cofactors of the elements of A can be obtained linearly in terms of the (first-order) cofactors of the elements of A_{nn}. Hence there are only $(n-1)^2$ independent second-order cofactors for A. The chief implication of this result is that, in the analysis of linear networks, there is a perfect duality between the admittance and impedance descriptions of networks, whatever be their complexity. For detailed treatment on this, the reader is referred to SHARPE and SPAIN [1960].

As an illustration, let us again consider the equicofactor matrix A of (4.4). The following second-order cofactors are computed:

$$\begin{aligned} A_{12,44} &= G_1 (G_2 + y_c), \\ A_{33,44} &= G_1 y_c, \\ A_{13,44} &= -G_1 (g_m - y_c), \\ A_{32,44} &= G_1 y_c, \\ A_{12,33} &= G_1 (g_m + G_2). \end{aligned} \tag{4.19}$$

Thus, from (4.18) we have

$$A_{12,33} = A_{12,44} + A_{33,44} - A_{13,44} - A_{32,44}. \tag{4.20}$$

§ 2. The associated directed graph

In the analysis of linear systems, equicofactor matrix of the following type is obtained (see Problem 4.68):

$$Y = \begin{bmatrix} \sum_{k=2}^{n} y_{1k} & -y_{12} & -y_{13} & \cdots & -y_{1n} \\ -y_{21} & \sum_{\substack{k=1 \\ k \neq 2}}^{n} y_{2k} & -y_{23} & \cdots & -y_{2n} \\ \vdots & \vdots & \vdots & \vdots & \vdots \\ -y_{n1} & -y_{n2} & -y_{n3} & \cdots & \sum_{\substack{k=1 \\ k \neq n}}^{n} y_{nk} \end{bmatrix}. \tag{4.21}$$

Y is also referred to as the *indefinite-admittance matrix* because it arises when the nodal system of equations is used in analyzing a linear network. The submatrix Y_{ii} is known as the *node-admittance matrix* with reference node i, the reason being that the node i in the corresponding network is chosen as the reference potential. (See ch. 2, § 3.3 and also Problem 2.21.)

In the expanded form the cofactor Y_{ij} of each element of Y has all its terms positive, and on that account such determinants or cofactors have been termed *unisignants* by SYLVESTER [1855]. By putting $y_{ij}=1$ for all i and j, he also showed that the maximum number of positive terms in the final expansions of these cofactors is n^{n-2}. It is apparent that the expansion of these cofactors by conventional procedures is an extremely time-consuming operation because the maximum number of terms due to the product of the principal diagonal elements alone is $(n-1)^{n-1}$. Furthermore, as remarked earlier, the elements of Y are usually given in *literal* rather than in numerical form, so that they cannot be combined in advance. For a fourth-order matrix Y with all elements present, for example, the cofactor Y_{44} of the (4, 4)-element of Y has the following form:

$$Y_{44} = \begin{vmatrix} y_{12}+y_{13}+y_{14} & -y_{12} & -y_{13} \\ -y_{21} & y_{21}+y_{23}+y_{24} & -y_{23} \\ -y_{31} & -y_{32} & y_{31}+y_{32}+y_{34} \end{vmatrix}. \tag{4.22}$$

The total number of terms obtained by usual expansion before cancellation would be 38, of which 22 would be canceling one another:

$$\begin{aligned}
Y_{44} &= y_{12}y_{21}y_{31} + y_{12}y_{21}y_{32} + y_{12}y_{21}y_{34} + y_{12}y_{23}y_{31} + y_{12}y_{23}y_{32} \\
&\quad + y_{12}y_{23}y_{34} + y_{12}y_{24}y_{31} + y_{12}y_{24}y_{32} + y_{12}y_{24}y_{34} + y_{13}y_{21}y_{31} \\
&\quad + y_{13}y_{21}y_{32} + y_{13}y_{21}y_{34} + y_{13}y_{23}y_{31} + y_{13}y_{23}y_{32} + y_{13}y_{23}y_{34} \\
&\quad + y_{13}y_{24}y_{31} + y_{13}y_{24}y_{32} + y_{13}y_{24}y_{34} + y_{14}y_{21}y_{31} + y_{14}y_{21}y_{32} \\
&\quad + y_{14}y_{21}y_{34} + y_{14}y_{23}y_{31} + y_{14}y_{23}y_{32} + y_{14}y_{23}y_{34} + y_{14}y_{24}y_{31} \\
&\quad + y_{14}y_{24}y_{32} + y_{14}y_{24}y_{34} - y_{21}y_{32}y_{13} - y_{12}y_{23}y_{31} - y_{13}y_{31}y_{21} \\
&\quad - y_{13}y_{31}y_{23} - y_{13}y_{31}y_{24} - y_{12}y_{21}y_{31} - y_{12}y_{21}y_{32} - y_{12}y_{21}y_{34} \\
&\quad - y_{23}y_{32}y_{12} - y_{23}y_{32}y_{13} - y_{23}y_{32}y_{14} \\
&= y_{12}y_{23}y_{34} + y_{12}y_{24}y_{31} + y_{12}y_{24}y_{32} + y_{12}y_{24}y_{34} + y_{13}y_{21}y_{34} \\
&\quad + y_{13}y_{23}y_{34} + y_{13}y_{32}y_{24} + y_{13}y_{24}y_{34} + y_{14}y_{21}y_{31} + y_{14}y_{21}y_{32} \\
&\quad + y_{14}y_{21}y_{34} + y_{14}y_{23}y_{31} + y_{14}y_{23}y_{34} + y_{14}y_{24}y_{31} + y_{14}y_{24}y_{32} \\
&\quad + y_{14}y_{24}y_{34}. \quad (4.23)
\end{aligned}$$

It was pointed out by Sylvester that the terms in the expansion of a symmetric unisignant are simply the "tree products". Sylvester's work was contemporaneous with that of KIRCHHOFF [1847], but he did not seem to be aware of the connection with the determinant of the node-admittance matrix of an *RLC* network which as discussed in ch. 2, § 6 is simply the tree-admittance products. Recently, CHEN [1965d] generalized the works of Sylvester and Kirchhoff and showed that the terms in the expansion of an asymmetric unisignant are simply the "directed-tree products" of the associated directed graph of *Y*. When the directed graph is symmetric, it reduces to the case of Sylvester and Kirchhoff.

In this section, we shall examine the close relationships between the terms in the expansions of the first- and second-order cofactors of the elements of a matrix of the type as shown in (4.21) and certain types of subgraphs in the associated directed graph of the matrix. We shall also show how these subgraphs are related to the subgraphs of its corresponding Coates graph discussed in the preceding chapter. Finally, the application of these techniques to the solution of network problems will be demonstrated.

2.1. *Directed-trees and first-order cofactors*

The basic idea of the correspondence between the number of terms in the expansion of a unisignant and the number of certain type of subgraphs called "directed trees" in the associated directed graph was first exploited by TUTTE [1948] in his study of dissection of equilateral triangles into equilateral triangles. BOTT and MAYBERRY [1954] applied this result to the analysis of economic activities. Their work essentially remained unknown to the engineers until a similar procedure was independently developed by CHEN [1965d] for the analysis of general electrical networks. Chen's approach is more general and systematic, and the development given in this section will follow closely that of CHEN [1965d, 1967b] but it will be presented in a different form.

DEFINITION 4.3: *Associated directed graph.* For an equicofactor matrix Y of (4.21), the *associated directed graph*, denoted by the symbol $G(Y)$ or simply G if Y is clearly understood or is not explicitly given, is an n-node, weighted labeled, directed graph. The nodes are labeled by the integers from 1 to n such that if $y_{ij} \neq 0$, $i \neq j$, there is an edge directed from node i to node j with associated weight y_{ij} for $i, j = 1, 2, \ldots, n$.

It should be pointed out that, in contrast with the Coates and Mason graphs discussed in the preceding chapter, the diagonal elements of Y have no direct effect on $G(Y)$. The definition shows how one goes from the matrix to the associated directed graph. The reverse process, to obtain the *associated equicofactor matrix*, denoted by $Y(G)$, from a given G is very simple and straightforward. As an illustration, consider the equicofactor matrix

$$Y = \begin{bmatrix} G_b + g_b & -g_b & -G_b & 0 \\ -g_b & g_b + g_e + g_c - ag_e & -g_c & -g_e + ag_e \\ -G_b & -g_c + ag_e & G_b + g_c + G_d & -G_d - ag_e \\ 0 & -g_e & -G_d & G_d + g_e \end{bmatrix}, \quad (4.24)$$

which is the indefinite-admittance matrix of a transistor feedback amplifier as shown in fig. 4.1. The associated directed graph $G(Y)$ of the matrix Y is presented in fig. 4.2. Conversely, if fig. 4.2 is given, the associated matrix $Y(G)$ can easily be obtained. Thus, there is a one-to-one correspondence between the matrix and its associated directed graph.

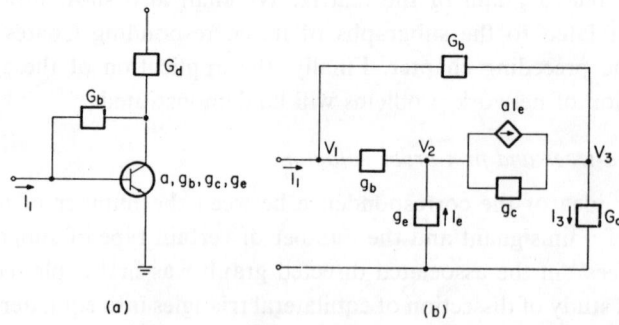

Fig. 4.1. (a) A transistor feedback amplifier and (b) its equivalent network.

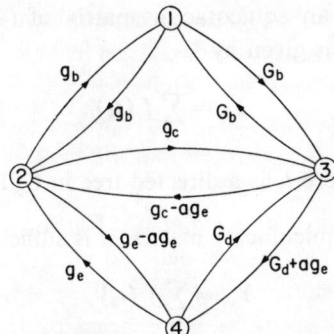

Fig. 4.2. The associated directed graph of the equicofactor matrix (4.24).

DEFINITION 4.4: *Directed tree*. A subgraph, denoted by the symbol t_i, of a directed graph G is said to be a *directed tree* of G with reference node i if, and only if, (1) it is a tree of G (Definition 2.2), and (2) the outgoing degree of each node of t_i is 1 except the node i which has outgoing degree 0.

The term directed tree is also called a *subtree* by TUTTE [1948], and a *rooted tree* by others. As an example, the set of directed trees t_4 with reference node 4 of the directed graph of fig. 4.2 is presented in fig. 4.3.

Like those symbols used in ch. 3, if g_s is a subgraph of a directed graph G, then by $f(g_s)$ we mean the product of the weights associated with the edges of g_s. In contrast to (3.1b), in this chapter it is convenient to define $f(\emptyset) = 1$ where $\emptyset$ is the null graph. Also, for a class of subgraphs g_s of G, the symbol $\sum_{g_s} f(g_s)$ denotes that the summation is taken over all possible g_s of the class.

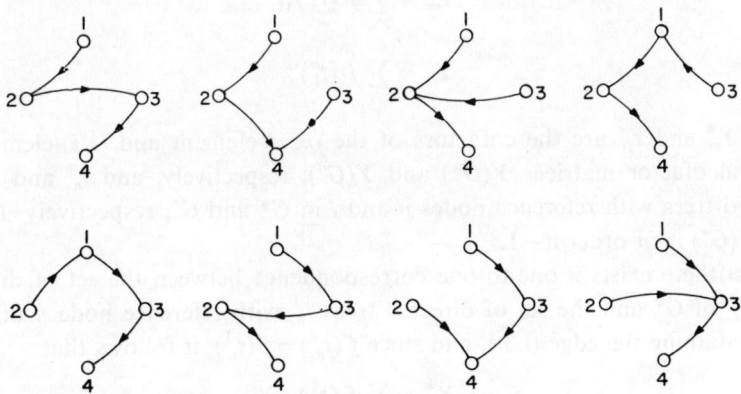

Fig. 4.3. The set of directed trees t_4 of the directed graph of fig. 4.2.

THEOREM 4.3: If Y is an equicofactor matrix of (4.21), then the cofactor Y_{ij} of (i,j)-element of Y is given by

$$Y_{ij} = \sum_{t_k} f(t_k) \qquad (4.25)$$

for $i, j, k = 1, 2, ..., n$, where t_k is a directed tree in $G(Y)$.

Proof. Since Y is an equicofactor matrix, it is sufficient to show that

$$Y_{nn} = \sum_{t_n} f(t_n). \qquad (4.26)$$

The theorem will be proved by induction over the number of edges of $G(Y)$. If $G(Y)$ has only one edge, the theorem is trivially satisfied. Assume that the theorem is true for any $G(Y)$ which has $b-1$ edges or fewer, where $b \geq 2$. To complete the induction we need to show that it is true for any $G(Y)$ that contains b edges.

Assume in $G(Y)$ there exists an edge (i, n) where $i \neq n$. For, if not, the theorem holds for this case since the sum of elements in every row of Y_{nn} is equal to zero. Let G^* be the directed graph obtained from $G(Y)$ first by removing the edge (i, n), and then by replacing the weight associated with the edge (n, i) by $(y_{ni} - y_{in})$. Thus, G^* is the associated directed graph of an equicofactor matrix. Also, let G'' be the directed graph obtained from $G(Y)$ first by identifying the nodes i and n, and then by removing all the self-loops thus produced. The combined node in G'' will be denoted by i. Thus, parallel edges may now appear in G''. Since G^* and G'' have fewer edges than $G(Y)$, the induction hypothesis can now be applied. Thus, we have

$$Y_{nn}^* = \sum_{t_n^*} f(t_n^*), \qquad (4.27)$$

$$Y_{ii}'' = \sum_{t_i''} f(t_i''), \qquad (4.28)$$

where Y_{nn}^* and Y_{ii}'' are the cofactors of the (n, n)-element and (i, i)-element of the equicofactor matrices $Y(G^*)$ and $Y(G'')$, respectively, and t_n^* and t_i'' are directed trees with reference nodes n and i in G^* and G'', respectively. Notice that $Y(G'')$ is of order $n-1$.

Since there exists a one-to-one correspondence between the set of directed trees t_n^* of G^* and the set of directed trees t_n^1 with reference node n of $G(Y)$ not containing the edge (i, n), and since $f(t_n^*) = f(t_n^1)$, it follows that

$$Y_{nn}^* = \sum_{t_n^1} f(t_n^1). \qquad (4.29)$$

Similarly, we can show that there exists a one-to-one correspondence between the set of directed trees t_i'' of G'' and the set of directed trees t_n^2 with reference node n of $G(Y)$ containing the edge (i, n). Since $y_{in} f(t_i'') = f(t_n^2)$, it follows that

$$y_{in} Y_{ii}'' = \sum_{t^2_n} f(t_n^2). \tag{4.30}$$

Thus, we have

$$\begin{aligned} Y_{nn} &= Y_{nn}^* + y_{in} Y_{ii}'' \\ &= \sum_{t^1_n} f(t_n^1) + \sum_{t^2_n} f(t_n^2) \\ &= \sum_{t_n} f(t_n), \end{aligned} \tag{4.31}$$

which completes the proof of the theorem.

In case the equicofactor matrix Y is symmetric, the associated directed graph $G(Y)$ can be greatly simplified.

DEFINITION 4.5: *Associated graph.* For a symmetric equicofactor matrix Y of (4.21), the *associated graph*, denoted by the symbol $G_u(Y)$ or simply G_u if Y is clearly understood or is not explicitly given, is an n-node, weighted, labeled (undirected) graph. The nodes are labeled by the integers from 1 to n such that if $y_{ij} \neq 0$, $i > j$, there is an edge connected between the nodes i and j with associated weight y_{ij} for $i, j = 1, 2, \ldots, n$.

In other words, $G_u(Y)$ is actually the associated undirected graph of $G(Y)$ such that the weight associated with the edge connected between the nodes i and j in $G_u(Y)$ is y_{ij} or y_{ji} for all i and j.

COROLLARY 4.3: If Y is symmetric, then

$$Y_{ij} = \sum_t f(t), \tag{4.32}$$

for $i, j = 1, 2, \ldots, n$, where t is a tree of $G_u(Y)$.

The basic idea of (4.32) was first given by MAXWELL [1892]. He stated that the determinant of the node-admittance matrix of a linear passive network without mutual inductances is equal to the sum of "tree-admittance products" of the network. We remark that the formula (4.32) is more general than that originally stated by Maxwell. For any passive network without mutual inductances, its associated indefinite-admittance matrix Y is always symmetric. However, the converse in general is not true. For example, the indefinite-admittance matrix of a passive linear network with mutual couplings is also symmetric. Thus, the formula is applicable to any network so long as it possesses a symmetric indefinite-admittance matrix. It is significant to point out that, for a passive linear network without mutual inductances, the associated graph $G_u(Y)$ is actually the network itself. This is indeed one of the important features of topological analysis. Clearly, the identity (4.25) is not restricted to the nodal system of equations; it may be applied to the mesh system or any other system of equations as well, as long as the associated matrix of the system can be put in the form as shown in (4.21). Finally, we mention the fact that if all the elements y_{ij} in (4.21) are considered distinct, the right-hand side of (4.25) introduces no redundancies due to duplications. Thus, it may be used rather effectively for the evaluation of the nodal determinant of a network.

As in Coates and Mason graphs, it is advantageous to know the number of directed trees of $G(Y)$ in advance. The solution to this is not hard to find. In the following, we shall treat the problem in a slightly general form. We shall derive formulas for computing the number of directed trees in any given directed graph.

DEFINITION 4.6: *Directed-tree matrix*. The *directed-tree matrix*, denoted by the symbol $D(G)$, associated with an n-node directed graph G is a matrix of order n such that if $D(G)=[d_{ij}]$, then d_{ii} denotes the outgoing degree of node i in G, and $-d_{ij}$, $i \neq j$, denotes the number of edges directed from node i to node j in G for $i, j = 1, 2, ..., n$ after all the self-loops, if they exist, have been removed from G.

In other words, the definition includes the possibility that G contains parallel edges. The removal of self-loops from G is permissible since they can never appear in any of its directed trees. Clearly, the directed-tree matrix $D(G)$ is not necessarily an equicofactor matrix. However, the sum of the elements of every row of $D(G)$ is necessarily zero. A weighted directed graph G is the associated directed graph of some equicofactor matrix if, and only if, the sum of the weights of the edges terminating at each of its nodes equals the sum of the weights of the edges outgoing from that node. Thus, using the same process of association defined for an equicofactor matrix (Definition 4.3), the associated directed graph of $D(G)$ is actually the directed graph G itself with the weight of each of its edges taken to be unity. Obviously, the process may be reversed, i.e., to obtain $D(G)$ from G. For convenience, we shall also say that G is the *associated directed graph* of $D(G)$, and $D(G)$ is the *associated matrix* of G with the weight of each of its edges taken to be unity even though the matrix itself may not be an equicofactor matrix.

COROLLARY 4.4: For a given directed graph G, the value of the cofactor of an ith row element of the directed-tree matrix $D(G)$ is equal to the number of directed trees of G with reference node i.

Proof. The corollary is equivalent to saying that if $D_{ix}(G)$ is the cofactor of the (i, x)-element of $D(G)$, and if $N(t_i)$ denotes the number of directed trees of G with reference node i, then

$$N(t_i) = D_{ix}(G) \qquad (4.33)$$

for $i, x = 1, 2, ..., n$, where n is the number of nodes of G. Since $D(G)$ is a matrix with the property that the sum of the elements of every row equals zero, by Lemma 4.1 it is sufficient to show that

$$N(t_n) = D_{nn}(G). \qquad (4.34)$$

Assume that the weight associated with each edge of G is 1. Let G^* be the directed graph obtained from G by altering the weights associated with the edges (n, u), $u = 1, 2, ..., n-1$, of G in such a way that the associated matrix $Y(G^*)$ of G^* is an equicofactor matrix. This is always possible since edges not originally contained in G may be considered as the edges with zero weight. In other words, we may have to add edges (n, u) to G to make $Y(G^*)$ an equicofactor matrix.

Since edges (n, u) of G or G^* cannot not appear in any directed tree with reference node n, it follows that G and G^* have the same number of directed trees with reference node n. By Theorem 4.3, we have

$$Y_{nn}^* = \sum_{t_n^*} f(t_n^*) = N(t_n^*), \qquad (4.35)$$

where Y_{nn}^* is the cofactor of the (n, n)-element of $Y(G^*)$; t_n^* is a directed tree of G^* with reference node n; and $N(t_n^*)$ denotes the number of directed trees t_n^* in G^*. In (4.35), we have also used the fact $f(t_n^*) = 1$ for every directed tree t_n^* in G^*. Since $D_{nn}(G) = Y_{nn}^*$ and $N(t_n) = N(t_n^*)$, the corollary follows immediately. This completes the proof of the corollary.

Corollary 4.4 was first derived by TUTTE [1948] in his study of the dissection of equilateral triangles into equilateral triangles. BOTT and MAYBERRY [1954] used this result for the analysis of economic activities.

We shall illustrate the above results by the following example.

Example 4.1: Consider the transistor feedback amplifier as shown in fig. 4.1. The indefinite-admittance matrix Y of the amplifier is given in (4.24), and its associated directed graph $G(Y)$ was obtained and is presented in fig. 4.2. The number of directed trees of $G(Y)$ can be computed from the directed-tree matrix $D(G)$, where $G=G(Y)$. From fig. 4.2, we obtain

$$D(G) = \begin{bmatrix} 2 & -1 & -1 & 0 \\ -1 & 3 & -1 & -1 \\ -1 & -1 & 3 & -1 \\ 0 & -1 & -1 & 2 \end{bmatrix}. \tag{4.36}$$

Suppose that we wish to compute the number $N(t_4)$ of directed trees with reference node 4 in $G(Y)$. By Corollary 4.4, we have

$$N(t_4) = D_{44}(G) = \begin{vmatrix} 2 & -1 & -1 \\ -1 & 3 & -1 \\ -1 & -1 & 3 \end{vmatrix} = 8. \tag{4.37}$$

Because of the special form of the directed-tree matrix, its determinant is not computationally difficult to evaluate. For example, (4.37) can easily be evaluated by simple row or column operations. For higher-order determinants of this type, an efficient iterative procedure will be discussed in § 5.2. Alternatively, the number $N(t_4)$ of directed trees may also be computed from any of the other three cofactors $D_{41}(G)$, $D_{42}(G)$, and $D_{43}(G)$. Thus, there are eight directed trees with reference node 4 in $G(Y)$. Since Y is an equicofactor matrix, all the cofactors of the elements of Y are equal. Thus, by Theorem 4.3 we have

$$\begin{aligned} Y_{ij} &= \sum_{t_4} f(t_4) \\ &= g_b g_c (G_d + a g_e) + g_b (g_e - a g_e)(G_d + a g_e) + g_b (g_c - a g_e)(g_e - a g_e) \\ &\quad + G_b g_b (g_e - a g_e) + g_b G_b (G_d + a g_e) + G_b (g_c - a g_e)(g_e - a g_e) \\ &\quad + G_b (G_d + a g_e)(g_e - a g_e) + g_c G_b (G_d + a g_e) \\ &= G_d [(1-a) g_e (G_b + g_b) + g_b (g_c + G_b) + g_c G_b] + g_e G_b (g_b + g_c) \\ &\quad + g_e g_b g_c \end{aligned} \tag{4.38}$$

for $i, j = 1, 2, 3, 4$, where the set of directed trees with reference node 4 of $G(Y)$ is given in fig. 4.3. The result is of course the same as that given in (3.25).

Let us make a similar computation for the number of directed trees with reference node 3 in $G(Y)$.

$$N(t_3) = D_{33}(G) = \begin{vmatrix} 2 & -1 & 0 \\ -1 & 3 & -1 \\ 0 & -1 & 2 \end{vmatrix} = 8, \tag{4.39}$$

which is expected since $D(G)$ is an equicofactor matrix. Thus, there are again

§2 The associated directed graph 239

eight directed trees with reference node 3 in $G(Y)$; they are as shown in fig. 4.4.
By Theorem 4.3 we have

$$Y_{ij} = \sum f(t_3)$$
$$= g_b g_c G_d + g_b(g_e - ag_e) G_d + g_e g_b G_b + g_b G_b G_d + (g_e - ag_e) G_d G_b$$
$$+ g_e g_c G_b + g_e g_b g_c + g_c G_b G_d. \quad (4.40)$$

A comparison of (4.40) with (4.38) shows that the redundant terms due to cancellations in (4.38) are avoided in (4.40). In other words, a proper choice of the reference node will greatly simplify the evaluation of the cofactors.

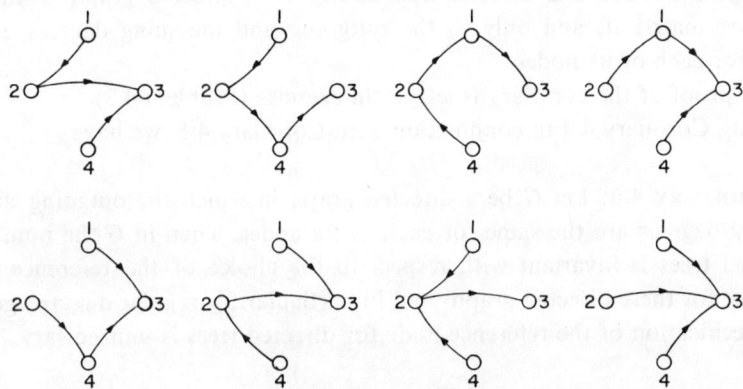

Fig. 4.4. The set of directed trees t_3 of the directed graph of fig. 4.2.

Earlier, we have indicated that if the equicofactor matrix Y is symmetric, its associated directed graph $G(Y)$ can be greatly simplified. The simplified graph is undirected and is denoted by the symbol $G_u(Y)$. A similar simplification may be applied to the directed-tree matrix $D(G)$ associated with a symmetric directed graph G. Since an (undirected) graph can be transformed into a symmetric directed graph by representing each (undirected) edge by a pair of directed edges with opposite directions, and *vice versa*, the problem may be stated in terms of the corresponding undirected graph of G. [It is not the associated undirected graph of G (Definition 1.25).]

DEFINITION 4.7: *Tree matrix.* The *tree matrix*, denoted by the symbol $D(G_u)$, associated with an n-node graph G_u is a symmetric matrix of order n such that if $D(G_u) = [d_{ij}]$, then d_{ii} denotes the degree of node i in G_u, and $-d_{ij}$, $i \neq j$, denotes the number of edges connected between the nodes i and j in G_u

for $i, j = 1, 2, ..., n$ after all the self-loops, if they exist, have been removed from G_u.

Thus, the tree matrix of a graph is an equicofactor matrix. It is also the indefinite-admittance matrix of an n-node resistive network with all the conductances taken to be unity. It is apparent that the directed-tree matrix associated with a symmetric directed graph is an equicofactor matrix. However, the converse is in general not true. The class of directed graphs that will yield equicofactor directed-tree matrices is the class of directed graphs defined below.

COROLLARY 4.5: The directed-tree matrix of a directed graph is an equicofactor matrix if, and only if, the outgoing and incoming degrees are the same for each of its nodes.

The proof of the corollary is left as an exercise (Problem 4.3).

Using Corollary 4.4 in conjunction with Corollary 4.5, we have

COROLLARY 4.6: Let G be a directed graph in which the outgoing and incoming degrees are the same for each of its nodes. Then in G the number of directed trees is invariant with respect to the choice of the reference node.

Thus, for these directed graphs and in particular for regular directed graphs, the specification of the reference node for directed trees is unnecessary.

COROLLARY 4.7: The number of trees in an (undirected) graph is equal to the value of the cofactor of an element of its tree matrix.

DEFINITION 4.8: *Complete directed graph.* A directed graph G having no self-loops is said to be *complete* and of *order k* if, for each pair of nodes of G, there are k and only k parallel edges in each direction. A complete directed graph of order 1 is simply called a *complete directed graph*.

Because of the special form of the directed-tree matrix associated with a complete directed graph of order k, the cofactors (numbers of directed trees) of the elements of its directed-tree matrix are not computationally difficult to evaluate. In order to eliminate the necessity of evaluating these cofactors, formulas for the numbers of directed trees in these graphs and their variants will be derived. Since a complete directed graph of any order is also symmetric (thus regular), we have

COROLLARY 4.8: The number of directed trees in the n-node complete directed graph of order k is given by $k(nk)^{n-2}$, $n \geq 2$.

Proof. Let $D_{nn}(G)$ be the submatrix obtained from $D(G)$ by deleting the nth row and the nth column, where G is the given complete directed graph. Since $D(G)$ is an equicofactor matrix, it is sufficient to evaluate the cofactor $D_{nn}(G)$ of the (n, n)-element of $D(G)$.

Let $D_{nn}(G) = [d_{ij}]$. Then, $d_{ii} = k(n-1)$, and for $i \neq j$, $d_{ij} = -k$. To evaluate $D_{nn}(G)$, we add all succeeding columns of $D_{nn}(G)$ to the first column, and then add the new first column to each of the remaining columns. The resulting matrix is now triangular, and its determinant is readily evaluated as $k(nk)^{n-2}$, which completes the proof of the corollary.

Similarly, the following two results can be derived. Since the derivations are straightforward, they are omitted here and are left as exercises (Problems 4.8 and 4.9).

COROLLARY 4.9: Let G be the directed graph obtained from the n-node complete directed graph of order k by removing k parallel edges from each of the p different ordered pairs of nodes. If all the nodes of the pairs are distinct and different from the reference node i, then the number $N(t_i)$ of directed trees with reference node i in G is given by

$$N(t_i) = k^{n-1}(n-1)^p n^{n-p-2}. \tag{4.41}$$

COROLLARY 4.10: Let G be the directed graph derived from the n-node complete directed graph of order k by removing all the edges (i_1, x), $x = i_2, i_3, \ldots, i_{q+1}$, $q < n$. If $i_1, i_2, \ldots, i_{q+1}$ are all distinct, then the number $N(t_i)$ of directed trees with reference node i in G is given by

$$N(t_i) = (n-q) n^{n-3} k^{n-1} \tag{4.42}$$

for $i \neq i_u$, $u = 1, 2, \ldots, q+1$, and

$$N(t_i) = (n-q-1) n^{n-3} k^{n-1} \tag{4.43}$$

for $i = i_u$, $u = 2, 3, \ldots, q+1$.

As an illustration, consider the 6-node complete directed graph of order 2. Let G^* be the directed graph obtained from G by the removal of all the edges $(1, 2)$ and $(3, 4)$. Then from (4.41) the number of directed trees t_5 or t_6 of G^* is

$$N(t_5) = N(t_6) = 2^5 \times 5^2 \times 6^2 = 28\,800.$$

From Corollary 4.8, there are 41 472 directed trees in G. Now suppose that we wish to use the formulas to find the number $N(t_i)$ of directed trees in the directed graph G of fig. 4.2. Since G is symmetric, the number of its directed trees is invariant with respect to the choice of the reference node. Thus, let node 4 be chosen as the reference node. If we add an edge (4, 1) to G to form a new directed graph G'', it is obvious that G and G'' have the same number of directed trees with reference node 4. Since G'' can be obtained from the 4-node complete graph by removing the edge (1, 4), it follows from (4.43) that

$$N(t_i) = (4-1-1) \times 4^{4-3} \times 1^{4-1} = 8$$

for $i = 1, 2, 3, 4$. On the other hand, the number $N(t_i'')$ of directed trees with reference node 2 or 3 in G'' can be computed either from (4.41) or (4.42):

$$N(t_2'') = N(t_3'') = 1^{4-1} \times (4-1)^1 \times 4^{4-1-2}$$
$$= (4-1) \times 4^{4-3} \times 1^{4-1} = 12.$$

Obviously, these results can easily be applied to (undirected) graphs. However, before we do this we need the following definitions.

DEFINITION 4.9: *Associated symmetric directed graph*. For an (undirected) graph G_u, the *associated symmetric directed graph* G_s of G_u is the directed graph obtained from G_u by representing each (undirected) edge of G_u by a pair of directed edges with opposite directions. If G_u is a weighted graph, then the weight associated with each edge of a pair of directed edges in G_s is the same as the weight of the corresponding edge in G_u.

With this definition, the following corollary is seen to be true.

COROLLARY 4.11: If G_s is the associated symmetric directed graph of a weighted graph G_u, then there exists a one-to-one correspondence between the trees t in G_u and directed trees t_i in G_s and such that $f(t) = f(t_i)$.

In the above corollary, the choice of the reference node i is arbitrary since G_s is symmetric.

DEFINITION 4.10: *Complete graph*. A self-loopless graph is said to be *complete* and of *order* k if for every pair of nodes there are k and only k edges connected between them. A complete graph of order 1 is simply called a *complete graph*.

Thus, there are $k(nk)^{n-2}$ trees in the n-node complete graph of order k. For $k = 1$, n^{n-2} is also the maximum number of positive terms in the expansion of the cofactor of an element of an equicofactor matrix of the type as shown in (4.21).

Example 4.2: Fig. 4.5 shows a three-degrees-of-freedom mass-spring-damper system driven by a forced excitation $F(t)$. It contains three masses m_1, m_2, and m_3, five springs with stiffnesses k_0, k_1, and k_2 as labeled, and one viscous damper with damping coefficient c. The system is assumed to vibrate only in the horizontal direction, where x_1, x_2, and x_3 denote the generalized coordinates. The associated equicofactor matrix Y of the system is given by

$$Y = \begin{bmatrix} m_1 s^2 + K & -k_0 & -k_2 & -m_1 s^2 - k_1 \\ -k_0 & m_2 s^2 + cs + 2k_0 & -k_0 & -m_2 s^2 - cs \\ -k_2 & -k_0 & m_3 s^2 + K & -m_3 s^2 - k_1 \\ -m_1 s^2 - k_1 & -m_2 s^2 - cs & -m_3 s^2 - k_1 & Ms^2 + 2k_1 + cs \end{bmatrix},$$

(4.44)

where $K = k_0 + k_1 + k_2$ and $M = m_1 + m_2 + m_3$. Since Y is symmetric, its associated graph $G_u(Y)$ is presented in fig. 4.6. The number $N(t)$ of trees of $G_u(Y)$ is equal to the cofactor $D_{ij}(G_u)$ of an element of $D(G_u)$, where

$$D(G_u) = \begin{bmatrix} 3 & -1 & -1 & -1 \\ -1 & 3 & -1 & -1 \\ -1 & -1 & 3 & -1 \\ -1 & -1 & -1 & 3 \end{bmatrix}.$$

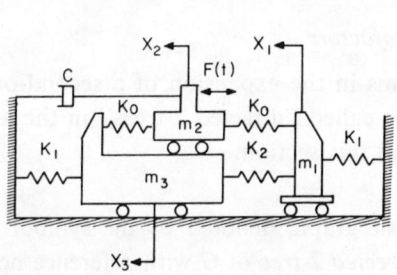
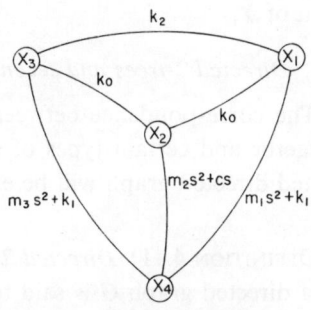

Fig. 4.5. A mass-spring-damper system.

Fig. 4.6. The associated graph of (4.44).

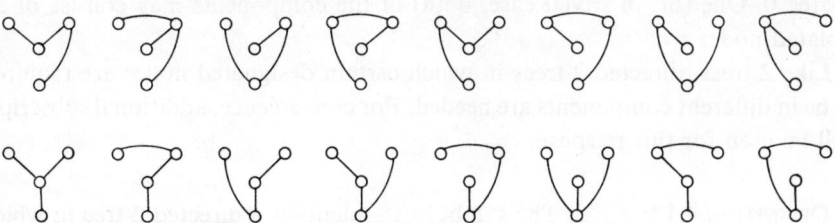

Fig. 4.7. The set of trees of the graph of fig. 4.6.

Thus, we have $N(t) = D_{ij}(G_u) = 16$. Since $G_u(Y)$ is also a complete graph, $N(t)$ is simply $4^{4-2} = 16$. The set of sixteen trees of $G_u(Y)$ is presented in fig. 4.7. By Corollary 4.3, the value of the cofactors Y_{ij} of the elements of Y is obtained as follows:

$$\begin{aligned}
Y_{ij} &= \sum f(t) \\
&= k_0^2(m_1 s^2 + k_1) + 2k_0 k_2(m_1 s^2 + k_1) + 2k_0(m_1 s^2 + k_1)(m_3 s^2 + k_1) \\
&\quad + 2k_0 k_2(m_3 s^2 + k_1) + k_0^2(m_3 s^2 + k_1) + k_0^2(m_2 s^2 + cs) \\
&\quad + 2k_0 k_2(m_2 s^2 + cs) + k_0(m_2 s^2 + cs)(m_3 s^2 + k_1) \\
&\quad + k_0(m_2 s^2 + cs)(m_1 s^2 + k_1) + (m_2 s^2 + cs)(m_1 s^2 + k_1) k_2 \\
&\quad + (m_1 s^2 + k_1)(m_2 s^2 + cs)(m_3 s^2 + k_1) + k_2(m_3 s^2 + k_1)(m_2 s^2 + cs)
\end{aligned}$$
(4.45)

for $i, j = 1, 2, 3, 4$. Had we expanded the cofactor Y_{ij} by the conventional techniques, we would have produced the following eleven additional positive and negative terms:

$$m_1 k_0^2 s^2, \; m_3 k_0^2 s^2, \; m_2 k_2^2 s^2, \; ck_2^2 s, \; k_0^2 k_1, \; k_0 k_2^2, \; k_0^2 k_3, \; k_0^2 k_2, \; k_0^2 k_2, \; k_0^3, \; k_0^3,$$

which would cancel one another, and thus would not appear in the final expansion of Y_{ij}.

2.2. *Directed 2-trees and second-order cofactors*

The correspondence between the terms in the expansion of a second-order cofactor and certain types of subgraphs called "directed 2-trees" in the associated directed graph will be exploited in this section.

DEFINITION 4.11: *Directed 2-tree*. A subgraph, denoted by the symbol $t_{i,j}$, of a directed graph G is said to be a *directed 2-tree* of G with reference nodes i and j if, and only if, (1) it is a 2-tree (Definition 2.32) of G, and (2) the outgoing degree of each node of $t_{i,j}$ is 1 except the nodes i and j which have outgoing degree 0. One (or, in trivial case, both) of the components may consist of an isolated node.

Like 2-trees, directed 2-trees in which certain designated nodes are required to be in different components are needed. For convenience, additional subscripts will be used for this purpose.

DEFINITION 4.12: $t_{ab,cd}$. The symbol $t_{ab,cd}$ denotes a directed 2-tree in which nodes a and b are in one component, and the nodes c and d in the other, where

§2 The associated directed graph

the nodes a and c (the first subscripts) are the reference nodes of the components.

As an example, consider the directed graph G given in fig. 4.2. The set of directed 2-trees $t_{13,4}$ is presented in fig. 4.8. In each of these directed 2-trees, the nodes 1 and 3 are in one component, and node 4 in the other. The nodes 1 and 4 are the reference nodes of the components. The subgraph consisting of the edges (2, 1) and (3, 4) is a directed 2-tree $t_{1,4}$ but it is not a directed 2-tree $t_{13,4}$ since nodes 1 and 3 are not contained in the same component.

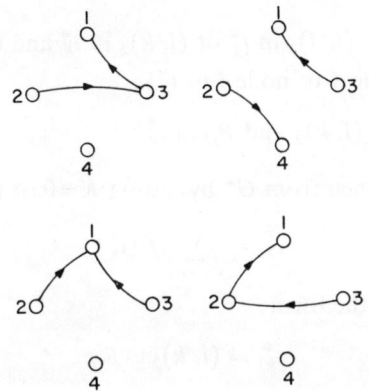

Fig. 4.8. The set of directed 2-trees $t_{13,4}$ of the directed graph of fig. 4.2.

THEOREM 4.4: If Y is an equicofactor matrix of (4.21), then

$$Y_{ij,kk} = \sum_{t_{ij,k}} f(t_{ij,k}) \tag{4.46}$$

for $i, j, k = 1, 2, \ldots, n$ and $i, j \neq k$, where $t_{ij,k}$ is a directed 2-tree in $G(Y)$.

Proof. There are two cases to consider: $i=j$ and $i \neq j$. The proofs are similar, and so let $i \neq j$.

Let G^* be the directed graph obtained from $G(Y)$ by altering the weights of the edges (i, j), (i, k), and (k, j) from y_{ij}, y_{ik}, and y_{kj} to $y_{ij}+K$, $y_{ik}-K$, and $y_{kj}-K$, respectively. Also, let Y^* be the associated equicofactor matrix of G^*. For our purpose, let the edges (i, j) and (i, k) in G^* be represented by parallel edges $(i, j)_1$ and $(i, j)_2$, and $(i, k)_1$ and $(i, k)_2$, respectively, such that

$$f((i, j)_1) = y_{ij}, \quad f((i, j)_2) = K, \tag{4.47a}$$

$$f((i, k)_1) = y_{ik}, \quad f((i, k)_2) = -K. \tag{4.47b}$$

It follows directly from Theorem 4.3 and Problem 4.2 that the cofactor Y^*_{kk}

of the (k, k)-element of Y^* is given by

$$Y_{kk}^* = \sum_{t_k^*} f(t_k^*), \qquad (4.48)$$

where t_k^* is a directed tree with reference node k in G^*.

Next, let us decompose the set S of all directed trees t_k^* of G^* into following three disjoint subsets S_1, S_2, and S_3 whose union is S:

$S_1 = \{t_k^*; t_k^* \text{ in } S, \text{ neither } (i, j)_2 \text{ nor } (i, k)_2 \text{ in } t_k^*\}$,

$S_2 = \{t_k^*; t_k^* \text{ in } S, \text{ either } (i, j)_2 \text{ in } t_k^* \text{ or } (i, k)_2 \text{ in } t_k^* \text{ and there exists no directed path } P_{ji} \text{ from node } j \text{ to node } i \text{ in } t_k^*\}$,

$S_3 = \{t_k^*; t_k^* \text{ in } S, \text{ both } (i, k)_2 \text{ and } P_{ji} \text{ in } t_k^*\}$.

Since G can be obtained from G^* by letting $K=0$, it follows that

$$Y_{kk} = \sum_{t_k^* \in S_1} f(t_k^*). \qquad (4.49)$$

For any t_{k1}^* in S_2 of the form

$$t_{k1}^* = (i, k)_2 \cup R, \qquad (4.50)$$

where R is the subgraph obtained from t_{k1}^* by removing the edge $(i, k)_2$, there always exists a unique t_{k2}^* in S_2 such that

$$t_{k2}^* = (i, j)_2 \cup R \qquad (4.51a)$$

and

$$f(t_{k1}^*) = -f(t_{k2}^*) \qquad (4.51b)$$

and vice versa. Thus, we have

$$\sum_{t_k^* \in S_2} f(t_k^*) = 0. \qquad (4.52)$$

Observe also the one-to-one correspondence between the elements t_k^* of S_3 and the directed 2-trees $t_{ij, k}$ in $G(Y)$ and such that

$$t_k^* - (i, k)_2 = t_{ij, k} \qquad (4.53a)$$

and

$$f(t_k^*) = -K f(t_{ij, k}). \qquad (4.53b)$$

Finally, we have the following relationship for the cofactors Y_{kk}^* and Y_{kk}:

$$Y_{kk}^* = Y_{kk} - K Y_{ij, kk}. \qquad (4.54)$$

§2 The associated directed graph

Thus, from (4.48), (4.49), (4.52), and (4.53) we have

$$Y_{kk}^* = \sum_{t^*_k \in S_1} f(t_k^*) + \sum_{t^*_k \in S_2} f(t_k^*) + \sum_{t^*_k \in S_3} f(t_k^*)$$
$$= Y_{kk} + 0 - K \sum_{t_{ij,k}} f(t_{ij,k}). \tag{4.55}$$

Comparing (4.54) with (4.55), the theorem follows immediately.

A direct consequence of this theorem and Lemma 4.2 is the following which may be considered as a generalization of Theorem 4.4.

COROLLARY 4.12: For $p \neq r$ and $q \neq s$,

$$Y_{pq,rs} = \sum_{t_{pq,rs}} f(t_{pq,rs}) - \sum_{t_{ps,rq}} f(t_{ps,rq}). \tag{4.56}$$

Proof.

$$Y_{pq,rs} = Y_{pq,rr} - Y_{ps,rr}$$
$$= \sum_{t_{pq,r}} f(t_{pq,r}) - \sum_{t_{ps,r}} f(t_{ps,r})$$
$$= \sum_{t_{pqs,r}} f(t_{pqs,r}) + \sum_{t_{pq,rs}} f(t_{pq,rs}) - \sum_{t_{psq,r}} f(t_{psq,r}) - \sum_{t_{ps,rq}} f(t_{ps,rq})$$
$$= \sum_{t_{pq,rs}} f(t_{pq,rs}) - \sum_{t_{ps,rq}} f(t_{ps,rq}), \tag{4.57}$$

which completes the proof of the corollary.

In particular, if Y is symmetric, the formulas (4.46) and (4.56) can be simplified, and $Y_{pq,rs}$ can be evaluated by the set of 2-trees of $G_u(Y)$.

COROLLARY 4.13: If Y is symmetric, then

$$Y_{pq,rs} = \sum_{t_{pq,rs}} f(t_{pq,rs}) - \sum_{t_{ps,rq}} f(t_{ps,rq}), \tag{4.58}$$

where $p \neq r$ and $q \neq s$, and $t_{pq,rs}$ and $t_{ps,rq}$ are 2-trees in the associated graph $G_u(Y)$.

COROLLARY 4.14: If Y is symmetric, and if $t_{pq,r}$ is a 2-tree in $G_u(Y)$, then

$$Y_{pq,rr} = \sum_{t_{pq,r}} f(t_{pq,r}). \tag{4.59}$$

Again, the basic idea of (4.59) was first given by MAXWELL [1892], and it was rigorously proved by MAYEDA and SESHU [1957]. In electrical network theory (ch. 2, § 6), it simply states that the cofactor of the (p, q)-element of the node-admittance matrix of a linear passive network without mutual inductances is equal to the sum of 2-tree-admittance products of all the 2-trees $t_{pq,r}$ of the network, where r is the point of its reference potential. Like the formula (4.32), it is obvious that the formula (4.59) is more general than that originally stated by Maxwell; it applies to any symmetric equicofactor matrix of the type as shown in (4.21). We remark that if all the elements y_{ij} of (4.21) are considered distinct, the formulas presented above do not introduce any redundancies due to duplications. Thus, they provide a short-cut for the evaluation of the cofactors of the elements of the node-admittance matrix.

COROLLARY 4.15: *For a given directed graph G, the value of the second-order cofactor $D_{ij,kk}(G)$ of the (i,j)- and (k,k)-elements of the directed-tree matrix $D(G)$ is equal to the number $N(t_{ij,k})$ of directed 2-trees $t_{ij,k}$ of G.*

We shall illustrate the above results by the following example.

Example 4.3: Consider the problem given in Example 4.1. Suppose that we wish to compute the number $N(t_{13,4})$ of directed 2-trees $t_{13,4}$ in $G(Y)$, as shown in fig. 4.2. The directed-tree matrix $D(G)$ is given in (4.36), where $G = G(Y)$. By Corollary 4.15, we have

$$N(t_{13,4}) = D_{13,44}(G)$$
$$= \text{sgn}(1-4)\,\text{sgn}(3-4)\,(-1)^{1+3+4+4} \begin{vmatrix} -1 & 3 \\ -1 & -1 \end{vmatrix}$$
$$= 4. \qquad (4.60)$$

Thus, there are four directed 2-trees $t_{13,4}$ in $G(Y)$, and they are presented in fig. 4.8. By Theorem 4.4, we have

$$Y_{13,44} = \sum_{t_{13,4}} f(t_{13,4})$$
$$= g_c G_b + G_b(g_e - ag_e) + g_b G_b + (g_c - ag_e)g_b$$
$$= G_b(g_c + g_e - ag_e + g_b) + g_b(g_c - ag_e). \qquad (4.61)$$

If the second-order cofactors $Y_{12,43}$ and $Y_{11,44}$ are needed, they may be computed in a similar fashion. Thus, from Corollary 4.12 we have

$$Y_{12,43} = \sum_{t_{12,43}} f(t_{12,43}) - \sum_{t_{13,42}} f(t_{13,42})$$
$$= g_b(G_d + ag_e) - G_b(g_e - ag_e),$$

§2   The associated directed graph   249

$$Y_{11,44} = \sum_{t_{1,4}} f(t_{1,4})$$
$$= g_c G_b + G_b(g_e - ag_e) + g_b G_b + (g_c - ag_e)g_b + g_b(G_d + ag_e)$$
$$+ g_c(G_d + ag_e) + (g_c - ag_e)(g_e - ag_e) + (g_e - ag_e)(G_d + ag_e)$$
$$= (g_c + g_e - ag_e + g_b)(G_b + G_d) + g_c(g_b + g_e). \tag{4.62}$$

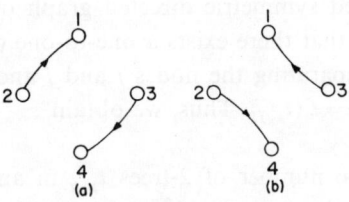

Fig. 4.9. The directed 2-trees $t_{12,43}$ and $t_{13,42}$ of the directed graph of fig. 4.2.

There is only one $t_{12,43}$ as shown in fig. 4.9(a), and one $t_{13,42}$ as shown in fig. 4.9(b). There are eight $t_{1,4}$, which are presented in figs. 4.8, 4.9(a), and 4.10. To show that these are the only directed 2-trees $t_{1,4}$, we need only compute the second-order cofactor $D_{11,44}(G)$. From (4.36), we get

$$N(t_{1,4}) = \text{sgn}(1-4)\,\text{sgn}(1-4)\,(-1)^{1+1+4+4}\begin{vmatrix} 3 & -1 \\ -1 & 3 \end{vmatrix}$$
$$= 8. \tag{4.63}$$

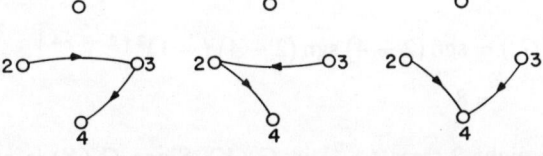

Fig. 4.10. Some directed 2-trees $t_{1,4}$ of the directed graph of fig. 4.2.

We shall now turn our attention to the derivation of formulas for the numbers of various types of directed 2-trees in a complete directed graph. Also we shall consider the number of 2-trees in an (undirected) graph.

COROLLARY 4.16: *The number of directed 2-trees $t_{ij,k}$ in the n-node complete directed graph of order p is given by*

$$N(t_{ij,k}) = p(pn)^{n-3} \tag{4.64}$$

if $i \neq j \neq k$, and if $i \neq k$

$$N(t_{i,k}) = 2p(pn)^{n-3}. \tag{4.65}$$

The proof of these formulas is similar to that of Corollary 4.8. A generalized version of them is described in Problems 4.22 and 4.23.

If G_s is the associated symmetric directed graph of a weighted graph G_u, it is not difficult to see that there exists a one-to-one correspondence between the 2-trees $t''_{i,j}$ of G_u separating the nodes i and j and directed 2-trees $t_{i,j}$ of G_s and such that $f(t''_{i,j}) = f(t_{i,j})$. Thus, we obtain

COROLLARY 4.17: *The number of 2-trees $t_{p,q}$ in an (undirected) graph G_u is equal to the value of the second-order cofactor $D_{pp,qq}(G)$ of the (p,p)- and (q,q)-elements of the tree matrix $D(G_u)$.*

Thus, there are $2p(pn)^{n-3}$ 2-trees separating any two designated nodes in a complete graph of order p. For $p=1$, $2n^{n-3}$ is also the maximum number of positive terms in the expansion of a second-order cofactor of the elements of a symmetric equicofactor matrix (4.21).

Example 4.4: Consider the problem given in Example 4.2. Suppose that we wish to evaluate the second-order cofactors $Y_{22,44}$, $Y_{21,44}$, and $Y_{23,44}$. Since Y is symmetric, we consider the associated graph $G_u(Y)$ of fig. 4.6. The number $N(t_{x_2,x_4})$ of 2-trees t_{x_2,x_4} of $G_u(Y)$ is equal to the value of the second-order cofactor $D_{22,44}(G_u)$ of the elements of the tree matrix $D(G_u)$ given in Example 4.2:

$$D_{22,44}(G_u) = \mathrm{sgn}(2-4)\,\mathrm{sgn}(2-4)(-1)^{2+2+4+4} \begin{vmatrix} 3 & -1 \\ -1 & 3 \end{vmatrix}$$
$$= 8. \tag{4.66}$$

Thus, there are eight 2-trees t_{x_2,x_4} in $G_u(Y)$. Since $G_u(Y)$ is also a complete graph, $N(t_{x_2,x_4})$ is simply $2 \cdot 4^{4-3} = 8$. The set of eight 2-trees t_{x_2,x_4} of $G_u(Y)$ is presented in fig. 4.11. From Corollary 4.14, we have

$$\begin{aligned}Y_{22,44} &= \sum f(t_{x_2,x_4}) \\ &= (m_1 s^2 + k_1)(m_3 s^2 + k_1) + (m_1 s^2 + k_1)k_0 + k_0(m_3 s^2 + k_1) \\ &\quad + k_0^2 + 2k_0 k_2 + k_2(m_1 s^2 + k_1) + k_2(m_3 s^2 + k_1) \\ &= (m_1 s^2 + K)(m_3 s^2 + K) - k_2^2, \end{aligned} \tag{4.67}$$

where $K = k_0 + k_1 + k_2$. Similarly, we can evaluate $Y_{21,44}$ and $Y_{23,44}$. The sets of desired 2-trees $t_{x_2 x_1, x_4}$ and $t_{x_2 x_3, x_4}$ of $G_u(Y)$ can be found in fig. 4.11, and

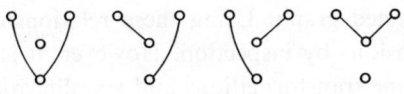

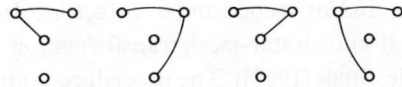

Fig. 4.11. The set of 2-trees t_{x_2,x_4} of the graph of fig. 4.6.

are as shown in figs. 4.12 and 4.13, respectively. Thus,

$$Y_{21,44} = \sum f(t_{x_2x_1,x_4}) = k_0(m_3s^2 + k_1) + k_0^2 + 2k_0k_2 \quad (4.68)$$

and

$$Y_{23,44} = \sum f(t_{x_2x_3,x_4}) = (m_1s^2 + k_1)k_0 + k_0^2 + 2k_0k_2. \quad (4.69)$$

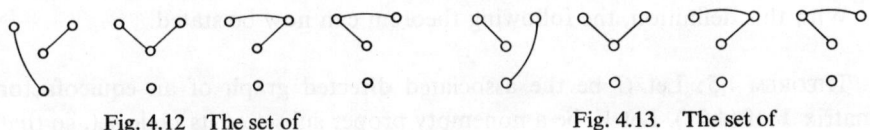

Fig. 4.12 The set of 2-trees $t_{x_2x_1,x_4}$.

Fig. 4.13. The set of 2-trees $t_{x_2x_3,x_4}$.

Finally, let us evaluate $Y_{12,43}$. From (4.58) and fig. 4.14 we obtain

$$Y_{12,43} = \sum f(t_{x_1x_2,x_4x_3}) - \sum f(t_{x_1x_3,x_4x_2})$$
$$= k_0(m_3s^2 + k_1) - k_2(m_2s^2 + cs). \quad (4.70)$$

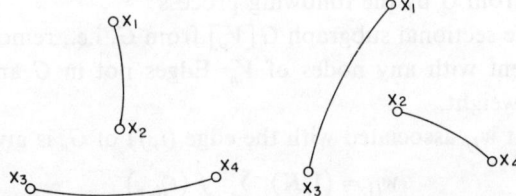

Fig. 4.14. The set of 2-trees used for the evaluation of $Y_{12,34}$ in (4.70).

§ 3. Equivalence and transformations

In the foregoing, we have presented the intimate relationships between the terms in the expansions of the first- and second-order cofactors of the elements of an equicofactor matrix and the sets of directed trees and directed 2-trees of its associated directed graph. Using these relationships, it is possible to write down the expansions by inspection. However, it is sometimes advantageous to carry out some transformations and simplifications of the associated directed graph G before applying the derived formulas. In the present section, we shall show how to eliminate some of the nodes of G without effecting the ratios of the first-order and/or second-order cofactors. In particular, the technique reduces to the well-known star-mesh transformation in electrical network theory (see, for example, SHEN [1947]). The procedure is similar to that for the Coates and Mason graphs discussed in ch. 3, and was first given by CHEN [1967b].

For a given directed graph G, let V_p be a nonempty proper subset of its node set. Let $G^*[V_p]$ be the directed graph obtained from G first by identifying all the nodes *not* contained in V_p, and then by removing all the self-loops thus produced. The combined node in $G^*[V_p]$ is denoted by p. Notice that the symbol $G^*[V_p]$ is similar to that used for a sectional subgraph $G[V_p]$ of G (Definition 1.15). The distinction is the asterisk used in the present case. This should not create any difficulty.

With this definition, the following theorem can now be stated.

THEOREM 4.5: Let G be the associated directed graph of an equicofactor matrix Y of (4.21). Let V_p be a nonempty proper subset of its node set, so that the cofactor K of an element of the associated equicofactor matrix of $G^*[V_p]$ is nonzero. If V_p contains less than $n-1$ nodes, $n \geq 3$, then

$$Y_{uv} = \sum_{t_k} f(t_k) = K \sum_{t^r{}_k} f(t_k^r) \qquad (4.71)$$

for u, v, $k = 1, 2, \ldots, n$ and k is not in V_p, where t_k and t_k^r are directed trees of G and G_r with reference node k, respectively, and G_r is the reduced directed graph obtained from G by the following process:

(i) Remove the sectional subgraph $G[V_p]$ from G, i.e., remove all the nodes and edges incident with any nodes of V_p. Edges not in G are considered as edges with zero weight.

(ii) The weight w_{ij} associated with the edge (i, j) of G_r is given by

$$w_{ij} = (1/K) \sum_{t^*{}_{ji,p}} f(t^*_{ji,p}) \qquad (4.72)$$

for all i and j not in V_p and $i \neq j$, and $w_{ii} = 0$ for all i, where $t^*_{ji,p}$ is a directed 2-tree of the type $t_{ji,p}$ in $G^*[V_p^{ij}]$, and V_p^{ij} is the set union of the sets V_p and $\{i, j\}$.

Proof. Since a relabeling of the nodes of G corresponds to simultaneous interchanges of the rows and the corresponding columns of Y, without loss

§ 3 Equivalence and transformations

of generality we may assume that the matrix Y can be partitioned in such a way that C_{11} corresponds to the node set V_p, which has h nodes:

$$Y = [c_{ij}] = \begin{bmatrix} C_{11} & C_{12} \\ C_{21} & C_{22} \end{bmatrix}, \qquad (4.73)$$

where C_{11} and C_{22} are square submatrices of orders h and $n-h$, respectively, and C_{11} is nonsingular. (Note that C_{ij} no longer represents the submatrix obtained from C by striking out the ith row and jth column. This should not create any confusion since it will only appear in this proof.).

It is easy to show that the identities (Problem 4.28)

$$\begin{bmatrix} U_h & 0 \\ -C_{21}C_{11}^{-1} & U_{n-h} \end{bmatrix} \begin{bmatrix} C_{11} & C_{12} \\ C_{21} & C_{22} \end{bmatrix} = \begin{bmatrix} C_{11} & C_{12} \\ 0 & C_{22} - C_{21}C_{11}^{-1}C_{12} \end{bmatrix} \qquad (4.74)$$

and

$$Y_{uv} = (-1)^{u+v} (\det C_{11}) \det (C_{22} - C_{21}C_{11}^{-1}C_{21})_{uv} \qquad (4.75)$$

for $h < u$, $v \le n$, hold where $(C_{22} - C_{21}C_{11}^{-1}C_{12})_{uv}$ denotes the submatrix obtained from $(C_{22} - C_{21}C_{11}^{-1}C_{12})$ by deleting the uth row and vth column, and U_h is the identity matrix of order h.

Let

$$C_{kt} = [c_{ij}^{kt}] \qquad (4.76)$$

for $k, t = 1, 2$, where the subscripts i and j are the row and column indices,

$$[w_{ij}''] = (\det C_{11})(C_{22} - C_{21}C_{11}^{-1}C_{12}) \qquad (4.77)$$

$$A_i = [c_{i1} \quad c_{i2} \quad \ldots \quad c_{ih}] \qquad (4.78)$$

for $i = h+1, h+2, \ldots, n$, and

$$B_j' = [c_{1j} \quad c_{2j} \quad \ldots \quad c_{hj}] \qquad (4.79)$$

for $j = h+1, h+2, \ldots, n$, where the prime indicates the matrix transpose. It follows from (4.76) and (4.77) that

$$\begin{aligned}
w_{ij}'' &= (\det C_{11}) c_{ij}^{22} - \sum_{k_1=1}^{h} \sum_{k_2=1}^{h} (-1)^{k_1+k_2} c_{ik_1}^{21} M_{k_2 k_1}^{11} c_{k_2 j}^{12} \\
&= c_{ij}^{22} (\det C_{11}) + \sum_{k_1=1}^{h} (-1)^{k_1+h+1} c_{ik_1}^{21} \left(\sum_{k_2=1}^{h} (-1)^{k_2+h} c_{k_2 j}^{12} M_{k_2 k_1}^{11} \right) \\
&= c_{ij}^{22} (\det C_{11}) + \sum_{k_1=1}^{h} (-1)^{k_1+h+1} c_{ik_1}^{21} \det [C_{11} \quad B_{j+h}]_{k_1} \\
&= c_{ij}^{22} (\det C_{11}) + \sum_{k_1=1}^{h} (-1)^{k_1+h+1} c_{ik_1}^{21} \det \begin{bmatrix} C_{11} & B_{j+h} \\ A_{i+h} & c_{ij}^{22} \end{bmatrix}_{(h+1) k_1} \\
&= \det \begin{bmatrix} C_{11} & B_{j+h} \\ A_{i+h} & c_{ij}^{22} \end{bmatrix}, \qquad (4.80)
\end{aligned}$$

where $M^{11}_{k_2 k_1}$ is the k_2th row, k_1th column minor of C_{11}; $[C_{11} \quad B_{j+h}]_{k_1}$ is the matrix obtained from the matrix $[C_{11} \quad B_{j+h}]$ by striking out the column k_1; and

$$\begin{bmatrix} C_{11} & B_{j+h} \\ A_{i+h} & c^{22}_{ij} \end{bmatrix}_{(h+1)k_1} \tag{4.81}$$

is the matrix obtained from the matrix

$$\begin{bmatrix} C_{11} & B_{j+h} \\ A_{i+h} & c^{22}_{ij} \end{bmatrix} \tag{4.82}$$

by striking out the $(h+1)$th row and the k_1th column. Thus,

$$\sum_{j=1}^{n-h} w''_{ij} = \sum_{j=1}^{n-h} \det \begin{bmatrix} C_{11} & B_{j+h} \\ A_{i+h} & c^{22}_{ij} \end{bmatrix}$$

$$= \det \begin{bmatrix} C_{11} & B_{h+1} + B_{h+2} + \cdots + B_n \\ A_{i+h} & c^{22}_{i1} + c^{22}_{i2} + \cdots + c^{22}_{i(n-h)} \end{bmatrix}$$

$$= \det \begin{bmatrix} C_{11} & 0 \\ A_{i+h} & 0 \end{bmatrix} = 0. \tag{4.83}$$

The last equation is obtained by adding the first h columns to the $(h+1)$th column, and then using the fact that Y is an equicofactor matrix.

Similarly, it can be shown that

$$\sum_{i=1}^{n-h} w''_{ij} = 0. \tag{4.84}$$

Thus, $[w''_{ij}]$ of (4.77) is an equicofactor matrix of order $(n-h)$ and is of the type of (4.21).

Next, let us consider the equicofactor matrix

$$\begin{bmatrix} C_{11} & B_{i+h} & B_{j+h} & \sum B_x \\ A_{i+h} & c^{22}_{ii} & c^{22}_{ij} & \sum c^{22}_{ix} \\ A_{j+h} & c^{22}_{ji} & c^{22}_{jj} & \sum c^{22}_{jx} \\ \sum A_x & \sum c^{22}_{xi} & \sum c^{22}_{xj} & c \end{bmatrix}, \tag{4.85a}$$

in which $i \neq j$, and

$$-c = \sum \left[\left(\sum_{q=1}^{h} c_{xq} \right) + c^{22}_{xi} + c^{22}_{xj} \right] \tag{4.85b}$$

(see Problem 4.65), where the unmarked summations are taken from $x = h+1$ to $x = n$ but $x \neq i+h$ and $x \neq j+h$.

Now it is clear that the associated directed graph of (4.85a) is actually $G^*[V_p^{ij}]$ with $(h+1)$th and $(h+2)$th rows and columns of (4.85a) corresponding to nodes

§3 Equivalence and transformations

i and j of $G^*[V_p^{ij}]$, respectively, and the $(h+3)$th row and column to node p. Since $-w_{ij}''$ is the second-order cofactor of the elements c_{ji}^{22} and c of (4.85a), it follows from Theorem 4.4 that

$$-w_{ij}'' = \sum_{t^*_{ji,p}} f(t^*_{ji,p}). \tag{4.86}$$

Thus, from (4.72), $w_{ij}'' = -Kw_{ij}$. Since $K = \det C_{11}$, it follows from (4.77) that G_r is actually the associated directed graph of the equicofactor matrix $(1/K) \times [w_{ij}'']$. Since the cofactor of the (u,v)-element of $(1/K)[w_{ij}'']$ is related to Y_{uv} through (4.75) and (4.77), the identity (4.71) follows directly from Theorem 4.3. So the theorem is proved.

Fortunately, the process outlined in (4.72) is easier to visualize than it is to describe in words, as a moment's study of its topological structure will verify. It amounts to saying that the weight associated with the edge (i,j), $i \neq j$, in the reduced directed graph G_r is equal to a constant $(1/K)$ times the sum of products of the weights associated with the edges of directed 2-trees of the type $t_{ji,p}$ in $G^*[V_p^{ij}]$.

COROLLARY 4.18: If $V_p = \{k\}$, then

$$w_{ij} = y_{ij} + y_{ik}y_{kj} \Big/ \Big(\sum_{\substack{x=1 \\ x \neq k}}^{n} y_{kx}\Big) \tag{4.87}$$

for all $i, j \neq k$, where y_{ij} is the weight associated with the edge (i,j) of G.

In electrical network theory, Corollary 4.18 may be considered as the star-mesh transformation for an active network with mutual inductances. The usefulness of Theorem 4.5 lies in the following.

THEOREM 4.6: For i, j, and k not in V_p, the second-order cofactor $Y_{ij,kk}$ is given by

$$Y_{ij,kk} = \sum_{t_{ij,k}} f(t_{ij,k}) = K \sum_{t^r_{ij,k}} f(t^r_{ij,k}), \tag{4.88}$$

where $t_{ij,k}$ and $t^r_{ij,k}$ are directed 2-trees of the type $t_{ij,k}$ in G and G_r, respectively.

The theorem follows from the fact that G_r is actually the associated directed graph of the equicofactor matrix $(1/K)[w_{ij}'']$. The details of its proof are left as an exercise (Problem 4.34).

Thus, from Theorems 4.5 and 4.6, we obtain the following result:

COROLLARY 4.19: The ratio of a first-order cofactor and a second-order cofactor and the ratio of any two second-order cofactors remain the same in both G and G_r for all i, j, and k not in V_p.

We shall illustrate the above results by an example.

Example 4.5: Consider a directed graph G as shown in fig. 4.15 which represents the associated directed graph of a general potential-feedback amplifier. Assume that $V_p = \{4\}$. The corresponding reduced directed graph G_r obtained from G by the removal of node 4 using the procedure outlined in Theorem 4.5 is presented in fig. 4.16. The weights w_{ij} associated with the edges (i, j) of G_r are obtained from (4.87), and are given by

$$w_{12} = w_{21} = y_1, \tag{4.89a}$$

$$w_{23} = y_f, \tag{4.89b}$$

$$w_{32} = (y_f - g_m) + (y_p + g_m) g_m/K, \tag{4.89c}$$

$$w_{35} = y_2 + (y_p + g_m) y_k/K, \tag{4.89d}$$

$$w_{53} = y_2 + y_k y_p/K, \tag{4.89e}$$

$$w_{25} = y_g, \tag{4.89f}$$

$$w_{52} = y_g + y_k g_m/K, \tag{4.89g}$$

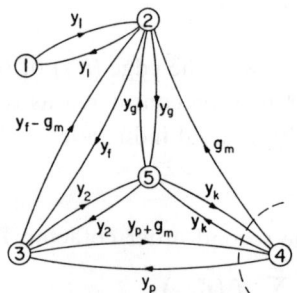

Fig. 4.15. The associated directed graph of a feedback amplifier.

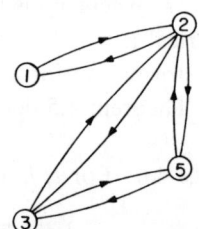

Fig. 4.16. The reduced directed graph obtained by removing the node 4.

and $w_{ij} = 0$ otherwise, where $K = y_p + g_m + y_k$. Remember that edges not contained in G are considered as edges with zero weight. For example, had there been an edge (1, 4) with weight y_{14} in G, there would have been an edge (1, 3) in G_r with weight $w_{13} = 0 + y_{14} y_p/K = y_{14} y_p/K$.

Suppose that we wish to remove the nodes 3 and 4 at the same time, using the procedure outlined in Theorem 4.5. Then $V_p = \{3, 4\}$, and the reduced directed graph G_r is presented in fig. 4.17. For illustrative purposes, we shall compute the weights w_{ij} of the edges of G_r.

The constant K is equal to the value of the cofactor of an element of the associated equicofactor matrix of the directed graph $G^*[3, 4]$ as shown in fig. 4.18. To avoid later confusion, use K' to denote this constant.

$$K' = \sum_{t^*_p} f(t^*_p) = \sum_{t^*_4} f(t^*_4) = \sum_{t^*_3} f(t^*_3)$$
$$= (y_f + y_2) y_p + (y_k + g_m)(y_f + y_2) + y_k y_p, \quad (4.90)$$

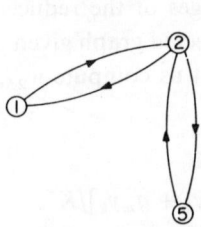
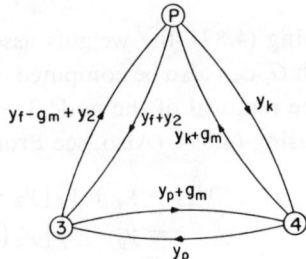

Fig. 4.17. The reduced directed graph obtained by removing the nodes 3 and 4.

Fig. 4.18. The directed graph $G^*[3, 4]$.

where directed trees t^*_3 of $G^*[3, 4]$ were used in the computation of (4.90) to avoid cancellations. Since G has only five nodes, it follows that $G^*[V^{ij}_p] = G$ for $i, j = 1, 2,$ and 5. Thus, we have

$$K' w_{12} = \sum_{t_{21, 5}} f(t_{21, 5})$$
$$= y_1(y_p + g_m) y_k + y_1 y_p y_2 + y_1 y_2 y_k + y_1(y_p + g_m) g_m + y_1 g_m y_2$$
$$+ y_1(y_f - g_m) y_p + y_1(y_f - g_m) g_m + y_1(y_f - g_m) y_k$$
$$= K' y_1, \quad (4.91)$$

$$K' w_{21} = \sum_{t_{12, 5}} f(t_{12, 5}) = \sum_{t_{21, 5}} f(t_{21, 5})$$
$$= K' y_1, \quad (4.92)$$

$$K' w_{25} = \sum_{t_{52, 1}} f(t_{52, 1}) = \sum_{t''_1} f(t''_1) = \sum_{t''_3} f(t''_3)$$
$$= g_m y_f y_g + g_m y_f y_2 + g_m y_f y_k + y_f y_p y_g + y_f y_p y_2 + y_f y_p y_k$$
$$+ y_f y_k y_g + y_f y_k y_2 + g_m y_g y_2 + y_g y_p y_2 + y_g y_p y_k + y_g y_k y_2$$
$$= K' y_g + y_f [y_2(y_p + y_k + g_m) + y_p y_k + g_m y_k], \quad (4.93)$$

where t''_i is a directed tree with reference node i in the sectional subgraph $G[2, 3, 4, 5]$. Since the associated matrix of $G[2, 3, 4, 5]$ is an equicofactor matrix, the choice of the reference node i is arbitrary. However, in (4.93) we choose $i=3$ for its evaluation so that redundancies due to cancellations are avoided. Finally, we have

$$K'w_{52} = \sum_{t_{25,1}} f(t_{25,1}) = \sum_{t''_i} f(t''_i) = K'w_{25}, \tag{4.94a}$$

$$K'w_{15} = \sum_{t_{51,2}} f(t_{51,2}) = 0, \tag{4.94b}$$

$$K'w_{51} = \sum_{t_{15,2}} f(t_{15,2}) = 0. \tag{4.94c}$$

Using (4.87), the weights associated with the edges of the reduced directed graph G_r can also be computed directly from the directed graph given in fig. 4.16 by the removal of the node 3. As an illustration, let us compute w_{25}, w_{12}, and w_{15} using (4.89). (Also, see Problem 4.66.)

$$\begin{aligned}w_{25} &= y_g + y_f[y_2 + y_k(y_p + g_m)/K]/K'' \\ &= y_g + y_f[y_2(y_p + y_k + g_m) + y_p y_k + g_m y_k]/K', \end{aligned} \tag{4.95a}$$

$$w_{12} = y_1, \tag{4.95b}$$

$$w_{15} = 0, \tag{4.95c}$$

where

$$K'' = (y_f - g_m) + g_m(y_p + g_m)/K + y_2 + y_k(y_p + g_m)/K. \tag{4.96}$$

Thus, from (4.71) and (4.88), the first-order cofactor Y_{ij} and the second-order cofactors $Y_{11,55}$ and $Y_{12,55}$ can be obtained from fig. 4.17:

$$Y_{ij} = K' \sum_{t^r_k} f(t^r_k) = K'w_{12}w_{25}, \tag{4.97}$$

$$Y_{11,55} = K' \sum_{t^r_{1,5}} f(t^r_{1,5}) = K'(w_{21} + w_{25}), \tag{4.98}$$

$$Y_{12,55} = K' \sum_{t^r_{12,5}} f(t^r_{12,5}) = K'w_{21}. \tag{4.99}$$

The ratio of these first- and second-order cofactors remains unaltered in both G and G_r.

Like the cases discussed earlier, when the associated equicofactor matrix of G is symmetric, the directed graphs G and G_r can both be simplified, and reduce to undirected graphs.

COROLLARY 4.20: If Y of (4.21) is symmetric, then

$$Y_{uv} = \sum_t f(t) = K \sum_{t^r} f(t^r), \tag{4.100}$$

$$Y_{ij,kk} = \sum_{t_{ij,k}} f(t_{ij,k}) = K \sum_{t^r_{ij,k}} f(t^r_{ij,k}), \qquad (4.101)$$

where t and t^r are trees in the associated graph $G_u(Y)$ and the reduced graph G_{ur}, respectively; $t_{ij,k}$ and $t^r_{ij,k}$ are 2-trees of the type $t_{ij,k}$ in G_u and G_{ur}, respectively; K is the cofactor of an element of the associated equicofactor matrix of $G_u^*[V_p]$ where V_p is a nonempty subset of the node set of G_u containing less than $n-1$ nodes; the symbol $G_u^*[V_p]$ again denotes the graph obtained from G_u first by identifying all the nodes not in V_p, with the combined node being represented by p, and then by removing all the self-loops thus produced; and G_{ur} is the reduced graph obtained from G_u by the following process:

(i) Remove the sectional subgraph $G_u[V_p]$. Edges not in G_u are considered as edges with zero weight.

(ii) The weight w_{ij} associated with the edge (i,j) or (j,i) of G_{ur} is given by

$$w_{ij} = (1/K) \sum_{t^*_{ij,p}} f(t^*_{ij,p}) \qquad (4.102)$$

for i and j not in V_p and $i \neq j$, and $w_{ii}=0$ for all i, where $t^*_{ij,p}$ is a 2-tree of the type $t_{ij,p}$ in $G_u^*[V_p^{ij}]$, and V_p^{ij} is the set union of the sets V_p and $\{i,j\}$.

Note that the order of the subscripts in the 2-trees of (4.101) and (4.102) is immaterial.

COROLLARY 4.21: If $V_p = \{k\}$, then

$$w_{ij} = y_{ij} + y_{ik}y_{kj} / \left(\sum_{\substack{x=1 \\ x \neq k}}^{n} y_{kx} \right) \qquad (4.103)$$

for all $i, j \neq k$, where y_{ij} is the weight associated with the edge (i,j) of G_u.

If Y is the indefinite-admittance matrix of a linear passive network without mutual couplings, formula (4.103) is the well-known star-mesh transformation in network theory, and was first given by SHEN [1947]. However, (4.103) is also applicable to networks with mutual couplings. Thus, Corollary 4.20 may be considered as a generalized star-mesh transformation, and Theorem 4.5, in turn, is a generalization of Corollary 4.20 for active networks. The reason for this is that network functions, as we shall see in § 6, can always be expressed as the ratios of the first- and/or second-order cofactors.

We shall illustrate the above procedure by the following example.

Example 4.6: Consider the network or graph G_u as shown in fig. 4.19. Assume that $V_p = \{1, 2\}$. The corresponding reduced graph G_{ur} obtained from G_u by the removal of the nodes 1 and 2 using the procedure outlined in Corollary

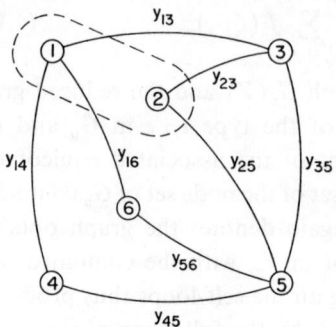
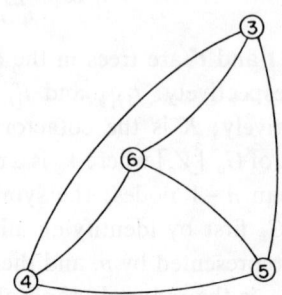

Fig. 4.19. An electrical network.   Fig. 4.20. The reduced network obtained by removing the nodes 1 and 2.

4.20 is presented in fig. 4.20. The weights w_{ij} associated with the edges of G_{ur} are computed from (4.102). For illustrative purposes, the weights ω_{34} and ω_{35} will be evaluated in detail.

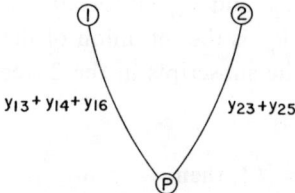

Fig. 4.21. The graph $G_u^*[1, 2]$ of the network of fig. 4.19.

The constant K is the cofactor of an element of the associated equicofactor matrix of the graph $G_u^*[1, 2]$ of fig. 4.21. Using (4.32) we have

$$K = \sum_{t^*} f(t^*)$$
$$= (y_{13} + y_{14} + y_{16})(y_{23} + y_{25}), \qquad (4.104)$$

where t^* is a tree of $G_u^*[1, 2]$. From figs. 4.22 and 4.23, we obtain

$$Kw_{34} = \sum_{t^*_{34,p}} f(t^*_{34,p})$$
$$= y_{13}y_{14}y_{25} + y_{13}y_{14}y_{23}, \qquad (4.105)$$
$$Kw_{35} = \sum_{t^*_{35,p}} f(t^*_{35,p})$$
$$= y_{35}y_{25}y_{13} + y_{35}y_{13}y_{23} + y_{35}y_{25}(y_{14} + y_{16}) + y_{35}y_{23}(y_{14} + y_{16})$$
$$\quad + y_{23}y_{25}(y_{14} + y_{16}) + y_{23}y_{25}y_{13}$$
$$= (y_{35}y_{25} + y_{35}y_{23} + y_{23}y_{25})(y_{13} + y_{14} + y_{16}), \qquad (4.106)$$

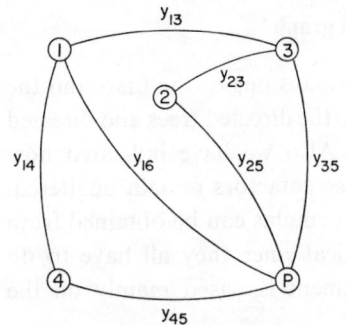
Fig. 4.22. The graph $G_u^*[1, 2, 3, 4]$.

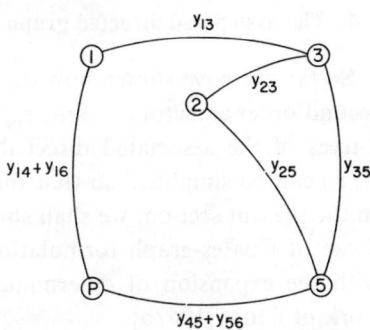
Fig. 4.23. The graph $G_u^*[1, 2, 3, 5]$.

where $t_{34,p}^*$ and $t_{35,p}^*$ are 2-trees in the graphs $G_u^*[1, 2, 3, 4]$ and $G_u^*[1, 2, 3, 5]$, respectively. Similarly, we get

$$Kw_{36} = y_{16}y_{13}y_{25} + y_{16}y_{13}y_{23}, \tag{4.107}$$

$$\begin{aligned}Kw_{45} &= y_{45}y_{25}y_{14} + y_{45}y_{25}(y_{13} + y_{16}) + y_{45}y_{23}y_{14} + y_{45}y_{23}(y_{13} + y_{16}) \\ &= (y_{13} + y_{14} + y_{16})(y_{45}y_{25} + y_{45}y_{23}),\end{aligned} \tag{4.108}$$

$$Kw_{46} = y_{14}y_{16}y_{23} + y_{14}y_{16}y_{25}, \tag{4.109}$$

$$\begin{aligned}Kw_{56} &= y_{56}y_{16}y_{25} + y_{56}y_{23}(y_{13} + y_{14}) + y_{56}y_{16}y_{23} + y_{56}y_{25}(y_{13} + y_{14}) \\ &= (y_{13} + y_{14} + y_{16})(y_{56}y_{23} + y_{56}y_{25}).\end{aligned} \tag{4.110}$$

Thus, from (4.100) and (4.101), the first-order cofactor Y_{ij} and the second-order cofactors $Y_{33,55}$ and $Y_{34,55}$ of the elements of the indefinite-admittance matrix of the network are given by

$$\begin{aligned}Y_{ij} &= K\sum_{t^r} f(t^r) \\ &= K[w_{36}(w_{34} + w_{46})(w_{35} + w_{56}) + w_{36}w_{45}(w_{34} + w_{46}) \\ &\quad + w_{36}w_{45}(w_{35} + w_{56}) + w_{34}w_{46}(w_{35} + w_{45} + w_{56}) \\ &\quad + w_{34}w_{45}w_{56} + w_{35}w_{45}w_{46} + w_{35}w_{56}(w_{34} + w_{46} + w_{45})],\end{aligned} \tag{4.111}$$

$$\begin{aligned}Y_{33,55} &= K\sum_{t^r_{3,5}} f(t^r_{3,5}) \\ &= K[(w_{34} + w_{45})(w_{36} + w_{56}) + w_{46}(w_{34} + w_{45}) + w_{46}(w_{36} + w_{56})],\end{aligned} \tag{4.112}$$

$$\begin{aligned}Y_{34,55} &= K\sum_{t^r_{34,5}} f(t^r_{34,5}) \\ &= K[w_{36}w_{46} + w_{34}w_{36} + w_{34}w_{46} + w_{34}w_{56}].\end{aligned} \tag{4.113}$$

§ 4. The associated directed graph and the Coates graph

So far, we have shown how the terms in the expansions of the first- and the second-order cofactors Y_{ij} and $Y_{pq,rs}$ are related to the directed trees and directed 2-trees of the associated directed graph $G(Y)$. Also we have indicated how $G(Y)$ can be simplified so that the ratios of these cofactors remain unaltered. In the present section, we shall show how these formulas can be obtained from those of Coates-graph formulation. This is logical since they all have to do with the expansion of determinants. Our treatment is based mainly on the work of CHEN [1967b].

4.1. *Directed trees, 1-factors, and semifactors*

Since Y is an equicofactor matrix of the type as shown in (4.21), it is sufficient to consider the cofactor Y_{nn}.

Let G_c be the associated Coates graph of the submatrix Y_{nn} of Y. Obviously, G_c has $n-1$ nodes, and the weights associated with the edges (i, j) of G_c are given by

$$f(i, j) = - y_{ji} \qquad (4.114)$$

for $i \neq j$, and

$$f(i, i) = \sum_{\substack{x=1 \\ x \neq i}}^{n} y_{ix}, \qquad (4.115)$$

where $i, j = 1, 2, ..., n-1$. For our purpose, each of the self-loops (i, i) of G_c will be represented by $n-1$ self-loops, denoted by $(i, i)_k$, $k=1, 2, ..., n$ but $k \neq i$, so that

$$f((i, i)_k) = y_{ik} \qquad (4.116)$$

for all i.

As an example, consider an equicofactor matrix Y of order 4 of (4.21) with all elements present. The associated Coates graph G_c of Y_{44}, being drawn in a slightly different form as outlined above, is as shown in fig. 4.24. The self-loops at node 1 are denoted by $(1, 1)_2$, $(1, 1)_3$, and $(1, 1)_4$ with weights y_{12}, y_{13}, and y_{14}, respectively.

Using the Coates-graph technique discussed in ch. 3, the cofactor Y_{nn} is evaluated by the formula given in Theorem 3.1;

$$Y_{nn} = \sum_{h} (-1)^q f(h), \qquad (4.117)$$

where h is a 1-factor of G_c, and q is the number of even components in h. As mentioned in ch. 3, § 3, the expansion of (4.117) is not very efficient because of the existence of a large number of cancellation terms. For example, the expansion of the cofactor Y_{44} of (4.22) by the formula (4.117) results in 38 terms of which 22 would be canceling one another. In the following, we shall first characterize those 1-factors of G_c which will not appear in the final expansion of Y_{nn}, and then show how the remaining 1-factors are related to the directed trees of $G(Y)$.

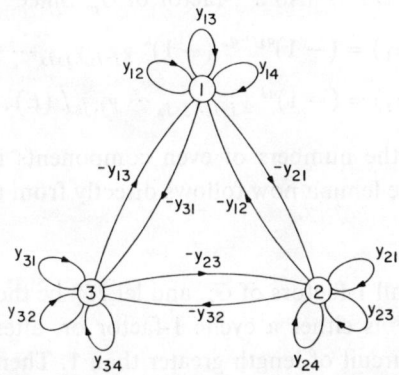

Fig. 4.24. The associated Coates graph of a fourth-order equicofactor matrix of (4.21).

DEFINITION 4.13: *Cyclic 1-factor*. A 1-factor of a Coates graph is said to be *cyclic* if it contains only self-loops of the form

$$(i_1, i_1)_{k_1}, (i_2, i_2)_{k_2}, \ldots, (i_j, i_j)_{k_j}, \quad (4.118)$$

so that there exists a set of indices $k_{t_1}, k_{t_2} \ldots k_{t_u}$, $u \geq 2$, of k's and a corresponding set of indices $i_{t_1}, i_{t_2} \ldots i_{t_u}$ of i's with $i_{t_{v+1}} = k_{t_v}$, $v = 1, 2, \ldots, u-1$, and $i_{t_1} = k_{t_u}$, where j and u are positive integers.

For example, in the Coates graph G_c of fig. 4.24, the 1-factor consisting of the self-loops $(1, 1)_2$, $(2, 2)_3$, and $(3, 3)_1$ of G_c is cyclic because there exists a set of indices 2, 3, and 1 and a corresponding set of indices 1, 2, and 3 with the above property. The 1-factor $(1, 1)_3 \cup (2, 2)_1 \cup (3, 3)_1$ is also cyclic, but $(1, 1)_4 \cup (2, 2)_1 \cup (3, 3)_2$ is not.

LEMMA 4.3: Let h_1 be a 1-factor of G_c containing at least one directed circuit of length greater than 1. Then there exists a 1-factor h_2 in G_c, $h_2 \neq h_1$, such that

$$-(-1)^{q_1} f(h_1) = (-1)^{q_2} f(h_2), \quad (4.119)$$

where q_1 and q_2 are the numbers of even components in h_1 and h_2, respectively.

Proof. Let L_1 be a directed circuit of length k, $k \geq 2$, in h_1. Let E be the complement of L_1 in h_1. Then $h_1 = L_1 \cup E$. Also let

$$L_1 = (j_1, j_2) \cup (j_2, j_3) \cup \cdots \cup (j_k, j_1), \tag{4.120a}$$

$$L_2 = (j_2, j_2)_{j_1} \cup (j_3, j_3)_{j_2} \cup \cdots \cup (j_1, j_1)_{j_k}. \tag{4.120b}$$

It follows that $h_2 = L_2 \cup E$ is also a 1-factor of G_c. Since

$$(-1)^{q_1} f(h_1) = (-1)^{q_L + q_e} (-1)^k y_{j_2 j_1} y_{j_3 j_2} \cdots y_{j_1 j_k} f(E), \tag{4.121}$$

$$(-1)^{q_2} f(h_2) = (-1)^{q_e} y_{j_2 j_1} y_{j_3 j_2} \cdots y_{j_1 j_k} f(E), \tag{4.122}$$

where q_L and q_e are the numbers of even components in the subgraphs L_1 and E, respectively, the lemma now follows directly from the fact that $(-1)^{k-1} = (-1)^{q_L}$.

Let H be the set of all 1-factors of G_c, and let H^* be the subset of H so that, for each h^* in H^*, h^* is either a cyclic 1-factor or, alternatively, it contains at least one directed circuit of length greater than 1. Then we have

THEOREM 4.7:

$$\sum_{h^*} (-1)^{q^*} f(h^*) = 0, \tag{4.123}$$

where h^* is an element of H^*; q^* is the number of even components in h^*.

Proof. Let $h_i = L \cup E$ be an element of H^* such that there exists no $h'_i = L' \cup E'$, also in H^*, which has the property that L is a proper subgraph of L', where L and L' are subgraphs of h_i and h'_i consisting of all the directed circuits of length ≥ 2 in h_i and h'_i, respectively; and E and E' are the complements of L and L' in h_i and h'_i, respectively.

For each edge (i, j), $i \neq j$, in G_c there exists a unique corresponding self-loop $(j, j)_i$ also in G_c. If L''_x is defined as the directed graph obtained from a directed circuit L_x of length ≥ 2 in G_c by replacing edges in L_x with their corresponding unique self-loops, then, for each choice of h_i, there are 2^m 1-factors of H^* that can be obtained by this type of operations where m is the number of directed circuits in L of h_i. In other words, if S_i is the subset of 1-factors of H^* that can be generated from h_i by the above operations, then

$$S_i = \{h; h = L''_{i_1} \cup \cdots \cup L''_{i_x} \cup L_{j_1} \cup \cdots \cup L_{j_{m-x}} \cup E\}, \tag{4.124}$$

where $L_1, L_2, \ldots,$ and L_m are m directed circuits of L, and $i_1 i_2 \ldots i_x$ and $j_1 j_2 \ldots j_{m-x}$ are the complementary indices of the integers $1, 2, \ldots, m$. It follows that half of the elements of S_i contains L_1, say; the other half does not. By replacing L_1 by L_1'' in each of the elements of the former half, we obtain the latter half. Thus, by Lemma 4.3 we have

$$\sum_{h \in S_i} (-1)^q f(h) = 0, \qquad (4.125)$$

where h is an element of S_i, and q is the number of even components in h.

If we repeat this process for all h_i in H^*, it is obvious that the elements of H^* can be partitioned into a set of disjoint subsets S_i with the above property. The theorem follows directly from here.

A direct consequence of the above theorem is the following which shows how the uncanceled 1-factors of H are related to the directed trees in the associated directed graph $G(Y)$ and the semifactors in the modified Coates graph G_c' of the Coates graph G_c. For convenience, let T_n and S be the sets of directed trees t_n and semifactors R of $G(Y)$ and G_c', respectively. Then, we have

THEOREM 4.8: There exists a one-to-one correspondence between the elements h of $(H-H^*)$ and the elements t_n of T_n (or R of S), and such that

$$f(h) = f(t_n) = (-1)^{q_R} f(R), \qquad (4.126)$$

where q_R is the number of even components in R.

Proof. If we associate each edge $(i, i)_j$ in G_c with an edge (i, j) in $G(Y)$ for all $i \neq j$, it follows immediately that the corresponding graph t_n of h in $G(Y)$ is a directed tree with reference node n. For, if not, t_n must contain a directed circuit of length ≥ 2. This would imply that h is a cyclic 1-factor of G_c, so it cannot appear in $(H-H^*)$, a contradiction. Thus, there exists a one-to-one correspondence between the elements of $(H-H^*)$ and T_n. Since $f((i, i)_j) = y_{ij}$ in G_c and $f(i, j) = y_{ij}$ in $G(Y)$, it follows that $f(h) = f(t_n)$. Similarly, we can prove the case for the semifactors in G_c'. However, the details are left as an exercise (Problem 4.33). This completes the proof of the theorem.

Thus, the cofactors Y_{ij} of the elements of Y of the type as shown in (4.21) can be expanded by any one of the following three formulas:

$$Y_{ij} = \sum_h (-1)^q f(h) \qquad (4.127a)$$

$$= \sum_{t_k} f(t_k) \qquad (4.127b)$$

$$= \sum_R (-1)^{q_R} f(R) \qquad (4.127c)$$

for $i, j, k = 1, 2, \ldots, n$, where h, t_k, and R are a 1-factor, a directed tree, and a semifactor in G_c, $G(Y)$, and G_c', respectively; q and q_R are the numbers of even components in h and R, respectively.

As an illustration, consider an equicofactor matrix Y of order 4 of (4.21) with all elements present. As mentioned above, if the cofactor Y_{44} is expanded by the formula (4.127a) above, 38 terms will be generated before cancellations, of which 22 would be canceling one another. The remaining 16 terms are in one-to-one correspondence with the directed trees t_4 of the associated directed graph $G(Y)$ of fig. 4.25 or with the semifactors of the modified Coates graph G_c' of fig. 4.26.

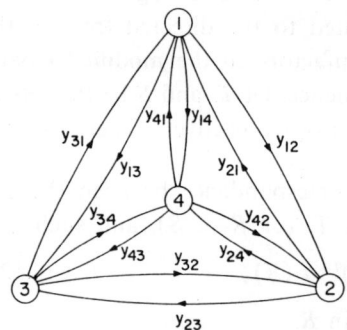
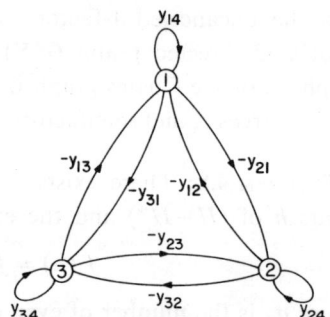

Fig. 4.25. The associated directed graph of a fourth-order equicofactor matrix of (4.21).

Fig. 4.26. The modified Coates graph of the Coates graph of fig. 4.24.

In other words, topological formulas using directed trees and semifactors are actually the same as those for the Coates graph (hence the Mason graph) except that the former look only for the uncanceled 1-factors of the latter. Consequently, insofar as the equicofactor matrix of (4.21) is concerned, the last two formulas of (4.127) are optimum for the evaluation of the cofactors of its elements, since the usual cancellations inherent in the expansions of these cofactors are avoided. Hence, they can be employed effectively for the analysis of linear systems in general and electrical networks in particular.

4.2. Directed 2-trees, 1-factorial connections, and 1-semifactors

Like the previous case, the directed 2-trees of $G(Y)$ and 1-semifactors of G_c' are intimately related to the 1-factorial connections of G_c.

DEFINITION 4.14: *Cyclic 1-factorial connection.* A 1-factorial connection H_{ij} from node i to node j of a Coates graph G_c is said to be *cyclic* if, after removing the nodes of its directed path from i to j from H_{ij}, the remaining subgraph becomes a cyclic 1-factor for some sectional subgraph of G_c.

Let F be the set of all 1-factorial connections H_{ij} from node i to node j in the associated Coates graph G_c of Y_{nn}, and let F^* be the subset of F so that, for each H_{ij}^* in F^*, H_{ij}^* is either a cyclic 1-factorial connection or, alternatively, it contains at least one directed circuit of length not less than 2. Then we have

THEOREM 4.9: For $i \neq j$,

$$\sum_{H^*_{ij}} (-1)^{q^*_H} f(H_{ij}^*) = 0, \qquad (4.128)$$

where H_{ij}^* is an element of F^*, and q_H^* is the number of even components in H_{ij}^*.

Proof. Let P_{ij} be a directed path from node i to node j in G_c. For each P_{ij}, let $F^*(P_{ij})$ be the subset of F^* in which each of its elements contains P_{ij}. Also, let $F_1^*(P_{ij})$ be the set obtained from $F^*(P_{ij})$ by removing P_{ij} from each of its elements. Clearly, $F_1^*(P_{ij})$ is the set of 1-factors, each of which is either a cyclic 1-factor or, alternatively, contains at least one directed circuit of length not less than 2, in the Coates graph $G_c(P_{ij})$ obtained from G_c by the removal of all the nodes and edges of P_{ij}. Thus, from Theorem 4.7 we obtain

$$\sum_{H^*_{ij} \in F^*(P_{ij})} (-1)^{q^*_H} f(H_{ij}^*) = \pm f(P_{ij}) \Big[\sum_{h^* \in F^*_1(P_{ij})} (-1)^{q^*_h} f(h^*) \Big] = 0, \qquad (4.129)$$

where h^* is a 1-factor of $G_c(P_{ij})$ in $F_1^*(P_{ij})$, and q_h^* is the number of even components in h^*.

Since the elements of the set F^* can be partitioned into a set of disjoint subsets $F^*(P_{ij})$ with the above property, the theorem follows from here.

Let $T_{ij,n}$ and S_{ij} be the sets of directed 2-trees $t_{ij,n}$ and 1-semifactors $R(ij)$ of $G(Y)$ and G'_c, respectively. Then we have

THEOREM 4.10: There exists a one-to-one correspondence between the elements H_{ij} of $(F - F^*)$ and the elements $t_{ij,n}$ of $T_{ij,n}$ (or $R(ij)$ of S_{ij}) and such that

$$(-1)^{q_H - 1} f(H_{ij}) = f(t_{ij,n}) = (-1)^{q_{ij} - 1} f(R(ij)), \qquad (4.130)$$

where q_H and q_{ij} are the numbers of even components in H_{ij} and $R(ij)$, respectively.

The proof of the above theorem is similar to that of Theorem 4.8, and is left as an exercise (Problem 4.36).

Thus, the second-order cofactors $Y_{ij,nn}$ can be evaluated by any one of the following three formulas:

$$Y_{ij,nn} = \sum_{H_{ij}} (-1)^{q_H - 1} f(H_{ij}) \tag{4.131a}$$

$$= \sum_{t_{ij,n}} f(t_{ij,n}) \tag{4.131b}$$

$$= \sum_{R(ij)} (-1)^{q_{ij} - 1} f(R(ij)) \tag{4.131c}$$

where H_{ij}, $t_{ij,n}$, and $R(ij)$ are a 1-factorial connection from i to j, a directed 2-tree, and a 1-semifactor in G_c, $G(Y)$, and G'_c, respectively; and q_H and q_{ij} are the numbers of even components in H_{ij} and $R(ij)$, respectively.

As an illustration, let us again consider the equicofactor matrix Y of order 4. Suppose that we wish to evaluate the second-order cofactor $Y_{22,44}$,

$$Y_{22,44} = \det \begin{bmatrix} y_{12} + y_{13} + y_{14} & -y_{13} \\ -y_{31} & y_{31} + y_{32} + y_{34} \end{bmatrix}$$

$$= y_{12}y_{31} + y_{12}y_{32} + y_{12}y_{34} + y_{13}y_{31} + y_{13}y_{32} + y_{13}y_{34}$$
$$+ y_{14}y_{31} + y_{14}y_{32} + y_{14}y_{34} - y_{13}y_{31}$$
$$= y_{12}y_{31} + y_{12}y_{32} + y_{12}y_{34} + y_{13}y_{32} + y_{13}y_{34} + y_{14}y_{31}$$
$$+ y_{14}y_{32} + y_{14}y_{34}. \tag{4.132}$$

Thus, if $Y_{22,44}$ is expanded by the formula (4.131a), ten terms will be generated before cancellations, of which two would be canceling each other. The remaining eight terms are in one-to-one correspondence with the directed 2-trees $t_{2,4}$ of $G(Y)$ of fig. 4.25 or with the 1-semifactors $R(2)$ of G'_c of fig. 4.26. Similarly, using (4.131) we can evaluate

$$Y_{12,44} = y_{21}y_{31} + y_{21}y_{32} + y_{21}y_{34} + y_{23}y_{31}, \tag{4.133}$$

in which the terms are in one-to-one correspondence with the 1-factorial connections H_{12} of G_c of fig. 4.24, or the directed 2-trees $t_{12,4}$ of $G(Y)$ of fig. 4.25, or the 1-semifactors $R(12)$ of G_c of fig. 4.26.

Again, topological rules using directed 2-trees or 1-semifactors for the evaluation of the second-order cofactors are actually the same as those for the Coates graph (hence the Mason graph) except that the former look only for those uncanceled 1-factorial connections of the latter. Thus, the last two formulas of (4.131) are also optimum for the evaluation of the second-order cofactors of the elements of Y of (4.21).

§ 5. Generation of directed trees and directed 2-trees

The efficiency with which the first- and second-order cofactors of the elements of a high-order equicofactor matrix may be analyzed by a digital computer depends largely upon the efficiency with which the sets of directed trees and directed 2-trees in the associated directed graph are generated. In this section, we shall introduce three techniques for their enumeration. The first one is similar to that for the generation of semifactors and k-semifactors discussed in ch. 3, § 4.2. The formulation is simple but duplications necessarily occur, and thus it is not very efficient. The second and third ones are iterative procedures which will not introduce any redundancies, duplications, or cancellations. The three techniques are all readily adaptable for digital computers. The development of the first two procedures will follow closely that of CHEN [1966a, 1968a]. The third one was implicitly given by FEUSSNER [1902, 1904], and has been used extensively by MASON [1957]. Other methods of tree generation are also available, and they will be treated in great detail in ch. 5.

5.1. *Algebraic formulation*

Let G be an n-node directed graph in which every edge has been assigned an edge-designation symbol e_k or e_{ij} where e_{ij}, if it is used, denotes the edge directed from node i to node j of G. For simplicity, we assume that G has no parallel edges and self-loops. However, the extension to the general case is straightforward, so it is left as an exercise (Problem 4.37).

Like those symbols used in ch. 3, § 4.1, let S_i^+ be the set consisting of all the edges of G having node i as their initial node. Using the Wang-product operation defined in (3.118) and (3.119), we can now state the result of this section.

THEOREM 4.11: In an n-node symmetric directed graph G, let

$$T_k = S_1^+ \text{@} \cdots \text{@} S_{k-1}^+ \text{@} S_{k+1}^+ \text{@} \cdots \text{@} S_n^+ . \tag{4.134}$$

Then the elements in the final expansion of T_k correspond to the set of directed trees t_k of G, provided that the edges e_{ij} and e_{ji} are considered as identical elements in all the operations.

Proof. It is not difficult to see that each element in the expansion of T_k corresponds to a subgraph of G, each of its nodes being of outgoing degree 1 except the node k which is of outgoing degree 0. Thus, every element of T_k is either a directed tree t_k or a subgraph containing a directed circuit. Also, it is clear that every directed tree t_k of G is contained in T_k. It remains to be shown that if an element of T_k which is not a directed tree t_k of G then it must appear an even number times. In the present case, we shall show that it will appear exactly twice in the expansion of the Wang product, and consequently it will not appear in the final expansion (4.134).

Let x be one of such elements. Since G has no self-loops and since the edges e_{ij} and e_{ji} are considered identical in all the operations, x corresponds to a subgraph of G containing a directed circuit L_1 of length u, $u \geq 3$. Let L_1 and E be the complementary subgraphs of x. Then $x = L_1 \cup E$. Since G is symmetric, the subgraph L_2 obtained from L_1 by reversing the direction of each of its edges is also a directed circuit of length u in G. It follows that the element corresponding to the subgraph $L_2 \cup E$ of G is also in the Wang product of (4.134). Thus, x will not appear in the final expansion of T_k. So the theorem is proved.

Since an (undirected) graph can always be transformed into a symmetric directed graph by representing each of its undirected edges by a pair of directed ones with opposite directions (Definition 4.9), the technique outlined above may also be used to generate the set of trees of any graph.

Like the case for the generation of semifactors, if G is not symmetric, it is always possible to make it symmetric. Let G_s be the symmetric directed graph obtained from G by adding a minimum number of edges to G. In order that we may distinguish the added edges from those already in G, let the primes be used for this purpose. For example, if (i, j) is in G and (j, i) in G_s but not in G, then in G_s let (i, j) be labeled by e_{ij} and (j, i) by e'_{ji}.

For an arbitrary directed graph G, suppose that the edges of its corresponding G_s have been labeled according to the rules just adopted. The set of directed trees of G_s can then be generated by the procedure outlined in Theorem 4.11 provided that e_{ij} and e_{ji} or e_{uv} and e'_{vu} are considered as identical elements in all the operations. If we eliminate all those elements of the set which contain at least one primed letter, then the remaining elements are the desired directed trees of G.

We remark that the expansion of the Wang product of the sets S_i^+ of (4.134) amounts to forming all combinations of edges without duplication of subscripts (disregarding the order of the subscripts), and then eliminating all those elements in which the same subscripts appear elsewhere as a permutation. We emphasize that even though the order of the subscripts is immaterial in all the operations, it is important in forming the weight product of a subgraph.

Since the Wang-product operation is associative, the number of combinations to be examined may be greatly reduced by eliminating some of the elements at an earlier stage; for example, if in $S_1^+ @ S_2^+$ there is an element with duplicated subscripts, then the element can be eliminated from $S_1^+ @ S_2^+$ before we form the Wang-product operation with the remaining sets. This is similarly valid if two elements have the same subscripts; they should be eliminated before other operations are performed.

We shall illustrate this by the following example.

Example 4.7: Consider the directed graph G of fig. 4.15. Suppose that we wish to generate the set T_5 of directed trees t_5 of G. The corresponding G_s of G is presented in fig. 4.27. For our purpose, let the edges (i, j) of G_s be labeled by the symbols e_{ij} for $i, j = 1, 2, 3, 4, 5$ except the edge $(2, 4)$ which is labeled by e'_{24}. From Theorem 4.11, we obtain

$$T'_5 = S_1^+ @ S_2^+ @ S_3^+ @ S_4^+$$
$$= \{e_{12}\} @ \{e_{21}, e_{23}, e'_{24}, e_{25}\} @ \{e_{32}, e_{34}, e_{35}\} @ \{e_{42}, e_{43}, e_{45}\}$$
$$= \{e_{12}e_{23}, e_{12}e'_{24}, e_{12}e_{25}\} @ \{e_{32}e_{42}, e_{32}e_{43}, e_{32}e_{45}, e_{34}e_{42}, e_{34}e_{45},$$
$$e_{35}e_{42}, e_{35}e_{43}, e_{35}e_{45}\}$$
$$= \{e_{12}e_{23}e_{34}e_{45}, e_{12}e_{23}e_{35}e_{42}, e_{12}e_{23}e_{35}e_{43}, e_{12}e_{23}e_{35}e_{45},$$
$$e_{12}e'_{24}e_{32}e_{45}, e_{12}e'_{24}e_{34}e_{45}, e_{12}e'_{24}e_{35}e_{43}, e_{12}e'_{24}e_{35}e_{45},$$
$$e_{12}e_{25}e_{32}e_{42}, e_{12}e_{25}e_{32}e_{43}, e_{12}e_{25}e_{32}e_{45}, e_{12}e_{25}e_{34}e_{42},$$
$$e_{12}e_{25}e_{34}e_{45}, e_{12}e_{25}e_{35}e_{42}, e_{12}e_{25}e_{35}e_{43}, e_{12}e_{25}e_{35}e_{45}\}. \quad (4.135)$$

Since T'_5 corresponds to the set of directed trees t'_5 of G_s, the set of directed trees t_5 of G can easily be obtained from T'_5 by eliminating those elements which contain at least one primed letter:

$$T_5 = \{e_{12}e_{23}e_{34}e_{45}, e_{12}e_{23}e_{35}e_{42}, e_{12}e_{23}e_{35}e_{43}, e_{12}e_{23}e_{35}e_{45},$$
$$e_{12}e_{25}e_{32}e_{42}, e_{12}e_{25}e_{32}e_{43}, e_{12}e_{25}e_{32}e_{45}, e_{12}e_{25}e_{34}e_{42},$$
$$e_{12}e_{25}e_{34}e_{45}, e_{12}e_{25}e_{35}e_{42}, e_{12}e_{25}e_{35}e_{43}, e_{12}e_{25}e_{35}e_{45}\}. \quad (4.136)$$

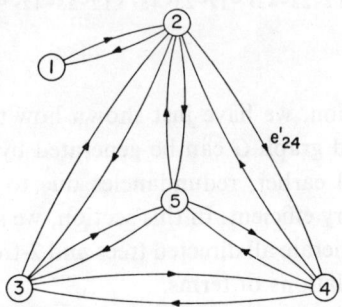

Fig. 4.27. The symmetric directed graph derived from the directed graph of fig. 4.15 by adding an edge e'_{24}.

With simple modifications, the above procedure can also be used to generate the set of directed 2-trees of G. We shall only state the result, and leave the details of the justification as exercises (Problems 4.38 and 4.54).

THEOREM 4.12: In an n-node symmetric directed graph G, let $T_{i,j}$ denote the Wang product of the sets S_k^+ for $k=1, 2, ..., n$ but $k \neq i, j$. Then the elements in the final expansion of $T_{i,j}$ correspond to the set of directed 2-trees $t_{i,j}$ of G, provided that the edges e_{uv} and e_{vu} are considered as identical elements in all the operations.

Like the case for the generation of directed trees, if G is not symmetric, the set of directed 2-trees of G_s can first be generated by the procedure outlined above. If we eliminate all those elements of the set which contain at least one primed letter, the remaining elements correspond to the set of desired directed 2-trees of G. We shall illustrate this by the following example.

Example 4.8: Consider the directed graph G as shown in fig. 4.15. Suppose that we wish to generate the set $T_{3,5}$ of directed 2-trees $t_{3,5}$ of G. The corresponding G_s of G is given in fig. 4.27. From Theorem 4.12, the set $T'_{3,5}$ of directed 2-trees $t'_{3,5}$ of G_s can be generated as follows:

$$T'_{3,5} = \{e_{12}\} @ \{e_{21}, e_{23}, e'_{24}, e_{25}\} @ \{e_{42}, e_{43}, e_{45}\}$$
$$= \{e_{12}e_{23}, e_{12}e'_{24}, e_{12}e_{25}\} @ \{e_{42}, e_{43}, e_{45}\}$$
$$= \{e_{12}e_{23}e_{42}, e_{12}e_{23}e_{43}, e_{12}e_{23}e_{45}, e_{12}e'_{24}e_{43}, e_{12}e'_{24}e_{45}, e_{12}e_{25}e_{42},$$
$$e_{12}e_{25}e_{43}, e_{12}e_{25}e_{45}\}. \quad (4.137a)$$

The set $T_{3,5}$ is then obtained from $T'_{3,5}$ by eliminating all those elements which contain at least one primed letter:

$$T_{3,5} = \{e_{12}e_{23}e_{42}, e_{12}e_{23}e_{43}, e_{12}e_{23}e_{45}, e_{12}e_{25}e_{42}, e_{12}e_{25}e_{43}, e_{12}e_{25}e_{45}\}.$$
$$(4.137b)$$

5.2. Iterative procedure

In the preceding section, we have just shown how the sets of directed trees and 2-trees of a directed graph G can be generated by an algebraic technique. However, as mentioned earlier, redundancies due to duplications necessarily occur. Thus, it is not very efficient. In this section, we shall describe an efficient procedure which will generate all directed trees and 2-trees without duplications, redundancies, or cancellations of terms.

Let G be an n-node directed graph, and also let e_{ij} be the symbol denoting the edge (i, j) directed from node i to node j in G. Without loss of generality,

§ 5 Directed trees and directed 2-trees

we may assume that G contains no self-loops and parallel edges, and is connected. Also, for convenience, we assume that the node n will be used as the reference node for the directed trees or one of the reference nodes of the directed 2-trees of G.

DEFINITION 4.15: *Unisignant.* The *unisignant*, denoted by the symbol $U(G)$, of an n-node labeled directed graph G is the determinant of the *unisignant matrix*

$$U(G) = \begin{bmatrix} \sum_{k=2}^{n} e_{1k} & -e_{12} & -e_{13} & \cdots & -e_{1(n-1)} \\ -e_{21} & \sum_{\substack{k=1 \\ k \neq 2}}^{n} e_{2k} & -e_{23} & \cdots & -e_{2(n-1)} \\ \vdots & \vdots & \vdots & & \vdots \\ -e_{(n-1)1} & -e_{(n-1)2} & -e_{(n-1)3} & \cdots & \sum_{\substack{k=1 \\ k \neq n-1}}^{n} e_{(n-1)k} \end{bmatrix}. \quad (4.138)$$

If the edge-designation symbols e_{ij} of G are considered as the weights associated with the edges of G, the unisignant $U(G)$ is actually the cofactor of the (n, n)-element of the associated matrix of G. The only difference is that the associated matrix may not be an equicofactor matrix. Note that $U(G) = \det U(G)$. In the following, we shall discuss a technique that modifies the column expansion for $U(G)$, which will iteratively remove all the cancellation terms in the expansion of $U(G)$, and which therefore reduces considerably the labor in making the expansion. Since the terms in the final expansion of $U(G)$ are in one-to-one correspondence with the directed trees t_n of G, the process also generates all the directed trees t_n of G.

The iterative removal of all the cancellation terms in the expansion of $U(G)$ is accomplished as follows: replace the first column of $U(G)$ by the sum of all its columns. This yields a matrix whose determinant is the same as that of $U(G)$:

$$\begin{bmatrix} e_{1n} & -e_{12} & -e_{13} & \cdots & -e_{1(n-1)} \\ e_{2n} & \sum_{\substack{k=1 \\ k \neq 2}}^{n} e_{2k} & -e_{23} & \cdots & -e_{2(n-1)} \\ \vdots & \vdots & \vdots & & \vdots \\ e_{(n-1)n} & -e_{(n-1)2} & -e_{(n-1)3} & \cdots & \sum_{\substack{k=1 \\ k \neq n-1}}^{n} e_{(n-1)k} \end{bmatrix}. \quad (4.139)$$

For convenience, we shall say that (4.139) is the *modified form* of $U(G)$ or simply that $U(G)$ is in modified form. Now, we shall show that, for a given unisignant matrix $U(G)$ in modified form, its determinant can be expressed as the weighted sum of the determinants of the unisignant matrices, again in modified form, of order one less than that of the original $U(G)$. Thus, the process may be repeated until all the unisignant matrices are of order 1.

Now, if we expand (4.139) along its first column, we have

$$U(G) = \sum_{i=1}^{n-1} e_{in} U_{i1}, \qquad (4.140)$$

where U_{ij} is the cofactor of the (i,j)-element of $U(G)$. Let $\boldsymbol{U}_{ij}$ be the submatrix obtained from $U(G)$ by deleting the ith row and jth column. For $i=1$, $\boldsymbol{U}_{i1}$ can be put in modified form simply by adding all the succeeding columns to the first column of $\boldsymbol{U}_{11}$. For $i \neq 1$, $\boldsymbol{U}_{i1}$ can be put in modified form simply by moving its $(i-1)$th column to the position of its first column, while keeping all other columns in their original order. In doing so, a total of $(i-2)$ column interchanges are needed. Since the elements of the first column of the resulting matrix all carry a negative sign, let the signs of all the elements of this column be changed to positive. Denote this derived matrix by $\boldsymbol{M}_{i1}$. It is not difficult to see that $\boldsymbol{M}_{i1}$ is $\boldsymbol{U}_{i1}$ in modified form, and

$$U_{i1} = (-1)^{i+1}(-1)^{i-2}(-1)M_{i1} = M_{i1}, \qquad (4.141a)$$

$$U(G) = \sum_{i=1}^{n-1} e_{in} M_{i1}, \qquad (4.141b)$$

where $M_{11} = \det \boldsymbol{U}_{11}$ and $M_{x1} = \det \boldsymbol{M}_{x1}$ for $x = 2, 3, \ldots, n-1$.

We shall illustrate the above procedure by the following examples.

Example 4.9: Consider the directed graph G as shown in fig. 4.28. $U(G)$ together with $U(G)$ in modified form and its expansion into unisignant matrices of lower orders, again in modified form, are given in determinantal form as

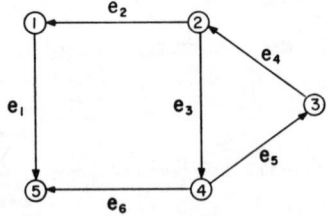

Fig. 4.28. A directed graph illustrating the generation of directed trees.

follows:

$$U(G) = \begin{vmatrix} e_1 & 0 & 0 & 0 \\ -e_2 & e_2+e_3 & 0 & -e_3 \\ 0 & -e_4 & e_4 & 0 \\ 0 & 0 & -e_5 & e_5+e_6 \end{vmatrix}$$

$$= \begin{vmatrix} e_1 & 0 & 0 & 0 \\ 0 & e_2+e_3 & 0 & -e_3 \\ 0 & -e_4 & e_4 & 0 \\ e_6 & 0 & -e_5 & e_5+e_6 \end{vmatrix}$$

$$= e_1 \begin{vmatrix} e_2 & 0 & -e_3 \\ 0 & e_4 & 0 \\ e_6 & -e_5 & e_5+e_6 \end{vmatrix} + e_6 \begin{vmatrix} 0 & 0 & 0 \\ e_3 & e_2+e_3 & 0 \\ 0 & -e_4 & e_4 \end{vmatrix}$$

$$= e_1 \left(e_2 \begin{vmatrix} e_4 & 0 \\ e_6 & e_5+e_6 \end{vmatrix} + e_6 \begin{vmatrix} e_3 & 0 \\ 0 & e_4 \end{vmatrix} \right)$$

$$= e_1 e_2 e_4 (e_5 + e_6) + e_1 e_6 e_3 e_4. \tag{4.142}$$

There are three terms in its final expansion, each of which corresponds to a directed tree of G with reference node 5. Observe that all the terms are given in factored form. From an application viewpoint, this indeed is the most desirable form since we are usually interested in the weight products of the subgraphs rather than the subgraphs themselves. Thus, there is no need to carry out the final expansion.

Example 4.10: Consider the associated undirected graph G_u of the directed graph G of fig. 4.28. The graph G_u is as shown in fig. 4.29. If the set of trees of G_u is needed, the same procedure may be employed. The only difference is that we use the unisignant $U(G_s)$ of the associated symmetric directed graph G_s of G_u, where a pair of oppositely directed edges in G_s are assigned the same edge-designation symbol, or, alternatively, we may consider the edge-designa-

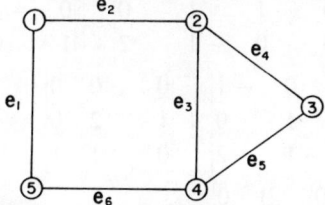

Fig. 4.29. The associated undirected graph of the directed graph of fig. 4.28.

tion symbols of G_u as the associated weights of the edges of G_u:

$$U(G_s) = \begin{vmatrix} e_1+e_2 & -e_2 & 0 & 0 \\ -e_2 & e_2+e_3+e_4 & -e_4 & -e_3 \\ 0 & -e_4 & e_4+e_5 & -e_5 \\ 0 & -e_3 & -e_5 & e_3+e_5+e_6 \end{vmatrix}$$

$$= \begin{vmatrix} e_1 & -e_2 & 0 & 0 \\ 0 & e_2+e_3+e_4 & -e_4 & -e_3 \\ 0 & -e_4 & e_4+e_5 & -e_5 \\ e_6 & -e_3 & -e_5 & e_3+e_5+e_6 \end{vmatrix}$$

$$= e_1 \begin{vmatrix} e_2 & -e_4 & -e_3 \\ 0 & e_4+e_5 & -e_5 \\ e_6 & -e_5 & e_3+e_5+e_6 \end{vmatrix} + e_6 \begin{vmatrix} 0 & -e_2 & 0 \\ e_3 & e_2+e_3+e_4 & -e_4 \\ e_5 & -e_4 & e_4+e_5 \end{vmatrix}$$

$$= e_1 \left(e_2 \begin{vmatrix} e_4 & -e_5 \\ e_3+e_6 & e_3+e_5+e_6 \end{vmatrix} + e_6 \begin{vmatrix} e_3 & -e_4 \\ e_5 & e_4+e_5 \end{vmatrix} \right)$$

$$+ e_6 \left(e_3 \begin{vmatrix} e_2 & 0 \\ e_4 & e_4+e_5 \end{vmatrix} + e_5 \begin{vmatrix} 0 & -e_2 \\ e_4 & e_2+e_3+e_4 \end{vmatrix} \right)$$

$$= e_1 e_2 (e_3 e_4 + e_4 e_5 + e_4 e_6 + e_3 e_5 + e_5 e_6) + e_1 e_6 (e_3 e_4 + e_3 e_5 + e_4 e_5)$$
$$+ e_6 e_3 (e_2 e_4 + e_2 e_5) + e_6 e_5 e_2 e_4. \tag{4.143}$$

There are eleven terms in its final expansion, each of which corresponds to a tree of G_u.

It is obvious that, if we let $e_{ij}=1$ for all i and j in G, then $U(G)$ becomes the cofactor of the (n, n)-element of the directed-tree matrix $D(G)$. This is similarly valid for the tree matrix $D(G_u)$. Thus, these cofactors represent the numbers of directed trees t_n of G and trees of G_u. Because of their special form, the technique discussed above may be effectively employed for their evaluation.

As an illustration, consider the unisignant $U(G)$ of Example 4.9. If we let all the edge-designation symbols be 1, then

$$U(G) = \begin{vmatrix} 1 & 0 & 0 & 0 \\ -1 & 2 & 0 & -1 \\ 0 & -1 & 1 & 0 \\ 0 & 0 & -1 & 2 \end{vmatrix} = \begin{vmatrix} 1 & 0 & 0 & 0 \\ 0 & 2 & 0 & -1 \\ 0 & -1 & 1 & 0 \\ 1 & 0 & -1 & 2 \end{vmatrix}$$

$$= \begin{vmatrix} 1 & 0 & -1 \\ 0 & 1 & 0 \\ 1 & -1 & 2 \end{vmatrix} + \begin{vmatrix} 0 & 0 & 0 \\ 1 & 2 & 0 \\ 0 & -1 & 1 \end{vmatrix}$$

$$= \begin{vmatrix} 1 & 0 \\ 1 & 2 \end{vmatrix} + \begin{vmatrix} 1 & 0 \\ 0 & 1 \end{vmatrix} = 3. \tag{4.144}$$

Thus, there are three directed trees t_5 in G. Similarly, if the number of trees in G_u of fig. 4.29 is needed, the same procedure may be used:

$$U(G_s) = \begin{vmatrix} 2 & -1 & 0 & 0 \\ -1 & 3 & -1 & -1 \\ 0 & -1 & 2 & -1 \\ 0 & -1 & -1 & 3 \end{vmatrix} = \begin{vmatrix} 1 & -1 & 0 & 0 \\ 0 & 3 & -1 & -1 \\ 0 & -1 & 2 & -1 \\ 1 & -1 & -1 & 3 \end{vmatrix}$$

$$= \begin{vmatrix} 1 & -1 & -1 \\ 0 & 2 & -1 \\ 1 & -1 & 3 \end{vmatrix} + \begin{vmatrix} 0 & -1 & 0 \\ 1 & 3 & -1 \\ 1 & -1 & 2 \end{vmatrix}$$

$$= \begin{vmatrix} 1 & -1 \\ 2 & 3 \end{vmatrix} + \begin{vmatrix} 1 & -1 \\ 1 & 2 \end{vmatrix} + \begin{vmatrix} 1 & 0 \\ 1 & 2 \end{vmatrix} + \begin{vmatrix} 0 & -1 \\ 1 & 3 \end{vmatrix} = 11. \quad (4.145)$$

So there are eleven trees in G_u.

With only minor modifications, the same procedure may also be used for the generation of directed 2-trees or 2-trees. Suppose that we wish to generate the set of directed 2-trees $t_{ij,n}$ in G using the iterative procedure discussed above. This is equivalent to the evaluation of the cofactor U_{ij}. For example, if $i>j$, let the $(i-1)$th column of U_{ij} be moved to the position of its jth column, while keeping all other columns in their original order. By doing so, a total of $(i-j-1)$ column interchanges are needed. Since the elements of the jth column of the resulting matrix all carry a negative sign, let the signs of all the elements of this column be changed to positive. For convenience, let this derived matrix M_{ij}'' be called the matrix U_{ij} in *modified form*. Then

$$U_{ij} = (-1)^{i+j}(-1)^{i-j-1}(-1) M_{ij}'' = M_{ij}'', \quad (4.146)$$

where $M_{ij}'' = \det M_{ij}''$. Also it is not difficult to see that any U_{ij} in modified form can be transformed into a unisignant matrix in modified form by simultaneous row and column interchanges. Thus, it will not change the value of its determinant. Similarly, this is valid for $i<j$. We have already considered the case for $i=j$ in the expansion of $U(G)$. Thus, the determinant of U_{ij} in modified form can be expressed as the weighted sum of the determinants of the unisignant matrices, again in modified form, of order one less than that of U_{ij}.

As an illustration, consider the cofactor U_{13} of the $(1, 3)$-element of $U(G)$ given in Example 4.9. U_{13} together with U_{13} in modified form and its expansion into unisignant matrices of lower orders, again in modified form, are given in determinantal form as follows:

$$U_{13} = (-1)^{1+3} \begin{vmatrix} -e_2 & e_2+e_3 & -e_3 \\ 0 & -e_4 & 0 \\ 0 & 0 & e_5+e_6 \end{vmatrix}$$

$$= \begin{vmatrix} e_2+e_3 & e_2 & -e_3 \\ -e_4 & 0 & 0 \\ 0 & 0 & e_5+e_6 \end{vmatrix} = \begin{vmatrix} 0 & -e_4 & 0 \\ e_2 & e_2+e_3 & -e_3 \\ 0 & 0 & .e_5+e_6 \end{vmatrix}$$

$$= e_2 \begin{vmatrix} e_4 & 0 \\ 0 & e_5+e_6 \end{vmatrix} = e_2 e_4 (e_5+e_6). \tag{4.147a}$$

Thus, there are two directed 2-trees $t_{13,5}$ in G. Similarly, if 2-trees $t_{13,5}$ in the graph G_u as shown in fig. 4.29 are needed, the same procedure may be employed. For illustrative purposes, we shall consider the cofactor U_{31} rather than U_{13} since the order of the subscripts in $t_{13,5}$ is immaterial,

$$U_{31} = \begin{vmatrix} -e_2 & 0 & 0 \\ e_2+e_3+e_4 & -e_4 & -e_3 \\ -e_3 & -e_5 & e_3+e_5+e_6 \end{vmatrix}$$

$$= \begin{vmatrix} 0 & -e_2 & 0 \\ e_4 & e_2+e_3+e_4 & -e_3 \\ e_5 & -e_3 & e_3+e_5+e_6 \end{vmatrix}$$

$$= e_4 \begin{vmatrix} e_2 & 0 \\ e_3 & e_3+e_5+e_6 \end{vmatrix} + e_5 \begin{vmatrix} 0 & -e_2 \\ e_3 & e_2+e_3+e_4 \end{vmatrix}$$

$$= e_4 e_2 (e_3+e_5+e_6) + e_5 e_2 e_3. \tag{4.147b}$$

Thus, there are four terms in the expansion of U_{31}, each of which corresponds to a tree $t_{13,5}$ or $t_{31,5}$ of G_u.

Like the directed-tree case, if we let all the edge-designation symbols be unity, the process outlined above can also be used rather effectively for the computation of the number of directed 2-trees in G or 2-trees in G_u. For example, the number $N(t_{13,5})$ of trees $t_{13,5}$ in G_u can be evaluated as follows:

$$N(t_{13,5}) = \begin{vmatrix} -1 & 0 & 0 \\ 3 & -1 & -1 \\ -1 & -1 & 3 \end{vmatrix} = \begin{vmatrix} 0 & -1 & 0 \\ 1 & 3 & -1 \\ 1 & -1 & 3 \end{vmatrix}$$

$$= \begin{vmatrix} 1 & 0 \\ 1 & 3 \end{vmatrix} + \begin{vmatrix} 0 & -1 \\ 1 & 3 \end{vmatrix} = 4. \tag{4.148}$$

As mentioned earlier in this section, the unisignant matrix $U(G)$ is very similar to the node-admittance matrix of an electrical network. Thus, the above procedure may also be employed for the evaluation of the nodal determinant without the necessity of first setting up the associated directed graph.

5.3. Partial factoring

The fundamental principle of the present approach is rather simple. The technique is based on the fact that if an edge e terminating at node k of a directed graph G is distinguished, then the set T_k of directed trees t_k of G can be classified as those $T_k(e)$ containing the edge e, and those $T_k(\bar{e})$ which do not. If the symbol $G(g_1; g_2)$ is used to denote the directed graph obtained from G first by shorting all the edges contained in g_1 (removing all the self-loops thus produced) and then by removing all the edges contained in g_2, where g_1 and g_2 are two edge-disjoint subgraphs of G, then the elements of $T_k(e)$ are in one-to-one correspondence with the directed trees t_k in $G(e; \emptyset)$, and $T_k(\bar{e})$ is the set of directed trees t_k in $G(\emptyset; e)$, where $\emptyset$ denotes the null graph. If we repeat the process for $G(e; \emptyset)$ and $G(\emptyset; e)$, we will eventually generate all the directed trees t_k of G without duplications. The procedure was implicitly given by FEUSSNER [1902, 1904] for undirected graphs, and has since been used explicitly by PERCIVAL [1953] for undirected graphs and by MASON [1957] for directed graphs. A simple generalization of the procedure is the following.

Let C_k^- be the subgraph of G consisting of all the edges that are terminated at node k. Let E and $\bar{E}$ be the complementary subgraphs of C_k^-. By $t_k(E)$ we mean a directed tree of G with reference node k and such that

$$t_k(E) \cap C_k^- = E.$$

THEOREM 4.13: For each directed tree with reference node k in $G(E; \bar{E})$, there corresponds a unique directed tree $t_k(E)$ in G and vice versa.

A different version of the theorem is given in Problem 4.63.

The procedure outlined above may also be used to generate the set of directed 2-trees of G without duplications as we can see from the following corollary.

COROLLARY 4.22: Let G^* be the directed graph derived from G by identifying the nodes i and j, the combined node being denoted by i. Then for each directed 2-tree $t_{i,j}$ of G, there corresponds a unique directed tree with reference node i in G^* and vice versa.

Instead of shorting an edge of G, it is possible to generate the set of directed trees by shorting the edges of a directed path. The technique is referred to as an *expansion in (directed) paths* or an *expansion on a node-pair* by MASON and ZIMMERMANN [1960], and is based on the following theorem.

THEOREM 4.14: A spanning subgraph t_k of a directed graph G without self-loops is a directed tree with reference node k if, and only if, either one of the following two conditions is satisfied:

(1) For each node $i \neq k$ there exists a unique directed path from node i to node k in t_k.

(2) There are $n-1$ edges in t_k in which, for each node $i \neq k$, there exists a directed path from node i to node k, where n is the number of nodes of G.

The proof of the theorem is straightforward, and is left as an exercise (Problem 4.44). The technique has been implemented on a digital computer. A computer program was written in ALGOL by BROWNELL [1968].

Thus, for a labeled directed graph G if each of its subgraphs is denoted by the "product" or by juxtaposition of its edge-designation symbols, then for each choice of the nodes $i \neq k$ in G, the set T_k of directed trees t_k of G can be uniquely represented by the expression

$$T_k = \bigcup_{(m)} \{P_{ik}^m\} \times T_k^*, \qquad (4.149)$$

where P_{ik}^m is the mth directed path from node i to node k in G; T_k^* is the set of directed trees with reference node k in $G(P_{ik}^m; \emptyset)$; the symbol "$\times$" denotes the Cartesian product of two sets; and the union is taken over all P_{ik}^m in G. Remember that the elements of the sets, though they are subgraphs, are assumed to have been represented by juxtaposition of their edge-designation symbols. Also, the combined node in $G(P_{ik}^m; \emptyset)$ is labeled by k. In other words, the set of directed trees of G can be expressed in a partially factored form. From an application viewpoint, there is no need to carry out the expansion.

As an illustration, consider the directed graph G as shown in fig. 4.15. The set T_5 of directed trees t_5 of G can now be generated by an expansion on a node-pair, say, nodes 1 and 5:

$$T_5 = \{e_{12}e_{25}\} \times \{e_{32}e_{45}, e_{32}e_{42}, e_{35}e_{45}, e_{35}e_{42}, e_{43}e_{32}, e_{43}e_{35}, e_{34}e_{45}, e_{34}e_{42}\}$$
$$\cup \{e_{12}e_{23}e_{35}\} \times \{e_{42}, e_{43}, e_{45}\} \cup \{e_{12}e_{23}e_{34}e_{45}\}. \qquad (4.150)$$

Obviously, if $G(P_{ik}^m; \emptyset)$ is still complicated, the same process may be repeated until all the terms can be obtained by inspection. The amount of work achieved by an expansion in directed paths will depend, of course, upon the structure of the particular directed graph under consideration and upon the choice of node-pairs for the expansion and subexpansions.

Similarly, if trees in the graph G_u of fig. 4.19 are needed, they may be generated in the same manner. The added flexibility is that we have freedom to choose the reference node. Thus, the set T of trees of G_u can be generated by an expansion on a node-pair, say, nodes 2 and 4. For illustrative purposes, we shall also use the subexpansions:

$$\begin{aligned}
T = \{e_{25}e_{45}\} &\times (\{e_{13}e_{23}\} \times \{e_{16}, e_{56}\} \cup \{e_{13}e_{35}\} \times \{e_{16}, e_{56}\} \cup \{e_{14}\} \\
&\times \{e_{56}e_{23}, e_{56}e_{35}, e_{56}e_{13}, e_{16}e_{23}, e_{16}e_{35}, e_{16}e_{13}\} \cup \{e_{16}e_{56}\} \\
&\times \{e_{13}, e_{23}, e_{35}\}) \cup (\{e_{25}e_{56}e_{16}e_{14}\} \times \{e_{13}, e_{23}, e_{35}\}) \cup (\{e_{23}e_{13}e_{14}\} \\
&\times \{e_{16}e_{25}, e_{16}e_{35}, e_{16}e_{45}, e_{56}e_{25}, e_{56}e_{35}, e_{56}e_{45}, e_{16}e_{56}\}) \\
\cup\, (\{e_{23}e_{13}e_{16}e_{56}e_{45}\}) &\cup (\{e_{23}e_{35}e_{45}\} \\
&\times \{e_{56}e_{16}, e_{56}e_{13}, e_{56}e_{14}, e_{16}e_{13}, e_{16}e_{14}\}) \cup (\{e_{23}e_{35}e_{56}e_{16}e_{14}\}) \\
\cup\, (\{e_{25}e_{35}e_{13}e_{14}\} &\times \{e_{16}, e_{56}\}),
\end{aligned} \tag{4.151}$$

where e_{ij} or e_{ji} denotes the edge connected between the nodes i and j of G_u.

§ 6. Direct analysis of electrical networks

Since the attempts of early pioneers in linear-circuit analysis to formalize the approach to circuit problems, very little seems to have been achieved towards developing rapid methods of network analysis. The techniques discussed in the preceding sections permit a quick analysis of a general linear network. In this section, we shall show how these techniques may be used to write down the network functions by inspection. The treatment here is similar to that given by MASON [1957] and BOISVERT and ROBICHAUD [1956]. However, their approach, viewpoint, and interpretation are quite different, but the essential results are the same. This is no surprise since they all have to do with topological analysis of linear networks. We shall show that their approach is actually a variant of the directed-tree formulation, and thus their results can be derived in a unified and systematic way.

6.1. *Open-circuit transfer-impedance and voltage-gain functions*

Let N be an n-node linear network composed of an arbitrary number of active and passive elements connected in any manner whatsoever. Let v_1, $v_2, \ldots, v_n$ be the potentials measured between nodes $1, 2, \ldots, n$ and some arbitrary but unspecified reference point, respectively. If $i_1, i_2, \ldots, i_n$ are the source currents entering the nodes $1, 2, \ldots, n$, respectively, from outside the network, then the nodal potentials and source currents are related by the system of linear algebraic equations

$$I = YV, \tag{4.152}$$

where I is an n-vector representing the source currents; V is an n-vector representing nodal voltages; and Y is the indefinite-admittance matrix of order n associated with the network N.

Since the indefinite-admittance matrix Y is also an equicofactor matrix of the type of (4.21) (see, for example, SHARPE and SPAIN [1960]), the sum of the elements of each of its rows equals zero. Hence, the system (4.152) can be put in the form

$$I = YV'', \qquad (4.153)$$

where $V'' = V - V_0$, and V_0 is an n-vector, all whose elements are v_m, the nodal voltage of the node m. Thus, the element v''_{jm} in the jth row of V'' denotes the potential rise from node m to node j.

In order to compute the network functions, let the network N be constrained so that the source current i_{rs} flows into the node r and out of the node s and no other source currents entering the network. (That is, if i_x is the xth row element of I, then $i_x = 0$, $x \neq r, s$, and $i_r = -i_s = i_{rs}$.)

Since Y is an equicofactor matrix, the equations of (4.153) are linearly dependent and so one equation is superfluous. Let us suppress the sth from (4.153). Then by Cramer's rule, it is not difficult to see that the solution of the system (4.153) is

$$v''_{km} = i_{rs} Y_{rk,sm}/Y_{uv}, \qquad (4.154)$$

for $k = 1, 2, ..., n$ but $k \neq m$, where v''_{km} is the kth row element of V'', and Y_{uv} and $Y_{rk,sm}$ are the first- and second-order cofactors of the elements of the indefinite-admittance matrix Y. The determinant of Y is of course identically zero. Note that the choice of the subscripts u and v in Y_{uv} is arbitrary.

The network N may be symbolically described by a "black box" as that shown in fig. 4.30. The voltages across the node pairs rs and pq are designated as v_{rs} and v_{pq}, respectively, with reference plus at the nodes r and p, and the output terminals are assumed to be open-circuited. This is quite general, since any load can be considered as part of the network N. Network functions can be meaningfully defined only when the black boxes such as fig. 4.30 do *not* contain any *independent* sources. For the present, we shall only make this

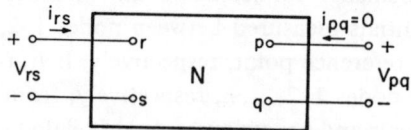

Fig. 4.30. A black-box description of the open-circuit transfer-impedance and voltage-gain functions.

assumption for N (in addition to lumped, linear, time-invariant components). However, N may contain *dependent* sources. Also, we assume that all the current and voltage functions are given in the Laplace-transform form.

DEFINITION 4.16: *Open-circuit transfer-impedance function.* The (*open-circuit*) *transfer-impedance function*, denoted by the symbol $z_{rp,sq}$, between the node pairs rs and pq of the network as shown in fig. 4.30 is defined as the ratio of the voltage v_{pq} to the current i_{rs}, with all initial conditions and independent sources in the network being set to zero and the output terminals being open-circuited.

DEFINITION 4.17: *Driving-point impedance function.* When $r=p$ and $s=q$, the transfer-impedance function $z_{rp,sq}$ becomes the *driving-point impedance function* $z_{rr,ss}$ between the node pair rs of the network of fig. 4.30.

Thus, from (4.154) we obtain

$$z_{rp,sq} = Y_{rp,sq}/Y_{uv}, \qquad (4.155)$$

$$z_{rr,ss} = Y_{rr,ss}/Y_{uv}. \qquad (4.156)$$

We remark that in the expressions for transfer and driving-point impedance functions, the order of the subscripts is as follows: r, the current injecting node; s, the current extracting node; p, the voltage measurement node; q, the voltage reference node. Nodes s and q form a sort of double datum, s for current and q for voltage; and nodes r and p then designate the input and output transfer measurement. It also helps to remember that, in any expression for an impedance function, the first subscript in a subscript pair of $Y_{rp,sq}$ has to do with current and the second subscript with voltage.

Let $G(Y)$ be the associated directed graph of the indefinite-admittance matrix Y. Then from (4.56) we have

$$Y_{rp,sq} = \sum_{t_{rp,sq}} f(t_{rp,sq}) - \sum_{t_{rq,sp}} f(t_{rq,sp}), \qquad (4.157)$$

for $r \neq s$ and $p \neq q$, where $t_{rp,sq}$ and $t_{rq,sp}$ are directed 2-trees of $G(Y)$.

The formula (4.157) can be written in an intuitive fashion by following a similar representation first given by PERCIVAL [1953] for passive networks without mutual inductances, and is as shown in fig. 4.31. The representation

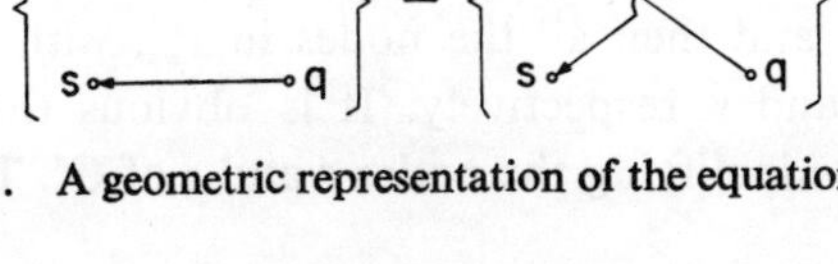

Fig. 4.31. A geometric representation of the equation (4.157).

shows the two types of directed 2-trees involved in the formula, including the reference nodes.

Now we shall show that the formula (4.157) can be further simplified with the aid of a few more definitions.

DEFINITION 4.18: *Proper path.* A path, denoted by the symbol P^k, of a directed graph is said to be *proper with respect to the node k* if for each node j in P^k, $j \neq k$, there exists a directed path from node k to node j in P^k.

As an example, consider the directed graph G as shown in fig. 4.28. The path P^4 denoted by the edge-designation symbols $e_2 e_4 e_5 e_6$ is proper with respect to the node 4, since for each node j, $j = 1, 2, 3, 5$, there exists a directed path from node 4 to node j in P^4. However, it is not proper with respect to the node 3 since there exists no directed path from node 3 to node 4 or node 5 in the path.

Let G_a be the directed graph obtained from $G(Y)$ by inserting an edge e_{pq} of weight 1, directed from node p to node q, in $G(Y)$. Also, let the symbol P_{rs}^p be used to represent a proper path with respect to the node p, connected between the nodes r and s, and containing the edge e_{pq} of G_a. For convenience, in $G_a(P_{rs}^p; \emptyset)$ let the combined node be denoted by p. Now, we decompose the proper path P_{rs}^p into the edge-disjoint union of three directed paths as follows:

$$P_{rs}^p = P_{pr} \cup e_{pq} \cup P_{qs} \tag{4.158}$$

or

$$P_{rs}^p = P_{ps} \cup e_{pq} \cup P_{qr}, \tag{4.159}$$

where P_{pr}, P_{qs}, P_{ps}, and P_{qr} are directed paths from p to r, q to s, p to s, and q to r in P_{rs}^p, respectively.

In the following, we shall only consider the case where P_{rs}^p is of the type as shown in (4.158). The other case may be treated in an entirely similar manner.

LEMMA 4.4: If P_{rs}^p is of the type as shown in (4.158), then there exists a one-to-one correspondence between the directed 2-trees $t_{rp,sq}$ of $G(Y)$ containing the directed paths P_{pr} and P_{qs}, and the directed trees t_p^* with reference node p in $G_a(P_{rs}^p; \emptyset)$ and such that

$$f(P_{rs}^p) f(t_p^*) = f(t_{rp,sq}). \tag{4.160}$$

Proof. Let G'' be the directed graph obtained from $G(Y)$ by first identifying all the nodes in P_{pr}, and then all the nodes in P_{qs}, with the combined nodes being denoted by r and s, respectively. It is obvious that $G_a(P_{rs}^p; \emptyset)$ can be obtained from G'' by identifying the nodes r and s of G''. Thus, from Corollary

4.22 we conclude that directed trees t_p^* of $G_a(P_{rs}^p; \emptyset)$ are in one-to-one correspondence with directed 2-trees $t_{r,s}''$ of the type $t_{r,s}$ in G'', and such that

$$f(t_p^*) = f(t_{r,s}''). \tag{4.161}$$

Also, it is not difficult to see that there exists a one-to-one correspondence between directed 2-trees $t_{r,s}''$ of G'' and directed 2-trees $t_{rp,sq}$ of $G(Y)$ containing the directed paths P_{pr} and P_{qs}, and such that

$$f(P_{pr} \cup P_{qs}) f(t_{r,s}'') = f(t_{rp,sq}). \tag{4.162}$$

Since $f(e_{pq}) = 1$, the lemma follows after making a few simple substitutions. So the lemma is proved.

LEMMA 4.5: Let P_{rs}^p be a proper path of the type of (4.158) in $G(Y)$. Then for $r \neq s$ and $p \neq q$ we have

$$\sum_{t_{rp,sq}} f(t_{rp,sq}) = \sum_{PP_{rs}} \left[f(P_{rs}^p) \sum_{t_p^*} f(t_p^*) \right], \tag{4.163}$$

where $t_{rp,sq}$ is a directed 2-tree of $G(Y)$, and t_p^* is a directed tree with reference node p in $G_a(P_{rs}^p; \emptyset)$.

The proof of the lemma is straightforward, and is left as an exercise (Problem 4.46). Similarly, we can show

LEMMA 4.6: Let P_{rs}^p be a proper path of the type of (4.159) in $G(Y)$. Then for $r \neq s$ and $p \neq q$ we have

$$\sum_{t_{rq,sp}} f(t_{rq,sp}) = \sum_{PP_{rs}} \left[f(P_{rs}^p) \sum_{t_p^*} f(t_p^*) \right], \tag{4.164}$$

where $t_{rq,sp}$ is a directed 2-tree in $G(Y)$, and t_p^* is a directed tree with reference node p in $G_a(P_{rs}^p; \emptyset)$.

As yet, we have said nothing about the algebraic sign associated with the second term in (4.157). This can easily be taken care of by assigning an orientation to a circuit. The definition of an oriented circuit was defined in ch. 2 (Definition 2.5), and we shall use this concept to fix the signs for the terms in (4.157). However, before we do this let us refresh this concept by an example.

In fig. 4.32, the circuit consisting of the edges a, c, and d can be oriented as (1, 2, 4, 1) or as (1, 4, 2, 1). Also, we can represent the orientation pictorially by an arrowhead. For our purpose, the orientations of an edge of a circuit and the circuit "coincide" if the nodes of the edge appear in the same order both in the ordered-pair representation of the edge and in the ordered-node representation of the circuit. Otherwise, they are "opposite". Pictorially, the

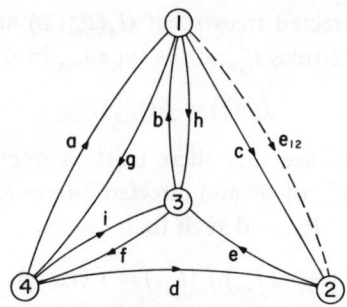

Fig. 4.32. A directed graph illustrating the circuit orientation.

meaning is obvious. For example, the orientation of the edge (1, 2) coincides with that of the circuit (1, 2, 4, 1), and the orientation of the edge (4, 2) is opposite to that of the circuit. Note that a circuit of a directed graph need not be a directed circuit.

Using (4.163) and (4.164) in conjunction with the above definition, the formula (4.157) can now be simplified as follows:

THEOREM 4.15: Let $G(Y)$ be the associated directed graph of an equicofactor matrix Y of the type as shown in (4.21). Then for $r \neq s$ and $p \neq q$ we have

$$Y_{rp,sq} = \sum_{PP_{rs}^p} \left[(-1)^\alpha f(P_{rs}^p) \sum_{t_p^*} f(t_p^*) \right], \qquad (4.165)$$

where P_{rs}^p is a proper path with respect to the node p, connected between the nodes r and s, and containing the edge e_{pq} of G_a; $\alpha = 0$ if the orientations of the edges e_{pq} and e_{sr} are both opposite to or coincident with that of the oriented circuit $e_{sr} \cup P_{rs}^p$, where e_{sr} is an edge directed from node s to node r, and $\alpha = 1$ otherwise; and t_p^* is a directed tree with reference node p in $G_a(P_{rs}^p; \emptyset)$, the combined node being denoted by p, where $G_a = G \cup e_{pq}$ and $f(e_{pq}) = 1$.

Proof. Since the corresponding subgraph of P_{rs}^p and t_p^* is either $t_{rp,sq}$ or $t_{rq,sp}$ in $G(Y)$, and vice versa, the theorem follows directly from (4.157), (4.163) and (4.164).

COROLLARY 4.23: If $G(Y)$ is the associated directed graph of the indefinite-admittance matrix Y of a linear network N, then

$$z_{rp,sq} = \frac{Y_{rp,sq}}{Y_{uv}} = \frac{\sum_{PP_{rs}^p} \left[(-1)^\alpha f(P_{rs}^p) \sum_{t_p^*} f(t_p^*) \right]}{\sum_{t_k} f(t_k)} \qquad (4.166)$$

for $r \neq s$ and $p \neq q$, where t_k is a directed tree of $G(Y)$.

§6 Direct analysis of electrical networks

For $r=p$ and $s=q$, the transfer-impedance function $z_{rp,sq}$ becomes the driving-point impedance function $z_{rr,ss}$, and (4.166) can be further simplified.

COROLLARY 4.24:
$$z_{rr,ss} = \frac{Y_{rr,ss}}{Y_{uv}} = \sum_{t''_m} f(t''_m) / \sum_{t_k} f(t_k), \qquad (4.167)$$

where t''_m is a directed tree with reference node m in the directed graph obtained from $G(Y)$ by identifying the nodes r and s.

We remark that in (4.163)–(4.167) the choice of the reference nodes p of $G_a(P^p_{rs}; \emptyset)$, k of $G(Y)$, and m of (4.167) is arbitrary since the associated matrices of these directed graphs are equicofactor matrices.

Example 4.11: Consider the directed graph G as shown in fig. 4.32. Let us compute all $f(t_{41,22})$ in G. So we add an edge e_{12} (dashed line) with weight 1 to G. From (4.163) we obtain

$$\sum_{t_{41,22}} f(t_{41,22}) = \sum_{P^1_{42}} \left[f(P^1_{42}) \sum_{t^*_1} f(t^*_1) \right] = g(b+f) + hf(1). \qquad (4.168)$$

Recall that $f(\emptyset) = 1$ as defined in §2.1.

Example 4.12: Consider the transistor feedback amplifier as shown in fig. 4.1. The associated directed graph $G(Y)$ of the indefinite-admittance matrix Y of the amplifier is given in fig. 4.2. Suppose that we wish to compute the transfer impedance function $z_{13,44}$. So let us add an edge e_{34} of weight 1 to $G(Y)$.

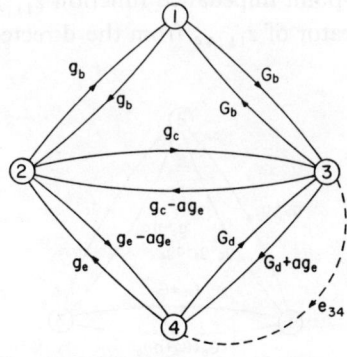

Fig. 4.33. The associated directed graph of the feedback amplifier of fig. 4.1 in which an edge e_{34} with unit weight is added.

The resulting graph is presented in fig. 4.33. From (4.166) we have

$$z_{13,44} = \frac{\sum_{P^3_{14}} [(-1)^\alpha f(P^3_{14}) \sum_{t^*_3} f(t^*_3)]}{\sum_{t_k} f(t_k)}$$

$$= \frac{G_b[g_b + (g_e - ag_e) + g_c] + g_b(g_c - ag_e)(1)}{\sum_{t_k} f(t_k)}, \quad (4.169)$$

where the denominator of (4.169) is given in (4.38). The numerator of (4.169) is of course the same as that given in (4.61). Note that there exists no proper path of the type as shown in (4.159). Since the current-gain function, which is defined as the ratio of i_3 to i_1, of the amplifier is related to the transfer impedance function $z_{13,44}$ by

$$i_3/i_1 = G_d z_{13,44}, \quad (4.170)$$

i_3/i_1 can be obtained by a simple substitution.

For illustrative purposes, let us compute the transfer impedance function $z_{12,43}$. So let us add an edge e_{23} of weight 1 to $G(Y)$. From (4.166) we have

$$z_{12,43} = \frac{\sum_{P^2_{14}} [(-1)^\alpha f(P^2_{14}) \sum_{t^*_2} f(t^*_2)]}{\sum_{t_k} f(t_k)}$$

$$= \frac{g_b(G_d + ag_e)(1) - G_b(g_e - ag_e)(1)}{\sum_{t_k} f(t_k)}, \quad (4.171)$$

where the denominator is given in (4.38).

Finally, if the driving-point impedance function $z_{11,44}$ is needed, we need only compute the numerator of $z_{11,44}$ from the directed graph G'' as shown in fig. 4.34:

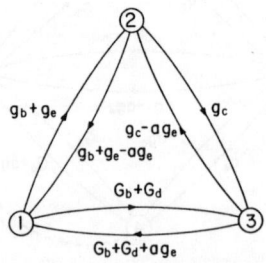

Fig. 4.34. The directed graph derived from that of fig. 4.33 by identifying the nodes 1 and 4.

$$\sum_{t''_m} f(t''_m) = (g_b + g_e - ag_e)(G_b + G_d + ag_e) + (G_b + G_d + ag_e)g_c$$
$$+ (g_b + g_e - ag_e)(g_c - ag_e)$$
$$= (G_b + G_d)g_c + (g_b + g_e)g_c + (g_b + g_e - ag_e)(G_b + G_d), \quad (4.172)$$

where the first two lines are obtained by using node 1 of G'' as the reference node, and the third line by using node 3 as the reference node.

DEFINITION 4.19: *Open-circuit voltage-gain or transfer-voltage-ratio function.* The *(open-circuit) voltage-gain or transfer-voltage-ratio function*, denoted by the symbol $g_{rp,sq}$, between the node pairs rs and pq of the network as shown in fig. 4.30 is defined as the ratio of the voltage v_{pq} to the voltage v_{rs}, with all initial conditions and independent sources in the network N being set to zero and the output terminals being open-circuited.

Thus, from (4.155) and (4.156) we get

$$g_{rp,sq} = z_{rp,sq}/z_{rr,ss} = Y_{rp,sq}/Y_{rr,ss}. \quad (4.173)$$

Using (4.166) and (4.167), the voltage-gain function $g_{rp,sq}$ of N can similarly be evaluated.

In fact, the formula (4.155) for the transfer impedance function $z_{rp,sq}$ is all that we need to remember. The other two expressions (4.156) and (4.173), for the driving-point impedance and voltage-gain functions, can always be deduced easily from the first. The formula (4.156) is a special case of the first, in which $r=p$ and $s=q$. The formula (4.173) is simply the quotient of the other two. The results presented in these formulas are, of course, quite general, since the letters $r, p, s,$ and q may refer to the labels of any four nodes in any network.

6.2. *Short-circuit transfer-admittance and current-gain functions*

In addition to the three network functions defined in the foregoing, there are three dual ones which can be treated in a similar manner.

DEFINITION 4.20: *Short-circuit current-gain or transfer-current-ratio function.* The *short-circuit current-gain or transfer-current-ratio function*, denoted by the

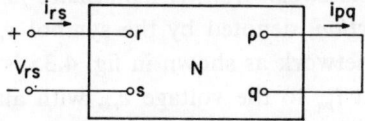

Fig. 4.35. A black-box description of the short-circuit current-gain and transfer-admittance functions.

symbol $\alpha_{rp,sq}$, between the node pairs rs and pq of the network as shown in fig. 4.35 is defined as the ratio of the short-circuit current i_{pq} to the current i_{rs}, with all initial conditions and independent sources in the network N being set to zero. For $r=p$ and $s=q$, $\alpha_{rp,sq}$ is defined to be unity.

COROLLARY 4.25: For $r \neq s$ and $p \neq q$, we have

$$\alpha_{rp,sq} = \frac{Y_{rp,sq}}{Y_{pp,qq}} = \frac{\sum_{pp_{rs}} [(-1)^\alpha f(P_{rs}^p) \sum_{t^*_p} f(t^*_p)]}{\sum_{t''_k} f(t''_k)}, \qquad (4.174)$$

where t''_k is a directed tree with reference node k in the directed graph obtained from $G(Y)$ by identifying the nodes p and q, and all other symbols are defined the same as in (4.165).

Proof. Let N'' be the network obtained from the given network N by connecting a branch of admittance y between the nodes p and q. For simplicity, let us use double prime to indicate the similar quantities defined for N''. For example, Y'', $z''_{rp,sq}$, and v''_{pq} denote the indefinite-admittance matrix, the transfer impedance function, and the voltage rise from q to p in N'', respectively. Thus, from (4.155) we have

$$yz''_{rp,sq} = yv''_{pq}/i''_{rs} = yY''_{rp,sq}/Y''_{qq}. \qquad (4.175)$$

Now, if we let y approach to infinity, the left-hand side of (4.175) is the short-circuit current-gain function $\alpha_{rp,sq}$ in N, and Y''_{qq} approaches to $yY_{pp,qq}$. Since $Y''_{rp,sq} = Y_{rp,sq}$, the identity (4.174) follows immediately, proving the corollary. (Also, see Problem 4.67.)

Thus, if the indefinite-admittance matrix Y of a network N is symmetric, it follows from (4.173) and (4.174) that the open-circuit voltage-gain function from one node-pair to another is equal to the short-circuit current-gain function in the opposite direction between the same node-pairs. An electrical network with symmetric indefinite-admittance matrix is referred to as *reciprocal*. More specifically, for a reciprocal network we have $g_{rp,sq} = \alpha_{pr,qs}$.

DEFINITION 4.21: *Short-circuit transfer-admittance function.* The *short-circuit transfer-admittance function*, denoted by the symbol $y_{rp,sq}$, between the node pairs rs and pq of the network as shown in fig. 4.35 is defined as the ratio of the short-circuit current i_{pq} to the voltage v_{rs}, with all initial conditions and independent sources in the network N being set to zero. For $r=p$ and $s=q$, $y_{rp,sq}$ is not defined.

Thus, if we use (4.174) and follow a similar argument as that for the proof of (4.174), we obtain (Problem 4.48)

COROLLARY 4.26:
$$y_{pp, sq} = Y_{pp, sq}/Y_{pp, ss, qq} \quad (4.176a)$$
for $s \neq q \neq p$, and
$$y_{rp, ss} = Y_{rp, ss}/Y_{rr, pp, ss} \quad (4.176b)$$
for $r \neq p \neq s$, and
$$y_{rp, sq} = Y_{rp, sq}/(Y_{rr, ss, pp} + Y_{rr, ss, qq} - Y_{rr, ss, pq} - Y_{rr, ss, qp}) \quad (4.176c)$$
for $r \neq p \neq q \neq s$, where the third-order cofactor $Y_{rr, ss, pq}$ is defined by
$$Y_{rr, ss, pq} = \operatorname{sgn}(r - p)\operatorname{sgn}(r - q)\operatorname{sgn}(s - p)\operatorname{sgn}(s - q)(-1)^{p+q} M_{rr, ss, pq}, \quad (4.177)$$
and $M_{rr, ss, pq}$ is the determinant of the submatrix obtained from the indefinite-admittance matrix Y by deleting the rows r, s, p, and columns r, s, q.

In order to express the terms in $Y_{rr, ss, pq}$ topologically, we need the following definition which is a simple extension of the concept of a directed 2-tree.

DEFINITION 4.22: *Directed 3-tree*. A subgraph, denoted by the symbol $t_{i,j,k}$, of an n-node directed graph G is said to be a *directed 3-tree* of G with reference nodes i, j, and k if, and only if, (1) it has $n-3$ edges and does not contain any circuit (not necessarily directed circuit); and (2) the outgoing degree of each of its nodes is 1 except the nodes i, j, and k which have outgoing degree 0.

Thus, a directed 3-tree is a spanning subgraph consisting of three circuitless components, each of which is a directed tree for some sectional subgraph of G. One, two, or three of these components may each consist of an isolated node. Obviously, the concept can be generalized to *directed k-trees* of G, and the techniques outlined in § 5 can also be extended for their generation. The details of these extensions are left as exercises (Problems 4.13 and 4.54).

Like directed 2-trees, sometimes certain specified nodes are required to appear in different components of a directed 3-tree. For convenience, we shall use additional subscripts for this purpose. For example, the symbol $t_{ab,c,def}$ is used to denote a directed 3-tree in which the nodes a and b are in one component, the node c in one component, and the nodes d, e, and f in one component, where the nodes a, c, and d (the first subscripts) are the reference nodes of the components.

With these definitions, we are now in a position to state the topological formulas for $y_{rp,sq}$. For simplicity, let

$$U_{r,s,pq} = \sum_{t_{r,s,pq}} f(t_{r,s,pq}), \qquad (4.178)$$

where $t_{r,s,pq}$ is a directed 3-tree of $G(Y)$. We see at once, by arguments similar to those of Theorem 4.4, that (Problem 4.52)

$$Y_{rr,ss,pq} = U_{r,s,pq}. \qquad (4.179)$$

Since directed 3-trees $t_{r,s,pq}$ occur both in $U_{r,s,pq}$ and $U_{r,s,p}$, such terms will not appear in the final expansion of the denominator of (4.176c). Similarly, directed 3-trees $t_{r,s,qp}$ also will not appear in the final expansion. Therefore,

$$Y_{rr,ss,pp} + Y_{rr,ss,qq} - Y_{rr,ss,pq} - Y_{rr,ss,qp}$$
$$= U_{r,s,p} + U_{r,s,q} - U_{r,s,pq} - U_{r,s,qp}$$
$$= U_{rq,s,p} + U_{r,sq,p} + U_{rp,s,q} + U_{r,sp,q}. \qquad (4.180)$$

It is significant to note that cancellations have been avoided in the above expansion. Using these, the formulas (4.176) can now be combined into one, and expanded topologically.

COROLLARY 4.27: For $r \neq s$ and $p \neq q$, we have

$$y_{rp,sq} = \frac{Y_{rp,sq}}{U_3} = \frac{\sum_{pp_{rs}} [(-1)^\alpha f(P_{rs}^p) \sum_{t_p^*} f(t_p^*)]}{U_3}, \qquad (4.181)$$

where $U_3 = U_{rq,s,p} + U_{r,sq,p} + U_{rp,s,q} + U_{r,sp,q}$.

Proof. The corollary follows directly from (4.176), (4.165), and (4.180).

In the following, we shall show that the term U_3 can be further simplified. It is equivalent to the sum of the weight products of the directed trees in the directed graph obtained from $G(Y)$ by identifying the nodes p and q, and r and s, respectively.

COROLLARY 4.28:

$$U_3 = U_{rq,s,p} + U_{r,sq,p} + U_{rp,s,q} + U_{r,sp,q} = \sum_{t_k''} f(t_k''), \qquad (4.182)$$

where t_k'' is a directed tree with reference node k in G''; and G'' is the directed graph obtained from $G(Y)$ by first identifying the nodes p and q, and then by identifying the nodes r and s. The choice of the reference node k for t_k'' of G'' is arbitrary.

§ 6   Direct analysis of electrical networks   293

Proof. Let G' be the directed graph obtained from $G(Y)$ by identifying the nodes p and q of $G(Y)$ with the combined node being denoted by p. Also in G'' let the combined nodes of r and s, and of p and q, be denoted by r and p, respectively. Since the associated matrix of G'' is an equicofactor matrix, we can let $k=r$.

$$\begin{aligned}\sum f(t''_r) &= \sum f(t'_{r,s}) \\ &= \sum f(t'_{rp,s}) + \sum f(t'_{r,sp}) \\ &= \sum f(t_{rp,s,q}) + \sum f(t_{rq,s,p}) + \sum f(t_{r,sp,q}) + \sum f(t_{r,sq,p}) \\ &= U_{rp,s,q} + U_{rq,s,p} + U_{r,sp,q} + U_{r,sq,p},\end{aligned} \quad (4.183)$$

where the prime indicates that directed 2-trees are taken from G', and the summations are taken over all the desired directed 2-trees and directed 3-trees in their respective directed graphs. This completes the proof of the corollary.

Finally, we mention that the reciprocal of a driving-point impedance function is called a *driving-point admittance function*, which is quite different from the short-circuit transfer-admittance function. (Also see ch. 2, § 6.3.)

Example 4.13: Consider the associated directed graph $G(Y)$ of fig. 4.33. Suppose that we wish to compute the short-circuit transfer-admittance function $y_{13,44}$. Then from (4.181) we have

$$\begin{aligned}y_{13,44} &= \frac{1}{U_3} \sum_{P^3_{14}} [(-1)^\alpha f(P^3_{14}) \sum_{t^*_3} f(t^*_3)] \\ &= \frac{1}{U_3} \{G_b[g_b + (g_e - ag_e) + g_c] + g_b(g_c - ag_e)\},\end{aligned} \quad (4.184a)$$

where

$$\begin{aligned}U_3 &= U_{14,4,3} + U_{1,4,3} + U_{13,4,4} + U_{1,43,4} \\ &= U_{1,4,3} = g_b + g_c + (g_e - ag_e).\end{aligned} \quad (4.184b)$$

From (4.182), the terms in (4.184b) are also directed trees in the directed graph G'' obtained from $G(Y)$ by identifying the nodes 1, 3, and 4. In G'' we can choose any node as the reference node, and (4.182) will all reduce to that of (4.184b).

If the short-circuit current-gain function $\alpha_{13,44}$ of the amplifier is needed, then from (4.174) we need only compute the denominator term $Y_{33,44}$ which is given by

$$(g_b + G_b)(g_c + g_e - ag_e) + g_b G_b.$$

6.3. *Open-circuit impedance and short-circuit admittance matrices*

From the discussion in the foregoing, it is clear that the formulas (4.166) and (4.181) can be applied directly to the evaluation of the open-circuit impedance parameters or the short-circuit admittance parameters of a two-port network (CHEN [1969c]). More specifically, a two-port network of fig. 4.36 is often described, independent of its load, by means of a system of equations

$$\begin{bmatrix} i_{rs} \\ i_{pq} \end{bmatrix} = \begin{bmatrix} y'_{11} & y'_{12} \\ y'_{21} & y'_{22} \end{bmatrix} \begin{bmatrix} v_{rs} \\ v_{pq} \end{bmatrix}. \tag{4.185}$$

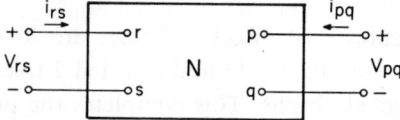

Fig. 4.36. A general two-port network.

The coefficient matrix of (4.185) is the *short-circuit admittance matrix* of the two-port, and is denoted by the symbol Y_{sc}. The entries of Y_{sc} are called the *short-circuit admittance parameters* or the *y parameters* for short. If Y_{sc} is nonsingular, the inverse of Y_{sc}, denoted by the symbol Z_{oc}, is the *open-circuit impedance matrix* of the two-port, and the entries of Z_{oc} are called the *open-circuit impedance parameters* or simply *z parameters*. The reason for these names "short-circuit" and "open-circuit" becomes obvious on noting that these parameters can be obtained by short-circuiting or open-circuiting the proper terminals of the two-port. For example, to get y'_{12} we short-circuit terminals *r* and *s* and connect a voltage generator across the terminals *p* and *q*. Then the ratio of the (Laplace) transform of the current at the short-circuited end to the transform of the voltage input is y'_{12}, which is also the negative of $y_{pr,qs}$. Similarly, this is valid for y'_{11}, y'_{21}, and y'_{22}. On the other hand, if $Z_{oc} = [z'_{ij}]$, then to find z'_{21} we leave the terminals *p* and *q* open and connect a current source at the terminals *r* and *s*. Then, the ratio of the transform of the voltage at the open-circuited end to the transform of the input current is z'_{21}. Similarly, z'_{11} and z'_{22} are the driving-point impedance functions at the respective ports when the other port is open-circuited, and z'_{12} is the open-circuit transfer-impedance function $z_{pr,qs}$.

For convenience, we use the shorthand symbol

$$W_{rp,sq} = \sum_{t_{rp,sq}} f(t_{rp,sq}), \tag{4.186}$$

where $t_{rp,sq}$ is a directed 2-tree of $G(Y)$.

COROLLARY 4.29: Let $G(Y)$ be the associated directed graph of a two-port network as shown in fig. 4.36. Then its short-circuit admittance matrix Y_{sc} is given by

$$Y_{sc} = \frac{1}{U_3}\begin{bmatrix} W_{p,q} & W_{qr,ps} - W_{pr,qs} \\ W_{rq,sp} - W_{rp,sq} & W_{r,s} \end{bmatrix}, \qquad (4.187)$$

where $U_3 = U_{rq,s,p} + U_{r,sq,p} + U_{rp,s,q} + U_{r,sp,q} \neq 0$.

Also, for convenience, let

$$V_k = \sum_{t_k} f(t_k), \qquad (4.188)$$

where t_k is a directed tree of $G(Y)$.

COROLLARY 4.30: If $V_k \neq 0$, the open-circuit impedance matrix Z_{oc} of a two-port network of fig. 4.36 is given by

$$Z_{oc} = \frac{1}{V_k}\begin{bmatrix} W_{r,s} & W_{pr,qs} - W_{qr,ps} \\ W_{rp,sq} - W_{rq,sp} & W_{p,q} \end{bmatrix}. \qquad (4.189)$$

Since Z_{oc} is the inverse of Y_{sc}, the determinant of Y_{sc} is the reciprocal of that of Z_{oc}. Using this in conjunction with (4.187) and (4.189), we have

COROLLARY 4.31:
$$\det Y_{sc} = 1/(\det Z_{oc}) = V_k/U_3. \qquad (4.190)$$

Thus, in particular if the two-port network N is passive and contains no mutual inductances, the associated directed graph $G(Y)$ becomes the two-port network itself. All the formulas derived in this and the preceding two sections can be applied directly to the two-port network itself with only a minor modification. For example, a proper path P_{rs}^p becomes a path in N connecting the nodes r and s and containing the added edge e_{pq}, and a directed 3-tree becomes a "3-tree". These terms are self-evident, and there is no point to go through all over again. Therefore, we shall be satisfied by simply pointing out this possibility, and leave the details as obvious. (See ch. 2, § 6.3.)

Using the intuitive representation of fig. 4.31, the off-diagonal elements of (4.187) and (4.189) can be stated in the following descriptive fashion:

$$W_{qr,ps} - W_{pr,qs} = \begin{pmatrix} r & & p \\ & \times & \\ s & & q \end{pmatrix} - \begin{pmatrix} r \longrightarrow p \\ s \longrightarrow q \end{pmatrix}, \qquad (4.191a)$$

$$W_{rq,sp} - W_{rp,sq} = \begin{pmatrix} r & & p \\ & \times & \\ s & & q \end{pmatrix} - \begin{pmatrix} r \longrightarrow p \\ s \longleftarrow q \end{pmatrix}, \qquad (4.191b)$$

which illustrated the two types of directed 2-trees involved in the formulas

Example 4.14: Consider the active two-port network N as shown in fig. 4.37(a). Its associated directed graph $G(Y)$ is presented in fig. 4.37(b). Suppose that we wish to compute the short-circuit admittance and open-circuit impedance parameters of N. Based on (4.187) and (4.189), the following terms are computed:

$$W_{24,31} = 0,$$
$$W_{21,34} = y_1 y_3 y_6 y_4;$$
$$W_{13,42} = g_m y_1 y_6 g_m,$$
$$W_{12,43} = y_1 (y_3 - g_m) y_6 (y_4 - g_m);$$
$$\begin{aligned} V_k'' = V_6'' &= (y_1 + y_2) y_3 (y_4 - g_m) + (y_1 + y_2) y_3 g_m \\ &+ (y_1 + y_2) y_3 (y_5 + y_6) + y_4 (y_5 + y_6)(y_1 + y_2) \\ &+ y_4 (y_5 + y_6) y_3 + y_4 g_m y_3, \end{aligned}$$

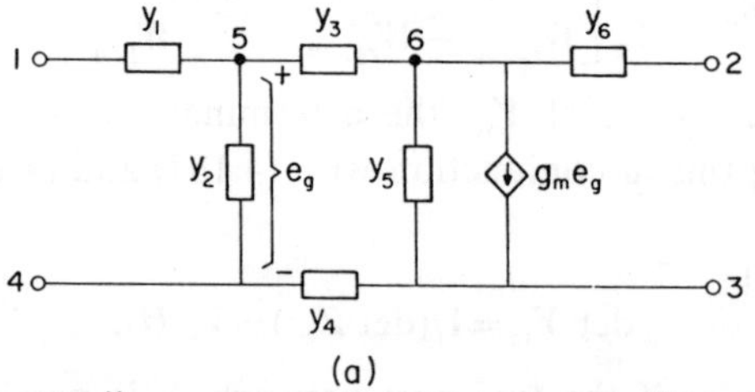

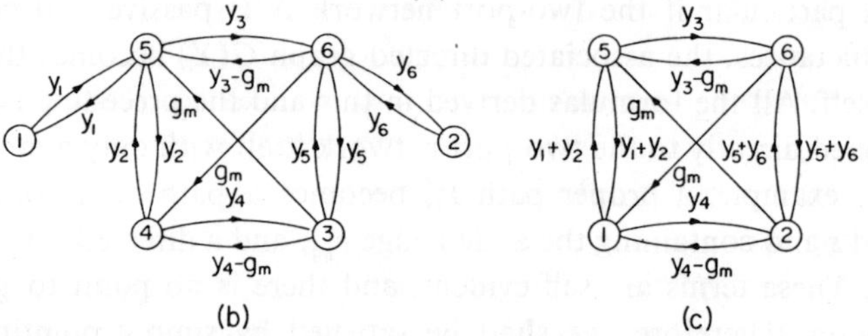

Fig. 4.37. (a) An active two-port network, (b) its associated directed graph, and (c) the directed graph derived from that of (b) by identifying the nodes 1 and 4, and 2 and 3, respectively.

where V_6'' is equal to the sum of directed-tree-admittance products with reference node 6 in the directed graph obtained from $G(Y)$ by identifying the nodes 1 and 4, and 2 and 3, respectively. The resulting graph is presented in fig. 4.37(c). The node 6 is chosen as the reference node to reduce the number of directed trees and also the number of cancellations.

Similarly, we can compute

$$\begin{aligned} W_{2,3} &= y_4 (y_5 + y_6) y_1 y_2 + y_4 (y_5 + y_6) y_1 y_3 + y_2 y_3 (y_4 - g_m) y_1 \\ &+ y_2 y_3 y_1 (y_5 + y_6) + y_4 g_m y_3 y_1 + y_2 y_3 y_1 g_m, \end{aligned}$$

which is the same as the sum of directed-tree-admittance products with reference node 6 in the directed graph obtained from $G(Y)$ by identifying the nodes 2 and 3. Again, the choice of the reference node in the resulting graph is arbitrary, but the number of terms in general will be different.

$$W_{1,4} = (y_1 + y_2) y_3 (y_4 - g_m) y_6 + (y_1 + y_2) y_3 g_m y_6 + (y_1 + y_2) y_3 y_5 y_6 \\ + y_4 y_5 y_6 (y_1 + y_2) + y_4 g_m y_3 y_6 + y_4 y_5 y_6 y_3,$$

which is identical to the sum of directed-tree-admittance products with reference node 6 in the directed graph obtained from $G(Y)$ by identifying the nodes 1 and 4.

The required network functions can now be computed by (4.187) and (4.189).

6.4. *The physical significance of the associated directed graph*

So far, we have formulated the network problems in terms of the indefinite-admittance matrix of a given network. The reason for this is that all the passive branch admittances enter the indefinite-admittance matrix in a beautifully symmetric form. Each branch admittance appears in a square pattern of four element positions centered about the main diagonal. The two rows and two columns associated with the square pattern have the same numbers as the nodes to which the branch in question is connected to the network. The branch admittance enters with a positive sign on the main diagonal and with a negative sign in the off-diagonal corners of the square pattern. Also, we find that the nonreciprocal element, such as the transconductance of a transistor or vacuum tube or any dependent source, will enter the indefinite-admittance matrix in a rectangular pattern of four element positions, but the pattern will no longer be centered on the main diagonal. The usefulness of this formulation lies in the fact that it facilitates the formulation of the open-circuit transfer-impedance and voltage-gain functions or the short-circuit admittance and current-gain functions from *any* pair of nodes in the network to *any* other pair of nodes directly without the necessity of referring to a specific node as the reference potential point.

This freedom of choice of the reference-potential point is reflected in the indefinite-admittance matrix of the network in that it is an equicofactor matrix. It also shows up in the associated directed graph of the matrix in that the choice of the reference node for the directed trees is arbitrary. These statements are really three different ways of saying the same thing. The approach emphasizes the fact that in solving an electrical network problem, it is immaterial as to which node is taken as the voltage reference or which equation we consider superfluous and suppress it from the given set of linear equations, since the solutions must be the same in every case. It states in effect that there is no meaning in absolute potential.

Consider a dependent current source, the equivalent network of which is presented in fig. 4.38(a). The indefinite-admittance matrix of the network is given by

$$\begin{array}{c} \\ i_a \\ i_b \\ i_c \\ i_d \end{array} \begin{array}{cccc} v_a & v_b & v_c & v_d \\ \left[\begin{array}{cccc} 0 & 0 & 0 & 0 \\ 0 & 0 & 0 & 0 \\ y & -y & 0 & 0 \\ -y & y & 0 & 0 \end{array} \right] \end{array}, \quad (4.192)$$

where v's and i's are the nodal voltages and entering currents of the network. As mentioned earlier, the transconductance y of a unilateral network element enters the indefinite-admittance matrix in a rectangular pattern which is not necessarily centered upon the main diagonal. This asymmetry is evidence of the violation of the reciprocity condition. Observe that the transconductance is actuated by the voltages at nodes a and b that correspond to the first two columns of the matrix. Similarly, the transconductance affects the currents of nodes c and d which are associated with the third and fourth rows of the matrix. Thus, the associated directed graph of the matrix is as shown in fig. 4.38(b).

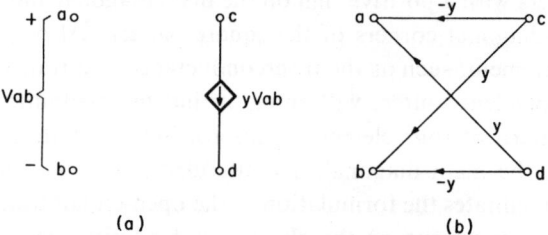

Fig. 4.38. (a) A voltage-controlled current source and (b) its associated directed graph.

This representation is of course quite general. The nodes a, b, c, and d need not be distinct. For example, if b and d are identical and if $y = g_m$, the transconductance of a vacuum tube, then, except for the passive elements, fig. 4.38 is the model for a tube. Similarly, if $b = d$ and $y = -ag_e$, then it is the model for a transistor in addition to the passive elements, where a is the current amplification factor and g_e the emitter conductance.

For the model of a gyrator, see Problem 4.56 and fig. 4.55.

Another element which frequently appears in the network models is the ideal transformer. The ideal transformer is characterized by a single real constant

n, which relates the currents and voltages of the two windings, as shown in fig. 4.39(a). The ideal transformer by itself has infinite short-circuit admittances and infinite open-circuit impedances, and hence cannot be formally characterized on the nodal basis. However, in practice a transformer usually appears in series with some other network element having a finite admittance. If we associate this element, say y_0, with the transformer, then the elements of the indefinite-admittance matrix become finite. The matrix is given by

$$\begin{bmatrix} y_0 & -y_0 & -ny_0 & ny_0 \\ -y_0 & y_0 & ny_0 & -ny_0 \\ -ny_0 & ny_0 & n^2y_0 & -n^2y_0 \\ ny_0 & -ny_0 & -n^2y_0 & n^2y_0 \end{bmatrix}. \qquad (4.193)$$

Since the matrix is symmetric, the associated (undirected) graph of the matrix is obtained in fig. 4.39(b).

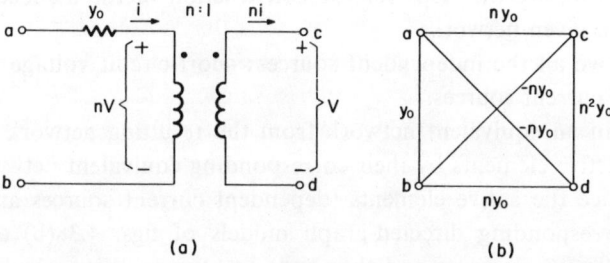

Fig. 4.39. (a) An ideal transformer having a finite series admittance and (b) its associated graph.

For the model of two imperfectly-coupled coils, see Problem 4.53 and fig. 4.49.

Thus, if these elements are imbedded in a network, the indefinite-admittance matrix of the network is simply the sum of the indefinite-admittance matrices of the subnetworks which form the complete network when combined in parallel by connecting together their corresponding nodes. The reason for this is that in parallel connection (superimposing the corresponding nodes) it implies equality of voltages and addition of currents at each node, and the addition of the indefinite-admittance matrices is a direct result. This would lead us to the direct construction of the associated directed graph from the given network.

In the preceding sections, we have shown how the network functions can be computed directly from the associated directed graph of the network by inspection. The technique thus depends upon the efficiency with which the associated directed graph of a network can be constructed. In the following, we shall present a simple procedure for constructing these graphs for a large class of the most commonly used networks. The networks to be considered are those consisting of resistors, capacitors, inductors (including ideal transformers in series with some finite admittances), gyrators, transistors, vacuum tubes and dependent sources. We assume that the dependent sources are dependent current sources, and they depend only upon voltages of other branches or node-pairs. This assumption is quite general since all other types of dependent sources can easily be transformed into the above type (use Blakesley transformation if necessary. See, for example, SESHU and BALABANIAN [1959]). It is possible, of course, to extend the technique to a very general class of linear networks. The reason for not doing this is that the rules needed here are few, and these rules are easy to remember and simple to operate. In case the network consists of other types of elements, it is simpler to draw the associated directed graph directly from the matrix.

The following are the steps for the construction of the associated directed graph from a given network:

(i) Remove all the independent sources: short-circuit voltage sources and open-circuit current sources.

(ii) Obtain an equivalent network from the resulting network of Step (i): replace all active elements by their corresponding equivalent networks.

(iii) Replace the active elements (dependent current sources and gyrators) by their corresponding directed-graph models of figs. 4.38(b) and 4.55(b): remove the active elements and then superimpose on the equivalent network already obtained in Step (ii) by their corresponding directed-graph models.

Similarly, we do the same for the ideal transformers and mutual couplings. The only difference is that we use the associated symmetric directed graphs of the graphs of figs. 4.39(b) and 4.49(b).

(iv) Replace each passive element by a pair of oppositely directed edges. The weight of each edge in the pair is the admittance of the passive element.

(v) Combine the parallel edges if necessary: the edges in the same direction between the same pair of nodes are combined into one with weight equal to the sum of the weights of the edges in the parallel connection.

There is a simple physical interpretation for the associated directed graph $G(Y)$ of the indefinite-admittance matrix Y of a network N. Since the nodes of $G(Y)$ are associated with the columns or rows of Y, it is convenient to assign the variables $v_1, v_2, \ldots,$ and v_n to the nodes $1, 2, \ldots,$ and n of $G(Y)$, in that order. These associated weights of the nodes are considered as the nodal potentials measured between nodes $1, 2, \ldots, n$ and some arbitrary but unspecified reference point. If v_{jk} denotes the potential difference between the nodes j and k, so that

§ 6   Direct analysis of electrical networks

$v_{jk} = v_j - v_k$, the source current I_j entering node j is given by

$$I_j = \sum_{\substack{x=1 \\ x \neq j}}^{n} f(j, x) v_{jx} \qquad (4.194)$$

for $j = 1, 2, \ldots, n$. This is exactly the nodal system of equations of the network given in (4.152). In other words, each edge of the associated directed graph $G(Y)$ may be considered as a basic network element called a *distor*, as shown in fig. 4.40(a). The current i_{jk} in a distor can only flow in the direction of the arrow, and is proportional to the potential difference of its two terminal potentials v_j and v_k. The proportionality constant is called the distor *admittance* y_{jk}.

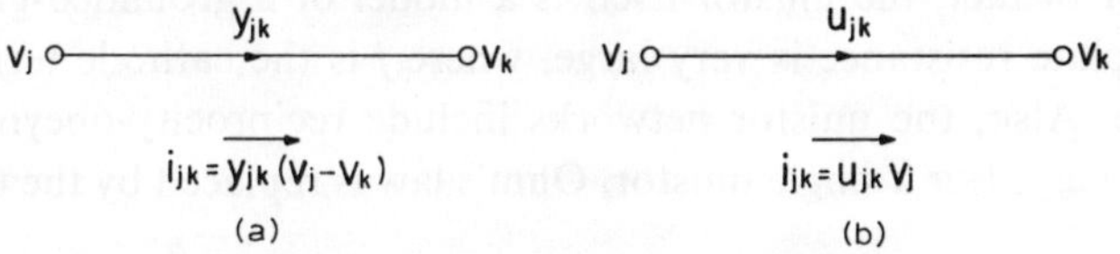

Fig. 4.40.   (a) A distor and (b) a unistor.

Actually, the distor itself is a model of a series connection of a passive branch with admittance y_{jk} and an ideal diode. Obviously, the distor does not obey reciprocity since it is a unilateral element. However, distor networks include reciprocity-obeying networks as a special case. To see this, it is only necessary to observe that two distors in parallel, pointing in opposite directions and having equal admittances, are equivalent to a single two-terminal branch admittance of the usual Ohm's-law variety. For a single distor, Ohm's law is replaced by the transmission law

$$i_{jk} = y_{jk}(v_j - v_k). \qquad (4.195)$$

This, together with Kirchhoff's current law, gives us the general nodal system of equations of (4.152).

A controlled current source, appearing in a conventional network representation such as a transistor or a vacuum tube, has a distor representation of fig. 4.38(b). The model probably appears to be strange and unphysical at first sight, in comparison with the more familiar controlled-current representation given in fig. 4.38(a). It can be argued however that the model is actually closer to physical reality. If v_a, v_b, v_c, and v_d denote the nodal potentials measured between nodes a, b, c, and d and some arbitrary but unspecified reference point, respectively, then the amount of current flowing out of node c is precisely $y(v_a - v_b) [= -y(v_c - v_a) + y(v_c - v_b)]$. Similarly, there is a current $y(v_a - v_b)$

flowing into node d, a current $y(v_c - v_d)$ flowing out of node a, and a current $y(v_c - v_d)$ flowing into node b. If we identify the nodes b and d and let $y = g_m$, the transconductance of a vacuum tube, then fig. 4.38(b) represents the model for a vacuum tube (in addition to the passive elements). The above argument corresponds to the flow of a positive current in a vacuum tube, for the positive current actually flows through the grid on its journey from plate to cathode.

There is another possibility as suggested by MASON [1957]. Here, we may consider each edge of $G(Y)$ as a basic network element called a *unistor*, as shown in fig. 4.40(b). The unistor has an admittance u_{jk}, and its current can only flow in the direction of the arrow, and is proportional to *one* of its two terminal potentials v_j and is independent of the other terminal potential v_k. As a matter of fact, the unistor itself is a model of a grounded-grid triode in which the plate resistance is very large, where j is the cathode terminal and k is the plate. Also, the unistor networks include reciprocity-obeying networks as a special case. For a single unistor, Ohm's law is replaced by the transmission law

$$i_{jk} = u_{jk} v_j. \tag{4.196}$$

Thus, in $G(Y)$ if we let $f(k, j) = u_{jk}$, (4.196) together with Kirchhoff's current law,

$$\sum_x i_{jx} = \sum_x i_{xj}, \tag{4.197}$$

gives us the general nodal system of equations of a unistor network,

$$v_j \sum_x u_{jx} = \sum_x u_{xj} v_x, \tag{4.198}$$

where $j, x = 1, 2, \ldots, n$ but $x \neq j$. Since

$$\sum_x u_{jx} = \sum_x u_{xj}, \tag{4.199}$$

(4.198) can be written in a slightly different form

$$\sum_x u_{xj}(v_j - v_x) = 0, \tag{4.200}$$

which is precisely the nodal system of equations given in (4.152). Thus, any linear network can also be fully represented by an interconnection of unistors.

6.5. *Direct analysis of the associated directed graph*

In the foregoing, we have developed topological formulas for various types of network functions of a general linear electrical network. These formulas express the network functions in terms of the various types of topological quantities in the associated directed graph of the network. We have also indicated how to construct these graphs for a large class of most commonly used networks, and provided for them a few simple physical interpretations. Thus, the associated directed graph $G(Y)$ itself may be considered as a physical model of the network. In this section, we shall show how the various topological formulas of a general network can be represented in a systematic and simple way.

Like many other topics discussed in this chapter, the basic idea of the present approach is not new, and was first exploited by KIRCHHOFF [1847] for the mesh system of equations, and by MAXWELL [1892] for the nodal system. The treatment to be presented here is similar to that given by MASON [1957]. However, as mentioned earlier, his approach, viewpoint, and interpretation are quite different, but the essential results are the same.

DEFINITION 4.23: *Meter branch.* A *meter branch* is a directed edge used to represent either an *am*meter, denoted by the symbol *am*, or a *vol*tmeter, denoted by the symbol *vm*. The current reference is given by the direction of the edge, and the voltage-reference plus is at the tail of the current-reference arrow.

DEFINITION 4.24: *Source branch.* A *source branch* is a directed edge used to represent either an independent *c*urrent *s*ource, denoted by the symbol *cs*, or an independent *v*oltage *s*ource, denoted by the symbol *vs*. The current reference is given by the direction of the edge, and the voltage-reference plus is at the *head* of the current-reference arrow.

Since the associated directed graph $G(Y)$ of a network may be considered as its physical model, the connection of a source branch and a meter branch to $G(Y)$ causes the meter to register. If it is an ammeter, then the reading indicates the amount of current flowing through the meter branch; if it is a voltmeter, then the reading indicates the voltage drop across the meter branch.

DEFINITION 4.25: *Transmission.* Let a meter branch directed from node p to node q and a source branch directed from s to r be inserted in $G(Y)$. The *transmission*, denoted by the symbol $H_{rp,sq}$, between the node pairs rs and pq of $G(Y)$ is defined as the ratio of meter reading to source value.

In other words, all the network functions defined in the preceding sections are special cases of transmission $H_{rp,sq}$. For example, if the meter branch denotes an ammeter, and if the source branch represents a voltage source, then the transmission $H_{rp,sq}$ represents the short-circuit transfer admittance $y_{rp,sq}$. On the other hand, if the meter branch denotes a voltmeter and if the source branch represents a current source, then the transmission $H_{rp,sq}$ represents the open-circuit transfer impedance $z_{rp,sq}$. This is similarly valid for other types of network functions. For convenience, various possible combinations are presented in Table 4.1.

Table 4.1

		meter branch	
		ammeter	voltmeter
source branch	current source	$\alpha_{rp,sq}$	$z_{rp,sq}$
source branch	voltage source	$y_{rp,sq}$	$g_{rp,sq}$

The above results are, of course, quite general, since the letters r, s, p, and q may refer to the labels of any four nodes in any linear network. In general, we also include the case where a meter branch and a source branch are first connected in series, and then inserted in $G(Y)$.

With these definitions, it is not difficult to see that all the topological formulas derived in the preceding sections can be formulated and represented in the following way.

THEOREM 4.16: Let G be the associated directed graph of the indefinite-admittance matrix Y of a general linear network. Let a meter branch directed from node p to node q and a source branch directed from s to r be inserted in G. Then for $r \neq s$ and $p \neq q$ the transmission $H_{rp,sq}$ of G is given by

$$H_{rp,sq} = \frac{\sum\limits_{pp_{rs}} (-1)^{\alpha} f(P_{rs}^{p}) V_{k}^{*}}{V_{j}''}, \qquad (4.201)$$

where

$$V_{k}^{*} = \sum_{t_{k}^{*}} f(t_{k}^{*}), \qquad (4.202a)$$

$$V_{j}'' = \sum_{t_{j}''} f(t_{j}''); \qquad (4.202b)$$

P_{rs}^{p} is a proper path with respect to the node p, connected between the nodes r and s and containing the meter branch of G, which is treated as an edge with unit weight; $\alpha = 0$ if the orientations of the meter and the source branches are both opposite to or coincident with that of the oriented circuit formed by P_{rs}^{p}

and the source branch, and $\alpha = 1$ otherwise; t_k^* and t_j'' are directed trees with reference nodes k and j in $G(P_{rs}^p; \emptyset)$ and G'', respectively; and G'' is the directed graph obtained from G by removing the current-source and voltmeter branches and by shorting the voltage-source and ammeter branches.

As an alternative to the sign problem, α can also be determined intuitively as follows: $\alpha = 0$ if the current in the positive direction of the source will cause the meter to register a positive reading in the circuit formed by the proper path and the meter branch, and $\alpha = 1$ otherwise.

We remark that in the theorem the choice of the reference nodes k and j is arbitrary since the associated matrices of the directed graphs $G(P_{rs}^p; \emptyset)$ and G'' are equicofactor matrices. Also, it is reasonable to expect that the voltage-source branch vs and the ammeter branch am should be short-circuited because they have zero internal impedance; that the current-source branch cs and the voltmeter branch vm should be open-circuited (removed) because they have infinite internal impedance.

A similar formulation of (4.201) based on unistors has also been given by MASON [1957]. However, the arguments that are followed in the derivations are heuristic, and sometimes quite devious. A formal proof of the topological rules for a unistor graph was given by NATHAN [1962] based on the Binet-Cauchy expansion of the determinant of the product of two matrices. The only difference between the rules given above and the ones given by Mason is that in the proper path P_{rs}^p all its edges are directed toward the node p instead of away from node p. The reason for this is not difficult to find. The unistor admittance associated with the unistor directed from node i to node j is the element y_{ji} of (4.21). In other words, the associated unistor graph of (4.21) is actually the associated directed graph of the transpose of the matrix (4.21). It is thus not surprising that the formulas of the two approaches are nearly identical. This fact has also been pointed out by TALBOT [1966]. In addition to the formulations discussed above, other approaches are also available. See, for example, PERCIVAL [1955], COATES [1958], MAYEDA [1958], NATHAN [1965], TALBOT [1965] and NUMATA and IRI [1973].

Example 4.15: The network diagram of a general potential-feedback amplifier and its equivalent network are given in fig. 4.41. The associated directed graph G of the amplifier can easily be obtained from the equivalent network using the procedures outlined in § 6.4, and is presented in fig. 4.42. Suppose that we wish to compute the transfer impedance function $z_{rp,kk}$ of the amplifier. So let us insert a voltmeter branch vm from node p to node k, and a current-source branch cs from node k to node r in G. They are denoted by the dashed

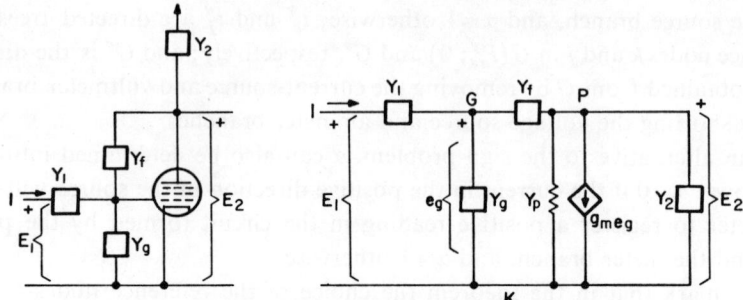

Fig. 4.41. A potential-feedback amplifier and its equivalent network.

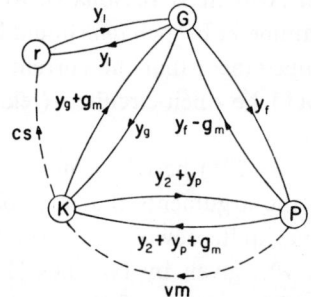

Fig. 4.42. The associated directed graph of the amplifier of fig. 4.41.

lines in fig. 4.42. From (4.201) we have $H_{rp,kk} = z_{rp,kk}$ and

$$z_{rp,kk} = \frac{1}{V_j''} \sum_{PP_{rk}} (-1)^\alpha f(P_{rk}^p) V_k^* = \frac{y_1(y_f - g_m)(1)}{V_j''}, \tag{4.203}$$

where

$$V_j'' = V_k'' = y_1 y_g [(y_f - g_m) + (y_2 + y_p + g_m)] + y_1 y_f (y_2 + y_p + g_m)$$
$$= y_1 [y_f(y_p + y_2 + g_m) + y_g(y_f + y_p + y_2)], \tag{4.204}$$

and the set of directed trees t_k'' of G'' needed in (4.204) is obtained by an expansion on the pair of nodes r and k in G'', and G'' is the directed graph obtained from G by removing the source branch and the meter branch.

Similarly, if the input admittance function of the amplifier is needed, a meter branch am and a source branch vs are inserted as shown in fig. 4.43(a). From (4.201) we have $H_{xx,kr} = (1/z_{rr,kk})$ and

$$\frac{1}{z_{rr,kk}} = \frac{1}{V_j''} \sum_{P^x_{xk}} (-1)^\alpha f(P_{xk}^x) V_k^*$$
$$= \frac{y_1 y_f(y_2 + y_p + g_m)(1) + y_1 y_g[(y_f - g_m) + (y_2 + y_p + g_m)]}{(y_1 + y_g)(y_2 + y_p) + (y_1 + y_g + g_m) y_f + (y_2 + y_p) y_f}, \tag{4.205}$$

where the terms in the denominator correspond to directed trees t_p'' in G'' as shown in fig. 4.43(b). The choice of node p as the reference node for the directed trees t_p'' of G'' eliminates redundancies due to duplications. Observe that the terms in the numerator are actually the same as those given in (4.204).

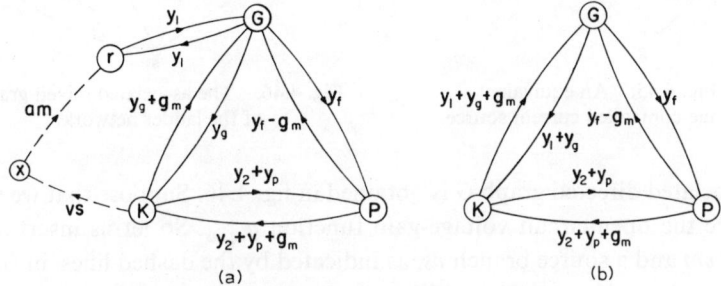

Fig. 4.43. The directed graphs used to compute the driving-point admittance function of the amplifier of fig. 4.41.

From the example that we have just worked out, it is obvious that the associated directed graph $G(Y)$ can be further simplified if we agree that an undirected edge stands for a pair of oppositely directed edges such that the weight associated with each directed edge in the pair is the same as that of the undirected edge. With this simplification, it is clear that the resulting $G(Y)$ is a mixed graph (ch. 1, § 6), which can be obtained directly from the equivalent network simply by superimposing the corresponding directed graphs of the unilateral elements, or the associated graphs of the ideal transformers and mutual couplings of the coils, upon the subnetwork consisting of resistors, capacitors, and (self-)inductors of the original network. We shall illustrate this with the following example.

Example 4.16: Consider the ladder network as shown in fig. 4.44. The dependent voltage source $z_m I_2$ can be converted into an equivalent dependent current source as shown in fig. 4.45. Following the procedures outlined in § 6.4,

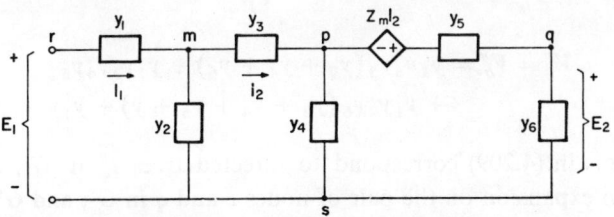

Fig. 4.44. An active ladder network.

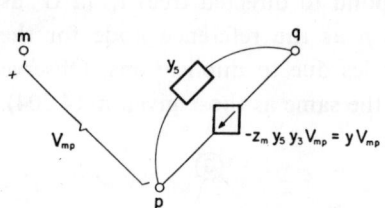
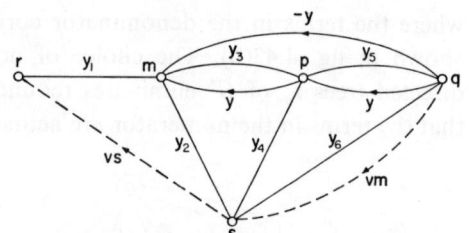

Fig. 4.45. An equivalent voltage-controlled current source.

Fig. 4.46. The associated mixed graph of the ladder network.

the associated directed graph G is obtained in fig. 4.46. Suppose that we wish to compute the open-circuit voltage-gain function $g_{rq,ss}$. So let us insert a meter branch vm and a source branch vs, as indicated by the dashed lines, in G. From (4.201) we have $H_{rq,ss} = g_{rq,ss}$ and

$$g_{rq,ss} = \frac{1}{V_j''} \sum_{Pq_{rs}} (-1)^\alpha f(P_{rs}^q) V_k^*$$
$$= [y_1(y_3+y)(y_5+y) + (-y)y_1(y_3+y_4+y_5+y)]/V_j'', \quad (4.206)$$

where

$$V_j'' = V_q'' = y_3 y_5 (y_1+y_2+y_4+y_6) + (y_1+y_2) y_4 y_5$$
$$+ (y_1+y_2) y_6 (y_3+y_4+y_5+y) + y_3 y_4 y_6, \quad (4.207)$$

and the set of directed trees t_q'' of G'' needed in V_q'' is obtained by an expansion on the pair of nodes m and q in G'', and G'' is given in fig. 4.47.

Suppose now that the open-circuit transfer impedance function $z_{rp,sq}$ of the network is needed. A meter branch vm and a source branch cs are inserted as shown in fig. 4.48. Again from (4.201) we have $H_{rp,sq} = z_{rp,sq}$ and

$$z_{rp,sq} = \frac{1}{V_j''} \sum_{Pp_{rs}} (-1)^\alpha f(P_{rs}^p) V_k^*$$
$$= \frac{y_1(y_3+y) y_6 - (-y) y_1 y_4}{V_j''}, \quad (4.208)$$

where

$$V_j'' = V_q'' = y_1 y_3 y_5 (y_2+y_4+y_6) + y_1 y_3 y_4 y_6$$
$$+ y_1 y_2 y_6 (y_3+y_4+y_5+y) + y_1 y_2 y_4 y_5, \quad (4.209)$$

and the terms in (4.209) correspond to directed trees t_q'' in G'', and are obtained by an expansion on the pair of nodes r and q in G'', and G'' is obtained from G by removing the meter and source branches.

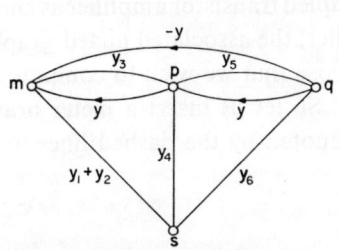
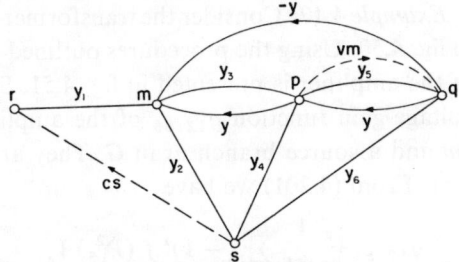

Fig. 4.47. The mixed graph derived from that of fig. 4.46 by removing vm and shorting vs.

Fig. 4.48. The mixed graph used to compute $z_{rp,sq}$.

Now, let us consider the model for two imperfectly-coupled coils of fig. 4.49(a). Following an argument similar to those given in § 6.4 for the dependent current sources and ideal transformers, the associated graph for the two imperfectly-coupled coils is presented in fig. 4.49(b) (Problem 4.53), where

$$y_1 = L_{11}/s(L_{11}L_{22} - M_{12}^2),$$
$$y_2 = L_{22}/s(L_{11}L_{22} - M_{12}^2), \qquad (4.210)$$
$$y = M/s(L_{11}L_{22} - M_{12}^2).$$

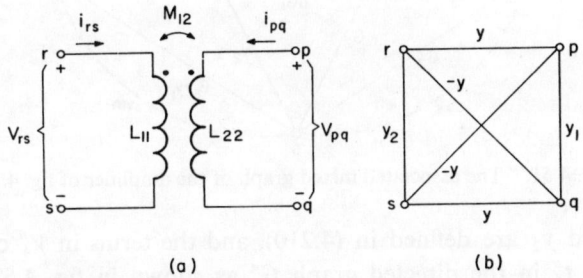

Fig. 4.49. (a) Two imperfectly-coupled coils and (b) their associated graph.

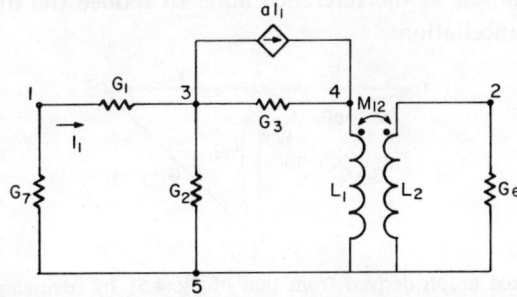

Fig. 4.50. A transformer-coupled transistor amplifier.

Example 4.17: Consider the transformer-coupled transistor amplifier as shown in fig. 4.50. Using the procedures outlined earlier, the associated mixed graph G of the amplifier is presented in fig. 4.51. Suppose that we wish to compute the voltage-gain function $g_{12,55}$ of the amplifier. So let us insert a meter branch vm and a source branch vs in G. They are denoted by the dashed lines in fig. 4.51. From (4.201) we have

$$g_{12,55} = \frac{1}{V_j''} \sum_{P_{15}^2} (-1)^\alpha f(P_{15}^2) V_k^*$$
$$= [G_1 G_3 y + (aG_1) y (G_1 + G_2 + G_3 - aG_1) + G_1(-aG_1) y$$
$$+ (-aG_1) G_3 y + (-aG_1)(-aG_1) y]/V_j''$$
$$= y(G_1 G_3 + aG_1 G_2)/V_j'', \tag{4.211}$$

where

$$V_j'' = V_4'' = (y_2 - y)[G_3 + (G_1 + G_2) - aG_1][y + (G_6 + y_1 - y)]$$
$$+ (G_1 + G_2) G_3 [y + (G_6 + y_1 - y)]$$
$$+ (G_6 + y_1 - y) y [G_3 + (G_1 + G_2) - aG_1], \tag{4.212}$$

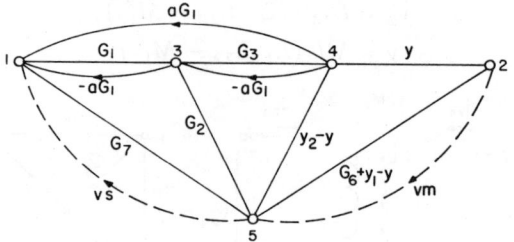

Fig. 4.51. The associated mixed graph of the amplifier of fig. 4.50.

and y, y_1, and y_2 are defined in (4.210), and the terms in V_4'' correspond to directed trees t_4'' in the directed graph G'' as shown in fig. 4.52. The set of directed trees t_4'' of G'' is obtained by an expansion on the pair of nodes 5 and 4. The node 4 is chosen as the reference node to reduce the number of redundancies due to cancellations.

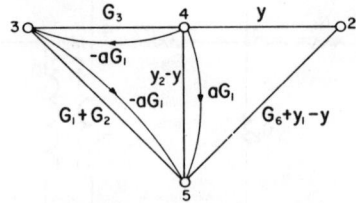

Fig. 4.52. The mixed graph derived from that of fig. 4.51 by removing the meter branch and shorting the source branch.

Finally, if the transfer impedance function $z_{12,55}$ of the amplifier is needed, it may be computed in a similar manner. All we have to do is to replace the voltage-source branch vs by a current-source branch cs in G. Thus, the numerator of $z_{12,55}$ is the same as that of $g_{12,55}$ given in (4.211). The terms in the denominator V_j'' of $z_{12,55}$ correspond to directed trees in the mixed graph obtained from G by removing the meter branch and the source branch from G, which, in fact, is the associated mixed graph of the indefinite-admittance matrix of the amplifier itself. In other words, V_j'' is simply the determinant of the node-admittance matrix of the amplifier. For convenience, let the set of directed trees be generated by an expansion on the pair of nodes 5 and 4:

$$\begin{aligned}V_4'' &= (y_2 - y)(G_6 + y_1)[(G_3 + G_2)(G_1 + G_7) + (G_1 - aG_1)G_7] \\ &+ G_2 G_3 (G_1 + G_7)(G_6 + y_1) + G_7 G_1 G_3 (G_6 + y_1) \\ &+ (G_6 + y_1 - y) y [(G_2 + G_3)(G_1 + G_7) + (G_1 - aG_1)G_7] \\ &= (y_1 + G_6)[(G_1 + G_7)(G_2 G_3 + G_2 y_2 + G_3 y_2) + G_1 G_7 (G_3 + y_2) \\ &- aG_1 G_7 y_2] + aG_1 G_7 y^2 - y^2 (G_2 + G_3)(G_1 + G_7) - y^2 G_1 G_7. \quad (4.213)\end{aligned}$$

The required transfer impedance function $z_{12,55}$ can now be computed.

We mention that, with the exception that all the directions of the edges are reversed, a proper path P_{rs}^p in (4.201) is also called a *proper transmission path* by MASON [1957]. The product of the weights associated with a proper transmission path is referred to as the *proper transmission-path value*, i.e., $f(P_{rs}^p)$. The term V_k^* is defined as the *cofactor* of the proper transmission path P_{rs}^p because it is the determinant of the node-admittance matrix of the network remaining when all the nodes in P_{rs}^p are identified.

§ 7. Conclusions

In this chapter, we have been concerned with a special class of system matrices known as the indefinite-admittance matrix. It arises when the nodal system of equations is used in analyzing a general linear electrical network. It is also called the equicofactor matrix because all the first-order cofactors of its elements are equal. We have examined some of the fundamental properties, and also shown how the second-order cofactors of its elements are related.

The basic relationships between the terms in the expansions of the first- and second-order cofactors of the elements of an equicofactor matrix of the type of (4.21) and certain types of subgraphs known as directed trees and directed 2-trees in the associated directed graph of the matrix have been demonstrated.

It has been pointed out that these subgraphs are related to the subgraphs in the corresponding Coates graph. A technique has been given which shows how to eliminate some of the nodes of the associated directed graph without affecting the ratios of the first-order and/or second-order cofactors. In particular, it reduces to the star-mesh transformation in electrical network theory.

Three methods for generating directed trees and directed 2-trees have been introduced. The first one is similar to that for the generation of semifactors and k-semifactors discussed in the preceding chapter. Its formulation is simple but duplications necessarily occur, and thus it is not very efficient. The other two are iterative procedures which would not introduce any redundancies, duplications, or cancellations of terms. Furthermore, they are all readily adaptable for a digital computer.

An application of the directed-tree approach to the direct analysis of electrical networks has been discussed. It has been shown that the technique permits a quick analysis of a general network. The network functions can be written down directly from the associated directed graph by inspection. Procedures for setting up the associated directed graph directly from the network diagram have also been outlined. We mention that a computer routine based on this technique was prepared by MCCLENAHAN and CHAN [1972].

The topological approach provides a great insight into the properties of the networks, and a knowledge of such methods can give much intuitive aid to the network designer or analyst even in those many cases where it is more convenient to formulate and solve the problem by formal matrix and determinantal methods. The rules presented in the chapter are an invaluable means of checking the reasonableness of any doubtful terms in the answer or of estimating the error in an approximation, since directed trees and directed paths are easy to identify. In a practical problem involving a rather complicated network, if the numerical values of the various branch admittances are known, we can search about in the network for those directed paths and directed trees which will lead to numerically dominant terms. The negligible terms never need be written down at all. This is a real saving!

Problems

4.1. Prove Corollary 4.1.
4.2. Show that the identity (4.25) is still valid if some of the edges of $G(Y)$ are represented by sets of parallel edges such that the sum of the weights associated with the parallel edges of each set is equal to the weight associated with the original edge of $G(Y)$.
4.3. Prove Corollary 4.5.

Problems

4.4. Prove Corollary 4.6.

4.5. Show that if Y is a matrix with the property that the sum of the elements of every row equals zero, and if it is of the type of (4.21), then

$$Y_{ix} = \sum_{t_i} f(t_i) \qquad (4.214)$$

for $i, x = 1, 2, \ldots, n$ where t_i is a directed tree in $G(Y)$.

4.6. Show that $Y_{ab,cd} = Y_{cd,ab}$.

4.7. Prove the identity (4.32).

4.8. Prove Corollary 4.9.

4.9. Prove Corollary 4.10.

4.10. Derive a formula similar to that of (4.41) for the case where the nodes of the pairs are not different from the reference node i.

4.11. Prove Corollary 4.11.

4.12. Show that the identity (4.46) is still valid if Y is a matrix with the property that the sum of the elements of every row equals zero, and is of the type of (4.21).

4.13. Extend the concept of directed 2-trees and directed 3-trees to "directed k-trees".

4.14. Using the concept of "directed k-trees" defined in Problem 4.13, show that if $i_1, j_1 < i_2, \ldots, i_k$ or $i_1, j_1 > i_2, \ldots, i_k$, then

$$(-1)^{i_1+j_1} \det Y_{i_1j_1, i_2i_2, \ldots, i_ki_k} = \sum f(t_{i_1j_1, i_2, \ldots, i_k}) \qquad (4.215)$$

for $k \geq 2$, where $Y_{i_1j_1, i_2i_2, \ldots, i_ki_k}$ is the submatrix obtained from the matrix Y by striking out the rows $i_1, i_2, \ldots, i_k$ and the columns $j_1, i_2, \ldots, i_k$; $t_{i_1j_1, i_2, \ldots, i_k}$ is a directed k-tree with reference nodes $i_1, i_2, \ldots,$ and i_k in $G(Y)$; Y is an equicofactor matrix of the type of (4.21); and the summation is taken over all $t_{i_1j_1, i_2, \ldots, i_k}$ in $G(Y)$.

4.15. Show that the identity (4.215) is still valid if Y is a matrix of (4.21) which has the property that the sum of the elements of every row equals zero.

4.16. Extend the identity (4.215) for any $i_1, j_1, i_2, \ldots,$ and i_k. (*Hint*: Extend the Definition 4.2.)

4.17. How many directed trees are there in a symmetric directed tetrahedron?

4.18. Show that if G_u is the graph that can be obtained from the n-node complete graph by removing p edges with different endpoints, then the number $N(t)$ of trees in G_u is given by

$$N(t) = (n-2)^p n^{n-p-2}, \qquad (4.216)$$

where $n > 1$ and $n \geq 2p$ (WEINBERG [1958]).

4.19. Show that if G_u is the graph that can be obtained from the n-node complete graph by removing p edges incident at a node of G_u, then the number $N(t)$ of trees of G_u is given by

$$N(t) = (n - p - 1)(n - 1)^{p-1} n^{n-p-2}, \qquad (4.217)$$

where $n > 1$ and $p < n$ (WEINBERG [1958]).

4.20. Compute the number of trees in each of the graphs given in fig. 4.53.

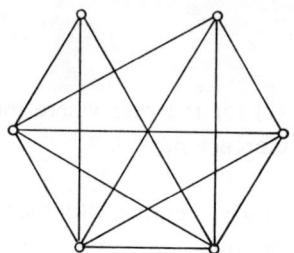 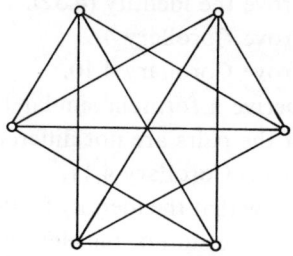

Fig. 4.53. Two incomplete graphs.

4.21. Prove Theorem 4.4 for the case where $i = j$.

4.22. Using the concept of "directed k-trees" defined in Problem 4.13, show that the number $N(t_{i_1 j_1, i_2, \ldots, i_k})$ of directed k-trees $t_{i_1 j_1, i_2, \ldots, i_k}$ with reference nodes $i_1, i_2, \ldots,$ and i_k, $i_1 \neq j_1$, of the n-node complete directed graph of order p is given by

$$N(t_{i_1 j_1, i_2, \ldots, i_k}) = p(np)^{n-k-1}, \qquad (4.218)$$

where $n > k \geq 2$ (CHEN [1966d]).

4.23. Show that in (4.218) if $i_1 = j_1$, then

$$N(t_{i_1, i_2, \ldots, i_k}) = kp(np)^{n-k-1}. \qquad (4.219)$$

4.24. Prove Corollary 4.13.

4.25. Show that if G_s is the associated symmetric directed graph of a weighted graph G_u, then there exists a one-to-one correspondence between the 2-trees $t^u_{i,j}$ of G_u and directed 2-trees $t^s_{i,j}$ of G_s and such that

$$f(t^u_{i,j}) = f(t^s_{i,j}). \qquad (4.220)$$

4.26. A set T_i of directed trees t_i is said to be *realizable* if there exists a directed graph whose directed trees t_i are all in T_i and there is no directed tree t_i in T_i which is not a directed tree of the directed graph. Show that a necessary and sufficient condition that a given set T_i of directed trees

t_i be realizable is that the sum graph of the directed trees t_i of T_i does not produce any directed trees t_i other than those contained in T_i.

4.27. Prove Corollary 4.15.

4.28. Prove the identity (4.75).

4.29. Prove the identity (4.84).

4.30. Prove Corollary 4.17.

4.31. Let t_n be a directed tree of a directed graph G and let (i,j) be an edge in G but not in t_n. Then for $i \neq n$ the operation

$$t_n \cup (i, j) - (i, k) = t'_n, \qquad (4.221)$$

where (i, k) is an edge of t_n, is called an *elementary transformation* on t_n if t'_n is a directed tree with reference node n in G. Show that any directed tree t_n of G can be obtained from any other one by a finite sequence of elementary transformations.

4.32. Let t_n and t'_n be two directed trees with reference node n of a directed graph. Show that if $t_n \neq t'_n$ there exists at least one directed edge (i, u), $i \neq n$, in t'_n but not in t_n and such that the sum graph $t_n \cup (i, u)$ is acyclic.

4.33. Prove the identity (4.126) for the semifactors.

4.34. Prove Theorem 4.6.

4.35. Prove Corollary 4.20.

4.36. Prove Theorem 4.10.

4.37. Extend Theorems 4.11 and 4.12 also to include directed graphs containing parallel edges.

4.38. Prove Theorem 4.12.

4.39. Realize the following set T of directed trees:

$$T = \{agec, aged, agid, ahid, ahic, agic\}.$$

If the realization is not unique, can you realize the set with two non-isomorphic directed graphs?

4.40. A subgraph of a directed graph is called a *subgraph product* if it is denoted by the product of its edge-designation symbols. Show that a given set of directed 2-tree products is realizable if, and only if, the set is realizable as the set of directed-tree products of a directed graph. (See Problem 4.26.)

4.41. Show that a necessary and sufficient condition that a set T_i of directed-tree products and a set $T_{i,j}$ of directed 2-tree products be realizable simultaneously as a directed graph is that $T_i \cup (\{e_0\} @ T_{i,j})$ be realizable as the set of directed-tree products of a directed graph, where i is the reference node for the directed trees and i and j are the reference nodes for the directed 2-trees, and e_0 is an arbitrarily chosen edge not contained in any element of $T_{i,j}$ (CHEN [1966b]).

4.42. Applying the condition given in Problem 4.41, find a simultaneous realization of the following given sets of directed-tree and directed 2-tree products:

$$T_i = \{e_1e_2e_3, e_1e_3e_5, e_1e_4e_5\},$$
$$T_{i,j} = \{e_3e_5, e_2e_3, e_4e_5, e_3e_6, e_4e_6\}.$$

4.43. Prove Theorem 4.13.
4.44. Prove Theorem 4.14.
4.45. Prove Corollary 4.22.
4.46. Prove Lemmas 4.5 and 4.6.
4.47. Consider the electrical network as shown in fig. 4.54, which represents a general potential-feedback amplifier. Compute the following network functions:
   (i) The output impedance function looking from the points p and r with the input terminals open-circuited.
   (ii) The voltage-gain function of the amplifier.
   (iii) The transfer voltage ratio V_{kr}/E_1, where V_{kr} is the potential drop from k to r.

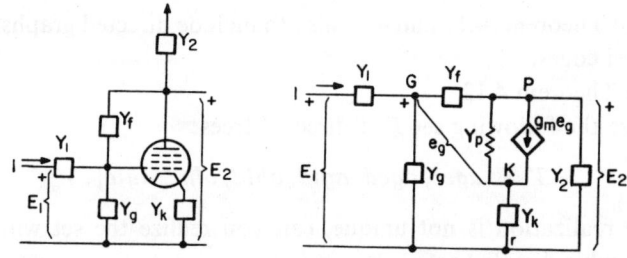

Fig. 4.54. A general potential-feedback amplifier and its equivalent network.

4.48. Prove Corollary 4.26.
4.49. In fig. 4.54, compute $z_{GP,rr}$ and $y_{GP,rr}$.
4.50. Derive the identities (4.187) and (4.189).
4.51. In (4.187), show that

$$W_{p,q} = W_{p,sq} + W_{q,sp}, \tag{4.222a}$$

$$W_{qr,ps} - W_{pr,qs} = W_{qr,s} - W_{pr,s}. \tag{4.222b}$$

(*Hint*: Use the identities (4.9) and (4.56).)

4.52. Show that the identity (4.179) is valid.

4.53. Justify that the model for the two imperfectly-coupled coils of fig. 4.49(a) is fig. 4.49(b).

4.54. Extend the procedure discussed in § 5.1 for the generation of directed 2-trees to the generation of directed k-trees defined in Problem 4.13.

4.55. Using the topological identities given in (4.187) and (4.189), show that det $Y_{sc}=1/(\det Z_{oc})=V_k/U_3$.

4.56. Show that the model for a gyrator of fig. 4.55(a) is fig. 4.55(b).

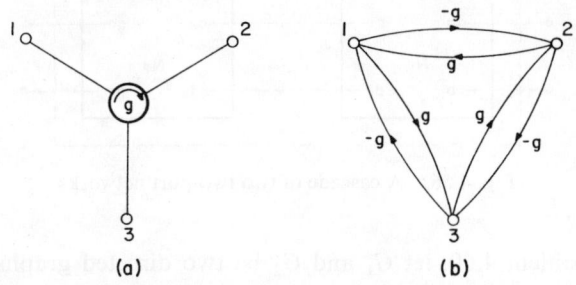

Fig. 4.55. (a) A gyrator and (b) its associated directed graph.

4.57. Consider the sum

$$\sum_{x=1}^{k} Y_{pi_x, ri_{x+1}}. \quad (4.223)$$

Show that the sum vanishes if $i_1 = i_{k+1}$. (*Hint*: Use the identity (4.59).)

4.58. Derive the following relationship:

$$-\sum_{\substack{x=1 \\ x \neq q}}^{n} Y_{pq, rx} = \sum_{\substack{x=1 \\ x \neq q}}^{n} Y_{px, rq}. \quad (4.224)$$

4.59. Let P_{ab} and P_{cd} be two node-disjoint directed paths from node a to node b and from node c to node d in a given directed graph G, respectively. Show that the set $T_{ba, dc}$ of directed 2-trees $t_{ba, dc}$ of G is given by

$$T_{ba, dc} = \bigcup \{P_{ab}P_{cd}\} \times T_{b, d}^*, \quad (4.225)$$

where the subgraphs of G are represented by the products of their edge-designation symbols; $T_{b, d}^*$ denotes the set of directed 2-trees $t_{b, d}^*$ with reference nodes b and d in G^*; G^* is obtained from G by identifying the nodes in P_{ab} and P_{cd}, respectively, with the combined nodes being denoted by b and d; and the set union is taken over all P_{ab} and P_{cd} of G.

4.60. Consider the cascade connection of two two-ports as shown in fig. 4.56. Show that the second-order cofactor $Y_{ag,bh}$ of the elements of the indefinite-admittance matrix of the over-all network is equal to the product of the second-order cofactors $Y^1_{ac,bd}$ and $Y^2_{eg,fh}$ of those of the individual networks N_1 and N_2, respectively:

$$Y_{ag,bh} = Y^1_{ac,bd} Y^2_{eg,fh}. \qquad (4.226)$$

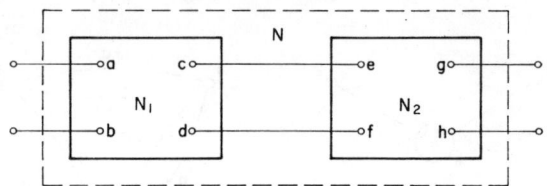

Fig. 4.56. A cascade of two two-port networks.

4.61. Like Problem 4.60, let G_1 and G_2 be two directed graphs connected in cascade. Assume that all the subgraphs of the directed graphs are represented by the products of their edge-designation symbols. Show that

$$T_{ag,bh} \cup T_{ah,bg} = (T^1_{ac,bd} \cup T^1_{ad,bc}) \times (T^2_{cg,dh} \cup T^2_{ch,dg}), \qquad (4.227)$$

where $T^1_{ab,cd}$, $T^2_{ab,cd}$, and $T_{ab,cd}$ are the sets of directed 2-trees $t_{ab,cd}$ in G_1, G_2, and $G_1 \cup G_2$, respectively; and the nodes c and d are the common nodes between the directed graphs G_1 and G_2.

4.62. Assume that all the subgraphs of a directed graph G are represented by their edge-designation symbols. Suppose that a, b, and c are the only edges terminated at node i of G. Show that the set T_i of directed trees t_i of G is given by

$$T_i = \{a\} \times T(a;bc) \cup \{b\} \times T(b;ac) \cup \{c\} \times T(c;ab) \cup \{ab\} \times T(ab;c)$$
$$\cup \{ac\} \times T(ac;b) \cup \{bc\} \times T(bc;a) \cup \{abc\} \times T(abc;\emptyset), \qquad (4.228)$$

where $T(a;bc)$ is the set of directed trees with reference node i in $G(a;bc)$; $G(a;bc)$ is defined in § 5.3 with the combined node being denoted by i; and all other terms are defined in a similar fashion. The technique is referred to as a *node expansion* by MASON and ZIMMERMANN [1960] for undirected graphs.

4.63. Generalize the identity (4.228) to accommodate any number of edges terminated at node i of G. For example, if C_i is the subgraph consisting of all the edges terminated at node i, then

$$T_i = \bigcup \{E_q\} \times T(E_q; \bar{E}_q), \qquad (4.229)$$

where E_q is a subgraph of C_i; $\bar{E}_q$ denotes the complement of E_q in C_i; $T(E_q; \bar{E}_q)$ is the set of directed trees with reference node i in $G(E_q; \bar{E}_q)$ in which i is the combined node and the set union is taken over all non-null subgraphs E_q of C_i.

4.64. Using the directed-tree technique, derive formulas for the solution of a system of linear nonhomogeneous equations. What is the associated directed graph of the system? Is the technique efficient? If not, give your reasons. (*Hint*: Use the identity (3.97).)

4.65. Show that in (4.85a)

$$-c = \sum_{\substack{x=h+1 \\ x \neq i+h \\ x \neq j+h}}^{n} \left[\left(\sum_{q=1}^{h} c_{qx} \right) + c_{ix}^{22} + c_{jx}^{22} \right]. \qquad (4.230)$$

4.66. Show that using (4.89) the weight w_{52} of (4.94a) can also be obtained from the directed graph of fig. 4.16 by the removal of node 3. Compare your result with (4.95a).

4.67. Using Thévenin's or Norton's theorem, prove the identity (4.176).

4.68. For a given network G, let A_a and Y_b be its incidence matrix (including all the nodes of G) and branch-admittance matrix. Show that $A_a Y_b A_a'$ is the indefinite-admittance matrix of G where the prime denotes the matrix transpose. Also show that $A_a Y_b A_a'$ is an equicofactor matrix of the type of (4.21). (*Hint*: Let E be a vector, all of its elements being unity. Then $A_a Y_b A_a' E = 0$.)

4.69. Let A be a square matrix of order n, $n > 1$, at least one of its cofactors being nonzero. Show that A is an equicofactor matrix if, and only if, all of its cofactors are equal. Is this generally true if all of the cofactors of the elements of A are zero?

4.70. Show that if any two parallel lines of an equicofactor matrix A are interchanged, the cofactors of the elements of the resulting matrix are equal to the negative of those of A.

CHAPTER 5

TREES AND THEIR GENERATION

The tree is perhaps one of the most important subgraphs in graph theory, insofar as engineering applications are concerned. In chs. 2 and 4 we have established a number of properties in electrical network theory that can be related to the concept of a tree. For example, the number of independent Kirchhoff's equations, the methods of choosing independent equations, and the topological formulas for network functions are all stated in terms of the single concept of a tree. In addition to these, the trees have been used successfully in chemical identification, scheduling and distributing problems, and a variety of other applications (see, for example, BUSACKER and SAATY [1965] and KAUFMANN [1967]). As a matter of fact, the term tree was first given by CAYLEY [1897] in his successful effort in applying graph theory to chemistry to solve the chemical identification problem.

In the topological analysis of a linear system, the problem ultimately reduces to that of finding the set of trees or directed trees in the associated graph. Three methods of efficient generation of these trees were presented in the preceding chapter. The first objective of this chapter is to present some of the fundamental properties of a tree, and to show that there exist intimate relationships among trees, cotrees, circuits and cuts of a graph. Based on these relationships, our second objective is to provide a unified summary of various existing tree-generating techniques and some of the related results.

Since trees in a directed graph are defined the same as those in an (undirected) graph, the present chapter is mainly concerned with the finite graphs.

Like those symbols used in ch. 2, throughout this chapter we denote the numbers of edges, nodes, and components of a graph G or a directed graph G_d by b, n, and c, respectively. Also we denote its rank and nullity by r and m, respectively.

§ 1. The characterizations of a tree

As simple consequences of the definition (Definition 2.2), trees can be characterized in many other ways, and we shall in the future make use of these equivalences given in the following theorem (Problem 5.1).

THEOREM 5.1: A graph G is a tree if, and only if, any one of the following properties is true:

(1) G is connected and does not contain any circuits;
(2) G has $n-1$ edges and does not contain any circuits;
(3) G is connected and has $n-1$ edges;
(4) G is connected and is of nullity $m=0$;
(5) G is connected but loses this property if any edge is removed;
(6) there exists a unique path between any two of its nodes.

THEOREM 5.2: A graph G contains a tree if, and only if, it is connected.

Proof. If G is not a tree, G must contain a circuit. By removing an edge of the circuit we obtain a spanning subgraph that is connected and contains fewer circuits than G. Repeated application of this procedure yields a tree of G, by virtue of the property (1) of Theorem 5.1.

COROLLARY 5.1: Let $d(x)$ denote the degree of the node x in a tree. Then

$$\sum_x [d(x) - 2] = -2, \qquad (5.1)$$

where the summation is taken over all the nodes x of the tree.

Proof.

$$\sum_x [d(x) - 2] = 2b - 2n = 2(b - n + 1) - 2 = 2m - 2 = -2.$$

A simple generalization of (5.1) is stated as a problem, and its proof is left as an exercise (Problem 5.2).

COROLLARY 5.2: A tree contains at least two nodes of degree 1.

Proof. By property (6) of Theorem 5.1, a tree contains a path whose length is maximal. The two endpoints of the path must be of degree 1 in the tree. For, if not, the path is not maximal. So the corollary is proved.

As a matter of fact, if a graph is nonseparable, any node can be made a node of degree 1 in a tree of the graph. This will be shown in the following theorem.

THEOREM 5.3: In a nonseparable graph G containing at least two edges, there exist at least two trees in which an arbitrary node of G is of degree 1 in these trees.

Proof. Let G^* be the graph obtained from G by removing an arbitrary node x and all the edges incident at x. Since G is nonseparable, it follows from Theorem 1.10 that G^* is a connected subgraph of G. By Theorem 5.2 there exists a tree t^* in G^*. If e_i is an edge of G incident at the node x, it is evident that the subgraph $t^* \cup e_i$ is a desired tree of G. Since G is nonseparable, there are at least two edges incident at x. Thus, there are at least two trees with the above property. This completes the proof of the theorem.

Earlier (Problems 2.43 and 2.44) we have indicated that a subgraph can be made part of a tree if, and only if, it contains no circuits, and that a subgraph can be made part of a cotree if, and only if, it contains no cuts. Then what are the necessary and sufficient conditions that a subgraph can be made part of a tree and another subgraph is contained in the corresponding cotree? This problem is characterized in the following theorem.

If t is a tree of a graph G, by $\bar{t}$ we mean the cotree with respect to t in G.

THEOREM 5.4: Let g_1 and g_2 be two edge-disjoint subgraphs of a connected graph G. If there exist a tree t_1 and a cotree $\bar{t}_2$ in G such that g_1 and g_2 are the subgraphs of t_1 and $\bar{t}_2$, respectively, then there exists a tree t of G for which g_1 is a subgraph of t and g_2 is a subgraph of $\bar{t}$.

Proof. Let G^* be the graph obtained from G by removing all the edges of g_2. Since the tree t_2 is contained in G^*, G^* is connected and contains all the nodes of G. Evidently, g_1 is also contained in G^* since g_1 and g_2 are edge-disjoint subgraphs of G. Since, by hypothesis, g_1 is contained in t_1, g_1 has no circuits, and hence by Problem 2.43 it can be made part of a tree of G^*. Let this tree be t. Then t is also a tree of G, and satisfies the conditions imposed in the theorem.

A different version of the theorem is the following.

COROLLARY 5.3: Let g_1 and g_2 be two edge-disjoint subgraphs of a connected graph G. Then there exists a tree t of G for which g_1 is a subgraph of t and g_2 is a subgraph of $\bar{t}$ if, and only if, g_1 contains no circuits and g_2 no cuts.

The result is of considerable practical interest. For example, in the state-space formulation of the electrical network problem, one is usually asked to construct a tree called a *normal tree* so that certain designated edges are tree branches and certain others are chords. (See, for example, BALABANIAN and BICKART [1969].)

In addition to these properties, in ch. 2, § 1 we have shown that a tree can also be characterized by its cut matrix or in particular by its incidence matrix. For an (undirected) graph G, the terms *cut*, *f-cutset*, and *f-circuit* are defined the same as those given in ch. 2, § 1 for a directed graph except that we ignore all the orientations associated with these subgraphs. The complete *incidence matrix* A_a, the complete *cut matrix* Q_a, and the complete *circuit matrix* B_a of G are similarly defined except that we replace all the -1's by 1's and use the algebra field of integers modulo 2. In this way, all the results in ch. 2, § 1 for directed graphs can be carried over to undirected graphs, except that all the negative signs disappear, as we now have the field modulo 2 to work with. There is hardly any point in going over the same ground once again, and so we shall blandly state the results, leaving the details as obvious. For convenience, again by a *basis incidence matrix* A of G we mean a submatrix of order $r \times b$ and of rank r of A_a. This is similarly valid for a *basis cut matrix* Q and a *basis circuit matrix* B of G. The choice of the symbolism here is guided by the fact that the essential structure of these matrices is the same for the directed and undirected graphs. Again, the prime denotes the matrix transpose.

THEOREM 5.5: The rank of the complete circuit matrix B_a of a graph G is equal to the nullity of G.

THEOREM 5.6: A square submatrix of a basis circuit matrix of a connected graph G is a major submatrix if, and only if, the columns of this submatrix correspond to the chords of a cotree of G.

THEOREM 5.7: The rank of the complete cut matrix Q_a of a graph G is equal to the rank of G.

THEOREM 5.8: A square submatrix of a basis cut matrix of a connected graph G is a major submatrix if, and only if, the columns of this submatrix correspond to the branches of a tree of G.

THEOREM 5.9: If the columns of the matrices B_a and Q_a of a graph are arranged in the same edge order, then

$$Q_a B'_a = 0 \quad \text{and} \quad B_a Q'_a = 0. \tag{5.2}$$

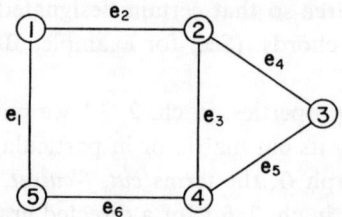

Fig. 5.1. A graph for illustrative purpose.

THEOREM 5.10: If the columns of A, B_f, and Q_f of a graph G are arranged in the order of chords and branches for the tree t defining the f-circuit matrix B_f and the f-cutset matrix Q_f such that

$$A = [A_{11} \quad A_{12}], \quad B_f = [U_m \quad B_{f12}], \quad \text{and} \quad Q_f = [Q_{f11} \quad U_r], \quad (5.3a)$$

then we have

$$B_f = [U_m \quad A'_{11}A'^{-1}_{12}] \quad \text{and} \quad Q_f = A^{-1}_{12}A = [B'_{f12} \quad U_r]. \quad (5.3b)$$

Example 5.1: In fig. 5.1, let $g_1 = e_1 e_3 e_4$ and $g_2 = e_2 e_5$. By Corollary 5.3 there exists a tree $t = e_1 e_3 e_4 e_6$ such that g_1 is a subgraph of t and g_2 is a subgraph of $\bar{t} = e_2 e_5$. The basis incidence matrix A of G with reference node 5 is given by

$$A = \begin{array}{c} \\ 1 \\ 2 \\ 3 \\ 4 \end{array} \begin{array}{c} e_1 \quad e_2 \quad e_3 \quad e_4 \quad e_5 \quad e_6 \\ \begin{bmatrix} 1 & 1 & 0 & 0 & 0 & 0 \\ 0 & 1 & 1 & 1 & 0 & 0 \\ 0 & 0 & 0 & 1 & 1 & 0 \\ 0 & 0 & 1 & 0 & 1 & 1 \end{bmatrix} \end{array}. \quad (5.4a)$$

Similarly, the f-cutset matrix of G with respect to the tree t is given by

$$Q_f = \begin{array}{c} e_1 \quad e_3 \quad e_4 \quad e_6 \quad e_2 \quad e_5 \\ \begin{bmatrix} 1 & 0 & 0 & 0 & 1 & 0 \\ 0 & 1 & 0 & 0 & 1 & 1 \\ 0 & 0 & 1 & 0 & 0 & 1 \\ 0 & 0 & 0 & 1 & 1 & 0 \end{bmatrix} \end{array}. \quad (5.4b)$$

The set T of trees of G is obtained as follows:

$$T = \{e_1 e_2 e_3 e_4, e_1 e_2 e_3 e_5, e_1 e_2 e_4 e_5, e_1 e_2 e_4 e_6, e_1 e_2 e_5 e_6, e_1 e_3 e_4 e_6,$$
$$e_1 e_3 e_5 e_6, e_1 e_4 e_5 e_6, e_2 e_3 e_4 e_6, e_2 e_3 e_5 e_6, e_2 e_4 e_5 e_6\}. \quad (5.4c)$$

Now, it is easy to check that a square submatrix of order 4 in A or Q_f is nonsingular (over the field mod 2) if, and only if, the columns of this submatrix correspond to the branches of a tree in T.

The f-circuit matrix of G with respect to t is given by

$$B_f = \begin{bmatrix} 1 & 1 & 0 & 1 & 1 & 0 \\ 0 & 1 & 1 & 0 & 0 & 1 \end{bmatrix}, \quad (5.5)$$

where the columns of B_f are arranged in the same order as those of Q_f of (5.4b). Again it is easy to check that a square submatrix of order 2 in B_f is nonsingular if, and only if, the columns of this submatrix correspond to the chords of a cotree in G. Furthermore, we have $B_f Q_f' = 0$.

§ 2. The codifying of a tree-structure

The discussions up to this point have been entirely in terms of the abstract tree; its geometric structure such as that shown in fig. 5.2 has served merely as an illustration of the theory. In a 3-dimensional Euclidean space, a geometric structure can always be associated with each abstract tree. In the preceding section, a tree is also expressed by the product of its branch-designation symbols.

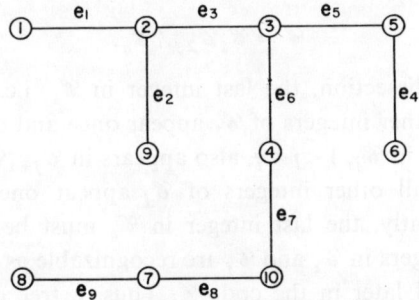

Fig. 5.2. The geometric structure of a tree.

However, this description is inadequate if one wants to recover the (abstract) tree or its geometric structure from this product. In this section, we shall show how the structure of a tree with n nodes can be encoded in a row of $n-2$ symbols so that it is recoverable from this description. The coding technique was originally proposed by NEVILLE [1953]. A different approach, based on a recurrence relation, was suggested by CHEN [1972c].

Let the nodes of a tree t be numbered by the integers $1, 2, ..., n$, $n \geq 2$. One of the nodes is required to play a special role in the codification; any node of the tree t will serve, and in the exposition it is convenient to suppose that this is node n. We emphasize that this node n has been implicitly specified for each of the codes to be discussed below.

2.1. Codification by paths

Let $d(x)$ denote the degree of the node x in t, and let

$$P_1 = (i_1, i_2) \cup (i_2, i_3) \cup \ldots \cup (i_{k-1}, i_k), \tag{5.6}$$

where $k \geq 2$ and $i_1 \neq n$, be a path in t such that $d(i_1)=1$, $d(i_u)=2$ for $u=2, 3, \ldots, k-1$, and $i_k = n$ or $d(i_k) \geq 3$ if $i_k \neq n$. The existence of P_1 in t follows directly from Corollary 5.2. Obviously, P_1 can be encoded by a row of $k-1$ integers

$$\mathscr{C}_1 = (i_2, i_3, \ldots, i_k). \tag{5.7}$$

Next, consider the graph $t - P_1$. If $t - P_1$ is not empty, it must be a tree, and the above process of dissection may be repeated. Let this path and its code be denoted by P_2 and $\mathscr{C}_2$, respectively. Now consider the graph $t - P_1 - P_2$. If we continue on this process, the tree t will eventually be decomposed into a set of edge-disjoint paths $P_1, P_2, \ldots, P_q$. Let $\mathscr{C}_1, \mathscr{C}_2, \ldots, \mathscr{C}_q$ be the corresponding codes of these paths. If we arrange these codes in a row with inner parentheses removed, we obtain a code $\mathscr{C}$ for t,

$$\mathscr{C} = \mathscr{C}_1 \mathscr{C}_2 \ldots \mathscr{C}_q. \tag{5.8}$$

By our process of dissection, the last integer in $\mathscr{C}_1$, i.e., i_k, also appears in $\mathscr{C}_2 \mathscr{C}_3 \ldots \mathscr{C}_q$ but all other integers of $\mathscr{C}_1$ appear once and only once in $\mathscr{C}$. Similarly, the last integer in $\mathscr{C}_j$, $1 < j < q$, also appears in $\mathscr{C}_{j+1} \mathscr{C}_{j+2} \ldots \mathscr{C}_q$ but not in $\mathscr{C}_1 \mathscr{C}_2 \ldots \mathscr{C}_{j-1}$, and all other integers of $\mathscr{C}_j$ appear once and only once in $\mathscr{C}_j \mathscr{C}_{j+1} \ldots \mathscr{C}_q$. Evidently, the last integer in $\mathscr{C}_q$ must be n. Conversely, if $\mathscr{C}$ is given, the last integers in $\mathscr{C}_1$ and $\mathscr{C}_j$ are recognizable as the first terms whose integers are repeated later in the code $\mathscr{C}$. Thus, a tree can be reconstructed from its code. Since the number of integers in the code of a path is equal to the length of the path, it follows that there are $n-1$ integers in $\mathscr{C}$, but if $\mathscr{C}$ is wanted only as a code for the tree t, the last integer in $\mathscr{C}$, i.e. n, is superfluous, and may be deleted. Let this simplified code be denoted by $\mathscr{C}^*$. Hence a tree with n nodes is expressible by a row of $n-2$ integers.

Finally, it should be pointed out that if an integer occurs k times in $\mathscr{C}^*$ then the corresponding node in the tree is of degree $k+1$. Also, the nodes of t which do not occur in $\mathscr{C}$ are nodes of degree 1 in t. The justifications for these are left as exercises (Problems 5.4 and 5.7).

As an example, consider the tree given in fig. 5.2. The tree may be decomposed into four edge-disjoint paths: $P_1 = (8, 7) \cup (7, 10)$, $P_2 = (9, 2)$, $P_3 = (1, 2) \cup (2, 3)$, and $P_4 = (6, 5) \cup (5, 3) \cup (3, 4) \cup (4, 10)$. The codes associated with these paths are $\mathscr{C}_1 = (7, 10)$, $\mathscr{C}_2 = (2)$, $\mathscr{C}_3 = (2, 3)$, and $\mathscr{C}_4 = (5, 3, 4, 10)$. Thus, the code associated with the tree is given by

$$\mathscr{C} = (7, 10, 2, 2, 3, 5, 3, 4, 10).$$

Conversely, if $\mathscr{C}$ is given the tree can be reconstructed as follows: Since the integers 10, 2, and 3 occur more than once in $\mathscr{C}$, the code $\mathscr{C}$ can be decomposed as (7, 10), (2), (2, 3), and (5, 3, 4, 10), each of which corresponds to the code of a path. These paths can easily be reconstructed from their codes, and are given by $P_1 = (w, 7) \cup (7, 10)$, $P_2 = (x, 2)$, $P_3 = (y, 2) \cup (2, 3)$, and $P_4 = (z, 5) \cup (5, 3) \cup (3, 4) \cup (4, 10)$. The tree can then be reconstructed from these paths. On the other hand, if $\mathscr{C}^* = (7, 10, 2, 2, 3, 5, 3, 4)$ is given, $\mathscr{C}$ can easily be reconstructed from $\mathscr{C}^*$ since the last integer n is equal to the number of integers in $\mathscr{C}^*$ plus 2. Now, if we count the number of times that an integer occurs in $\mathscr{C}^*$, we find that the integers 2 and 3 each occur twice, and that 7, 10, 5 and 4 each occur once. Thus, the nodes 2 and 3 are of degree 3 in t, and the nodes, 7, 10, 5 and 4 are of degree 2. All other nodes are of degree 1.

Note that the nodes w, x, y and z of the tree cannot be determined uniquely except that they are of degree 1. Suppose that we wish to encode a labeled tree so that it can be recovered completely, not just to within isomorphism. This can easily be accomplished as follows: At each stage in the process of dissecting a tree into paths, we always choose a path whose initial node has smallest value among all the nodes of degree 1. In this way, a tree can be completely recovered from its code. As an illustration, again consider the graph as shown in fig. 5.2. The tree is decomposed uniquely into four edge-disjoint paths: $P_1 = (1, 2)$, $P_2 = (6, 5) \cup (5, 3)$, $P_3 = (8, 7) \cup (7, 10)$, and $P_4 = (9, 2) \cup (2, 3) \cup (3, 4) \cup (4, 10)$. The associated codes are $\mathscr{C}_1 = (2)$, $\mathscr{C}_2 = (5, 3)$, $\mathscr{C}_3 = (7, 10)$, and $\mathscr{C}_4 = (2, 3, 4, 10)$. Thus, the code for the tree is given by

$$\mathscr{C} = (2, 5, 3, 7, 10, 2, 3, 4, 10).$$

Conversely, if $\mathscr{C}$ is given, the tree can be completely recovered as follows: Since the nodes 1, 6, 8 and 9 do not occur in $\mathscr{C}$, they must be the initial nodes of the paths. These nodes will appear in the paths P_1, P_2, P_3 and P_4 in the order of 1, 6, 8 and 9. Since the integers 2, 3 and 10 occur more than once in $\mathscr{C}$, $\mathscr{C}$ can be decomposed as (2), (5, 3), (7, 10), and (2, 3, 4, 10). Thus, the paths can be completely recovered from their codes.

As another example, let us encode the tree given in fig. 5.3. The code is given below:

$$\mathscr{C}^* = (15, 8, 3, 12, 32, 5, 1, 25, 29, 23, 25, 27, 17, 29, 29,$$
$$21, 1, 10, 21, 13, 13, 9, 6, 21, 32, 19, 23, 10, 15, 19). \tag{5.9}$$

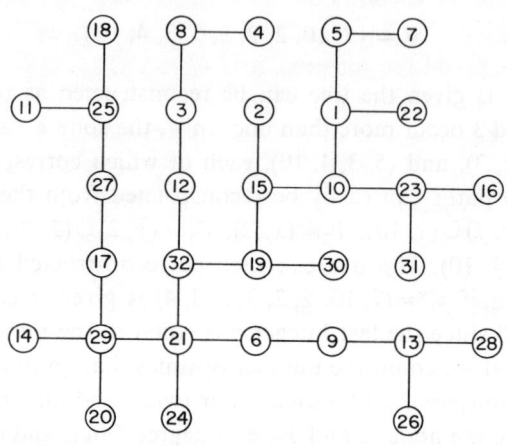

Fig. 5.3. A geometric tree whose code is given in (5.9).

2.2. Codification by terminal edges

In the foregoing, we have discussed a technique of codifying a tree by dissecting it into paths. In the present section, we shall show how a tree can be encoded by the endpoints of the terminal edges. (Also see Problem 5.9.)

DEFINITION 5.1: *Terminal edge*. An edge of a graph is called a *terminal edge* if one of its endpoints is of degree 1 in the graph.

For a given tree t, let (i_x, j_x), $x = 1, 2, ..., k$, be the terminal edges of t such that $d(i_x) = 1$ and $j_x = n$ or $d(j_x) \geq 2$. Furthermore, we require that $i_u < i_{u+1}$ for $u = 1, 2, ..., k-1$. Obviously, the nodes i_x are all distinct, but any number of j_x may coincide. Let

$$\mathscr{U}_1 = (i_1, i_2, ..., i_k), \tag{5.10a}$$

$$\mathscr{D}_1 = (j_1, j_2, ..., j_k). \tag{5.10b}$$

Also let E_1 be the subgraph consisting only of the terminal edges of t. Consider the graph $t - E_1$. If $t - E_1 \neq \emptyset$, then it is a tree and the above process may be repeated. Let $\mathscr{U}_2$ and $\mathscr{D}_2$ be the corresponding rows of integers thus generated. If we continue on this process, we will eventually generate two sequences of codes $\mathscr{U}_1, \mathscr{U}_2, ..., \mathscr{U}_q$ and $\mathscr{D}_1, \mathscr{D}_2, ..., \mathscr{D}_q$, each of which has the above property. If the latter sequence is arranged in a row with inner parentheses omitted, we obtain yet another code $\mathscr{D}$ for the tree t,

$$\mathscr{D} = \mathscr{D}_1 \mathscr{D}_2 \ldots \mathscr{D}_q, \qquad (5.11)$$

in which the elements of $\mathscr{U}_1$ are the nodes of t which do not occur in $\mathscr{D}$. If $\mathscr{D}$ is given, the number k of elements in $\mathscr{U}_1$, and thus $\mathscr{D}_1$, can be determined. The integers in $\mathscr{U}_j$, $1 < j < q$, are the integers which occur among the integers in $\mathscr{D}_{j-1}$ but not among the integers in $\mathscr{D}_j \mathscr{D}_{j+1} \ldots \mathscr{D}_q$. Since *all* the available nodes (integers) are taken at each stage, the number of elements in $\mathscr{D}_2$ or $\mathscr{U}_2$ is implicit in the composition of $\mathscr{D}$, when k is known, in the relation of the first k terms of $\mathscr{D}$ to the remainder of the row. The number of elements in $\mathscr{D}_3$ or $\mathscr{U}_3$ is implicit when the numbers in $\mathscr{D}_1$ and $\mathscr{D}_2$ are found, and so on. Finally, the elements of $\mathscr{U}_q$ are integers which do not occur in $\mathscr{U}_1, \mathscr{U}_2, \ldots,$ and $\mathscr{U}_{q-1}$ but less than n. Thus, a row of $n-1$ integers subject only to the condition that the last one is n can be punctuated as the row $\mathscr{D}$ of a tree, and the tree can then be reconstructed from its code $\mathscr{D}$.

As an illustration, consider the tree t given in fig. 5.2. The set of terminal edges of t consists of $(1, 2)$, $(6, 5)$, $(8, 7)$, and $(9, 2)$. The codes for these edges are $\mathscr{U}_1 = (1, 6, 8, 9)$ and $\mathscr{D}_1 = (2, 5, 7, 2)$. Similarly, the other codes are $\mathscr{U}_2 = (2, 5, 7)$ and $\mathscr{D}_2 = (3, 3, 10)$, $\mathscr{U}_3 = (3)$ and $\mathscr{D}_3 = (4)$, and $\mathscr{U}_4 = (4)$ and $\mathscr{D}_4 = (10)$. Thus, the code for the tree is given by

$$\mathscr{D} = (2, 5, 7, 2, 3, 3, 10, 4, 10).$$

Conversely, if $\mathscr{D}$ is given, the tree can be completely reconstructed as follows: Since 1, 6, 8 and 9 are the nodes in t which do not occur in $\mathscr{D}$, they must belong to $\mathscr{U}_1$, i.e., $\mathscr{U}_1 = (1, 6, 8, 9)$. Thus, we have $k = 4$ and $\mathscr{D}_1$ consists of the first four terms of $\mathscr{D}$, i.e., $\mathscr{D}_1 = (2, 5, 7, 2)$. By pairing the corresponding nodes in $\mathscr{U}_1$ and $\mathscr{D}_1$, we obtain the set of terminal edges of t. They are $(1, 2)$, $(6, 5)$, $(8, 7)$, and $(9, 2)$. Since the elements in $\mathscr{U}_2$ are the integers which occur among the integers in $\mathscr{D}_1$ but not among the integers in $\mathscr{D}_2 \mathscr{D}_3 \mathscr{D}_4 = (3, 3, 10, 4, 10)$, we find that $\mathscr{U}_2 = (2, 5, 7)$. Thus, $\mathscr{D}_2 = (3, 3, 10)$. The elements in $\mathscr{U}_3$ are those integers which occur in $\mathscr{D}_2$ but not in $\mathscr{D}_3 \mathscr{D}_4 = (4, 10)$, and we have $\mathscr{U}_3 = (3)$ and $\mathscr{D}_3 = (4)$. Since the elements in $\mathscr{U}_4$ are integers which do not occur in $\mathscr{U}_1$, $\mathscr{U}_2$, and $\mathscr{U}_3$ and which are less than 10 ($=n$), we have $\mathscr{U}_4 = (4)$ and $\mathscr{D}_4 = (10)$. Thus, the other edges of the tree are $(2, 3)$, $(5, 3)$, $(7, 10)$, $(3, 4)$, and $(4, 10)$. The tree is then completely recovered from these edges.

As another example, let us encode the tree as shown in fig. 5.3. The code for the tree is given below:

$$\mathscr{D} = (15, 8, 5, 25, 29, 23, 25, 29, 1, 21, 13, 13, 19, 23, 1,$$
$$3, 9, 10, 27, 10, 12, 6, 17, 21, 15, 32, 29, 19, 21, 32, 32). \qquad (5.12)$$

Like the code $\mathscr{C}$ discussed in the preceding section, if $\mathscr{D}$ is wanted only as a code for the tree t, the last integer which is always n is superfluous, and may be deleted. Let the simplified code be denoted by $\mathscr{D}^*$. Hence, a tree with n nodes is again expressible by a row of $n-2$ integers. Also, if an integer occurs k times in $\mathscr{D}^*$ then the corresponding node in t is of degree $k+1$. For example, the code $\mathscr{D}^*$ for the tree of fig. 5.2 is given by

$$\mathscr{D}^* = (2, 5, 7, 2, 3, 3, 10, 4).$$

Since the integers 2 and 3 each occur twice in $\mathscr{D}^*$, nodes 2 and 3 in t are nodes of degree 3. This is similarly valid for the other integers in $\mathscr{D}^*$.

§ 3. Decomposition into paths

In § 2.1 we have demonstrated how to encode a tree by dissecting it into paths. In the following, we shall consider the problem of decomposing a tree into the minimal number of edge-disjoint paths, and then determine the number of such decompositions. The work follows that of KOTZIG [1967], and is based on the fact that the number of nodes of odd degree in a graph is always even (Corollary 1.2).

THEOREM 5.11: Let $2k$, $k>0$, be the number of nodes of odd degree in a tree t. Then t can be decomposed into k edge-disjoint paths and any decomposition of t into edge-disjoint paths contains at least k paths.

Proof. From Corollary 1.2 we have $k \geq 1$. Let

$$P_1 = (i_1, i_2) \cup (i_2, i_3) \cup \cdots \cup (i_u, i_{u+1}) \tag{5.13}$$

be a path of t such that the nodes i_1 and i_{u+1} are of odd degree in t and the nodes i_z, $z=2, 3, ..., u$, are of even degree. Consider the graph $t-P_1$. Obviously, there are $2(k-1)$ nodes of odd degree in $t-P_1$. If $k-1 \neq 0$ the same process may be repeated for each of the components of $t-P_1$. Thus, t can be decomposed into k edge-disjoint paths.

Suppose that a decomposition of t into edge-disjoint paths contains less than k paths. Then the union of these paths will produce a tree t containing less than $2k$ nodes of odd degree, since the union of a graph and a path (edge-disjoint with the graph) will increase the number of nodes of odd degree at most by 2. Thus, any decomposition contains at least k edge-disjoint paths, which completes the proof of the theorem.

THEOREM 5.12: Let n_i be the number of nodes of degree i in a tree t, and

§3 Decomposition into paths

let $2k$ be the number of nodes of odd degree in t. Then the number β of different decompositions of t into k edge-disjoint paths is given by

$$\beta = \prod_{i=1}^{n-1} f(i)^{n_i}, \qquad (5.14)$$

where $f(2i-1) = f(2i) = 1 \cdot 3 \cdot 5 \cdot \ldots \cdot (2i-1)$.

Proof. For a set R, let $\alpha(R)$ denote the number of its elements. Let S_u, $u = 1, 2, \ldots, n$, be the set consisting of all the edges of t incident at the node u. Partition the elements of S_u into a set R_u of disjoint subsets r_{ux} with the following property: Let $R_u = \{r_{ux}\}$. If $\alpha(S_u)$ is even, then $\alpha(r_{ux}) = 2$ for all x; and if $\alpha(S_u)$ is odd, then $\alpha(r_{u1}) = 1$ and $\alpha(r_{ux}) = 2$ for $x \geq 2$. It is not difficult to see that if the node u is of degree i then there are $f(i)$ different partitions of the elements of S_u into R_u.

Consider a system $\mathscr{R} = \{R_1, R_2, \ldots, R_n\}$ of partitions with the above property. In the following, we shall show that for each choice of $\mathscr{R}$ there corresponds a unique decomposition of t into k edge-disjoint paths, and vice versa. The dissection process amounts to repeatedly generating a path according to $\mathscr{R}$ and then removing it. The rules for generating a path according to $\mathscr{R}$ are outlined as follows:

(1) We start at a node w of odd degree and then proceed along the edge contained in r_{w1} of R_w.

(2) If we arrive at an edge (u, v) in r_{vx} of R_v, then for $x \neq 1$ we proceed along the other edge contained in r_{vx}, and for $x = 1$ a path is constructed.

It is evident that the above procedure will decompose t into k edge-disjoint paths. Conversely, to each decomposition there corresponds a system of partitions with the above property. It follows that the number β of different decompositions is equal to the number of different systems $\mathscr{R}$ with the required property. Thus, we have

$$\beta = \prod_{x=1}^{n} f(d(x)) = \prod_{i=1}^{n-1} f(i)^{n_i}, \qquad (5.15)$$

which completes the proof of the theorem.

As an illustration, consider the tree t as shown in fig. 5.2. We have $n_1 = 4$, $n_2 = 4$, $n_3 = 2$, and $n_i = 0$ for $i = 4, 5, \ldots, 9$. Since there are six nodes of odd degree in t, i.e., $k = 3$, it follows that any decomposition of t into edge-disjoint paths contains at least 3 paths, and in fact there are

$$\beta = (1)^4 (1)^4 (3)^2 = 9$$

different decompositions possible for t. They are given by

$$\{e_1e_2, e_3e_5e_4, e_6e_7e_8e_9\}, \{e_1e_2, e_3e_6e_7e_8e_9, e_5e_4\}, \{e_1, e_2e_3e_5e_4, e_6e_7e_8e_9\},$$
$$\{e_1, e_2e_3e_6e_7e_8e_9, e_5e_4\}, \{e_2, e_1e_3e_5e_4, e_6e_7e_8e_9\}, \{e_2, e_1e_3e_6e_7e_8e_9, e_5e_4\},$$
$$\{e_1e_3, e_2, e_4e_5e_6e_7e_8e_9\}, \{e_2e_3, e_1, e_4e_5e_6e_7e_8e_9\}, \{e_3, e_1e_2, e_4e_5e_6e_7e_8e_9\}.$$

§ 4. The Wang-algebra formulation

In many engineering applications of graph theory, it is necessary that we list all the trees and cotrees of a graph. For example, the topological formulas for network functions are stated in terms of tree or cotree products. Thus, the efficiency with which a complex linear system can be analyzed by a digital computer using topological formulas as a tool depends largely upon the efficiency with which the trees or cotrees of the associated graph are generated.

The simplest way to generate trees of an n-node connected graph is to examine all possible combinations of $n-1$ edges. If b is the total number of its edges, we must examine $\binom{b}{n-1}$ sets. For a moderately large graph, the number of these sets is too large even for a high-speed digital computer. For example, for a graph with 21 nodes and 40 edges the number of such sets to be examined is 137846528820.

The main difficulties in generating trees by direct approach lie in the fact that the number of operations increases exponentially with the number of nodes. What is needed is a method of generating all trees which requires that the number of operations vary approximately linearly with the number of nodes of the graph, and which does not need to check every new generated tree against all the previously found ones in order to eliminate duplications, because the exhaustive checking procedure is an extremely time-consuming process as the number of trees, as the above example shows, could be in the *millions*. Apart from slowing down the process, this fact also limits the sizes of the systems that can be analyzed by a digital computer, since one must retain the set of all trees in computer memory and check each new tree against the list for possible duplications.

In this section, we shall demonstrate the intimate relationships among trees, cotrees, multi-trees, multi-cotrees, circuits, paths, and cuts, and then show how these relationships can be used to generate trees and cotrees. The entire section is based on the recent work of CHEN and MARK [1969], BERGER and NATHAN [1968], and CHEN [1969d–h, 1971c].

4.1. The Wang algebra

The Wang algebra was first suggested by WANG [1934] in an attempt to formulate a set of rules to obtain a short-cut for expanding the network determinants. Because of the intimate relationships between trees or cotrees of a network (graph) and the uncancelled terms of the network determinant, the algebra has been widely used in enumerating trees or cotrees of a graph.

For a given graph G, let $\mathscr{G}$ be the set of all subgraphs containing no isolated nodes except the null graph $\emptyset$ which is also in $\mathscr{G}$. Each of these subgraphs is represented by the product of its edge-designation symbols. Consider the set $\mathscr{S}$ of all subsets of $\mathscr{G}$ for which an *equivalence relation* "$=$" is defined in the usual sense. In $\mathscr{S}$ we define two binary operations. One of these binary operations is called the *ring sum* (or *symmetric difference*) and is denoted by the symbol $\oplus$, and the second is called the *Wang product* and is denoted by @. These operations are required to satisfy the following rules: For any S_1 and S_2 in $\mathscr{S}$, we have

$$S_1 \oplus S_2 = \{g; g \text{ is in } S_1 \text{ or in } S_2 \text{ but not in both}\}; \qquad (5.16)$$

and

$$S_1 @ S_2 = \{w_1\} \oplus \{w_2\} \oplus \cdots \oplus \{w_{\alpha(W)}\}, \qquad (5.17a)$$

where $\alpha(W)$ denotes the number of elements in the set $W = \{w_i\}$, and

$$W = \{g_1 \cup g_2; g_1 \text{ is in } S_1, g_2 \text{ in } S_2, \text{ and } g_1 \cap g_2 = \emptyset\}. \qquad (5.17b)$$

In other words, the Wang product of the sets S_1 and S_2 is the set of subgraphs of G obtained by taking the ring-sum operations of all the sets, each of which contains only one element obtained by taking the union of two edge-disjoint subgraphs, one from S_1 and one from S_2.

It is not difficult to check that the set $\mathscr{S}$ forms a commutative ring with a unity (Problem 5.13): For any S_1, S_2, and S_3 in $\mathscr{S}$, we have

A. Ring sum

1. commutativity: $S_1 \oplus S_2 = S_2 \oplus S_1$,
2. associativity: $S_1 \oplus (S_2 \oplus S_3) = (S_1 \oplus S_2) \oplus S_3$,
3. closure: $S_1 \oplus S_2$ in $\mathscr{S}$,
4. zero element: $S_0 \oplus S_1 = S_1$,
5. negative element: $S_1 \oplus S_1 = S_0$,

where S_0 denotes the empty set of $\mathscr{S}$;

B. Wang product

1. commutativity: $S_1 @ S_2 = S_2 @ S_1$,
2. associativity: $S_1 @ (S_2 @ S_3) = (S_1 @ S_2) @ S_3$,
3. closure: $S_1 @ S_2$ in $\mathscr{S}$,
4. unity: $\{\emptyset\} @ S_1 = S_1$;

C. Distributivity

$$S_1 @ (S_2 \oplus S_3) = S_1 @ S_2 \oplus S_1 @ S_3.$$

In addition to these properties, we have

$$S_0 @ S_1 = S_0,$$
$$S_1 @ S_1 = S_0 \text{ if } \emptyset \text{ is not in } S_1$$
$$= \{\emptyset\} \text{ if } \emptyset \text{ is in } S_1.$$

As an illustration, consider the sets of subgraphs of the graph G as shown in fig. 5.1:

$$S_1 = \{e_1 e_2, e_3 e_4 e_5, e_5 e_6\},$$
$$S_2 = \{e_1 e_2, e_1 e_3 e_4, e_1 e_6\}.$$

Then

$$S_1 \oplus S_2 = \{e_3 e_4 e_5, e_5 e_6, e_1 e_3 e_4, e_1 e_6\},$$
$$S_1 @ S_2 = \{e_1 e_2 e_3 e_4 e_5, e_1 e_2 e_5 e_6\}.$$

4.2. *Linear dependence*

Let g_x, $x = 1, 2, \ldots, k$, be the subgraphs of a graph G. If for some set of constants $c_x = 1$ or 0, not all of which are zero, we have

$$c_1 g_1 \oplus c_2 g_2 \oplus \cdots \oplus c_k g_k = \emptyset, \tag{5.18}$$

where $1 g_x = g_x$ and $0 g_x = \emptyset$, then the subgraphs $g_1, g_2, \ldots, g_k$ are said to be *linearly dependent*. If however (5.18) is satisfied only when all the constants c_x are zero, the subgraphs are said to be *linearly independent*. In particular, if for some i and j, $i \neq j$ and $i, j \leq k$, we have $g_i = g_j$ or $g_i = \emptyset$, then the subgraphs g_x are always linearly dependent.

Alternatively, the linear dependence of subgraphs can be interpreted in terms of linear dependence of vectors. The linear vector space associated with the graph G consists of the set $\mathscr{G}$ of all subgraphs of G. The field $\mathscr{F}$ over which the subgraphs of G constitute a linear vector space is the field of integers mod 2, and the addition of vectors is the ring-sum operation. It is not difficult to verify that $\mathscr{G}$ constitutes a linear vector space $\mathscr{V}(G)$ of dimension b. If the edges of G are labeled by $e_1, e_2, \ldots, e_b$, any of its subgraphs can be expressed as a b-tuple of 1's and 0's. The elementary b-vectors, each of which consists of a single edge of G, form a basis of $\mathscr{V}(G)$. (Also see Problems 5.14 and 5.70–5.74 for some of the related properties.) Then the subgraphs $g_1, g_2, \ldots,$ and g_k of G are linearly dependent (independent) if, and only if, their corresponding b-vectors are linearly dependent (independent) over $\mathscr{F}$. In particular, the columns of the transposes of the complete incidence matrix A_a, the complete circuit matrix B_a, and the complete cut matrix Q_a are elements in $\mathscr{V}(G)$. Thus, from our discussions in chs. 2, §1 and 5, §1, we have (Problem 5.15)

COROLLARY 5.4: Let $\{L_x\}$, $x = 1, 2, \ldots, m$, be a set of linearly independent circuits or edge-disjoint unions of circuits of a graph G. Then any circuit or edge-disjoint union of circuits of G can be expressed as a linear combination (ring sums) of these L_x.

COROLLARY 5.5: Let $\{C_x\}$, $x = 1, 2, \ldots, r$, be a set of linearly independent cuts of a graph G. Then any cut of G can be expressed as a linear combination of the cuts C_x.

As an illustration, consider the set of circuits of the graph G as shown in fig. 5.1: $L_1 = e_1 e_2 e_3 e_6$, $L_2 = e_3 e_4 e_5$, and $L_3 = e_1 e_2 e_4 e_5 e_6$. Since

$$L_1 \oplus L_2 \oplus L_3 = \emptyset,$$

it follows that the circuits L_1, L_2, and L_3 are linearly dependent. However, any two of these three circuits are linearly independent. Thus, any one can be expressed as a linear combination of the other two. For example, L_1 can be expressed as the rung sum of L_2 and L_3.

Similarly, we can define the linear dependence of the elements of $\mathscr{S}$. Let S_x, $x = 1, 2, \ldots, k$, be elements of $\mathscr{S}$. If for some set of constants $c_x = 1$ or 0, not all of which are zero, we have

$$c_1 S_1 \oplus S_2 c_2 \oplus \cdots \oplus c_k S_k = S_0, \tag{5.19}$$

where $1 S_x = S_x$ and $0 S_x = S_0$, then the sets $S_1, S_2, \ldots, S_k$ are said to be *linearly dependent*. If however (5.19) is satisfied only when all the constants c_x are zero, the sets are said to be *linearly independent*. Like the case for $\mathscr{G}$, we can also show that the set $\mathscr{S}$ is a linear vector space over $\mathscr{F}$ with respect to the ring-sum operation (Problem 5.68).

For a non-null subgraph g, it is sometimes convenient to represent this subgraph by a set $H(g)$ in $\mathscr{S}$ consisting only of its edges,

$$H(g) = \{e; e \text{ is an edge of } g\}. \tag{5.20}$$

For example, if $L_1 = e_1 e_2 e_3 e_6$, then $H(L_1) = \{e_1, e_2, e_3, e_6\}$.

Direct consequences of Corollaries 5.4 and 5.5 are the following. Their proofs are left as a problem (Problem 5.17).

COROLLARY 5.6: Let L be a circuit or edge-disjoint union of circuits of G. Then $H(L)$ can be expressed as a linear combination of the sets $H(L_x)$, $x = 1, 2, \ldots, m$.

COROLLARY 5.7: Let C be a cut. Then $H(C)$ can be expressed as a linear combination of the sets $H(C_x)$, $x = 1, 2, \ldots, r$.

As an example, consider the cuts of the graph G given in fig. 5.1: $C_1 = e_2 e_6$, $C_2 = e_2 e_3 e_5$, $C_3 = e_3 e_4 e_6$, and $C_4 = e_1 e_2$. Then we have

$$H(C_1) = \{e_2, e_6\}, \qquad H(C_2) = \{e_2, e_3, e_5\},$$
$$H(C_3) = \{e_3, e_4, e_6\}, \quad H(C_4) = \{e_1, e_2\}.$$

It is not difficult to check that these sets are linearly independent. Consider the cut $C = e_1 e_3 e_4$ and its corresponding set $H(C) = \{e_1, e_3, e_4\}$. This set can be expressed as a linear combination of the above four sets, namely,

$$H(C) = H(C_1) \oplus H(C_3) \oplus H(C_4).$$

THEOREM 5.13: Let $\{L_x\}$ and $\{L_x^*\}$, $x = 1, 2, \ldots, m$, be two sets of linearly independent circuits or edge-disjoint unions of circuits of a graph. Then we have

$$H(L_1) @ H(L_2) @ \cdots @ H(L_m) = H(L_1^*) @ H(L_2^*) @ \cdots @ H(L_m^*). \tag{5.21}$$

Proof. Since L_x^* are linearly independent, it follows from Corollary 5.6 that we have
$$H(L_x) = c_{x1} H(L_1^*) \oplus c_{x2} H(L_2^*) \oplus \cdots \oplus c_{xm} H(L_m^*), \tag{5.22}$$

where $c_{xi} = 1$ or 0 and $i = 1, 2, \ldots, m$. Using (5.22) and the fact that the Wang-product operation is distributive over the ring-sum operations, we have

$$H(L_1) @ H(L_2) @ \cdots @ H(L_m)$$
$$= \bigoplus_{(j_1 j_2 \cdots j_m)} c_{1j_1} H(L_{j_1}^*) @ c_{2j_2} H(L_{j_2}^*) @ \cdots @ c_{mj_m} H(L_{j_m}^*)$$
$$= \bigoplus_{(j_1 j_2 \cdots j_m)} c_{1j_1} c_{2j_2} \cdots c_{mj_m} [H(L_{j_1}^*) @ H(L_{j_2}^*) @ \cdots @ H(L_{j_m}^*)]$$
$$= (\det M) \, H(L_1^*) @ H(L_2^*) @ \cdots @ H(L_m^*), \tag{5.23}$$

§4 The Wang-algebra formulation 337

where $j_1 j_2 \ldots j_m$ is a permutation of $12 \ldots m$; M is a matrix of order m over the field mod 2 whose ith row and jth column element is c_{ij}; and the ring sum is taken over all possible permutations $j_1 j_2 \ldots j_m$.

Now, we shall show that M is nonsingular, i.e., det $M = 1$. Assume that M is singular. Consider the transpose M' of M which is also singular. Then there exists a nontrivial solution X over $\mathscr{F}$ for the system

$$M'X = 0, \tag{5.24}$$

where the transpose of X is given by $[x_1, x_2, \ldots, x_m]$. Thus, we have

$$\sum_{i=1}^{m} c_{ij} x_i = 0 \tag{5.25}$$

for $j = 1, 2, \ldots, m$. Since from (5.22) and (5.25)

$$\sum_{i=1}^{m} x_i H(L_i) = \sum_{i=1}^{m} \sum_{j=1}^{m} x_i c_{ij} H(L_j^*)$$

$$= \sum_{j=1}^{m} \sum_{i=1}^{m} x_i c_{ij} H(L_j^*)$$

$$= \sum_{j=1}^{m} 0 H(L_j^*) = S_0, \tag{5.26}$$

where the summations denote the ring sums, and since not all x_i are zero, it follows that $H(L_1), H(L_2), \ldots, H(L_m)$ are linearly dependent. This is impossible by our hypothesis. Thus M is nonsingular over $\mathscr{F}$, which completes the proof of the theorem.

Similarly, we can state the dual result for the cuts of G. Since the proof can be carried out in an entirely similar manner, the details are left as a problem (Problem 5.19).

THEOREM 5.14: Let $\{C_x\}$ and $\{C_x^*\}$, $x = 1, 2, \ldots, r$, be two sets of linearly independent cuts of a graph. Then we have

$$H(C_1) @ H(C_2) @ \cdots @ H(C_r) = H(C_1^*) @ H(C_2^*) @ \cdots @ H(C_r^*). \tag{5.27}$$

We shall illustrate the above results by the following example.

Example 5.2: Consider the graph G as shown in fig. 5.1. Since $L_1 = e_1 e_2 e_3 e_6$ and $L_2 = e_3 e_4 e_5$, and $L_1^* = e_1 e_2 e_3 e_6$ and $L_2^* = e_1 e_2 e_4 e_5 e_6$ are two sets of linearly independent circuits, we have

$$H(L_1) @ H(L_2) = \{e_1, e_2, e_3, e_6\} @ \{e_3, e_4, e_5\}$$
$$= \{e_1e_3, e_1e_4, e_1e_5, e_2e_3, e_2e_4, e_2e_5, e_3e_4, e_3e_5, e_6e_3, e_6e_4, e_6e_5\},$$

$$H(L_1^*) @ H(L_2^*) = \{e_1, e_2, e_3, e_6\} @ \{e_1, e_2, e_4, e_5, e_6\}$$
$$= \{e_1e_2\} \oplus \{e_1e_4\} \oplus \{e_1e_5\} \oplus \{e_1e_6\} \oplus \{e_2e_1\} \oplus \{e_2e_4\}$$
$$\oplus \{e_2e_5\} \oplus \{e_2e_6\} \oplus \{e_3e_1\} \oplus \{e_3e_2\} \oplus \{e_3e_4\} \oplus \{e_3e_5\}$$
$$\oplus \{e_3e_6\} \oplus \{e_6e_1\} \oplus \{e_6e_2\} \oplus \{e_6e_4\} \oplus \{e_6e_5\}$$
$$= \{e_1e_4, e_1e_5, e_2e_4, e_2e_5, e_3e_1, e_3e_2, e_3e_4, e_3e_5, e_3e_6, e_6e_4, e_6e_5\}$$
$$= H(L_1) @ H(L_2),$$

which verifies the identity (5.21). Similarly, we can show that the cuts $C_1 = e_2e_6$, $C_2 = e_2e_3e_5$, $C_3 = e_3e_4e_6$, and $C_4 = e_1e_2$, and the cuts $C_1^* = e_1e_2$, $C_2^* = e_2e_3e_4$, $C_3^* = e_4e_5$, and $C_4^* = e_1e_6$ are two sets of linearly independent cuts. Then

$$H(C_1) @ H(C_2) @ H(C_3) @ H(C_4)$$
$$= \{e_2, e_6\} @ \{e_2, e_3, e_5\} @ \{e_3, e_4, e_6\} @ \{e_1, e_2\}$$
$$= \{e_2e_3, e_2e_5, e_6e_2, e_6e_3, e_6e_5\} @ \{e_3e_1, e_3e_2, e_4e_1, e_4e_2, e_6e_1, e_6e_2\}$$
$$= \{e_2e_3e_4e_1, e_2e_5e_3e_1, e_2e_5e_4e_1, e_2e_5e_6e_1, e_6e_2e_4e_1, e_6e_3e_4e_1,$$
$$e_6e_3e_4e_2, e_6e_5e_3e_1, e_6e_5e_3e_2, e_6e_5e_4e_1, e_6e_5e_4e_2\},$$

$$H(C_1^*) @ H(C_2^*) @ H(C_3^*) @ H(C_4^*)$$
$$= \{e_1, e_2\} @ \{e_2, e_3, e_4\} @ \{e_4, e_5\} @ \{e_1, e_6\}$$
$$= \{e_1e_6, e_2e_6, e_2e_1\} @ \{e_2e_4, e_2e_5, e_3e_4, e_3e_5, e_4e_5\}$$
$$= \{e_1e_6e_2e_4, e_1e_6e_2e_5, e_1e_6e_3e_4, e_1e_6e_3e_5, e_1e_6e_4e_5, e_2e_6e_3e_4,$$
$$e_2e_6e_3e_5, e_2e_6e_4e_5, e_2e_1e_3e_4, e_2e_1e_3e_5, e_2e_1e_4e_5\}$$
$$= H(C_1) @ H(C_2) @ H(C_3) @ H(C_4),$$

which verifies the identity (5.27). Note that in the above operations we have applied the associative and commutative properties of the Wang products.

4.3. *Trees and cotrees*

In this section, the relationships between trees and cotrees and independent cuts and circuits or edge-disjoint unions of circuits will be formulated in terms of the Wang algebra. The idea is not new, and was first suggested by WANG [1934]. However, Wang's rules are applicable only to planar graphs. TING [1935] generalized Wang's rules, to make it applicable to any graph. The dual procedure of Wang and Ting, based on the incidence cuts of a graph, was first

§4 The Wang-algebra formulation

formulated by TSAI [1939], and generalized to any set of independent cuts by DUFFIN [1959].

THEOREM 5.15: Let $\{L_x\}$, $x = 1, 2, \ldots, m$, be a set of linearly independent circuits or edge-disjoint unions of circuits of a connected graph G. Then the set $\bar{T}$ of cotrees of G is given by

$$\bar{T} = H(L_1) \,@\, H(L_2) \,@\, \cdots \,@\, H(L_m). \tag{5.28}$$

Proof. Without loss of generality, from Theorem 5.13 we may assume that $\{L_x\}$ is a set of f-circuits with respect to a tree t. We shall first show that every cotree of G is contained in $H = H(L_1) \,@\, \cdots \,@\, H(L_m)$. Assume otherwise, i.e., there exists a cotree that is not contained in H. Let this cotree $\bar{t} = e_1 e_2 \cdots e_m$ be the complement of the tree t in G. Since e_i, $i = 1, 2, \ldots, m$, is in $H(L_i)$ but not in any other sets $H(L_x)$, $x \neq i$, it follows that $\bar{t}$ is in H, a contradiction. Thus, every cotree of G is contained in H.

Next, we have to show that every term in H is a cotree. Since every circuit L of G may be included in a set of m linearly independent circuits, let $\{L_x^*\}$ be a set of m linearly independent circuits including L. Then for each element h^* in $H^* = H(L_1^*) \,@\, H(L_2^*) \,@\, \cdots \,@\, H(L_m^*)$ we have $L \cap h^* \neq \emptyset$. Since by Theorem 5.13, $H = H^*$, it follows that for each h in H we have

$$L \cap h \neq \emptyset. \tag{5.29}$$

Assume that there exists an element h in H that is not a cotree. Then the complement $\bar{h}$ of h in G must be a subgraph containing at least one circuit. Let this circuit be L. Evidently, $L \cap h = \emptyset$, since L is contained in $\bar{h}$. This is impossible by (5.29). Thus, every term in H must be a cotree. This completes the proof of the theorem.

The dual result based on cuts is given as

THEOREM 5.16: Let $\{C_x\}$, $x = 1, 2, \ldots, r$, be a set of linearly independent cuts of a connected graph G. Then the set T of trees of G is given by

$$T = H(C_1) \,@\, H(C_2) \,@\, \cdots \,@\, H(C_r). \tag{5.30}$$

The proof is analogous to that of Theorem 5.15 and is omitted here (Problem 5.31).

From Theorems 5.15 and 5.16, we conclude that the terms in $H(L_1) \,@\, H(L_2)$ and $H(C_1) \,@\, H(C_2) \,@\, H(C_3) \,@\, H(C_4)$ given in Example 5.2 correspond to the cotrees and trees of the graph as shown in fig. 5.1. The correctness of the statement may be verified by the reader.

4.4. *Multi-trees and multi-cotrees*

The relationships between multi-trees and multi-cotrees and independent cuts and circuits will be presented in this section. The concept of a multi-tree or multi-cotree is a simple extension of that of a 2-tree or a 2-cotree defined in ch. 2, § 6.2 (Definitions 2.32 and 2.34).

DEFINITION 5.2: *k-tree*. A spanning subgraph of a graph G is said to be a *k-tree*, $k \geq 1$, if, and only if, it has k components and contains no circuits. Many of the components may each consist of an isolated node. A k-tree is also referred to as a *multi-tree* or a *forest* if k is not specifically given.

DEFINITION 5.3: *k-cotree*. For a graph G, the complement of a *k-tree* in G is called a *k-cotree* of G. A *k*-cotree is also referred to as a *multi-cotree* or a *coforest* if k is not specifically given.

Very often, k-trees in which certain designated nodes are required to be in different components are used. For convenience, we shall use the semicolons to separate the groups of nodes required to appear in different components†. For example, the symbol $t(i_1, i_2; i_3, i_4, i_5; i_6, i_7)$ or simply $t(I)$ where $I = \{i_1, i_2; i_3, i_4, i_5; i_6, i_7\}$, denotes a 3-tree in which the nodes i_1 and i_2 are in one component, the nodes i_3, i_4, and i_5 in one component, and the nodes i_6 and i_7 in one component. In particular, if I consists only of one element, we define $t(I) = t$, a tree, i.e., a 1-tree is also a tree. The extension of this to general case is obvious. We denote the set of all possible multi-trees $t(I)$ of G by $T(I)$. Similarly, $\bar{t}(I)$ and $\bar{T}(I)$ denote a multi-cotree [the complement of a multi-tree $t(I)$] and the set of all possible multi-cotrees $\bar{t}(I)$ in G, respectively. Obviously, if I consists only of one element, then $\bar{t}(I) = \bar{t}$, a cotree, i.e., a 1-cotree is also a cotree. For simplicity, let T and $\bar{T}$ be the sets of trees and cotrees of G, respectively.

For example, in fig. 5.4 the sets of 3-trees $t(1, 2; 4; 3, 6)$ and 3-cotrees $\bar{t}(1, 2; 4; 3, 6)$ are presented in figs. 5.5 and 5.6, respectively.

For a given graph G, let $G(I)$ be the graph obtained from G by identifying the nodes contained in I. The operation may produce self-loops and we assume that these self-loops are *not* removed. Since trees (cotrees) in $G(I)$ are in one-

---

† In order to avoid double subscripts and to simplify our notation, the symbol $t(i; j)$ for a 2-tree defined here is slightly different from that $t_{i,j}$ given in ch. 2, § 6.2. This should not create any difficulty.

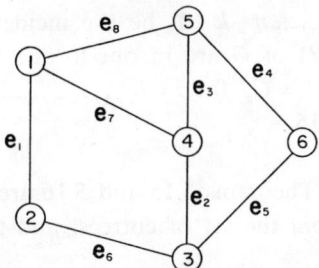

Fig. 5.4. A graph illustrating the multi-trees and multi-cotrees.

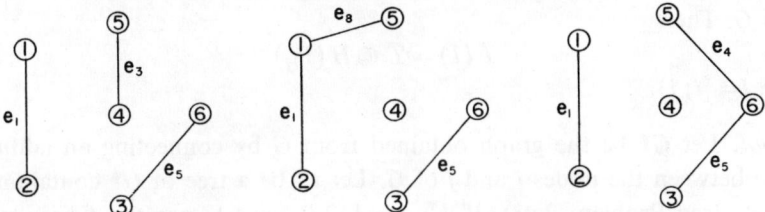

Fig. 5.5. The set of 3-trees $t(1, 2; 4; 3, 6)$ of the graph of fig. 5.4.

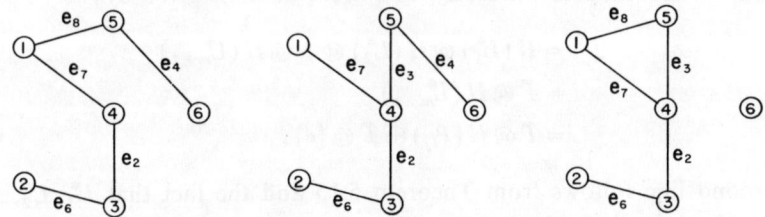

Fig. 5.6. The set of 3-cotrees $\bar{t}(1, 2; 4; 3, 6)$ of the graph of fig. 5.4.

to-one correspondence with the multi-trees $t(I)$ (multi-cotrees $\bar{t}(I)$) of G, Theorems 5.15 and 5.16 can also be used to generate these subgraphs (Problems 5.22–5.25). For our purpose, we shall only state the case based on incidence cuts (see CHEN [1966a]).

THEOREM 5.17: Let C_x, $x = 1, 2, \ldots, n$, be the incidence cuts of the nodes x of a connected graph G. Then the set of k-trees $t(I)$ of G, up to the null graph, is given by

$$T(I) = H(C_{i_{k+1}}) @ H(C_{i_{k+2}}) @ \cdots @ H(C_{i_n}), \tag{5.31}$$

where $I = \{i_1; i_2; \ldots; i_k\}$ and $\{i_{k+1}; i_{k+2}; \ldots; i_n\}$ are the complementary sets of the set of integers $\{1; 2; \ldots; n\}$.

Proof. Let $C^*_{j_y}$, $y=1, 2, \ldots, n-k+1$, be the incidence cuts of the nodes j_y of $G(I)$. Since k-trees $t(I)$ of G are in one-to-one correspondence with the trees of $G(I)$, and since $C_z = C^*_z$ for $z = i_{k+1}, i_{k+2}, \ldots, i_n$, the corollary follows directly from Theorem 5.16.

Other consequences of Theorems 5.15 and 5.16 are that the set of 2-cotrees of G can be obtained from the set of cotrees, and the set of trees from the set of 2-trees.

THEOREMS 5.18: Let P_{ij} be a path connecting the nodes i and j in a connected graph G. Then
$$\bar{T}(I) = \bar{T} @ H(P_{ij}), \tag{5.32}$$
where $I = \{i; j\}$.

Proof. Let G^* be the graph obtained from G by connecting an additional edge e between the nodes i and j of G. Let t^* be a tree of G^* containing the path P_{ij} (see Problem 2.43). If L^*_x, $x = 1, 2, \ldots, m+1$, are the f-circuits with respect to the tree t^* in G^*, and if L^*_{m+1} is the f-circuit defined by the chord e, then from Theorem 5.15 the set $\bar{T}^*$ of cotrees of G^* is given by†

$$\begin{aligned}\bar{T}^* &= H(L^*_1) @ H(L^*_2) @ \cdots @ H(L^*_{m+1}) \\ &= \bar{T} @ H(L^*_{m+1}) \\ &= \bar{T} @ H(P_{ij}) \oplus \bar{T} @ \{e\}.\end{aligned} \tag{5.33}$$

The second line follows from Theorem 5.15 and the fact that $L^*_1, L^*_2, \ldots, L^*_m$ are also the f circuits with respect to t^* in G, and the third line is obtained after expressing $H(L^*_{m+1}) = H(P_{ij}) \oplus \{e\}$.

Since cotrees in G^* can be classified into those $\bar{T}^*(e)$ which contain the edge e, and those $\bar{T}^*(\bar{e})$ which do not, it follows that we have

$$\begin{aligned}\bar{T}^* &= \bar{T}^*(\bar{e}) \oplus \bar{T}^*(e) \\ &= \bar{T}(I) \oplus \bar{T} @ \{e\},\end{aligned} \tag{5.34}$$

since cotrees in $\bar{T}^*(\bar{e})$ are in one-to-one correspondence with 2-cotrees in G, and cotrees in $\bar{T}^*(e)$ are in one-to-one correspondence with cotrees in G. Equating (5.33) and (5.34), and using the fact that, for every S_1 in $\mathscr{S}$, $S_1 \oplus S_1 = S_0$, we obtain the desired result. So the theorem is proved.

---

† In the absence of parentheses, we assume the following order of operations: the Wang product, the ring sum, and the union.

THEOREM 5.19: Let C_{ij} be a cutset separating the node i from the node j in a connected graph G. Then

$$T = T(I) @ H(C_{ij}), \tag{5.35}$$

where $I = \{i; j\}$.

Proof. Let G^* be the graph obtained from G by connecting an additional edge e between the nodes i and j of G. Let t^* be a tree of G^* containing the edge e but not any of the edges of C_{ij}. This is always possible since $G^* - C_{ij}$ is connected. If C_x^*, $x = 1, 2, \ldots, r$, are the f-cutsets of G^* defined by the tree t^* with C_r^* being the f-cutset formed by the edge e, then the corresponding cutsets C_y'' of C_y^*, $y = 1, 2, \ldots, r-1$, in $G^*(I)$ are the f-cutsets with respect to the corresponding tree t'' of t^* in $G^*(I)$. Furthermore, we have $t'' = t^* - e$ and $C_y'' = C_y^*$. By Theorem 5.16, the set T^* of trees of G^* may be expressed in terms of the set T'' of trees of $G^*(I)$:

$$\begin{aligned} T^* &= H(C_1^*) @ H(C_2^*) @ \cdots @ H(C_r^*) \\ &= H(C_1'') @ H(C_2'') @ \cdots @ H(C_{r-1}'') @ H(C_{ij} \cup \{e\}) \\ &= T'' @ [H(C_{ij}) \oplus \{e\}] \\ &= T(I) @ H(C_{ij}) \oplus T'' @ \{e\}. \end{aligned} \tag{5.36}$$

The last line follows from the fact that trees in $G^*(I)$ are in one-to-one correspondence with 2-trees $t(I)$ in G.

Since trees in G^* can be classified into those $T^*(e)$ which contain the edge e, and those $T^*(\bar{e})$ which do not, it follows that

$$\begin{aligned} T^* &= T^*(e) \oplus T^*(\bar{e}) \\ &= T(I) @ \{e\} \oplus T \\ &= T'' @ \{e\} \oplus T. \end{aligned} \tag{5.37}$$

(5.35) now follows directly from (5.36) and (5.37). This completes the proof of the theorem.

Example 5.3: Suppose that we wish to generate all the 3-trees $t(1; 3; 5)$ of the graph G as shown in fig. 5.4. Then from Theorem 5.17, we have $I = \{1; 3; 5\}$, and

$$\begin{aligned} T(I) &= H(C_2) @ H(C_4) @ H(C_6) \\ &= \{e_1, e_6\} @ \{e_2, e_3, e_7\} @ \{e_4, e_5\} \\ &= \{e_1 e_2 e_4, e_1 e_2 e_5, e_1 e_3 e_4, e_1 e_3 e_5, e_1 e_4 e_7, e_1 e_5 e_7, e_2 e_4 e_6, \\ &\quad e_2 e_5 e_6, e_3 e_4 e_6, e_3 e_5 e_6, e_5 e_6 e_7, e_4 e_6 e_7\}. \end{aligned}$$

Example 5.4: Consider the graph G given in fig. 5.7. We shall first generate the set of 2-trees $t(1; 3)$ of G by (5.31). Then $I = \{1; 3\}$, and

$$T(I) = H(C_2) @ H(C_4)$$
$$= \{e_2, e_3, e_6\} @ \{e_1, e_3, e_5\}$$
$$= \{e_1e_2, e_1e_3, e_1e_6, e_2e_3, e_3e_6, e_2e_5, e_3e_5, e_5e_6\}.$$

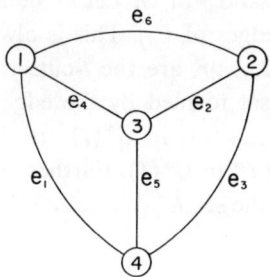

Fig. 5.7. A graph illustrating the generation of trees, 2-trees, and 2-cotrees.

Now, we use (5.35) to generate the set of trees of G. Let $C_{13} = e_1e_4e_6$.

$$T = T(I) @ \{e_1, e_4, e_6\}$$
$$= \{e_1e_2e_3, e_1e_2e_5, e_1e_3e_5, e_1e_5e_6, e_1e_2e_4, e_1e_3e_4, e_1e_4e_6, e_2e_3e_4, e_3e_4e_6,$$
$$e_2e_4e_5, e_3e_4e_5, e_4e_5e_6, e_1e_2e_6, e_2e_3e_6, e_2e_5e_6, e_3e_5e_6\}.$$

Finally, we shall use (5.32) to generate the set of 2-cotrees $\bar{t}(I)$ of G. Let $P_{13} = e_4$. Then

$$\bar{T}(I) = \bar{T} @ \{e_4\}$$
$$= \{e_4e_5e_6, e_3e_4e_6, e_2e_4e_6, e_2e_3e_4, e_3e_5e_6, e_2e_5e_6, e_2e_3e_5, e_1e_5e_6, e_1e_2e_5,$$
$$e_1e_3e_6, e_1e_2e_6, e_1e_2e_3, e_3e_4e_5, e_1e_4e_5, e_1e_3e_4, e_1e_2e_4\} @ \{e_4\}$$
$$= \{e_3e_4e_5e_6, e_2e_4e_5e_6, e_2e_3e_4e_5, e_1e_4e_5e_6, e_1e_2e_4e_5, e_1e_3e_4e_6,$$
$$e_1e_2e_4e_6, e_1e_2e_3e_4\}.$$

It is easy to check that each of the terms in $\bar{T}(I)$ is the complement of a term in $T(I)$. We remark that no duplicate terms are generated in the above process. However, had we chosen $P_{13} = e_1e_5$, we would have generated each of the following terms twice during the process: $e_1e_3e_5e_6$, $e_1e_2e_5e_6$, $e_1e_2e_3e_5$, and $e_1e_3e_4e_5$. Of course, the end result would be the same. Thus, we may conclude that a proper choice of P_{ij} in (5.32) may greatly reduce the number of redundancies due to duplications. This is similarly valid for all other formulas based on the Wang-algebra formulation (Problems 5.26 and 5.27).

4.5. Decomposition

In the foregoing, we have shown how the trees or cotrees can be generated by independent cuts or circuits. However, these techniques are not efficient if the given graph is complicated, because a large number of cancellation terms will usually be generated in the process.

In order to reduce the computational difficulties, in the following we shall show how to decompose the original graph into several components, determine the appropriate subgraphs in each component, and then combine these subgraphs to obtain the desired results. The formulas, however, still involve redundancies due to duplications, but the number will be greatly reduced. In the following section, we shall discuss the conditions under which duplications may be avoided.

Let G be the connected graph formed by superimposing the nodes $i'_1, i'_2, \ldots, i'_k$ of a graph G' upon the nodes $i''_1, i''_2, \ldots, i''_k$ of another graph G'', respectively, with the combined nodes in G being denoted by $i_1, i_2, \ldots, i_k$, where $0 < k-1 \leq \min(r', r'')$, and r' and r'' are the ranks of G' and G'', respectively. If T, $T'(I')$, and $T''(I'')$ are the set of trees and the sets of k-trees of the forms $t(I')$ and $t(I'')$ in G, G', and G'', respectively, where $I' = \{i'_1; i'_2; \ldots; i'_k\}$ and $I'' = \{i''_1; i''_2; \ldots; i''_k\}$, then we have

THEOREM 5.20:
$$T = T'(I') \,@\, T''(I'') \,@\, H(C_{i_1}) \,@\, H(C_{i_2}) \,@\, \cdots \,@\, H(C_{i_{k-1}}), \quad (5.38)$$
where C_{i_x}, $x = 1, 2, \ldots, k-1$, are the incidence cuts of the nodes i_x in G.

Proof. Without loss of generality, let $i_x = x$ for $x = 1, 2, \ldots, k$. Also, let the nodes of G be labeled in the following way: Those contained in G' but not in $I = \{1; 2; \ldots; k\}$ are labeled by $k+1, k+2, \ldots, k+n'$; and those contained in G'' but not in I are labeled by $k+n'+1, k+n'+2, \ldots, k+n'+n''$; where $k+n'$ and $k+n''$ are the numbers of nodes in G' and G'', respectively. Thus, if C_u, C'_u, and C''_u are the incidence cuts of the nodes u in G, G', and G'', respectively, then $C_u = C'_u$ for $u = k+1, \ldots, k+n'$; $C_u = C''_u$ for $u = k+n'+1, \ldots, k+n'+n''$; and C_u is the edge-disjoint union of C'_u and C''_u for $u = 1, 2, \ldots, k$. By Theorem 5.16, we have

$$\begin{aligned}
T &= H(C_1) \,@\, \cdots \,@\, H(C_{k-1}) \,@\, H(C_{k+1}) \,@\, \cdots \,@\, H(C_{k+n'+n''}) \\
&= H(C_1) \,@\, \cdots \,@\, H(C_{k-1}) \,@\, H(C'_{k+1}) \,@\, \cdots \,@\, H(C'_{k+n'}) \,@\, H(C''_{k+n'+1}) \\
&\qquad\qquad\qquad\qquad\qquad\qquad\qquad @\, \cdots \,@\, H(C''_{k+n'+n''}) \\
&= H(C_1) \,@\, \cdots \,@\, H(C_{k-1}) \,@\, T'_s \,@\, T''_s \\
&= H(C_1) \,@\, \cdots \,@\, H(C_{k-1}) \,@\, T'(I') \,@\, T''(I''), \quad (5.39)
\end{aligned}$$

where T'_s and T''_s are the sets of trees in $G'(I')$ and $G''(I'')$, respectively. The theorem follows from the fact that k-trees $t(I')$ and $t(I'')$ in G' and G'' are in one-to-one correspondence with trees in $G'(I')$ and $G''(I'')$, respectively. So the theorem is proved.

Similarly, if $\bar{T}$, $\bar{T}'$, and $\bar{T}''$ are the sets of cotrees in G, G', and G'', respectively, and if $P'_{\alpha\beta}$ and $P''_{\alpha\beta}$ are the paths connecting the nodes α and β in G' and G'', respectively, where $\alpha = i_x$ and $\beta = i_{x+1}$ for $x = 1, 2, \ldots, k-1$, then we have

THEOREM 5.21: If G' and G'' are connected, then

$$\bar{T} = \bar{T}' \ @\ \bar{T}'' \ @\ H(P'_{i_1 i_2} \cup P''_{i_1 i_2}) \ @ \cdots @\ H(P'_{i_{k-1} i_k} \cup P''_{i_{k-1} i_k}). \tag{5.40}$$

Proof. Without loss of generality, let $i_x = x$. Also, let

$$L_x = P'_{x(x+1)} \cup P''_{x(x+1)}. \tag{5.41}$$

It is not difficult to see that the set of circuits L_x, $x = 1, 2, \ldots, k-1$, are linearly independent. Let $\{L'_u\}$, $u = 1, 2, \ldots, m'$, and $\{L''_v\}$, $v = 1, 2, \ldots, m''$, be the sets of linearly independent circuits in G' and G'', respectively, such that L'_u, L_x, and L''_v constitute $(m' + m'' + k - 1)$ linearly independent circuits in G, where m' and m'' are the nullities of G' and G'', respectively. Since the nullity m of G is equal to $(m' + m'' + k - 1)$, it follows from Theorem 5.15 that

$$\begin{aligned}\bar{T} &= H(L'_1) \ @ \cdots @\ H(L'_{m'}) \ @\ H(L''_1) \ @ \cdots @\ H(L''_{m''}) \ @\ H(L_1) \ @ \cdots @\ H(L_{k-1}) \\ &= \bar{T}' \ @\ \bar{T}'' \ @\ H(L_1) \ @ \cdots @\ H(L_{k-1}). \end{aligned} \tag{5.42}$$

The theorem follows from here. (See Problem 5.29 for a generalization.)

We shall illustrate the above results by the following example.

Example 5.5: Consider the graph G of fig. 5.1. The graph may be decomposed into two subgraphs G' and G'' through the nodes 2 and 4 as shown in fig. 5.8. Then $I = \{2; 4\}$, $I' = \{2'; 4'\}$, and $I'' = \{2''; 4''\}$. The incidence cut C_2 of the node 2 in G is $e_2 e_3 e_4$. Thus, according to Theorem 5.20 we have

$$\begin{aligned}T &= T'(I') \ @\ T''(I'') \ @\ H(C_2) \\ &= \{e_1 e_2, e_1 e_6, e_2 e_6\} \ @\ \{e_4, e_5\} \ @\ \{e_2, e_3, e_4\} \\ &= \{e_1 e_2 e_3 e_4, e_1 e_2 e_3 e_5, e_1 e_2 e_4 e_5, e_1 e_2 e_4 e_6, e_1 e_3 e_4 e_6, e_1 e_2 e_5 e_6, \\ &\qquad e_1 e_3 e_5 e_6, e_1 e_4 e_5 e_6, e_2 e_3 e_4 e_6, e_2 e_3 e_5 e_6, e_2 e_4 e_5 e_6\},\end{aligned}$$

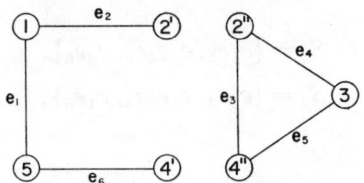

Fig. 5.8. A decomposition of the graph of fig. 5.1.

which agrees with that given in Example 5.2. Alternatively, we may use C_4 instead of C_2, and the result is given by

$$T = T'(I') @ T''(I'') @ H(C_4)$$
$$= \{e_1e_2, e_1e_6, e_2e_6\} @ \{e_4, e_5\} @ \{e_3, e_5, e_6\},$$

which yields the same result as above.

Now suppose that we wish to generate the set of cotrees of G by (5.40). Let $P'_{24} = e_1e_2e_6$ and $P''_{24} = e_3$. Then

$$\bar{T} = \bar{T}' @ \bar{T}'' @ H(P'_{24} \cup P''_{24})$$
$$= \{\emptyset\} @ \{e_3, e_4, e_5\} @ \{e_1, e_2, e_6, e_3\}$$
$$= \{e_3e_1, e_4e_1, e_5e_1, e_3e_2, e_4e_2, e_5e_2, e_3e_6, e_4e_6, e_5e_6, e_4e_3, e_5e_3\},$$

which is the same as that given in Example 5.2. Also, we may use $P''_{24} = e_4e_5$ and obtain the same result as shown above.

We will now proceed to show that Theorem 5.20 may be further simplified, and can be expressed in terms of the ring-sum operations. The result is of particular interest to us because it serves as the starting point on the problem of generating trees by decomposition without duplications to be discussed in the next section.

In order to simplify the notation, from here on by $T'(I_k)$ or $T''(I_k)$ we mean $T'(I')$ or $T''(I'')$, where $I_k = \{i_1; i_2; \ldots; i_k\}$. This is similarly valid for the sets $\bar{T}'(I_k)$ and $\bar{T}''(I_k)$ of k-cotrees in G' and G'', respectively.

In the following, we define the *Cartesian-product* operation of any two sets S_1 and S_2 in $\mathscr{S}$, denoted by the symbol $S_1 \times S_2$, as

$$S_1 \times S_2 = \{g_i \cup g_j; g_i \text{ is in } S_1 \text{ and } g_j \text{ in } S_2\}, \quad (5.43)$$

and $S_1 \times S_0 = S_0 \times S_1 = S_0$, i.e., the set of all subgraphs that are obtained as the result of the union of all possible combinations of a subgraph in S_1 and a subgraph in S_2.

For example, if
$$S_1 = \{e_1e_2, e_3e_4e_5, e_5e_6\},$$
$$S_2 = \{e_1e_2, e_1e_3e_4, e_1e_6\},$$
then
$$S_1 \times S_2 = \{e_1e_2, e_1e_2e_3e_4, e_1e_2e_6, e_1e_2e_3e_4e_5, e_1e_3e_4e_5,$$
$$e_1e_3e_4e_5e_6, e_1e_2e_5e_6, e_1e_3e_4e_5e_6, e_1e_5e_6\}.$$

We remark that not all the elements in $S_1 \times S_2$, as defined above, are necessarily distinct. This definition deviates somewhat from the usual sense of the Cartesian product. Perhaps a different name should be used, but we do not find it necessary. In the above example, $e_1e_3e_4e_5e_6$ appears twice in the product.

LEMMA 5.1: Let C be a cut separating the node k from the nodes in I_{k-1} in G. Then for $k \geq 2$ we have

$$T(I_{k-1}) = T(I_k) \, @ \, H(C). \tag{5.44}$$

The lemma follows directly from Theorem 5.19, and may be considered as a generalization of (5.35). Its proof is straightforward, and is left as an exercise (Problem 5.30).

LEMMA 5.2: If S_1 and S_2 are the sets of subgraphs of G' and G'', respectively, then

$$S_1 \times S_2 = S_1 \, @ \, S_2. \tag{5.45}$$

THEOREM 5.22: Let J be a subset of the set I_{k-1}, and also let $\bar{J}$ be the complement of J in I_{k-1}. Then for $k \geq 1$ we have†

$$T = \oplus \, T'(J \cup \{i_k\}) \times T''(\bar{J} \cup \{i_k\}) \tag{5.46a}$$
$$= \oplus \, T'(\bar{J} \cup \{i_k\}) \times T''(J \cup \{i_k\}), \tag{5.46b}$$

where $I_0 = S_0$ and the ring sums are taken over all subsets J of I_{k-1}.

Proof. Without loss of generality, let $i_x = x$ for $x = 1, 2, ..., k > 1$. Let C_x, C'_x, and C''_x be the incidence cuts of the node x in G, G', and G'', respectively. Obviously, C_x is the edge-disjoint union of C'_x and C''_x. Thus, we may write

---

† In the absence of parentheses, we assume the following order of operations: the Wang product or Cartesian product, the ring sum, and the union.

§4 The Wang-algebra formulation

$C_x = C'_x \oplus C''_x$. We shall prove the theorem by induction over the Wang-product operations as follows: From (5.38), (5.44) and (5.45), we have

$$T = H(C_{k-1}) @ \cdots @ H(C_1) @ T'(I_k) @ T''(I_k)$$
$$= H(C_{k-1}) @ \cdots @ H(C_2) @ [H(C'_1) @ T'(I_k) @ T''(I_k) \oplus H(C''_1)$$
$$@ T'(I_k) @ T''(I_k)]$$
$$= H(C_{k-1}) @ \cdots @ H(C_2) @ [T'(I_k - \{1\}) \times T''(I_k) \oplus T'(I_k)$$
$$\times T''(I_k - \{1\})]$$
$$= H(C_{k-1}) @ \cdots @ H(C_2) @ [\bigoplus_{J \in I_1} T'(I_k - J) \times T''(I_k - J)]$$
$$= \cdots$$
$$= H(C_{k-1}) @ [\bigoplus_{J \in I_{k-2}} T'(I_k - J) \times T''(I_k - J)]$$
$$= [H(C'_{k-1}) \oplus H(C''_{k-1})] @ [\bigoplus_{J \in I_{k-2}} T'(I_k - J) @ T''(I_k - J)]$$
$$= \bigoplus_{J \in I_{k-2}} [H(C'_{k-1}) @ T'(I_k - J) @ T''(I_k - J) \oplus H(C''_{k-1})$$
$$@ T'(I_k - J) @ T''(I_k - J)]$$
$$= \bigoplus_{J \in I_{k-2}} [H(C'_{k-1}) @ T'(J \cup \{k-1, k\}) @ T''(J \cup \{k-1, k\})$$
$$\oplus H(C''_{k-1}) @ T'(J \cup \{k-1, k\}) @ T''(J \cup \{k-1, k\})]$$
$$= \bigoplus_{J \in I_{k-2}} [T'(J \cup \{k\}) @ T''(J \cup \{k-1, k\}) \oplus T'(J \cup \{k-1, k\})$$
$$@ T''(J \cup \{k\})]$$
$$= \bigoplus_{J \in I_{k-1}} T'(J \cup \{k\}) \times T''(J \cup \{k\})$$
$$= \bigoplus_{J \in I_{k-1}} T'(J \cup \{k\}) \times T''(J \cup \{k\}),$$

which completes the proof of the theorem.

Since for every tree in a connected graph there corresponds a unique cotree, a different version of the above theorem is the following.

THEOREM 5.23: Let J be a subset of the set I_{k-1}, and also let $\bar{J}$ be the complement of J in I_{k-1}. Then for $k \geq 1$ we have

$$\bar{T} = \oplus \, \bar{T}'(J \cup \{i_k\}) \times \bar{T}''(J \cup \{i_k\}) \tag{5.47a}$$
$$= \oplus \, \bar{T}'(\bar{J} \cup \{i_k\}) \times \bar{T}''(\bar{J} \cup \{i_k\}), \tag{5.47b}$$

where $I_0 = S_0$ and the ring sums are taken over all subsets J of I_{k-1}.

COROLLARY 5.8: For $k=2$, we have

$$T = T' \times T''(I_2) \cup T'(I_2) \times T'', \tag{5.48a}$$

$$\bar{T} = \bar{T}' \times \bar{T}''(I_2) \cup \bar{T}'(I_2) \times \bar{T}''. \tag{5.48b}$$

The above corollary was originally given by PERCIVAL [1953]. In particular, if we let $G'' = e$, a single edge of G, (5.48a) reduces to the well-known result of FEUSSNER [1902, 1904]:

$$T = T(\bar{e}) \cup T(e), \tag{5.49}$$

where $T(e)$ and $T(\bar{e})$ denote the sets of trees of G containing the edge e and not containing the edge e, respectively. They correspond to the sets of trees in the graphs obtained from G by identifying the two endpoints of the edge e and by removing the edge e, respectively. Thus, if we repeat this process for the resulting graphs, we will eventually generate all the trees of G without duplications. The procedure has been implemented on a digital computer by MCILROY [1969]. In § 7, we shall show that $T(\bar{e})$ can be obtained from $T(e)$ without duplications. This result should simplify the implementation of the technique on a digital computer.

So far, we have only discussed the partitioning of a graph into two parts (not necessarily two components). In the following, we shall show how the above results may be used to decompose a graph into several components.

Let G_{k+1} be a subgraph of a connected graph G whose removal from G decomposes G into k components $G_1, G_2, ..., $ and G_k. Let V and V_x be the node sets of G and G_x, $x = 1, 2, ..., k+1$, respectively.

THEOREM 5.24: Let $T_x(J_x)$, $x = 1, 2, ..., k+1$, be the set of multi-trees $t(J_x)$ of G_x. Then the set T of trees of G is given by

$$T = \oplus \, T_1(J_1) \times T_2(J_2) \times \cdots \times T_{k+1}(J_{k+1}), \tag{5.50}$$

where J_u is a nonempty subset of $V_{k+1} \cap V_u$, $u = 1, 2, ..., k-1$; J_k is a subset of $V_{k+1} \cap V_k$ such that an arbitrarily designated node i is always contained in J_k; J_{k+1} is the complement of the set $(J_1 \cup J_2 \cup \cdots \cup J_k) - \{i\}$ in $(V_1 \cup V_2 \cup \cdots \cup V_k) \cap V_{k+1}$; and the ring sum is taken over all possible such subsets J_x.

Proof. The theorem follows directly from Theorem 5.22 by letting $G' = G_1 \cup G_2 \cup \cdots \cup G_k$, $G'' = G_{k+1}$, and $i = i_k$.

Let us illustrate the above results by the following examples.

Example 5.6: Suppose that we wish to generate the set of trees of the graph

G of fig. 5.9 by the formula (5.46a). The graph G may be decomposed into two subgraphs G' and G'', as shown in fig. 5.10, through the nodes 1, 2, and 3. According to (5.46a), we have

$$T = T'(3) \times T''(1;2;3) \oplus T'(1;3) \times T''(2;3) \oplus T'(2;3) \times T''(1;3)$$
$$\oplus T'(1;2;3) \times T''(3).$$

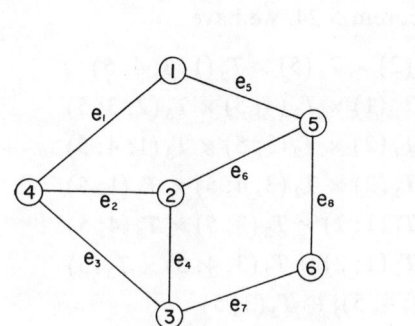

Fig. 5.9. A graph G.

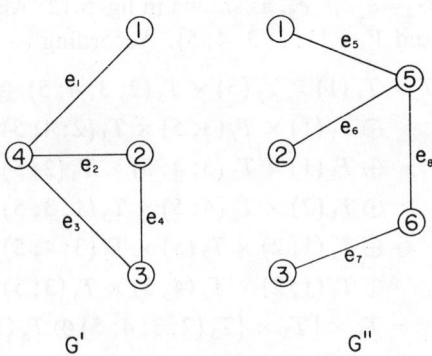

Fig. 5.10. A decomposition of G into G' and G''.

The needed trees and multi-trees of G' and G'' are given as follows:

$$T' = T'(3) = \{e_1e_2e_3, e_1e_2e_4, e_1e_3e_4\},$$
$$T'' = T''(3) = \{e_5e_6e_7e_8\},$$
$$T'(1;3) = \{e_2e_3, e_2e_4, e_3e_4, e_1e_2, e_1e_4\},$$
$$T''(1;3) = \{e_5e_6e_7, e_5e_6e_8, e_6e_7e_8\},$$
$$T'(2;3) = \{e_1e_2, e_1e_3\},$$
$$T''(2;3) = \{e_5e_6e_7, e_5e_6e_8, e_5e_7e_8\},$$
$$T'(1;2;3) = \{e_1, e_2, e_3\},$$
$$T''(1;2;3) = \{e_5e_8, e_6e_8, e_7e_8, e_6e_7, e_5e_7\}.$$

Substituting these into the formula, we have

$$T = \{e_1e_2e_3e_5e_8, e_1e_2e_3e_6e_8, e_1e_2e_3e_7e_8, e_1e_2e_3e_6e_7, e_1e_2e_3e_5e_7, e_1e_2e_4e_5e_8,$$
$$e_1e_2e_4e_6e_8, e_1e_2e_4e_7e_8, e_1e_2e_4e_6e_7, e_1e_2e_4e_5e_7, e_1e_3e_4e_5e_8, e_1e_3e_4e_6e_8,$$
$$e_1e_3e_4e_7e_8, e_1e_3e_4e_6e_7, e_1e_3e_4e_5e_7, e_2e_3e_5e_6e_7, e_2e_3e_5e_6e_8, e_2e_3e_5e_7e_8,$$
$$e_2e_4e_5e_6e_7, e_2e_4e_5e_6e_8, e_2e_4e_5e_7e_8, e_3e_4e_5e_6e_7, e_3e_4e_5e_6e_8, e_3e_4e_5e_7e_8,$$
$$e_1e_2e_5e_7e_8, e_1e_4e_5e_6e_7, e_1e_4e_5e_6e_8, e_1e_4e_5e_7e_8, e_1e_2e_6e_7e_8, e_1e_3e_5e_6e_7,$$
$$e_1e_3e_5e_6e_8, e_1e_3e_6e_7e_8, e_1e_5e_6e_7e_8, e_2e_5e_6e_7e_8, e_3e_5e_6e_7e_8\}.$$

There are 35 terms in T, each of which is a tree of G. The terms $e_1e_2e_5e_6e_7$ and $e_1e_2e_5e_6e_8$ are each generated twice during the process, and thus do not appear in the final expansion.

Example 5.7: Consider the graph G as shown in fig. 5.11. We shall generate the set of trees by Theorem 5.24. Let $G_3 = e_6e_7e_8$ and $i=5$. Then $G_1 = e_1$ and $G_2 = e_2e_3e_4e_5$, as shown in fig. 5.12. Also we have $V_1 = \{1; 2\}$, $V_2 = \{3; 4; 5; 6\}$, and $V_3 = \{1; 2; 3; 4; 5\}$. According to Theorem 5.24, we have

$$\begin{aligned}
T &= T_1(1) \times T_2(5) \times T_3(2;3;4;5) \oplus T_1(2) \times T_2(5) \times T_3(1;3;4;5) \\
&\oplus T_1(1) \times T_2(3;5) \times T_3(2;4;5) \oplus T_1(1) \times T_2(4;5) \times T_3(2;3;5) \\
&\oplus T_1(1) \times T_2(3;4;5) \times T_3(2;5) \oplus T_1(2) \times T_2(3;5) \times T_3(1;4;5) \\
&\oplus T_1(2) \times T_2(4;5) \times T_3(1;3;5) \oplus T_1(2) \times T_2(3;4;5) \times T_3(1;5) \\
&\oplus T_1(1;2) \times T_2(5) \times T_3(3;4;5) \oplus T_1(1;2) \times T_2(3;5) \times T_3(4;5) \\
&\oplus T_1(1;2) \times T_2(4;5) \times T_3(3;5) \oplus T_1(1;2) \times T_2(3;4;5) \times T_3(5) \\
&= T_1 \times [T_2 \times \{T_3(2;3;4;5) \oplus T_3(1;3;4;5)\} \oplus T_2(3;5) \\
&\times \{T_3(2;4;5) \oplus T_3(1;4;5)\} \oplus T_2(4;5) \times \{T_3(2;3;5) \oplus T_3(1;3;5)\} \\
&\oplus T_2(3;4;5) \times \{T_3(2;5) \oplus T_3(1;5)\}] \oplus T_1(1;2) \\
&\times [T_2 \times T_3(3;4;5) \oplus T_2(3;5) \times T_3(4;5) \oplus T_2(4;5) \times T_3(3;5) \\
&\oplus T_2(3;4;5) \times T_3] \\
&= \{e_1\} \times [\{e_2e_3e_4, e_3e_4e_5, e_2e_4e_5, e_2e_3e_5\} \times (\{e_7, e_8\} \oplus \{e_6\}) \\
&\oplus \{e_2e_4, e_3e_4, e_2e_5, e_3e_5\} \times (\{e_6e_7, e_6e_8\} \oplus S_0) \\
&\oplus \{e_2e_4, e_2e_5, e_4e_5\} \times (\{e_7e_8\} \oplus \{e_6e_7\}) \oplus \{e_4, e_5\} \times (\{e_6e_7e_8\} \oplus S_0)] \\
&\oplus \{\emptyset\} \times [\{e_2e_3e_4, e_3e_4e_5, e_2e_4e_5, e_2e_3e_5\} \times \{e_6e_7, e_6e_8\} \\
&\oplus \{e_2e_4, e_2e_5, e_3e_4, e_3e_5\} \times S_0 \oplus \{e_2e_4, e_2e_5, e_4e_5\} \times \{e_6e_7e_8\} \oplus S_0] \\
&= \{e_1\} \times [\{e_2e_3e_4, e_3e_4e_5, e_2e_4e_5, e_2e_3e_5\} \times \{e_6, e_7, e_8\} \cup \{e_2e_4e_6e_8, \\
&e_3e_4e_6e_8, e_2e_5e_6e_8, e_3e_5e_6e_8, e_3e_4e_6e_7, e_3e_5e_6e_7, e_4e_5e_6e_7, e_2e_4e_7e_8, \\
&e_2e_5e_7e_8, e_4e_5e_7e_8, e_4e_6e_7e_8, e_5e_6e_7e_8\}] \cup [\{e_2e_3e_4, e_3e_4e_5, e_2e_4e_5, e_2e_3e_5\} \\
&\times \{e_6e_7, e_6e_8\}] \cup \{e_2e_4e_6e_7e_8, e_2e_5e_6e_7e_8, e_4e_5e_6e_7e_8\},
\end{aligned}$$

where T_1, T_2 and T_3 are the sets of trees in G_1, G_2 and G_3, respectively. There are 35 terms in the final expansion of the Cartesian products, each corresponding to a tree of G. During the process, the terms $e_1e_2e_4e_6e_7$ and $e_1e_2e_5e_6e_7$ are each generated twice, and thus do not appear in the last equation given above.

From the example we have just worked out, one soon realizes that for a complicated graph the process of eliminating duplicate terms is indeed cumbersome. Apart from slowing down the process of generating trees, this fact

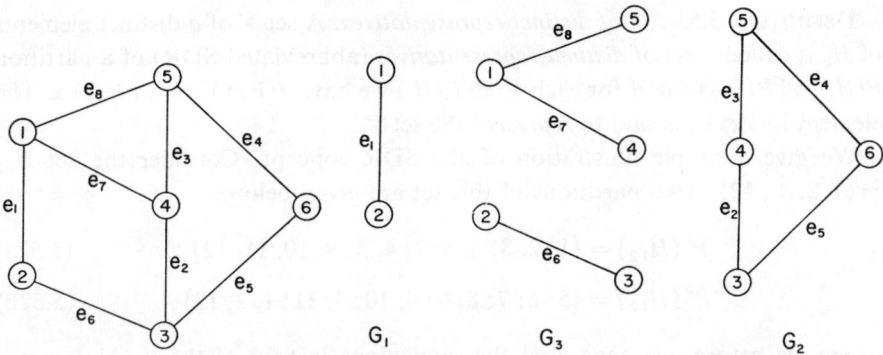

Fig. 5.11. A graph G.

Fig. 5.12. A decomposition of G into G_1, G_3, and G_2.

limits the sizes of the graphs that can be analyzed by a digital computer, since one must retain the set of all trees in a computer memory and check each new tree against the list for possible duplications. Thus, the problem of finding out the conditions under which the above procedure will not generate any duplicate terms is very important, and the results will be extremely useful. We shall consider this in greater detail in the following section.

§ 5. Generation of trees by decomposition without duplications

In this section, we shall first define "essential complementary partitions" of a set and then present a simple and efficient algorithm to test the essentiality of a pair of partitions. Finally, we shall show how they can be applied to Theorem 5.22 to generate trees without duplications.

The present section is based entirely on the work of CHEN [1969e, f].

5.1. *Essential complementary partitions of a set*

For a given set V, we denote the number of its elements by $\alpha(V)$. By $V \subseteq H$ we mean that V is a subset of the set H. Unless stated to the contrary, H_k denotes a set consisting of k elements.

DEFINITION 5.4: *Partition.* A *partition* of a set H_k, denoted by the symbol $P(H_k)$, is a collection of nonempty subsets V_x, $x = 1, 2, \ldots, q$, of H_k such that

$$H_k = V_1 \cup V_2 \cup \cdots \cup V_q \tag{5.51a}$$

and

$$V_u \cap V_w = S_0 \tag{5.51b}$$

for $1 \leq u < w \leq q$.

DEFINITION 5.5: *Set of distinct representatives*. A set V of q distinct elements of H_k is called a *set of distinct representatives* (abbreviated SDR) of a partition $P(H_k)$, $\alpha[P(H_k)] = q$, if for each V_x in $P(H_k)$ we have $\alpha(V \cap V_x) = 1$ for all x. The element in $V \cap V_x$ is said to *represent* the set V_x.

We give a simple illustration of the SDR concept. Consider the set $H_{12} = \{1, 2, \ldots, 12\}$. Two partitions of this set are given below:

$$P'(H_{12}) = \{1; 2, 8; 3, 6, 7; 4, 5; 9, 10, 11, 12\}, \quad (5.52a)$$

$$P''(H_{12}) = \{5; 6; 7; 8; 9; 4, 10; 3, 11; 1, 2, 12\}, \quad (5.52b)$$

where, as before, we have used the semicolons instead of the usual commas and curly brackets to separate the elements of a set of subsets. The SDRs of $P'(H_{12})$ and $P''(H_{12})$ are given by

$$V' = \{1; 2; 3; 4; 12\} \quad (5.53a)$$

and

$$V'' = \{5; 6; 7; 8; 9; 10; 11; 12\}, \quad (5.53b)$$

respectively. Obviously, for a given partition, there may be more than one SDR. For example, the set $\{1; 8; 6; 5; 10\}$ is also an SDR of $P'(H_{12})$.

DEFINITION 5.6: *Complementary partitions*. A pair of partitions $P'(H_k)$ and $P''(H_k)$, denoted by the symbol $C_j[P'(H_k), P''(H_k)]$, are said to be *complementary with respect to an element j* of H_k if there exists a $V \subseteq H_k$, j in V, and such that V and $\overline{V - \{j\}}$ are the SDRs of $P'(H_k)$ and $P''(H_k)$, respectively, where $\overline{V - \{j\}}$ denotes the complement of the set $V - \{j\}$ in H_k.

From the definition, it is evident that for a given H_k there always exist a pair of complementary partitions with respect to any element of H_k. For example, the pair of partitions of H_{12} of (5.52) are complementary with respect to 12, since in (5.53) $V'' = \overline{V' - \{12\}}$.

We mention that in general if $i \neq j$, $C_i[P'(H_k), P''(H_k)] \neq C_j[P'(H_k), P''(H_k)]$. However, $C_i[P'(H_k), P''(H_k)] = C_i[P''(H_k), P'(H_k)]$ for all i. Thus, for a pair of complementary partitions to be meaningful, the subscript i or j must be specified. Clearly, if $P'(H_k)$ has q elements, then $P''(H_k)$ must have $k - q + 1$ elements, and

$$\alpha\{P'(H_k)\} + \alpha\{P''(H_k)\} = k + 1. \quad (5.54)$$

DEFINITION 5.7: *Essential complementary partitions*. A pair of partitions $C_j[P'(H_k), P''(H_k)]$ are said to be *essential* if there exists no $C_j[P'(V), P''(V)]$

where V is a nonempty proper subset of H_k, and $P'(V) \subseteq P'(H_k)$ and $P''(V) \subseteq P''(H_k)$.

As an illustration, consider the set $H_{10} = \{1, 2, \ldots, 10\}$. The partitions

$$P'(H_{10}) = \{1, 2, 3; 4, 5; 6, 7, 8; 9, 10\} \tag{5.55a}$$

and

$$P''(H_{10}) = \{1; 2; 3, 4, 6; 5, 7; 8; 9; 10\} \tag{5.55b}$$

are complementary with respect to 10, since there exists a subset $J = \{3; 5; 6; 10\}$ of H_{10} such that J and $\overline{J - \{10\}}$ are the SDRs of $P'(H_{10})$ and $P''(H_{10})$, respectively. However, the pair are not essential since there exists a nonempty proper subset $V = \{9; 10\}$ of H_{10} such that the pair

$$P'(V) = \{9, 10\} \quad \text{and} \quad P''(V) = \{9; 10\}$$

are complementary with respect to 10.

In the following section, we shall present a simple algorithm for testing the essentiality of a pair of complementary partitions. However, before we do this, let us discuss some of the fundamental properties of essential complementary partitions.

THEOREM 5.25: *If a pair of complementary partitions of H_k are essential with respect to an element of H_k, they are essential with respect to any element of H_k.*

Proof. Assume that $C_i[P'(H_k), P''(H_k)]$ is essential. From Problem 5.76, the pair are complementary with respect to any j. If $C_j[P'(H_k), P''(H_k)]$, $j \neq i$, is not essential, then there exists a $C_j[P'(V), P''(V)]$, where V is a minimal nonempty proper subset of H_k, and $P'(V) \subseteq P'(H_k)$ and $P''(V) \subseteq P''(H_k)$. If i is contained in an element of $P'(V)$ or $P''(V)$, then there exists $C_i[P'(V), P''(V)]$, which is impossible by our hypothesis that $C_i[P'(H_k), P''(H_k)]$ is essential. Thus, i is contained in the complement $\overline{V}$ of V in H_k. Now let us consider the pair of partitions of $\overline{V}$:

$$P'(\overline{V}) = P'(H_k) - P'(V) \quad \text{and} \quad P''(\overline{V}) = P''(H_k) - P''(V).$$

Obviously, they are complementary with respect to i. In other words, there exists a $C_i[P'(\overline{V}), P''(\overline{V})]$ where $\overline{V}$ is a nonempty proper subset of H_k, which is again impossible since $C_i[P'(H_k), P''(H_k)]$ is essential by hypothesis. Thus, $C_j[P'(H_k), P''(H_k)]$ must be essential, which completes the proof of the theorem.

Thus, it may be concluded that the subscript i in a pair of essential complementary partitions $C_i[P'(H_k), P''(H_k)]$ is superfluous, and may be omitted. For simplicity, we denote it by $C[P'(H_k), P''(H_k)]$.

COROLLARY 5.9: $C[P'(H_k), P''(H_k)]$ is essential if, and only if, $C[P''(H_k), P'(H_k)]$ is essential.

5.2. *Algorithm*

In the present section, we shall first define two types of operators D' and D''. They will then be used as a tool for testing the essentiality of a pair of complementary partitions in an algorithm to be discussed below.

In $C_i[P'(H_k), P''(H_k)]$, let V and $\overline{V-\{i\}}$ be the SDRs of $P'(H_k)$ and $P''(H_k)$, respectively. If R' and R'' are the subsets of $P'(H_k)$ and $P''(H_k)$, respectively, then the operators D' and D'' are defined as follows:

$D'(R') = \{p''; p'' \in P''(H_k)$, and there exists an $r' \in R'$ such that

the intersection of p'' and $r'-(r'\cap V)$ is not empty$\}$, (5.56a)

and

$D''(R'') = \{p'; p' \in P'(H_k)$, and there exists an $r'' \in R''$ such that

the intersection of p' and $r''-r''\cap(\overline{V-\{i\}})$ is not empty$\}$. (5.56b)

Sometimes, operations of the above types may be repeated many times. For convenience, let $D_1(R') = D'(R')$ and

$$D_2(R') = D''(D'(R')),$$
$$D_3(R') = D'(D''(D'(R'))) = D'(D_2(R')),$$
$$D_4(R') = D''(D'(D''(D'(R')))) = D''(D_3(R')),$$

and so on. Similarly, let $D_1(R'') = D''(R'')$ and

$$D_2(R'') = D'(D''(R'')),$$
$$D_3(R'') = D''(D'(D''(R''))) = D''(D_2(R'')),$$
$$D_4(R'') = D'(D''(D'(D''(R'')))) = D'(D_3(R'')),$$

and so on. It is obvious that the sets $D_{2x}(R')$ and $D_{2x-1}(R'')$ are all subsets of $P'(H_k)$, and the sets $D_{2x-1}(R')$ and $D_{2x}(R'')$ are all subsets of $P''(H_k)$, where x is a positive integer.

As an example, consider the pair of complementary partitions of (5.52) together with the SDRs in (5.53). Let $R' = \{9, 10, 11, 12\}$ and $R'' = \{1, 2, 12\}$. Using the operators defined above, the following sets can easily be generated:

$$D_1(R') = D'(R') = \{9; 4, 10; 3, 11\},$$
$$D_2(R') = D''(D'(R')) = D''(9; 4, 10; 3, 11) = \{3, 6, 7; 4, 5\},$$
$$D_3(R') = D'(3, 6, 7; 4, 5) = \{5; 6; 7\}, \quad (5.57)$$
$$D_1(R'') = D''(1, 2, 12) = \{1; 2, 8\},$$
$$D_2(R'') = D'(D''(R'')) = D'(1; 2, 8) = \{8\}.$$

Note that since the elements of a set of subsets are separated by semicolons, while the elements of a subset are separated by commas, the absence of semicolons inside the parentheses or curly brackets indicates that the set consists only of one element. For example, there is only one element, a subset of H_{12}, inside the parentheses of D'' (1, 2, 12), while 1, 2, and 12 are elements of the subset.

LEMMA 5.3: Given $C_i[P'(H_k), P''(H_k)]$, there exists an element p either in $P'(H_k)$ or in $P''(H_k)$ such that $\alpha(p)=1$.

Proof. Suppose that the lemma is false, then $k \geq 2q$ and $k \geq 2(k-q+1)$ where $q = \alpha(P'(H_k))$, since $\alpha(P''(H_k)) = k - q + 1$. It follows that $k \geq k+1$, which is impossible, so the lemma is proved.

LEMMA 5.4: Given $C_i[P'(H_k), P''(H_k)]$, let p' and p'' be the elements, each containing i, of $P'(H_k)$ and $P''(H_k)$, respectively. Then

$$D_u(p') \cap D_v(p'') = S_0 \tag{5.58}$$

for all positive integers u and v.

Proof. We shall first show that if R_1 and R_2 are two disjoint subsets of $P'(H_k)$ or $P''(H_k)$, i.e., $R_1 \cap R_2 = S_0$, then $D'(R_1)$ and $D'(R_2)$ or $D''(R_1)$ and $D''(R_2)$ are also disjoint. Suppose that there is an element p in $P''(H_k)$ which is contained in both $D'(R_1)$ and $D'(R_2)$, then there exist r'_1 and r'_2 in R_1 and R_2, respectively, such that

$$p \cap [r'_1 - (r'_1 \cap V)] \neq S_0 \tag{5.59a}$$

and

$$p \cap [r'_2 - (r'_2 \cap V)] \neq S_0, \tag{5.59b}$$

where V is an SDR of $P'(H_k)$. It follows that there are at least two elements j_1 and j_2 in p. Obviously, j_1 and j_2 cannot be in V, so they must be in $\overline{V - \{i\}}$. This is impossible since p has only one distinct representative. Thus, $D'(R_1) \cap D'(R_2) = S_0$. Similarly, we can show that $D''(R_1) \cap D''(R_2) = S_0$. Obviously, the intersection of R_1 and $D'(R_2)$, or R_2 and $D'(R_1)$, or R_1 and $D''(R_2)$, or R_2 and $D''(R_1)$ is also empty.

Now, we shall complete our proof by induction over $w = u + v$. For $w = 2$, $D'(p')$ and $D''(p'')$ are obviously disjoint. For, if not, there is an element, say, p in both $D'(p')$ and $D''(p'')$. From the definition of operator D' we have $p \cap p' \neq S_0$. Since p is also in $D''(p'')$, it follows that p and p' are disjoint elements of $P'(H_k)$, i.e., $p \cap p' = S_0$. This is impossible, so $D'(p')$ and $D''(p'')$ are disjoint. Now, we assume that (5.58) is true for any $w-1$ or less where $w>2$. We shall show that it is also true for any $w=u+v$. For convenience, let $D_0(p') = \{p'\}$ and $D_0(p'') = \{p''\}$.

If u and v in (5.58) are of different parity, i.e., one is even and one is odd, then $D_{u-1}(p')$ and $D_{v-1}(p'')$ are, by induction hypothesis, disjoint subsets of $P'(H_k)$ or $P''(H_k)$. Thus, from our earlier argument, $D_u(p')$ and $D_v(p'')$ must be disjoint. If, on the other hand, u and v are of the same parity, and if there is an element p in both $D_u(p')$ and $D_v(p'')$, we may assume, without loss of generality, that $u>v$. Since p is in $D_u(p')$, there exists an r in $D_{u-1}(p')$ such that $r \cap p \neq S_0$. But r and p are both in $P'(H_k)$ or $P''(H_k)$; they must be disjoint, a contradiction. So the lemma is proved.

In Lemma 5.4, if we let u_m and v_m be the largest possible integers u and v that can be achieved in $D_u(p')$ and $D_v(p'')$, respectively, such that $D_{u_m}(p') \neq S_0$ and $D_{v_m}(p'') \neq S_0$, then the following lemma is obvious.

LEMMA 5.5: If p is an element of $D_{u_m}(p')$ or $D_{v_m}(p'')$, then $\alpha(p)=1$.
Let

$$F' = \{p'\} \cup \bigcup_{x=1}^{w'} D_{2x}(p') \cup \bigcup_{y=1}^{z'} D_{2y-1}(p''), \tag{5.60a}$$

where w' and z' are the largest integers not greater than $\tfrac{1}{2}u_m$ and $\tfrac{1}{2}(v_m+1)$, respectively, and

$$F'' = \{p''\} \cup \bigcup_{x=1}^{w''} D_{2x-1}(p') \cup \bigcup_{y=1}^{z''} D_{2y}(p''), \tag{5.60b}$$

where w'' and z'' are the largest integers not greater than $\tfrac{1}{2}(u_m+1)$ and $\tfrac{1}{2}v_m$, respectively. Then we have

THEOREM 5.26: Let p' and p'' be the elements, each containing i, of $P'(H_k)$ and $P''(H_k)$, respectively. Then $C_i[P'(H_k), P''(H_k)]$ is essential if, and only if, $F'=P'(H_k)$ or $F''=P''(H_k)$.

Proof. Necessity. If $F' \neq P'(H_k)$, then $P'(H_k) - F'$ is a nonempty proper subset of $P'(H_k)$. Let J' be the union of all the elements of F'. Also let J'' be the union of all the elements of $D'(F') \cup \{p''\}$. If $J' \neq J''$, let j be an integer in J'' but not in J'. Consider the element p in $D'(F') \cup \{p''\}$ that contains the integer j. Since p is also in $D_{2x-1}(p')$ or in $D_{2x}(p'')$ for some x, it follows that neither $D_{2x}(p')$ nor $D_{2x+1}(p'')$ can be empty, and one of the elements of $D_{2x}(p')$ or $D_{2x+1}(p'')$ must contain the integer j. Since $D_{2x}(p')$ and $D_{2x+1}(p'')$ are also in F', the integer j must be in J'. Thus, $J' = J''$. In other words, F' and $D'(F') \cup \{p''\}$ are a pair of complementary partitions of J' or J'' with respect to i. This contradicts to our assumption that $C_i[P'(H_k), P''(H_k)]$ is essential. So $F' = P'(H_k)$. Similarly, we can show the other case.

Sufficiency. Let $F' = P'(H_k)$. If $C_i[P'(H_k), P''(H_k)]$ is not essential, then there exists a $C_i[P'(J), P''(J)]$, where J is a nonempty proper subset of H_k and i is in J. It is not difficult to see that $F' \subseteq P'(J)$. Thus, $F' \ne P'(H_k)$. Since $F' = P'(H_k)$ by hypothesis, $C_i[P'(H_k), P''(H_k)]$ is essential. Similarly, we can prove the other case. This completes the proof of the theorem.

The theorem provides a simple and efficient way for testing the essentiality of a pair of complementary partitions. As an illustration, consider the pair of complementary partitions together with their SDRs of (5.52) and (5.53). Let $p' = \{9, 10, 11, 12\}$ and $p'' = \{1, 2, 12\}$. The desired sets are generated in (5.57). From (5.60a) we have

$$F' = \{p'\} \cup D_2(p') \cup D_1(p'')$$
$$= \{9, 10, 11, 12\} \cup \{3, 6, 7; 4, 5\} \cup \{1; 2, 8\}$$
$$= \{1; 2, 8; 3, 6, 7; 4, 5; 9, 10, 11, 12\} = P'(H_{12}).$$

Thus, according to Theorem 5.26, $C[P'(H_{12}), P''(H_{12})]$ is essential. We can also use (5.60b) for our test:

$$F'' = \{p''\} \cup D_1(p') \cup D_3(p') \cup D_2(p'')$$
$$= \{1, 2, 12\} \cup \{9; 4, 10; 3, 11\} \cup \{5; 6; 7\} \cup \{8\}$$
$$= \{5; 6; 7; 8; 9; 4, 10; 3, 11; 1, 2, 12\} = P''(H_{12}),$$

which also shows that $C[P'(H_{12}), P''(H_{12})]$ is essential.

As another example, consider the pair of complementary partitions with respect to 10 of (5.55). The SDRs are $\{3; 5; 6; 10\}$ and $\{1; 2; 4; 7; 8; 9; 10\}$. Let $p' = \{9, 10\}$ and $p'' = \{10\}$. Then we have $D_1(p') = \{9\}$, $D_2(p') = S_0$, and $D_1(p'') = S_0$. It follows from (5.60) that $F' = \{p'\} = \{9, 10\}$ and $F'' = \{p''\} \cup D_1(p') = \{9; 10\}$. Thus, according to Theorem 5.26, $C_{10}[P'(H_{10}), P''(H_{10})]$ is not essential since $F' \ne P'(H_{10})$ or $F'' \ne P''(H_{10})$.

5.3. Decomposition without duplications

We shall now apply the above results to Theorem 5.22 for generation of trees by decomposition without duplications. However, before we do this, it is necessary that we put (5.46) in a different but more convenient form.

THEOREM 5.27:

$$T = \oplus \, T'(P'(I_k)) \times T''(P''(I_k)), \qquad (5.61)$$

where the ring sum is taken over all $C_{i_k}[P'(I_k), P''(I_k)]$.

Proof. For each choice of J in (5.46), the nodes in J must belong to the components of a multi-tree $t'(J \cup \{i_k\})$ in $T'(J \cup \{i_k\})$, and the nodes in $\bar{J}$ to the components of a multi-tree $t''(\bar{J} \cup \{i_k\})$ in $T''(\bar{J} \cup \{i_k\})$. Thus, the components of $t'(J \cup \{i_k\})$ and $t''(\bar{J} \cup \{i_k\})$ partition the nodes in I_k into a pair of complementary partitions with respect to i_k with $J \cup \{i_k\}$ and $\bar{J} \cup \{i_k\}$ as their SDRs. In other words, the nodes of I_k in these components form a $C_{i_k}[P'(I_k), P''(I_k)]$. This means that for every term in (5.46a) there is a unique term in (5.61) before the ring-sum operations are taken. Conversely, for each $C_{i_k}[P'(I_k), P''(I_k)]$, the SDRs of $P'(I_k)$ and $P''(I_k)$ may be expressed as $J \cup \{i_k\}$ and $\bar{J} \cup \{i_k\}$, respectively, where $J \subseteq I_{k-1}$. Thus, for each term in (5.61) there corresponds a unique term in (5.46a) before the ring-sum operations are taken. So the theorem is proved.

Evidently, each term generated on the right-hand side of (5.61) is either a tree or a subgraph that contains at least one circuit. If it is a subgraph having circuits, such subgraphs must appear an even number of times. Thus, our problem of how to generate trees by Theorem 5.22 without duplications reduces to that of characterizing a class of complementary partitions which will not generate subgraphs having circuits. The class is, in fact, the class of essential complementary partitions of I_k discussed in the preceding section.

THEOREM 5.28:

$$T = \bigcup T'(P'(I_k)) \times T''(P''(I_k)), \qquad (5.62)$$

where the union is taken over all possible essential complementary partitions $C[P'(I_k), P''(I_k)]$.

Proof. By Theorem 5.25, it is sufficient to show that the theorem is valid for all essential complementary partitions with respect to i_k,

$$C_{i_k}[P'(I_k), P''(I_k)]. \qquad (5.63)$$

We shall first prove that if (5.63) is essential, then the corresponding term in (5.62) will not generate any subgraphs having circuits. Assume otherwise, i.e., let g be a subgraph having at least one circuit in the set

$$T'(P'(I_k)) \times T''(P''(I_k)). \qquad (5.64)$$

Let V_c be the node set of a circuit in g, and let

$$R' = \{x;\ x \text{ is in } P'(I_k) \quad \text{and} \quad x \cap V_c \neq S_0\} \tag{5.65a}$$

and

$$R'' = \{x;\ x \text{ is in } P''(I_k) \quad \text{and} \quad x \cap V_c \neq S_0\}. \tag{5.65b}$$

If p' and p'' are elements of $P'(I_k)$ and $P''(I_k)$ containing the integer (node) i_k, then none of the elements in R' or R'' can be contained in $D_u(p')$ or $D_u(p'')$ for any u. For, if not, say p_1 is an element of R' or R'' that is also contained in $D_u(p')$ for some u, then $p_1 \cap V_c = \{j_1, j_2, \ldots\}$ since every element in $R' \cup R''$ contains at least two nodes from V_c. It follows that there exists a p_2 in $R' \cup R''$ such that $p_2 \cap V_c = \{j_2, j_3, \ldots\}$. Obviously, p_2 is contained in $D_{u+1}(p')$. If we continue on this process, we will eventually include all the elements of $R' \cup R''$ and return to j_1 since the elements of V_c correspond to the nodes of a circuit. This would imply that u in $D_u(p')$ can be increased indefinitely. This is clearly impossible since it can easily be shown that the integers u_m and v_m in $D_{u_m}(p')$ and $D_{v_m}(p'')$ of Lemma 5.5 must be finite, so none of the elements of $R' \cup R''$ can be contained in $D_u(p')$ or $D_u(p'')$ for any u. Thus, the set F' defined in (5.60a) is a nonempty subset of $P'(I_k) - R'$. This is again impossible by Theorem 5.26, so all the terms generated in (5.64) are trees of the graph.

Next, we have to show that if (5.63) is not essential, then each element generated in (5.64) is a subgraph containing at least one circuit provided that (5.64) is not empty. Since (5.63) is not essential, there exists $C_{i_k}[P'(I), P''(I)]$ where I is a nonempty proper subset of I_k, and $P'(I) \subseteq P'(I_k)$ and $P''(I) \subseteq P''(I_k)$. It is not difficult to see that there are $\alpha(I) + 1$ elements in $P'(I)$ and $P''(I)$. Note that since not all the elements in $P'(I)$ and $P''(I)$ are necessarily distinct, in general we have $\alpha[P'(I) \cup P''(I)] \leq \alpha(I) + 1$. Consider the pair $M' = P'(I_k) - P'(I)$ and $M'' = P''(I_k) - P''(I)$. Obviously, there are $q = k - \alpha(I)$ elements in M' and M''. This is also the number of distinct representatives of (5.63) contained in M' and M''. For convenience, let j_x, $x = 1, 2, \ldots, q$, be the distinct representatives of the elements J_x in M' and M''. It follows that $I_k - I = \{j_1, j_2, \ldots, j_q\}$, and every j_x appears exactly in two different elements of M' and M''. Since j_1 represents J_1, i.e., j_1 is the distinct representative of J_1, there must be a set, say, J_2 containing j_1 but it cannot be represented by j_1. Thus, j_1 and j_2 are both in J_2. Since j_2 represents J_2, there must be a set, say, J_3 containing j_2 but it cannot be represented by j_2. Thus, j_2 and j_3 are both in J_3. If we continue on this argument, the process will generate a sequence of the form $j_1 j_2 \ldots j_s$ such that j_y and j_{y+1} are both in J_{y+1} for $y = 1, 2, \ldots, s-1 \leq q$. Since the number of elements in M' and M'' is finite, the process can only be continued indefinitely when, for some s and z, $j_s = j_z$ where $z < s$. Thus, if (5.64) is not empty, each element generated in (5.64) is a subgraph containing at least one circuit; the circuit is contained in the edge train formed by the paths of the components of $t'(P'(I_k))$ and $t''(P''(I_k))$, where $t'(P'(I_k))$ and $t''(P''(I_k))$ are elements of $T'(P'(I_k))$ and $T''(P''(I_k))$, respectively. Each of the paths is connected between two nodes contained in the sequence $j_z j_{z+1} \ldots j_s$.

Since two different pairs of complementary partitions generate two disjoint sets of subgraphs in (5.62), the theorem follows from here. This completes the proof of the theorem.

Now, we have achieved our original goal of generating trees by first decomposing the graph into two parts, determining appropriate multi-trees for each part, and then combining these multi-trees in certain ways to obtain the desired result without duplications. However, the price we paid in achieving this is that we have to enumerate the set of all possible essential complementary partitions. This is not a high price since these partitions are independent of the structure of a graph, and thus they can be tabulated. Actually, we do not need to list them all; half would be sufficient. This follows directly from Corollary 5.9.

As an illustration, consider $I_3 = \{i_1; i_2; i_3\}$. The following pairs of complementary partitions $C_{i_3}[P'(I_3), P''(I_3)]$ are essential; their essentiality can be tested by a procedure outlined in § 5.2 (also see Problem 5.75 for I_4):

$$\begin{array}{cc} P'(I_3) & P''(I_3) \\ \{i_1; i_2; i_3\} & \{i_1, i_2, i_3\} \\ \{i_1; i_2, i_3\} & \{i_1, i_3; i_2\} \\ \{i_1; i_2, i_3\} & \{i_1, i_2; i_3\} \\ \{i_1, i_2; i_3\} & \{i_1, i_3; i_2\} \end{array}$$

By interchanging the roles of $P'(I_3)$ and $P''(I_3)$, we obtain four additional pairs of essential complementary partitions. The only pair of complementary partitions which are not essential are $C_{i_3}[\{i_1, i_2; i_3\}, \{i_1, i_2; i_3\}]$.

We shall illustrate Theorem 5.28 by the following examples.

Example 5.8: Consider the graph G of fig. 5.11. The graph may be decomposed into two subgraphs G' and G'' along the nodes 3, 4, and 5. Such a decomposition is as shown in fig. 5.13. Using the essential complementary partitions listed above, the desired multi-trees are found as follows:

$$T'(3; 4; 5) = \{e_1 e_6, e_1 e_7, e_1 e_8, e_6 e_7, e_6 e_8\},$$
$$T'(3; 4, 5) = \{e_1 e_7 e_8, e_6 e_7 e_8\},$$
$$T'(3, 4; 5) = \{e_1 e_6 e_7\},$$

$$T' = T'(3, 4, 5) = \{e_1 e_6 e_7 e_8\},$$
$$T'(3, 5; 4) = \{e_1 e_6 e_8\};$$
$$T'' = T''(3, 4, 5) = \{e_2 e_3 e_4, e_2 e_3 e_5, e_2 e_4 e_5, e_3 e_4 e_5\},$$
$$T''(3, 5; 4) = \{e_4 e_5\},$$
$$T''(3, 4; 5) = \{e_2 e_4, e_2 e_5\},$$
$$T''(3; 4; 5) = \{e_4, e_5\},$$
$$T''(3; 4, 5) = \{e_3 e_4, e_3 e_5\}.$$

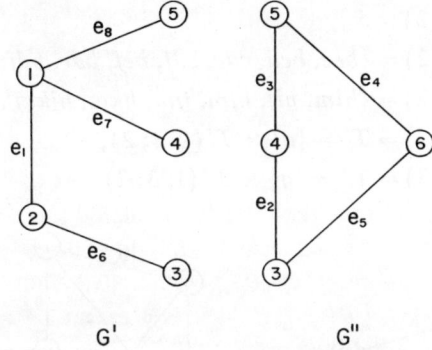

Fig. 5.13. A decomposition of the graph of fig. 5.11.

According to Theorem 5.28, the set of trees of G is given by

$$T = T'(3; 4; 5) \times T'' \cup T''(3; 4; 5) \times T' \cup T'(3, 4, 5) \times T''(3, 5; 4) \cup$$
$$T''(3; 4, 5) \times T'(3, 5; 4) \cup T'(3, 4, 5) \times T''(3, 4; 5) \cup$$
$$T''(3; 4, 5) \times T'(3, 4; 5) \cup T'(3, 4; 5) \times T''(3, 5; 4) \cup$$
$$T''(3, 4; 5) \times T'(3, 5; 4)$$
$$= \{e_1 e_6, e_1 e_7, e_1 e_8, e_6 e_7, e_6 e_8\} \times \{e_2 e_3 e_4, e_2 e_3 e_5, e_2 e_4 e_5, e_3 e_4 e_5\} \cup$$
$$\{e_1 e_4 e_6 e_7 e_8, e_1 e_5 e_6 e_7 e_8\} \cup \{e_4 e_5\} \times \{e_1 e_7 e_8, e_6 e_7 e_8, e_1 e_6 e_7\} \cup$$
$$\{e_1 e_3 e_4 e_6 e_8, e_1 e_3 e_5 e_6 e_8\} \cup \{e_2 e_4, e_2 e_5\} \times \{e_1 e_7 e_8, e_6 e_7 e_8, e_1 e_6 e_8\} \cup$$
$$\{e_1 e_3 e_4 e_6 e_7, e_1 e_3 e_5 e_6 e_7\},$$

which is of course the same as that obtained in Example 5.7.

It is significant to point out that in the present analysis the set of trees of G appears in the form of unions of Cartesian products of classes of subgraphs. For network-theoretic purpose, this is indeed the most desirable form and there is no need to carry out the Cartesian products, since we are really interested in computing the tree-admittance products.

Example 5.9: The graph G given in fig. 5.14 may be decomposed into two subgraphs G' and G'' through the nodes 1, 2, and 3. Such a decomposition is as shown in fig. 5.15. The desired multi-trees are found as follows:

$$T'(1;2;3) = \{bc, bf, be, cd, cf, de, df, ef\},$$
$$T''(1;2;3) = \{hi, hk, hm, ij, ik, jk, jm, km\},$$
$$T'(1;2,3) = S_0,$$
$$T''(1;2,3) = \{g\} \times T''(1;2;3),$$
$$T'(1,2;3) = \{a\} \times T'(1;2;3),$$
$$T''(1,2;3) = S_0,$$
$$T'(1,3;2) = \{bce, bcd, cde, cdf, bef, bde, bdf, cef\},$$
$$T''(1,3;2) = \{him, hij, hjm, ijm, hkm, hjk, ikm, ijk\},$$
$$T'(1,2,3) = T' = \{a\} \times T'(1,3;2),$$
$$T''(1,2,3) = T'' = \{g\} \times T''(1,3;2).$$

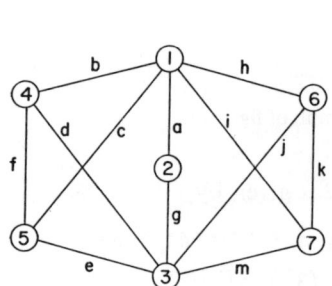

Fig. 5.14. A graph G.

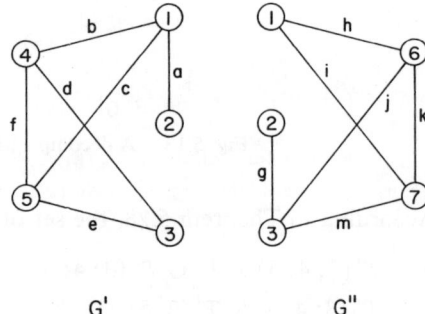

Fig. 5.15. A decomposition of G.

According to Theorem 5.28, the set of trees of G is given by

$$T = T'(1;2;3) \times \{g\} \times T''(1,3;2) \cup S_0 \cup S_0 \cup \{a\} \times T'(1;2;3)$$
$$\times T''(1,3;2) \cup \{a\} \times T'(1,3;2) \times T''(1;2;3) \cup T'(1,3;2) \times \{g\}$$
$$\times T''(1;2;3) \cup \{a\} \times T'(1;2;3) \times \{g\} \times T''(1;2;3) \cup S_0$$
$$= \{bc, bf, be, cd, cf, de, df, ef\} \times \{ghim, ghij, ghjm, gijm, aijm, ghkm,$$
$$ghjk, gikm, gijk, ahim, ahij, ahjm, ahkm, ahjk, aikm, aijk, aghi,$$
$$aghk, aghm, agij, agik, agjk, agjm, agkm\} \cup \{hi, hk, hm, ij, ik,$$
$$jk, jm, km\} \times \{bce, bcd, cde, cdf, bef, bde, bdf, cef\} \times \{a, g\}.$$

There are 320 terms in the expansion of the Cartesian products, each of which corresponds to a tree of G. Again, for network-theoretic purpose, there is no need to carry out the Cartesian products.

Thus, we may conclude that the formula given in Theorem 5.28 is very useful from practical viewpoint for computer analysis of very large networks when the topological approach is seen at its best. The advantages of decomposition are clear. Not only is the quantity of trees sharply reduced, but they are "smaller" trees requiring less manipulation for the evaluation of each. The fact that in many applications we are not really concerned with the set of trees but the sum of the tree-admittance products makes them more significant. As an illustration, the number of terms in T obtained in Example 5.9 before expansion is 50 in comparison with 320 terms that we would have obtained by techniques other than decomposition. Furthermore, the maximum number of symbols contained in each of the terms of T is four, while each of the 320 terms would have contained six symbols. This is of course particularly significant when the graph under consideration has high rank and nullity. Finally, we mention that, based on (5.62), generation formulas for a cascade of multi-terminal networks were derived by CHEN [1969f].

§ 6. The matrix formulation

Since trees (cotrees) of a graph or directed graph are in one-to-one correspondence with the major submatrices of a basis cut (circuit) matrix, the problem of tree generation is equivalent to that of enumerating all the major submatrices (see ch. 2, § 1.2 and § 1.3, ch. 5, § 1, and Definition 2.4). In this section, we shall first present an algorithm that enumerates all the major submatrices of a matrix without duplications, and then apply this result to the generation of trees, cotrees, directed trees, and directed 2-trees. The present section is based on the work of MAYEDA et al. [1968] and CHEN [1968b].

6.1. *The enumeration of major submatrices of an arbitrary matrix*

Let F be an arbitrary matrix of order $p \times b$ and of rank p. Like those used in ch. 2, § 4.2, $F(I_u, J_v)$ denotes the submatrix of F consisting of the rows and columns corresponding to the integers contained in $I_u = \{i_1, i_2, ..., i_u\}$ and $J_v = \{j_1, j_2, ..., j_v\}$, respectively. Also, by $F(J_p)$ we mean $F(J_p) = F(I_p, J_p)$ and by $\bar{I}_u$ we mean the complement of I_u in $\{1, 2, ..., b\}$.

DEFINITION 5.8: *Matrix in proper form.* An arbitrary matrix F of order $p \times b$ containing a diagonal major submatrix is referred to as a *matrix in proper form*. The defining diagonal major submatrix is denoted by $F(K_p)$ where K_p

$= \{k_1, k_2, ..., k_p\}$. If $F(K_p)$ is the identity matrix, then F is said to be in *normal form*.

In the following discussion, we assume that all the given matrices are in proper form. The assumption is not a serious restriction, as it can easily be shown that to every matrix F there corresponds a matrix in proper form which can be obtained from F by using only the following row operations:
(1) addition to one row of a multiple of another row,
(2) permutation of rows.

Since the matrix thus obtained generates the same linear row-vector space as F does, they would have the same set of major submatrices (consisting of the same columns) whose corresponding determinants differ at most by a sign. Furthermore, the relative signs of these determinants are preserved.

For convenience, we assume that the columns of F have been labeled by the integers $1, 2, ..., b$, and the rows of F are labeled in such a way that the ith row of F is denoted by k_i for $i = 1, 2, ..., p$, where k_i is in K_p. For example, if the defining diagonal major submatrix $F(K_3)$ consists of the columns 5, 6, and 7, then the rows of F are labeled by 5, 6, and 7 in that order. Throughout this and the following two sections, the term matrix means a matrix which has been labeled according to the rules just outlined. Also, we assume that all the integers in the parentheses appear in their natural order.

If $I_u \subseteq K_p$, by $K_p J_v / I_u$ we mean the set $(K_p - I_u) \cup J_v$. For example, if $K_3 = \{1, 5, 6\}$, $J_2 = \{2, 8\}$, and $I_2 = \{1, 6\}$, then $(K_3 J_2 / I_2) = \{2, 5, 8\}$. It follows that $F(K_p J_u / I_u)$, $1 \leq u \leq p$, is a major submatrix of F other than $F(K_p)$ if, and only if, $F(I_u, J_u)$ is a nonsingular submatrix of $F(\bar{K}_p)$. Consequently, the following theorem is obvious (Problem 5.33).

THEOREM 5.29: Let F be a matrix of order $p \times b$ and of rank p in proper form. Then the set M of all major submatrices of F is given by

$$M = \bigcup_{q=0}^{r} \bigcup_{(I_q)} \bigcup_{(J_q)} \{F(K_p J_q / I_q)\}, \tag{5.66}$$

where r is the rank of $F(\bar{K}_p)$; $F(I_q, J_q)$ is a nonsingular submatrix of $F(\bar{K}_p)$; and $F(K_p J_0 / I_0) = F(K_p)$ by definition.

The problem of enumerating all the major submatrices of F is now reduced to that of enumerating all the nonsingular submatrices of orders $1, 2, ..., $ and r in $F(\bar{K}_p)$. The nonsingular submatrices of order 1 are very easy to find; they are the nonzero elements in $F(\bar{K}_p)$. For the nonsingular submatrices of order greater than 1, the whole process outlined above may be repeated. For example, consider the submatrix $F(I_q, J_{b-p})$, $2 \leq q \leq r$, where K_p and J_{b-p} are sets of

§6   The matrix formulation   367

complementary indices of the integers 1, 2, ..., b. This matrix is first put in proper form, and, by Theorem 5.29, all the major submatrices of this matrix can be obtained. If all the submatrices of $F(\bar{K}_p)$ of the above type are considered, the totality of all these major submatrices is the set of all nonsingular submatrices of order q in $F(\bar{K}_p)$. Note that in each of the above steps, the same process may again be repeated until all the nonsingular submatrices can be obtained by inspection or are of order 1. Because of the iterative nature of the technique, the algorithm can readily be programmed for a digital computer.

In order to simplify the notation, in the following we shall omit all the commas separating the elements of a set together with its curly brackets, and use the semicolons to separate the elements of different sets.

Example 5.10: Find the set of major submatrices of the matrix

$$F = \begin{array}{c} \\ 2 \\ 5 \\ 7 \end{array} \begin{array}{c} 1\ 2\ 3\ 4\ 5\ 6\ 7 \\ \left[\begin{array}{ccccccc} 1 & 1 & 2 & 3 & 0 & 0 & 0 \\ 1 & 0 & 3 & 0 & 8 & 2 & 0 \\ 0 & 0 & 0 & 6 & 0 & 5 & 9 \end{array}\right] \end{array},$$

where $r=3$ and $F(K_3)=F(257)$. Let W_i, $i=1, 2, 3$, be the sets of nonsingular submatrices of orders i in $F(\bar{K}_3) = F(1346)$. Then

$W_1 = \{F(2; 1), F(2; 3), F(2; 4), F(5; 1), F(5; 3), F(5; 6), F(7; 4), F(7; 6)\}$,

$W_2 = \{F(25; 13), F(25; 14), F(25; 16), F(25; 34), F(25; 36), F(25; 46),$
$\quad\quad F(57; 14), F(57; 16), F(57; 34), F(57; 36), F(57; 46), F(27; 14),$
$\quad\quad F(27; 16), F(27; 34), F(27; 36), F(27; 46)\}$,

$W_3 = \{F(257; 346), F(257; 146), F(257; 134), F(257; 136)\}$,

where W_3 is obtained by considering the matrix $F(1346)$ as a given matrix, and then by repeating the whole process again. The sets M_i of major submatrices of F corresponding to the sets W_i are obtained from (5.66), and are given by

$M_1 = \{F(257\ 1/2) = F(157), F(357), F(457), F(127), F(237), F(267),$
$\quad\quad F(245), F(256)\}$,

$M_2 = \{F(257\ 13/25) = F(137), F(147), F(167), F(347), F(367), F(467),$
$\quad\quad F(124), F(126), F(234), F(236), F(246), F(145), F(156), F(345),$
$\quad\quad F(356), F(456)\}$,

$M_3 = \{F(257\ 346/257) = F(346), F(146), F(134), F(136)\}$.

The set M of all possible distinct major submatrices of F is then obtained by taking the set union of the sets M_i, $i=0, 1, 2, 3$, where $M_0 = \{F(257)\}$.

6.2. Trees and cotrees

The application of Theorem 5.29 to the generation of trees and cotrees of a graph without duplications is straightforward. Before considering specific results, let us discuss the general problem. Suppose that F is a matrix over the field of integers mod 2 which is a basis cut matrix (circuit matrix) of a graph G. Assign arbitrary orientations to the edges of G and consider it as a directed graph G_d. Construct the basis cut matrix (circuit matrix) F_d of G_d for the same cuts (circuits) as in F, retaining the column ordering as well. As discussed in chs. 2, § 1 and 5, § 1, since the major submatrices of F and F_d correspond to trees (cotrees) of G, major submatrices of F and F_d correspond. Thus, it does not matter which matrix is used to generate the set of trees. The only thing that we should remember is that if F is used then all the operations are over the field of integers mod 2, and if F_d is used then the operations are over the real field. Thus, we will not specifically state the field over which the operations are to be taken.

COROLLARY 5.10: If Q is a basis cut matrix in proper form of a connected graph G, then the set of trees of G is given by

$$T = \bigcup_{q=0}^{r} T^{(q)}, \qquad (5.67a)$$

where

$$T^{(q)} = \{t; t = e_{x_1} e_{x_2} \ldots e_{x_r} \text{ and } (x_1 x_2 \ldots x_r) = (K_r J_q / I_q)\}; \qquad (5.67b)$$

$Q(I_q, J_q)$ is a nonsingular submatrix of $Q(\bar{K}_r)$; t_0 is the tree corresponding to the columns of $Q(K_r)$; and $T^{(0)} = \{t_0\}$.

COROLLARY 5.11: If B is a basis circuit matrix in proper form of a connected graph G, then the set of cotrees of G is given by

$$\bar{T} = \bigcup_{q=0}^{m} \bar{T}^{(q)}, \qquad (5.68a)$$

where

$$\bar{T}^{(q)} = \{\bar{t}; \bar{t} = e_{x_1} e_{x_2} \ldots e_{x_m} \text{ and } (x_1 x_2 \ldots x_m) = (K_m J_q / I_q)\}; \qquad (5.68b)$$

$B(I_q, J_q)$ is a nonsingular submatrix of $B(\bar{K}_m)$; $\bar{t}_0$ is the cotree corresponding to the columns of $B(K_m)$; and $\bar{T}^{(0)} = \{\bar{t}_0\}$.

One of the main difficulties in computer generation of trees is that it requires a large memory. The matrix technique presented above provides an effective solution to the problem; it amounts to repeated normalization of the submatrices of a matrix. Because of the special nature of the matrices Q and B, their normalization can easily be achieved. One simple way to accomplish this is the following: Suppose that we wish to normalize a matrix F over the field of integers mod 2. First we move some row of F which has a 1 in the first column to the first row, and then remove all other 1's, if they exist, in the first column by simple row additions (mod 2). Let the resulting matrix be F_1. In F_1 move any row, other than the first, which has a 1 in the second column to the second row, and then remove all other 1's, if they exist, in the second column by simple row additions. Continuing this process, we will eventually arrive at a diagonal major submatrix, or possibly it could not be continued. If the latter happens, we simply ignore this column, and proceed to the next one. If the process fails to achieve a diagonal major submatrix, then F has no major submatrix at all. This is of course similarly valid for a matrix F_d with entries 1, -1 and 0 over the real field.

Example 5.11: Suppose that we wish to generate the set of trees of the graph G as shown in fig. 5.16. A basis cutset matrix of G is given by

$$Q = \begin{array}{c} \\ 1 \\ 4 \\ 7 \end{array} \begin{array}{c} 1\ 2\ 3\ 4\ 5\ 6\ 7 \\ \left[\begin{array}{ccccccc} 1 & 1 & 1 & 0 & 1 & 1 & 0 \\ 0 & 1 & 0 & 1 & 1 & 1 & 0 \\ 0 & 0 & 1 & 0 & 1 & 1 & 1 \end{array}\right] \end{array},$$

where $r=3$ and $Q(K_3)=Q(147)$. The sets W_i, $i=1, 2, 3$, of nonsingular submatrices of orders i in $Q(2356)$ are given by

$W_1 = \{Q(1;2), Q(1;3), Q(1;5), Q(1;6), Q(4;2), Q(4;5),$
$\qquad Q(4;6), Q(7;3), Q(7;5), Q(7;6)\}$,
$W_2 = \{Q(14;23), Q(14;35), Q(14;36), Q(47;23), Q(47;25), Q(47;26),$
$\qquad Q(47;35), Q(47;36), Q(17;25), Q(17;26), Q(17;23)\}$,
$W_3 = \{Q(147;235), Q(147;236)\}$.

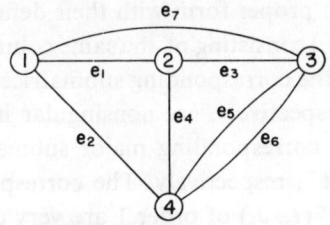

Fig. 5.16. A graph illustrating the generation of trees by a matrix technique.

The sets $T^{(i)}$ of trees corresponding to the sets W_i are obtained from (5.67b), and are given by

$$T^{(1)} = \{e_2e_4e_7, e_3e_4e_7, e_4e_5e_7, e_4e_6e_7, e_1e_2e_7, e_1e_5e_7, e_1e_6e_7,$$
$$e_1e_3e_4, e_1e_4e_5, e_1e_4e_6\},$$
$$T^{(2)} = \{e_2e_3e_7, e_3e_5e_7, e_3e_6e_7, e_1e_2e_3, e_1e_2e_5, e_1e_2e_6, e_1e_3e_5, e_1e_3e_6,$$
$$e_2e_4e_5, e_2e_4e_6, e_2e_3e_4\},$$
$$T^{(3)} = \{e_2e_3e_5, e_2e_3e_6\}.$$

The set of trees is then obtained by taking the set union of the sets $T^{(i)}$, $i=0$, 1, 2, 3, where $T^{(0)} = \{e_1e_4e_7\}$.

6.3. *Directed trees and directed 2-trees*

The application of the present technique to the generation of directed trees and 2-trees of a directed graph G_d will now be discussed. If A is a basis incidence matrix of G_d, by A^+ we mean the matrix obtained from A by replacing entries -1's by 0's. Since in forming A we have deleted a row from the complete incidence matrix of G_d, without loss of generality let this deleted row be n, and our problem is to generate all the directed trees t_n of G_d.

LEMMA 5.6: $A^+(I_r)$ is a major submatrix of A^+ if, and only if, the edges corresponding to the columns of $A^+(I_r)$ constitute a subgraph, each of its nodes being of outgoing degree 1 except node n which is of outgoing degree 0.

Since the trees of G_d are in one-to-one correspondence with the major submatrices of A, it follows from Lemma 5.6 that we have

THEOREM 5.30: $A(I_r)$ and $A^+(I_r)$ are both nonsingular if, and only if, the edges corresponding to the columns of $A(I_r)$ constitute a directed tree t_n in G_d.

Thus, our problem reduces to that of enumerating all the corresponding major submatrices of A and A^+ efficiently. The technique discussed earlier can easily be modified for our purpose. For convenience, we assume that both A and A^+ have been put in proper form with their defining diagonal major submatrices $A(K_r)$ and $A^+(K_r)$ consisting of the same columns (by interchanging the rows if necessary). Then the corresponding submatrices $A(I_q, J_q)$ and $A^+(I_q, J_q)$ in $A(\bar{K}_r)$ and $A^+(\bar{K}_r)$, respectively, are nonsingular if, and only if, $A(K_rJ_q/I_q)$ and $A^+(K_rJ_q/I_q)$ are the corresponding major submatrices, other than $A(K_r)$ and $A^+(K_r)$, in A and A^+, respectively. The corresponding nonsingular submatrices $A(I_q, J_q)$ and $A^+(I_q, J_q)$ of order 1 are very easy to find; they are the corresponding nonzero elements in $A(\bar{K}_r)$ and $A^+(\bar{K}_r)$, respectively. For the corresponding nonsingular submatrices of order greater than one, the whole process may be repeated. Thus, it again amounts to repeatedly putting the corresponding matrices in proper form.

Alternatively, for a given directed graph G_d we can consider it as the associated directed graph of an equicofactor matrix Y of the type as shown in (4.21). The edge-designation symbols of G_d are considered as weights associated with the edges. From our discussion in ch. 2, § 3.3, it is not difficult to see that the cofactor of the (n, n)-element of Y is, in fact, the node-admittance matrix Y_n with node n chosen as the reference point. For a given network, the node-admittance matrix Y_n can easily be computed from the relation $Y_n = AY_bA'$, where A is the basis incidence matrix of the network with reference node n, and Y_b is the branch-admittance matrix. However, if Y_n is given, the branch-admittance matrix Y_b is usually hard to find, since it is in general not symmetric. In the following, we shall show how Y_n can be decomposed in another convenient form.

THEOREM 5.31: Let Y be the equicofactor matrix of the type of (4.21) associated with a directed graph G_d. Then the cofactor Y_{nn} of the (n, n)-element of Y can be decomposed as
$$Y_{nn} = \det A^+DA', \qquad (5.69)$$
where A is the basis incidence matrix of G_d with reference node n, and D is a $b \times b$ diagonal matrix whose iith entry is the edge-designation symbol of the edge corresponding to the ith column of A.

The proof of the theorem is straightforward, and is left as an exercise (Problem 5.38). For a related treatment, see CHEN [1972d].

LEMMA 5.7: All the nonsingular submatrices of A^+ have their determinant 1 or -1.

Now, if we apply the Binet-Cauchy theorem to the matrix triple product A^+DA' in (5.69) and make use of the fact that D is diagonal, we conclude that the positive terms in the final expansion of Y_{nn} are in one-to-one correspondence with the corresponding major submatrices of A^+ and A. From Theorem 4.3, we also know that these positive terms correspond to directed trees of G_d. Thus, the technique discussed earlier is also applicable to the generation of directed trees.

For convenience, let A_{-i} be the matrix obtained from A by deleting the row i. Then from Theorem 5.31 we have

COROLLARY 5.12: The second-order cofactor $Y_{ij,\,nn}$ of the elements of Y can be expressed as

$$Y_{ij,nn} = (-1)^{i+j} \det A_{-i}^{+} D (A_{-j})'. \tag{5.70}$$

Thus, from Theorem 4.4 in conjunction with (5.70), we conclude that directed 2-trees $t_{ij,n}$ of G_d are in one-to-one correspondence with the corresponding major submatrices of A_{-i}^{+} and A_{-j}. In other words, we can use the above procedure for the generation of directed 2-trees.

Since 2-trees $t_{ij,n}$ of G_d are in one-to-one correspondence with the corresponding major submatrices of A_{-i} and A_{-j}, the above technique can also be used for their enumeration.

Example 5.12: In fig. 5.17, suppose that we wish to generate the set of directed trees t_4 of the directed graph G_d. The basis incidence matrix A with reference node 4 is given by

$$A = \begin{array}{c} \\ 2 \\ 4 \\ 5 \end{array} \begin{array}{c} 1 \quad 2 \quad\; 3 \quad 4 \quad 5 \quad 6 \quad\;\; 7 \\ \begin{bmatrix} -1 & 1 & 0 & 0 & 0 & 0 & 1 \\ 1 & 0 & -1 & 1 & 0 & 0 & 0 \\ 0 & 0 & 1 & 0 & 1 & 1 & -1 \end{bmatrix} \end{array},$$

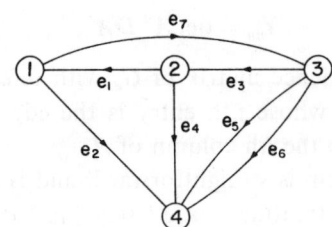

Fig. 5.17. A directed graph illustrating the generation of directed trees and 2-trees by a matrix technique.

which is already in normal form with $A(K_3) = A(245)$. Then

$$A^{+} = \begin{array}{c} 2 \\ 4 \\ 5 \end{array} \begin{bmatrix} 0 & 1 & 0 & 0 & 0 & 0 & 1 \\ 1 & 0 & 0 & 1 & 0 & 0 & 0 \\ 0 & 0 & 1 & 0 & 1 & 1 & 0 \end{bmatrix},$$

where $A^{+}(K_3) = A^{+}(245)$. The sets W_i, $i = 1, 2, 3$, of corresponding nonsingular submatrices of orders i of $A(1367)$ and $A^{+}(1367)$ are given by

$$W_1 = \{A(2;7), A(4;1), A(5;3), A(5;6)\},$$
$$W_2 = \{A(24;17), A(25;37), A(25;67), A(45;13), A(45;16)\},$$
$$W_3 = \{A(245;167)\}.$$

The sets M_i of corresponding major submatrices of A and A^+ corresponding to the sets W_i are obtained as follows:

$$M_1 = \{A(245\ 7/2) = A(457), A(125), A(234), A(246)\},$$
$$M_2 = \{A(245\ 17/24) = A(157), A(347), A(467), A(123), A(126)\},$$
$$M_3 = \{A(167)\}.$$

Thus, the set of directed trees t_4 of G_d can easily be obtained from M_i, $i=0$, 1, 2, 3, where $M_0 = \{A(245)\}$, and is given by

$$T_4 = \{e_4 e_5 e_7, e_1 e_2 e_5, e_2 e_3 e_4, e_2 e_4 e_6, e_1 e_5 e_7, e_3 e_4 e_7, e_4 e_6 e_7,$$
$$e_1 e_2 e_3, e_1 e_2 e_6, e_1 e_6 e_7, e_2 e_4 e_5\}.$$

Similarly, if directed 2-trees $t_{13,4}$ of G_d are needed, we use the matrices

$$A^+_{-1} = \begin{matrix} \\ 1 \\ 3 \end{matrix} \begin{matrix} 1 & 2 & 3 & 4 & 5 & 6 & 7 \\ \left[\begin{matrix} 1 & 0 & 0 & 1 & 0 & 0 & 0 \\ 0 & 0 & 1 & 0 & 1 & 1 & 0 \end{matrix}\right] & & & & & & \end{matrix}$$

and

$$A_{-3} = \begin{bmatrix} -1 & 1 & 0 & 0 & 0 & 0 & 1 \\ 1 & 0 & -1 & 1 & 0 & 0 & 0 \end{bmatrix}.$$

Since A_{-3} is not in proper form with respect to the columns 1 and 3, let Z be the matrix A_{-3} in proper form with $Z(K_2) = Z(13)$:

$$Z = \begin{matrix} 1 \\ 3 \end{matrix} \begin{bmatrix} -1 & 1 & 0 & 0 & 0 & 0 & 1 \\ 0 & 1 & -1 & 1 & 0 & 0 & 1 \end{bmatrix}.$$

Since there exist no corresponding nonsingular submatrices in $A^+_{-1}(24567)$ and $Z(24567)$, there is only one directed 2-tree $t_{13,4} = e_1 e_3$ in G_d.

Finally, we mention that a computer routine based on this technique was written for the generation of trees, 2-trees, directed trees, and directed 2-trees. It was written in FORTRAN IV for the IBM System 360 Model 44 digital computer, and the details of the program may be found in CHEN and LI [1973].

§ 7. Elementary transformations

In this section, we shall show that trees of a graph can be generated by a series of elementary tree transformations. The idea was originally suggested by WHITNEY [1935], and has since been exploited by many others (see, for example, WATANABE [1960] and MAYEDA and SESHU [1965]).

DEFINITION 5.9: *Distance.* Let g_1 and g_2 be two subgraphs containing the same number of edges of a graph. The *distance* between g_1 and g_2 is the number of edges in one but not in the other.

In particular, g_1 and g_2 could be trees, cotrees, 2-trees, or 2-cotrees. For example, in fig. 5.14 let us consider the trees $t' = bcghim$, $t'' = bfghij$, $t^* = abcekm$. Then, t' is of distance 2 from t'', and t'' is of distance 5 from t^*. Using this definition, it is obvious that the sets $T^{(q)}$ and $\bar{T}^{(q)}$ of (5.67b) and (5.68b) are the sets of trees and cotrees of distance q from the reference tree t_0 and cotree $\bar{t}_0$, respectively.

DEFINITION 5.10: *Elementary tree transformation.* Let t be a tree of a graph G and also let e_1 be an edge in G but not in t. Then the operation

$$e_1 \cup t - e_2 = t^* \tag{5.71}$$

is called an *elementary tree transformation* if t^* is a tree of G and e_2 is an edge of t.

The following result, which is originally due to WHITNEY [1935], is stated as a theorem. Although his proof is considerably more elegant and general, the one given here serves our purpose best.

THEOREM 5.32: Let t be a tree of a connected graph G. Then every tree of G can be obtained from t by a finite sequence of elementary tree transformations.

Proof. Let t' be a tree of G. If $t' \neq t$, there exists a branch e_1 in t but not in t'. Since not all the edges of the unique circuit of $t' \cup e_1$ can be in t, there is a branch e_2 in t' but not in t. Consider the tree $t^* = t' \cup e_1 - e_2$. Clearly, t^* is closer to t than t'. Thus, if $t^* \neq t$, the process may be repeated. Since t is finite, we will eventually arrive at t by a finite sequence of elementary tree transformations. So the theorem is proved.

Obviously, we can obtain similar results for cotrees, 2-trees, and directed trees. The details are left as exercises (Problems 5.40, 5.41 and 4.31).

Let $t_0 = e_1 e_2 \ldots e_r$ be a reference tree of G, and let

$$E_q = e_{i_1} e_{i_2} \ldots e_{i_q} \tag{5.72}$$

be a subgraph of t_0. By $\bar{E}_q$ we mean the complement of E_q in t_0. The symbol $t(E_q)$ denotes a tree of G such that

$$t(E_q) \cap t_0 = E_q, \tag{5.73}$$

where $q = 0, 1, \ldots, r$ and $E_0 = \emptyset$. We denote by $T(E_q)$ the set of trees $t(E_q)$ of G.

Evidently, $t(E_q)$ is a tree of distance $r-q$ from t_0, and $T(E_q)$ is a set of trees of distance $r-q$ from t_0 in G. Since

$$T(E_u) \cap T(E'_v) = S_0 \tag{5.74}$$

if, and only if, $E_u \neq E'_v$ where E_u and E'_v are two subgraphs of t_0, the set T of trees of G can be expressed as

$$T = \bigcup_{q=0}^{r} \bigcup_{(I_q)} T(E_q), \tag{5.75}$$

where $i_1 < i_2 < \cdots < i_q$, and the second union is taken over all subsets $I_q = \{i_1, i_2, \ldots, i_q\}$ of $\{1, 2, \ldots, r\}$.

Using the elementary tree transformations defined in (5.71), it is not difficult to see that the set $T(E_k)$ of trees of distance $r-k$ from t_0 can be obtained from the set $T(E_{k+1})$ of trees of distance $r-k-1$ from t_0 for $k=0, 1, \ldots, r-1$. However, the principal disadvantage of this procedure is that duplications may occur, although they arise within a class $T(E_k)$. We shall next discuss how to avoid duplications. The technique was first proposed by MAYEDA and SESHU [1965], and in the following we shall present a simple proof and interpretation of their result based on the work of CHEN [1969d].

For a given graph G, let $Q_t(e)$ be the f-cutset defined by the branch e of the tree t in G, with respect to t, and let Q_{i_1} be the incidence cut of the node i_1 in G. As mentioned in ch. 4, § 5.3 and also in (5.49), if an edge e of G is distinguished, the trees can be classified into those $T(e)$ which contain the edge e and those $T(\bar{e})$ which do not. In ch. 4, § 5.3, we use the symbol $G(g_1; g_2)$ to denote the graph obtained from G by first shorting all the edges in g_1, and then by removing all the edges in g_2, where g_1 and g_2 are edge-disjoint subgraphs of G. Then the elements of $T(e)$ are in one-to-one correspondence with the trees in $G(e; \emptyset)$, and $T(\bar{e})$ is the set of trees in $G(\emptyset; e)$. If we repeat this process for $G(e; \emptyset)$ and $G(\emptyset; e)$, we will eventually generate all of the trees of G without duplications. In the following, we shall first show that $T(\bar{e})$ can actually be obtained from $T(e)$ without duplications, and then use this result to derive the Mayeda–Seshu's algorithm.

THEOREM 5.33: If e is an edge incident at the node i_1 of G, then

$$T(\bar{e}) = \{t; t = (t' \oplus e) \cup e_x, t' \text{ is in } T(e), e_x \text{ is in } Q_{t'}(e) \cap Q_{i_1},$$
$$\text{and} \quad e_x \neq e\}. \tag{5.76}$$

Proof. The proof consists of two parts: First, that we can generate every tree of $T(\bar{e})$ by this process, and second, that every term generated by this process is a tree in $T(\bar{e})$ and no duplications will arise.

Suppose that there exists a tree t in $T(\bar{e})$ which cannot be generated by this process. Consider the subgraph $t \cup e$ which must contain a unique circuit L. Then the operation $(t \cup e) \oplus e' = t'$ will produce a tree that is an element of $T(e)$ where $e' \neq e$ is an edge of L which is incident with the node i_1. Since e' is also in $Q_{t'}(e) \cap Q_{i_1}$, this means that t can be obtained from t' by the above process, a contradiction. Thus, every tree of $T(\bar{e})$ can be generated by this process.

It is obvious that every term generated by the above process is a tree in $T(\bar{e})$. However, we have to show that no duplications will arise. Suppose that t is a tree in $T(\bar{e})$ which appears more than once by the above process, then there exist at least two different edges e_1 and e_2 in t with node i_1 as one of their endpoints such that $(t \cup e) \oplus e_1$ and $(t \cup e) \oplus e_2$ are different elements of $T(e)$. This would imply that both e_1 and e_2 are contained in the unique circuit of $t \cup e$, which is obviously impossible. Thus the theorem is proved.

Since a tree $t_0 = e_1 e_2 \ldots e_r$ is a connected subgraph without circuits, it follows that the branches of t_0 can always be labeled in such a way that every subgraph $e_1 e_2 \ldots e_x$, $x = 1, 2, \ldots, r$, of t_0 is connected (Problem 5.43).

THEOREM 5.34: If the branch labeling of a reference tree $t_0 = e_1 e_2 \ldots e_r$ is such that every subgraph $e_1 e_2 \ldots e_x$, $x = 1, 2, \ldots, r$, is connected, and if $E_q = e_{i_1} e_{i_2} \ldots e_{i_q}$, $i_1 < i_2 < \cdots < i_q$, is a subgraph of t_0, then

$$T(\bar{E}_q) = \{t; t = (t' \oplus e_{i_q}) \cup e, \ t' \text{ is in } T(\bar{E}_{q-1}),$$
$$e \text{ is in } Q_{t'}(e_{i_q}) \cap Q_{t_0}(e_{i_q}), \text{ and } e \neq e_{i_q}\} \quad (5.77)$$

and

$$T(\bar{E}_q) \cap T(\bar{E}'_p) = S_0 \quad (5.78)$$

if, and only if, E_q and E'_p are two different subgraphs of t_0, where $p, q = 1, 2, \ldots, r$.

Proof. Without loss of generality, we assume that each edge of G is a circuit edge. For, if not, we can consider the graph G^* obtained from G by shorting all its noncircuit edges. Since noncircuit edges must belong to every tree of G, it is clear that if (5.77) is valid for G^* then it is also valid for G. This assumption is necessary to ensure the connectedness of $G(\bar{E}_q; E_{q-1})$ since we have to talk about its f-cutsets. (Otherwise, we must use the concept of a forest and then define its f-cutsets.) For convenience, let the trees of $G(\bar{E}_q; E_{q-1})$ be classified into those $T''(e_\alpha)$, $\alpha = i_q$, which contain the edge e_α and those $T''(\bar{e}_\alpha)$ which do not. Let C_{i_1} be the incidence cut of the node i_1 in $G(\bar{E}_q; E_{q-1})$, and let $C_{t''}(e)$

be the f-cutset defined by the branch e of t'' and with respect to the tree t'' in $G(\bar{E}_q; E_{q-1})$. Since from Theorem 4.13 there exists a one-to-one correspondence between trees t' in $T(\bar{E}_{q-1})$ and those t'' in $T''(e_\alpha)$, it follows that $t' = g \cup \bar{E}_q$, where g is the corresponding subgraph of t'' in G. In other words, t'' can be obtained from t' by shorting the edges in $\bar{E}_q$. Since $e_1 e_2 \ldots e_\alpha$ is a connected subgraph of t_0, and since E_q is contained in $e_1 e_2 \ldots e_\alpha$, one of the two components of $t' - e_\alpha$ is either an isolated node or a subgraph of $\bar{E}_q$. Let i_1 be the node in this component which is also one of the two endpoints of e_α. But by shorting the edges in $\bar{E}_q$ or by removing the edges in E_{q-1} the f-cutset $Q_{t_0}(e_\alpha)$ remains unaltered. Thus, we have $C_{i_1} = Q_{t_0}(e_\alpha)$. Similarly, the f-cutsets $C_{t''}(e_\alpha)$ and $Q_{t'}(e_\alpha)$ differ only by those edges contained in E_q. We conclude that

$$C_{i_1} \cap C_{t''}(e_\alpha) = Q_{t_0}(e_\alpha) \cap Q_{t'}(e_\alpha). \tag{5.79}$$

From Theorem 5.33 we have

$$T''(\bar{e}_\alpha) = \{t^*; t^* = (t'' \oplus e_\alpha) \cup e_x, t'' \text{ is in } T''(e_\alpha),$$
$$e_x \text{ is in } C_{t''}(e_\alpha) \cap C_{i_1}, \text{ and } e_x \neq e_\alpha\}. \tag{5.80}$$

Since there also exists a one-to-one correspondence between the elements in $T(\bar{E}_q)$ and those in $T''(\bar{e}_\alpha)$ such that the corresponding elements differ only by those in $\bar{E}_q$, (5.77) follows directly from (5.80) after a simple substitution.

The proof of (5.78) is straightforward, and is omitted.

This is a very useful result in enumerating trees of a graph by a digital computer, since it provides the theoretical foundation for an iterative tree-generation procedure which will not introduce any duplications. The procedure requires that one generate a reference tree, which is easy to do. In fact, if the starting tree t_0 is generated by a computer, it is naturally generated with the property that every subgraph $e_1 e_2 \ldots e_x$ is connected. The procedure also requires that one compute the f-cutsets for a given tree, which is again easy to do. The f-cutsets for a given tree are simply obtained from any basis cutset matrix by normalizing the matrix with respect to the columns corresponding to the branches of the tree. The algorithm seems to imply that one must examine 2^r replacement sets. However, if any replacement yields an empty set, the following replacements are not examined. Thus, the number of sets to be examined is not 2^r in general but it is not the absolute minimum required. When the network of a graph is active, an additional advantage of generating trees by elementary tree transformations is the ease of determining the sign of each tree relative to the previous tree. A computer routine based on the above algorithm was written as a portion of a larger linear-network fault-diagnosis program by SESHU and WAXMAN [1966]. It was originally written in FAP for IBM System 7094 computer, and a second version, now in use, was written in FORTRAN IV for the IBM System 7094 computer. The procedure has also been extended to the generation of directed trees by PAUL [1967] (Problem 5.65).

Example 5.13: In fig. 5.16, the set of trees containing the edge e_1 is given by
$$T(e_1) = \{e_1e_2e_7, e_1e_5e_7, e_1e_6e_7, e_1e_3e_4, e_1e_4e_5, e_1e_4e_6, e_1e_2e_3,$$
$$e_1e_2e_5, e_1e_2e_6, e_1e_3e_5, e_1e_3e_6, e_1e_4e_7\}.$$

Suppose that we wish to generate the set $T(\bar{e}_1)$ from the set $T(e_1)$ by the procedure outlined in Theorem 5.33. Let $i_1 = 1$. Then $Q_1 = e_1e_2e_7$. If $W(t')$ denotes the subset of $T(\bar{e}_1)$ that is generated by t' in $T(e_1)$, then from (5.76) we have

$$W(e_1e_2e_7) = \{t; t = (e_1e_2e_7 \oplus e_1) \cup e_x, e_x \text{ is in } e_1e_3e_4 \cap e_1e_2e_7 = e_1,$$
$$\text{and} \quad e_x \neq e_1\}$$
$$= S_0,$$
$$W(e_1e_5e_7) = W(e_1e_6e_7) = W(e_1e_2e_5) = W(e_1e_2e_6) = S_0,$$
$$W(e_1e_3e_4) = \{t; t = (e_1e_3e_4 \oplus e_1) \cup e_x, e_x \text{ is in } e_1e_2e_7 \cap e_1e_2e_7 = e_1e_2e_7,$$
$$\text{and} \quad e_x \neq e_1\}$$
$$= \{e_2e_3e_4, e_3e_4e_7\},$$
$$W(e_1e_4e_5) = \{t; t = e_4e_5 \cup e_x, e_x \text{ is in } e_2e_7\}$$
$$= \{e_2e_4e_5, e_4e_5e_7\},$$
$$W(e_1e_4e_6) = \{t; t = e_4e_6 \cup e_x, e_x \text{ is in } e_2e_7\}$$
$$= \{e_2e_4e_6, e_4e_6e_7\},$$
$$W(e_1e_2e_3) = \{t; t = e_2e_3 \cup e_x, e_x \text{ is in } e_7\}$$
$$= \{e_2e_3e_7\},$$
$$W(e_1e_3e_5) = \{t; t = e_3e_5 \cup e_x, e_x \text{ is in } e_2e_7\}$$
$$= \{e_2e_3e_5, e_3e_5e_7\},$$
$$W(e_1e_3e_6) = \{t; t = e_3e_6 \cup e_x, e_x \text{ is in } e_2e_7\}$$
$$= \{e_2e_3e_6, e_3e_6e_7\},$$
$$W(e_1e_4e_7) = \{t; t = e_4e_7 \cup e_x, e_x \text{ is in } e_2\}$$
$$= \{e_2e_4e_7\}.$$

The set $T(\bar{e}_1)$ is then obtained by taking the set union of the sets $W(t')$, and is given by

$$T(\bar{e}_1) = \{e_2e_3e_4, e_3e_4e_7, e_2e_4e_5, e_4e_5e_7, e_2e_4e_6, e_4e_6e_7, e_2e_3e_7,$$
$$e_2e_3e_5, e_3e_5e_7, e_2e_3e_6, e_3e_6e_7, e_2e_4e_7\}.$$

Example 5.14: Consider the graph G as shown in fig. 5.16. The set of trees of G will now be generated by the procedure outlined in Theorem 5.34. Let $t_0 = e_1 e_4 e_7$ which has the property that e_1 and $e_1 e_4$ are connected subgraphs of t_0. Then we have

$$T(e_4 e_7) = \{t; t = (t_0 \oplus e_1) \cup e, e \text{ is in } Q_{t_0}(e_1) = e_1 e_2 e_3 e_5 e_6, e \neq e_1\}$$
$$= \{e_2 e_4 e_7, e_3 e_4 e_7, e_4 e_5 e_7, e_4 e_6 e_7\},$$
$$T(e_1 e_7) = \{t; t = e_1 e_7 \cup e, e \text{ is in } Q_{t_0}(e_4) = e_2 e_4 e_5 e_6, e \neq e_4\}$$
$$= \{e_1 e_2 e_7, e_1 e_5 e_7, e_1 e_6 e_7\},$$
$$T(e_1 e_4) = \{t; t = e_1 e_4 \cup e, e \text{ is in } Q_{t_0}(e_7) = e_3 e_5 e_6 e_7, e \neq e_7\}$$
$$= \{e_1 e_3 e_4, e_1 e_4 e_5, e_1 e_4 e_6\},$$
$$T(e_7) = \{t; t = (t' \oplus e_4) \cup e, t' \text{ is in } T(e_4 e_7), e \text{ in }$$
$$Q_{t'}(e_4) \cap e_2 e_4 e_5 e_6, \text{ and } e \neq e_4\}$$
$$= \{e_2 e_3 e_7, e_3 e_5 e_7, e_3 e_6 e_7\},$$
$$T(e_4) = \{t; t = (t' \oplus e_7) \cup e, t' \text{ is in } T(e_4 e_7), e \text{ in }$$
$$Q_{t'}(e_7) \cap e_3 e_5 e_6 e_7, e \neq e_7\}$$
$$= \{e_2 e_3 e_4, e_2 e_4 e_5, e_2 e_4 e_6\},$$
$$T(e_1) = \{t; t = (t' \oplus e_7) \cup e, t' \text{ is in } T(e_1 e_7), e \text{ in }$$
$$Q_{t'}(e_7) \cap e_3 e_5 e_6 e_7, e \neq e_7\}$$
$$= \{e_1 e_2 e_3, e_1 e_2 e_5, e_1 e_2 e_6, e_1 e_3 e_5, e_1 e_3 e_6\},$$
$$T(\emptyset) = \{t; t = (t' \oplus e_7) \cup e, t' \text{ is in } T(e_7), e \text{ in }$$
$$Q_{t'}(e_7) \cap e_3 e_5 e_6 e_7, e \neq e_7\}$$
$$= \{e_2 e_3 e_5, e_2 e_3 e_6\}.$$

Thus, the set of trees of G is obtained by taking the set union of the above sets and $T(t_0) = \{e_1 e_4 e_7\}$.

§ 8. Hamilton circuits in directed-tree graphs

In the preceding section, we have shown that any tree of a graph G can be obtained from any other one by a finite sequence of elementary tree transformations. In other words, the trees of G containing at least two trees can be ordered in such a way that the successive elements in the ordering are related by an elementary tree transformation if each tree is allowed to appear any finite number of times in the ordering. In this section, we shall prove a stronger requirement in showing that the condition that a tree be allowed to appear a finite number of times in an ordering may be removed, and an ordering with the above property exists. This result was first pointed out by CUMMINS [1966], and has been extended to acyclic directed graphs by CHEN [1967c].

DEFINITION 5.11: *Elementary directed-tree transformation.* Let t_n be a directed tree with reference node n in a directed graph G_d, and also let $e_{ij}=(i,j)$ be an edge in G_d but not in t_n. Then for $i \neq n$ the operation

$$e_{ij} \cup t_n - e_{ik} = t_n^* \tag{5.81}$$

is called an *elementary directed-tree transformation* if t_n^* is a directed tree with reference node n in G_d, where e_{ik} is an edge of t_n.

DEFINITION 5.12: *Directed-tree graph.* The *directed-tree graph* of a directed graph G_d, denoted by the symbol $T_n(G_d)$, is an (undirected) graph in which each node corresponds to a directed tree t_n of G_d, and each edge corresponds to an elementary directed-tree transformation between directed trees such that there is an edge between the nodes i and j if, and only if, the corresponding directed trees of these nodes are related by an elementary directed-tree transformation. In particular, if G_d is symmetric, the directed-tree graph is simply referred to as the *tree graph*, and is denoted by $T(G_d)$.

As an example, consider the directed graph G_d as shown in fig. 5.18. The set of directed trees t_3 is given by

$$T_3 = \{efc, bef, aec, dec, abe, bde\}. \tag{5.82}$$

The directed-tree graph $T_3(G_d)$ is presented in fig. 5.19.

DEFINITION 5.13: *Hamilton circuit.* A *Hamilton circuit* of a graph G (or a directed graph) is a circuit which contains all the nodes of G.

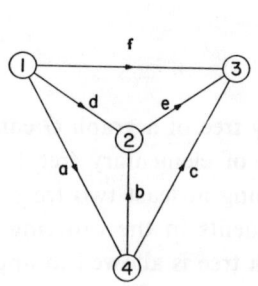

Fig. 5.18. A directed graph G_d.

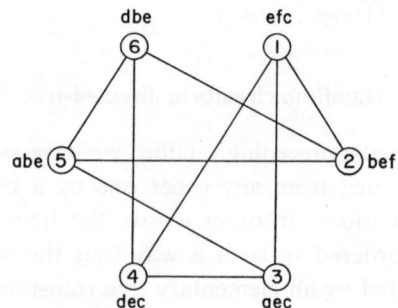

Fig. 5.19. The directed-tree graph $T_3(G_d)$.

For example, in fig. 5.19 the following are two Hamilton circuits H_1 and H_2 of $T_3(G_d)$:

$$H_1 = e_{21}e_{14}e_{43}e_{35}e_{56}e_{62}, \tag{5.83a}$$

$$H_2 = e_{21}e_{13}e_{34}e_{46}e_{65}e_{52}. \tag{5.83b}$$

Now, we shall proceed to show that there exists a Hamilton circuit in a class of directed-tree graphs. For convenience, we denote by t_n^i the corresponding directed tree of the node i in $T_n(G_d)$. The following lemma is obvious.

LEMMA 5.8: In a directed graph, any two of its directed trees with the same reference node can be obtained from each other by a finite sequence of elementary directed-tree transformations.

THEOREM 5.35: For an acyclic or symmetric directed graph G_d containing at least three directed trees t_n, any edge of its directed-tree graph $T_n(G_d)$ can be made part of a Hamilton circuit.

Proof. We shall only prove the case where G_d is acyclic by induction over the number of edges of G_d; the other case can be proved in a similar manner.

Suppose that G_d is acyclic. Clearly, the theorem is trivially satisfied if G_d has three edges. Assume that the assertion is true for any G_d which has $k-1$, $k \geq 4$, edges or less. We must show that it is also true if G_d has k edges.

Let G_1 be the directed graph obtained from G_d by removing the edge e_{kn}, and G_2 be the directed graph obtained from G_1 by identifying the nodes k and n. Let the combined node in G_2 be denoted by n. It is evident that $T_n(G_1)$ and $T_n(G_2)$ are two node-disjoint sectional subgraphs of $T_n(G_d)$ which together include all the nodes of $T_n(G_d)$. Note that from the graph-theoretic viewpoint, $T_n(G_2)$ is a sectional subgraph of $T_n(G_d)$, even though the nodes of $T_n(G_2)$ do not represent the directed trees t_n of G_d. As a matter of fact, if t_{2n}^m is the directed tree in G_2 corresponding to the node m in $T_n(G_2)$, then $t_n^m = e_{kn} \cup t_{2n}^m$ is the directed tree of G_d corresponding to the node m of $T_n(G_d)$. Now we shall complete the proof of the theorem by the following lemmas. In order to avoid possible confusion and to simplify our notation, we denoted by E_{ij} an undirected edge connected between the nodes i and j in a directed-tree graph.

LEMMA 5.9: If $T_n(G_1)$ and $T_n(G_2)$ contain at least two nodes each, then for every edge E_{ij}, i in $T_n(G_1)$ and j in $T_n(G_2)$, in $T_n(G_d)$, there exist two nodes u and v in $T_n(G_d)$ where u is in $T_n(G_1)$ and v in $T_n(G_2)$ such that E_{jv}, E_{vu}, and E_{ui} are all in $T_n(G_d)$ with distinct $i, j, u,$ and v.

Proof. Since $T_n(G_2)$ contains at least two nodes, from Lemma 5.8 there exists a t_{2n}^x in G_2 which can be obtained from t_{2n}^j by an elementary directed-tree transformation. Let e_{qs} be an edge in t_{2n}^x but not in t_{2n}^j and e_{qv} be the edge in t_{2n}^j but not in t_{2n}^x. Obviously, $q \neq k$, and $t_n^x = e_{kn} \cup t_{2n}^x$ and $t_n^j = e_{kn} \cup t_{2n}^j$ in G_d are related by an elementary directed-tree transformation.

Since E_{ij} is an edge of $T_n(G_d)$, it follows that t_n^i and t_n^j are related by an elementary directed-tree transformation such that e_{kn} is an edge in t_n^j but not in t_n^i and e_{kw} is the edge in t_n^i but not in t_n^j. Then the subgraph t_n^u obtained by the following operation

$$t_n^u = t_n^j \cup e_{qs} e_{kw} - e_{qv} e_{kn}$$

is a directed tree of G_d such that t_n^u and t_n^i, and t_n^u and t_n^x, respectively, are related by an elementary directed-tree transformation. Consequently, E_{jx}, E_{xu}, E_{ui}, and E_{ij} are edges of $T_n(G_d)$. This completes the proof of the lemma.

LEMMA 5.10: If $T_n(G_1)$ and $T_n(G_2)$ contain at least three nodes each, then $T_n(G_d)$ has the property stated in the theorem.

Proof. Since G_1 and G_2 have fewer edges than G_d, by induction hypothesis any edge of $T_n(G_y)$, $y = 1, 2$, can be made part of a Hamilton circuit of $T_n(G_y)$. Three different cases are considered:

Case 1: Let E_{ij}, i in $T_n(G_1)$ and j in $T_n(G_2)$, be an edge of $T_n(G_d)$. By Lemma 5.9, there exist two nodes u and v in $T_n(G_d)$ where u is in $T_n(G_1)$ and v in $T_n(G_2)$ such that E_{jv}, E_{vu}, and E_{ui} are in $T_n(G_d)$. If H_1 and H_2 are Hamilton circuits containing the edges E_{ui} and E_{jv} in $T_n(G_1)$ and $T_n(G_2)$, respectively, then the subgraph of $T_n(G_d)$ formed by the subgraphs $H_1 - E_{ui}$, E_{ij}, $H_2 - E_{jv}$, and E_{vu} is a Hamilton circuit of $T_n(G_d)$ containing the edge E_{ij}.

Case 2: Let E_{jv} be an edge of $T_n(G_2)$. Consider $t_n^j = e_{kn} \cup t_{2n}^j$ and $t_n^v = e_{kn} \cup t_{2n}^v$. Since $T_n(G_1)$ contains more than one node, by Lemma 5.8 there exists a directed tree t_n^i in $T_n(G_1)$ such that t_n^i and t_n^j are related by an elementary directed-tree transformation. Thus, there is an edge e_{kw} in t_n^i other than e_{kn} such that

$$t_n^i = t_n^j \cup e_{kw} - e_{kn}.$$

Similarly, if e_{qs} is an edge in t_n^v but not in t_n^j, and if e_{qz} is the edge in t_n^j but not in t_n^v, then

$$t_n^u = t_n^v \cup e_{kw} - e_{kn}$$

is a directed tree in G_1. Furthermore, the directed trees t_n^j and t_n^v, t_n^v and t_n^u, t_n^j and t_n^i, and t_n^i and t_n^u, respectively, are related by an elementary directed-tree transformation. By an argument similar to that of Case 1, we obtain a desired Hamilton circuit.

Case 3: Let E_{iu} be an edge of $T_n(G_1)$. The proof of this case is identical to that of Case 2, and therefore is omitted.

LEMMA 5.11: If $T_n(G_1)$ contains exactly two nodes, and if $T_n(G_2)$ contains at least three nodes, or if $T_n(G_2)$ contains two nodes and $T_n(G_1)$ at least three nodes, then $T_n(G_d)$ has the property stated in the theorem.

The proof of this lemma is similar to that of Lemma 5.10, and is omitted.

LEMMA 5.12: If $T_n(G_1)$ and $T_n(G_2)$ contain exactly two nodes each, then $T_n(G_d)$ has the property stated in the theorem.

Proof. Since there exists at least one edge E_{ij}, i in $T_n(G_1)$ and j in $T_n(G_2)$, in $T_n(G_d)$, the lemma follows immediately from Lemma 5.9.

LEMMA 5.13: If either $T_n(G_1)$ or $T_n(G_2)$ contains only one node, then $T_n(G_d)$ has the property stated in the theorem.

The proof of this lemma is omitted, and is left as an exercise (Problem 5.45).

Since the decomposition of $T_n(G_d)$ into $T_n(G_1)$ and $T_n(G_2)$ can only be one of the above types, the theorem is thus proved.

COROLLARY 5.13: Let G_d be an acyclic or symmetric directed graph containing at least two directed trees t_n. Then the directed trees t_n of G_d can be arranged in a sequence (row) in which two successive elements are related by an elementary directed-tree transformation. Furthermore, any two of its directed trees t_n that are related by an elementary directed-tree transformation can be made as the successive elements of some sequence.

We remark that the first and the last elements in a sequence are considered as the successive elements. Clearly, all major submatrices of any matrix can also be arranged in a similar fashion (Problem 5.67).

Since an (undirected) graph may be transformed into a symmetric directed graph by representing each undirected edge by a pair of directed edges with opposite directions, and *vice versa*, and since, as pointed out in Corollary 4.11, there exists a one-to-one correspondence between the trees of a graph and the directed trees of its associated symmetric directed graph, a graph and its associated directed graph may be considered synonymous. Thus, two nodes in a tree graph are connected by an edge if, and only if, their corresponding trees are related by an elementary tree transformation. This leads to the following well-known result.

COROLLARY 5.14: Any edge of a tree graph containing at least three nodes can be made part of a Hamilton circuit.

In a similar manner, we can order the directed 2-trees or 2-trees of G_d (Problem 5.46). Thus, the problem of generating the set of trees of a graph is equivalent to that of enumerating a Hamilton circuit or a tree in the tree graph. KAMAE [1967a] and KISHI and KAJITANI [1968] have shown how to generate a Hamilton circuit in a complete subgraph of a tree graph, and then to combine these local Hamilton circuits to yield a Hamilton circuit of the tree graph. This is of course similarly valid for directed trees, directed 2-trees, or 2-trees.

As an illustration, the directed-tree graph $T_3(G_d)$ of fig. 5.19 contains at least two Hamilton circuits as shown in (5.83). In terms of directed trees t_3 of G_d of fig. 5.18, the corresponding sequences of H_1 and H_2 are given by

$$bef - efc - dec - aec - abe - dbe -,$$

and

$$bef - efc - aec - dec - dbe - abe -,$$

respectively. Since every edge of $T_3(G_d)$ is contained either in H_1 or in H_2, any edge of $T_3(G_d)$ can be made part of a Hamilton circuit. Similarly, the set of directed 2-trees $t_{2,3}$ of G_d is given by

$$T_{2,3} = \{ab, ac, db, dc, fb, fc\},$$

whose elements can be ordered in at least the two following ways:

$$ab - ac - fc - dc - db - fb -,$$

and

$$ab - db - fb - fc - dc - ac -,$$

respectively.

§ 9. Directed trees and directed Euler lines

In this section, we shall discuss some of the fundamental relationships between directed trees and directed Euler lines and the number of such lines for a class of directed graphs. These relationships were first pointed out by VAN AARDENNE-EHRENFEST and DE BRUIJN [1951].

DEFINITION 5.14: *Euler line*. A closed edge train containing all the edges of a graph or a directed graph is called an *Euler line* of the graph or directed graph.

THEOREM 5.36: A connected graph is an Euler line if, and only if, the degree of each of its nodes is even.

Proof. The necessary part is evident. We shall only prove that the condition is also sufficient. Let G be a connected graph with the desired property. We begin a closed edge train E at some arbitrary node i in G and continue as far as possible, always through new edges. Since the number of edges incident at each node is even, this process can only be terminated by returning to i. If $G - E = G'$ is not empty, then the degree of each of its nodes is also even. Since G is connected, there must be some node j in E such that (j, k) is in G'. From j we can construct a new closed edge train E' containing only edges of G'. Again E' must terminate by returning to j. But E and E' can be combined to form a new closed edge train E''. If $G - E''$ is not empty, the process may be repeated. Thus, eventually we will generate an Euler line of G, which completes the proof of the theorem.

COROLLARY 5.15: A connected graph is an Euler line if, and only if, it is an edge-disjoint union of circuits.

COROLLARY 5.16: A connected directed graph is an Euler line if, and only if, the sum of the incoming and the outgoing degrees of each of its node is even.

In fig. 5.20, since each of the nodes of the complete pentagon is of degree 4, the graph has an Euler line. Using the technique outlined in the proof of Theorem 5.36, a sequence of circuits is generated: *aed, ghf, ijcb*. Thus, the subgraph *edafhijcbg* is an Euler line of the complete pentagon.

Since the number of nodes of odd degree is always even, we have

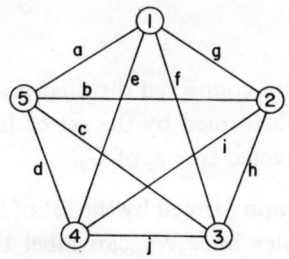

Fig. 5.20. The 5-node complete graph.

COROLLARY 5.17: Let G be a connected graph with $2k$, $k \geq 1$, nodes of odd degree. Then G can be decomposed into an edge-disjoint union of k open edge trains, k being the minimal number of such edge trains in any decomposition.

Proof. Since each node of odd degree must be the endpoint of at least one of the open edge trains, it follows that there are at least k such open edge trains in any decomposition. To show that G can be decomposed by this number, let G' be the graph obtained from G by adding to it k node-disjoint edges, each connecting a pair of distinct nodes of odd degree. Then by Theorem 5.36, G' has an Euler line E'. When the added edges are removed from E', E' is decomposed into k open edge trains. So the corollary is proved.

DEFINITION 5.15: *Directed Euler line.* An Euler line in a directed graph is said to be *directed* if for each pair of successive edges (i, j) and (u, v) in the Euler line we have either $j = u$ or $i = v$.

Following Definition 1.31, let $d^+(i)$ and $d^-(i)$ denote the outgoing and the incoming degrees of the node i of G_d. An argument similar to that used in the proof of Theorem 5.36 leads to the complete characterization of the directed Euler lines given below (Problem 5.47).

THEOREM 5.37: A connected directed graph is a directed Euler line if, and only if, $d^+(i) = d^-(i)$ for all i.

COROLLARY 5.18: In a connected graph, it is always possible to construct a closed edge sequence passing through each edge once and only once in each direction.

Let G_d be a directed graph with the property that $d^+(i) = d^-(i)$ for all i. Choose a fixed edge (n, u) and consider an arbitrary directed Euler line E_d of G_d. Now, if we start at the edge (n, u) and traverse along E_d, each node k of G_d will be visited $d^+(k)$ or $d^-(k)$ times. The edge by which we leave the node k after having visited it for the $d^-(k)$th time will be called the *last exit edge* of the node k.

THEOREM 5.38: Let G_d be a connected directed graph in which $d^+(i) = d^-(i)$ for all i. Then the subgraph formed by the set of last exit edges of the nodes k, $k = 1, 2, \ldots, n-1$, is a directed tree t_n of G_d.

Proof. Let g be the subgraph formed by the set of last exit edges of the nodes. Let E_d be the directed Euler line. We can label the edges of G_d according to the order in E_d with $e_1, e_2, \ldots, e_b$; the edge (n, u) in E_d gets the label e_1.

If e_i and e_j are both last exit edges in G_d, and if they are successive in G_d with e_i being followed by e_j, then $i<j$, since e_{i+1} and e_j must have the same initial node and since e_j is the edge whose index j is the highest among all the indices of the edges with this initial node. Consequently, g cannot contain any directed circuits. For, if not, the indices k of the edges e_k corresponding to a directed circuit in g would increase indefinitely. The theorem follows from the fact that g has $n-1$ edges and each of its nodes is of outgoing degree 1 except the node n which is of outgoing degree 0.

THEOREM 5.39: Let G_d be a connected directed graph in which $d^+(i)=d^-(i)$ for all i. Then for a given edge (n, u) and for a given directed tree t_n there are exactly

$$\prod_{i=1}^{n} (d_i - 1)! \tag{5.84}$$

directed Euler lines whose set of last exit edges coincides with t_n, where $d_i = d^+(i) = d^-(i)$.

Proof. At each node of G_d, we shall label the edges outgoing from the node with following restrictions: At node n, let $e_1 = (n, u)$; at node k, $k<n$, the edge in t_n gets the highest possible index. The number of ways in which the labeling process is possible is expressed in (5.84). It remains to be shown that, for each labeling of the above type, there exists exactly one directed Euler line corresponding to this labeling.

We begin a directed Euler line at e_1 in G_d and continue as far as possible, always through the edge e_k which has not been traversed before and which has the lowest index k among all the edges having the same initial node as e_k. Since G_d is assumed to be finite, the process can only be terminated by returning to node n. Now we have to show that the process has used up all the edges of G_d. Assume that an edge, say, (i, j) has not been traversed. Then there is an edge (j, v) that also has not been traversed. In particular, the last exit edge (j, x) of the node j has not been traversed. It follows that the last exit edge (x, y) cannot be traversed, and so on. Since there exists a unique directed path from node j to node n in t_n, the above process will eventually arrive at n, and we find that there is an edge outgoing from n which has not been traversed. This contradicts to our assumption that we have continued this process as far as possible. This completes the proof of the theorem.

In the theorem, we consider two directed Euler lines as being identical if they differ only in the starting points. From Theorems 5.38 and 5.39, we obtain the following useful identity (Problem 5.69).

THEOREM 5.40: Let G_d be a connected directed graph in which $d^+(i) = d^-(i) = d_i$ for all i. If $N(t_k)$ and $N(E_d)$ denote the numbers of directed trees t_k and directed Euler lines E_d of G_d, respectively, then

$$N(E_d) = N(t_k) \prod_{i=1}^{n} (d_i - 1)! \qquad (5.85)$$

for $k = 1, 2, \ldots, n$.

The theorem shows that the number of directed trees of G_d is independent of the choice of the reference node, a fact that was pointed out in Corollary 4.6.

Example 5.15: Consider the directed graph G_d as shown in fig. 5.21. We have $d^+(i) = d^-(i) = d_i$, $i = 1, 2, 3, 4$, where $d_1 = d_3 = 1$ and $d_2 = d_4 = 2$. By

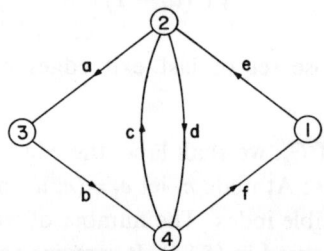

Fig. 5.21. A directed graph illustrating the relationships between its directed Euler lines and directed trees.

Theorem 5.37, there exists a directed Euler line in G_d. One of the directed Euler lines is given by *abcdfe*. Next, let $t_4 = abe$ and $(n, u) = c$. Then according to Theorem 5.39 there is

$$\prod_{i=1}^{4} (d_i - 1)! = (1-1)!\,(2-1)!\,(1-1)!\,(2-1)! = 1$$

directed Euler line whose set of last exit edges coincides with t_4, i.e., $E_d = cdfeab$. Actually, this number is independent of the choice of t_4 in G_d. Since there are only two directed trees *abe* and *bde* in G_d, according to Theorem 5.40 we have only two directed Euler lines in G_d. They are *fedcab* and *feabcd*. Note that we do not count two such directed Euler lines as different if they differ only in the starting points. Now if we choose node 1 as the reference node for the directed trees, we still have only two directed trees *abf* and *bdf*. The corresponding directed Euler lines are *edcabf* and *eabcdf*. They are of course identical to the two given earlier.

As another example, consider the 5-node complete directed graph G_d. By

Corollary 4.8, there are 125 directed trees. Since each node is of degree 4, according to Theorem 5.40 there are

$$N(E_d) = 125 \cdot (3!)^5 = 972\,000$$

directed Euler lines in G_d. In other words, in the complete pentagon as shown in fig. 5.20 there are 972 000 closed edge sequences, each of which passes through every edge of the complete pentagon once and only once in each direction.

§ 10. Conclusions

We began the present chapter by discussing some of the fundamental properties of a tree, and we have shown how the structure of a tree with n nodes can be encoded in a row of $n-2$ symbols so that it is recoverable from this description. Two codes of this type have been given. We have also considered the problem of decomposing a tree into the minimal number of edge-disjoint paths and of determining the number of such decompositions.

In the topological analysis of a linear system, the problem ultimately reduces to that of listing the set of trees or directed trees in the associated graph. Thus, the efficiency with which a complex linear system can be analyzed by a digital computer using topological formulas as a tool depends largely upon the efficiency with which the trees or directed trees are generated. For this reason, we have devoted a major portion of this chapter to the problem of efficient generation of trees of a graph. Several techniques are available, and they have been presented according to the nature of the techniques in one of the following four formulations: the Wang-algebra formulation, decomposition, the matrix formulation, and the elementary transformations. For a complicated graph, the techniques based on the Wang-algebra formulation are not efficient in that a large number of redundant terms are generated. Apart from slowing down the process of generating trees, this fact limits the sizes of the graphs that can be analyzed by a digital computer, since one must retain the set of generated trees in a computer memory and check each new tree against the list for possible duplications. However, this formulation leads naturally to the techniques of generating trees by decomposition without duplications. The advantages of decomposition are clear. Not only is the quantity of trees sharply reduced, but they are "smaller" trees requiring less manipulation for the evaluation of each. The fact that in many applications we are not really concerned with the set of trees but the sum of the tree products makes them more significant. However, in the decomposition, we assume that trees and multi-trees in the component graphs are available. The other two formulations are useful in that they provide the theoretical foundations for iterative tree-generation procedures which will not introduce any redundancies due to duplications. The introduction of these various techniques allows one to choose a particular approach or combination of approaches, so as to solve the problem at hand in the simplest and most satisfying manner.

Finally, we have discussed the existence of a Hamilton circuit in a directed-tree graph and some of the fundamental relationships between directed trees and directed Euler lines for a class of directed graphs. A formula for the number of directed Euler lines in these graphs has also been presented.

Problems

5.1. Prove Theorem 5.1.
5.2. Let $d(i)$ be the degree of the node i of a graph G. Show that

$$\sum_i [d(i) - 2] = 2(m - c). \qquad (5.86)$$

5.3. Show that a graph is a forest (multi-tree) if, and only if, the nullity of the graph is zero.
5.4. Show that if an integer occurs k times in the code $\mathscr{C}^*$, then its corresponding node in the tree is of degree $k+1$ where $k \geq 0$.
5.5. Using the code $\mathscr{C}$ of (5.8), show that the number of different trees with n given nodes is n^{n-2}.
5.6. Given the code $\mathscr{C}^*$ of (5.9), find, in the following order, (1) the degrees of the nodes, (2) the codes for the paths, (3) the path structures, and (4) the structure of the tree.
5.7. Show that the nodes that do not appear in the code $\mathscr{C}$ of (5.8) are nodes of degree 1 in the tree.
5.8. Suppose that the tree structure of fig. 5.3 is not given. Reconstruct the tree from the code given in (5.12).
5.9. The simplest rule for dissecting a tree with a view to encoding its structure is to remove the edges one at a time, choosing always the edge whose endpoint of degree 1 has the smallest value. Based on this rule, construct a code like codes $\mathscr{C}$ and $\mathscr{D}$ for a tree structure.
5.10. In a connected graph containing at least two trees, show that the distance of a pair of maximally distant trees is not greater than the rank or the nullity of the graph.
5.11. If the *distance* of two nodes in a connected graph G is defined as the length of a shortest path connecting these two nodes, show that for some

fixed node k there exists a tree t in G such that the distance from k to any node is the same in t as in G (see Definition 1.23).

5.12. Find a system $\mathscr{R}$ of partitions with the property specified in the proof of Theorem 5.12 for the tree of fig. 5.2.

5.13. Show that the operations defined in (5.16) and (5.17) are associative and commutative, and the set $\mathscr{S}$ forms a commutative ring with a unity with respect to these operations.

5.14. Show that the set $\mathscr{G}$ of all subgraphs containing no isolated nodes of a graph G constitutes a linear vector space over the field of integers mod 2, the addition of vectors being the ring-sum operations.

5.15. Prove Corollaries 5.4 and 5.5.

5.16. Let t be a tree of a graph G, and let k be the number of components in $G-t$. Show that there exists a tree of distance $n-k$ from t in G, and any tree is of distance not greater than $n-k$ from t in G.

5.17. Prove Corollaries 5.6 and 5.7.

5.18. Show that if a graph G has no parallel edges and self-loops with $m \leq 4$ and $c=1$, then $m \leq r$.

5.19. Prove Theorem 5.14.

5.20. A spanning subgraph L^* of a graph G is said to be a *generalized circuit* of G if, and only if, each component of L^* is of nullity 1. L^* is said to be *odd* if the circuit contained in each of its components is of odd length. Show that if $n \geq 3$ and $m \geq 1$, then a square submatrix of order n of the complete incidence matrix A_a of G over the real field is nonsingular if, and only if, the columns of this submatrix correspond to the edges of an odd generalized circuit L^* of G. Note that the matrix A_a consists only of 1's and 0's. Also show that the determinant of the submatrix is equal to $\pm 2^q$ where q is the number of circuits in L^* (CHEN [1970c]).

5.21. In Problem 5.20, show that the permanent of the submatrix over the real field is nonzero if, and only if, the columns of this submatrix correspond to the edges of a generalized circuit L^* of G, and its value is again equal to 2^q (CHEN [1970c]).

5.22. Show that for a given graph G, trees and cotrees in $G(I)$ are in one-to-one correspondence with the multi-trees $t(I)$ and the multi-cotrees $\bar{t}(I)$ of G, respectively, where I is a subset of the node set of G and $G(I)$ is defined in § 4.4.

5.23. In Problem 5.22, let $\{L_x\}$, $x=1, 2, \ldots, m_s$, be a set of linearly independent circuits or edge-disjoint unions of circuits of $G(I)$, where m_s is the nullity of $G(I)$. Show that if $G(I)$ is connected, then the set of multi-cotrees of G, up to the null graph, is given by

$$\bar{T}(I) = H(L_1) \,@\, H(L_2) \,@\cdots@\, H(L_{m_s}). \tag{5.87}$$

5.24. In Problem 5.22, let $\{C_x\}$, $x=1, 2,\ldots, r_s$, be a set of linearly independent cuts of $G(I)$, where r_s is the rank of $G(I)$. Show that if $G(I)$ is connected, then the set of trees of G, up to $\emptyset$, is given by

$$T(I) = H(C_1) \,@\, H(C_2) \,@\cdots@\, H(C_{r_s}). \tag{5.88}$$

5.25. Extend the formulas (5.87) and (5.88) to unconnected graphs.

5.26. Give an example to show that the number of redundancies due to duplications is different for different choice of circuits in (5.28).

5.27. Repeat Problem 5.26 for (5.30) and (5.31). Can you give a set of rules so that the redundancies can be reduced?

5.28. In fig. 5.7, let $P_{ij} = e_2 e_6$. Use (5.32) to generate the set $\bar{T}(1;3)$.

5.29. Extend Theorem 5.21 to the case where G' or G'' is unconnected.

5.30. Prove Lemma 5.1.

5.31. Prove Theorem 5.16.

5.32. If every edge of a connected graph G without parallel edges is a circuit edge, and if $m \leq \min(5, r)$, then m is the distance of two maximally distant trees of G (BARON and IMRICH [1968]).

5.33. Prove Theorem 5.29.

5.34. Prove Lemma 5.6.

5.35. Let A_a^+ be the matrix obtained from the complete incidence matrix A_a of a directed graph G_d by replacing entries -1's by 0's. Show that a submatrix $A_a^+(I_q, J_q)$ is nonsingular if, and only if, each node of the subgraph formed by the nodes in I_q and the edges corresponding to the columns in $A_a^+(I_q, J_q)$ is of outgoing degree 1 (ASH [1959]).

5.36. If A^- and A_a^- are the matrices obtained from A and A_a, respectively, by replacing entries 1's by 0's, state the dual results of Problems 5.34 and 5.35.

5.37. From Problems 5.35 and 5.36, show that the corresponding submatrices $A_a^+(I_q, J_q)$ and $A_a^-(I_q, J_q)$ are both nonsingular if, and only if, the edges corresponding to the columns of the submatrices form a circuit or node-disjoint union of circuits (ASH [1959]).

5.38. Prove Theorem 5.31.

5.39. Prove Lemma 5.7.

5.40. Define an elementary cotree transformation, and then show that every cotree of a connected graph G can be obtained from any other one by a finite sequence of elementary cotree transformations. (*Hint*: Follow (5.71).)

5.41. Repeat Problem 5.40 for 2-trees and directed 2-trees.
5.42. Prove Lemma 5.8.
5.43. Let $t = e_1 e_2 \ldots e_r$ be a tree. Show that the edges of t can always be labeled in such a way that $e_1 e_2 \ldots e_x$, $x = 1, 2, \ldots, r-1$, are connected subgraphs of t.
5.44. Let G_d^* be the directed graph obtained from a given directed graph G_d by replacing each edge of G_d by k parallel edges with the same direction, where k is a positive integer. Show that the number of directed trees t_n^* of G_d^* with reference node n is equal to k^{n-1} times the number of directed trees t_n of G_d.
5.45. Prove Lemmas 5.11 and 5.13.
5.46. Define the directed 2-tree graph for the directed 2-trees $t_{i,j}$ of a directed graph. Show that there exists a Hamilton circuit in the directed 2-tree graph of any acyclic directed graph containing at least three $t_{i,j}$.
5.47. Prove Theorem 5.37.
5.48. Show that a graph is an open edge train if, and only if, it is connected and contains exactly two nodes of odd degree.
5.49. A spanning subgraph $t^*(I_k)$ of a graph G is said to be a *generalized k-tree* of G if, and only if, $t^*(I_k)$ has k components of nullity 0 with the nodes in $I_k = \{i_1; i_2; \ldots; i_k\}$ appearing in different components, and all other components, if they exist, are of nullity 1. $t^*(I_k)$ is said to be *odd* if it is a k-tree or all the circuits contained in its components of nullity 1 are of odd length. Let $(A_a)_{I_k}$ be the matrix obtained from the complete incidence matrix A_a by deleting the rows corresponding to the integers in I_k. Note that A_a is a matrix consisting only of 1's and 0's. Show that a square submatrix of $(A_a)_{I_k}$ of order $n-k$ over the real field is nonsingular if, and only if, the columns of this submatrix correspond to the edges of an odd generalized k-tree $t^*(I_k)$ in G, and its determinant is equal to $\pm 2^q$ where q is the number of circuits contained in $t^*(I_k)$ (CHEN [1970c]).
5.50. In Problem 5.49, show that the permanent of the submatrix is nonzero if, and only if, the columns of this submatrix correspond to the edges of a generalized k-tree $t^*(I_k)$ in G, and its value is equal to 2^q (CHEN [1970c]).
5.51. A spanning subgraph $t_d^*(I_k)$ of a directed graph G_d is said to be a *generalized directed k-tree with reference nodes* $i_1, i_2, \ldots, i_k$ in G_d if, and only if, $d^+(x) = 1$ for each node x in $t_d^*(I_k)$ but not in I_k, and $d^+(x) = 0$ for each x in I_k, where $I_k = \{i_1; i_2; \ldots; i_k\}$. Let $(A_a^+)_{I_k}$ be the submatrix obtained

from A_a^+ by deleting the rows corresponding to the integers in I_k. Then any square submatrix of order $n-k$ of $(A_a^+)_{I_k}$ is nonsingular if, and only if, the columns of this submatrix correspond to the edges of a generalized directed k-tree $t_d^*(I_k)$ in G_d, where A_a^+ is defined in Problem 5.35 (CHEN [1970c]).

5.52. For a graph G, let g be an element of $\mathscr{G}$ and let e be an edge of G. Define

$$\partial g/\partial e = g \oplus e \text{ if } e \text{ is in } g$$
$$= \emptyset \text{ if } e \text{ is not in } g. \tag{5.89a}$$

If S_1 is an element of $\mathscr{S}$, and if $g = e_1 e_2 \cdots e_k$, define

$$\partial S_1/\partial g = \partial S_1/\partial e_1 \oplus \partial S_1/\partial e_2 \oplus \cdots \oplus \partial S_1/\partial e_k, \tag{5.89b}$$

where

$$\partial S_1/\partial e = \{\partial s/\partial e; s \text{ is in } S_1 \text{ and } e \text{ in } s\}. \tag{5.89c}$$

Using these definitions, show that if $G = e_1 e_2 \cdots e_b$ then

$$\partial^k \{G\}/\partial g_1 \, \partial g_2 \cdots \partial g_k = \{G \oplus g_\alpha; g_\alpha \text{ is in } W\}, \tag{5.90a}$$

where $g_1, g_2, \ldots,$ and g_k are elements of $\mathscr{G}$;

$$\partial^2 \{G\}/\partial g_1 \, \partial g_2 = \partial (\partial \{G\}/\partial g_2)/\partial g_1, \tag{5.90b}$$

and it is similarly defined for the general case; and

$$W = H(g_1) @ H(g_2) @ \cdots @ H(g_k). \tag{5.90c}$$

5.53. Using (5.90), show that (5.28) and (5.30) can also be expressed as

$$T = \partial^m \{G\}/\partial L_1 \, \partial L_2 \cdots \partial L_m, \tag{5.91a}$$

$$\bar{T} = \partial^r \{G\}/\partial C_1 \, \partial C_2 \cdots \partial C_r. \tag{5.91b}$$

(*Hint*: For (5.91a), use (5.28), and for (5.91b), use (5.30).) (CHEN and MARK [1969].)

5.54. Using (5.91), show that the identities (5.38) and (5.40) can also be expressed as

$$T = \frac{\partial^{k-1} T' \times T''}{\partial (P'_{i_1 i_2} \cup P''_{i_1 i_2}) \, \partial (P'_{i_2 i_3} \cup P''_{i_2 i_3}) \cdots \partial (P'_{i_{k-1} i_k} \cup P''_{i_{k-1} i_k})}, \tag{5.92a}$$

$$\bar{T} = \frac{\partial^{k-1} \bar{T}'(I') \times \bar{T}''(I'')}{\partial C_{i_1} \, \partial C_{i_2} \cdots \partial C_{i_{k-1}}} \tag{5.92b}$$

(CHEN and MARK [1969]).

5.55. Show that if $C_{ij}=e_1e_2\ldots e_k$ is a cutset separating the nodes i and j in a graph G, and if $Y(\bar{e}_q)$ is the set of 2-trees $t(i;j)$ of G, each of which does not contain the edge e_q, $q=1, 2, \ldots, k$, then the set of trees of G is given by

$$T = Z(e_1) \oplus Z(e_2) \oplus \cdots \oplus Z(e_k), \tag{5.93}$$

where $Z(e_q) = \{e_q\} \times Y(\bar{e}_q)$.

5.56. Use the formulas given in (5.91) and (5.92) to generate the sets of trees and cotrees of the graphs as shown in figs. 5.14 and 5.16.

5.57. Let $P(H) = \{E_1, E_2, \ldots, E_q\}$ be a partition of the edge set H of a tree in a graph G. Let $G(\bar{E}_x)$, $x=1, 2, \ldots, q$, be the graph obtained from G by shorting all the edges in H but not in E_x. Show that if T_x denotes the set of trees of $G(\bar{E}_x)$, then the set of trees of G is given by

$$T = T_1 \text{ @ } T_2 \text{ @} \cdots \text{@ } T_q. \tag{5.94}$$

What is the dual result of (5.94)? (*Hint*: Use (5.30) and the fact that by shorting a branch of a tree the *f*-cutsets defined by the other branches of the tree are not effected.) (BERGER and NATHAN [1968].)

5.58. Consider the n-node complete graph G_c, each of its edges being assigned a nonnegative weight called the *length* of the edge. The *length* of a tree is defined as the sum of the lengths of its branches. For simplicity, we assume that there are no trees of G_c of equal length. Show that the longest edge of a circuit never belongs to the shortest tree, and that every tree except the shortest one contains at most one branch that does not belong to the union of the shorter trees. (*Hint*: Use elementary tree transformation.) (DIJKSTRA [1960].)

5.59. Using the results obtained in Problem 5.58, give an algorithm to construct a shortest tree of G_c.

5.60. In Problem 5.58, let all the trees of G_c be arranged in order of increasing length, i.e., $t_0, t_1, t_2, \ldots$. If q branches of the shortest tree t_0 are contained in the tree t_k, show that the index k must satisfy the inequality: $k \geq 2^q - 1$ (DIJKSTRA [1960]).

5.61. Show that every cutset of a graph contains at least k edges if, and only if, every set of $r-1$ edges of any tree is in at least k different trees.

5.62. Let a positive number be assigned to each branch of a tree t. The *distance* d_{ij} between the two nodes i and j of t is the sum of the weights of the branches in the unique path connected between the nodes i and j of t. (With simple modifications, the present definition is somewhat more general than that given in Problem 5.11.) The matrix D whose elements are d_{ij} is called the *distance matrix* of t with d_{ii} taken as zero. Show that a necessary and sufficient condition for a given $n \times n$ nonnegative symmetric matrix D to be the distance matrix of a tree is that $d_{ii}=0$ for all i and

$$d_{ij} \leq d_{ik} + d_{kj} \tag{5.95}$$

for all i, j, and k (HAKIMI and YAU [1964]).

5.63. In Problem 5.62, if $n \geq 4$, show that D is realizable if, and only if, each of its 4×4 principal submatrices is realizable as a tree (PEREIRA [1969]).

5.64. Using the result given in Problem 5.11, extend Problems 5.62 and 5.63 to any graph.

5.65. For a given directed graph G_d, let t_n^0 be a reference directed tree with reference node n. By $t_n(E_q)$ we mean that it is a directed tree with reference node n in G_d such that $t_n(E_q) \cap t_n^0 = E_q$ where E_q is a subgraph of t_n^0. Denote by $T_n(E_q)$ the set of all directed trees $t_n(E_q)$ of G_d. Show that if the branch labeling of $t_n^0 = e_1 e_2 \cdots e_r$ is such that every subgraph $e_1 e_2 \cdots e_x$, $x=1, 2, \ldots, r$, is connected, and if $E_q = e_{i_1} e_{i_2} \cdots e_{i_q}$ with $i_1 < i_2 < \cdots < i_q$, then

$$T_n(\bar{E}_q) = \{t_n; t_n = (t_n' \oplus e_{i_q}) \cup e, t_n' \text{ is in } T_n(\bar{E}_{q-1}), e \text{ is}$$
$$\text{in } Q_{t_n'}(e_{i_q}) \cap \xi(e_{i_q}), e_{i_q} \text{ in } t_n' \cap t_n^0, \text{ and } e \neq e_{i_q} \} \tag{5.96}$$

and

$$T_n(\bar{E}_q) \cap T_n(\bar{E}_p') = S_0 \tag{5.97}$$

if, and only if, E_q and E_p' are two different subgraphs of t_n^0, where $\bar{E}_q$ denotes the complement of E_q in t_n^0, and $\xi(e_{i_q})$ denotes the subgraph of G_d consisting of all the edges having the same initial node as edge e_{i_q}, and $p, q=1, 2, \ldots, r$. (*Hint*: See Theorem 5.34.) (PAUL [1967].)

5.66. The *tree matrix* $T=[t_{ij}]$ of a connected graph G is defined as a matrix whose columns correspond to the edges e_j of G, and whose rows correspond to the trees t^i of G with $t_{ij}=1$ if e_j is in t^i and $t_{ij}=0$ otherwise. Show that if T^* is the augmented matrix obtained from T by adding a column of 1's placed at the $(b+1)$th column, then the rank of T^* is $b-p+1$ if, and only if, G contains p separable subgraphs (HAKIMI [1961]).

5.67. Let F be a matrix containing at least two major submatrices. Show that the major submatrices of F can be arranged in a sequence (row) in which two successive elements in the sequence differ only by one column (KAMAE [1967b]). (For a matroid, see HOLZMANN and HARARY [1972].)

5.68. Show that the set $\mathscr{S}$ of all subsets of the set $\mathscr{G}$ constitutes a linear vector space over the field of integers mod 2.

5.69. Prove Theorem 5.40.

5.70. In Problem 5.14, let $\mathscr{V}(C)$, $\mathscr{V}(L)$, and $\mathscr{V}(G)$ be the vector spaces associated with the sets of cuts, circuits or edge-disjoint unions of circuits, and subgraphs not containing isolated nodes of G, respectively. Give an example to show that $\mathscr{V}(C)$ and $\mathscr{V}(L)$, in general, do not constitute orthogonal complements of $\mathscr{V}(G)$. Can you justify this?

5.71. Using the symbols defined in Problem 5.70, show that if G is connected then a necessary and sufficient condition for the subspaces $\mathscr{V}(C)$ and $\mathscr{V}(L)$ to be the orthogonal complements of $\mathscr{V}(G)$ over the field of integers mod 2 is that the graph G contain an odd number of trees or cotrees. Extend this to the case where G is not connected (CHEN [1971a]).

5.72. Using the results of Problem 5.71, show that if G contains an odd number of trees, then any subgraph of G is uniquely expressible as the ring sum of a cut and a circuit or edge-disjoint union of circuits of G.

5.73. Using the symbols defined in Problem 5.70, show that a b-vector over the field of integers mod 2 is in the intersection of the subspaces $\mathscr{V}(C)$ and $\mathscr{V}(L)$ if, and only if, it is in the null space of the matrix formed by the rows of a basis cut matrix and the rows of a basis circuit matrix of G (CHEN [1971a]).

5.74. In Problem 5.70, show that the vector corresponding to the graph G itself is in the sum of the subspaces $\mathscr{V}(C)$ and $\mathscr{V}(L)$. Use this result to show that G can always be expressed as the ring sum of a cut and a circuit or edge-disjoint union of circuits of G (CHEN [1971a]).

5.75. Let $H_4 = \{1, 2, 3, 4\}$. Show that each of the following is a pair of essential complementary partitions of H_4 and that they are the only ones for H_4:

$[\{1,2,3,4\}, \{1;2;3;4\}]$ $[\{1,2,3;4\}, \{1,4;2;3\}]$ $[\{1,2,3;4\}, \{1;2,4;3\}]$
$[\{1,2,3;4\}, \{1;2;3,4\}]$ $[\{1;2,3,4\}, \{1,3;2;4\}]$ $[\{1;2,3,4\}, \{1,2;3;4\}]$
$[\{1;2,3,4\}, \{1,4;2;3\}]$ $[\{1,4;2,3\}, \{1,3;2;4\}]$ $[\{1,4;2,3\}, \{1,2;3;4\}]$
$[\{1,4;2,3\}, \{1;2,4;3\}]$ $[\{1,4;2,3\}, \{1;2;3,4\}]$ $[\{1,3,4;2\}, \{1;2,3;4\}]$
$[\{1,3,4;2\}, \{1,2;3;4\}]$ $[\{1,3,4;2\}, \{1;2,4;3\}]$ $[\{1,3;2,4\}, \{1,2;3;4\}]$
$[\{1,3;2,4\}, \{1,2;3;4\}]$ $[\{1,3;2,4\}, \{1,4;2;3\}]$ $[\{1,3;2,4\}, \{1;2;3,4\}]$
$[\{1,2,4;3\}, \{1;2,3;4\}]$ $[\{1,2,4;3\}, \{1,3;2;4\}]$ $[\{1,2,4;3\}, \{1;2;3,4\}]$
$[\{1,2;3,4\}, \{1;2,3;4\}]$ $[\{1,2;3,4\}, \{1,3;2;4\}]$ $[\{1,2;3,4\}, \{1,4;2;3\}]$
$[\{1,2;3,4\}, \{1;2,4;3\}]$.

(For H_5 see CHEN and GOYAL [1971].)

5.76. Show that if $C_i[P'(H_k), P''(H_k)]$ is essential, then it is complementary with respect to any j of H_k.

CHAPTER 6

THE REALIZABILITY OF DIRECTED GRAPHS WITH PRESCRIBED DEGREES

As implied by the title, the main objective of this chapter is to present general conditions that are necessary and sufficient for the existence of a directed graph with prescribed incoming and outgoing degrees. In particular, we shall consider the realizability conditions of a set of nonnegative integers as the degrees of the nodes of a graph. Other related problems such as the realizability of a set of nonnegative integer pairs as various types of directed graphs, the unique realizability of a set of nonnegative integers, and the existence of a weighted directed graph will also be considered.

One motivation for considering these problems lies in the fact that the existence problem for a directed graph having prescribed degrees is equivalent to that for a square matrix of nonnegative integers having prescribed row and column sums. The other reason is the potential applications of these results. For example, in enumerating the isomers of the saturated hydrocarbons C_nH_{2n+2}, with a given number of carbon atoms, the problem is equivalent to listing all the trees in which every node has degree 1 or 4. As another example, consider the chemical compound N_2O_3. To find all isomers of this compound, it is necessary to generate all connected graphs realizing the set of integers 2, 2, 2, 3, 3 as the degrees of their nodes. For interested readers, we refer to SENIOR [1951a] for a more detailed treatment. As a matter of fact, it is the phenomenon of isomerism among organic chemical compounds that leads SENIOR [1951b] to the investigation of this problem.

Like many other chapters in this book, this chapter is mainly based on a few published papers of fundamental importance, which are referred to in the appropriate sections. In general, we follow the recent work of CHEN [1966e].

§ 1. Existence and realization as a (p, s)-digraph

For an n-node directed graph G_d, let $d^+(i)$ and $d^-(i)$ be the outgoing and the incoming degrees of the node i, respectively. The nonnegative integer pair $[d^+(i), d^-(i)]$ is called the *degree pair* of the node i. If, on the other hand, a set of nonnegative integer pairs (a_i, b_i), $i=1, 2, ..., n$, is given, how can we tell whether or not there exists an n-node directed graph G_d whose nodes i have degree pairs (a_i, b_i)? If such a G_d exists, we say that the set of nonnegative integer pairs (a_i, b_i) is *realizable*, or the directed graph G_d realizes the set. It is obvious that not every set is realizable. For example, the set of nonnegative integer pairs (1, 1) and (1, 2) is not realizable. In this section, we present some general conditions that are necessary and sufficient for the existence of a class of directed graphs called (p, s)-*digraphs* with prescribed degrees. The simple constructive proof of the main theorem by means of a directed bipartite graph suggests a simple procedure for constructing such a (p, s)-digraph from a given set of realizable nonnegative integer pairs.

Throughout this chapter, by $\{(a_i, b_i)\}$ we mean a set of n nonnegative integer pairs (a_i, b_i) for $i = 1, 2, \ldots, n$. Similarly, $\{a_i\}$ and $\{b_i\}$ denote the sets of n nonnegative integers a_i and b_i, respectively. If G_d has parallel edges from node i to node j, pick one of them arbitrarily and denote it by (i, j). Thus, (i, j) denotes any one of the parallel edges from node i to node j in G_d. In particular, for $i = j$, (i, i) denotes any one of the self-loops at the node i of G_d. For a finite set S, the number of its elements is denoted by $\alpha(S)$. Similarly, we denote by $\alpha(i, j)$ the number of parallel edges directed from node i to node j in G_d.

DEFINITION 6.1: (p, s)-*digraph*. A (p, s)-digraph is a directed graph in which $\alpha(i, j) \leq p$ for all $i \neq j$, and $\alpha(i, i) \leq s$ for all i where p and s are given nonnegative integers. When $p = s$, a (p, s)-digraph is simply called a p-*digraph*.

The letters p and s stand for the words *p*arallel and *s*elf-loop, respectively.

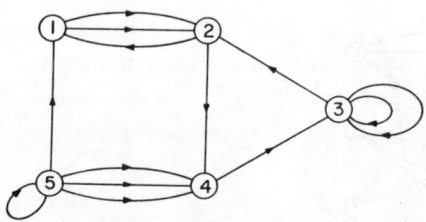

Fig. 6.1. A directed graph illustrating its degree pairs.

In fig. 6.1, the directed graph is a (3, 2)-digraph, since the maximum number of parallel edges is 3 and the maximum number of self-loops is 2. Evidently, it is also a (p, s)-digraph for any $p \geq 3$ and $s \geq 2$. The set of degree pairs of the directed graph is given by

$$\{(2, 2), (2, 3), (3, 3), (1, 4), (5, 1)\}. \tag{6.1}$$

1.1. *Directed graphs and directed bipartite graphs*

For any directed graph G_d with node set V, there is an equivalent representation as a *directed bipartite graph* B_d: To the node set V of G_d, we construct a replica V' whose elements are in a one-to-one correspondence with those of V. The edge (i, j'), i in V and j' in V', is an edge of B_d if, and only if, there exists an edge (i, j) in G_d, each of the parallel edges being considered individually. In other words, in G_d and B_d we have $\alpha(i, j) = \alpha(i, j')$. The directed graph B_d is called the *corresponding directed bipartite graph* of G_d. Conversely, it is evident that in a directed bipartite graph B_d with $\alpha(V) = \alpha(V')$ there exists a one-to-one correspondence between the nodes in V and V'. If these corresponding nodes in V and V' of B_d are identified, we obtain a *corresponding directed graph* of B_d. Obviously in B_d, we have $d^-(i) = 0$ for all i in V and $d^+(i') = 0$ for all i' in V'. Furthermore, the outgoing and the incoming degrees of G_d at node i are the same as the outgoing degree of B_d at i in V and the incoming degree of B_d at i' in V', respectively. Thus, the problem of realizing a directed graph with prescribed outgoing and incoming degrees is equivalent to that of finding a directed bipartite graph with specified outgoing degrees of the nodes in V and incoming degrees of the nodes in V'. The correspondence between B_d and G_d makes it possible to translate the results from one to the other.

As an illustration, consider the directed graph G_d given in fig. 6.1. The corresponding directed bipartite graph B_d of G_d is presented in fig. 6.2. where 1, 2, 3, 4, 5 and 1', 2', 3', 4', 5', respectively, are the corresponding nodes of the sets V and V' of B_d. Conversely, if B_d is given and if the corresponding nodes

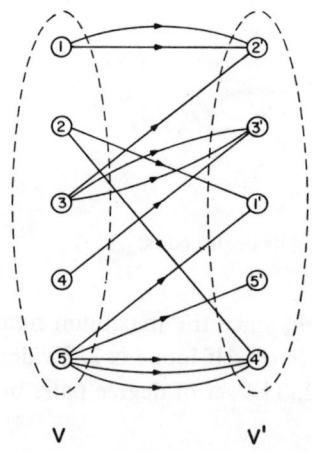
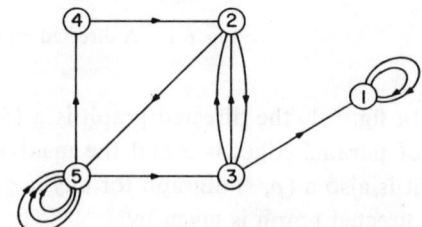

Fig. 6.2. The corresponding directed bipartite graph of the directed graph of fig. 6.1.

Fig. 6.3. A corresponding directed graph of the directed bipartite graph of fig. 6.2.

are specified, its corresponding directed graph G_d can be obtained in a unique and straightforward manner, that is to within isomorphism. We remark that, for a different choice of the node correspondence, we may obtain a different G_d. For example, in fig. 6.2 if 1, 2, 3, 4, 5 and 2', 3', 1', 5', 4', respectively, are considered as the corresponding nodes of V and V', the corresponding directed graph as shown in fig. 6.3 is a (2, 3)-digraph, which is certainly not isomorphic to the one given in fig. 6.1. In order to obtain a unique association, in the following we assume that all the nodes of B_d are labeled in such a way that i and i' are the corresponding nodes of B_d.

1.2. Existence

The realizability of a set of n nonnegative integer pairs as the degree pairs of the nodes of an n-node (p, s)-digraph is stated in

THEOREM 6.1: The necessary and sufficient conditions for a set $\{(a_i, b_i)\}$ of n nonnegative integer pairs to be realizable as the degree pairs of the nodes of an n-node (p, s)-digraph are

$$\sum_{i=1}^{n} a_i = \sum_{j=1}^{n} b_j, \tag{6.2a}$$

$$\sum_{a_i \in S_A} \min\left[a_i, \alpha(S_A) p\right] + \sum_{a_j \in S_A} \min\{a_j, [\alpha(S_A) - 1] p + s\} \geq \sum_{b_k \in S_B} b_k \tag{6.2b}$$

for all nonempty subsets S_A of $\{a_i\}$, where $\bar{S}_A$ is the complement of the set S_A in $\{a_i\}$, and S_B is the corresponding set of S_A in $\{b_i\}$ which has the property that b_x is in S_B if and only if a_x is in S_A.

Proof. The necessity of these two conditions is easy to establish. Suppose that B_d is the corresponding directed bipartite graph with node sets V and V' of a realized (p, s)-digraph G_d. It is obvious that the sum of the outgoing degrees of the nodes in V is equal to the sum of the incoming degrees of the nodes in V', therefore, (6.2a) holds.

Let V_A and V'_B be the corresponding node sets of the nonempty subsets S_A and S_B in V and V', respectively. Then the maximum number of edges of the form (i, j'), $i \in V_A$ and $j' \in V'_B$, in B_d is given by

$$\sum_{i \in V_A} \min\{d^+(i), [\alpha(V'_B) - 1] p + s\}.$$

Similarly the maximum number of edges of the form (i, j'), $i \in \bar{V}_A$ and $j' \in V'_B$, in B_d is given by

$$\sum_{i \in \bar{V}_A} \min[d^+(i), \alpha(V'_B) p],$$

where $\bar{V}_A$ denotes the complement of V_A in V. It follows that

$$\sum_{i \in \bar{V}_A} \min\left[d^+(i), \alpha(V_B')\, p\right] + \sum_{j \in V_A} \min\{d^+(j), [\alpha(V_B') - 1]\, p + s\}$$

$$\geq \sum_{k' \in V_B'} d^-(k').$$

Since

$$d^+(i) = a_i \quad \text{for all } i \text{ in } V,$$

$$d^-(k') = b_k \quad \text{for all } k' \text{ in } V',$$

and $\alpha(V_B') = \alpha(V_A)$, (6.2b) follows immediately from the preceding inequality.

The proof of sufficiency of these two conditions is much more involved and will be given in the next section.

Facilitated results of the theorem can now be obtained. If we let p and s be arbitrarily large, the second condition of the theorem is always satisfied. Hence we have

COROLLARY 6.1: A necessary and sufficient condition for a set $\{(a_i, b_i)\}$ to be realizable as the degree pairs of the nodes of an n-node directed graph is

$$\sum_{i=1}^{n} a_i = \sum_{j=1}^{n} b_j. \tag{6.3}$$

If $s = 0$ and $p = \infty$, we obtain a result of HAKIMI [1965].

COROLLARY 6.2: The necessary and sufficient conditions for a set $\{(a_i, b_i)\}$ to be realizable as the degree pairs of the nodes of an n-node directed graph, $n \geq 2$, having no self-loops, are

$$\sum_{i=1}^{n} a_i = \sum_{j=1}^{n} b_j \tag{6.4a}$$

and

$$\sum_{\substack{i=1 \\ i \neq j}}^{n} a_i \geq b_j \quad \text{for } j = 1, 2, \ldots, n. \tag{6.4b}$$

If the elements of the set $\{(a_i, b_i)\}$ are ordered such that

$$a_i + b_i \geq a_{i+1} + b_{i+1} \tag{6.5}$$

for $i=1, 2, \ldots, n-1$, then (6.4b) is satisfied if and only if

$$\sum_{i=2}^{n} a_i \geq b_1. \tag{6.6}$$

(See Problem 6.13 for a special case of the corollary.)

By putting $s=p$, (6.2b) reduces to

$$\sum_{i=1}^{n} \min[a_i, \alpha(S_A)p] \geq \sum_{b_k \in S_B} b_k \tag{6.7}$$

for each choice of S_A. If b's are ordered such that $b_i \geq b_{i+1}$ for $i=1, 2, \ldots, n-1$ then the inequality (6.7) is satisfied if and only if

$$\sum_{i=1}^{n} \min[a_i, kp] \geq \sum_{j=1}^{k} b_j \tag{6.8}$$

for $k=1, 2, \ldots, n$. This establishes a theorem of BERGE [1962].

COROLLARY 6.3: Consider a set $\{(a_i, b_i)\}$ in which, by altering the indices if necessary, $b_i \geq b_{i+1}$ for $i=1, 2, \ldots, n-1$, $n \geq 2$. Then the set is realizable as the degree pairs of the nodes of an n-node p-digraph if, and only if,

$$\sum_{i=1}^{n} a_i = \sum_{j=1}^{n} b_j \tag{6.9a}$$

and

$$\sum_{i=1}^{n} \min[a_i, kp] \geq \sum_{j=1}^{k} b_j \quad \text{for} \quad k=1, 2, \ldots, n. \tag{6.9b}$$

For the case in which $s=p=1$, Corollary 6.3 can be further simplified. It is possible to reformulate the result so that it can be tested more rapidly. This result was first established by RYSER [1957] and GALE [1957] in an attempt to obtain the existence conditions of a matrix of zeros and ones having prescribed row and column sums. This equivalent matrix problem will be treated in a later section.

COROLLARY 6.4: In a given set $\{(a_i, b_i)\}$ alter the indices of the b_i so that $b_i \geq b_{i+1}$ for $i=1, 2, \ldots, n-1$, $n \geq 2$. The necessary and sufficient conditions for the set to be realizable as the degree pairs of the nodes of an n-node 1-digraph are

$$\sum_{i=1}^{n} a_i = \sum_{j=1}^{n} b_j \tag{6.10a}$$

and

$$\sum_{i=1}^{k} a_i^* \geq \sum_{j=1}^{k} b_j \tag{6.10b}$$

for all integers k where a_i^* denotes the number of the a_j greater than or equal to the integer i, and the sequences b_j and a_i^* are assumed to have been brought to the same length by adding an appropriate number of zeros if necessary.

Proof. From Corollary 6.3, it is sufficient to show

$$\sum_{i=1}^{n} \min[a_i, k] = \sum_{j=1}^{k} a_j^* \qquad (6.11)$$

for all integers k. We shall prove the above identity by induction over k. For $k=1$, it is clear from the definition that

$$\sum_{i=1}^{n} \min[a_i, 1] = a_1^*.$$

Assume that (6.11) is true for any k. We need to show that it is also true for $k+1$. Since

$$\min[a_i, k+1] = \min[a_i, k] \quad \text{for } a_i \leq k$$

and

$$\min[a_i, k+1] = \min[a_i, k] + 1 \quad \text{for } a_i \geq k+1,$$

it follows that

$$\sum_{i=1}^{n} \min[a_i, k+1] = a_{k+1}^* + \sum_{i=1}^{n} \min[a_i, k]$$
$$= a_{k+1}^* + \sum_{j=1}^{k} a_j^*.$$

The second line follows from the induction hypothesis. So the corollary is proved.

There is a simple pictorial representation of the number a_j^*: Let each integer a_i be represented by a row of a_i dots, and arrange these rows in a vertical array so that a_{i+1} lies under a_i. Then the number a_j^* is simply the number of dots in the jth column of the array. For example, let $a_1 = 5$, $a_2 = 4$, $a_3 = 4$, $a_4 = 3$, and $a_5 = 1$. The array corresponding to these integers is as shown below:

$$
\begin{array}{c|ccccccc}
 & a_1^* & a_2^* & a_3^* & a_4^* & a_5^* & a_6^* & a_7^* \\
a_1 & \cdot & \cdot & \cdot & \cdot & \cdot & & \\
a_2 & \cdot & \cdot & \cdot & \cdot & & & \\
a_3 & \cdot & \cdot & \cdot & \cdot & & & \\
a_4 & \cdot & \cdot & \cdot & & & & \\
a_5 & \cdot & & & & & & \\
\end{array}
\qquad (6.12)
$$

Thus, we have $a_1^*=5$, $a_2^*=4$, $a_3^*=4$, $a_4^*=3$, $a_5^*=1$, and $a_i^*=0$ for $i \geq 6$; they correspond to the numbers of dots in the columns of the array. Obviously, they are independent of the ordering of the elements of $\{a_i\}$. As a matter of fact, if for a given set $\{a_i\}$ we define

$$\{a_i\}^* = \{a_j^*; j = 1, 2, \ldots\}, \quad (6.13)$$

then it is clear that $\{a_i\}^*$ determines $\{a_i\}$ since the integer a_i occurs exactly $a_x^* - a_{x+1}^*$, $x = a_i$, times in $\{a_i\}$.

The set $\{a_i\}^*$ is called the *dual set* of $\{a_i\}$. In fact, the correspondence between $\{a_i\}$ and $\{a_i\}^*$ is completely dual in the following sense:

$$\{\{a_i\}^*\}^* = \{a_i\}. \quad (6.14)$$

The validity of (6.14) is quite obvious in view of the pictorial representation discussed above. This result will not be needed in the sequel and its formal proof is left as an exercise (Problem 6.3).

After this digression into a discussion of the dual set, let us return to Theorem 6.1, and consider other special cases. One of these cases was considered by FULKERSON [1960] who varied the problem slightly and looked for conditions under which there is a self-loopless directed graph on n-nodes having given outgoing and incoming degrees but no parallel edges. This is a special case of Theorem 6.1 in which $s=0$ and $p=1$. Hence, we have

COROLLARY 6.5: *The necessary and sufficient conditions for a set $\{(a_i, b_i)\}$ to be realizable as the degree pairs of the nodes of an n-node $(1, 0)$-digraph are*

$$\sum_{i=1}^{n} a_i = \sum_{j=1}^{n} b_j \quad (6.15a)$$

and

$$\sum_{a_j \in S_A} \min[a_j, k-1] + \sum_{a_i \in S_A} \min[a_i, k] \geq \sum_{b_t \in S_B} b_t \quad (6.15b)$$

for all nonempty subsets S_A of $\{a_i\}$, where $k = \alpha(S_A)$ and $n \geq 2$.

Like the main theorem, the existence conditions of (6.15b) consist of a set of $2^n - 1$ inequalities. It would be of great practical interest if these can be reduced to a set of n inequalities. This is possible if the integers a_i and b_j are properly ordered. The result is given as follows:

COROLLARY 6.6: *Consider a set $\{(a_i, b_i)\}$ in which, by altering the indices if necessary, $b_i \geq b_{i+1}$ and, furthermore, if $b_i = b_{i+1}$ then $a_i \geq a_{i+1}$ for $i = 1, 2, \ldots, n-1$, $n \geq 2$. The set is realizable as the degree pairs of the nodes of an*

n-node (1, 0)-digraph if, and only if,

$$\sum_{i=1}^{n} a_i = \sum_{j=1}^{n} b_j, \qquad (6.16\text{a})$$

$$\sum_{i=1}^{n} \min[a_i; n-1] \geq \sum_{j=1}^{n} b_j, \qquad (6.16\text{b})$$

and

$$\sum_{i=1}^{k} \min[a_i, k-1] + \sum_{j=k+1}^{n} \min[a_j, k] \geq \sum_{i=1}^{k} b_i \qquad (6.16\text{c})$$

for $k=1, 2, \ldots, n-1$. [See Problem 6.68 for an equivalent statement of (6.16b).]

Proof. Necessity. (6.16b) and (6.16c) follow directly from (6.15b) by taking $S_B = \{b_i\}$ and $S_B = \{b_i; i=1, 2, \ldots, k\}$, respectively.

Sufficiency. It is clear that (6.16b) implies (6.15b) for $S_B = \{b_i\}$. To show that (6.16c) implies (6.15b) for all nonempty proper subsets S_B of $\{b_i\}$, let

$$a_t^*(S_A) = \alpha(\{a_i; a_i \text{ in } S_A \quad \text{and} \quad a_i \geq t\}) \qquad (6.17)$$

for $t=1, 2, \ldots$. Evidently, we have

$$a_t^* = a_t^*(\{a_i\}). \qquad (6.18)$$

In other words, we retain the functional notation only when proper subsets of $\{a_i\}$ are being used. Then the left-hand side of (6.15b) can be written as

$$a_k^*(\bar{S}_A) + \sum_{t=1}^{k-1} a_t^*,$$

where the second term is defined to be zero for $k=1$. If we can show that the right-hand side of the equation

$$\sum_{t=1}^{k-1} a_t^* \geq \sum_{b_i \in S_B} b_i - a_k^*(\bar{S}_A)$$

is maximized, over all S_A (or S_B) with $\alpha(S_B) = k$, by letting $S_B = \{b_i; i=1, 2, \ldots, k\}$, then the proof of sufficiency is accomplished. This is indeed the case as we shall show in the following:

Let S_B be any proper subset of $\{b_i\}$ with $\alpha(S_B) = k$. Assume that $S_B \neq \{b_i; i=1, 2, \ldots, k\}$. Let b_x be an element in $\{b_i; i=1, 2, \ldots, k\}$ but not in S_B, and also let b_y be an element in S_B but not in $\{b_i; i=1, 2, \ldots, k\}$. By virtue of the ordering of the b_i, it is clear that $x < y$. Consider the set $S_B'' = (S_B - \{b_y\}) \cup \{b_x\}$ and its corresponding set S_A'' in $\{a_i\}$. It is apparent that $a_k^*(\bar{S}_A'')$ and $a_k^*(\bar{S}_A)$

differ at most by 1 where $\bar{S}_A''$ denotes the complement of S_A'' in $\{a_i\}$. Thus, for $b_x \neq b_y$ we have

$$\sum_{b_i \in S''_B} b_i - a_k^*(\bar{S}_A'') \geq \sum_{b_i \in S_B} b_i - a_k^*(\bar{S}_A), \tag{6.19}$$

since

$$\sum_{b_i \in S''_B} b_i > \sum_{b_i \in S_B} b_i. \tag{6.20}$$

For $b_x = b_y$, (6.20) is satisfied with the equal sign, and by our ordering of the a_i, $a_x \geq a_y$, it follows that

$$a_k^*(\bar{S}_A'') \leq a_k^*(\bar{S}_A). \tag{6.21}$$

So (6.19) is still valid for this case. Since S_B'' has one more element in common with $\{b_i; i=1, 2, ..., k\}$ than S_B, it is obvious that $\{b_i; i=1, 2, ..., k\}$ can be obtained from S_B by repeated application of this process. Thus, from (6.16c)

$$\sum_{t=1}^{k-1} a_t^* \geq \sum_{i=1}^{k} b_i - a_k^*(\{a_j; j = k+1, ..., n\})$$

$$\geq \sum_{b_i \in S_B} b_i - a_k^*(\bar{S}_A)$$

for all nonempty proper subsets S_B of $\{b_i\}$. Hence we obtain the desired result. This completes the proof of the corollary.

Before we consider other special cases, we shall illustrate the above results by the following examples.

Example 6.1: Consider the following set of nonnegative integer pairs:

$$\{(a_i, b_i); i = 1, 2, 3, 4\} = \{(3, 1), (2, 5), (4, 1), (0, 2)\}. \tag{6.22}$$

Suppose that we wish to realize the set as the degree pairs of the nodes of a 4-node (2, 1)-digraph. From Theorem 6.1, we can check if the problem is realizable: (6.2a) is satisfied since

$$\sum_{i=1}^{4} a_i = 3 + 2 + 4 + 0 = 9$$

and

$$\sum_{j=1}^{4} b_j = 1 + 5 + 1 + 2 = 9.$$

(6.2b) consists of 15 ($=2^4-1$) inequalities, as follows:
  For $S_A = \{a_1\}$, we have

$$\min(2, 2) + \min(4, 2) + \min(0, 2) + \min(3, 1) = 2 + 2 + 0 + 1 = 5 \geq 1.$$

Similarly, for $S_A = \{a_2\}$, $\{a_3\}$, and $\{a_4\}$ we have

$$2 + 2 + 0 + 1 = 5 \geq 5,$$
$$2 + 2 + 0 + 1 = 5 \geq 1,$$
$$2 + 2 + 2 + 0 = 6 \geq 2.$$

For $S_A = \{a_1, a_2\}$, we have

$$\min(4, 4) + \min(0, 4) + \min(3, 3) + \min(2, 3)$$
$$= 4 + 0 + 3 + 2 = 9 \geq 1 + 5 = 6.$$

Similarly, for $S_A = \{a_1, a_3\}$, $\{a_1, a_4\}$, $\{a_2, a_3\}$, $\{a_2, a_4\}$, and $\{a_3, a_4\}$, we have

$$2 + 0 + 3 + 3 = 8 \geq 1 + 1 = 2,$$
$$2 + 4 + 3 + 0 = 9 \geq 1 + 2 = 3,$$
$$3 + 0 + 2 + 3 = 8 \geq 5 + 1 = 6,$$
$$3 + 4 + 2 + 0 = 9 \geq 5 + 2 = 7,$$
$$3 + 2 + 3 + 0 = 8 \geq 1 + 2 = 3.$$

For $S_A = \{a_1, a_2, a_3\}$, we have

$$\min(0, 6) + \min(3, 5) + \min(2, 5) + \min(4, 5)$$
$$= 0 + 3 + 2 + 4 = 9 \geq 1 + 5 + 1 = 7.$$

Similarly, for $S_A = \{a_1, a_2, a_4\}$, $\{a_1, a_3, a_4\}$, and $\{a_2, a_3, a_4\}$ we have

$$4 + 3 + 2 + 0 = 9 \geq 1 + 5 + 2 = 8,$$
$$2 + 3 + 4 + 0 = 9 \geq 1 + 1 + 2 = 4,$$
$$3 + 2 + 4 + 0 = 9 \geq 5 + 1 + 2 = 8.$$

Finally, for $S_A = \{a_1, a_2, a_3, a_4\}$ we have

$$\min(3, 7) + \min(2, 7) + \min(4, 7) + \min(0, 7)$$
$$= 3 + 2 + 4 + 0 = 9 \geq 1 + 5 + 1 + 2 = 9.$$

Thus, the problem possesses a solution which may be found by the algorithm outlined in the next section. One of the realizations is as shown in fig. 6.4.

Next, suppose that we wish to know if the set is realizable as a self-loopless directed graph. Then from Corollary 6.2 it is only necessary to test the following four inequalities:

$$2 + 4 + 0 = 6 \geq 1,$$
$$3 + 4 + 0 = 7 \geq 5,$$
$$3 + 2 + 0 = 5 \geq 1,$$
$$3 + 2 + 4 = 9 \geq 2.$$

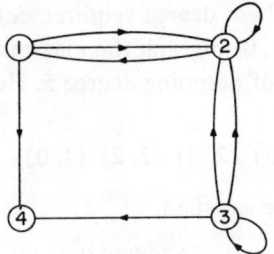
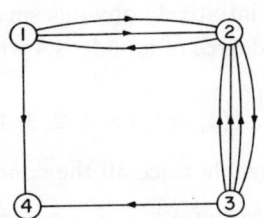

Fig. 6.4. A (2, 1)-digraph which realizes the set (6.22).

Fig. 6.5. A self-loopless realization of the set (6.22)

Thus, there exists a 4-node self-loopless directed graph whose degree pairs of the nodes are those given in (6.22). One of such realizations is presented in fig. 6.5.

Finally, let us consider the realizability of the set as a 1-digraph. In order that we may apply the result in Corollary 6.4, it is necessary that we alter the indices of the b_i so that $b_i \geq b_{i+1}$ for $i=1, 2, 3$. A new sequence is given as follows:

$$\{(a_i, b_i); i = 1, 2, 3, 4\} = \{(2, 5), (0, 2), (4, 1), (3, 1)\}. \qquad (6.23)$$

The dual set of the set $\{2, 0, 4, 3\}$ is given by

$$\{a_i\}^* = \{3, 3, 2, 1\}.$$

Using (6.10b), we have

$$3 \not\geq 5,$$
$$3 + 3 = 6 \not\geq 5 + 2 = 7,$$
$$3 + 3 + 2 = 8 \geq 5 + 2 + 1 = 8,$$
$$3 + 3 + 2 + 1 = 9 \geq 5 + 2 + 1 + 1 = 9.$$

Since not all the inequalities are satisfied, there exists no 1-digraph which realizes the set as the degree pairs of its nodes.

Example 6.2: Consider the (3, 2)-digraph G_d of fig. 6.1. We wish to know if there exists a 5-node (1, 0)-digraph whose degree pairs of the nodes are the same as those of G_d. The solution to this is given by Corollary 6.6. Our first step is to arrange the outgoing and the incoming degrees of the nodes of G_d according to the rules given in the corollary, and the result is given by

$$\{(a_i, b_i); i = 1, 2, 3, 4, 5\} = \{(1, 4), (3, 3), (2, 3), (2, 2), (5, 1)\}.$$

It is easy to check that (6.16a) is satisfied. However, (6.16b) is violated since

$$1 + 3 + 2 + 2 + 4 = 12 \not\geq 4 + 3 + 3 + 2 + 1 = 13.$$

Thus, we cannot find any $(1, 0)$-digraph with these degree requirements. This is also intuitively obvious since in any 5-node $(1, 0)$-digraph the maximum outgoing degree of a node is 4 while G_d has a node of outgoing degree 5. However, the set

$$\{(a_i, b_i); i = 1, 2, 3, 4, 5\} = \{(1, 4), (3, 3), (2, 3), (2, 2), (4, 0)\}$$

is realizable since all the conditions in (6.16) are satisfied.

(6.16a): $\quad 1 + 3 + 2 + 2 + 4 = 12 = 4 + 3 + 3 + 2 + 0 = 12,$

(6.16b): $\quad 1 + 3 + 2 + 2 + 4 = 12 \geq 4 + 3 + 3 + 2 + 0 = 12,$

(6.16c): $\quad 0 + 1 + 1 + 1 + 1 = 4 \geq 4,$

$\quad\quad\quad\quad 1 + 1 + 2 + 2 + 2 = 8 \geq 4 + 3 = 7,$

$\quad\quad\quad\quad 1 + 2 + 2 + 2 + 3 = 10 \geq 4 + 3 + 3 = 10,$

$\quad\quad\quad\quad 1 + 3 + 2 + 2 + 4 = 12 \geq 4 + 3 + 3 + 2 = 12.$

One of such realizations is given in fig. 6.6.

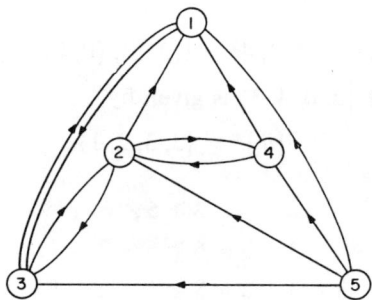

Fig. 6.6. A $(1,0)$-digraph realization.

We shall now proceed to consider other special cases. The reason for doing this is that the conditions obtained for these special cases are much the simplest, and they are sufficiently important to be considered explicitly. However, before we proceed, let us first show that the conditions given in Corollary 6.6 can be put in another form.

For a given set $\{a_i\}$ and for $k = 1, 2, \ldots,$ let

$$a_k'' = \alpha(I_k) + \alpha(J_k), \tag{6.24a}$$

where

$$I_k = \{i; i < k \quad \text{and} \quad a_i \geq k - 1\}, \tag{6.24b}$$

$$J_k = \{i; i > k \quad \text{and} \quad a_i \geq k\}. \tag{6.24c}$$

Like the number a_k^*, the number a_k'' also has a simple pictorial representation, again in terms of rows of dots, except that this time no dots are placed on the main diagonal of the array. For example, the array corresponding to the set $\{4, 4, 3, 5, 2\}$ is as shown below:

$$
\begin{array}{c|cccccc}
 & a_1'' & a_2'' & a_3'' & a_4'' & a_5'' & a_6'' \\
a_1 & & \cdot & \cdot & \cdot & & \cdot \\
a_2 & \cdot & & \cdot & \cdot & & \cdot \\
a_3 & \cdot & \cdot & & & & \cdot \\
a_4 & \cdot & \cdot & & & \cdot & \cdot \\
a_5 & \cdot & \cdot & & & & \\
\end{array}
$$

Then the number a_k'' is the number of dots in the kth column of the diagonally restricted array of dots. Thus, $a_1''=4$, $a_5''=3$, and so on. Using this interpretation, we see immediately that the left-hand sides of (6.16b) and (6.16c) are just $\sum_{t=1}^{n} a_t''$ and $\sum_{t=1}^{k} a_t''$, respectively. Thus, the conditions of (6.16) are simply that the partial sums of the sequence b_i be dominated by those of the sequence a_i''.

This result was first pointed out by FULKERSON [1960] in connection with his study of the existence of a square matrix of zeros and ones having prescribed row and column sums and zero trace.

COROLLARY 6.7: Consider a set $\{(a_i, b_i)\}$ in which, by altering the indices if necessary, $b_i \geq b_{i+1}$ and, furthermore, if $b_i = b_{i+1}$ then $a_i \geq a_{i+1}$ for $i = 1, 2, \ldots, n-1$, $n \geq 2$. The set is realizable as the degree pairs of the nodes of a $(1, 0)$-digraph if, and only if,

$$\sum_{i=1}^{n} a_i = \sum_{j=1}^{n} b_j \tag{6.25a}$$

and

$$\sum_{i=1}^{k} a_i'' \geq \sum_{j=1}^{k} b_j \quad \text{for} \quad k = 1, 2, \ldots, n. \tag{6.25b}$$

Following a similar argument, as in the proof of Corollary 6.6, a generalized version of the above corollary may be stated (Problems 6.7 and 6.11).

COROLLARY 6.8: In a given set $\{(a_i, b_i)\}$ if the indices of a_i and b_i can be altered such that $a_i \geq a_{i+1}$ and $b_i \geq b_{i+1}$ for $i = 1, 2, \ldots, n-1$, $n \geq 2$, then the set is realizable as the degree pairs of the nodes of a (p, s)-digraph, $p \geq s$ if, and only if,

$$\sum_{i=1}^{n} a_i = \sum_{j=1}^{n} b_j, \tag{6.26a}$$

$$\sum_{i=1}^{n} \min\left[a_i, (n-1)p + s\right] \geq \sum_{j=1}^{n} b_j, \qquad (6.26b)$$

$$\sum_{i=1}^{k} \min\left[a_i, (k-1)p + s\right] + \sum_{t=k+1}^{n} \min\left[a_t, kp\right] \geq \sum_{j=1}^{k} b_j \qquad (6.26c)$$

for $k = 1, 2, \ldots, n-1$.

We now study the realizability conditions for a more restricted class of directed graphs called semi-regular directed graphs.

DEFINITION 6.2: *Semi-regular directed graph*. A directed graph is said to be *semi-regular with incoming (outgoing) degree* k if $d^-(i)$ $(d^+(i))$ is equal to k for each node i in the directed graph.

COROLLARY 6.9: In a given set $\{(a_i, b_i)\}$ in which, by altering the indices if necessary, $b_i \geq b_{i+1}$ for $i = 1, 2, \ldots, n-1$, $n \geq 2$, then the set is realizable as the degree pairs of the nodes of a semi-regular $(1, 0)$-digraph with outgoing degree k if, and only if, $a_j = k$ and $b_j \leq n-1$ for $j = 1, 2, \ldots, n$, and

$$\sum_{t=1}^{n} b_t = kn. \qquad (6.27)$$

COROLLARY 6.10: The set in the above corollary is realizable as the degree pairs of the nodes of a regular $(1, 0)$-digraph of degree k if, and only if, $a_i = b_i = k \leq (n-1)$.

The proof of these two corollaries is straightforward, and is left as an exercise (Problem 6.9).

Example 6.3: Consider the directed graph G_d as shown in fig. 6.7. Suppose that we wish to construct a 4-node $(2, 1)$-digraph which has the same degree requirements as G_d. This amounts to realizing the set

$$\{(a_i, b_i); i = 1, 2, 3, 4\} = \{(5, 6), (5, 4), (4, 4), (4, 4)\}$$

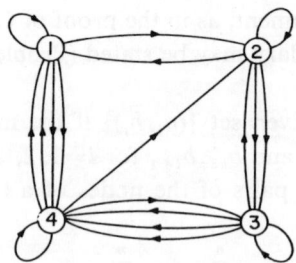

Fig. 6.7. A $(3,1)$-diagram whose degree pairs of the nodes satisfy (6.26) with $p = 2$ and $s = 1$.

as the degree pairs of the nodes of a 4-node (2, 1)-digraph. Since the elements of the set have already been ordered properly, using Corollary 6.8 we need to test the following:

(6.26a): $\quad 5 + 5 + 4 + 4 = 18 = 6 + 4 + 4 + 4,$
(6.26b): $\quad 5 + 5 + 4 + 4 = 18 \geq 6 + 4 + 4 + 4 = 18,$
(6.26c): $\quad 1 + 2 + 2 + 2 = 7 \geq 6,$
$\quad\quad\quad\quad 3 + 3 + 4 + 4 = 14 \geq 6 + 4 = 10,$
$\quad\quad\quad\quad 5 + 5 + 4 + 4 = 18 \geq 6 + 4 + 4 = 14.$

Thus, there is a 4-node (2, 1)-digraph with the desired degree requirements.

Example 6.4: Consider the set

$$\{(a_i, b_i); i = 1, 2, \ldots, 8\}$$
$$= \{(2, 3), (2, 3), (2, 3), (2, 3), (2, 1), (2, 1), (2, 1), (2, 1)\}.$$

The problem is to check if there exists an 8-node self-loopless directed graph having no parallel edges whose degree pairs of the nodes are those given in the set. Since from Corollary 6.9, $a_j = 2$ and $b_j \leq 7$ for $j = 1, 2, \ldots, 8$ and

$$8 \cdot 2 = 16 = 3 + 3 + 3 + 3 + 1 + 1 + 1 + 1 = 16,$$

the problem therefore possesses a solution which may be found by applying the algorithm to be described in the next section.

1.3. *A simple algorithm for the realization*

The usefulness of the results described so far depends upon the ability to construct a directed graph if the given specifications satisfy the constraints stated in Theorem 6.1 or their equivalences. This constructive procedure also serves as a proof of sufficiency of Theorem 6.1.

Let B_0 be a $2n$-node directed bipartite graph with node sets V and V', $\alpha(V) = \alpha(V') = n$, such that its corresponding directed graph is a (p, s)-digraph and, furthermore,

$$d^+(i) \leq a_i \quad \text{and} \quad d^-(j') \leq b_j$$

for all nodes $i \in V$ and $j' \in V'$, respectively, where i and i' are the corresponding nodes of B_0.

Step 1. If in B_0, $d^+(i) < a_i$ and $d^-(j') < b_j$ for some $i \in V$ and $j' \in V'$, and such that either

$$\alpha(i, j') < s \quad \text{for} \quad i = j \quad\quad\quad (6.28a)$$

or

$$\alpha(i, j') < p \quad \text{for} \quad i \neq j \quad\quad\quad (6.28b)$$

is satisfied, construct a new directed bipartite graph B_1 from B_0 by inserting an additional edge (i, j') to B_0. Repeat this process until no more edges can be added. Let the final directed graph thus obtained be denoted by B_{β_1}, where β_1 is the number of edges added in the above process. Evidently, B_t has one more edge than B_{t-1} for $t = 1, 2, ..., \beta_1$.

Step 2. If in B_{β_1}, there exists a sequence of nodes $i_1, j'_1, i_2, j'_2, ..., i_q, j'_q$ $(q \geq 2)$ where $i_u \in V$ and $j'_u \in V'$ and

$$d^+(i_1) < a_{i_1} \quad \text{and} \quad d^-(j'_q) < b_{j_q}, \tag{6.29a}$$

and such that either

$$\alpha(i_u, j'_u) < p \quad \text{for} \quad i_u \neq j_u \tag{6.29b}$$

or

$$\alpha(i_u, j'_u) < s \quad \text{for} \quad i_u = j_u, \tag{6.29c}$$

$u = 1, 2, ..., q$, is satisfied, and

$$\alpha(i_{x+1}, j'_x) > 0 \tag{6.29d}$$

for $x = 1, 2, ..., q-1$, construct a new directed bipartite graph B_{β_1+1} from B_{β_1} by removing the edges (i_{x+1}, j'_x), $x = 1, 2, ..., q-1$, and by adding the new edges (i_u, j'_u), $u = 1, 2, ..., q$. Repeat this process until no more edges can be added. Let the final directed graph thus obtained be denoted by B_β where $\beta = \beta_1 + \beta_2$ and β_2 is the number of edges added in Step 2. Again, B_t has one more edge than B_{t-1} for $t = 1, 2, ..., \beta$.

The problem remaining is that if the set $\{(a_i, b_i)\}$ satisfies the constraints imposed in Theorem 6.1 and if the processes of Steps 1 and 2 cannot be repeated for B_β, then the corresponding directed graph of B_β is a realization of the set. The proof is given as follows:

By the nature of our constructive processes described above, it is apparent that B_β is a (p, s)-digraph and such that $d^+(i) \leq a_i$, $i \in V$, and $d^-(j') \leq b_j$, $j' \in V'$. Since a_i and b_j are assumed to be finite, β_1 and β_2 must also be finite, and the processes will be terminated at some point. If we can show that in the resulting B_β the equalities $d^+(i) = a_i$, $i \in V$, and $d^-(j') = b_j$, $j' \in V'$, hold then B_β would be a realization.

In B_β let

$$V_1 = \{i; i \in V \quad \text{and} \quad d^+(i) < a_i\}, \tag{6.30}$$

and define the subsets X of V and X' of V' recursively as follows:

(1) V_1 is a subset of X.

(2) If i is in X and if for some $j \in V$ either $\alpha(i, j') < p$ for $i \neq j$ or $\alpha(i, j') < s$ for $i = j$, $j' \in V'$, is satisfied, then j' is in X'.
(3) If j' is in X' and if for some $i \in V$, $\alpha(i, j') > 0$, then i is in X.

The proof that B_β is a realization is contained in the following lemmas.

LEMMA 6.1: If V_1 is empty then B_β is a realization.

Proof. From (6.2a) it follows that no $j' \in V'$ exists such that $d^-(j') < b_j$. Hence, the lemma holds.

LEMMA 6.2: For each $j' \in X'$, $d^-(j') = b_j$.

Proof. Assume otherwise. It follows from our definition of the subsets X and X' that there exists a sequence of nodes with the properties described in Step 1 or 2. Since by hypothesis no such sequence can exist, the lemma follows.

Using Lemma 6.2 in conjunction with (6.2a), the following results are evident (Problem 6.12).

LEMMA 6.3: If V_1 is not empty, then there exists at least one node $j' \in (V' - X')$ such that $d^-(j') < b_j$.

LEMMA 6.4: If $X \neq V$, then for each node $i \in (V - X)$, $d^+(i) = a_i$, and furthermore, there is no edge of the form (i, j'), $j' \in X'$.

LEMMA 6.5: If $X' \neq V'$, then for each $i \in X$ and $j' \in (V' - X')$ we have

$$\alpha(i, j') = p \quad \text{for} \quad i \neq j, \tag{6.31a}$$

$$\alpha(i, j') = s \quad \text{for} \quad i = j, \tag{6.31b}$$

where j is in V.

For each subset V_x of V, let $S_A(V_x)$ be the subset of $\{a_i\}$ such that a_i is in $S_A(V_x)$ if and only if i is in V_x. Similarly, for each subset V'_x of V', let $S_B(V'_x)$ be the subset of $\{b_i\}$ such that b_j is in $S_B(V'_x)$ if and only if j' is in V'_x. From Lemmas 6.4 and 6.5 the following result is obvious (Problem 6.14).

LEMMA 6.6: If $X' \neq V'$, the total number N of edges (i, j'), $i \in V$ and $j' \in (V' - X')$, in B_β is given by

$$N = \sum_{a_i \in S_1} a_i + \sum_{a_j \in S_2} a_j + \sum_{a_i \in S_3} \{[\alpha(V_2) - 1] p + s\} + \sum_{a_j \in S_4} \alpha(V_2) p, \tag{6.32}$$

where
$$S_1 = S_A[V - (V_2 \cup X)],$$
$$S_2 = S_A[V_2 \cap (V - X)],$$
$$S_3 = S_A[V_2 \cap X],$$
$$S_4 = S_A[V - \{V_2 \cup (V - X)\}],$$

and $V_2' = V' - X'$ and V_2 is the corresponding node set of $V_2' \in V$.

Using Lemma 6.6 in conjunction with Lemma 6.3, the lemma below is obvious.

LEMMA 6.7: If V_1 is not empty then

$$N < \sum_{b_j \in S_B(V'_2)} b_j. \tag{6.33}$$

However, by hypothesis, (6.2b) must be satisfied for every nonempty subset S_B. In particular, if we let $S_B = S_B(V_2')$ then $S_A = S_A(V_2)$, and (6.2b) can be written in a slightly different form:

$$\sum_{a_i \in S_1} \min[a_i, \alpha(S_A)\,p] + \sum_{a_j \in S_2} \min\{a_j, [\alpha(S_A) - 1]\,p + s\}$$
$$+ \sum_{a_i \in S_3} \min\{a_i, [\alpha(S_A) - 1]\,p + s\} + \sum_{a_j \in S_4} \min[a_j, \alpha(S_A)\,p]$$
$$= \sum_{a_i \in S_1} a_i + \sum_{a_j \in S_2} a_j + \sum_{a_i \in S_3} \{[\alpha(V_2) - 1]\,p + s\} + \sum_{a_j \in S_4} \alpha(V_2)\,p$$
$$\geq \sum_{b_t \in S_B(V'_2)} b_t.$$

Thus, from (6.32) and (6.33) V_1 must be empty. By Lemma 6.1, we conclude that B_β is a realization of $\{(a_i, b_i)\}$ which completes the proof of sufficiency.

We shall illustrate the procedures outlined in Steps 1 and 2 by the following examples.

Example 6.5: Consider the set

$\{(a_i, b_i); i = 1, 2, \ldots, 8\}$
$$= \{(2, 1), (2, 1), (2, 3), (2, 3), (2, 3), (2, 3), (2, 1), (2, 1)\},$$

which, by Example 6.4, is known to be realizable as the degree pairs of the nodes of an 8-node (1, 0)-digraph. Using the algorithm outlined in this section, we shall realize a (1, 0)-digraph with the above properties.

§1   Existence and realization   417

In fig. 6.8, the subgraph consisting of the edges corresponding to the solid lines is a 16-node directed bipartite graph B_0 which has the property that its corresponding directed graph is a (1, 0)-digraph.

Step 1. In B_0 add the following edges one at a time:

$$\{(1, 3'), (2, 4'), (3, 5'), (4, 6'), (5, 3'), (6, 4'), (7, 5'), (8, 6')\}.$$

The order in adding these edges is immaterial. The final directed bipartite graph B_8 is presented in fig. 6.8 with the added edges being represented by the dashed lines.

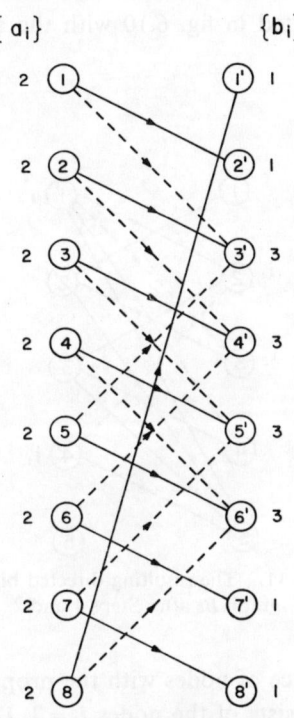

Fig. 6.8. A 16-node directed bipartite graph.

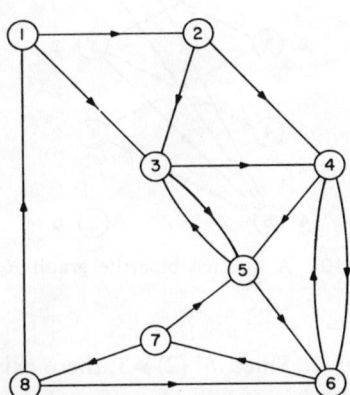

Fig. 6.9. The corresponding (1,0)-digraph of the directed bipartite graph of fig. 6.8.

Since there is no node i in V with $d^+(i) < a_i$, Step 2 is not necessary, and the corresponding directed graph, as shown in fig. 6.9, of B_8 is a realization of the set.

Example 6.6: The set

$$\{(a_i, b_i); i = 1, 2, 3, 4, 5\} = \{(1, 4), (3, 3), (2, 3), (2, 2), (4, 0)\}$$

is known to be realizable (see Example 6.2) as the degree pairs of the nodes of a 5-node (1, 0)-digraph. One of such realizations is given in fig. 6.6. We wish to construct another one which has the same properties.

In fig. 6.10, the spanning subgraph consisting of the edges corresponding to the solid lines is a 10-node directed bipartite graph B_0 which has the property that its corresponding directed graph is a (1, 0)-digraph.

Step 1. In B_0 add the following edges one at a time:

$$\{(2, 1'), (3, 2'), (4, 3'), (4, 1')\}.$$

The final directed bipartite graph B_4 is presented in fig. 6.10 with the added edges being represented by the dashed lines.

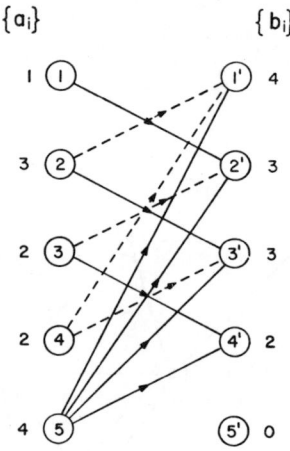

Fig. 6.10. A directed bipartite graph B_0.

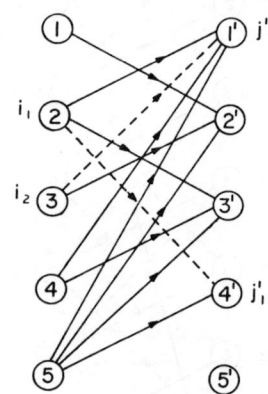

Fig. 6.11. The resulting directed bipartite graph B_5 after Steps 1 and 2.

Step 2. Since $d^+(2) \neq 3$, there exists a sequence of nodes with the properties described in (6.29). One of such sequences consists of the nodes $i_1 = 2$, $j_1' = 4'$, $i_2 = 3$, and $j_2' = 1'$. In B_4 if we remove edge (3, 4') and then add edges (2, 4') and (3, 1'), we obtain B_5 which is given in fig. 6.11 with the added edges being represented by the dashed lines. Since there exists no node i in V with $d^+(i) < a_i$, Step 2 has now been completed. The corresponding directed graph, as shown in fig. 6.12, of B_5 is another realization of the set. It is interesting to note that the two realizations are not isomorphic since one has three circuits of length 2 (fig. 6.6) while the other has only two (fig. 6.12).

Finally, it should be pointed out that the node sequences described in Step 2 are not really too difficult to enumerate; it is equivalent to generating an alternating path P with endpoints i_1 and j'_q in the directed bipartite graph B_d which is based on the same node sets as B_t and which has the properties that $\alpha(i, j') = p$ for $i \neq j$ and $\alpha(i, i') = s$. Clearly, B_t is a subgraph of B_d. If $\bar{B}_t$ denotes the complement of B_t in B_d, then by an *alternating path* with respect to B_t in B_d we mean a path P (not directed path) in which the edges belong alternately to B_t and $\bar{B}_t$, beginning in an edge (i_1, j'_1) in $\bar{B}_t$ and also terminating in an edge (i_q, j'_q) in $\bar{B}_t$. For example, the path $P = (2, 4') \cup (3, 4') \cup (3, 1')$ is an alternating path with respect to B_4 of fig. 6.10 in B_d.

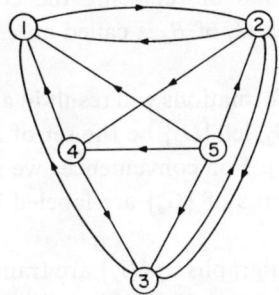

Fig. 6.12. The corresponding $(1, 0)$-digraph of the directed bipartite graph of fig. 6.11.

1.4. *Degree invariant transformations*

It is apparent from the previous discussion that if a set of nonnegative integer pairs is realizable as the degree pairs of the nodes of a (p, s)-digraph, the realization in general is not unique, and an example is given in Example 6.6. Suppose that $\{G_d\}$ represents the set of all possible realizations, the problem of generating $\{G_d\}$ from any one of its elements is interesting, and will be considered in this section.

Unless stated to the contrary, in this chapter two directed graphs G_1 and G_2 are considered identical when they are isomorphic and may be denoted by $G_1 = G_2$.

DEFINITION 6.3: *d-invariant.* Two (p, s)-digraphs G_1 and G_2 are said to be *d-invariant* if there exists a one-to-one correspondence between the nodes of G_1 and G_2 such that the corresponding nodes have the same degree pair.

DEFINITION 6.4: *Elementary (p, s) d-invariant transformation.* If in a directed bipartite graph B_d with node sets V and V', there exist two edges (i, j') and

(k, t'), i, $k \in V$ and j', $t' \in V'$ and $i \neq k$, $j' \neq t'$, such that

$$\alpha(i, t') < p \quad \text{if} \quad i \neq t, \tag{6.34a}$$
$$\alpha(i, t') < s \quad \text{if} \quad i = t, \tag{6.34b}$$
$$\alpha(k, j') < p \quad \text{if} \quad k \neq j, \tag{6.34c}$$
$$\alpha(k, j') < s \quad \text{if} \quad k = j, \tag{6.34d}$$

the operation of replacing the edges (i, j') and (k, t') by the edges (i, t') and (k, j') is called an *elementary (p, s) d-invariant transformation* in B_d. Similarly, for $\alpha(V) = \alpha(V')$ the operation of replacing the corresponding edges in the corresponding directed graph G_d of B_d is called an *elementary (p, s) d-invariant transformation* in G_d.

Clearly, this kind of transformations will result in a d-invariant (p, s)-digraph. For a given (p, s)-digraph G_d, let $\{G_d\}$ be the set of all possible (p, s)-digraphs that are d-invariant from G_d. For convenience, we assume that all the corresponding nodes of the elements of $\{G_d\}$ are labeled by the same integer.

THEOREM 6.2: Two (p, s)-digraphs of $\{G_d\}$ are transformable from each other by a finite sequence of elementary (p, s) d-invariant transformations. (For the minimal number of such transformations required, see Problems 6.59–6.61.)

Proof. As in Definition 5.9, we first introduce a distance between pairs of (p, s)-digraphs in $\{G_d\}$. Let G_α and G_β be any two non-isomorphic (p, s)-digraphs of $\{G_d\}$. If $\alpha(i, j)$ and $\beta(i, j)$ denote the numbers of edges directed from node i to node j in G_α and G_β, respectively, then the *distance* between G_α and G_β is defined as

$$\tfrac{1}{2} \sum_{i=1}^{n} \sum_{j=1}^{n} |\alpha(i, j) - \beta(i, j)|,$$

where n is the number of nodes of G_d.

Let $\{G_\alpha\}$ be the set of all (p, s)-digraphs into which G_α is transformable by finite sequences of elementary (p, s) d-invariant transformations, and let $\{G_\beta\}$ be the corresponding set arising from G_β. Let G_1 and G_2 be the elements of $\{G_\alpha\}$ and $\{G_\beta\}$, respectively, such that the distance between them is the minimum distance between the (p, s)-digraphs in $\{G_\alpha\}$ and $\{G_\beta\}$. If the distance between G_1 and G_2 is zero or if G_1 and G_2 are isomorphic, we are finished. If $p = 0$ then all the elements of $\{G_d\}$ are isomorphic. So let us assume that $p \neq 0$ and G_1 and G_2 are not isomorphic and of positive distance.

We denote by $G_1 \oplus G_2$ the n-node directed graph obtained from G_1 and G_2 such that (i, j) is in $G_1 \oplus G_2$ if and only if (i, j) is either in G_1 or in G_2 but not in both. Now in $G_1 \oplus G_2$ we introduce some notation. If the edge directed from node i to node j is in G_1 but not in G_2 we shall write $(i,j)_1$. If the edge directed from node i to node j is in G_2 but not in G_1 we shall write $(i,j)_2$. Since G_1 and G_2 are not the same, there must exist at least one pair of distinct nodes i_1 and i_2 such that $(i_1, i_2)_1$ is in $G_1 \oplus G_2$. Since the incoming degree of i_2 is the same in both G_1 and G_2, there must exist a node i_3, $i_3 \neq i_1$, such that $(i_3, i_2)_2$ is in $G_1 \oplus G_2$. Continuing this way, the process can be repeated indefinitely only by generating a closed edge train. Thus, for some positive integer k, $k \geq 2$, there exists a sequence of nodes

$$i_1, i_2, \ldots, i_{2k-1}, i_{2k}, i_1 \tag{6.35a}$$

such that

$$(i_1, i_2)_1, (i_3, i_2)_2, \ldots, (i_{2k-1}, i_{2k})_1, (i_1, i_{2k})_2. \tag{6.35b}$$

The subgraph formed by the edges of (6.35b) is a closed edge train in $G_1 \oplus G_2$. Note that not all the nodes in (6.35a) are necessarily distinct. However, nodes i_1 and i_2 are distinct. We first examine the case where $k=2$. We then have

$$(i_1, i_2)_1, (i_3, i_2)_2, (i_3, i_4)_1, (i_1, i_4)_2.$$

Hence, an elementary (p, s) d-invariant transformation which replaces edges (i_3, i_2) and (i_1, i_4) by edges (i_1, i_2) and (i_3, i_4) in G_2 would result in a (p, s)-digraph in $\{G_\beta\}$ which is closer to G_1 than G_2, violating our assumption on the minimality of the distance between G_1 and G_2. Thus, $k > 2$. We shall now prove that no sequence of nodes of type (6.35) can exist by induction over k. Assume that the assertion is true for any sequence (6.35) of length (number of edges) $2q$, $q < k$. We shall prove the non-existence of such a sequence of length $2k$, $k > 2$. We now consider two cases.

Case 1: $s \neq 0$. If $\alpha(i_1, i_4) < p$ for $i_1 \neq i_4$ or $\alpha(i_1, i_4) < s$ for $i_1 = i_4$, then the operation of replacing the edges (i_3, i_4) and (i_1, i_2) by the edges (i_1, i_4) and (i_3, i_2) in G_1 is an elementary (p, s) d-invariant transformation and would result in a (p, s)-digraph in $\{G_\alpha\}$ which is closer to G_2 than G_1, a contradiction. Hence $\alpha(i_1, i_4) \geq 1$. If $\beta(i_1, i_4)$ of G_2 is not zero, then an elementary (p, s) d-invariant transformation of replacing the edges (i_1, i_4) and (i_3, i_2) by the edges (i_3, i_4) and (i_1, i_2) in G_2 would result in a (p, s)-digraph in $\{G_\beta\}$ which is closer to G_1 than G_2. Hence $\beta(i_1, i_4) = 0$. Now, consider the sequence of nodes i_1, i_4, $i_5, \ldots, i_{2k-1}, i_{2k}, i_1$. This sequence has the same property as (6.35) except its length is shorter than $2k$. By induction hypothesis, no such sequence can exist. Thus, G_1 and G_2 are either isomorphic or of distance zero, a contradiction.

Case 2: $s = 0$. Then nodes i_1, i_2, and i_3 are distinct. Suppose that there are at least four distinct nodes in the sequence (6.35). Then we may assume, without loss of generality, that i_1, i_2, i_3, and i_4 are distinct nodes of (6.35), and that the edges $(i_1, i_2)_1$, $(i_3, i_2)_2$, and $(i_3, i_4)_1$ are in $G_1 \oplus G_2$. Following an argument similar to that of Case 1, we conclude that no such sequence can exist. Hence there are exactly three distinct nodes in (6.35). Without loss of generality, let these nodes be i_x, $x = 1, 2, 3$. This means that the closed edge train of (6.35b) consists of at least two directed circuits of length 3 oriented in the opposite directions. If in G_1 or G_2 there exists a node j, which is distinct from i_x, and such that, say, (i_1, j) or (j, i_1) is in G_1 or G_2 and $\alpha(i_y, j) < p$ or $\alpha(j, i_y) < p$ for $y = 2$ or 3, then an elementary $(p, 0)$ d-invariant transformation involving the nodes i_1, i_2, i_3, and j would yield a $(p, 0)$-digraph G_3 in $\{G_\alpha\}$ or $\{G_\beta\}$ such that the distance between G_3 and G_2 or G_3 and G_1 is the same as that between G_1 and G_2. Furthermore, in $G_3 \oplus G_2$ or $G_3 \oplus G_1$ there exists a sequence of nodes of type (6.35) containing at least four distinct nodes, so that either the Case 1 argument applies or our assumption on the minimality of the distance between G_1 and G_2 is violated. Thus, in G_1 (G_2) either $\alpha(i_x, j) = p$ or $\alpha(j, i_x) = p$ or no such node j exists. In any case, G_1 and G_2 would be isomorphic, again a contradiction. So the theorem is proved.

Thus, for any G_α and G_β of $\{G_d\}$ there exists a finite sequence of (p, s)-digraphs G_x, $x = 1, 2, \ldots, k$ such that

$$G_\alpha = G_1, G_2, \ldots, G_k = G_\beta,$$

where G_i and G_{i+1}, $i = 1, 2, \ldots, k-1$, are related by an elementary (p, s) d-invariant transformation.

COROLLARY 6.11: Every element of $\{G_d\}$ is isomorphic with G_d if, and only if, each elementary (p, s) d-invariant transformation would result in a directed graph which is isomorphic with G_d.

As an example, consider the d-invariant $(1, 0)$-digraphs of figs. 6.6 and 6.12. If in fig. 6.12 we replace the edges $(1, 2)$ and $(4, 3)$ by $(1, 3)$ and $(4, 2)$, we obtain a $(1, 0)$-digraph which is isomorphic to the one given in fig. 6.6.

For simple directed graphs, these transformations can usually be obtained by inspection. However, for complicated graphs, a systematic procedure is necessary. The process outlined in the proof of Theorem 6.2 would serve for this purpose.

1.5. *Realizability as a connected (p, s)-digraph*

Necessary and sufficient conditions for the realization of a set of nonnegative integer pairs as the degree pairs of the nodes of a connected (p, s)-digraph, an acyclic (p, s)-digraph, and a tree will be presented in this section. The development follows the work of HAKIMI [1965] and CHEN [1966e].

LEMMA 6.8: Let G_d be a (p, s)-digraph containing k components with $p \neq 0$ and $k \geq 2$. If G_d has no isolated nodes and if one of its components contains a circuit (not necessarily directed), then there exists a (p, s)-digraph which is d-invariant from G_d but has $k-1$ components.

Proof. Let (i, j) be a circuit edge of a component g_α. Let (u, v) be an edge in another component g_β. If we replace the edges (i, j) and (u, v) by the edges (i, v) and (u, j), we obtain a (p, s)-digraph which has the desired property. So the lemma is proved.

THEOREM 6.3: The necessary and sufficient conditions for a set $\{(a_i, b_i)\}$ to be realizable as the degree pairs of the nodes of a connected (p, s)-digraph, $n \geq 2$, are

(a) the set is realizable as the degree pairs of the nodes of a

$$(p, s)\text{-digraph (it satisfies (6.2))}, \tag{6.36a}$$

(b) $\qquad a_i + b_i \neq 0 \quad \text{for} \quad i = 1, 2, ..., n,$ (6.36b)

(c) $\qquad p \neq 0,$ (6.36c)

(d) $\qquad \displaystyle\sum_{i=1}^{n} a_i \geq (n-1).$ (6.36d)

Proof. Necessity. Let G_d be a connected (p, s)-digraph realization of the set. The first three conditions are obviously true. The fourth condition follows from the fact that the number of edges in any connected (p, s)-digraph is at least $(n-1)$, and therefore

$$\sum_{i=1}^{n} a_i = \sum_{j \in V} d^+(j) \geq (n-1),$$

where V is the node set of G_d.

Sufficiency. From (6.36a), we may assume that G_d is a (p, s)-digraph realization of the set. The problem now is equivalent to proving that there exists an element in $\{G_d\}$ which is connected. If G_d is connected, there is no problem; therefore, let us assume that G_d has k, $k \geq 2$, components. Without loss of generality, we may further assume that G_d contains no circuits; for, if not, there is an element in $\{G_d\}$ which has $k-1$ components as suggested by (6.36b), (6.36c), and Lemma 6.8, and we can continue this operation until all the components are circuitless or we find a connected (p, s)-digraph in $\{G_d\}$. If all the components of G_d are circuitless, then the number of edges in G_d is equal to $n-k$. By (6.36d) this would imply that

$$\sum_{i=1}^{n} a_i = (n-k) \geq (n-1).$$

Since $k \geq 2$, this is clearly impossible, hence the theorem.

The following corollary can easily be established as a simple consequence of this theorem, and its proof is left as an exercise (Problem 6.19).

COROLLARY 6.12: The necessary and sufficient conditions for a set $\{(a_i, b_i)\}$ to be realizable as the degree pairs of the nodes of a tree are that $(a_i + b_i) \neq 0$ for $i = 1, 2, \ldots, n$, and

$$\sum_{i=1}^{n} a_i = \sum_{j=1}^{n} b_j = n - 1. \tag{6.37}$$

This result was also given by MENON [1964]. If we further stipulate that $a_n = 0$ and $a_k = 1$ for $k = 1, 2, \ldots, n-1$, the set is also realizable as the degree pairs of the nodes of a directed tree t_n (Problem 6.20).

LEMMA 6.9: A connected directed graph G_d is strongly connected if, and only if, every edge of G_d is contained in at least one directed circuit.

Proof. The necessary part is self-evident. To prove sufficiency, we assume that in G_d there exist two nodes i and j such that there is no directed path from node i to node j, and then show that it is impossible. Since G_d is connected, there exists a path P_{ij} (not directed path) connected between the nodes i and j. If P_{ij} is of length 1, then there is a directed path from node i to node j in G_d since P_{ij} is an edge which by assumption must be contained in some directed circuit. Assume that the assertion is correct for any P_{ij} which is of length $k-1$ or less, $k \geq 2$. To complete the induction, we assume that P_{ij} is of length k. If the last edge in P_{ij} is (x, y) where $x = j$ or $y = j$, by induction hypothesis there exists a directed path from node i to node x or y in G_d. Since by assumption every edge of G_d is contained in some directed circuit, there exists a directed path from node y to node x and a directed path from node x to node y in G_d, the latter being the edge (x, y). Thus, there is a directed path from node i to node j in G_d, a contradiction. So the lemma is proved.

COROLLARY 6.13: The necessary and sufficient conditions for a set $\{(a_i, b_i)\}$

to be realizable as the degree pairs of the nodes of a strongly-connected directed graph without self-loops are

(a) the set is realizable as the degree pairs of the nodes of a self-loopless directed graph (it satisfies (6.4)), (6.38a)

(b) $\min(a_i, b_i) \neq 0$ for $i = 1, 2, \ldots, n, n \geq 2$. (6.38b)

Proof. The necessary part is obvious. To prove sufficiency, we first note that if the set satisfies (6.38) then it is realizable as the degree pairs of the nodes of a connected self-loopless directed graph, for the set satisfies all four conditions (6.36) in Theorem 6.3. Let G_d be one of such realizations. If every edge of G_d is contained in some directed circuit, by Lemma 6.9 G_d is a desired realization. Suppose that there exists an edge (i, j) in G_d which is not in any directed circuit. Since by (6.38b), $d^+(j) \neq 0$, there exists an edge (j, k), and similarly there is an edge (k, t). Continuing this process, we will generate a directed-edge train. Since G_d is finite, the argument can be continued indefinitely only by generating a directed circuit, say, L_1. Using a similar argument, we can establish the existence of another directed-edge train which terminates in an edge (u, i) since $d^-(v) \neq 0$ for all nodes v of G_d. The second directed-edge train must also contain a directed circuit, say, L_2. If the two directed circuits intersect each other at some node, then the edge (i, j) will be contained in some directed circuit. This is impossible by assumption, so we have two node-disjoint directed circuits. Let (i_1, i_2) and (j_1, j_2) be the edges in L_1 and L_2, respectively. Since G_d has no self-loops, $i_1 \neq i_2$ and $j_1 \neq j_2$. If we replace the edges (i_1, i_2) and (j_1, j_2) by the edges (i_1, j_2) and (j_1, i_2), we obtain a new directed graph G_1 which is d-invariant from G_d and which does not contain any self-loops. Furthermore, it is clear that every edge that was in some directed circuit in G_d remains to be so in G_1, and the edge (i, j) is in a directed circuit. Obviously, we can continue this process until we obtain a directed graph G_β which is d-invariant from G_d and in which every edge is in some directed circuit. So by Lemma 6.9 the corollary is proved.

We should point out that the condition (6.38b) is not necessary for the realization of a strongly-connected (p, s)-digraph. For example, consider the $(1, 0)$-digraph G_d as shown in fig. 6.13. There exists no strongly-connected $(1, 0)$-digraph in $\{G_d\}$. The reason for this is that by Theorem 6.2 any such realization can be obtained from G_d by a finite sequence of elementary $(1, 0)$ d-invariant transformations. This is clearly impossible for G_d. On the other hand, if (6.38b) is satisfied for at most one i, then all the realizations of the set have no directed circuits (BROWNLEE [1968]). (Also see Problems 6.66 and 6.67 for related results.)

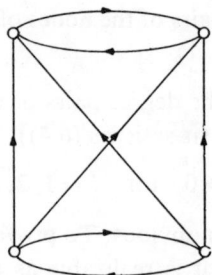

Fig. 6.13. A (1,0)-digraph which cannot be transformed into a strongly-connected (1,0)-digraph by any finite sequence of elementary (1,0) d-invariant transformations.

COROLLARY 6.14: Let the set $\{(a_i, b_i)\}$ be realizable as the degree pairs of the nodes of a self-loopless directed graph. A necessary and sufficient condition for all the self-loopless realizations of the set to be acyclic is that min $(a_i, b_i) \neq 0$ for at most one i where $i = 1, 2, ..., n$.

Proof. Necessity. Assume that there exists an acyclic realization G_d which has the property that $\min [d^+(x), d^-(x)] \neq 0$ for at least two nodes x of G_d. Let i and j, $i \neq j$, be two such nodes. Then there exist edges (u, i), (i, v), (y, j), and (j, z) in G_d. Since G_d is acyclic, nodes u, i, and v are all distinct. Similarly, all the nodes y, j, and z are distinct. If we replace the edges (u, i), (i, v), (y, j), and (j, z) by the edges (u, v), (i, i), (j, j), and (y, z), we obtain a directed graph G_1 which has two self-loops and which is d-invariant from G_d. If in G_1 we replace the two self-loops by the edges (i, j) and (j, i), we obtain a directed graph G_2 which is d-invariant from G_d and which has at least one directed circuit and has no self-loops. This is contrary to our assumption that all the self-loopless realizations of the set are acyclic. So the condition is necessary.

Sufficiency. Let G_d be a self-loopless realization of the set with the property that there exists at most one node i such that $\min [d^+(i), d^-(i)] \neq 0$. Since a directed circuit of length greater than 1 contains at least two nodes x with $\min [d^+(x), d^-(x)] \neq 0$, it is evident that G_d must be acyclic. Hence the condition is sufficient.

Note that the corollary does not guarantee the existence of an acyclic realization of a set of nonnegative integer pairs as the degree pairs of its nodes; it merely characterizes a class of acyclic realizations of the set.

Example 6.7: Consider the (3, 2)-digraph G_d of fig. 6.1. The corresponding set of degree pairs of its nodes is given by

$$\{[d^+(i), d^-(i)]; i = 1, 2, 3, 4, 5\} = \{(2, 2), (2, 3), (3, 3), (1, 4), (5, 1)\}.$$

§2 Realizability as a symmetric (p, s)-digraph

Suppose that we wish to construct a strongly-connected $(p, 0)$-digraph, $p < \infty$, which is d-invariant from the $(3, 2)$-digraph G_d. The problem would possess a solution if the set satisfies the conditions given in (6.4b) and (6.38b). (6.38b) is clearly satisfied since all the integers are positive. From (6.5) and (6.6), (6.4b) is equivalent to the following:

$$2 + 2 + 3 + 1 + 5 = 13 \geq 5 + 1 = 6.$$

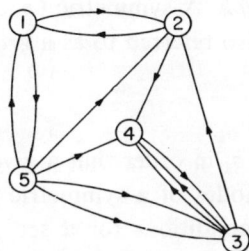

Fig. 6.14. A strongly-connected $(2, 0)$-digraph which is d-invariant from the $(3, 2)$-digraph of fig. 6.1.

Thus, there is a strongly-connected $(p, 0)$-digraph which realizes the set, and one of such realizations is presented in fig. 6.14. We remark that the original $(3, 2)$-digraph is not strongly connected since there exists no directed path from node 1 to node 5. The minimum value of p's that can be achieved for this set is 2, since in Example 6.2 we have shown that there exists no $(1, 0)$-digraph which realizes the set.

§2. Realizability as a symmetric (p, s)-digraph

In this section, we shall present necessary and sufficient conditions for a set of n nonnegative integer pairs to be the degree pairs of the nodes of an n-node symmetric (p, s)-digraph. Then from this general characterization we deduce the realizability conditions for a set of n nonnegative integers to be the degrees of the nodes of an n-node (undirected) graph.

The term "symmetric directed graph" was defined earlier (Definition 1.35). For present purpose, we need a slight modification. Perhaps we should use a different name, but we do not find it necessary, since only the modified definition will be used throughout the remainder of this chapter.

DEFINITION 6.5: *Symmetric (p, s)-digraph*. A (p, s)-digraph G_d is said to be *symmetric* if for each edge (i, j) in G_d, the edge (j, i) is also in G_d such that

$\alpha(i,j) = \alpha(j,i)$ for $i \neq j$ and $\alpha(i,j) = 2e$ for $i = j$ where e is a nonnegative integer.

By a *symmetric subgraph* of a (p, s)-digraph we mean a subgraph which is also a symmetric (p, s)-digraph.

Since an undirected graph may be transformed into a symmetric directed graph by representing each undirected edge by a pair of directed edges with opposite directions, and *vice versa*, it is convenient to use a simpler term.

DEFINITION 6.6: (p, s)-*graph*. A symmetric $(p, 2s)$-digraph is called a (p, s)-*graph*. A (∞, ∞)-graph is also referred to as a *graph*.

2.1. Existence

As it is necessary for $a_i = b_i$ in order that a given set $\{(a_i, b_i)\}$ be realizable as the degree pairs of the nodes of a symmetric directed graph, it is simpler to specify the realizability conditions for a set $\{a_i\}$ of nonnegative integers instead of a set of nonnegative integer pairs.

There is one difficulty in treating the (p, s)-graph as a special case of a (p, s)-digraph: Let G_d be a (p, s)-digraph realization of a set $\{(a_i, b_i)\}$ with $a_i = b_i$ for all i. The conditions $a_i = b_i$ are not sufficient to guarantee that there exists a (p, s)-graph realization. For example, there exists no $(1, 0)$-graph which is d-invariant from a directed circuit of length 3. In other words, the realizability conditions for a (p, s)-graph are not merely the conditions given in Theorem 6.1 and the constraints $a_i = b_i$ for all i. However, as we shall see, additional restrictions must be satisfied.

THEOREM 6.4: Let $\{G_d\}$ be the set of all possible $(p, 0)$-digraph realizations of a given realizable set $\{(a_i, b_i)\}$. Then there exists a $(p, 0)$-graph in $\{G_d\}$ if, and only if, $a_i = b_i$ for all i and

$$\sum_{j=1}^{n} a_j = 2e, \qquad (6.39)$$

where e is a nonnegative integer.

In other words, the theorem states that if a set is realizable as the degree pairs of the nodes of a $(p, 0)$-digraph, then it is realizable as the degrees of the nodes of a $(p, 0)$-graph provided the conditions of the theorem are satisfied.

The necessary part of the theorem is self-evident. The proof of sufficiency of these conditions is much more involved and will be given in the next section.

Following an argument similar to that given in § 2.2, we arrive at the following conclusions (Problem 6.24).

THEOREM 6.5: Assume that the set $\{(a_i, b_i)\}$ is realizable as the degree pairs of the nodes of a (p, s)-digraph. If $a_i = b_i$ for all i and if $\sum_{j=1}^{n} a_j$ is even, then the set is also realizable as the degrees of the nodes of a (p, h)-graph if $p \neq 0$ and a $(1, h)$-graph if $p = 0$ where h denotes the largest integer $\leq \frac{1}{2}(s+1)$.

From Corollary 6.8 and Theorem 6.4, we obtain the following result (Problem 6.27).

COROLLARY 6.15: The necessary and sufficient conditions for a set $\{a_i\}$ ($a_i \geq a_{i+1}$ for $i = 1, 2, \ldots, n-1$, $n \geq 2$) to be realizable as the degrees of the nodes of a $(p, 0)$-graph are

$$\sum_{j=1}^{n} a_j = 2e, \tag{6.40a}$$

where e is a nonnegative integer, and

$$\sum_{i=1}^{n} \{\min[a_i, (n-1)p] - a_i\} \geq 0, \tag{6.40b}$$

$$\sum_{i=k+1}^{n} \min[a_i, kp] \geq \sum_{j=1}^{k} \{a_j - \min[a_j, (k-1)p]\} \tag{6.40c}$$

for $k = 1, 2, \ldots, n-1$.

By putting $p = \infty$ in the above corollary, a basic theorem of SENIOR [1951b] is obtained.

COROLLARY 6.16: The necessary and sufficient conditions for a set $\{a_i\}$ ($a_i \geq a_{i+1}$ for $i = 1, 2, \ldots, n-1$) to be realizable as the degrees of the nodes of a graph without self-loops are

$$\sum_{j=1}^{n} a_j = 2e, \tag{6.41a}$$

$$\sum_{j=1}^{n} a_j \geq 2a_1, \tag{6.41b}$$

where e is a nonnegative integer and $n \geq 2$.

Proof. The conditions (6.40b) and (6.40c) are always satisfied for n, $k \geq 2$. For $k = 1$, (6.41b) follows immediately.

COROLLARY 6.17: A set $\{a_i\}$ is realizable as the degrees of the nodes of a graph if, and only if, $\sum_{j=1}^{n} a_j$ is even.

The proof of this corollary is straightforward, and the details are left as an exercise (Problem 6.30).

COROLLARY 6.18: The necessary and sufficient conditions for a set $\{a_i\}$ ($a_i \geq a_{i+1}$ for $i = 1, 2, \ldots, n-1$) to be realizable as the degrees of the nodes of a (1, 0)-graph are

$$\sum_{j=1}^{n} a_j = 2e, \tag{6.42a}$$

$$\sum_{i=1}^{k} a_i'' \geq \sum_{i=1}^{k} a_i \tag{6.42b}$$

for $k = 1, 2, \ldots, n$, $n \geq 2$, where e is a nonnegative integer and the a_i'' is defined in (6.24). (See Problems 6.35 and 6.47 for the equivalent conditions of (6.42b).)

The corollary follows directly from Theorem 6.4 and Corollary 6.7.

Another existential answer to this problem was given by HAVEL [1955] and HAKIMI [1962]. Their theorem is iterative in nature and can be used as an algorithm for constructing a (1, 0)-graph with prescribed degrees. We first give a lemma and then prove their result.

For a given graph G, we denote by $d(i)$ the degree of the node i of G.

LEMMA 6.10: Consider an n-node (1, 0)-graph G in which, by altering the node labelings if necessary, $d(i) \geq d(i+1)$ for $i = 1, 2, \ldots, n-1$, $n \geq 2$. Then there exists a (1, 0)-graph G_1 which is d-invariant from G and which has the property that the edges $(1, x)$, $x = 2, 3, \ldots, d(1)+1$, are all in G_1.

Proof. In G we have $d(i) \leq n-1$ for all i. Let V be the node set of G and let V_α be the subset of V consisting of the nodes from 2 to $d(1)+1$. If the edges $(1, x)$, $x \in V_\alpha$, are all in G, then there is no problem. Assume that $(1, y)$, $y \in V_\alpha$, is not in G. Then there exists a node z in $(V - V_\alpha)$ such that $(1, z)$ is in G. Since $d(y) \geq d(z)$, it is not difficult to see that there exists a node w in V, $w \neq z$, such that (w, y) is in G but (w, z) is not. If in G we replace the edges (w, y) and $(1, z)$ by the edges $(1, y)$ and (w, z), we obtain a (1, 0)-graph G_2 which is d-invariant from G and which has the property that the number of edges $(1, x)$, $x \in V_\alpha$, in G_2 is one more than that in G. Repeating this process results in a (1, 0)-graph G_1 in which the node 1 has the desired property. So the lemma is proved.

The graph G_1 obtained in the above lemma may be considered as the (1, 0)-graph G in *canonical form*. Thus, every (1, 0)-graph can be transformed into its canonical form by a finite sequence of elementary (1, 0) d-invariant transformations.

THEOREM 6.6: The necessary and sufficient conditions for a set $\{a_i\}$ ($a_i \geq a_{i+1}$

for $i=1, 2, ..., n-1$, $n\geq 2$) to be realizable as the degrees of the nodes of a $(1, 0)$-graph are that $n-1\geq a_1$ and the modified set

$$\{a_2 - 1, a_3 - 1, ..., a_{a_1+1} - 1, a_{a_1+2}, ..., a_n\} \tag{6.43}$$

be realizable as the degrees of the nodes of a $(1, 0)$-graph.

Proof. Necessity. Let G be a $(1, 0)$-graph realization of the set. Since there are at most $n-1$ edges incident at a node of G, a_1 cannot exceed $n-1$. By Lemma 6.10, there exists a $(1, 0)$-graph G_1 which is d-invariant from G and which has the property that the edges $(1, x)$, $x=2, 3, ..., d(1)+1$, are all in G_1. Thus, the removal of node 1 and all the edges incident with this node results in a $(1, 0)$-graph with the degrees of its nodes given by the set (6.43).

Sufficiency. If the set (6.43) is realizable as the degrees of the nodes of a $(1, 0)$-graph, then we can construct a new $(1, 0)$-graph G by adding a new node y and also the edges connected between y and the nodes of degrees a_2-1, $a_3-1, ..., a_{z+1}-1$ where $z=a_1$. It is clear that G is a realization of the set $\{a_i\}$. This completes the proof of the theorem.

In addition to the complete characterizations presented in (6.42) and (6.43), there are other necessary conditions that a set must satisfy in order that it be realizable. Since these conditions can usually be checked by inspection, we could avoid the more complicated test of (6.42b) or (6.43) unless it passes the inspection test. Thus, the question of whether a set is realizable may be answered more quickly. We discuss two conditions which were originally given by BEHZAD and CHARTRAND [1967].

THEOREM 6.7: There exists no $(1, 0)$-graph in which no two of its nodes have the same degree.

Proof. Let G be an n-node $(1, 0)$-graph in which $d(i)>d(i+1)$ for $i=1, 2, ..., n-1$, $n\geq 2$. If no two of its nodes have the same degree, then since $0\leq d(j)\leq n-1$, it follows that $d(j)=n-j$ for $j=1, 2, ..., n$. This implies, however, that n is an isolated node while edges $(1, x)$, $x=2, 3, ..., n$, are all in G, which is clearly impossible. Hence, no such $(1, 0)$-graph G exists.

Since no sequence of distinct nonnegative integers can be realized, we consider those sequences in which exactly two integers are equal.

COROLLARY 6.19: Let G be an n-node $(1, 0)$-graph, $n\geq 3$, having precisely two nodes with the same degree. Then G has either exactly one node of degree $n-1$ or exactly one of degree 0.

Proof. Since all degrees lie among the n integers $0, 1, \ldots, n-1$, the degrees of the nodes of G take on all these values except one. Hence, at least one of 0 and $n-1$ is the degree of a node of G. If G contains two nodes of degree 0 or two of degree $n-1$, then the deletion of one of these together with its incident edges, if they exist, results in a $(1, 0)$-graph with all distinct degrees of its nodes, which is contrary to Theorem 6.7. Similary, G cannot contain both a node of degree 0 and one of degree $n-1$. This completes the proof of the corollary.

Example 6.8: Obtain a $(2, 0)$-graph realization of the set

$$\{a_i; i = 1, 2, 3, 4, 5\} = \{5, 4, 4, 3, 2\},$$

in which the integers have been arranged in a non-increasing order. The problem possesses a solution since from Corollary 6.15:

(6.40a): $\quad 5 + 4 + 4 + 3 + 2 = 2 \cdot 9$,

(6.40b): $\quad 0 + 0 + 0 + 0 + 0 \geq 0$,

(6.40c): $\quad 2 + 2 + 2 + 2 = 8 \geq 5 - 0 = 5$,

$\qquad\qquad 4 + 3 + 2 = 9 \geq (5 - 2) + (4 - 2) = 5$,

$\qquad\qquad 3 + 2 = 5 \geq (5 - 4) + (4 - 4) + (4 - 4) = 1$,

$\qquad\qquad 2 \geq (5 - 5) + (4 - 4) + (4 - 4) + (3 - 3) = 0$.

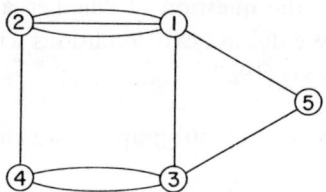

Fig. 6.15. A $(3,0)$-graph realization.

Fig. 6.15 is a $(3, 0)$-graph realization, which can be transformed into a $(2, 0)$-graph by a d-invariant transformation. Since a_1 is greater than 4 ($=n-1$), there exists no $(1, 0)$-graph which is d-invariant from the graph of fig. 6.15. Thus, $p=2$ is the minimum integer that can be achieved for the set with $s=0$.

Suppose that we substract 1 from the integers a_1, a_2, a_4, and a_5, we obtain a new set

$$\{b_i; i = 1, 2, 3, 4, 5\} = \{4, 4, 3, 2, 1\}.$$

Since from Corollary 6.19 there are two nodes of degree 4 ($=n-1$), the set cannot be realized as the degrees of the nodes of a $(1, 0)$-graph. However, there exists a (2, 0)-graph which realizes the set since the conditions (6.40) are satisfied for $p=2$. (For related results, also see Problems 6.45, 6.54, 6.56, and 6.63.)

Example 6.9: Consider the set
$$\{a_i; i = 1, 2, 3, 4, 5\} = \{4, 4, 3, 3, 2\},$$
in which the integers have been arranged in a non-increasing order. We wish to construct a (1, 0)-graph whose degrees are those given in the set.

Using (6.24), we find $a_i'' = a_i$ for $i = 1, 2, 3, 4, 5$, and the conditions (6.42) are trivially satisfied. Thus, there is a (1, 0)-graph which realizes the set. Let us construct a (1, 0)-graph using the procedure outlined in Theorem 6.6. After two successive reductions, the corresponding modified sets are $\{3, 2, 2, 1\}$ and $\{1, 1, 0\}$. The set $\{1, 1, 0\}$ can easily be realized as the degrees of the nodes of a graph consisting of an edge and an isolated node (edge (3, 4) and node 5 in fig. 6.16). If we add a node (node 2) and three edges to this graph, we obtain a realization for the set $\{3, 2, 2, 1\}$ (heavy and dashed lines). Finally, if we add another node (node 1) and four more edges, we obtain a desired realization for the original set. Clearly, this procedure is valid in general.

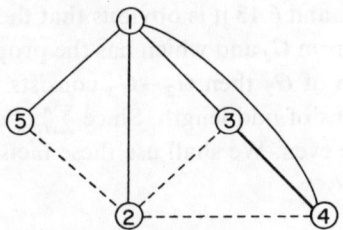

Fig. 6.16. The realization of a (1,0)-graph by the iterative procedure outlined in Theorem 6.6.

2.2. *Realization*

A proof of sufficiency of Theorem 6.4 will be given here. The proof is constructive in nature, and may be used to generate a $(p, 0)$-graph from a given $(p, 0)$-digraph which satisfies the constraints stated in the theorem.

Let G_d be a $(p, 0)$-digraph realization of the set $\{(a_i, b_i)\}$ in which $a_i = b_i$ for all i and $\sum_{j=1}^{n} a_j$ is even. Let G_d' be a maximal symmetric subgraph of G_d. Consider $G_d - G_d'$ and assume that $p \neq 0$.

LEMMA 6.11: *If the edge* (i, j) *is in* $G_d - G_d'$, *then there exists a closed directed-edge train of length* ≥ 3 *in* $G_d - G_d'$.

Proof. We begin a directed-edge train at the node i, and follow the edge (i,j) and continue as far as possible, always through new edges in $G_d - G'_d$. Since at each node k in $G_d - G'_d$, $d^+(k) = d^-(k)$, this process can only come to an end by returning to i, hence the lemma.

LEMMA 6.12: In $G_d - G'_d$, if there is a closed directed-edge train of length $2k$, $k \geq 2$, then there exists a $(p, 0)$-digraph G_1 which is d-invariant from G_d and which has the property that G'_d is a proper subgraph of a maximal symmetric subgraph of G_1.

Proof. Let the closed directed-edge train be

$$Q = (i_1, i_2) \cup (i_2, i_3) \cup \cdots \cup (i_{2k}, i_1). \tag{6.44}$$

The operation of replacing the edges (i_{2t-1}, i_{2t}) by the edges (i_{2t+1}, i_{2t}) for $t = 1, 2, \ldots, k$ in G_d where $i_{2k+1} = i_1$ generates a desired $(p, 0)$-digraph G_1, hence the lemma.

LEMMA 6.13: In $G_d - G'_d$ if Q_1 and Q_2 are two edge-disjoint, closed directed-edge trains of odd length such that they have at least one node in common, then there exists a closed directed-edge train of even length in $Q_1 \cup Q_2$.

By Lemmas 6.11, 6.12, and 6.13 it is obvious that there exists a $(p, 0)$-digraph G_2 which is d-invariant from G_d and which has the property that if G'_2 is a maximal symmetric subgraph of G_2 then $G_2 - G'_2$ consists of a set of node-disjoint closed directed-edge trains of odd length. Since $\sum_{j=1}^{n} a_j$ is even, the number of such edge trains must be even. We shall use these facts to deduce the following result.

LEMMA 6.14: In $G_2 - G'_2$ let Q_1 and Q_2 be two closed directed-edge trains. If in G_2 there exists an edge (i, j), $i \in Q_1$ and $j \in Q_2$, then there exists a $(p, 0)$-digraph G_3 which is d-invariant from G_2 and which has the property that the number of directed circuits of length 2 in G_3 is at least three more than that in G_2.

Proof. Let

$$Q_1 = (i_1, i_2) \cup (i_2, i_3) \cup \cdots \cup (i_{2u-1}, i_1), \tag{6.45a}$$

$$Q_2 = (j_1, j_2) \cup (j_2, j_3) \cup \cdots \cup (j_{2v-1}, j_1), \tag{6.45b}$$

where $u, v \geq 2$. For convenience, let $i = i_1$ and $j = j_1$. The operation of replacing the edges (i_{2t}, i_{2t+1}) by the edges (i_{2t}, i_{2t-1}) for $t = 1, 2, \ldots, u-1$; and (i_1, j_1) by (i_1, i_{2u-1}); and (j_{2x}, j_{2x+1}) by (j_{2x}, j_{2x-1}) for $x = 1, 2, \ldots, v-1$; and (j_1, i_1)

by (j_1, j_{2v-1}), respectively, in G_2 generates a $(p, 0)$-digraph G_3 with the desired properties. Remember that if (i, j) is in G_2 then (j, i) is also in G_2. This completes the proof of the lemma.

LEMMA 6.15: If in Lemma 6.14 there exists no edge (i, j), $i \in Q_1$ and $j \in Q_2$, in G_2, then the result described in Lemma 6.14 still holds.

Proof. The operation of replacing the edges (i_1, i_2) and (j_1, j_2) by the edges (i_1, j_2) and (j_1, i_2), $(i_1, i_2) \in Q_1$ and $(j_1, j_2) \in Q_2$, generates a closed directed-edge train of even length. Thus, from Lemma 6.12 the lemma follows.

By Lemmas 6.14 and 6.15 and the fact that $G_2 - G_2'$ consists of an even number of node-disjoint closed directed-edge trains, it is apparent that there exists a symmetric $(p, 0)$-digraph which is d-invariant from the $(p, 0)$-digraph G_d. This completes the proof of the theorem.

Example 6.10: Consider the $(2,0)$-digraph G_d as shown in fig. 6.17. Since $d^+(i) = d^-(i)$ for $i = 1, 2, ..., 8$ and since $\sum_{j=1}^{8} d^+(i)$ is even, by Theorem 6.4 there exists a symmetric $(2,0)$-digraph which is d-invariant from G_d. Using the procedures outlined in the above lemmas, it is not difficult to decompose G_d

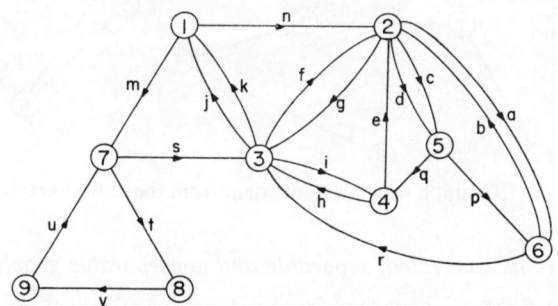

Fig. 6.17. A $(2,0)$-digraph that will be transformed into a $(2,0)$-graph.

into a set of edge-disjoint subgraphs consisting of a maximal symmetric subgraph G_d' and some closed directed-edge trains Q_i. One of such decompositions is as follows: $G_d' = abfghi$, $Q_1 = edq$, $Q_2 = cprjmskn$, and $Q_3 = tuv$. In G_d if we move the edges c, r, m, and k of Q_2 to the new positions as indicated by the dashed lines in fig. 6.18, we generate a $(2,0)$-digraph which is d-invariant from G_d. Since Q_1 and Q_3 are of odd length, an elementary $(2,0)$ d-invariant transformation in G_d will result in a closed directed-edge train of length 6, as indicated in fig. 6.18. Finally, if in fig. 6.18 we replace the edges $e=(4,2)$, $q=(5,9)$, and $t=(7,8)$ by the edges (5,2), (7,9), and (4,8), we obtain a symmetric (2,0)-digraph or simply a (2,0)-graph as shown in fig. 6.19. Actually, the graph of fig. 6.19 is a (1,0)-graph.

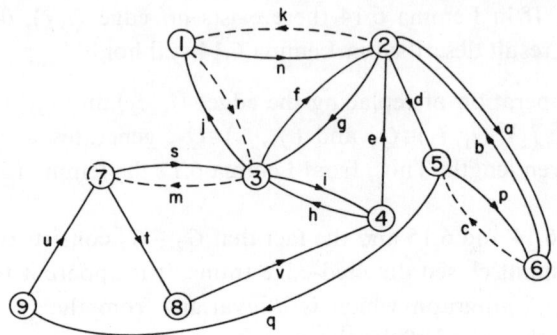

Fig. 6.18. A (2,0)-digraph derived from that of fig. 6.17 by applying sequences of elementary (2,0) d-invariant transformations.

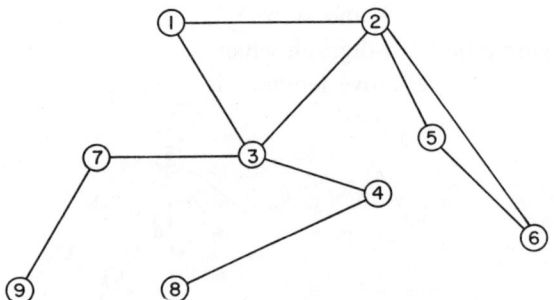

Fig. 6.19. A (1,0)-graph which is d-invariant from the (2,0)-digraph of fig. 6.17.

2.3. *Realizability as connected, separable and nonseparable graphs*

The realizability of a set of nonnegative integers as the degrees of the nodes of a connected (p, s)-graph, a separable graph and a nonseparable graph will be discussed in this section. The development of separable and nonseparable graphs follows that of HAKIMI [1962].

THEOREM 6.8: A set $\{a_i\}$, $n \geq 2$, is realizable as the degrees of the nodes of a connected (p, s)-graph if, and only if, the set is realizable as the degrees of the nodes of a (p, s)-graph with $p \neq 0$ and $a_i \neq 0$ for all i and

$$\sum_{j=1}^{n} a_j \geq 2(n-1). \tag{6.46}$$

The proof of the theorem is similar to that of Theorem 6.3, and is left as an exercise (Problem 6.34). In the light of Theorem 6.8 and Corollaries 6.16 and 6.17, the following two corollaries can easily be established. They were first given by SENIOR [1951b].

COROLLARY 6.20: The necessary and sufficient conditions for a set $\{a_i\}$ ($a_i \geq a_{i+1}$ for $i=1, 2, \ldots, n-1$, $n \geq 2$) to be realizable as the degrees of the nodes of a connected graph without self-loops are that $a_n \neq 0$ and

$$\sum_{j=1}^{n} a_j = 2e, \tag{6.47a}$$

$$\sum_{j=1}^{n} a_j \geq 2 \max(a_1, n-1), \tag{6.47b}$$

where e is a positive integer. (Also see Problems 6.40 and 6.41.)

COROLLARY 6.21: The necessary and sufficient conditions for a set $\{a_i\}$ to be realizable as the degrees of the nodes of a connected graph are that $a_i \neq 0$ for all i and the sum of the integers a_i is even and not less than $2(n-1)$.

Thus, from Corollaries 6.16, 6.20, and 6.21 we conclude that if, for a given set, neither the class of connected-graph realizations nor the class of self-loopless-graph realizations is the null class, then their intersection cannot be null.

COROLLARY 6.22: A set $\{a_i\}$ is realizable as the degrees of the nodes of a tree if, and only if, $a_i \neq 0$ for all i and $\sum_{j=1}^{n} a_j = 2(n-1)$.

THEOREM 6.9: The necessary and sufficient conditions for a set $\{a_i\}$ ($a_i \geq a_{i+1}$ for $i=1, 2, \ldots, n-1$, $n > 2$) to be realizable as the degrees of the nodes of a nonseparable graph without self-loops are that $a_i \geq 2$ for all i and

$$\sum_{j=1}^{n} a_j = 2e, \tag{6.48a}$$

$$\sum_{j=1}^{n} a_j \geq 2(n + a_1 - 2), \tag{6.48b}$$

where e is a positive integer.

Proof. We shall only prove that (6.48b) is necessary; the other conditions are self-evident. Let G be a nonseparable realization of the set, and let G_1 be the graph obtained from G by removing the node of degree a_1 and all the edges incident at this node. Then for the nodes i of G_1 we have

$$\sum_{i=2}^{n} d(i) = \sum_{j=1}^{n} a_j - 2a_1. \tag{6.49}$$

If (6.48b) is not satisfied, (6.49) would be less than $2(n-2)$. From Corollary 6.21 this would imply that G_1 is not connected. In other words, the removed node is a cutpoint (Definition 1.18), and the graph is separable which is contrary to our assumption.

To prove sufficiency, consider the set $\{c_j\}$ of nonnegative integers $c_j = a_j - 2$, $1 \leq j \leq n$. We now see that the set $\{c_j\}$ is realizable as the degrees of the nodes of a graph without self-loops, for applying Corollary 6.16, we have

$$\sum_{j=1}^{n} c_j = \sum_{j=1}^{n} a_j - 2n \geq 2(a_1 - 2) = 2c_1.$$

The right-hand side follows from (6.48b). Let this realization be denoted by G_2. If in G_2 we superimpose a Hamilton circuit of length n, we obtain a graph G without self-loops which realizes the original set $\{a_i\}$. Since the graph contains a Hamilton circuit, G is also nonseparable, which completes the proof.

THEOREM 6.10: A set $\{a_i\}$, $n \geq 3$, is realizable as the degrees of the nodes of a separable but connected graph without self-loops if, and only if, it is realizable as the degrees of the nodes of a connected graph without self-loops and any one of the following three conditions is satisfied:

   (i) If $n = 3$, and $a_1 \geq a_2 \geq a_3$, then $a_1 = a_2 + a_3$.
   (ii) If $n = 4$, there exist a_i and a_j in $\{a_i\}$ such that $a_i \neq a_j$.
   (iii) If $n > 4$, there exists an a_j in $\{a_i\}$ such that $a_j \neq 2$.

Proof. Proofs for conditions (i) and (ii) are straightforward, and the details are left as an exercise (Problem 6.36). In the following we shall only prove the case where $n > 4$.

The necessity of (iii) is proved merely by observing that a connected graph with each of its nodes being of degree 2 is a circuit, which is nonseparable. To prove sufficiency, two cases are considered.

Case 1: $a_1 = a_2 = \cdots = a_n$, $n > 4$. Suppose that a_1 is even. Since $a_1 > 2$, let $a_1 = 2 + a_{11}$. Consider the following two sets of integers 2, a_2, a_3 and a_{11}, a_4, $a_5, \ldots, a_n$. Clearly, both of these sets satisfy the conditions of Corollary 6.20; hence they are realizable as the degrees of the nodes of connected graphs without self-loops. If we superimpose the nodes of lowest degree in these graphs, we obtain a separable but connected graph without self-loops which realizes $\{a_i\}$. Now suppose that a_1 is odd. Then n must be even and $a_1 \neq 1$, for otherwise the conditions of Corollary 6.20 would not be satisfied. Let $a_1 = 2 + a_{12}$. Consider the following two sets of integers 2, a_2, a_3 and a_{12}, a_4, a_5, ..., a_n. Again, using the same argument, we obtain a separable but connected graph without self-loops which realizes $\{a_i\}$.

Case 2: Assume that $a_i \geq a_{i+1}$ for $i = 1, 2, ..., n-1$, and $a_1 \neq a_n$. Consider the sets of integers a_n, a_n and $(a_1 - a_n)$, a_2, a_3, ..., a_{n-1}. The first set is obviously realizable as the degrees of the nodes of a connected graph without self-loops. To show that the second set satisfies Corollary 6.20, observe that

$$\sum_{i=1}^{n-1} a_i - a_n = \sum_{j=1}^{n} a_j - 2a_n$$
$$\geq 2 \max(a_1, n-1, a_2 + a_n, \tfrac{1}{2}na_n) - 2a_n$$
$$= 2 \max[a_1 - a_n, n - a_n - 1, a_2, \tfrac{1}{2}(n-2)a_n]$$
$$\geq 2 \max(a_1 - a_n, n-2, a_2).$$

In the second line we have used the condition that $\{a_i\}$ satisfies (6.47b). Thus, there exist connected graphs without self-loops which realize the two given sets. Let these graphs be G_1 and G_2. If we superimpose a node of G_1 upon the node of G_2 corresponding to the integer $a_1 - a_n$, we obtain a desired realization of the original set, which completes the proof of the theorem.

Example 6.11: Obtain a separable but connected graph without self-loops which is d-invariant from the graph as shown in fig. 6.16. Since the degrees of the given graph satisfy all the conditions specified in Theorem 6.10, the problem possesses a solution. Using the procedures outlined in the proof of Theorem 6.10, let us consider the sets of integers 2, 2, and 2, 4, 3, 3. These sets can be realized as the degrees of the nodes of (2,0)-graphs, as shown in fig. 6.20 with dashed and solid lines. The final realization is obtained by superimposing the corresponding nodes of degree 2.

Next, consider the set of integers

$$\{a_i; i = 1, 2, ..., 8\} = \{5, 4, 4, 3, 3, 3, 2, 2\}.$$

We wish to realize a nonseparable graph without self-loops. Since the integers satisfy all the requirements of Theorem 6.9, the problem is solvable. Using the technique outlined in the proof of Theorem 6.9, let us consider the set $\{3, 2, 2, 1, 1, 1, 0, 0\}$. This set can easily be realized as the degrees of the nodes of a

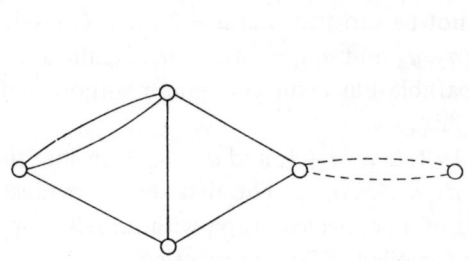
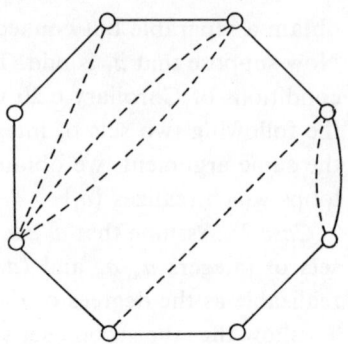

Fig. 6.20. A separable but connected (2,0)-graph which is d-invariant from the graph of fig. 6.16.

Fig. 6.21. A nonseparable (2,0)-graph.

(2,0)-graph, as shown by the dashed lines in fig. 6.21. If we superimpose upon this graph a Hamilton circuit of length 8 (solid lines), we obtain a desired realization.

§ 3. Unique realizability of graphs without self-loops

In this section, we shall consider the problem of unique realizability of a set of nonnegative integers as the degrees of the nodes of a graph or connected graph without self-loops. The problem is of considerable interest in structural organic chemistry (see, for example, SENIOR [1951a]). The solutions were first given by SENIOR [1951b] for connected graphs, and by HAKIMI [1963] for any graph. Also, they were independently obtained by BÄBLER [1953] in a different way. Our viewpoint here is that of Hakimi.

We consider two isomorphic graphs G_1 and G_2 as identical, and write $G_1 = G_2$. By a *unique realization* of a set of nonnegative integers we mean that all the realizations of the set, if they exist, are within isomorphism. There are two difficulties that one is faced with this problem. The first one is that it is not easy to examine two unlabeled graphs for isomorphism, and the second is that a graph which permits an elementary $(\infty, 0)$ d-invariant transformation does not necessarily prove that the graph cannot be a unique realization of a set of nonnegative integers. Therefore, we must make use of those elementary $(\infty, 0)$ d-invariant transformations that produce a fundamental change in the graph, such as increasing or decreasing the length of a path or circuit of maximum length, increasing or decreasing the maximum number of parallel edges, etc.

Unless stated to the contrary, throughout this section by a graph we mean a graph without self-loops.

3.1. Preliminary considerations

For a given graph G, let $\{G\}$ be the set of all connected graphs that are d-invariant from G. We shall discuss conditions under which there exists a graph G_1 in $\{G\}$ such that $G \neq G_1$. To this end we appeal to Theorem 6.2 which states that any element of $\{G\}$ can be obtained from any other one by a finite sequence of elementary $(\infty, 0)$ d-invariant transformations. However, the main complication is that we must choose only those transformations that do not destroy the connectivity of the graph.

LEMMA 6.16: For an n-node tree G, $n \geq 3$, let k be the number of its nodes of degree ≥ 2. Then there exists a connected graph G_1 in $\{G\}$ such that $G_1 \neq G$ if either one of the following two conditions is satisfied:

(i) $4 \leq k < n-2$,
(ii) $k=3$ and the three highest degrees of the nodes do not have the same value.

Proof. (i). In G let

$$P_{ij} = (i, t_1) \cup (t_1, t_2) \cup \cdots \cup (t_q, j), q > 1, \qquad (6.50)$$

be a path of maximum length. If in G there exists a node x which is distinct from the nodes in P_{ij} and such that $d(x) \neq 1$ and $q \geq 3$, then there exists a path $(y_1, y_2) \cup (y_2, t_z)$, $2 \leq z \leq q-1$, in $G - P_{ij}$. If in G we replace the edges (i, t_1) and (y_2, t_z) by (i, t_z) and (t_1, y_2), we generate a graph G_1 in $\{G\}$ and such that $G_1 \neq G$ since G_1 contains a path of length $> q+1$. If, on the other hand, no such node x exists in G, then we may assume that $q \geq 4$ and that all the nodes of degree ≥ 2 are contained in P_{ij}. Since $k < n-2$ by hypothesis, there exists at least one node of degree > 2 in P_{ij}. Without loss of generality, let us assume that t_w, $1 \leq w \leq \frac{1}{2}(q+1)$, is one of such nodes; for otherwise we can start from node j rather than from node i in P_{ij}. Since $d(t_w) > 2$, there exists a node v which is distinct from the nodes in P_{ij} and such that (v, t_w) is in G. If $w \neq 1$, let G_2 be the graph obtained from G by replacing the edges (v, t_w) and (t_{q-1}, t_q) by the edges (v, t_{q-1}) and (t_w, t_q). If $w=1$ let G_3 be the graph obtained from G by replacing the edges (v, t_1), (t_{q-1}, t_q), (i, t_1), and (t_{q-2}, t_{q-1}) by the edges (v, t_{q-1}), (t_1, t_q), (i, t_{q-2}), and (t_1, t_{q-1}). Clearly, G_2 and G_3 are in $\{G\}$. Since in G_2 or G_3 there are three nodes of degree ≥ 2 which are adjacent to the node t_w and since there are at most two nodes of degree ≥ 2 which are adjacent to any node in G, it follows that $G \neq G_2$ or G_3.

(ii). Let G be a tree with $k=3$, and let P_{ij} of (6.50) be a path of maximum length in G. Then $q=3$, and all the nodes of degree ≥ 2 are in P_{ij}. By hypothesis, we may assume, without loss of generality, that $d(t_1) \neq d(t_2)$. If in G we replace the edges (i, t_1) and (t_2, t_3) by the edges (t_1, t_3) and (i, t_2), we generate a new connected graph G_1 in $\{G\}$ and such that $G_1 \neq G$. This completes the proof of the theorem.

LEMMA 6.17: Let G be a connected graph containing a circuit of length at least 4. If in G there exists a node of degree $\neq 2$, then there exists a connected graph G_1 in $\{G\}$ such that $G_1 \neq G$.

Proof. Two cases are considered.

Case 1. G contains no parallel edges. Let L be a circuit of maximum length in G. By hypothesis, there exists an edge (i, j) in $G - L$ which is incident at the node j of L. If the node i is not in L, let us consider the edges (j, k) and (k, u) of L. If in G we replace the edges (i, j) and (k, u) by the edges (j, k) and (i, u), we generate a new graph G_1 in $\{G\}$ and such that $G_1 \neq G$ since G_1 has two parallel edges between the nodes j and k. If the node i is also in L, let us consider the edges (i, k) and (j, u) of L which belong to two distinct circuits formed by L and (i, j). Now we can construct a new graph G_2 in $\{G\}$ by transforming the edges (i, k) and (j, u) into (i, j) and (k, u). Since G_2 has a pair of parallel edges, $G_2 \neq G$.

Case 2: G contains some parallel edges. Let the maximum number of parallel edges be $p, p > 1$, connected between the nodes i and j, i.e., $\alpha(i, j) = p$. Let L be a circuit of maximum length in G. Four subcases are considered.

(i). (i,j) is in L. This includes the possibility that (i,j) is contained in a circuit of maximum length in G. Consider the edge (k, u) of L where k and u are distinct from i and j. This is always possible since L is of length at least 4. Let G_1 be the graph obtained from G by replacing the edges (i, j) and (k, u) by the edges (i, u) and (j, k). Clearly, G_1 is in $\{G\}$. If $G_1 = G$, then there must be (i, u) or (j, k) in G. In such a case, we consider (i, v) and (j, w), $v \neq j$ and $w \neq i$, of L. If in G we replace (i, v) and (j, w) by (v, w) and (i, j), we generate a graph G_2 in $\{G\}$ which contains an edge (i, j) with $\alpha(i, j) = p + 1$. Thus, $G_2 \neq G$.

(ii). (i, j) is not in L but i and j are in L. This is similar to Case 1, and the proof is omitted here.

(iii). Nodes i and j are not in L. A simple transformation involving an edge of L and (i, j) would result in a graph G_1 in $\{G\}$ in which a circuit of maximum length contains at least two more edges than L.

(iv). i is not in L but j is in L. The case can be proved by similar procedures as outlined in *Case* 1 and *Case* 2(i). The details of the proof are left as an exercise (Problem 6.37).

LEMMA 6.18: Let G be a connected graph containing a circuit L of maximum length. If there is an edge which is not incident at any of the nodes of L, then there exists a connected graph G_1 in $\{G\}$ such that $G_1 \neq G$.

3.2. *Unique realizability as a connected graph*

The necessary and sufficient conditions for a set of nonnegative integers to be realizable as the degrees of the nodes of one, and only one, connected graph without self-loops are discussed in this section.

THEOREM 6.11: Consider a set $\{a_i\}$ ($a_i \geq a_{i+1}$ for $i=1, 2, \ldots, n-1, n \geq 2$) which is realizable as the degrees of the nodes of a connected graph without self-loops. Then the realization as a connected graph without self-loops is unique if, and only if, the integers of the set satisfy at least one of the following conditions:

(1) $n \leq 3$, (6.51a)

(2) $a_1 = 2$, (6.51b)

(3) $n = 3a_1 - 1$ and $a_1 = a_3 > a_4 = 1$, (6.51c)

(4) $n = 4$ and $a_1 = a_3 > a_4 = 1$, (6.51d)

(5) $a_1 = a_2 + a_3 + \cdots + a_n$, (6.51e)

(6) $a_1 = (n-1)a_2 - 2$, $a_1 > a_2 = a_n$, and $n > 3$, (6.51f)

(7) $a_3 = 1$ and $n > 3$. (6.51g)

Proof. Since the set $\{a_i\}$ is realizable, condition (2) is completely equivalent to either one of the following two conditions (Problem 6.38):

$$a_1 = a_n = 2, \qquad (6.52a)$$

$$\sum_{j=1}^{n} a_j = 2(n-1) \text{ and } a_{n-2} > a_{n-1} = a_n = 1, n \geq 3. \qquad (6.52b)$$

Also, condition (3) is equivalent to

$$\sum_{j=1}^{n} a_j = 2(n-1) \text{ and } a_1 = a_3 > a_4 = 1. \qquad (6.53)$$

Necessity. Let G be a connected-graph realization of the set with $d(j) = a_j$ for $j = 1, 2, \ldots, n$. Then from Corollary 6.20 we have

$$\sum_{j=1}^{n} a_j \geq 2 \max(a_1, n-1). \qquad (6.54)$$

Assume that G is the unique realization. Four different cases are considered.

Case 1. *G contains no circuits.* Then *G* must be a tree. Let k be the number of its nodes of degree ≥ 2. If $k=1$, then the degrees of the nodes must satisfy condition (5). If $k=2$, then the degrees of the nodes must satisfy condition (7). If $k=3$ and $a_1=a_3$, then (6.53) is satisfied which is equivalent to the condition (3). If $k=n-2$, then from Corollary 6.22 the degrees of the nodes must satisfy (6.52b) which is a special case of condition (2) for a tree. Since $k \leq n-2$, it follows from Lemma 6.16 that all other possibilities of k cannot have a unique realization.

Case 2. *G contains a circuit of maximum length which is* 2. Let L be a circuit of maximum length in G, and let i and j be the nodes of L. Then by Lemma 6.18 every edge (u, v) of G is either incident at i or j. If all the nodes except i and j are of degree 1, then the condition (7) is satisfied. If, on the other hand, there exists a node u of degree ≥ 2, $u \neq i, j$, then all the edges incident at u must all be incident either at i or at j but not both, for otherwise G would contain a circuit of length 3. If the edges (u, j) are in G, then by Lemma 6.18 (x, i), $x \neq j, u$, cannot be in G. This shows that the condition (5) is satisfied. Similarly, this is true for the case where the edges (u, i) are in G.

Case 3. *G contains a circuit of maximum length which is* 3. Let $L=(i, j) \cup (j, k) \cup (k, i)$ be a circuit of G. Then by Lemma 6.18 every edge of G must be incident at a node of L. Assume that the number of noncircuit edges incident at node i is at least as large as that of j or k. Furthermore, we may assume that if (u, v), $u \neq i, j, k$, is in G, then there are no (x, y), $x \neq i, j, k, u$ and $y \neq v$ with $\alpha(x, y) \geq 2$ in G; for if not an elementary $(\infty, 0)$ *d*-invariant transformation would produce a graph containing a circuit of length 4. If in G there are edges (x, i) and (y, j) where x and y are not in L, then the transformation of replacing the edges (i, k) and (j, y) by the edges (i, y) and (j, k) would result in a graph G_1 in $\{G\}$ and such that $G_1 \neq G$ since the number of noncircuit edges incident at the node i of G_1 is one more than that in G. So we shall only consider following types of edges (x, y) of G: (i) x and y are both in L, and (ii) if x is not in L then $y=i$. These two types of edges are considered in three separate subcases. *Subcase* 1. There exists only one edge of the form (x, i) in G where x is not in L. Then G has exactly four nodes and one is of degree 1. If $d(i)=d(j)=d(k)$, then condition (4) is satisfied. Thus, let $d(i) \neq d(k)$. If in G we replace the edges (x, i) and (j, k) by the edges (x, k) and (i, j), we produce a new graph G_2 in $\{G\}$ and such that $G_2 \neq G$ since G_2 has a noncircuit edge incident at a node of different degree from the node i in G. *Subcase* 2. There are at least two edges of the form (x, i) in G where x is not in L. If (x, i) is a noncircuit edge, then by a transformation similar to the one discussed above we can generate a graph in $\{G\}$ which is distinct from G. Thus, we may assume that all (x, i) are circuit

§ 3   Unique realizability   445

edges, and furthermore there exist no parallel edges between the nodes k and j in G; for if there are parallel edges connected between the nodes k and j then the transformation of replacing the edges (x, i) and (k, j) by the edges (x, j) and (i, k) would generate a graph G_3 in $\{G\}$ and such that $G_3 \neq G$ since G_3 contains a circuit of length 4. If, in G, all the nodes except i have the same degree then condition (6) is satisfied. If there are at least two nodes of different degrees, other than i, then a simple transformation would result in a graph G_4 in $\{G\}$ and such that $G_4 \neq G$. *Subcase* 3. No edges of type (ii) are in G. Then condition (1) is satisfied.

Case 4. *G contains a circuit of length at least* 4. If all the nodes of G are of degree 2, then (6.52a) is satisfied which is a special case of condition (2); the other condition (6.52b) is given in Case 1. If there exists a node of degree $\neq 2$, by Lemma 6.17 the realization cannot be unique. This completes the proof of the necessary part of the theorem.

Sufficiency. If the set $\{a_i\}$ is realizable as the degrees of the nodes of a connected graph, and if it satisfies any one of the seven conditions, we shall show that the realization is unique.

(1) If $n \leq 3$, the realization is obviously unique since the application of any elementary $(\infty, 0)$ d-invariant transformation that does not destroy the connectivity of the graph would result in a graph which is isomorphic to the original graph (Corollary 6.11).

(2) If $a_1 = a_n$, then the realization is a circuit of length n which is unique. If (6.52b) is satisfied, the only realization is a path of length $n-1$. Thus, from (6.52) the condition (2) is sufficient.

(3) Condition (3) is equivalent to (6.53). Any connected-graph realization of the set must be a tree (Corollary 6.22), and furthermore every tree contains a path of length 4. Let $P = (i_1, i_2) \cup (i_2, i_3) \cup (i_3, i_4) \cup (i_4, i_5)$ be one of such paths. Then $d(i_x) = a_1 = a_3$ for $x = 2, 3, 4$ and $d(y) = 1$ for all other nodes in the tree. Thus, there are exactly $a_1 - 2$ edges of the tree, not contained in the path P, incident at each of the nodes i_2, i_3, and i_4. This type of trees is obviously unique, to within isomorphism.

(4) If $n = 4$ and $a_1 = a_3 > a_4 = 1$, the realization is obviously unique. The proof is similar to that of (1).

(5) If $a_1 = a_2 + a_3 + \cdots + a_n$, then all the edges of a realization are incident at the node corresponding to the integer a_1. Thus, it is unique.

(6) If $a_1 = (n-1) a_2 - 2$, $a_1 > a_2 = a_n$, and $n > 3$, then all but one of the edges of a realization of the set are incident at the node corresponding to the integer a_1. Clearly, if $a_2 = a_n$, no elementary $(\infty, 0)$ d-invariant transformation will generate a graph which is distinct from the original one. Thus, the realization is unique.

(7) If $a_3 = 1$ and $a_2 \neq 1$, any realization must contain a path of length 3. Let $P = (i_1, i_2) \cup (i_2, i_3) \cup (i_3, i_4)$ be one of such paths. Without loss of generality, let us assume that $d(i_2) = a_1$ and $d(i_3) = a_2$. Since $a_k = 1$ for $k = 3, 4, \ldots, n$, the realization must have $\frac{1}{2}(a_1 + a_2 - n + 2)$ parallel edges connected between the nodes i_2 and i_3. All other edges are incident at i_2 or i_3, until the degree requirement of these two nodes are fulfilled. Thus, the realization is unique. If $a_2 = 1$, the realization is obviously unique. This completes the proof of the theorem.

Evidently, the seven conditions given in the theorem are not mutually exclusive. For example, the set $\{2, 1, 1\}$ satisfies the conditions (1), (2), and (5). On the other hand, there are sets that satisfy one and only one of these conditions: the set $\{3, 3, 2\}$ is met only by condition (1), $\{2, 2, 2, 2\}$ only by (2), $\{3, 3, 3, 1, 1, 1, 1, 1\}$ only by (3), $\{3, 3, 3, 1\}$ only by (4), $\{6, 3, 2, 1\}$ only by (5), $\{4, 2, 2, 2\}$ only by (6), and $\{4, 4, 1, 1\}$ only by (7). Thus, we have demonstrated that no one of the seven conditions may be omitted, and each of these sets is realizable as the degrees of the nodes of one, and only one, graph without self-loops. These realizations are presented in fig. 6.22.

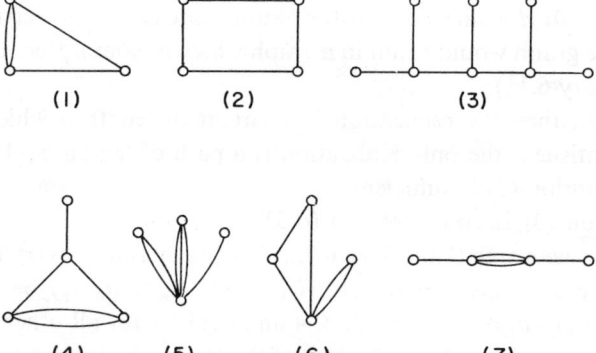

Fig. 6.22. The degrees of the nodes of each of these seven connected graphs satisfying one and only one of the conditions (6.51).

3.3. Unique realizability as a graph

The condition that a realization of a set be connected will be dropped in this section. This means that the unique realizability conditions of Theorem 6.11 can now be relaxed somewhat.

THEOREM 6.12: Consider a set $\{a_i\}$ ($a_i \geq a_{i+1}$ for $i = 1, 2, \ldots, n-1, n \geq 2$) which is realizable as the degrees of the nodes of a graph without self-loops. Then the realization as a graph without self-loops is unique if, and only if, the integers of the set satisfy at least one of the following conditions:

(1) $n \leq 3$, (6.55a)

(2) $n = 4$ and $a_1 = a_3 > a_4 = 1$, (6.55b)

(3) $a_1 = a_2 + a_3 + \cdots + a_n$, (6.55c)

(4) $a_1 = (n-1)a_2 - 2$, $a_1 > a_2 = a_n$, and $n > 3$, (6.55d)

(5) $a_2 = a_n = 1$. (6.55e)

Proof. Necessity. Let G be the unique realization of the set $\{a_i\}$. We shall show that if the integers of the set do not satisfy any one of the five conditions (6.55), then the realization is not unique. Without loss of generality, we may assume that G has no isolated nodes; for otherwise we can delete these nodes and the corresponding zeros in $\{a_i\}$, and treat the remaining graph and set as our problem. Two cases are considered.

Case 1. G is connected. Then for G to be unique the degrees of its nodes must satisfy at least one of the seven conditions (6.51) of Theorem 6.11. Considering each of these conditions separately and eliminating those conditions that allow an unconnected realization, we obtain the five conditions of the hypothesis. Note that condition (6.51g) permits an unconnected realization if $a_2 \neq 1$.

Case 2. G is not connected. Let $d(j) = a_j$ for all the nodes j in G. If the sum of all the integers a_j is not less than $2(n-1)$, then from Corollary 6.20 there exists a connected realization since the sum of the integers satisfies (6.41b). If G has a circuit, then from Lemma 6.8 there is a realization which is distinct from G. Thus, we may assume that G contains no circuits and that the sum of the integers $d(j)$ is less than $2(n-1)$. Suppose that k is the number of its nodes of degree ≥ 2. If $k \leq 1$, then condition (5) is satisfied. For $k > 1$, two subcases are considered. If these k nodes are in k different components of G, then by an elementary $(\infty, 0)$ d-invariant transformation, we can generate a graph G_1 in which these k nodes are in $k-1$ components of G_1. If these k nodes are in q, $q < k$, components of G, then again by an elementary $(\infty, 0)$ d-invariant transformation we can generate a graph G_2 in which these k nodes are in $q+1$ components of G_2.

Sufficiency. Using Corollary 6.11 and Theorem 6.11, the sufficiency of the five conditions (6.55) is obvious. So the theorem is proved.

As an illustration, consider the seven realizations as shown in fig. 6.22. It

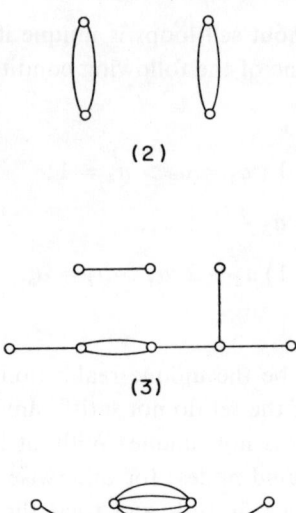

Fig. 6.23. The unconnected realizations of the graphs (2), (3), and (7) of fig. 6.22.

can easily be shown that the graphs (2), (3), and (7) of fig. 6.22 have unconnected realizations as well, and they are presented in fig. 6.23.

§ 4. Existence and realization of a (p, s)-matrix

In the preceding sections, we have discussed the necessary and sufficient conditions for the existence of a (p, s)-digraph with prescribed incoming and outgoing degrees. In this section we shall show that the existence of a (p, s)-digraph also establishes the existence of a class of matrices known as the (p, s)-matrices.

DEFINITION 6.7: (p, s)-*matrix*. For given nonnegative integers p and s, a square matrix of nonnegative integers is called a (p, s)-*matrix* if, and only if, all of its off-diagonal elements are bounded above by p and all of its diagonal elements are bounded above by s. A (p, p)-matrix is also referred to as a *p-matrix*.

For a given (p, s)-matrix M of order n, let the sum of row i of M be denoted by a_i and let the sum of column j of M be denoted by b_j for $i, j = 1, 2, ..., n$. It is clear that the sum of integers in M is given by

$$\sum_{i=1}^{n} a_i = \sum_{j=1}^{n} b_j. \tag{6.56}$$

§4 Existence and realization of a (p, s)-matrix

Our problem is to establish arithmetic conditions that are necessary and sufficient for the existence of a (p, s)-matrix having prescribed row sums a_i and column sums b_j. The problem can be interpreted in terms of the existence of a (p, s)-digraph with prescribed degree pairs (a_i, b_i) of its nodes: Let $M = [\alpha_{ij}]$. Then the *corresponding* (p, s)-*digraph* G_d of M is an n-node directed graph in which if $\alpha_{ij} \neq 0$ there are α_{ij} parallel edges directed from node i to node j for $i, j = 1, 2, \ldots, n$. Conversely, for a given (p, s)-digraph the process of constructing a corresponding (p, s)-matrix is straightforward, and needs no further elaboration. Thus, there exists a one-to-one correspondence between these two different representations. As an example, consider the $(2, 0)$-digraph as shown in fig. 6.14. The corresponding $(2,0)$-matrix M is given by

$$M = \begin{array}{c} \\ 2 \\ 2 \\ 3 \\ 1 \\ 5 \end{array} \overset{\begin{array}{ccccc} 2 & 3 & 3 & 4 & 1 \end{array}}{\begin{bmatrix} 0 & 1 & 0 & 0 & 1 \\ 1 & 0 & 0 & 1 & 0 \\ 0 & 1 & 0 & 2 & 0 \\ 0 & 0 & 1 & 0 & 0 \\ 1 & 1 & 2 & 1 & 0 \end{bmatrix}}.$$

It is now clear that all the results obtained so far for (p, s)-digraphs can be carried over for (p, s)-matrices. Since they are only trivially different, it is hardly necessary to repeat the same results all over again. However, we shall only point out the corresponding operation for a (p, s)-matrix of an elementary (p, s) d-invariant transformation for a (p, s)-digraph.

Consider the 2×2 submatrices of M of the types

$$M_1 = \begin{bmatrix} a & b \\ c & d \end{bmatrix} \quad \text{and} \quad M_2 = \begin{bmatrix} b & a \\ d & c \end{bmatrix}, \tag{6.57}$$

where $\min(a, d) \neq 0$, and $b, c < p$ if b or c is an off-diagonal element of M and $b, c < s$ if b or c is a diagonal element of M. Let

$$M_3 = \begin{bmatrix} a-1 & b+1 \\ c+1 & d-1 \end{bmatrix} \quad \text{and} \quad M_4 = \begin{bmatrix} b+1 & a-1 \\ d-1 & c+1 \end{bmatrix}. \tag{6.58}$$

A (p, s)-*interchange* is defined as a transformation of the elements of M that changes a specified submatrix of type M_1 into M_3, or else a submatrix of type M_2 into M_4, and leaves all other elements of M unaltered. Suppose that we apply a finite number of (p, s)-interchanges to M. Then by the nature of the operation, the resulting matrix has the same row sums and the same column sums. Then from Theorem 6.2 if $\{M\}$ is the set of all possible (p, s)-matrices having the same row and column sums as those of M, every element of $\{M\}$ can be obtained from any other one by a finite sequence of (p, s)-interchanges.

As an illustration, consider the two (1,0)-matrices H_1 and H_2

$$H_1 = \begin{array}{c} \\ 2 \\ 2 \\ 3 \\ 1 \\ 4 \end{array} \begin{array}{c} 2\ 3\ 3\ 4\ 0 \\ \left[\begin{array}{ccccc} 0 & 0 & 1 & 1 & 0 \\ 0 & 0 & 1 & 1 & 0 \\ 1 & 1 & 0 & 1 & 0 \\ 0 & 1 & 0 & 0 & 0 \\ 1 & 1 & 1 & 1 & 0 \end{array}\right] \end{array},$$

$$H_2 = \begin{array}{c} 2 \\ 2 \\ 3 \\ 1 \\ 4 \end{array} \left[\begin{array}{ccccc} 0 & 1 & 0 & 1 & 0 \\ 0 & 0 & 1 & 1 & 0 \\ 1 & 1 & 0 & 1 & 0 \\ 0 & 0 & 1 & 0 & 0 \\ 1 & 1 & 1 & 1 & 0 \end{array}\right].$$

If in H_2 we replace the submatrix $H_2(14; 23)$ by the corresponding submatrix $H_1(14; 23)$, we transform H_2 into H_1 by a (1,0)-interchange where $H_2(14; 23)$ denotes the submatrix consisting of the rows 1 and 4 and columns 2 and 3 of H_2. This is similarly valid for $H_1(14; 23)$.

Clearly, the discussions on (p, s)-matrices can easily be extended to any matrix of bounded nonnegative integers with prescribed row and column sums by the addition of an appropriate number of rows or columns consisting only of zeros. Since a matrix of 1's and 0's is of considerable practical interest, in the following we shall present a simple "n-step" method, first given by GALE [1957], for constructing a 1-matrix with prescribed row and column sums.

Suppose that we wish to construct a 1-matrix H having given row sums a_i and column sums b_i for $i=1, 2,..., n$, $n \geq 2$. Without loss of generality, let $a_j \geq a_{j+1}$ and $b_j \geq b_{j+1}$ for $j=1, 2,..., n-1$. Then the problem possesses a solution if, and only if, the integers satisfy the condition (6.10). Assume that these conditions are satisfied. The procedure for constructing a 1-matrix H is the following: Let $H = [h_{ij}]$. Since $b_1 \leq a_1^*$, it follows that $a_1, a_2, ..., a_{b_1} \geq 1$. Let $h_{i1} = 1$ for $i \leq b_1$ and $h_{i1} = 0$ for $i > b_1$. Now consider the new problem with the matrix H_1 of order n having the row sums $a_i' = a_i - h_{i1}$ and the column sums b_i' where $b_1' = 0$ and $b_j' = b_j$ for $j \geq 2$. Clearly, the first column of H_1 consists only of zeros. We assert that the new problem is again solvable so that by repeating the process we will eventually fill out the whole matrix H.

Let the b_j' be ordered as $b_2', b_3', ..., b_n', b_1'$. To show that the new problem is solvable we must prove that, for any k,

$$\sum_{j=1}^{k} (a'_j)^* \geq \sum_{j=2}^{k+1} b_j, \quad (6.59a)$$

where $b_j=0$ for $j>n$. Remember that the sequences b_j or b'_j and $(a'_j)^*$ are assumed to have been brought to the same length by adding an appropriate number of zeros, if necessary. From (6.11), the left-hand side of (6.59a) can be written as

$$\sum_{j=1}^{k} (a'_j)^* = \sum_{i=1}^{n} \min(a'_i, k)$$

$$= \sum_{i=1}^{b_1} \min(a_i - 1, k) + \sum_{i=b_1+1}^{n} \min(a_i, k), \quad (6.59b)$$

where the second summation is zero for $b_1=n$. If $a^*_{k+1} \geq b_1$, then $a_i - 1 \geq k$ for $i \leq b_1$, and hence from (6.11) and (6.10b), (6.59a) is satisfied, since $b_1 \geq b_{k+1}$. If $a^*_{k+1} < b_1$, then

$$a_i \geq k+1 \quad \text{for} \quad i \leq a^*_{k+1},$$

$$a_i \leq k \quad \text{for} \quad a^*_{k+1} < i \leq b_1.$$

Since

$$\min(a_i - 1, k) = \min(a_i, k) \quad \text{for} \quad a_i \geq k+1$$

and

$$\min(a_i - 1, k) = \min(a_i, k) - 1 \quad \text{for} \quad a_i \leq k,$$

the right-hand side of (6.59b) becomes

$$\sum_{i=1}^{n} \min(a'_i, k) = \sum_{i=1}^{n} \min(a_i, k) - (b_1 - a^*_{k+1})$$

$$= \sum_{i=1}^{k+1} a^*_i - b_1$$

$$\geq \sum_{j=1}^{k+1} b_j - b_1 = \sum_{j=2}^{k+1} b_j,$$

since (6.10b) is satisfied. This completes the proof of the problem.

We shall illustrate the above procedure by the following example.

Example 6.12: Suppose that we wish to construct a 1-matrix H with the following realizable sequences of row and column sums

$$\{a_i; i = 1, 2, 3, 4, 5\} = \{4, 3, 2, 2, 1\},$$
$$\{b_i; i = 1, 2, 3, 4, 5\} = \{4, 3, 3, 2, 0\},$$

respectively, which have already been arranged in non-increasing orders. Since $b_1 = 4$ and $a_1^* = 5$, the transpose of the first column of H is then given by

$$[1\ 1\ 1\ 1\ 0].$$

Next, consider the new sequences

$$\{a_i'; i = 1, 2, 3, 4, 5\} = \{3, 2, 1, 1, 1\},$$

$$\{b_i'; i = 1, 2, 3, 4, 5\} = \{0, 3, 3, 2, 0\}.$$

Since $b_2' = 3$, the transpose of the second column of H is given by

$$[1\ 1\ 1\ 0\ 0].$$

Now, consider the new sequences $\{2, 1, 0, 1, 1\}$ and $\{0, 0, 3, 2, 0\}$. The transpose of the third column of H is then given by

$$[1\ 1\ 0\ 1\ 0].$$

The next sequences are $\{1, 0, 0, 0, 1\}$ and $\{0, 0, 0, 2, 0\}$, and $\{0, 0, 0, 0, 0\}$ and $\{0, 0, 0, 0, 0\}$, respectively; they generate the last two columns of H. Thus, we have

$$H = \begin{bmatrix} 1 & 1 & 1 & 1 & 0 \\ 1 & 1 & 1 & 0 & 0 \\ 1 & 1 & 0 & 0 & 0 \\ 1 & 0 & 1 & 0 & 0 \\ 0 & 0 & 0 & 1 & 0 \end{bmatrix}.$$

§ 5. Realizability as a weighted directed graph

A weighted directed graph is a directed graph in which every edge has been assigned a weight. In Chapters 3 and 4, we have applied this concept to the analysis of electrical networks and other physical systems. It is, therefore, natural to extend our present discussion to the case of weighted directed graphs.

DEFINITION 6.8: *Weighted (p, s)-digraph.* Let G_w be a 1-digraph in which every edge has been assigned a nonnegative real number. The number associated with the edge (i, j) is denoted by $f(i, j)$. For given nonnegative real numbers p and s, the 1-digraph G_w is called a *weighted (p, s)-digraph* if

$$0 \leq f(i, j) \leq p$$

for all $i \neq j$, and

$$0 \leq f(i, i) \leq s$$

§ 5 Realizability as a weighted directed graph

for all i. A weighted (p,p)-digraph is also referred to as a *weighted p-digraph*.

In a weighted directed graph there is no need to consider parallel edges since they can be represented by a single edge with weight being equal to the sum of the weights of these parallel edges. For an n-node weighted (p, s)-digraph G_w, the pair (w_i, w'_i) of nonnegative real numbers

$$w_i = \sum_{j=1}^{n} f(i, j) \tag{6.60a}$$

and

$$w'_i = \sum_{j=1}^{n} f(j, i) \tag{6.60b}$$

for $i = 1, 2, \ldots, n$, are called the *weighted degree pair* of the node i of G_w. Our problem is to find arithmetic conditions that are necessary and sufficient for a set of n nonnegative real pairs (w_i, w'_i) to be the weighted degree pairs of the nodes of an n-node weighted (p, s)-digraph.

The realizability conditions turn out to be exactly the same as those for the (p, s)-digraphs given in (6.2) except that we use w_i and w'_i instead of a_i and b_i in (6.2), respectively. The reason for this is that the result can be proved in an entirely similar manner as that of Theorem 6.1. However, we shall present a simple interpretation of a weighted (p, s)-digraph in terms of a (kp, ks)-digraph.

For any (p, s)-digraph G_d, there is a unique representation as a weighted (p, s)-digraph G_w: (i, j) is in G_w if there exists an (i, j) in G_d, and $f(i, j)$ is equal to the number of parallel edges directed from node i to node j in G_d. Conversely, if p, s, and all the weights $f(i, j)$ of a weighted (p, s)-digraph G_w are rational, there exists a positive integer k such that $kf(i, j)$ are nonnegative integers for all i and j. Then its corresponding directed graph G_d is a (kp, ks)-digraph in which there are $kf(i, j)$ parallel edges directed from node i to node j for all i and j. Thus, the problem of realizing a set of nonnegative real pairs (w_i, w'_i) as the weighted degree pairs of the nodes of a weighted (p, s)-digraph is equivalent to that of realizing a set of nonnegative integer pairs (kw_i, kw'_i) as the degree pairs of the nodes of a (kp, ks)-digraph. Since kp, ks, kw_i, and kw'_i all have a common factor k, the integer k can be eliminated from the conditions (6.2) of Theorem 6.1. Thus, the realizability conditions for a weighted (p, s)-digraph are exactly the same as those for a (p, s)-digraph. In other words, most of the results discussed in the preceding sections remain valid, *mutatis mutandis*, as can be seen either from the above argument or by making slight changes in the proofs throughout.

As an illustration, consider the set of nonnegative real pairs

$$\{(w_i, w'_i); i = 1, 2, 3, 4, 5\} = \{(4, 13.5), (6, 4), (7, 2), (4, 5), (3.5, 0)\}.$$

The set is realizable as the weighted degree pairs of a weighted directed graph without self-loops since from (6.4), (6.5), and (6.6) we have

(6.4a): $\quad 4 + 6 + 7 + 4 + 3.5 = 24.5 = 13.5 + 4 + 2 + 5 + 0$,
(6.6): $\quad 6 + 7 + 4 + 3.5 = 20.5 \geq 13.5$.

One of such realizations is presented in fig. 6.24.

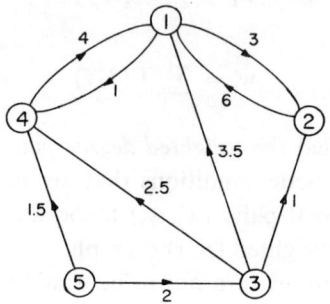

Fig. 6.24. A realization of a set of nonnegative real pairs as the weighted degree pairs of a weighted directed graph.

§ 6. Conclusions

In this chapter, we have presented the necessary and sufficient conditions for a set of nonnegative integer pairs to be the degree pairs of the nodes of a (p, s)-digraph. The simple constructive proof of the main theorem suggests a simple procedure for constructing such a (p, s)-digraph from a realizable set. The general existence conditions consist of a set of $2^n - 1$ inequalities where n is the number of elements of the set. These conditions in many cases can be reduced to a set of n inequalities under assumed constraints. We have shown that any two realizations of a set can be transformed into each other by a finite sequence of elementary (p, s) d-invariant transformations. Using this type of transformations, we have deduced the realizability conditions for a set of nonnegative integers to be the degrees of the nodes of a (p, s)-graph from those for a (p, s)-digraph. The realizability conditions for a set to be the degree pairs or degrees of the nodes of a connected (p, s)-digraph or a separable or non-separable graph without self-loops have also been given.

We have presented conditions that are necessary and sufficient for the unique realization of a set of nonnegative integers as the degrees of the nodes of a graph or a connected graph without self-loops. The problem is of considerable interest in structural organic chemistry because of its close relationship with the structural isomers of a given chemical compound.

The existence of a (p, s)-digraph also establishes the arithmetic conditions that are necessary and sufficient for the existence of a (p, s)-matrix having prescribed row and column sums. The well-known zero–one matrices become a special case.

Finally, we have extended our discussions to the existence of a weighted (p, s)-digraph with prescribed weighted degree pairs. The problem is interesting because the concept of a weighted graph is found to be extremely useful in the applications of graph theory to electrical networks and other physical systems.

Problems

6.1. Let $x_1, x_2, \ldots, x_n$ and $y_1, y_2, \ldots, y_n$ be two sets of real numbers such that

$$\sum_{i=1}^{n} x_i = \sum_{j=1}^{n} y_j. \qquad (6.61a)$$

Show that

$$\sum_{i=1}^{n-1} x_i \geq y_n \qquad (6.61b)$$

if, and only if,

$$\sum_{i=1}^{n-1} (x_i + y_i) \geq x_n + y_n. \qquad (6.61c)$$

6.2. Show that if (6.5) holds then (6.4b) and (6.6) are equivalent.

6.3. Show that for a set $\{a_i\}$ we have

$$\{\{a_i\}^*\}^* = \{a_i\}. \qquad (6.62)$$

6.4. Prove Corollary 6.2 by induction over n.

6.5. Is there a $(2,0)$-digraph which is d-invariant from the directed graph as shown in fig. 6.5?

6.6. Show that in $\{a_i\}$ if $a_i \geq a_{i+1}$ for $i = 1, 2, \ldots, n-1$, $n \geq 2$, then the sequence of integers a_j'' defined in (6.24) for $j = 1, 2, \ldots$ is either monotone nonincreasing or else, for some k, $1 \leq k < n$,

$$a_1'' \geq \cdots \geq a_k''; \ a_{k+1}'' = a_k'' + 1 \geq a_{k+2}'' \geq \cdots \geq a_n''. \qquad (6.63)$$

6.7. Prove Corollary 6.8.

6.8. What are the necessary and sufficient conditions for a set $\{(a_i, b_i)\}$ to be realizable as the degree pairs of the nodes of a semi-regular $(1,0)$-digraph with incoming degree k?

6.9. Prove Corollaries 6.9 and 6.10.
6.10. Extend Problem 6.8 to the realization of a semi-regular $(p, 0)$-digraph with incoming degree k.
6.11. Is Corollary 6.8 still valid for $p<s$? If not, give a counterexample.
6.12. Prove Lemmas 6.3, 6.4, and 6.5.
6.13. In Corollary 6.2, assume that (6.4b) is satisfied with the equal sign. What type of directed graphs will the realizations be?
6.14. Prove Lemma 6.6.
6.15. Assume that a directed graph G_d contains at most one node of zero incoming or outgoing degree. Show that for $n>1$, G_d has a directed circuit.
6.16. Let G_d be an acyclic directed graph containing at least two nodes. Show that there are at least two nodes of zero incoming or outgoing degree.
6.17. Prove Corollary 6.11.
6.18. Given a graph G without self-loops whose nodes are numbered from 1 to n. Let b_i be the nonnegative integers associated with the nodes i of G for $i=1, 2, ..., n$. Show that the directions can be assigned to the edges of G so that the incoming degrees of the nodes i of the resulting directed graph are b_i if, and only if, for each non-null subgraph g of G,

$$\sum_{i \in g} b_i - \alpha(g) \geq 0, \tag{6.64a}$$

$$\sum_{j=1}^{n} b_j = \alpha(G), \tag{6.64b}$$

where i is a node of g and $\alpha(g)$ denotes the number of edges of g (HAKIMI [1965]).
6.19. Prove Corollary 6.12.
6.20. Give necessary and sufficient conditions for a set $\{(a_i, b_i)\}$ to be realizable as the degree pairs of the nodes of a directed tree t_n. Justify your results. (*Hint*: Use (6.37).)
6.21. In Problem 6.18, assume that the assignment of directions is possible. Show that the assignment is unique if, and only if, the resulting directed graph is acyclic.
6.22. Let G_d be a planar $(1, 0)$-digraph. Show that if in G_d there is a node i with $d^+(i) \geq 3$, then there exists in G_d a directed path from node i to a node j with $d^+(j) \leq 2$. (*Hint*: $b \leq 3(n-2)$ where b and n are the numbers of edges and nodes of G, respectively.)
6.23. Show that every planar $(1, 0)$-digraph can be reoriented in such a way

that the outgoing degree of each node is not greater than 3. (*Hint*: Apply the result of Problem 6.22.) (LEMPEL [1968].)

6.24. Prove Theorem 6.5. Also show that h can be lowered by 1 if s is odd and $p=0$.

6.25. Show that every graph realization G of a set $\{a_i\}$ is of nullity $c-1$ where c is the number of components of G if, and only if, the sum of the nonnegative integers a_i is equal to $2(n-1)$.

6.26. Applying the result of Problem 6.25, prove Corollary 6.22.

6.27. Prove Corollary 6.15.

6.28. Determine if the sets $\{1, 1, 1, 1, 1, 1, 1, 1, 4, 4, 4\}$ and $\{2, 2, 2, 4\}$ are uniquely realizable as the degrees of the nodes of connected graphs without self-loops.

6.29. Show that if a set $\{a_i\}$ is realizable as the degrees of the nodes of a planar (1,0)-graph, then its nonnegative integers must satisfy the conditions of Theorem 6.6 and

$$\sum_{i=k+1}^{n} a_i - \sum_{j=1}^{k} a_j \leq 6(n-k-2) \tag{6.65}$$

for $k=0, 1, \ldots, n-3$, $n \geq 3$ where the second summation is zero for $k=0$. Also show that the conditions are sufficient for $n<6$. However, for $n \geq 6$ the conditions are not sufficient. Use the set $\{3, 4, 4, 4, 4, 5\}$ to justify the last statement. (*Hint*: The set has a unique realization.) (BÖTTGER and HARDERS [1964].) (For a related treatment, see OWENS [1971].)

6.30. Prove Corollary 6.17.

6.31. Let $a_1+a_2+\cdots+a_n=b_1+b_2+\cdots+b_n$. Show that the conditions for the existence of a p-matrix having prescribed row sums a_i and column sums b_i for $i=1, 2, \ldots, n$ are equivalent to those for the existence of a p-matrix whose row sums are bounded above by nonnegative integers a_i and whose columns sums are bounded below by nonnegative integers b_i.

6.32. Show that there is an $n \times n$ (1,0)-matrix whose column sums are bounded below by a nonnegative integer b, and whose ith row sum is bounded above by a nonnegative integer a_i, if and only if, for $n \geq 2$,

$$nb \leq \sum_{i=1}^{n} a_i'', \tag{6.66}$$

where a_i'' is defined in (6.24), and $a_j \geq a_{j+1}$ for $j=1, 2, \ldots, n-1$. (*Hint*: See Problem 6.31.) (FULKERSON [1960].)

6.33. Show that there is an $n \times n$ (1, 0)-matrix whose row sums are bounded above by a nonnegative integer a, and whose ith column sum is bounded below by a nonnegative integer b_i, if and only if $b_i \leq n-1$ and

$$na \geq \sum_{j=1}^{n} b_j, \qquad (6.67)$$

where $i=1, 2, \ldots, n$. (*Hint*: See Problem 6.31.) (FULKERSON [1960].)

6.34. Prove Theorem 6.8.

6.35. Show that (6.42b) holds for all k in $1 \leq k \leq n$ if, and only if,

$$\sum_{i=x+1}^{n} a_i + k(x-1) \geq \sum_{j=1}^{k} a_j \qquad (6.68)$$

holds for all k, x in $1 \leq k \leq x \leq n$ where the first summation is defined to be zero for $x=n$. (*Hint*: Use the identity

$$\sum_{j=1}^{k} a_j'' = \min_{0 \leq x \leq n} \left[\sum_{i=x+1}^{n} a_i + kx - \min(k, x) \right].) \qquad (6.69)$$

(FULKERSON et al. [1965] and CHEN [1973].)

6.36. Justify the conditions (i) and (ii) of Theorem 6.10.

6.37. Supply the details of the proofs of subcases (ii)–(iv) of Case 2 in the proof of Lemma 6.17.

6.38. Verify the equivalent conditions given in (6.52) and (6.53).

6.39. Show that a (1,0)-graph can always be transformed into a simple canonical form by a finite sequence of elementary (1,0) *d*-invariant transformations. What is the form of the canonical (1,0)-graph? (*Hint*: Use Lemma 6.10.)

6.40. Assume that a set $\{a_i\}$ is realizable as the degrees of the nodes of a (1,0)-graph. Show that there is a connected (1,0)-graph realization if, and only if, $a_i \neq 0$ for all i and the sum of the integers a_i is no less than $2(n-1)$.

6.41. Assume that for a given set $\{a_i\}$, $n \geq 2$, there is a (1,0)-graph realization. Show that, for the set $\{a_i\}$, there is a (1,0)-graph realization with the property that the removal of any fewer than k, $k \geq 2$, edges from the graph leaves a connected graph if, and only if, $a_i \geq k$ for all i (EDMONDS [1964]).

6.42. Show that all the (1,0)-graph realizations, if they exist, of a set consisting of four or less nonnegative integers are isomorphic.

6.43. Consider a non-null *n*-node (1,0)-graph. Let $n = a_1 + a_2$ where a_1 and a_2 are nonnegative integers. Show that there exists a partition $V = V_1 \cup V_2$ of the node set V of G such that V_1 has a_1 elements, V_2 has a_2 elements, and

$$\max[d(i)] + \max[d(j)] \leq \max[d(k)], \qquad (6.70)$$

where nodes i and j are in the sectional subgraphs $G[V_1]$ and $G[V_2]$, respectively, and nodes k are in G (LOVÁSZ [1966]).

6.44. Show that for every positive integer $n \geq 2$, there are exactly two non-isomorphic n-node (1,0)-graphs in each of which just two nodes have the same degree, and these graphs are complementary (BEHZAD and CHARTRAND [1967]).

6.45. Assume that the set $\{a_i\}$ is realizable as the degrees of the nodes of a graph G and a self-loopless graph G_1. Let $\{G\}$ be the set of graphs that are d-invariant from G, and let $\{G_1\}$ be the set of self-loopless graphs that are d-invariant from G_1. In $\{G\}$ let G' be an element containing fewest number of parallel edges and self-loops, and in $\{G_1\}$ let G'_1 be an element containing fewest number of parallel edges. Define $X_k = \{x_1, x_2, ..., x_k\}$. Show that if $x_j = 2$ for all j then the minimum k with which the set $\{a_i\} \cup X_k$ is realizable as the degrees of the nodes of a (1,0)-graph is equal to the minimal number of edges that must be removed from G'_1, so that the resulting graph is a (1,0)-graph. Similarly, show that if $x_j = 1$ for all j then the minimum k with which the set $\{a_i\} \cup X_{2k}$ is realizable as the degrees of the nodes of a (1,0)-graph is equal to the minimal number of edges that must be removed from G', so that the resulting graph is a (1,0)-graph (OWENS and TRENT [1967]).

6.46. Let G_1 be a (p, s)-graph containing k, $k \geq 1$, self-loops where $p \neq 0$. Show that there exists a (p, s)-graph G_2 which is d-invariant from G_1 and which contains k self-loops at the nodes of highest degrees.

6.47. Show that (6.42b) holds for all k in $1 \leq k \leq n$ if, and only if,

$$x(x-1) + \sum_{i=x+1}^{n} \min(x, a_i) \geq \sum_{j=1}^{x} a_j \qquad (6.71)$$

for $x = 1, 2, ..., n-1$ (ERDÖS and GALLAI [1960] and CHEN [1973]).

6.48. Let t be a tree with n labeled nodes in which $d(i) = a_i$ for each node i. Let $\{t\}$ be the set of trees which are based on the n labeled nodes of t and which are d-invariant from t. Show that for $n \geq 3$

$$\alpha\{t\} = (n-2; a_1 - 1, a_2 - 1, ..., a_n - 1), \qquad (6.72a)$$

where

$$(n; a_1, a_2, ..., a_k) = \frac{n!}{a_1! a_2! ... a_k!}. \qquad (6.72b)$$

(*Hint*: Use the identity

$$(n; a_1, a_2, ..., a_k) = \sum (n-1; a_1, ..., a_i - 1, ..., a_k), \qquad (6.72c)$$

where the summation is taken over all i for $a_i \geq 1$.)

6.49. In Problem 6.48, show that in $\{t\}$ the number of trees in which a given node is of degree k is given by

$$\binom{n-2}{k-1}(n-1)^{n-k-1}, \qquad (6.73)$$

where $n \geq 3$ (CLARKE [1958]).

6.50. Which of the following sets are realizable as the degrees of the nodes of a (1,0)-graph?

(a) $\{2, 2, 3, 3, 3, 3\}$.
(b) $\{4, 4, 4, 4, 3, 3, 2\}$.
(c) $\{3, 3, 3, 3, 3, 3, 3, 3, 3, 3\}$.
(d) $\{5, 5, 4, 4, 4, 3, 3\}$.

6.51. Let G be a $(p, 0)$-graph, $p \geq 2$, in which there are at least two nodes with p parallel edges connecting between them. Let k be a node of G of maximum degree. Show that there exists a $(p, 0)$-graph G_1 which is d-invariant from G and which has the property that, for some j, there are p parallel edges connecting the nodes k and j of G_1.

6.52. In Problem 6.51, show that if k_1 and k_2, $k_1 \neq k_2$, are two of the nodes of G with largest degrees, then there exists a $(p, 0)$-graph G_2 which is d-invariant from G and which has the property that there are p parallel edges connecting its nodes k_1 and k_2.

6.53. Assume that the set $\{a_i\}$ ($a_j \geq a_{j+1}$ for $j=1, 2, \ldots, n-1$, $n \geq 2$) is realizable as the degrees of the nodes of a graph G without self-loops. Let $\{G\}$ be the set of self-loopless graphs which are d-invariant from G. Show that if there exists no $(p, 0)$-graph in $\{G\}$ then the set

$$\{a_1 - p, a_2 - p, a_3, \ldots, a_n\} \qquad (6.74)$$

is also realizable as the degrees of the nodes of a self-loopless graph. (*Hint*: Use the result of Problem 6.52.)

6.54. In (6.74), let $p=1$. Assume that the reduction cycle has been carried out k times. Let k_m be the minimal k such that the set corresponding to the k_mth reduction cycle is realizable as the degrees of the nodes of a (1,0)-graph. Show that k_m is the minimum number of edges that must be removed from a graph of $\{G\}$ containing the fewest number of parallel edges, so that the resulting graph is a (1,0)-graph (OWENS [1970]).

6.55. Applying the result of Problem 6.54, show that $k_m = 3$ for the set

$$\{8, 7, 5, 5, 4, 3\}.$$

6.56. Assume that the set $\{a_i\}$ ($a_j \geq a_{j+1}$ for $j=1, 2, \ldots, n-1$, $n \geq 2$) is realizable as the degrees of the nodes of a graph G having no self-loops. Let $k_m(\{a_i\})$ be the integer k_m defined for the set $\{a_i\}$ in Problem 6.54. Let a be the sum of the integers of $\{a_i\}$. Show that the integer $k_m(\{a_i\})$ can also be determined by the following algorithm:

Case 1. If $a_n < n-1$ and $2(a_1 + a_n) < a$, then

$$k_m(\{a_i\}) = k_m(\{b_j\}), \qquad (6.75a)$$

where $\{b_j\}$ is the set consisting of the nonnegative integers

$$a_1 - 1, a_2 - 1, \ldots, a_u - 1, a_{u+1}, \ldots, a_{n-1},$$

$u = a_n$, arranged in a non-increasing order.

Case 2. If $a_n < n-1$ and $2(a_1 + a_n) \geq a$, then

$$k_m(\{a_i\}) = a_1 - n + 1. \qquad (6.75b)$$

Case 3. If $a_n \geq n-1$, then

$$k_m(\{a_i\}) = \max[(a_1 - n + 1), \tfrac{1}{2}a - \tfrac{1}{2}n(n-1)]. \qquad (6.75c)$$

(*Hint*: For Case 1, there is a graph of $\{G\}$ which has no parallel edges outgoing from node n. For Case 2, we can connect node n by one edge each to a_n nodes other than node 1, with $a = 2(a_1 + a_n)$, and have all other edges connecting node 1 and the other nodes. For Case 3, we note that there are $\tfrac{1}{2}n(n-1)$ edges in the n-node complete (1,0)-graph.) (KLEITMAN [1970].)

6.57. Using (6.75), show that

(a) $k_m(\{8, 8, 4, 2, 2\}) = 5$.
(b) $k_m(\{8, 8, 7, 7\}) = 9$.
(c) $k_m(\{5, 5, 5, 4, 3, 2\}) = 1$.
(d) $k_m(\{8, 8, 4, 1, 1, 1, 1, 1, 1\}) = 4$.

6.58. Let $\{G_d\}$ be the set of directed graphs that are d-invariant from an n-node (p, s)-digraph G_d. Let G_1 and G_2 be two elements of $\{G_d\}$. Denote by $G_1 \oplus G_2$ the n-node directed graph obtained from G_1 and G_2 such that (i, j) is in $G_1 \oplus G_2$ if, and only if, (i, j) is either in G_1 or in G_2 but not in both. Assume that the edges of $G_1 \oplus G_2$ have been properly labeled so that we can distinguish the edges of G_1 from those of G_2. An *alternating edge train* in $G_1 \oplus G_2$ is an edge train in which the edges, along the edge train, belong alternately to G_1 and G_2 and in which if we reverse the directions of those edges of G_1 (G_2) we obtain a directed edge train. Show that if $G_1 \oplus G_2 \neq \emptyset$, then it can be decomposed into an edge-disjoint union of closed alternating edge trains of even length ≥ 4 (CHEN [1971b]).

6.59. In Problem 6.58, let $s \neq 0$. Show that G_1 and G_2 can be transformed into each other by a sequence of

$$\tfrac{1}{2} E(G_1 \oplus G_2) - L(G_1 \oplus G_2) \qquad (6.76)$$

and no fewer elementary (p, s) d-invariant transformations, where $E(G_1 \oplus G_2)$ denotes the number of edges in $G_1 \oplus G_2$, and $L(G_1 \oplus G_2)$ is the maximum number of edge-disjoint closed alternating edge trains in $G_1 \oplus G_2$ (CHEN [1971b]).

6.60. In Problem 6.58, let $s = 0$. Show that G_1 and G_2 can be transformed into each other by a sequence of k and no fewer elementary $(p, 1)$ d-invariant transformations where k is the number defined by (6.76).

6.61. Show that the statement of Problem 6.59 is valid for any s if G_1 and G_2 are two d-invariant (p, s)-graphs.

6.62. Consider the (2,0)-digraphs G_1 and G_2 as shown in figs. 6.17 and 6.18, respectively. Using the result of Problem 6.60, show that G_1 and G_2 can be transformed into each other by a sequence of four elementary (2,0) d-invariant transformations and no fewer than four of such transformations are possible.

6.63. For an n-node self-loopless graph G, let G^* be a self-loopless graph which is d-invariant from G and which has the fewest number of parallel edges. Let G_1 be a maximal spanning subgraph of G^* containing no parallel edges. Let D^* and D_1 be the sets of degrees of the nodes of G^* and G_1, respectively. For a given set $\{a_i\}$, a set $\{b_i\}$ is said to be a *maximal set relative to* $\{a_i\}$ if the sum of the elements of $\{b_i\}$ is the largest possible such that $b_i \leq a_i$ for all i and $\{b_i\}$ is realizable as the degrees of the nodes of a (1,0)-graph. Clearly, D_1 is a maximal set relative to D^*. Now, let $\bar{G}_1$ be the spanning subgraph of G^* containing all the edges not in G_1 (this is slightly different from the complement of a subgraph defined in ch. 1, § 2.1). Consider $\bar{G}_1^* = (\bar{G}_1)^*$. Let G_2 be a maximal spanning subgraph of $\bar{G}_1^*$ containing no parallel edges, and let D_1^* and D_2 be the sets of degrees of the nodes of $\bar{G}_1^*$ and G_2, respectively. Again, it is clear that D_2 is a maximal set relative to D_1^*. Thus, by continuing this process, we will generate a sequence of sets D_i^* and D_i such that D_i is a maximal set relative to D_{i-1}^* for $i = 1, 2, ..., k$ with $D^* = D_0^*$, where k is the smallest positive integer such that $\bar{G}_{k-1}$ has no parallel edges. Therefore, $\bar{G}_{k-1} = (\bar{G}_{k-1})^* = G_k$. Let

$$D_i^* = \{d_i^*(x); x = 1, 2, ..., n\},$$
$$D_i = \{d_i(x); x = 1, 2, ..., n\}.$$

Then, for all x, we have

$$d_0^*(x) = d_1(x) + d_2(x) + \cdots + d_k(x), \qquad (6.77a)$$

$$d_i^*(x) = d_{i-1}^*(x) - d_i(x). \qquad (6.77b)$$

Thus, D_i^* can be obtained from D_{i-1}^* and D_i recursively, where D_i is a maximal set relative to D_{i-1}^*. Using this procedure, show that, for a given set $D_0^* = \{a_i\}$, if it can be decomposed into a sequence of sets D_i^* and D_i with the above properties, then the graph obtained by superimposing the corresponding nodes of the (1,0)-graph realizations of the sets D_i, $i=1, 2, \ldots, k$, is a self-loopless realization of $\{a_i\}$ containing the fewest number of parallel edges (AYOUB and FRISCH [1970]).

6.64. Extend Problem 6.63 to any graph.

6.65. Let

$$D_0^* = \{8, 4, 3, 1, 1, 1\}.$$

Using the results of Problem 6.63, show that it can be decomposed into the sets

$$D_1 = \{5, 2, 2, 1, 1, 1\},$$
$$D_1^* = \{3, 2, 1, 0, 0, 0\},$$
$$D_2 = \{2, 1, 1, 0, 0, 0\},$$
$$D_2^* = \{1, 1, 0, 0, 0, 0\}.$$

What is the corresponding realization of the set D_0^*?

6.66. Show that a set $\{(a_i, b_i)\}$ ($a_j \geq a_{j+1}$ for $j=1, 2, \ldots, n-1$) is realizable as the degree pairs of the nodes of a $(1, 0)$-digraph that is also a directed Euler line if, and only if, $a_i = b_i \leq n-1$ for all i and the conditions (6.42b) are satisfied. (*Hint:* Apply Theorem 5.37.)

6.67. Assume that the set $\{(a_i, b_i)\}$ ($a_j \geq a_{j+1}$ for $j=1, 2, \ldots, n-1$) is realizable as the degree pairs of the nodes of a $(1, 0)$-digraph. Show that there exists a strongly-connected $(1, 0)$-digraph realization if, and only if, $\min(a_i, b_i) \neq 0$ for all i and

$$\sum_{i=1}^{k} b_i + \sum_{j=k+1}^{n} \min(b_j, k) > \sum_{i=1}^{k} a_i \qquad (6.78)$$

for $k=1, 2, \ldots, n-1$ (BEINEKE and HARARY [1965]).

6.68. Show that if (6.16a) holds, then (6.16b) is satisfied if, and only if, $a_i \leq n-1$ for all i.

CHAPTER 7

STATE EQUATIONS OF NETWORKS

In chapter 2 we used the Laplace transform technique to formulate the network equations in the form of systems of linear algebraic equations. To obtain the final time-domain solution, we must take the inverse Laplace transformation, which is usually computationally difficult. As an alternative, we can formulate the network equations in the time-domain as a system of first-order differential equations which governs the dynamic behavior of the network. The advantages of representing the network equations in this form are numerous. First of all, such a system has been widely studied in mathematics and its solution, both analytical and numerical, is known and readily available. Secondly, the representation can easily and naturally be extended to time-varying and nonlinear networks. In fact, nearly all time-varying and nonlinear networks are characterized by this approach. Finally, the first-order differential equations are easily programmed for a digital computer or simulated on an analog computer.

In this chapter, we first present procedures for the systematic formulation of network equations in the form of a system of first-order differential equations known as the *state equations*, and then discuss the number of dynamically independent *state variables* required in the formulation of these state equations. We shall demonstrate how this number or its upper bounds can be determined from network topology alone. In the following discussion, we shall limit ourselves to linear lumped networks that may be passive or active, reciprocal or non-reciprocal, and time-invariant or time-varying.

§1. State equations in normal form

Our objective is to choose an appropriate set of real functions $x_1(t)$, $x_2(t)$, ..., $x_k(t)$ of the time t called the *state variables* so that the network equations can be put in the form of a system of k first-order differential equations as

$$\dot{x}_i(t) = \sum_{j=1}^{k} a_{ij}(t) x_j(t) + \sum_{j=1}^{h} b_{ij}(t) u_j(t), \qquad (i = 1, 2, \ldots, k) \tag{7.1}$$

$\dot{x}_j(t)$ being the time derivative of $x_j(t)$, or, in matrix form,

$$\dot{x}(t) = A x(t) + B u(t), \qquad (7.2)$$

where the k-vector $x(t)$ formed by the state variables $x_j(t)$ is called the *state vector*, the h-vector $u(t)$ formed by the h known forcing functions or excitations $u_j(t)$ is referred to as the *input vector*, and $\dot{x}(t)$ denotes the k-vector whose elements are the time derivatives of those of $x(t)$. The coefficient matrices A and B, depending only upon the network parameters, are of orders $k \times k$ and $k \times h$, respectively. For a linear time-invariant network, they are constant matrices, being independent of time t.

The state variables $x_j(t)$ may or may not be the desired output variables. For this reason, another matrix equation is required to express the desired output variables $y_j(t)$ ($j = 1, 2, \ldots, w$) in terms of the state variables $x_j(t)$ ($j = 1, 2, \ldots, k$) and the input excitations $u_j(t)$ ($j = 1, 2, \ldots, h$) as

$$y(t) = C x(t) + D u(t), \qquad (7.3)$$

where the w-vector $y(t)$ whose elements are the w output variables $y_j(t)$ is called the *output vector*, and the known coefficient matrices C and D, depending only on the network parameters, are of orders $w \times k$ and $w \times h$, respectively.

Equation (7.2) is referred to as the *state equation in normal form*. Equation (7.3) is called the *output equation*. Together they are known as the *state equations*.

Our immediate problem is to choose the network variables as the state variables in order to formulate the state equations. If we call the set of instantaneous values of all the branch currents and voltages as the 'state' of the network, then the knowledge of the instantaneous values of all these variables determines this instantaneous state. However, not all of these instantaneous values are required in order to determine the instantaneous state, since some can be calculated from the others. For example, the instantaneous voltage of a resistor can be obtained from its instantaneous current through Ohm's law. The question arises as to how many branch voltages and currents whose instantaneous values are sufficient to determine completely the instantaneous state of the network. This leads to the following definition.

DEFINITION 7.1: *Complete set of state variables.* In a given network, a minimal set of its branch variables is said to be a *complete set of state variables* if their instantaneous values are sufficient to determine completely the instantaneous values of all the branch variables.

In general, the complete set of state variables is not unique. In most situations, the voltages across the capacitors and the currents through the inductors will constitute a complete set of state variables for the network. However, this set is not always complete. For example, in a network containing two capacitors connected in parallel, the voltages of these two capacitors cannot be chosen independently, one being determined by the other. Hence, they cannot both be the members of a complete set, since any complete set must be minimal. For our purposes, we shall choose the capacitor voltages and inductor currents as the state variables in the case of linear time-invariant networks. For linear time-varying networks, it is convenient to select the capacitor charges and the inductor flux linkages as the state variables. They are, of course, not the only possible choices. In fact, any linear combinations of capacitor voltages or charges and any linear combinations of inductor currents or flux linkages may be chosen as the state variables. Methods for obtaining a complete set of state variables will be discussed in the later sections.

After selecting the state variables, we now proceed to develop the state equations. Our starting point is the time-domain counterparts of the three primary systems of equations (2.40)–(2.42), as given in §2 of chapter 2. Let N be a network represented by a directed graph G. Let Q and B be a basis cut matrix and a basis circuit matrix of G. As before, denote by n, b, r and m the number of nodes, number of edges, rank and nullity of G, respectively. For each edge of G, there associate two real functions $v(t)$ and $i(t)$ of time t, satisfying the following constraints:

1. Kirchhoff's current law:

$$QI(t) = 0, \qquad (7.4)$$

where the branch-current vector $I(t)$ is a b-vector denoting the currents $i(t)$ of the edges of G.

2. Kirchhoff's voltage law:

$$BV(t) = 0, \qquad (7.5)$$

where the branch-voltage vector $V(t)$ is a b-vector denoting the voltages $v(t)$ of the edges of G.

3. Element defining equations or generalized Ohm's law

§1 State equations in normal form 467

A *capacitive edge* obeys a differential equation like

$$i(t) = \frac{d}{dt}[C(t)v(t)] = \frac{dq(t)}{dt}, \tag{7.6a}$$

where $C(t)$ and $q(t)$ are the capacitance and charges of the corresponding network element. An *inductive edge* obeys a differential equation like

$$v(t) = \frac{d}{dt}[L(t)i(t)] = \frac{d\lambda(t)}{dt}, \tag{7.6b}$$

where $L(t)$ and $\lambda(t)$ are the inductance and flux linkages of the corresponding network element. A *resistive edge* obeys an algebraic equation like

$$v(t) = R_k(t)i_k(t) \tag{7.6c}$$

or

$$i(t) = G_k(t)v_k(t), \tag{7.6d}$$

where $i_k(t)$ and $v_k(t)$ denote the current and voltage of the edge e_k, which may or may not be the same edge whose current and voltage are $i(t)$ and $v(t)$. Thus, (7.6c) or (7.6d) represents either a controlled source whose controlling parameter is $R_k(t)$ or $G_k(t)$, or a one-port resistor whose resistance is $R_k(t)$ or whose conductance is $G_k(t)$.

To simplify our notation, from here on we shall drop the time variable t in writing the variables, vectors or matrices with the understanding that they are functions of time t. We shall include t for situations where confusion may arise or, occasionally, for emphasis.

Example 7.1: Consider the network of fig. 7.1 represented by the directed graph G shown in fig. 7.2. Let Q and B be the fundamental cutset and circuit matrices of G with respect to the tree $t = e_1 e_4 e_5 e_8$. (It is unlikely that this t will be confused with the time variable t. To be consistent with the symbols

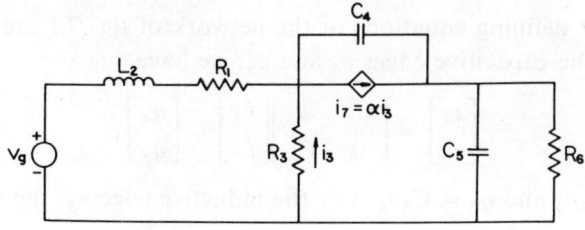

Fig. 7.1. The equivalent network of a transistor amplifier.

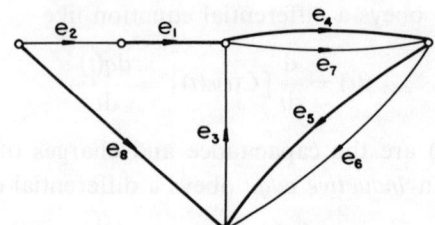

Fig. 7.2. The directed graph representing the network of fig. 7.1.

adopted in the preceding chapters, this practice will be continued throughout the remainder of the chapter.) Then the Kirchhoff's current and voltage laws become

$$
\begin{array}{c}
\begin{matrix} e_8 & e_4 & e_5 & e_1 & e_3 & e_6 & e_7 & e_2 \end{matrix} \\
\begin{bmatrix} 1 & 0 & 0 & 0 & 0 & 0 & 0 & 1 \\ 0 & 1 & 0 & 0 & -1 & 0 & 1 & -1 \\ 0 & 0 & 1 & 0 & -1 & 1 & 0 & -1 \\ 0 & 0 & 0 & 1 & 0 & 0 & 0 & -1 \end{bmatrix}
\begin{bmatrix} i_8 \\ i_4 \\ i_5 \\ i_1 \\ i_3 \\ i_6 \\ i_7 \\ i_2 \end{bmatrix} = \mathbf{0},
\end{array} \qquad (7.7\text{a})
$$

$$
\begin{bmatrix} 0 & 1 & 1 & 0 & 1 & 0 & 0 & 0 \\ 0 & 0 & -1 & 0 & 0 & 1 & 0 & 0 \\ 0 & -1 & 0 & 0 & 0 & 0 & 1 & 0 \\ -1 & 1 & 1 & 1 & 0 & 0 & 0 & 1 \end{bmatrix}
\begin{bmatrix} v_8 \\ v_4 \\ v_5 \\ v_1 \\ v_3 \\ v_6 \\ v_7 \\ v_2 \end{bmatrix} = \mathbf{0}. \qquad (7.7\text{b})
$$

The element defining equations of the network of fig. 7.1 are obtained as follows: For the capacitive edges e_4 and e_5, we have

$$
\begin{bmatrix} i_4 \\ i_5 \end{bmatrix} = \begin{bmatrix} C_4 & 0 \\ 0 & C_5 \end{bmatrix} \begin{bmatrix} \dot{v}_4 \\ \dot{v}_5 \end{bmatrix} = \begin{bmatrix} \dot{q}_4 \\ \dot{q}_5 \end{bmatrix}, \qquad (7.8\text{a})
$$

where $q_4 = C_4 v_4$ and $q_5 = C_5 v_5$. For the inductive edge e_2, the relation is

$$
v_2 = L_2 \dot{i}_2 = \dot{\lambda}_2, \qquad (7.8\text{b})
$$

§1 State equations in normal form

where $\lambda_2 = L_2 i_2$. For the resistive edges e_3, e_6, e_7 and e_1, we can write

$$\begin{bmatrix} i_3 \\ i_6 \\ i_7 \\ \hdashline v_1 \end{bmatrix} = \begin{bmatrix} 1/R_3 & 0 & 0 & \vdots & 0 \\ 0 & 1/R_6 & 0 & \vdots & 0 \\ \alpha/R_3 & 0 & 0 & \vdots & 0 \\ \hdashline 0 & 0 & 0 & \vdots & R_1 \end{bmatrix} \begin{bmatrix} v_3 \\ v_6 \\ v_7 \\ \hdashline i_1 \end{bmatrix}. \quad (7.8c)$$

To develop the state equations, we choose the capacitor voltages v_4 and v_5 and the inductor current i_2 as the state variables. The problem is now to eliminate the unwanted variables in equations (7.7) and (7.8), so that the final equations contain only the state variables v_4, v_5 and i_2 and their derivatives and the known forcing function $v_8 = v_g$. To this end, we expand equations (7.7) into

$$i_8 = -i_2, \quad (7.9a)$$

$$\begin{bmatrix} i_4 \\ i_5 \end{bmatrix} = -\begin{bmatrix} -1 & 0 & 1 \\ -1 & 1 & 0 \end{bmatrix} \begin{bmatrix} i_3 \\ i_6 \\ i_7 \end{bmatrix} - \begin{bmatrix} -1 \\ -1 \end{bmatrix} [i_2], \quad (7.9b)$$

$$i_1 = i_2, \quad (7.9c)$$

$$\begin{bmatrix} v_3 \\ v_6 \\ v_7 \end{bmatrix} = -\begin{bmatrix} 0 \\ 0 \\ 0 \end{bmatrix} [v_8] - \begin{bmatrix} 1 & 1 \\ 0 & -1 \\ -1 & 0 \end{bmatrix} \begin{bmatrix} v_4 \\ v_5 \end{bmatrix} - \begin{bmatrix} 0 \\ 0 \\ 0 \end{bmatrix} [v_1], \quad (7.9d)$$

$$[v_2] = -[-1][v_8] - [1 \; 1]\begin{bmatrix} v_4 \\ v_5 \end{bmatrix} - [1][v_1]. \quad (7.9e)$$

First, we eliminate the variables i_4, i_5 and v_2 by substituting (7.8a) and (7.8b) in (7.9b) and (7.9e), yielding

$$\begin{bmatrix} C_4 & 0 \\ 0 & C_5 \end{bmatrix} \begin{bmatrix} \dot{v}_4 \\ \dot{v}_5 \end{bmatrix} = \begin{bmatrix} 1 & 0 & -1 \\ 1 & -1 & 0 \end{bmatrix} \begin{bmatrix} i_3 \\ i_6 \\ i_7 \end{bmatrix} + \begin{bmatrix} 1 \\ 1 \end{bmatrix} [i_2], \quad (7.10a)$$

$$[L_2 \dot{i}_2] = [v_8] - [1 \; 1]\begin{bmatrix} v_4 \\ v_5 \end{bmatrix} - [v_1]. \quad (7.10b)$$

To continue the elimination process, it is necessary to express the unwanted variables i_3, i_6, i_7 and v_1 in terms of the state variables v_4, v_5 and i_2. For this purpose, we have available equations (7.8c), (7.9c) and (7.9d). First, we expand (7.8c) as

$$\begin{bmatrix} i_3 \\ i_6 \\ i_7 \end{bmatrix} = \begin{bmatrix} 1/R_3 & 0 & 0 \\ 0 & 1/R_6 & 0 \\ \alpha/R_3 & 0 & 0 \end{bmatrix} \begin{bmatrix} v_3 \\ v_6 \\ v_7 \end{bmatrix} + \begin{bmatrix} 0 \\ 0 \\ 0 \end{bmatrix} [i_1], \qquad (7.11a)$$

$$[v_1] = [0 \ 0 \ 0] \begin{bmatrix} v_3 \\ v_6 \\ v_7 \end{bmatrix} + [R_1][i_1]. \qquad (7.11b)$$

Then we substitute (7.9c) and (7.9d) in (7.11), resulting in

$$\begin{bmatrix} i_3 \\ i_6 \\ i_7 \end{bmatrix} = \begin{bmatrix} -1/R_3 & -1/R_3 \\ 0 & 1/R_6 \\ -\alpha/R_3 & -\alpha/R_3 \end{bmatrix} \begin{bmatrix} v_4 \\ v_5 \end{bmatrix}, \qquad (7.12a)$$

$$v_1 = R_1 i_2. \qquad (7.12b)$$

To complete the elimination process, we insert (7.12) in (7.10), giving

$$\begin{bmatrix} C_4 & 0 \\ 0 & C_5 \end{bmatrix} \begin{bmatrix} \dot{v}_4 \\ \dot{v}_5 \end{bmatrix} = \begin{bmatrix} (\alpha-1)/R_3 & (\alpha-1)/R_3 \\ -1/R_3 & -1/R_3 - 1/R_6 \end{bmatrix} \begin{bmatrix} v_4 \\ v_5 \end{bmatrix} + \begin{bmatrix} 1 \\ 1 \end{bmatrix} [i_2], \qquad (7.13a)$$

$$[L_2 \dot{i}_2] = -[1 \ 1] \begin{bmatrix} v_4 \\ v_5 \end{bmatrix} - [R_1][i_2] + [1][v_g]. \qquad (7.13b)$$

Writing these equations in normal form yields

$$\begin{bmatrix} \dot{i}_2 \\ \dot{v}_4 \\ \dot{v}_5 \end{bmatrix} = \begin{bmatrix} -R_1/L_2 & -1/L_2 & -1/L_2 \\ 1/C_4 & (\alpha-1)/C_4 R_3 & (\alpha-1)/C_4 R_3 \\ 1/C_5 & -1/C_5 R_3 & -1/C_5 R_3 - 1/C_5 R_6 \end{bmatrix} \begin{bmatrix} i_2 \\ v_4 \\ v_5 \end{bmatrix} + \begin{bmatrix} 1/L_2 \\ 0 \\ 0 \end{bmatrix} [v_g]. \qquad (7.14)$$

This is the state equation in normal form for the network of fig. 7.1.

Alternatively, we can choose the capacitor charges $q_4 = C_4 v_4$ and $q_5 = C_5 v_5$ and inductor flux linkage $\beta_2 = L_2 i_2$ as the state variables. In this case, equations (7.13) can be put in the form

$$\begin{bmatrix} \dot{\lambda}_2 \\ \dot{q}_4 \\ \dot{q}_5 \end{bmatrix} = \begin{bmatrix} -R_1/L_2 & -1/C_4 & -1/C_5 \\ 1/L_2 & (\alpha-1)/C_4 R_3 & (\alpha-1)/C_5 R_3 \\ 1/L_2 & -1/C_4 R_3 & -1/C_5 R_3 - 1/C_5 R_6 \end{bmatrix} \begin{bmatrix} \lambda_2 \\ q_4 \\ q_5 \end{bmatrix} + \begin{bmatrix} 1 \\ 0 \\ 0 \end{bmatrix} [v_g],$$

(7.15)

which is again in the form of (7.2).

We remark that, in developing the state equation for the network of fig. 7.1, we carried out the elimination process in great detail. In §3 we shall show that, with the exception of a few degenerate cases, the process is valid in general, and will yield the desired state equation. In fact, the procedures developed in §3 follow closely those outlined above. We shall refer to this example from time to time for illustration.

Suppose that the variables i_5, i_6 and v_2 are selected as the outputs. Then combining the second equation of (7.9b) with (7.12a) gives

$$[i_5] = [-1/R_3 \quad -1/R_3 - 1/R_6] \begin{bmatrix} v_4 \\ v_5 \end{bmatrix} + [1][i_2]. \quad (7.16a)$$

The current i_6 is contained in (7.12a) or

$$[i_6] = [0 \quad 1/R_6] \begin{bmatrix} v_4 \\ v_5 \end{bmatrix}. \quad (7.16b)$$

From (7.9e) and (7.12b), we have

$$[v_2] = [-1 \quad -1] \begin{bmatrix} v_4 \\ v_5 \end{bmatrix} - [R_1][i_2] + [1][v_g]. \quad (7.16c)$$

Writing these equations in matrix form yields the desired output equation

$$\begin{bmatrix} i_5 \\ i_6 \\ v_2 \end{bmatrix} = \begin{bmatrix} 1 & -1/R_3 & -1/R_3 - 1/R_6 \\ 0 & 0 & 1/R_6 \\ -R_1 & -1 & -1 \end{bmatrix} \begin{bmatrix} i_2 \\ v_4 \\ v_5 \end{bmatrix} + \begin{bmatrix} 0 \\ 0 \\ 1 \end{bmatrix} [v_g]. \quad (7.17)$$

In terms of the state variables q_4, q_5 and λ_2, the output equation becomes

$$\begin{bmatrix} i_5 \\ i_6 \\ v_2 \end{bmatrix} = \begin{bmatrix} 1/L_2 & -1/C_4 R_3 & -1/C_5 R_3 - 1/C_5 R_6 \\ 0 & 0 & 1/C_5 R_6 \\ -R_1/L_2 & -1/C_4 & -1/C_5 \end{bmatrix} \begin{bmatrix} \lambda_2 \\ q_4 \\ q_5 \end{bmatrix} + \begin{bmatrix} 0 \\ 0 \\ 1 \end{bmatrix} [v_g]. \quad (7.18)$$

Equations (7.14) and (7.17) or (7.15) and (7.18) are the state equations for the network of fig. 7.1.

§2. Procedures for writing the state equations

Following the steps outlined in the foregoing example, we now proceed to develop procedures for writing the state equations.

As shown in §7 of chapter 2, if a network has a circuit composed only of the independent voltage sources or a cut composed only of the independent current sources, then the network does not possess a unique solution, if one exists. Therefore, we shall exclude those networks from our considerations. Also, from Corollary 2.9 of chapter 2, we see that, with respect to a chosen tree, the tree-branch voltages determine all other voltages and cotree-chord currents determine all other currents. Thus, to develop state equations using inductor currents or flux linkages and capacitor voltages or charges as the state variables, we should place all the voltage sources and as many capacitors as possible in a tree, and all the current sources and as many inductors as possible in a cotree. This leads to the following definition.

DEFINITION 7.2: *Normal tree.* A *normal tree* of the connected directed graph representing a network is a tree that contains all the independent-voltage-source edges, the maximum number of capacitive edges, the minimum number of inductive edges, and none of the independent-current-source edges.

Notice that in this definition, we excluded the possibility of having unconnected networks. This can easily be taken care of by considering the *normal forest* composed of the normal trees of the components. Since a subgraph can be made part of a tree if and only if it does not contain any circuit, we see that, under the stipulated hypothesis, all the independent-voltage-source edges can be made part of a tree. Likewise, a subgraph can be made part of a cotree if and only if it does not contain any cut. This implies that all the independent-current-source edges can be made part of a cotree, being excluded from a tree. Appealing to Theorem 5.4 of chapter 5, we conclude that there exists a tree containing all the independent-voltage-source edges and none of the independent-current-source edges. The maximum number of capacitors and the minimum number of inductors that are included in such a tree will be discussed later.

Based on the observations made in the preceding section, we now outline steps for writing the state equations.

Step 1. From a given network, construct a directed graph G representing its interconnection and indicating the voltage and current references of its elements.

§2 Procedures for writing the state equations

Step 2. In G select a normal tree t, and choose either the capacitor voltages of t and the inductor currents of the cotree $\bar{t}$, the complement of t, or the capacitor charges of t and the inductor flux linkages of $\bar{t}$ as the state variables.

Step 3. Assign each branch of tree t a voltage symbol, and assign each chord of cotree $\bar{t}$ a current symbol.

Step 4. Using Kirchoff's current law, write each tree-branch current as a sum of cotree-chord currents, and indicate it in G if necessary.

Step 5. Using Kirchhoff's voltage law, write each cotree-chord voltage as a sum of tree-branch voltages, and indicate it in G if necessary.

Step 6. Apply the element defining equations (7.6) to each edge of G.

Step 7. Eliminate the non-state variables among the equations obtained in the preceding step.

Step 8. Write the resulting equations in normal form.

We shall illustrate the above steps by the following examples.

Example 7.2: Consider the network of fig. 7.1. We shall follow the above steps to develop the state equation.

Step 1. The directed graph G representing the network of fig. 7.1 was obtained earlier, and is given in fig. 7.2.

Step 2. As in Example 7.1, we see that the tree $t = e_1 e_4 e_5 e_8$ is a normal tree of G. We choose the tree capacitor voltages and cotree inductor current as the state variables.

Step 3. As indicated in fig. 7.3, the tree branches are assigned the voltage symbols v_1, v_4, v_5 and v_8 and the cotree chords are assigned the current symbols i_2, i_3, i_6 and i_7. Thus, the state variables are v_4, v_5 and i_2.

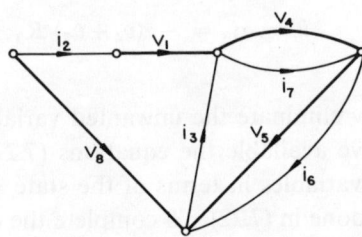

Fig. 7.3. The assignment of voltage symbols for the tree branches and the current symbols for the cotree chords for the directed graph of fig. 7.2.

Step 4. Using Kirchhoff's current law, the tree-branch currents can be expressed as the sums of the cotree-chord currents (the f-cutsets):

$$i_1 = i_2, \tag{7.19a}$$

$$i_4 = i_2 + i_3 - i_7, \tag{7.19b}$$

$$i_5 = i_2 + i_3 - i_6, \tag{7.19c}$$

$$i_8 = -i_2. \tag{7.19d}$$

Step 5. Using Kirchoff's voltage law, the cotree-chord voltages can be expressed as the sums of the tree-branch voltages (the *f*-circuits):

$$v_2 = v_8 - v_1 - v_4 - v_5, \tag{7.20a}$$

$$v_3 = -v_4 - v_5, \tag{7.20b}$$

$$v_6 = v_5, \tag{7.20c}$$

$$v_7 = v_4. \tag{7.20d}$$

Step 6. Apply the element defining equations (7.6) to each edge of G. From figs. 7.1 and 7.3 in conjunction with (7.19) and (7.20), we obtain

$$C_4 \dot{v}_4 = i_2 + i_3 - i_7, \tag{7.21a}$$

$$C_5 \dot{v}_5 = i_2 + i_3 - i_6, \tag{7.21b}$$

$$L_2 \dot{i}_2 = v_8 - v_1 - v_4 - v_5, \tag{7.21c}$$

$$v_1 = R_1 i_2, \tag{7.22a}$$

$$i_3 = -(v_4 + v_5)/R_3, \tag{7.22b}$$

$$i_6 = v_5/R_6, \tag{7.22c}$$

$$i_7 = \alpha i_3 = -\alpha(v_4 + v_5)/R_3, \tag{7.22d}$$

where $v_8 = v_g$.

Step 7. We must eliminate the unwanted variables v_1, i_3, i_6 and i_7 in (7.21). For this we have available the equations (7.22), from which we can express the unwanted variables in terms of the state variables i_2, v_4 and v_5. This already has been done in (7.22). To complete the elimination process, we substitute (7.22) in (7.21), giving

$$L_2 \dot{i}_2 = v_g - v_4 - v_5 - R_1 i_2, \tag{7.23a}$$

$$C_4 \dot{v}_4 = i_2 + (\alpha - 1)v_4/R_3 + (\alpha - 1)v_5/R_3, \tag{7.23b}$$

$$C_5 \dot{v}_5 = i_2 - v_4/R_3 - (1/R_3 + 1/R_6)v_5. \tag{7.23c}$$

Step 8. Writing equations (7.23) in matrix form yields the desired state equation

$$\dot{x} = Ax + Bu, \tag{7.24a}$$

where

$$A = \begin{bmatrix} -R_1/L_2 & -1/L_2 & -1/L_2 \\ 1/C_4 & (\alpha-1)/C_4 R_3 & (\alpha-1)/C_4 R_3 \\ 1/C_5 & -1/C_5 R_3 & -1/C_5 R_3 - 1/C_5 R_6 \end{bmatrix}, \tag{7.24b}$$

$$B = \begin{bmatrix} 1/L_2 \\ 0 \\ 0 \end{bmatrix}, \quad x = \begin{bmatrix} i_2 \\ v_4 \\ v_5 \end{bmatrix}, \quad u = [v_g]. \tag{7.24c}$$

Since the resistors, capacitors and inductor are time-invariant, the matrices A and B are constant matrices, being independent of the time t, as expected.

Example 7.3: To illustrate the procedure for a time-varying network, consider the network of fig. 7.1, in which we let

$$C_4(t) = \frac{1}{2+\sin 2t}, \tag{7.25a}$$

$$C_5(t) = \frac{1}{1+\tfrac{1}{2}\cos t}, \tag{7.25b}$$

$$R_1(t) = 2+\sin 3t. \tag{7.25c}$$

The first five steps are the same as those outlined in Example 7.2 except that, for convenience, we choose the capacitor charges

$$q_4 = C_4 v_4, \tag{7.26a}$$

$$q_5 = C_5 v_5, \tag{7.26b}$$

and inductor flux linkage

$$\lambda_2 = L_2 i_2 \tag{7.26c}$$

as the state variables.

Step 6. Applying the element defining equations yields

$$\dot{q}_4 = i_2 + i_3 - i_7, \tag{7.27a}$$

$$\dot{q}_5 = i_2 + i_3 - i_6, \tag{7.27b}$$

$$\dot{\lambda}_2 = v_8 - v_1 - v_4 - v_5, \tag{7.27c}$$

$$v_1 = R_1 i_2, \tag{7.28a}$$

$$i_3 = -(v_4 + v_5)/R_3, \tag{7.28b}$$

$$i_6 = v_5/R_6, \tag{7.28c}$$

$$i_7 = -\alpha(v_4 + v_5)/R_3, \tag{7.28d}$$

where $v_8 = v_g$.

Step 7. To eliminate the unwanted variables v_1, i_3, i_6 and i_7 in (7.27), we express them in terms of the state variables q_4, q_5 and λ_2 by making use of (7.26) and (7.28). This gives

$$v_1 = R_1 \lambda_2/L_2, \tag{7.29a}$$

$$i_3 = -q_4/C_4 R_3 - q_5/C_5 R_3, \tag{7.29b}$$

$$i_6 = q_5/C_5 R_6, \tag{7.29c}$$

$$i_7 = -\alpha q_4/C_4 R_3 - \alpha q_5/C_5 R_3. \tag{7.29d}$$

Substituting these in (7.27) in conjunction with (7.25) and (7.26) yields

$$\dot{\lambda}_2 = v_g - R_1 \lambda_2/L_2 - (2 + \sin 2t)q_4 - (1 + \tfrac{1}{2}\cos t)q_5, \tag{7.30a}$$

$$\dot{q}_4 = \lambda_2/L_2 + (2 + \sin 2t)(\alpha - 1)q_4/R_3 + (1 + \tfrac{1}{2}\cos t)(\alpha - 1)q_5/R_3, \tag{7.30b}$$

$$\dot{q}_5 = \lambda_2/L_2 - (2 + \sin 2t)q_4/R_3 - (1 + \tfrac{1}{2}\cos t)(1/R_3 + 1/R_6)q_5. \tag{7.30c}$$

Step 8. Writing the equations (7.30) in matrix form gives

$$\dot{x}(t) = A(t)x(t) + B(t)u(t), \tag{7.31a}$$

where

$$A(t) = \begin{bmatrix} -R_1/L_2 & -(2+\sin 2t) & -(1+\tfrac{1}{2}\cos t) \\ 1/L_2 & (2+\sin 2t)(\alpha-1)/R_3 & (1+\tfrac{1}{2}\cos t)(\alpha-1)/R_3 \\ 1/L_2 & -(2+\sin 2t)/R_3 & -(1+\tfrac{1}{2}\cos t)(R_3+R_6)/R_3 R_6 \end{bmatrix},$$

$$\tag{7.31b}$$

$$\boldsymbol{B} = \begin{bmatrix} 1 \\ 0 \\ 0 \end{bmatrix}, \quad \boldsymbol{x}(t) = \begin{bmatrix} \lambda_2 \\ q_4 \\ q_5 \end{bmatrix}, \quad \boldsymbol{u}(t) = [v_g(t)]. \tag{7.31c}$$

As expected, since the network is time-varying, the coefficient matrix $A(t)$ is a function of t.

To demonstrate that not all the capacitor voltages or charges and inductor currents or flux linkages can be members of a complete set of state variables, we insert an inductor L_{10} and a capacitor C_9 in the network of fig. 7.1. Also we add an independent current source $i_g(t)$. The resulting network is shown in fig. 7.4.

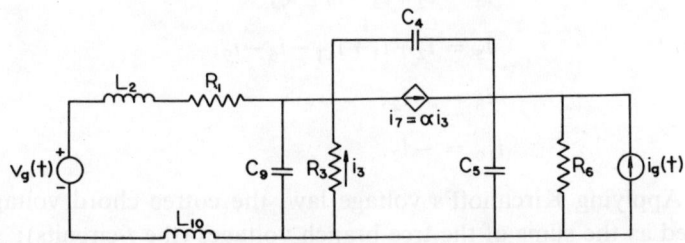

Fig. 7.4. An active network used to illustrate the formulation of its state equation.

Example 7.4: Following the eight steps developed in this section, obtain the state equation of the network of fig. 7.4.

Step 1. The directed graph G representing the network of fig. 7.4 is presented in fig. 7.5.

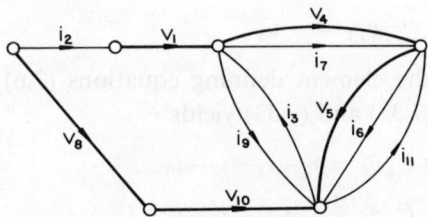

Fig. 7.5. The directed graph representing the active network of fig. 7.4.

Step 2. In G let e_k be the edge associated with the variable v_k or i_k. Since $e_4 e_5 e_9$ is a circuit, at most two capacitive edges can be in any normal tree.

Similarly, since $e_2 e_{10}$ is a cutset, at least one inductive edge will be in a normal tree. A possible normal tree $t = e_1 e_4 e_5 e_8 e_{10}$ is shown in heavy lines in fig. 7.5. Other possible choices are $e_1 e_4 e_8 e_9 e_{10}$, $e_1 e_5 e_8 e_9 e_{10}$, $e_1 e_2 e_4 e_5 e_8$, $e_1 e_2 e_4 e_8 e_9$ and $e_1 e_2 e_5 e_8 e_9$.

Step 3. As indicated in fig. 7.5, the tree branches are assigned the voltage symbols v_1, v_4, v_5, v_8 and v_{10} and the cotree chords are assigned the current symbols i_2, i_3, i_6, i_7, i_9 and i_{11}. The state variables are chosen to be v_4, v_5 and i_2.

Step 4. Using Kirchhoff's current law, the tree-branch currents can be expressed as the sums of the cotree-chord currents (the f-cutsets):

$$i_1 = i_2, \tag{7.32a}$$

$$i_4 = i_2 + i_3 - i_7 - i_9, \tag{7.32b}$$

$$i_5 = i_2 + i_3 + i_{11} - i_6 - i_9, \tag{7.32c}$$

$$i_8 = -i_2, \tag{7.32d}$$

$$i_{10} = -i_2. \tag{7.32e}$$

Step 5. Applying Kirchhoff's voltage law, the cotree chord voltages can be expressed as the sums of the tree-branch voltages (the f-circuits):

$$v_2 = v_8 + v_{10} - v_1 - v_4 - v_5, \tag{7.33a}$$

$$v_3 = -v_4 - v_5, \tag{7.33b}$$

$$v_6 = v_5, \tag{7.33c}$$

$$v_7 = v_4, \tag{7.33d}$$

$$v_9 = v_4 + v_5, \tag{7.33e}$$

$$v_{11} = -v_5. \tag{7.33f}$$

Step 6. Applying the element defining equations (7.6) to the edges of G in conjunction with (7.32) and (7.33) yields

$$C_4 \dot{v}_4 = i_2 + i_3 - i_7 - i_9, \tag{7.34a}$$

$$C_5 \dot{v}_5 = i_2 + i_3 - i_6 - i_9 + i_{11}, \tag{7.34b}$$

$$L_2 \dot{i}_2 = v_8 + v_{10} - v_1 - v_4 - v_5, \tag{7.34c}$$

$$v_1 = R_1 i_2, \tag{7.35a}$$

$$v_8 = v_g, \tag{7.35b}$$

§3 The explicit form of the state equation 479

$$v_{10} = -L_{10}\dot{i}_2, \tag{7.35c}$$

$$i_3 = -v_4/R_3 - v_5/R_3, \tag{7.35d}$$

$$i_6 = v_5/R_6, \tag{7.35e}$$

$$i_7 = \alpha v_3/R_3 = \alpha(-v_4 - v_5)/R_3, \tag{7.35f}$$

$$i_9 = C_9 \dot{v}_4 + C_9 \dot{v}_5, \tag{7.35g}$$

$$i_{11} = i_g. \tag{7.35h}$$

Step 7. We must now eliminate the unwanted variables v_1, v_8, v_{10}, i_3, i_6, i_7, i_9 and i_{11} in (7.34). This can easily be accomplished by substituting (7.35) in (7.34), giving

$$\begin{bmatrix} C_4+C_9 & C_9 \\ C_9 & C_5+C_9 \end{bmatrix} \begin{bmatrix} \dot{v}_4 \\ \dot{v}_5 \end{bmatrix} = \begin{bmatrix} (\alpha-1)/R_3 & (\alpha-1)/R_3 \\ -1/R_3 & -1/R_3 - 1/R_6 \end{bmatrix} \begin{bmatrix} v_4 \\ v_5 \end{bmatrix} + \begin{bmatrix} i_2 \\ i_2+i_g \end{bmatrix},$$
(7.36a)

$$(L_2 + L_{10})\dot{i}_2 = v_g - R_1 i_2 - v_4 - v_5. \tag{7.36b}$$

Step 8. Writing the equations (7.36) in normal form results in

$$\dot{x} = Ax + Bu, \tag{7.37a}$$

where

$$A = \frac{1}{\Delta} \begin{bmatrix} -R_1 \Delta/(L_2+L_{10}) & -\Delta/(L_2+L_{10}) & -\Delta/(L_2+L_{10}) \\ C_5 & (\alpha-1)(C_5+C_9)/R_3 & (\alpha-1)(C_5+C_9)/R_3 \\ & +C_9/R_3 & +C_9(1/R_3+1/R_6) \\ C_4 & -(\alpha C_9+C_4)/R_3 & -(C_4+C_9)(1/R_3+1/R_6) \\ & & -(\alpha-1)C_9/R_3 \end{bmatrix},$$
(7.37b)

$$B = \frac{1}{\Delta} \begin{bmatrix} 0 & \Delta/(L_2+L_{10}) \\ -C_9 & 0 \\ C_4+C_9 & 0 \end{bmatrix}, \tag{7.37c}$$

$$x = \begin{bmatrix} i_2 \\ v_4 \\ v_5 \end{bmatrix}, \quad u = \begin{bmatrix} i_g \\ v_g \end{bmatrix}, \tag{7.37d}$$

and
$$\Delta = C_4 C_5 + C_9(C_4 + C_5). \tag{7.37e}$$

§3. The explicit form of the state equation

In the preceding section, we introduced a procedure for obtaining the state equation in normal form for linear networks that may be passive or active, time-invariant or time-varying. In this section, we shall develop the explicit form of state equation in normal form for a general linear network. For the general procedure to succeed, we shall assume the nonsingularity of certain matrices. If it happens that the assumption is not valid for a particular network, then it will not be possible to obtain the desired form by the process that we shall describe. Due to its generality, the explicit form is extremely complicated, and for a particular network it is usually simpler to obtain the desired state equation by the procedure outlined in the preceding section. The objective of this section is to formalize the formulation of the state equation for a general network, and to exhibit its general form.

Our starting point is the three primary systems of network equations (7.4)–(7.6). As before, the first step is to choose a normal tree t in the directed graph G representing the network N. Let $\boldsymbol{Q}_f$ and $\boldsymbol{B}_f$ be the f-cutset and f-circuit matrices of G with respect to the tree t. Partition $\boldsymbol{Q}_f$ and $\boldsymbol{B}_f$ according to the branches and chords of the tree t, and partition the branch-current vector $\boldsymbol{I}$ and branch-voltage vector V accordingly. Then by appealing to Theorem 2.9, (7.4) and (7.5) can be written as

$$[U_r \ \ \boldsymbol{Q}_{\bar{t}}] \begin{bmatrix} I_t \\ I_{\bar{t}} \end{bmatrix} = \boldsymbol{0}, \tag{7.38}$$

$$[-\boldsymbol{Q}'_{\bar{t}} \ \ U_m] \begin{bmatrix} V_t \\ V_{\bar{t}} \end{bmatrix} = \boldsymbol{0}. \tag{7.39}$$

To facilitate our discussion, we must further partition the branch-current and branch-voltage vectors in accordance with the types of edges appearing in the normal tree t and its complement, the cotree $\bar{t}$. To this end, we adopt the following convention in representing various types of edges:

- e independent voltage sources;
- ct capacitive branches of the tree t;
- rt resistive branches of the tree t;

§3   The explicit form of the state equation

lt inductive branches of the tree t;
$c\bar{t}$ capacitive chords of the cotree $\bar{t}$;
$r\bar{t}$ resistive chords of the cotree $\bar{t}$;
$\bar{lt}$ inductive chords of the cotree $\bar{t}$;
j independent current sources.

Thus, I_{ct} denotes the current vector of the capacitor currents in the tree t, and $V_{r\bar{t}}$ represents the voltage vector of the resistor voltages in the cotree $\bar{t}$. The extension of these symbols to other types of elements is straightforward, and no further elaboration is necessary. Following this convention, write

$$V_t = \begin{bmatrix} V_e \\ V_{ct} \\ V_{rt} \\ V_{lt} \end{bmatrix} \text{ and } V_{\bar{t}} = \begin{bmatrix} V_{c\bar{t}} \\ V_{r\bar{t}} \\ V_{\bar{lt}} \\ V_j \end{bmatrix}, \quad (7.40\text{a})$$

$$I_t = \begin{bmatrix} I_e \\ I_{ct} \\ I_{rt} \\ I_{lt} \end{bmatrix} \text{ and } I_{\bar{t}} = \begin{bmatrix} I_{c\bar{t}} \\ I_{r\bar{t}} \\ I_{\bar{lt}} \\ I_j \end{bmatrix}. \quad (7.40\text{b})$$

Likewise, the submatrix $Q_{\bar{t}}$ can be written as

$$Q_{\bar{t}} = \begin{array}{c} \\ e \\ c \\ r \\ l \end{array} \overset{\begin{array}{cccc} c & r & l & j \end{array}}{\begin{bmatrix} Q_{ec} & Q_{er} & Q_{el} & Q_{ej} \\ Q_{cc} & Q_{cr} & Q_{cl} & Q_{cj} \\ Q_{rc} & Q_{rr} & Q_{rl} & Q_{rj} \\ Q_{lc} & Q_{lr} & Q_{ll} & Q_{lj} \end{bmatrix}}. \quad (7.41)$$

By virtue of the definition of a normal tree, we must have

$$Q_{rc} = 0, \quad Q_{lc} = 0, \quad Q_{lr} = 0. \quad (7.42)$$

For example, if $Q_{rc} \neq 0$, there exists an f-cutset defined by a resistive branch of the normal tree and containing a capacitive chord. Replacing this resistive branch by the capacitive chord in the normal tree yields a tree that contains one more capacitive edge than the "normal tree", all other types of edges being the same except the resistive edges. This contradicts to the definition of a normal tree, and, consequently, $Q_{rc} = 0$. In a similar fashion, we can show $Q_{lc} = 0$ and $Q_{lr} = 0$ (Problem 7.2).

Substituting these partitioned matrices into the Kirchhoff's equations (7.38) and (7.39) and expanding the resulting equations yield

$$I_e + Q_{ec}I_{c\bar{t}} + Q_{er}I_{r\bar{t}} + Q_{el}I_{l\bar{t}} + Q_{ej}I_j = 0, \tag{7.43a}$$

$$I_{ct} + Q_{cc}I_{c\bar{t}} + Q_{cr}I_{r\bar{t}} + Q_{cl}I_{l\bar{t}} + Q_{cj}I_j = 0, \tag{7.43b}$$

$$I_{rt} + Q_{rr}I_{r\bar{t}} + Q_{rl}I_{l\bar{t}} + Q_{rj}I_j = 0, \tag{7.43c}$$

$$I_{lt} + Q_{ll}I_{l\bar{t}} + Q_{lj}I_j = 0, \tag{7.43d}$$

$$V_{c\bar{t}} - Q'_{cc}V_{ct} - Q'_{ec}V_e = 0, \tag{7.43e}$$

$$V_{r\bar{t}} - Q'_{cr}V_{ct} - Q'_{rr}V_{rt} - Q'_{er}V_e = 0, \tag{7.43f}$$

$$V_{l\bar{t}} - Q'_{cl}V_{ct} - Q'_{rl}V_{rt} - Q'_{ll}V_{lt} - Q'_{el}V_e = 0, \tag{7.43g}$$

$$V_j - Q'_{cj}V_{ct} - Q'_{rj}V_{rt} - Q'_{lj}V_{lt} - Q'_{ej}V_e = 0. \tag{7.43h}$$

In addition to these equations, we have the element defining equations as shown in (7.6). For the capacitive edges, we have

$$\begin{bmatrix} I_{ct} \\ I_{c\bar{t}} \end{bmatrix} = \frac{d}{dt}\begin{bmatrix} C_{tt} & 0 \\ 0 & C_{\bar{t}\bar{t}} \end{bmatrix}\begin{bmatrix} V_{ct} \\ V_{c\bar{t}} \end{bmatrix} = \begin{bmatrix} \dot{q}_{ct} \\ \dot{q}_{c\bar{t}} \end{bmatrix}, \tag{7.44a}$$

where the capacitor charge vectors q_{ct} and $q_{c\bar{t}}$ are defined by the relation

$$\begin{bmatrix} q_{ct} \\ q_{c\bar{t}} \end{bmatrix} = \begin{bmatrix} C_{tt}V_{ct} \\ C_{\bar{t}\bar{t}}V_{c\bar{t}} \end{bmatrix}. \tag{7.44b}$$

For the inductive edges, we have the relation

$$\begin{bmatrix} V_{l\bar{t}} \\ V_{lt} \end{bmatrix} = \frac{d}{dt}\begin{bmatrix} L_{\bar{t}\bar{t}} & L_{\bar{t}t} \\ L_{t\bar{t}} & L_{tt} \end{bmatrix}\begin{bmatrix} I_{l\bar{t}} \\ I_{lt} \end{bmatrix} = \begin{bmatrix} \dot{\lambda}_{\bar{t}\bar{t}} + \dot{\lambda}_{\bar{t}t} \\ \dot{\lambda}_{t\bar{t}} + \dot{\lambda}_{tt} \end{bmatrix}, \tag{7.45a}$$

where the inductor self-flux-linkage vectors $\lambda_{\bar{t}\bar{t}}$ and λ_{tt} and mutual flux-linkage vectors $\lambda_{\bar{t}t}$ and $\lambda_{t\bar{t}}$ are defined by

$$\begin{bmatrix} \lambda_{\bar{t}\bar{t}} \\ \lambda_{tt} \\ \lambda_{\bar{t}t} \\ \lambda_{t\bar{t}} \end{bmatrix} = \begin{bmatrix} L_{\bar{t}\bar{t}}I_{l\bar{t}} \\ L_{tt}I_{lt} \\ L_{\bar{t}t}I_{lt} \\ L_{t\bar{t}}I_{l\bar{t}} \end{bmatrix}. \tag{7.45b}$$

We remark that, unlike the capacitance matrix defined in (7.44a), the inductance matrix of (7.45a) is not necessarily diagonal in order to take into consideration the possibility of mutual inductance. However, it is symmetric with $L_{\bar{t}\bar{t}} = L'_{\bar{t}\bar{t}}$, $L_{tt} = L'_{tt}$ and $L_{\bar{t}t} = L'_{t\bar{t}}$.

§3   The explicit form of the state equation   483

Finally, for the resistive edges, we assume that they can be characterized by the hybrid relation

$$\begin{bmatrix} I_{r\bar{t}} \\ V_{rt} \end{bmatrix} = \begin{bmatrix} H_{\bar{t}\bar{t}} & H_{\bar{t}t} \\ H_{\bar{t}\bar{t}} & H_{tt} \end{bmatrix} \begin{bmatrix} V_{r\bar{t}} \\ I_{rt} \end{bmatrix}. \tag{7.46}$$

Our objective is to eliminate, among the equations (7.43)–(7.46), all the unwanted vectors with the aim of retaining only the vectors V_{ct} and $I_{l\bar{t}}$, the source vectors, and their derivatives. To this end, we proceed with the elimination process in the following manner.

A. *Eliminating the vectors I_{ct}, $I_{c\bar{t}}$ and $V_{c\bar{t}}$*

This is accomplished by first substituting $V_{c\bar{t}}$ from (7.43e) in (7.44a) and then inserting I_{ct} and $I_{c\bar{t}}$ from the resulting (7.44a) in (7.43b), giving

$$\frac{d}{dt}(C_c V_{ct}) = -Q_{cr} I_{r\bar{t}} - Q_{cl} I_{l\bar{t}} - Q_{cj} I_j + \frac{d}{dt}(C_e V_e), \tag{7.47}$$

where

$$C_c = C_{tt} + Q_{cc} C_{\bar{t}\bar{t}} Q'_{cc}, \tag{7.48a}$$

$$C_e = -Q_{cc} C_{\bar{t}\bar{t}} Q'_{ec}. \tag{7.48b}$$

B. *Eliminating the vectors V_{lt}, $V_{l\bar{t}}$ and I_{lt}*

Substituting I_{lt} from (7.43d) in (7.45a) and then inserting $V_{l\bar{t}}$ and V_{lt} from the resulting (7.45a) in (7.43g), yielding

$$\frac{d}{dt}(L_l I_{l\bar{t}}) = Q'_{cl} V_{ct} + Q'_{rl} V_{rt} + Q'_{el} V_e + \frac{d}{dt}(L_j I_j), \tag{7.49}$$

where

$$L_l = L_{\bar{t}\bar{t}} - L_{\bar{t}t} Q_{ll} - Q'_{ll} L_{\bar{t}\bar{t}} + Q'_{ll} L_{tt} Q_{ll}, \tag{7.50a}$$

$$L_j = -Q'_{ll} L_{tt} Q_{lj} + L_{\bar{t}t} Q_{lj}. \tag{7.50b}$$

C. *Eliminating the vectors $I_{r\bar{t}}$ and V_{rt}*

Equations (7.47) and (7.49) contain all the desired vectors except $I_{r\bar{t}}$ and V_{rt} which must be eliminated. For this we still have the unused equations (7.43c), (7.43f) and (7.46). When the former two are substituted into the latter, we get

$$(U_{r\bar{t}} + H_{\bar{t}\bar{t}} Q_{rr}) I_{r\bar{t}} - H_{\bar{t}t} Q'_{rr} V_{rt} = F_1, \tag{7.51a}$$

$$H_{tt} Q_{rr} I_{r\bar{t}} + (U_{rt} - H_{\bar{t}\bar{t}} Q'_{rr}) V_{rt} = F_2, \tag{7.51b}$$

where $U_{r\bar{t}}$ and U_{rt} are identity matrices of appropriate orders, and

$$F_1 = H_{\bar{t}\bar{t}}Q'_{cr}V_{ct} - H_{\bar{t}\bar{t}}Q_{rl}I_{l\bar{t}} + H_{\bar{t}\bar{t}}Q'_{er}V_e - H_{\bar{t}\bar{t}}Q_{rj}I_j, \qquad (7.52a)$$

$$F_2 = H_{\bar{t}\bar{t}}Q'_{cr}V_{ct} - H_{tt}Q_{rl}I_{l\bar{t}} + H_{\bar{t}\bar{t}}Q'_{er}V_e - H_{tt}Q_{rj}I_j. \qquad (7.52b)$$

To simplify the notation, define

$$W_1 = U_{r\bar{t}} + H_{\bar{t}\bar{t}}Q_{rr}, \qquad (7.53a)$$

$$W_2 = U_{rt} - H_{\bar{t}\bar{t}}Q'_{rr}. \qquad (7.53b)$$

If W_1 and W_2 are nonsingular, (7.51) can be solved for $I_{r\bar{t}}$ and V_{rt}. When these are substituted back into (7.51) and the terms are rearranged, we obtain

$$M_1 I_{r\bar{t}} = F_{11}V_{ct} - F_{12}I_{l\bar{t}} + P_{11}V_e - P_{12}I_j, \qquad (7.54a)$$

$$M_2 V_{rt} = F_{21}V_{ct} - F_{22}I_{l\bar{t}} + P_{21}V_e - P_{22}I_j, \qquad (7.54b)$$

where

$$M_1 = W_1 + H_{\bar{t}\bar{t}}Q'_{rr}W_2^{-1}H_{tt}Q_{rr}, \qquad (7.55a)$$

$$M_2 = W_2 + H_{tt}Q_{rr}W_1^{-1}H_{\bar{t}\bar{t}}Q'_{rr}, \qquad (7.55b)$$

$$F_{11} = H_{\bar{t}\bar{t}}Q'_{cr} + H_{\bar{t}\bar{t}}Q'_{rr}W_2^{-1}H_{\bar{t}\bar{t}}Q'_{cr}, \qquad (7.56a)$$

$$F_{12} = H_{\bar{t}\bar{t}}Q_{rl} + H_{\bar{t}\bar{t}}Q'_{rr}W_2^{-1}H_{tt}Q_{rl}, \qquad (7.56b)$$

$$F_{21} = H_{\bar{t}\bar{t}}Q'_{cr} - H_{tt}Q_{rr}W_1^{-1}H_{\bar{t}\bar{t}}Q'_{cr}, \qquad (7.56c)$$

$$F_{22} = H_{tt}Q_{rl} - H_{tt}Q_{rr}W_1^{-1}H_{\bar{t}\bar{t}}Q_{rl}, \qquad (7.56d)$$

$$P_{11} = H_{\bar{t}\bar{t}}Q'_{er} + H_{\bar{t}\bar{t}}Q'_{rr}W_2^{-1}H_{\bar{t}\bar{t}}Q'_{er}, \qquad (7.57a)$$

$$P_{12} = H_{\bar{t}\bar{t}}Q_{rj} + H_{\bar{t}\bar{t}}Q'_{rr}W_2^{-1}H_{tt}Q_{rj}, \qquad (7.57b)$$

$$P_{21} = H_{\bar{t}\bar{t}}Q'_{er} - H_{tt}Q_{rr}W_1^{-1}H_{\bar{t}\bar{t}}Q'_{er}, \qquad (7.57c)$$

$$P_{22} = H_{tt}Q_{rj} - H_{tt}Q_{rr}W_1^{-1}H_{\bar{t}\bar{t}}Q_{rj}. \qquad (7.57d)$$

Assuming that M_1 and M_2 are nonsingular, (7.54) can be solved for $I_{r\bar{t}}$ and V_{rt}. When the solutions are substituted back into (7.47) and (7.49), the resulting equations can be put in the form

$$\begin{bmatrix} \dot{V}_{ct} \\ \dot{I}_{l\bar{t}} \end{bmatrix} = \begin{bmatrix} A_{11} & A_{12} \\ A_{21} & A_{22} \end{bmatrix} \begin{bmatrix} V_{ct} \\ I_{l\bar{t}} \end{bmatrix} + \begin{bmatrix} B_{11} & B_{12} & B_{13} & 0 \\ B_{21} & B_{22} & 0 & B_{24} \end{bmatrix} \begin{bmatrix} V_e \\ I_j \\ \dot{V}_e \\ \dot{I}_j \end{bmatrix}, \qquad (7.58)$$

where, assuming that C_c and L_l are nonsingular,

$$A_{11} = -C_c^{-1}(Q_{cr}M_1^{-1}F_{11}+\dot{C}_c), \tag{7.59a}$$

$$A_{12} = C_c^{-1}(Q_{cr}M_1^{-1}F_{12}-Q_{cl}), \tag{7.59b}$$

$$A_{21} = L_l^{-1}(Q'_{cl}+Q'_{rl}M_2^{-1}F_{21}), \tag{7.59c}$$

$$A_{22} = -L_l^{-1}(Q'_{rl}M_2^{-1}F_{22}+\dot{L}_l), \tag{7.59d}$$

$$B_{11} = C_c^{-1}(\dot{C}_e - Q_{cr}M_1^{-1}P_{11}), \tag{7.60a}$$

$$B_{12} = C_c^{-1}(Q_{cr}M_1^{-1}P_{12}-Q_{cj}), \tag{7.60b}$$

$$B_{13} = C_c^{-1}C_e, \tag{7.60c}$$

$$B_{21} = L_l^{-1}(Q'_{el}+Q'_{rl}M_2^{-1}P_{21}), \tag{7.60d}$$

$$B_{22} = L_l^{-1}(\dot{L}_j - Q'_{rl}M_2^{-1}P_{22}), \tag{7.60e}$$

$$B_{24} = L_l^{-1}L_j. \tag{7.60f}$$

Equation (7.58) is the desired state equation for a general network. We emphasize that it is valid only under the assumption that the matrices W_1, W_2, M_1, M_2, C_c and L_l are nonsingular.

Alternatively, we can choose the linear combinations of the capacitor charges and the linear combinations of the inductor flux linkages, defined by the equations

$$q = C_c V_{ct}, \tag{7.61a}$$

$$\lambda = L_l I_{lt}, \tag{7.61b}$$

as the state variables of the network. In this case, the state equation becomes

$$\begin{bmatrix} \dot{q} \\ \dot{\lambda} \end{bmatrix} = \begin{bmatrix} -Q_{cr}M_1^{-1}F_{11}C_c^{-1} & (Q_{cr}M_1^{-1}F_{12}-Q_{cl})L_l^{-1} \\ (Q'_{cl}+Q'_{rl}M_2^{-1}F_{21})C_c^{-1} & -Q'_{rl}M_2^{-1}F_{22}L_l^{-1} \end{bmatrix} \begin{bmatrix} q \\ \lambda \end{bmatrix}$$

$$+ \begin{bmatrix} \dot{C}_e-Q_{cr}M_1^{-1}P_{11} & Q_{cr}M_1^{-1}P_{12}-Q_{cj} & C_e & 0 \\ Q'_{el}+Q'_{rl}M_2^{-1}P_{21} & \dot{L}_j-Q'_{rl}M_2^{-1}P_{22} & 0 & L_j \end{bmatrix} \begin{bmatrix} V_e \\ I_j \\ \dot{V}_e \\ \dot{I}_j \end{bmatrix}. \tag{7.62}$$

We shall illustrate the above results by the following example.

Example 7.5: We wish to use the above formulas to obtain the state equation of the network of fig. 7.4 represented by the directed graph G as shown in fig. 7.5. As in Example 7.4, we select the normal tree $t = e_1 e_4 e_5 e_8 e_{10}$. The submatrix $Q_{\bar{t}}$ of the f-cutset matrix Q_f with respect to this tree is partitioned according to (7.41), and is given by

$$Q_{\bar{t}} = \begin{array}{c} \\ e \left\{ e_8 \right. \\ c \left\{ \begin{array}{l} e_4 \\ e_5 \end{array} \right. \\ r \; e_1 \\ l \; e_{10} \end{array} \begin{array}{c} \overbrace{}^{c} \quad \overbrace{}^{r} \quad \overbrace{}^{l} \quad \overbrace{\phantom{e_{11}}}^{j} \\ e_9 \quad\; e_3 \;\; e_6 \;\; e_7 \quad\; e_2 \quad\; e_{11} \\ \left[\begin{array}{c|ccc|c|c} 0 & 0 & 0 & 0 & 1 & 0 \\ \hline 1 & -1 & 0 & 1 & -1 & 0 \\ 1 & -1 & 1 & 0 & -1 & -1 \\ \hline 0 & 0 & 0 & 0 & -1 & 0 \\ \hline 0 & 0 & 0 & 0 & 1 & 0 \end{array} \right] \end{array} = \begin{bmatrix} Q_{ec} & Q_{er} & Q_{el} & Q_{ej} \\ Q_{cc} & Q_{cr} & Q_{cl} & Q_{cj} \\ Q_{rc} & Q_{rr} & Q_{rl} & Q_{rj} \\ Q_{lc} & Q_{lr} & Q_{ll} & Q_{lj} \end{bmatrix}.$$

(7.63)

The parameter matrices are obtained as

$$\begin{bmatrix} C_{tt} & 0 \\ 0 & C_{\bar{t}\bar{t}} \end{bmatrix} = \begin{bmatrix} C_4 & 0 & 0 \\ 0 & C_5 & 0 \\ \hline 0 & 0 & C_9 \end{bmatrix}, \tag{7.64a}$$

$$\begin{bmatrix} L_{\bar{t}\bar{t}} & L_{\bar{t}t} \\ L_{t\bar{t}} & L_{tt} \end{bmatrix} = \begin{bmatrix} L_2 & 0 \\ 0 & L_{10} \end{bmatrix}, \tag{7.64b}$$

§3 The explicit form of the state equation

$$\begin{bmatrix} H_{\bar{t}\bar{t}} & H_{\bar{t}t} \\ H_{t\bar{t}} & H_{tt} \end{bmatrix} = \left[\begin{array}{ccc|c} 1/R_3 & 0 & 0 & 0 \\ 0 & 1/R_6 & 0 & 0 \\ \alpha/R_3 & 0 & 0 & 0 \\ \hline 0 & 0 & 0 & R_1 \end{array}\right]. \quad (7.64c)$$

The other parameter matrices are computed from (7.48) and (7.50):

$$C_c = C_{tt} + Q_{cc} C_{\bar{t}\bar{t}} Q'_{cc} = \begin{bmatrix} C_4 & 0 \\ 0 & C_5 \end{bmatrix} + \begin{bmatrix} 1 \\ 1 \end{bmatrix} [C_9] [1 \ 1]$$

$$= \begin{bmatrix} C_4 + C_9 & C_9 \\ C_9 & C_5 + C_9 \end{bmatrix}, \quad (7.65a)$$

$$C_e = -Q_{cc} C_{\bar{t}\bar{t}} Q'_{ec} = -\begin{bmatrix} 1 \\ 1 \end{bmatrix} [C_9] [0] = \begin{bmatrix} 0 \\ 0 \end{bmatrix}, \quad (7.65b)$$

$$L_l = L_{\bar{t}\bar{t}} - L_{\bar{t}t} Q_{ll} - Q'_{ll} L_{t\bar{t}} + Q'_{ll} L_{tt} Q_{ll}$$
$$= [L_2] - [0][1] - [1][0] + [1][L_{10}][1]$$
$$= [L_2 + L_{10}], \quad (7.65c)$$

$$L_j = (L_{\bar{t}t} - Q'_{ll} L_{tt}) Q_{lj} = ([0] - [1][L_{10}])[0] = [0]. \quad (7.65d)$$

From (7.53), (7.55), (7.56) and (7.57), we get

$$W_1 = U_{\bar{r}\bar{t}} + H_{\bar{t}\bar{t}} Q_{rr} = U_3, \quad (7.66a)$$

$$W_2 = U_{rt} - H_{t\bar{t}} Q'_{rr} = [1], \quad (7.66b)$$

$$M_1 = W_1 + H_{\bar{t}t} Q'_{rr} W_2^{-1} H_{tt} Q_{rr} = U_3, \quad (7.67a)$$

$$M_2 = W_2 + H_{tt} Q_{rr} W_1^{-1} H_{\bar{t}t} Q'_{rr} = [1], \quad (7.67b)$$

$$F_{11} = H_{\bar{t}\bar{t}} Q'_{cr} + H_{\bar{t}t} Q'_{rr} W_2^{-1} H_{t\bar{t}} Q'_{cr} = \begin{bmatrix} -1/R_3 & -1/R_3 \\ 0 & 1/R_6 \\ -\alpha/R_3 & -\alpha/R_3 \end{bmatrix}, \quad (7.68a)$$

$$F_{12} = H_{\bar{t}\bar{t}} Q_{rl} + H_{\bar{t}t} Q'_{rr} W_2^{-1} H_{tt} Q_{rl} = \begin{bmatrix} 0 \\ 0 \\ 0 \end{bmatrix}, \quad (7.68b)$$

$$F_{21} = H_{\bar{t}\bar{t}}Q'_{cr} - H_{tt}Q_{rr}W_1^{-1}H_{\bar{t}\bar{t}}Q'_{cr} = [0 \quad 0], \tag{7.68c}$$

$$F_{22} = H_{tt}Q_{rl} - H_{tt}Q_{rr}W_1^{-1}H_{\bar{t}\bar{t}}Q_{rl} = -[R_1], \tag{7.68d}$$

$$P_{11} = H_{\bar{t}\bar{t}}Q'_{er} + H_{\bar{t}\bar{t}}Q'_{rr}W_2^{-1}H_{\bar{t}\bar{t}}Q'_{er} = \begin{bmatrix} 0 \\ 0 \\ 0 \end{bmatrix}, \tag{7.69a}$$

$$P_{12} = H_{\bar{t}\bar{t}}Q_{rj} + H_{\bar{t}\bar{t}}Q'_{rr}W_2^{-1}H_{tt}Q_{rj} = \begin{bmatrix} 0 \\ 0 \\ 0 \end{bmatrix}, \tag{7.69b}$$

$$P_{21} = H_{\bar{t}\bar{t}}Q'_{er} - H_{tt}Q_{rr}W_1^{-1}H_{\bar{t}\bar{t}}Q'_{er} = [0], \tag{7.69c}$$

$$P_{22} = H_{tt}Q_{rj} - H_{tt}Q_{rr}W_1^{-1}H_{\bar{t}\bar{t}}Q_{rj} = [0]. \tag{7.69d}$$

Finally, to obtain the state equation, we compute from (7.59) and (7.60)

$$A_{11} = -C_c^{-1}(Q_{cr}M_1^{-1}F_{11} + \dot{C}_c)$$
$$= \frac{1}{\Delta}\begin{bmatrix} C_5 + C_9 & -C_9 \\ -C_9 & C_4 + C_9 \end{bmatrix}\begin{bmatrix} (\alpha-1)/R_3 & (\alpha-1)/R_3 \\ -1/R_3 & -1/R_3 - 1/R_6 \end{bmatrix}, \tag{7.70a}$$

$$A_{12} = C_c^{-1}(Q_{cr}M_1^{-1}F_{12} - Q_{cl}) = \frac{1}{\Delta}\begin{bmatrix} C_5 \\ C_4 \end{bmatrix}, \tag{7.70b}$$

$$A_{21} = L_l^{-1}(Q'_{cl} + Q'_{rl}M_2^{-1}F_{21}) = \frac{1}{L_2 + L_{10}}[-1 \quad -1], \tag{7.70c}$$

$$A_{22} = -L_l^{-1}(Q'_{rl}M_2^{-1}F_{22} + \dot{L}_l) = \frac{-1}{L_2 + L_{10}}[R_1], \tag{7.70d}$$

$$B_{11} = C_c^{-1}(\dot{C}_e - Q_{cr}M_1^{-1}P_{11}) = \begin{bmatrix} 0 \\ 0 \end{bmatrix}, \tag{7.71a}$$

$$B_{12} = C_c^{-1}(Q_{cr}M_1^{-1}P_{12} - Q_{cj}) = \frac{1}{\Delta}\begin{bmatrix} -C_9 \\ C_4 + C_9 \end{bmatrix}, \tag{7.71b}$$

$$B_{13} = C_c^{-1}C_e = \begin{bmatrix} 0 \\ 0 \end{bmatrix}, \tag{7.71c}$$

$$B_{21} = L_l^{-1}(Q'_{el} + Q'_{rl}M_2^{-1}P_{21}) = \frac{1}{L_2 + L_{10}}[1], \tag{7.71d}$$

§3   The explicit form of the state equation

$$B_{22} = L_l^{-1}(\dot{L}_j - Q'_{rl}M_2^{-1}P_{22}) = [0], \tag{7.71e}$$

$$B_{24} = L_l^{-1}L_j = [0]. \tag{7.71f}$$

Finally, substituting these in (7.58) we obtain the desired state equation

$$\begin{bmatrix} \dot{v}_4 \\ \dot{v}_5 \\ \dot{i}_2 \end{bmatrix} = \frac{1}{\Delta} \begin{bmatrix} C_5+C_9 & -C_9 & 0 \\ -C_9 & C_4+C_9 & 0 \\ 0 & 0 & \dfrac{\Delta}{L_2+L_{10}} \end{bmatrix} \begin{bmatrix} \dfrac{\alpha-1}{R_3} & \dfrac{\alpha-1}{R_3} & 1 \\ -\dfrac{1}{R_3} & -\dfrac{1}{R_3}-\dfrac{1}{R_6} & 1 \\ -1 & -1 & -R_1 \end{bmatrix} \begin{bmatrix} v_4 \\ v_5 \\ i_2 \end{bmatrix}$$

$$+ \begin{bmatrix} 0 & -C_9/\Delta & 0 & 0 \\ 0 & (C_4+C_9)/\Delta & 0 & 0 \\ 1/(L_2+L_{10}) & 0 & 0 & 0 \end{bmatrix} \begin{bmatrix} v_g \\ i_g \\ \dot{v}_g \\ \dot{i}_g \end{bmatrix}. \tag{7.72}$$

This confirms (7.37).

Alternatively, as indicated in (7.61), we can choose charges and flux linkages as the state variables. In this case, the state variables are given by

$$q = \begin{bmatrix} q_\alpha \\ q_\beta \end{bmatrix} = C_c V_{ct} = \begin{bmatrix} C_4 v_4 + C_9(v_4+v_5) \\ C_5 v_5 + C_9(v_4+v_5) \end{bmatrix}, \tag{7.73a}$$

$$\lambda = [\lambda] = L_l I_{l\bar{l}} = (L_2+L_{10})[i_2]. \tag{7.73b}$$

Substituting the appropriate quantities in (7.62) yields

$$\begin{bmatrix} \dot{q}_\alpha \\ \dot{q}_\beta \\ \dot{\lambda} \end{bmatrix} = \frac{1}{\Delta} \begin{bmatrix} \dfrac{(\alpha-1)}{R_3} & \dfrac{(\alpha-1)}{R_3} & 1 \\ -\dfrac{1}{R_3} & -\dfrac{1}{R_3}-\dfrac{1}{R_6} & 1 \\ -1 & -1 & -R_1 \end{bmatrix} \begin{bmatrix} C_5+C_9 & -C_9 & 0 \\ -C_9 & C_4+C_9 & 0 \\ 0 & 0 & \dfrac{\Delta}{L_2+L_{10}} \end{bmatrix} \begin{bmatrix} q_\alpha \\ q_\beta \\ \lambda \end{bmatrix}$$

$$+ \begin{bmatrix} 0 & 0 & 0 & 0 \\ 0 & 1 & 0 & 0 \\ 1 & 0 & 0 & 0 \end{bmatrix} \begin{bmatrix} v_g \\ i_g \\ \dot{v}_g \\ \dot{i}_g \end{bmatrix}. \tag{7.74}$$

After simplification, (7.74) becomes

$$\begin{bmatrix} \dot{q}_\alpha \\ \dot{q}_\beta \\ \dot{\lambda} \end{bmatrix} = \frac{1}{\Delta} \begin{bmatrix} (\alpha-1)C_5/R_3 & (\alpha-1)C_4/R_3 & \Delta/(L_2+L_{10}) \\ C_9/R_6-C_5/R_3 & -(C_4+C_9)/R_6-C_4/R_3 & \Delta/(L_2+L_{10}) \\ -C_5 & -C_4 & -R_1\Delta/(L_2+L_{10}) \end{bmatrix} \begin{bmatrix} q_\alpha \\ q_\beta \\ \lambda \end{bmatrix}$$

$$+ \begin{bmatrix} 0 & 0 \\ 0 & 1 \\ 1 & 0 \end{bmatrix} \begin{bmatrix} v_g \\ i_g \end{bmatrix}. \tag{7.75}$$

We remark that the variables q_α, q_β and λ denote the linear combinations of capacitor charges or inductor flux linkages. To see this, we rewrite (7.73) as

$$q_\alpha = C_4 v_4 + C_9(v_4+v_5) = C_4 v_4 + C_9 v_9 = q_4 + q_9, \tag{7.76a}$$

$$q_\beta = C_5 v_5 + C_9(v_4+v_5) = C_5 v_5 + C_9 v_9 = q_5 + q_9, \tag{7.76b}$$

$$\lambda = L_2 i_2 + L_{10} i_2 = L_2 i_2 - L_{10} i_{10} = \lambda_2 - \lambda_{10}. \tag{7.76c}$$

§4. An alternative representation of the state equation

In the preceding section, we demonstrated that the state equation of a linear network has the general form

$$\dot{x} = Ax + B_1 u + B_2 \dot{u}, \tag{7.77}$$

$\dot{u}$ being the time derivative of the excitation u. It is possible that we can remove the derivative of the excitation by the transformation

$$x \to x + B_2 u, \tag{7.78}$$

which, when substituted in (7.77), yields

$$\dot{x} = Ax + Bu, \tag{7.79a}$$

where

$$B = B_1 + AB_2 - B_2. \tag{7.79b}$$

In the case of (7.62) where charges and flux linkages are chosen as the state variables, the vectors x and u in (7.77) can be identified as

$$x = \begin{bmatrix} q \\ \lambda \end{bmatrix} \quad \text{and} \quad u = \begin{bmatrix} V_e \\ I_j \end{bmatrix}. \tag{7.80}$$

After transformation (7.78), the new state vector x becomes

$$x = \begin{bmatrix} q - C_e V_e \\ \lambda - L_j I_j \end{bmatrix}. \tag{7.81}$$

Using (7.61) in conjunction with (7.48), (7.50), (7.43d) and (7.43e), we can write

$$\begin{aligned} x &= \begin{bmatrix} C_c V_{ct} - C_e V_e \\ L_l I_{l\bar{t}} - L_j I_j \end{bmatrix} \\ &= \begin{bmatrix} C_{tt} V_{ct} + Q_{cc} C_{\bar{t}\bar{t}}(Q'_{cc} V_{ct} + Q'_{ec} V_e) \\ L_{\bar{t}\bar{t}} I_{l\bar{t}} - Q'_{ll} L_{tt} I_{l\bar{t}} - L_{\bar{t}\bar{t}}(Q_{ll} I_{l\bar{t}} + Q_{lj} I_j) + Q'_{ll} L_{tt}(Q_{ll} I_{l\bar{t}} + Q_{lj} I_j) \end{bmatrix} \\ &= \begin{bmatrix} C_{tt} V_{ct} + Q_{cc} C_{\bar{t}\bar{t}} V_{c\bar{t}} \\ L_{\bar{t}\bar{t}} I_{l\bar{t}} - Q'_{ll} L_{\bar{t}\bar{t}} I_{l\bar{t}} + L_{\bar{t}\bar{t}} I_{lt} - Q'_{ll} L_{tt} I_{lt} \end{bmatrix} \\ &= \begin{bmatrix} q_{ct} + Q_{cc} q_{c\bar{t}} \\ \lambda_{\bar{t}\bar{t}} - Q'_{ll} \lambda_{\bar{t}\bar{t}} + \lambda_{\bar{t}\bar{t}} - Q'_{ll} \lambda_{tt} \end{bmatrix}, \end{aligned} \tag{7.82}$$

The last step follows from (7.44b) and (7.45b). Thus, the elements of x, being the state variables, denote the linear combinations of either the capacitor charges or the inductor flux linkages.

It would be pointless, in manual computations, to use these general formulas for obtaining state equations for simple, specific networks. One may therefore wonder whether these general formulas are of any value. The derivation here has two purposes. Firstly, it is valuable to have seen, at least once, the general formulation of state equation on a sufficiently firm basis. Secondly, the general formulas are suitable for the formulation of state equations when digital computers are available.

§5. Physical interpretations of the parameter matrices

In the preceding two sections, we have presented the explicit forms of the state equation in normal form for a general linear network. In this section, we discuss the physical meanings of the parameter matrices of the state equation. Since we shall consider the general time-varying case, it is convenient to choose the charges and flux linkages as the state variables.

Our starting point is (7.61). Combining (7.61a), (7.44) and (7.82) yields

$$\frac{d}{dt}(C_c V_{ct} - C_e V_e) = \dot{q} - C_e \dot{V}_e - \dot{C}_e V_e$$

$$= \frac{d}{dt}(C_{tt} V_{ct} + Q_{cc} C_{\bar{t}\bar{t}} V_{c\bar{t}})$$

$$= I_{ct} + Q_{cc} I_{c\bar{t}}. \qquad (7.83)$$

Similarly, from (7.61b), (7.45) and (7.82), we obtain

$$\frac{d}{dt}(L_l I_{l\bar{t}} - L_j I_j) = \dot{\lambda} - L_j \dot{I}_j - \dot{L}_j I_j$$

$$= \frac{d}{dt}(L_{\bar{t}\bar{t}} I_{l\bar{t}} + L_{\bar{t}t} I_{lt} - Q'_{ll} L_{tt} I_{lt} - Q'_{ll} L_{t\bar{t}} I_{l\bar{t}})$$

$$= V_{l\bar{t}} - Q'_{ll} V_{lt}. \qquad (7.84)$$

Substituting $\dot{q} - C_e \dot{V}_e$ from (7.83) and $\dot{\lambda} - L_j \dot{I}_j$ from (7.84) in (7.62) and rearranging the terms give a purely algebraic equation whose coefficient matrices can easily be interpreted physically:

$$\begin{bmatrix} I_{ct} \\ V_{l\bar{t}} \end{bmatrix} = \begin{bmatrix} -Q_{cr} M_1^{-1} F_{11} & Q_{cr} M_1^{-1} F_{12} - Q_{cl} \\ Q'_{cl} + Q'_{rl} M_2^{-1} F_{21} & -Q'_{rl} M_2^{-1} F_{22} \end{bmatrix} \begin{bmatrix} V_{ct} \\ I_{l\bar{t}} \end{bmatrix} + \begin{bmatrix} -Q_{cc} & 0 \\ 0 & Q'_{ll} \end{bmatrix} \begin{bmatrix} I_{c\bar{t}} \\ V_{lt} \end{bmatrix}$$

$$+ \begin{bmatrix} -Q_{cr} M_1^{-1} P_{11} & Q_{cr} M_1^{-1} P_{12} - Q_{cj} \\ Q'_{el} + Q'_{rl} M_2^{-1} P_{21} & -Q'_{rl} M_2^{-1} P_{22} \end{bmatrix} \begin{bmatrix} V_e \\ I_j \end{bmatrix}. \qquad (7.85)$$

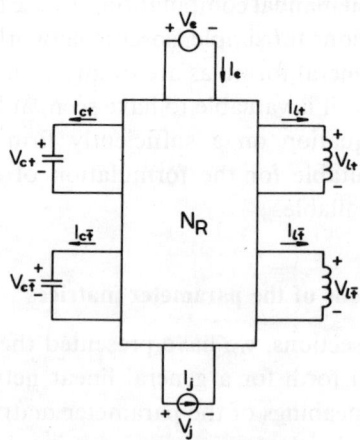

Fig. 7.6. Vector port representation of a general network, where each single vector port denotes a subnetwork of its class.

§5   Physical interpretations of the parameter matrices   493

Referring to the original network N which is redrawn symbolically as shown in fig. 7.6, where each single "vector port" represents a subnetwork of its class. For example, V_{ct} denotes the subnetwork composed of all the tree-branch capacitors of N, and V_e all the independent voltage sources. The references for the voltages and currents, as indicated in fig. 7.6, are consistent with those for which the equations were developed. By setting appropriate quantities equal to zero, the physical interpretations of the various submatrices of the preceding equation can be easily determined as follows:

$$-I_{ct} = Q_{cr} M_1^{-1} F_{11} V_{ct}|_{I_{\bar{l}t},\, I_{\bar{c}t},\, V_e \text{ and } I_j = 0}, \qquad (7.86a)$$

$$I_{ct} = (Q_{cr} M_1^{-1} F_{12} - Q_{cl}) I_{\bar{l}t}|_{V_{ct},\, I_{\bar{c}t},\, V_e \text{ and } I_j = 0}, \qquad (7.86b)$$

$$V_{\bar{l}t} = (Q'_{cl} + Q'_{rl} M_2^{-1} F_{21}) V_{ct}|_{I_{\bar{l}t},\, V_{lt},\, V_e \text{ and } I_j = 0}, \qquad (7.86c)$$

$$V_{\bar{l}t} = Q'_{rl} M_2^{-1} F_{22}(-I_{\bar{l}t})|_{V_{ct},\, V_{lt},\, V_e \text{ and } I_j = 0}. \qquad (7.86d)$$

Looking back at (7.43d) and (7.43e), we recognize that if we set $I_{\bar{l}t} = 0$ and $I_j = 0$, the tree-branch inductor-current vector I_{lt} must also be zero, and that if we set $V_{ct} = 0$ and $V_e = 0$, so also will the cotree-chord capacitor-voltage vector $V_{\bar{c}t}$ be zero. Recall that by removing an independent source, we mean short-circuiting a voltage source or open-circuiting a current source. Thus, we conclude that

$$Y_{cc} = Q_{cr} M_1^{-1} F_{11} \qquad (7.87)$$

is the short-circuit admittance matrix of the resistive multiport whose ports are formed by the terminals of the tree-branch capacitors, when all other capacitors and inductors are open-circuited and all independent sources are removed. The resistive multiport is depicted symbolically as shown in fig. 7.7. In a similar way, the submatrix

$$Z_{ll} = Q'_{rl} M_2^{-1} F_{22} \qquad (7.88)$$

is the open-circuit impedance matrix of the resistive multiport whose ports

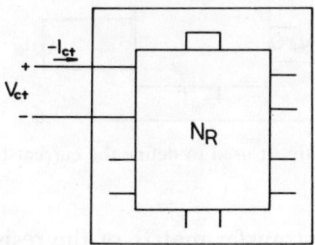

Fig. 7.7. The resistive multiport used to define the admittance matrix Y_{cc}.

are formed by the terminals of the cotree-chord inductors, when all other capacitors and inductors are short-circuited and all independent sources are removed, as depicted symbolically in fig. 7.8.

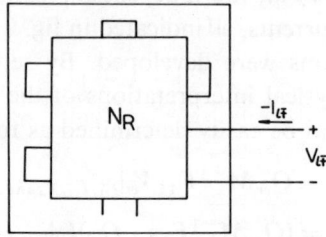

Fig. 7.8. The resistive multiport used to define the impedance matrix Z_{ll}.

To find the other two submatrices, we set

$$H_{lc} = Q_{cr}M_1^{-1}F_{12} - Q_{cl}, \qquad (7.89)$$

$$H_{cl} = Q'_{cl} + Q'_{rl}M_2^{-1}F_{21}. \qquad (7.90)$$

Thus, H_{lc} is the current-transfer matrix of the resistive multiport whose input ports are formed by the cotree-chord inductors and whose outputs are the currents in the short-circuited terminals of the tree-branch capacitors, when all the cotree-chord capacitors are open-circuited and all independent sources are removed. The tree-branch inductors are replaced by current sources whose values are determined by (7.43d) with $I_j = 0$. The resistive multiport is presented in fig. 7.9. Notice that, as dictated by Kirchhoff's voltage law, the voltages across the cotree-chord capacitors are always zero.

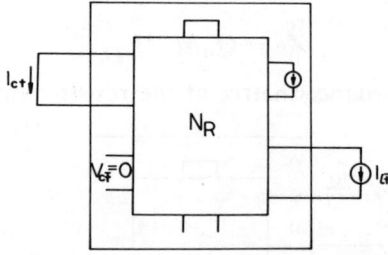

Fig. 7.9. The resistive multiport used to define the current-transfer matrix H_{lc}.

Finally, H_{cl} is the voltage-transfer matrix of the resistive multiport whose input ports are formed by the tree-branch capacitors and whose outputs are the open-circuit voltages across the terminals of the cotree-chord inductors, when all the tree-branch inductors are short-circuited and all independent sources are removed. From (7.43e) with $V_e = 0$, the voltages of the cotree chord capacitors are now known and, thus, they can be removed. The resistive multiport is depicted in fig. 7.10, in which the tree-branch inductor currents are always zero.

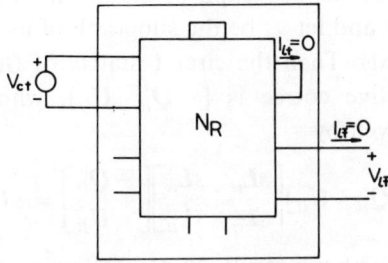

Fig. 7.10. The resistive multiport used to define the voltage-transfer matrix H_{cl}.

In a similar manner, the coefficient matrices of V_e and I_j in (7.85), as defined below, can be interpreted as follows:

$$-I_{ct} = Y_{ec}V_e = Q_{cr}M_1^{-1}P_{11}V_e|_{V_{ct}, I_{l\bar{t}}, I_{c\bar{t}} \text{ and } I_j = 0}, \qquad (7.91a)$$

$$I_{ct} = H_{jc}I_j = (Q_{cr}M_1^{-1}P_{12} - Q_{cj})I_j|_{V_{ct}, I_{l\bar{t}}, I_{c\bar{t}} \text{ and } V_e = 0}, \qquad (7.91b)$$

$$V_{l\bar{t}} = H_{el}V_e = (Q'_{el} + Q'_{rl}M_2^{-1}P_{21})V_e|_{V_{ct}, I_{l\bar{t}}, V_{lt} \text{ and } I_j = 0}, \qquad (7.91c)$$

$$V_{l\bar{t}} = -Z_{jl}I_j = -Q'_{rl}M_2^{-1}P_{22}I_j|_{V_{ct}, I_{l\bar{t}}, V_{lt} \text{ and } V_e = 0}. \qquad (7.91d)$$

As is evident from the previous discussions, the physical meanings of these submatrices are clear; they represent the transfer-admittance, current-transfer, voltage-transfer, and transfer-impedance matrices of the various associated multiports. The details of these interpretations are left as exercises (Problems 7.3 and 7.4).

Before we illustrate the above interpretations, we mention that the parameter matrices C_c and L_l also possess simple physical interpretations. As in §5.3 of chapter 4, let $G(g_1; g_2)$ denote the directed graph obtained from G first by shorting all the edges contained in subgraph g_1 (do not remove the self-loops thus produced) and then by removing all the edges contained in the subgraph g_2, g_1 and g_2 being two edge-disjoint subgraphs of G.

For C_c let g_1 be the subgraph of the normal tree containing all the non-capacitive branches and let g_2 be the subgraph of its cotree containing all the non-capacitive chords. Then the cutset matrix of $G(g_1; g_2)$ with respect to the reduced normal tree composed only of the capacitive tree branches is $[U_{cc} \ Q_{cc}]$, whose associated cutset-admittance matrix is given by

$$[U_{cc} \ Q_{cc}] \begin{bmatrix} sC_{tt} & 0 \\ 0 & sC_{\bar{t}\bar{t}} \end{bmatrix} \begin{bmatrix} U_{cc} \\ Q'_{cc} \end{bmatrix} = sC_c. \qquad (7.92)$$

Similarly, for L_l let g_3 be the subgraph of the normal tree containing all the non-inductive branches and let g_4 be the subgraph of its cotree containing all the non-inductive chords. Then the circuit matrix of $G(g_3, g_4)$ with respect to this reduced inductive cotree is $[-Q'_{ll} \ U_{ll}]$, whose associated loop-impedance matrix is given by

$$[-Q'_{ll} \ U_{ll}] \begin{bmatrix} sL_{tt} & sL_{t\bar{t}} \\ sL_{\bar{t}t} & sL_{\bar{t}\bar{t}} \end{bmatrix} \begin{bmatrix} -Q_{ll} \\ U_{ll} \end{bmatrix} = sL_l. \qquad (7.93)$$

We shall illustrate the above results by the following example.

Example 7.6: Suppose that we wish to use the above physical interpretations to obtain the state equation of the network of fig. 7.4 represented by the directed graph G as shown in fig. 7.5. As in Example 7.4, we select the normal tree $t = e_1 e_4 e_5 e_8 e_{10}$.

To compute Y_{cc} (7.87), we open-circuit the two inductors and the capacitor C_9, remove the independent current source, short-circuit the independent voltage source, and replace the two capacitors C_4 and C_5 with two voltage sources v_4 and v_5 having the same polarities as the capacitor voltages. The resulting network is shown in fig. 7.11. Our objective is to compute the currents $-i_4$ and $-i_5$ in terms of the voltage sources v_4 and v_5. This gives

$$\begin{bmatrix} -i_4 \\ -i_5 \end{bmatrix} = \begin{bmatrix} (1-\alpha)/R_3 & (1-\alpha)/R_3 \\ 1/R_3 & 1/R_3 + 1/R_6 \end{bmatrix} \begin{bmatrix} v_4 \\ v_5 \end{bmatrix}, \qquad (7.94)$$

the coefficient matrix being the short-circuit admittance matrix Y_{cc}, as defined in (7.87).

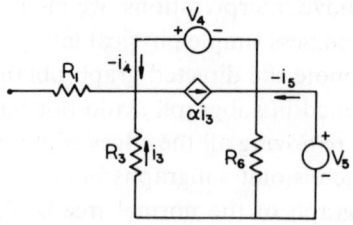

Fig. 7.11. The resistive network used to compute the admittance matrix Y_{cc}.

To find Z_{ll} (7.88), we short-circuit the three capacitors, the inductor L_{10} and the independent voltage source, remove the independent current source, and replace the inductor L_2 with a current source whose current reference is opposite to that of the inductor L_2. The final network is depicted in fig. 7.12.

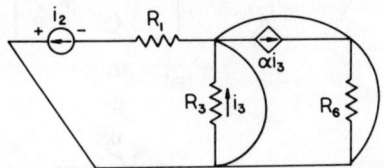

Fig. 7.12. The resistive network used to compute the impedance matrix Z_{ll}.

The input impedance looking into this current source is given by

$$Z_{ll} = [R_1]. \qquad (7.95)$$

Finally, to determine H_{lc} and H_{cl}, we use the networks of figs. 7.13 (a) and (b). The desired equations are obtained as follows:

$$\begin{bmatrix} i_4 \\ i_5 \end{bmatrix} = \begin{bmatrix} 1 \\ 1 \end{bmatrix} [i_2], \qquad (7.96a)$$

$$[v_2] = [-1 \quad -1] \begin{bmatrix} v_4 \\ v_5 \end{bmatrix}, \qquad (7.96b)$$

the coefficient matrices being H_{lc} and H_{cl}, respectively.

In a similar fashion, we can compute the coefficient matrices of (7.91). For the coefficient matrices Y_{ec} and Z_{jl} of (7.91a) and (7.91d), it is easy to verify that they are zero matrices. For (7.91b), we use the network of fig. 7.14, yielding

$$\begin{bmatrix} i_4 \\ i_5 \end{bmatrix} = \begin{bmatrix} 0 \\ 1 \end{bmatrix} [i_g] = H_{jc} I_j. \qquad (7.97)$$

Finally, for (7.91c) the coefficient matrix H_{el} is seen to be [1].

As indicated in (7.92) and (7.93), it is easy to verify that

$$sC_c = s \begin{bmatrix} C_4 + C_9 & C_9 \\ C_9 & C_5 + C_9 \end{bmatrix} \qquad (7.98)$$

is the cutset-admittance matrix of the capacitance network of fig. 7.15(a) with respect to the tree composed of the branches C_4 and C_5, and that

$$sL_l = s[L_2 + L_{10}] \qquad (7.99)$$

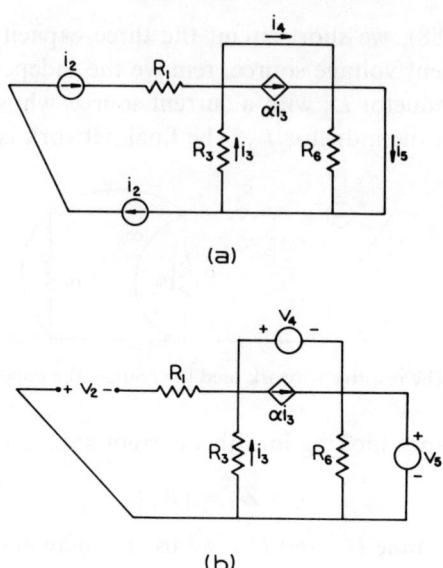

Fig. 7.13. (a) The resistive network used to compute the current-transfer matrix H_{lc}.
(b) The resistive network used to compute the voltage-transfer matrix H_{cl}.

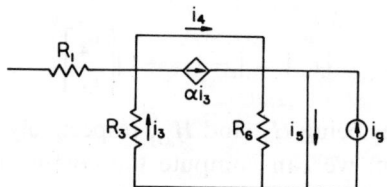

Fig. 7.14. The resistive network used to compute the current-transfer matrix H_{jc}.

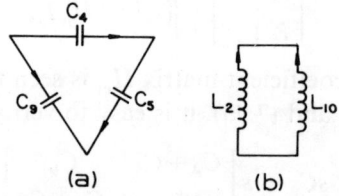

Fig. 7.15. (a) The capacitance network used to compute the matrix C_c.
(b) The inductance network used to compute the matrix L_l.

is the loop-impedance matrix of the inductance network of fig. 7.15(b) with respect to the cotree L_2.

§6 Order of complexity 499

In terms of the coefficient matrices of (7.86) and (7.91), the state equation (7.62) becomes

$$\begin{bmatrix} \dot{q} \\ \dot{\lambda} \end{bmatrix} = \begin{bmatrix} -Y_{cc}C_c^{-1} & H_{lc}L_l^{-1} \\ H_{cl}C_c^{-1} & -Z_{ll}L_l^{-1} \end{bmatrix} \begin{bmatrix} q \\ \lambda \end{bmatrix} + \begin{bmatrix} \dot{C}_e - Y_{ec} & H_{jc} & C_e & 0 \\ H_{el} & \dot{L}_j - Z_{jl} & 0 & L_j \end{bmatrix} \begin{bmatrix} V_e \\ I_j \\ \dot{V}_e \\ \dot{I}_j \end{bmatrix}$$

(7.100)

Substituting the appropriate quantities from (7.94)–(7.99) in (7.100) yields the desired state equation as shown in (7.75).

§6. Order of complexity

In the foregoing, we presented the explicit form of state equation in normal form for a general linear network. For the general procedure to succeed, we assume the existence of certain matrices. To complete our formulation, we must demonstrate that all the network variables can be expressed in terms of the state variables and the known sources. After this is done, we can claim that the output variables, being a part of the network variables, are expressible in the form of (7.3). That this is always the case follows directly from (7.43)–(7.46). For example, from (7.43d) the vector I_{lt} can be expressed in terms of $I_{l\bar{t}}$ and I_j. To compute V_{lt}, we use the second equation of (7.45a) in conjunction with (7.43d), yielding

$$V_{lt} = \frac{d}{dt}\left([L_{t\bar{t}} \;\; L_{tt}] \begin{bmatrix} I_{l\bar{t}} \\ -Q_{ll}I_{l\bar{t}} - Q_{lj}I_j \end{bmatrix}\right)$$

$$= (L_{t\bar{t}} - L_{tt}Q_{ll})\dot{I}_{l\bar{t}} + (\dot{L}_{t\bar{t}} - \dot{L}_{tt}Q_{ll})I_{l\bar{t}} - \dot{L}_{tt}Q_{lj}I_j - L_{tt}Q_{lj}\dot{I}_j$$

$$= (L_{t\bar{t}} - L_{tt}Q_{ll})A_{21}V_{ct} + [(L_{t\bar{t}} - L_{tt}Q_{ll})A_{22} + \dot{L}_{t\bar{t}} - \dot{L}_{tt}Q_{ll}]I_{l\bar{t}}$$

$$+ (L_{t\bar{t}} - L_{tt}Q_{ll})[B_{21}V_e + B_{22}I_j + B_{24}\dot{I}_j]$$

$$- \dot{L}_{tt}Q_{lj}I_j - L_{tt}Q_{lj}\dot{I}_j.$$

(7.101)

The last line follows directly from the state equation (7.58). Likewise, we can compute all other variables, and the details of their derivations are left as exercises (Problems 7.8–7.11). In other words, once the state variables are determined, the remaining network variables can be found purely algebraically. Since the state variables that we choose are capacitor voltages and inductor currents, each of their element-defining equations contains a derivative. Thus, unlike the variables associated with the resistive edges, which are *static*, these variables are *dynamical*. The question arises as to how many *dynamically independent* state variables there are, whose instantaneous values are sufficient to determine completely the instantaneous state of the network. This leads to the concept of the 'order of complexity' of a network.

An early solution to determine this number called the *degrees of freedom* was given by GUILLEMIN [1931], applicable only to time-invariant *RLC* networks that do not contain any all-capacitor or all-inductor circuits. The term originates in the study of mechanical systems, in which we can attach significance to the word 'position' or 'configuration'. Later, REZA [1955] gave the solution for networks containing only two types of elements and adopted the term 'order of complexity', which we shall follow in the following discussion. The complete solution for time-invariant *RLC* networks was obtained independently by BRYANT [1959, 1960], BERS [1959] and SESHU and REED [1961]. The extension to active networks was very recent and has been considered by PURSLOW and SPENCE [1967], TOW [1968], MARK [1971] and CHEN [1972a]. The main difficulty lies in the fact that, unlike the case for *RLC* networks, the number is not decided, in general, purely by the topological structure; it also depends on the values of the parameters of the network. This complicates the problem considerably. However, various upper bounds on this number for a general network are known. In this section, we shall present a unified summary on some of these known results. Unfortunately, no simple necessary and sufficient conditions are available, and they are left as an unsolved problem.

We first define precisely the term 'order of complexity' of a general network, and then show how it can be determined or its upper bounds obtained by inspection. We restrict ourselves to linear time-invariant networks composed of independent voltage and current sources, resistors, capacitors, self and mutual inductors possessing the positive-definite inductance matrix, and controlled current sources. With proper modeling, the controlled voltage sources, gyrators and impedance converters can also be included in the consideration. Each controlled source is assumed to have a real constant controlling parameter and is controlled by the current through an element having a finite impedance or by the node-pair voltage across a pair of nodes which may or may not have an element connected directly between them. Each controlled source is controlled by one element current or node-pair voltage. Each element is treated as a separate edge in the directed graph G representing the network N. As before, in order to seek meaningfully solutions, we further stipulate that the network N possess neither circuits composed only of independent and dependent voltage sources nor cuts composed only of independent and dependent current sources. Since independent sources and initial conditions do not affect our argument to be presented below, for simplicity and without loss of generality, we assume that all the independent sources and initial conditions have been removed from the network N.

In order to apply the topological formulas developed in chapter 4, we now convert all the current-controlled current sources to equivalent voltage-controlled current sources. In doing so, the resulting controlling parameters of the controlled sources may contain an integral and/or differential operator. For this, define the operational notation

$$px(t) = \frac{dx(t)}{dt} \quad \text{and} \quad \frac{1}{p}x(t) = \int_0^t x(\tau)d\tau. \tag{7.102}$$

Then the element-defining equations (7.44a), (7.45a) and (7.46), other than the controlled sources, may be combined into one matrix equation of the form

$$\mathbf{Z}_{LMR}(p)\mathbf{I}_\alpha(t) = \mathbf{Y}_C(p)\mathbf{V}_\alpha(t), \tag{7.103}$$

where $\mathbf{I}_\alpha(t)$ and $\mathbf{V}_\alpha(t)$ are vectors representing currents and voltages of the resistors, capacitors and inductors. The matrix $\mathbf{Y}_C(p)$ is diagonal, has units in those diagonal elements corresponding to inductors or resistors, and a term of the form Cp corresponding to a capacitor. Likewise, $\mathbf{Z}_{LMR}(p)$ is a symmetric matrix having units in those diagonal elements corresponding to the capacitors, a term of the form R in the diagonal corresponding to a resistor, and a term of the form Lp corresponding to an inductor. For the voltage-controlled current sources, they can be characterized by the matrix equation

$$\mathbf{I}_\beta(t) = \mathbf{Y}_\beta(p)\mathbf{V}_\alpha(t), \tag{7.104}$$

where $\mathbf{I}_\beta(t)$ denotes the vector of the currents in the controlled sources, and $\mathbf{Y}_\beta(p)$ is the matrix of the controlling parameters.

In addition to the element-defining equations, we have Kirchhoff's current and voltage laws. Let $\mathbf{Q}_f$ and $\mathbf{B}_f$ be the fundamental cutset and circuit matrices of G with respect to a chosen normal tree. Partitioning these matrices according to (7.103) and (7.104) yields

$$[\mathbf{Q}_\alpha \ \mathbf{Q}_\beta]\begin{bmatrix}\mathbf{I}_\alpha(t)\\ \mathbf{I}_\beta(t)\end{bmatrix} = \mathbf{0}, \tag{7.105a}$$

$$[\mathbf{B}_\alpha \ \mathbf{B}_\beta]\begin{bmatrix}\mathbf{V}_\alpha(t)\\ \mathbf{V}_\beta(t)\end{bmatrix} = \mathbf{0}, \tag{7.105b}$$

$V_\beta(t)$ being the vector of the voltages of the controlled sources. Combining (7.103)–(7.105) gives the following primary system of network equations:

$$\begin{bmatrix} Q_\alpha & Q_\beta & 0 & 0 & 0 \\ 0 & 0 & B_\alpha & B_\beta & B_\beta \\ -Z_{LMR}(p) & 0 & Y_C(p) & 0 & 0 \\ 0 & -U_\beta & Y_\beta(p) & 0 & 0 \end{bmatrix} \begin{bmatrix} I_\alpha(t) \\ I_\beta(t) \\ V_\alpha(t) \\ V_\beta(t) \end{bmatrix} = \mathbf{0} \qquad (7.106a)$$

or more compactly

$$H(p) \begin{bmatrix} I(t) \\ V(t) \end{bmatrix} = \mathbf{0}, \qquad (7.106b)$$

$I(t)$ and $V(t)$ being the branch-current and branch-voltage vectors of G. It should be emphasized that $H(p)$ contains the operators and must be handled with care.

With these preliminaries, we now proceed to define precisely the term 'natural frequency' of a general network, which in turn is used to define the order of complexity.

DEFINITION 7.3: *Natural frequency*. The *natural frequencies* of a network are defined as the roots of the determinantal polynomial of the operator matrix of the network equations when these are framed as a set of first-order differential equations and/or algebraic equations for the currents and voltages of the network elements.

DEFINITION 7.4: *Order of complexity*. The number of natural frequencies of a network is called the *order of complexity* of the network, counting each frequency according to its multiplicity.

Thus, the roots of the polynomial det $H(s)$ obtained from the operator matrix $H(p)$ of (7.106) are the natural frequencies of the network N, whose number is the order of complexity of N. As indicated earlier, in state-variable formulation of network equations, the order of complexity is the minimal number of dynamically-independent element currents and voltages whose instantaneous values are sufficient to determine completely the instantaneous state of the network, and hence is the dimension of the *state space*, a space spanned by the state vector $x(t)$, its coordinates being the components of $x(t)$. In mathematics, it is the number of arbitrary constants appearing in the general solution of the network equations. Physically, it is the maximum number of linearly independent energy-storing elements in N or, equivalently, the maximum number of linearly independent initial conditions that can be specified for N. These are the various ways in stating the same thing.

We shall illustrate the above results by the following example.

Example 7.7: Consider the active network N of fig. 7.16(a) represented by the directed graph G as shown in fig. 7.16(b). Choose $e_2 e_3$ as the normal tree.

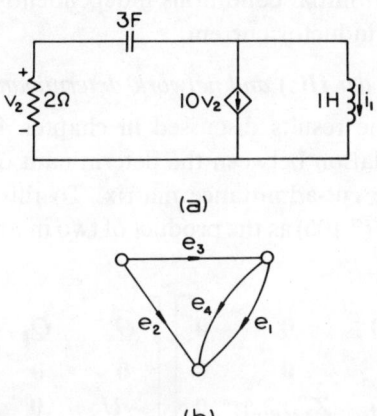

Fig. 7.16. (a) An active network illustrating the formulation of the operator matrix $H(p)$.
(b) The directed graph of the active network shown in (a).

Then the primary system of network equations (7.106) becomes

$$\begin{bmatrix} -1 & 0 & 1 & -1 & 0 & 0 & 0 & 0 \\ 1 & 1 & 0 & 1 & 0 & 0 & 0 & 0 \\ \hline 0 & 0 & 0 & 0 & 1 & -1 & 1 & 0 \\ 0 & 0 & 0 & 0 & 0 & -1 & 1 & 1 \\ \hline -p & 0 & 0 & 0 & 1 & 0 & 0 & 0 \\ 0 & -2 & 0 & 0 & 0 & 1 & 0 & 0 \\ 0 & 0 & -1 & 0 & 0 & 0 & 3p & 0 \\ \hline 0 & 0 & 0 & -1 & 0 & 10 & 0 & 0 \end{bmatrix} \begin{bmatrix} i_1(t) \\ i_2(t) \\ i_3(t) \\ \hline i_4(t) \\ \hline v_1(t) \\ v_2(t) \\ v_3(t) \\ \hline v_4(t) \end{bmatrix} = \mathbf{0}, \quad (7.107)$$

the coefficient matrix being the operator matrix $H(p)$. Thus, the natural frequencies are the roots of the polynomial det $H(s)$, which is given by

$$\det H(s) = -63s^2 - 6s - 1, \tag{7.108}$$

yielding

$$s_1, s_2 = -0.0476 \pm j\, 0.1166. \tag{7.109}$$

The order of complexity of the network of fig. 7.16(a) is therefore 2, meaning that we can specify two initial conditions independently, the initial capacitor voltage and the initial inductor current.

6.1. *Relations between det $H(s)$ and network determinant*

In order to apply the results discussed in chapter 4, we now proceed to examine the mutual relation between the determinant of the matrix $H(s)$ and the determinant of the cut-admittance matrix. To this end, we first rewrite the coefficient matrix of (7.106) as the product of two matrices and then evaluate its determinant:

$$\det H(s) = \det \begin{bmatrix} U_r & 0 & 0 & 0 \\ 0 & U_m & 0 & 0 \\ 0 & 0 & Z_{LMR}(s) & 0 \\ 0 & 0 & 0 & U_\beta \end{bmatrix} \begin{bmatrix} Q_\alpha & Q_\beta & 0 & 0 \\ 0 & 0 & B_\alpha & B_\beta \\ -U_\alpha & 0 & Z_{LMR}^{-1}(s)Y_C(s) & 0 \\ 0 & -U_\beta & Y_\beta(s) & 0 \end{bmatrix}$$

$$= [\det Z_{LMR}(s)] \det \begin{bmatrix} Q_f & 0 \\ 0 & B_f \\ -U_b & Y(s) \end{bmatrix}, \tag{7.110}$$

where

$$Y(s) = \begin{bmatrix} Z_{LMR}^{-1}(s)Y_C(s) & 0 \\ Y_\beta(s) & 0 \end{bmatrix} \tag{7.111}$$

is the branch-admittance matrix of the network N. We remark that the matrix $Z_{LMR}(s)$ is nonsingular since the inductance matrix is assumed to be positive definite.

THEOREM 7.1: Let Q be a basis cut matrix of the directed graph of a given network, whose branch-admittance matrix is $Y(s)$. Then

$$\det H(s) = \text{constant} \cdot [\det Z_{LMR}(s)][\det QY(s)Q'], \tag{7.112}$$

§6 Order of complexity

$Z_{LMR}(s)$ being the branch-impedance matrix of that part of the network corresponding to the resistors and self and mutual inductors.

Proof. Appealing to Theorem 2.20, we may assume, without loss of generality, that $Q = Q_f$, the f-cutset matrix with respect to a normal tree. Partitioning the matrices Q_f, B_f and $Y(s)$ of (7.110) according to the branches and chords of this normal tree yields

$$\det \begin{bmatrix} Q_f & 0 \\ 0 & B_f \\ -U_b & Y_s \end{bmatrix} = \det \begin{bmatrix} U_r & Q_{22} & 0 & 0 \\ 0 & 0 & -Q'_{22} & U_m \\ -U_r & 0 & Y_{tt} & Y_{t\bar{t}} \\ 0 & -U_m & Y_{\bar{t}t} & Y_{\bar{t}\bar{t}} \end{bmatrix}$$

$$= \det \begin{bmatrix} 0 & 0 & Y_{tt}+Q_{22}Y_{\bar{t}t} & Y_{t\bar{t}}+Q_{22}Y_{\bar{t}\bar{t}} \\ 0 & 0 & -Q'_{22} & U_m \\ -U_r & 0 & Y_{tt} & Y_{t\bar{t}} \\ 0 & -U_m & Y_{\bar{t}t} & Y_{\bar{t}\bar{t}} \end{bmatrix}$$

$$= \pm \det \begin{bmatrix} Y_{tt}+Q_{22}Y_{\bar{t}t} & Y_{t\bar{t}}+Q_{22}Y_{\bar{t}\bar{t}} \\ Q'_{22} & -U_m \end{bmatrix}$$

$$= \pm \det \begin{bmatrix} Y_{tt}+Q_{22}Y_{\bar{t}t}+Y_{t\bar{t}}Q'_{22}+Q_{22}Y_{\bar{t}\bar{t}}Q'_{22} & Y_{t\bar{t}}+Q_{22}Y_{\bar{t}\bar{t}} \\ 0 & -U_m \end{bmatrix}$$

$$= \pm \det (Y_{tt}+Q_{22}Y_{\bar{t}t}+Y_{t\bar{t}}Q'_{22}+Q_{22}Y_{\bar{t}\bar{t}}Q'_{22})$$

$$= \pm \det Q_f Y(s) Q'_f, \qquad (7.113)$$

where $Q_f = [U_r \quad Q_{22}]$, $B_f = [-Q'_{22} \quad U_m]$ after appealing to Theorem 2.9, and

$$Y(s) = \begin{bmatrix} Y_{tt} & Y_{t\bar{t}} \\ Y_{\bar{t}t} & Y_{\bar{t}\bar{t}} \end{bmatrix}. \qquad (7.114)$$

Substituting (7.113) in (7.110) yields the desired result.

As an example, we compute the node-admittance matrix $Y_n(s)$ of the network of fig. 7.16(a), which is given by

$$Y_n(s) = \begin{bmatrix} 3s+0.5 & -3s \\ 10-3s & 3s+1/s \end{bmatrix}, \qquad (7.115)$$

whose determinant is obtained as

$$\det Y_n(s) = (63s^2 + 6s + 1)/2s. \tag{7.116}$$

Comparing (7.116) with (7.108) yields

$$-\det H(s) = [\det Z_{LMR}(s)] \det Y_n(s), \tag{7.117}$$

where

$$Z_{LMR}(s) = \begin{bmatrix} s & 0 & 0 \\ 0 & 2 & 0 \\ 0 & 0 & 1 \end{bmatrix}. \tag{7.118}$$

This confirms (7.112).

Having established the desired relation (7.112), we can now apply the results of chapter 4 to determine the order of complexity or its upper bounds of a network N. To this end, let $\hat{Y}$ be the indefinite-admittance matrix of N, and let $G(\hat{Y})$ be the associated directed graph of $\hat{Y}$ according to Definition 4.3, which is different from the directed graph G representing N. As discussed in §6.5 of chapter 4, if we agree that an undirected edge stands for a pair of oppositely directed edges such that the weight (admittance) associated with each directed edge in the pair is the same as that of the undirected edge, the resulting $G(\hat{Y})$ is a mixed graph, which can be obtained directly from the equivalent network simply by superimposing the corresponding directed graphs of the unilateral elements or the associated graphs of the ideal transformers and mutual couplings of the coils upon the subnetwork consisting of resistors, capacitors and self-inductors of the original network. We shall further agree that the directed edges in the mixed graph $G(\hat{Y})$ will be called the *active edges*, and the undirected edges the *passive edges*.

Appealing to Theorems 2.20 and 4.3, (7.112) can be expressed as

$$\det H(s) = \text{constant} \cdot s^{B_l} \cdot \det Y_n(s)$$

$$= \text{constant} \cdot s^{B_l} \sum_{t_k} f(t_k), \tag{7.119}$$

where $Y_n(s)$ is the node-admittance matrix of N, B_l denotes the number of inductors in N, and t_k is a directed tree in $G(\hat{Y})$ with reference node k. We remark that since k on the right-hand side of (7.119) is arbitrary, we have freedom to choose any node as the reference node of the directed trees.

Observe, therefore, there are three classes of edges in $G(\hat{Y})$:

(i) *Inductive edges*: edges whose associated admittances are of the form $1/Ls$, L being a real constant.
(ii) *Capacitive edges*: edges whose associated admittances are of the form Cs, C being a real constant.
(iii) *Resistive edges*: edges whose associated admittances are real constants.

The inductive edges are due to self-inductors and/or inductor-current-controlled current sources, the capacitive edges are due to capacitors and/or capacitor-current-controlled current sources, and the resistive edges are due to resistors and/or voltage-controlled current sources whose controlling parameters are, by hypothesis, real constants or resistor-current-controlled current sources.

DEFINITION 7.5: *C-cut and L-cut*. A *C-cut* (*L-cut*) of $G(\hat{Y})$ is a cut composed only of capacitive (inductive) edges.

DEFINITION 7.6: *C-circuit and L-circuit*. A *C-circuit* (*L-circuit*) of $G(\hat{Y})$ is a circuit composed only of capacitive (inductive) edges.

The *L-subgraph* $G_L(\hat{Y})$ of $G(\hat{Y})$ is a subgraph consisting all of its inductive edges, being the subgraph obtained by removing all the non-inductive edges. Likewise, we define the *C-subgraph* $G_C(\hat{Y})$ of $G(\hat{Y})$ to be the subgraph consisting of all capacitive edges. Thus, an L-cut or an L-circuit is a cut or a circuit in the L-subgraph, but an L-cut in the L-subgraph is not necessarily an L-cut in $G(\hat{Y})$. Denote by $G_L^*(\hat{Y})$ the directed graph formed from $G(\hat{Y})$ by shorting all the non-inductive edges (do not remove the self-loops thus produced). In a similar manner, we define the directed graph $G_C^*(\hat{Y})$.

For a subgraph g of $G(\hat{Y})$, denote by $r(g)$ and $m(g)$ the rank and nullity of g, respectively. Also, by b_{lg} and b_{cg} we mean the numbers of the inductive edges and the capacitive edges of g, respectively. For $g = t_k$, b_{lt_k} represents the number of inductive edges of the directed tree t_k. In particular, for $g = G(\hat{Y})$ we write $b_c = b_{cg}$ and $b_l = b_{lg}$.

LEMMA 7.1: Let g be a subgraph of a connected graph G. Then the maximum number of edges of g contained in any tree of G is given by the rank of g, and the minimum number of edges of g contained in any tree of G is given by the rank of the graph obtained from G by shorting all the edges not contained in g.

Proof. The results are valid for both directed and undirected graphs G. In Problem 2.43, we have indicated that a subgraph can be made part of a tree if and only if it does not contain any circuit. The maximum number of edges contained in any circuitless subgraph of g is $r(g)$. Thus, $r(g)$ is the maximum number of edges of g contained in any tree of G.

To show the second part of the lemma, we appeal to Problem 2.44 which indicates that a subgraph can be made part of a cotree if and only if it does not contain any cut of G. Thus, the minimum number of edges of g contained in any tree of G is equal to the number of linearly independent cuts of G, contained in g. Each of these cuts remains a cut in the graph obtained from G by shorting all the edges not contained in g. Consequently, the number of such linearly independent cuts is equal to the rank of the resulting graph. The lemma follows from here. This completes the proof of the lemma.

With these preliminaries, we now proceed to deduce formulas for determining the order of complexity or its upper bounds. First, we consider the RLC networks, since the formulas for such networks are much the simplest.

6.2. RLC networks

In the case where N is an RLC network, $G(\hat{Y})$ and N are isomorphic, and the inductive, capacitive and resistive edges of $G(\hat{Y})$ become the inductors, capacitors and resistors of N, respectively. Therefore, we can use $G(\hat{Y})$ and N interchangeably. When we speak of a C-circuit in N, we mean a circuit composed only of capacitors. This is similarly valid for all other terms.

THEOREM 7.2: *The order of complexity of an RLC network is given by the number of reactive elements, less the number of linearly independent L-cuts and the number of linearly independent C-circuits.*

Proof. Write

$$\det Y_n(s) = a_k s^k + a_{k-1} s^{k-1} + \ldots + a_0 + a_{-1} s^{-1} + \ldots + a_{-q} s^{-q}, \quad (7.120)$$

where $a_k \neq 0$ and $\mathbf{a}_{-q} \neq 0$, k and q being integers to be determined. From (7.19) and (7.120), it is evident that the order of complexity of N is given by

$$\sigma = k + q + (b_l - q) = k + b_l. \quad (7.121)$$

Hence it suffices to determine k. Since $\det Y_n(s)$ is equal to the sum of all the tree-admittance products, it follows that

$$k = b_{ct} - b_{lt}, \quad (7.122)$$

t being a tree of N that maximizes the difference. We now show that it is possible simultaneously to maximize the number of capacitors and minimize the number of inductors in the tree t. Let t_1 be a tree of N containing the maximum number of capacitors, the subgraph corresponding to these capacitors being denoted by g_1. Let t_2 be a tree containing the minimum number of inductors, the subgraph corresponding to these inductors being denoted by $\bar{g}_2$. If g_2 is the complement of $\bar{g}_2$ in the L-subgraph, then the cotree $\bar{t}_2$ contains g_2. Appealing to Theorem 5.4, we conclude that there exists a tree t of N for which g_1 and $\bar{g}_2$ are subgraphs of t. From Lemma 7.1, we have $b_{ct} = r(N_C)$ and $b_{lt} = r(N_L^*)$, where $N_C = G_C(\hat{Y})$ and $N_L^* = G_L^*(\hat{Y})$. Substituting these in (7.121) in conjunction with (7.122) yields

$$\sigma = b_{ct} + b_l - b_{lt} = r(N_C) + b_l - r(N_L^*)$$
$$= b_c - m(N_C) + b_l - r(N_L^*). \tag{7.123}$$

Observe that the number of linearly independent C-circuits in N_C, being equal to its nullity $m(N_C)$, is the same as that in N, and that the number of linearly independent L-cuts in N_L^*, being equal to its rank $r(N_L^*)$, is the same as that in N. The theorem follows from here, and our proof is completed.

COROLLARY 7.1: *The number of nonzero natural frequencies of an RLC network is equal to its order of complexity, less the number of linearly independent L-circuits and the number of linearly independent C-cuts.*

Proof. From (7.119) and (7.120), it is evident that the number of zero natural frequencies is given by $b_l - q$. As in the proof of Theorem 7.2, we have $q = b_{l\hat{t}} - b_{c\hat{t}}$, $\hat{t}$ being a tree of N that maximizes the difference. Also, we can show that $b_{l\hat{t}} = r(N_L)$ and $b_{c\hat{t}} = r(N_C^*)$, where $N_L = G_L(\hat{Y})$ and $N_C^* = G_C^*(\hat{Y})$ (Problem 7.12). This leads to

$$b_l - q = b_l - b_{l\hat{t}} + b_{c\hat{t}} = b_l - r(N_L) + r(N_C^*)$$
$$= m(N_L) + r(N_C^*), \tag{7.124}$$

showing that the number of zero natural frequencies is given by $m(N_L)$, which is equal to the number of linearly independent L-circuits of N, plus $r(N_C^*)$ which is equal to the number of linearly independent C-cuts of N. So the corollary is proved.

Alternative expressions for the corollary are possible and are stated as

COROLLARY 7.2: *The number of nonzero natural frequencies of an RLC network is given by*

$$f = r(N_C) + r(N_L) - r(N_C^*) - r(N_L^*) \qquad (7.125a)$$

$$= m(N_C^*) + m(N_L^*) - m(N_C) - m(N_L) \qquad (7.125b)$$

$$= [m(N_L^*) - m(N_L)] + [r(N_C) - r(N_C^*)]. \qquad (7.125c)$$

COROLLARY 7.3: The number of nonzero natural frequencies of an LC network is given by

$$f = 2[m(N) - m(N_C) - m(N_L)]. \qquad (7.126)$$

In the case where the RLC network does not contain any L-circuit and C-circuit, (7.125b) reduces to $f = m(N_C^*) + m(N_L^*)$, giving an algorithm due to GUILLEMIN [1931]. Formula (7.126) was first stated by REZA [1955] who called f the *number of dynamically independent loops*.

Physically, the natural frequencies at the origin correspond to the constant currents that circulate around the circuits composed only of inductors or the constant voltages that apply across the cuts composed only of capacitors. These constant currents and voltages do not give rise to any additional algebraic constraints among the currents in the L-circuits or the voltages across the C-cuts, only on their rates of change. For example, an L-circuit gives rise to a differential relation of the form

$$L_1 \frac{di_1}{dt} + L_2 \frac{di_2}{dt} + \cdots + L_k \frac{di_k}{dt} = 0. \qquad (7.127)$$

Integration of this expression from 0 to t leads to

$$L_1 i_1 + L_2 i_2 + \cdots + L_k i_k = \text{constant}. \qquad (7.128)$$

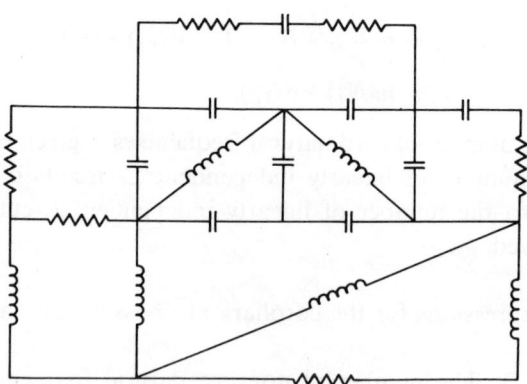

Fig. 7.17. An RLC network illustrating the determination of its order of complexity.

It might appear that this also represents a constraint on the inductor currents. However, the constant term on the right-hand side is not specified, whose determination requires an independent relationship. Similar statement can be made about voltages across the network elements of a C-cut.

We shall illustrate the above results by the following example.

Example 7.8: Consider the RLC network N of fig. 7.17. Using the symbols defined above gives

$$b_l = 6, \qquad b_c = 9, \tag{7.129a}$$

$$r(N_C) = 9-2 = 7, \qquad r(N_L) = 6-1 = 5, \tag{7.129b}$$

$$r(N_C^*) = 3-1 = 2, \qquad r(N_L^*) = 2-1 = 1, \tag{7.129c}$$

$$m(N_C) = 9-9+2 = 2, \qquad m(N_L) = 6-6+1 = 1, \tag{7.129d}$$

$$m(N_C^*) = 9-3+1 = 7, \qquad m(N_L^*) = 6-2+1 = 5. \tag{7.129e}$$

Now we can state the following:

The number of linearly independent C-cuts $= r(N_C^*) = 2,$ (7.130a)

The number of linearly independent L-cuts $= r(N_L^*) = 1,$ (7.130b)

The number of linearly independent C-circuits $= m(N_C) = 2,$ (7.130c)

The number of linearly independent L-circuits $= m(N_L) = 1.$ (7.130d)

Thus, the order of complexity of the RLC network is given by

$$\sigma = b_c + b_l - r(N_L^*) - m(N_C) = 9+6-1-2 = 12, \tag{7.131}$$

which is the maximum number of linearly independent initial conditions that can be specified for N or, equivalently, the total number of natural frequencies of N, the zero frequency and multiplicity included. The number of nonzero natural frequencies, according to Corollary 7.1, is given by

$$f = \sigma - m(N_L) - r(N_C^*) = 12-1-2 = 9, \tag{7.132}$$

which can also be computed directly from (7.125). The reader is urged to verify this.

6.3. *Active networks*

From (7.119), it is clear that the order of complexity of an active network N is equal to the number of inductors plus the highest index of s in det $Y_n(s)$. However, the highest index of s in det $Y_n(s)$ is not necessarily contained in a directed tree t_k of $(G(\hat{Y}))$ called the *normal directed tree* that maximizes the difference between the number of capacitive edges and the number of inductive edges contained in it, since, unlike the RLC case, cancellations may occur with other normal directed trees after forming the admittance products. The new term differs from the tree used for an RLC network in that it is not always possible to maximize the number of capacitive edges and minimize the number of inductive edges in the same directed tree.

THEOREM 7.3: An upper bound on the order of complexity σ of an active network N containing B_l inductors is given by

$$\sigma \leq \sigma_{\max} = B_l + b_{ct} - b_{lt}, \tag{7.133}$$

t being a normal directed tree of $G(\hat{Y})$, whose reference node is arbitrary. The equality is attained if and only if the sum of all the normal-directed-tree-admittance products is nonzero.

It is more convenient if the formula (7.133) is stated directly in terms of the subnetworks of N rather than the subgraphs of $G(\hat{Y})$. To this end, we define the following.

DEFINITION 7.7: *Passive tree of a network*. A *passive tree* of a network N is a tree composed only of resistors, inductors and capacitors of N.

THEOREM 7.4: In a network N composed of resistors, inductors, capacitors and current-controlled current sources whose controlling parameters are real constants, let t be a passive tree of N chosen so that the difference between the number of capacitors and the number of inductors is maximized. Then an upper bound on the order of complexity σ of N is given by the number of capacitors contained in t plus the number of inductors contained in the cotree $\bar{t}$, the complement of t in N, or

$$\sigma \leq \sigma_{\max} = b_{ct} + b_{l\bar{t}}. \tag{7.134}$$

Proof. Let $\hat{N}$ be the network obtained from N by converting all its current-controlled current sources into equivalent voltage-controlled current sources. In doing so, we introduce active resistive, capacitive or inductive edges in $G(\hat{Y})$. In $G(\hat{Y})$ let $\hat{t}_k$ be a directed tree (not necessarily a normal directed tree) that maximizes the difference $b_{c\hat{t}_k} - b_{l\hat{t}_k}$ and that does not result in complete cancellation in det $Y_n(s)$. Our objective is to show that

$$B_l + b_{c\hat{t}_k} - b_{l\hat{t}_k} = b_{ct} + b_{l\bar{t}}. \tag{7.135}$$

For this it is sufficient to demonstrate that we can construct a passive tree t of N that possesses the same index of s as $\hat{t}_k$. If $\hat{t}_k$ is composed only of passive edges, its corresponding elements in N is the desired tree t. Thus, assume that $\hat{t}_k$ contains at least one active edge. As stated in Problem 7.13, at most one of the four active edges of fig. 4.38(b) of a voltage-controlled current source can appear in $\hat{t}_k$ that will not result in cancellations in det $Y_n(s)$. Without loss of generality, let (c, b) be this active edge in $\hat{t}_k$. Appealing to the result stated in Problem 7.24, we may further stipulate that $\hat{t}_k - (c, b) = t_{ca, kbd}$ (see Definition 4.12). Then the operation $\hat{t}_k \cup (a, b) - (c, b) = t'$ yields a tree (not necessarily a directed tree) of $G(\hat{Y})$, where (a, b) is the passive edge corresponding to the controlling element (resistor, capacitor or inductor) of the controlled source. Since the admittances of (a, b) and (c, b) differ only by a multiplicative constant, t' and $\hat{t}_k$ possess the same index of s. If there are other active edges in t', clearly the process can be repeated. Thus, we can generate a tree t'' of $G(\hat{Y})$ composed only of the passive edges, whose corresponding subnetwork in N is a tree t possessing the same index of s as $\hat{t}_k$. Furthermore, we have $b_{ct} = b_{ct''} = b_{\hat{ct}_k}$ and $b_{lt} = b_{l\hat{t}_k}$. Substituting these in (7.135) and using the fact that $B_l - b_{lt} = t_{l\bar{t}}$ yield the desired formula (7.134). This completes the proof of the theorem.

The above theorem may be extended to the case of voltage-controlled current sources. In fact, following the same argument as given above, we can show that if N does not contain any cut composed only of inductors and controlled current sources such that the voltage of one of its inductors is a controlling element, then (7.134) remains valid (Problem 7.27).

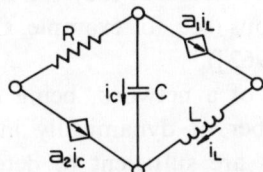

Fig. 7.18. An active network illustrating the determination of an upper bound on its order of complexity.

As an illustration, consider the network N of fig. 7.18. It is easy to verify that the elements R, C and L constitute the only complete tree of N. According to the theorem, an upper bound on its order of complexity is given by

$$\sigma \leq \sigma_{\max} = 1 + 0 = 1. \qquad (7.136)$$

It can be shown that if N possesses a unique solution, then $a_1 \neq 1$, $a_2 \neq -1$ and $\sigma = 1$ (Problem 7.14).

COROLLARY 7.4: The order of complexity of a network containing a gyrator (transistor or vacuum tube) is not less than that of the network (equivalent network) with the gyrator (controlled current source) removed.

Proof. By partitioning the set of directed trees of $G(\hat{Y})$ according to those that do not contain any active edge, and those that contain at least one, the nodal determinant can be put in two summations, each being nonnegative. Using this in conjunction with (7.119), the corollary follows immediately. The details of the derivations are left as exercises (Problems 7.15 and 7.16).

§7. Conclusions

We began this chapter by presenting procedures for the systematic formulation of network equations in the form of a system of first-order differential equations known as the state equations. We then developed the explicit form of state equation in normal form for a general linear network. For the explicit form to exist, we assume the nonsingularity of certain matrices. Due to its generality, the explicit form is extremely complicated, and for a particular network it is usually simpler to derive the desired state equation by the procedure outlined in §2. The objectives for presenting the general formulas were to show, at least once, the general formulation of state equation on a sufficiently firm basis, and to demonstrate that its derivation can be achieved by means of a digital computer. This is followed by a discussion of the physical meanings of the various parameter matrices of the state equation. However, the solutions of the state equation are not discussed; they can be found in any text on the theory of differential equations (see, for example, CODDINGTON and LEVINSON [1955] and PONTRYAGIN [1962]).

The order of complexity of a network, being the dimension of its state space, is the minimal number of dynamically independent state variables whose instantaneous values are sufficient to determine completely the instantaneous state of the network. It is also the maximum number of linearly independent initial conditions that can be specified for a network. In the theory of differential equations, it is the number of arbitrary constants appearing in the general solution of the network equations. These are different ways in stating the same thing.

For a network composed of resistors, inductors, capacitors and controlled current sources whose controlling parameters are real constants, an upper bound on its order of complexity can be associated with a passive tree of the network that maximizes the difference between the number of capacitors and the number of inductors in the passive tree. In general, networks can attain this bound. In the case of an *RLC* network, its order of complexity can be determined by its topology alone, being equal to the number of reactive elements, less the number of linearly independent *L*-cuts and the number of linearly independent *C*-circuits.

Problems

7.1. Determine the order of complexity of the network of fig. 7.19, and then derive its state equation.

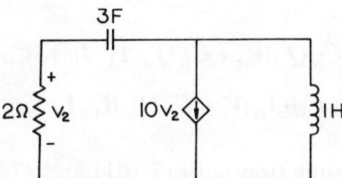

Fig. 7.19. A network containing a voltage-controlled current source.

7.2. In (7.41) show that $Q_{lc} = 0$ and $Q_{lr} = 0$.

7.3. Give a physical interpretation for the transfer-admittance matrix Y_{ec} of (7.91a) and the transfer-impedance matrix Z_{jl} of (7.91d).

7.4. Give a physical interpretation for the current-transfer, matrix H_{jc} of (7.91b) and the voltage-transfer, matrix H_{el} of (7.91c).

7.5. Consider the network of fig. 7.20, in which the current in the resistor R_4 and voltage across the resistor R_3 are chosen as the output variables. Derive the state equation and the output equation for the network.

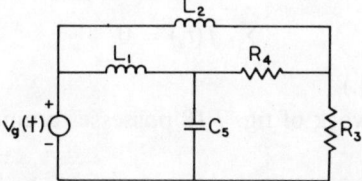

Fig. 7.20. An *RLC* network with an independent voltage generator.

7.6. Express all the element voltages and currents of the network of fig. 7.20 in terms of the state variables and the voltage source.

7.7. Choose a normal tree and derive the state equation for the network of fig. 7.21. Also express all the element voltages and currents in terms of

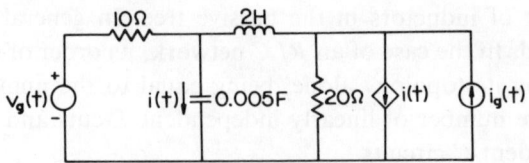

Fig. 7.21. An active network containing a current-controlled current source and two independent sources.

the state variables and the sources. What is the order of complexity of the network? Determine all its natural frequencies.

7.8. Show that

$$I_{c\hat{t}} = (C_{\overline{tt}}Q'_{cc}A_{11} + \dot{C}_{\overline{tt}}Q'_{cc})V_{ct} + C_{\overline{tt}}Q'_{cc}A_{12}I_{l\hat{t}} + (C_{\overline{tt}}Q'_{cc}B_{11} + \dot{C}_{\overline{tt}}Q'_{ec})V_e$$
$$+ (C_{\overline{tt}}Q'_{cc}B_{13} + C_{\overline{tt}}Q'_{ec})\dot{V}_e + C_{\overline{tt}}Q'_{cc}B_{12}I_j. \tag{7.137}$$

7.9. Using (7.85) in conjunction with (7.101) and (7.137), express I_{ct} and $V_{l\hat{t}}$ in terms of the state variables and source variables.

7.10. Using (7.43a) and (7.43h) in conjunction with (7.54), (7.101) and (7.137), express I_e and V_j in terms of the state variables and source variables.

7.11. From Problems 7.8–7.10 and (7.54) and (7.101), justify the statement that the branch-current vector and branch-voltage vector of a network can be expressed in terms of the state vector and the source vector, as in (7.3).

7.12. In an RLC network N, assume that $\hat{t}$ is a tree that maximizes the difference $b_{l\hat{t}} - b_{c\hat{t}}$. Show that $b_{l\hat{t}} = r(N_L)$ and $b_{c\hat{t}} = r(N_C^*)$.

7.13. Referring to fig. 4.38(b), assume that T_{k2} is the set of directed trees t_k of $G(\hat{Y})$, each of which contains two of the four active edges. Show that

$$\sum_{t_k \in T_{k2}} f(t_k) = 0. \tag{7.138}$$

(See CHEN [1972b].)

7.14. Show that the network of fig. 7.18 possesses a unique solution if $a_1 \neq 1$ and $a_2 \neq -1$.

7.15. Referring to fig. 4.38(b) and using the symbols defined in (4.186) and (4.188). Show that the nodal determinant can be expanded as

$$\det Y_n(s) = V_k = V_d = V_d^0 - yW_{cb,da}^0 + yW_{ca,db}^0, \tag{7.139}$$

in which the superscript 0 indicates that the various admittance products are to be evaluated in the subgraph obtained from $G(\hat{Y})$ by setting $y = 0$. Using (7.139) in conjunction with (7.140), prove Corollary 7.4.

Problems 517

7.16 Referring to fig. 4.55(b) and using the symbols defined in (4.178) and (4.188). Show that the nodal determinant can be expanded as

$$\det \mathbf{Y}_n(s) = V_k^0 + g^2 U_{a,b,c}, \qquad (7.140)$$

where V_k^0 denotes the nodal determinant of the network with the gyrator removed. (*Hint*: Directed trees containing two gyrator edges occur in triplets and one gyrator edge in pairs.)

7.17. Applying the formula (7.58), compute the parameter matrices of the state equation of the network as shown in fig. 7.20. Confirm the results by employing the multiport interpretation of (7.86) and (7.91).

7.18. Repeat Problem 7.17 for the network of fig. 7.21.

7.19. Repeat Problem 7.5 but this time assume that the capacitor is time-varying with capacitance $C_5 = 1 + \frac{1}{4} \sin t$.

7.20. Shorting the two resistors on top of the network of fig. 7.17, determine the order of complexity of the resulting network, and the number of its nonzero natural frequencies.

7.21. The natural frequencies of a network are conventionally defined as the zeros of the determinant of the loop-impedance matrix or the node-admittance matrix. This definition involves some ambiguities and inconsistencies. State these ambiguities and inconsistencies.

7.22. Repeat Problem 7.5 but this time a capacitor of capacitance C_6 is connected across the inductor L_1.

7.23. In fig. 7.17, replace the two resistors on top of the network by two inductors. Determine the order of complexity and the number of nonzero natural frequencies of the resulting network.

7.24. Show that (7.138) is satisfied if T_{k2} is replaced by $T_{k1} = \{t_k:$ one of the four active edges is in t_k and such that if (c, a) is in t_k then $t_k - (c, a) \neq t_{cb, kad}$; if (d, b) is in t_k then $t_k - (d, b) \neq t_{da, kbc}$; if (c, b) is in t_k then $t_k - (c, b) \neq t_{ca, kbd}$; and if (d, a) is in t_k then $t_k - (d, a) \neq t_{db, kca}\}$.

7.25. Show that the derivatives of the voltage sources will appear in the state equation only when there is a C-circuit, and that the derivatives of the current sources will appear only when there is a cut composed only of inductors and/or independent current sources.

7.26. Simplify the formulas (7.58)–(7.60) for an RLC network with independent sources.

7.27. Show that formula (7.134) remains valid if N does not contain any cut composed only of inductors and controlled current sources such that the voltage of one of its inductors is a controlling element.

CHAPTER 8

SQUARING RECTANGLES AND LAYOUTS

Most of the engineering problems are solved with the help of mathematics. In this chapter, we demonstrate that, in rare occasions, solutions to electrical network problems can also help resolve many mathematics puzzles. We show how the solution to an electrical resistive network problem can be used to dissect a rectangle of commensurable sides into a finite number of non-overlapping squares, and how the dissection of a rectangle can be applied to circuit layout in Very-Large-Scale-Integrated (VLSI) circuit design.

§ 1. Introduction

Like many mathematical problems, the problem of dividing a rectangle into a finite number of non-overlapping squares, no two of which are equal, started as a puzzle. For a small number of squares, the problem may not be that difficult for a skillful puzzlers. However, the problem to construct all rectangles with 10 or more squares will not be so easy.

Consider, for example, the rectangle shown in fig. 8.1, which is composed of nine squares with sides 2, 5, 7, 9, 16, 25, 28, 33, and 36, as indicated. Observe that the sides of all the squares and of the rectangle are parallel to two perpendicular lines. One is horizontal and the other is vertical.

DEFINITION 8.1: *Squared rectangle.* A *squared rectangle of order n* is a dissection of a rectangle into a finite number n of non-overlapping squares. The n squares are called the *elements* of the dissection.

The dissection shown in fig. 8.1 is an example of a squared rectangle of order 9, because it contains 9 squares as its elements. Observe that, in this case, all of its elements are unequal. This leads to the following term:

§ 1 Introduction

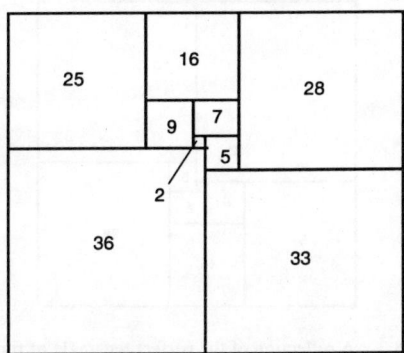

Fig. 8.1. The dissection of a rectangle into squares.

DEFINITION 8.2: *Perfect rectangle*. A *perfect rectangle* is a squared rectangle of more than one element, all the elements of which are unequal in size. Otherwise, it is *imperfect*.

A perfect rectangle is also referred to as a *perfect squared rectangle*. The rectangle in fig. 8.1 is perfect, because all of its elements are unequal. In fact, it was shown by BROOKS, SMITH, STONE and TUTTE [1940] that there are no perfect rectangles of order less than 9, and exactly two of order 9. One shown in fig. 8.1 was obtained by BROOKS et al. [1940], and the other given in fig. 8.2 was first discovered by MORON [1925]. When we speak of dissections, we exclude those obtained by trivial transformations such as reflections, rotations, or scaling.

For example, the perfect rectangle of fig. 8.3 is a reflection of that of fig. 8.1, and is, therefore, considered to be the same. Likewise, the perfect rectangle of fig. 8.4 is the same as that of fig. 8.2, because one can be obtained from the other by a scaling factor of 2.

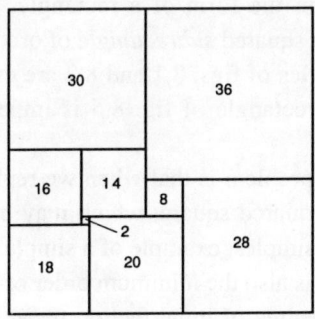

Fig. 8.2. A perfect rectangle of order 9.

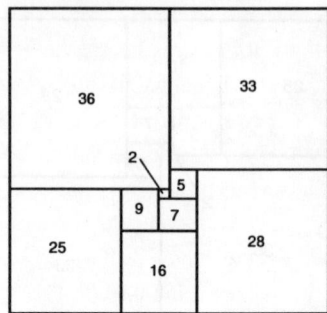

Fig. 8.3. A reflection of the perfect rectangle of fig. 8.1.

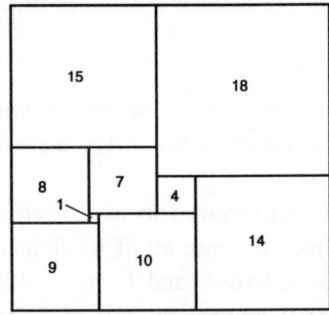

Fig. 8.4. A scaled dissection of the perfect rectangle in fig. 8.2.

DEFINITION 8.3: *Simple squared rectangle*. A squared rectangle is said to be *simple* if it contains no squared rectangle of order more than one. All other squared rectangles are *compound*.

In other words, a simple squared rectangle is a squared rectangle with no subset of its elements arranged in the form of a rectangle, and a compound squared rectangle always contains a squared *subrectangle* of order larger than one.

The two perfect rectangles of figs. 8.1 and 8.2 are examples of simple squared rectangles. The squared rectangle of fig. 8.5 is imperfect because some of its elements are equal in size.

A much more difficult problem is that when we replace the word rectangle by square, and try to find a squared square, which may be *perfect* or *imperfect* and *simple* or *compound*. The simplest example of a simple squared square of order 13 is shown in fig. 8.6, which is also the minimum order of such a square. Although it is simple, it is to a high degree of imperfection because it contains five pairs of elements of equal size. As before, a perfect squared square is simply called a *perfect square.* The dissection shown in fig. 8.7 is an example of a compound perfect square, because all of its elements are unequal in size and there is a subrectangle (unshaded area), which by itself is simple and perfect. The construction of a simple perfect square is difficult, let alone finding one with least order. DUIJVESTIJN [1962] showed that there is no simple perfect square of order less than 21. KAZARINOFF and WEITZENKAMP [1973a] complemented Duijvestijn's result by proving the nonexistence of a compound perfect square of order less than 22. Therefore, there is no perfect square of order less than 21. Using the computer as a tool, DUIJVESTIJN [1978] succeeded in finding the simple perfect square of lowest order 21. Finally, DUIJVESTIJN, FEDERICO and LEEUW [1982] demonstrated that there are no compound perfect squares of order less than 24 and there is one and only one compound perfect square of order 24. This lowest-order compound perfect square of order 24 was discovered by WILLCOCKS [1948].

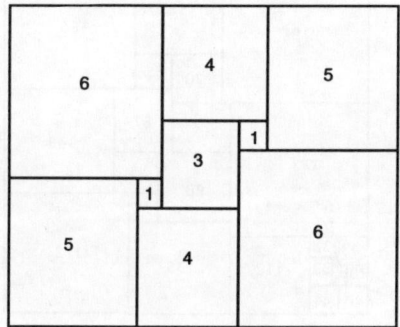

Fig. 8.5. An imperfect squared rectangle.

§ 2. The Bouwkamp Code

It is known that there are 2 simple perfect rectangles of order 9, 6 of order 10, 22 of order 11, 67 of order 12 and 214 of order 13. In addition, there are 1 simple imperfect rectangle of order 9, 0 of order 10, 0 of order 11, 9 of order 12 and 33 of order 13. Therefore, there are 354 simple squared rectangles of order less than 14, 311 of which are perfect and 43 imperfect. Even with these low orders, it is impractical to draw out all of them. The difficulty, however, can be overcome by a suitable coding system proposed by BOUWKAMP [1946] called the *Bouwkamp code*.

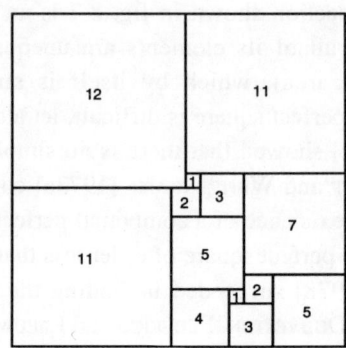

Fig. 8.6. A simple squared square of lease order 13.

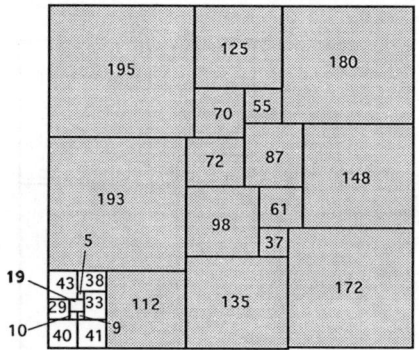

Fig. 8.7. A compound perfect square of order 25.

First assume that the rectangle is to be drawn in such a way that the longer sides are horizontal, and that the element in the upper-left corner is not smaller than the elements located at the remaining three corners. It was verified by BOUWKAMP [1946] that, except for the simple squared square of order 13 shown in fig. 8.6, the geometrical configuration of the aforementioned orientation for a squared rectangle is unique for orders not exceeding 13. This orientation will help identify and reconstruct a rectangle from its code, but it is not necessary in applying the Bouwkamp code.

Given an oriented squared rectangle with horizontal and vertical line segments, consider the group of elements with their upper horizontal sides in a common horizontal segment. The individual elements of this group are ordered from left to right. The various groups themselves are ordered from top to bottom in accordance with the horizontal level of the common horizontal segments. If two line segments are at the same horizontal level, they will be ordered from left to right too. Starting from the upper horizontal side of the given rectangle, we read the elements from left to right and from top to bottom. In the written code, the various groups are separated by parentheses, and the elements of a group by commas. The resulting code is referred to as the *Bouwkamp code*.

Consider the perfect rectangle of order 9, shown in fig. 8.1. After proper orientation, it is redrawn in fig. 8.8. The first group of elements touching the top horizontal side of the rectangle is denoted by (36, 33). The second group of elements with their upper sides in common with the second horizontal line segment is found to be (5, 28). Likewise, the third, fourth and fifth groups are given by (25, 9, 2), (7), and (16), respectively. Arranging these groups horizontally from left to right yields the Bouwkamp code below:

$$(36, 33)\ (5, 28)\ (25, 9, 2)\ (7)\ (16)$$

As another example, consider Moron's perfect rectangle of fig. 8.4, which is redrawn in fig. 8.9 after reorientation. Its Bouwkamp code reads as follows:

$$(18, 15)\ (7, 8)\ (14, 4)\ (10, 1)\ (9)$$

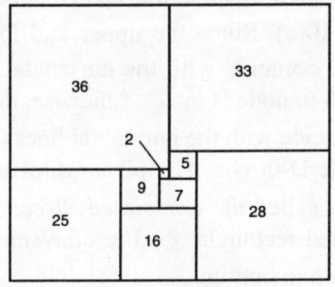

Fig. 8.8. An oriented perfect rectangle of order 9.

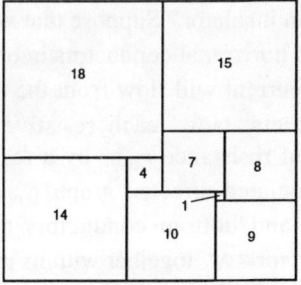

Fig. 8.9. An oriented Moron's perfect rectangle.

The Bouwkamp code for a compound squared rectangles is similar. For example, we give the code for a compound perfect square of order 25 shown in fig. 8.7 below:

(195, 125, 180) (70, 55) (87, 148) (193, 72) (98, 61)

(37, 172) (135) (43, 38, 112) (5, 33) (29, 19) (10, 9) (41) (40)

§ 3. The Electrical Network Associated with a Dissected Rectangle

In this section, we show that the problem of dissecting a rectangle or square into squares is closely related to the solution of a resistive network associated with the dissected rectangle.

Consider the dissected rectangle $\mathcal{R}$ of fig. 8.2, which is redrawn in fig. 8.10(b). This dissected rectangle can be represented by a planar directed graph G_p. Each node corresponds to a horizontal line (thick line) in fig. 8.10(b), and each element (square) of $\mathcal{R}$ is represented by an edge connecting the two nodes corresponding to the two horizontal lines that include upper and lower sides of the element. The dissected rectangle $\mathcal{R}$ has six horizontal lines corresponding to the six nodes A, B, C, D, E and F in fig. 8.10(a). Since the upper and lower sides of the element (shaded area) of sides 30 coincide with the horizontal lines A and B, there is a directed edge from node A to node B in G_p. Likewise, the upper and lower sides of the element of sides 8 coincide with the horizontal lines C and D, there is a directed edge from node C to node D in G_p. The planar directed graph G_p constructed in this way in fig. 8.10(a) is called the *associated directed graph* or simply the *p-digraph* G_p of the dissected rectangle $\mathcal{R}$. For convenience, the directions of the edges were chosen from top to bottom.

Imagine that each horizontal line is a perfect conductor, that each square is a uniform resistive sheet, and that each vertical line between the squares inside the rectangle is an infinitely thin insulator. Suppose that we apply a dc voltage source between the top and bottom horizontal conductors in order to create a voltage drop from top to bottom. A dc current will flow from the top to the bottom conductor through these resistive sheets. Now, each resistive sheet can be represented equivalently by a resistor of resistance r_k or by a resistive edge of conductance $g_k = 1/r_k$. Thus, the associated directed graph G_p of $\mathcal{R}$ is a resistive one-port network $\mathcal{N}_p$ with the top and bottom conductors corresponding to the input terminals. This one-port network $\mathcal{N}_p$ together with its input excitation is denoted by $\mathcal{N}_c$ and is shown in fig. 8.11.

§ 3   The electrical network associated with a dissected rectangle   525

Let G_c be the planar directed graph associated with the network $\mathcal{N}_c$ of fig. 8.11. It is also the directed graph obtained from the *p*-digraph G_p of $\mathcal{R}$ by inserting a directed edge from node F to node A corresponding to the excitation source E, as shown in fig. 8.12. BROOKS et al. [1940] call the planar one-port network $\mathcal{N}_p$ a *polar net* or simply *p-net*, and the complete planar network $\mathcal{N}_c$ a *completed net* or simply *c-net*.

In the associated directed graph G_c called the *c-digraph* of fig. 8.12, let i_k and v_k be the current and voltage of the branch e_k, respectively. If we set branch currents i_k equal to their corresponding sides of the squares in 8.10(b), as follows:

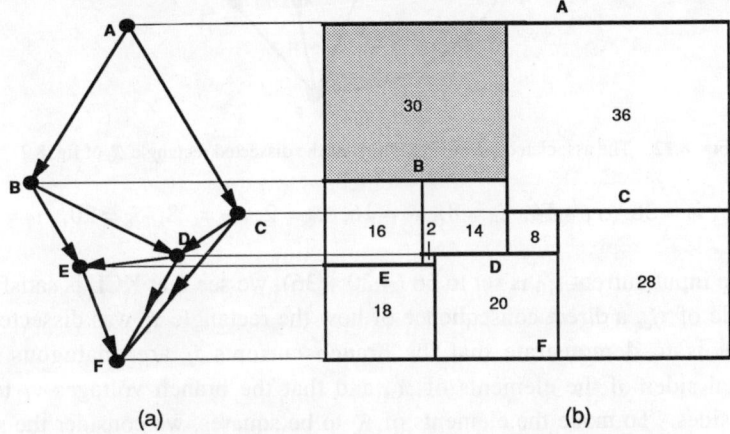

Fig. 8.10. The association of a directed graph with a dissected rectangle.

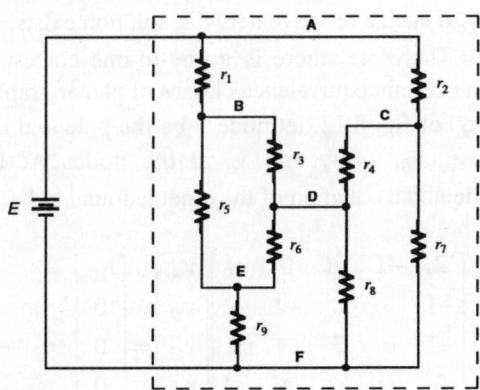

Fig. 8.11. The one-port network together with its excitation forming the network $\mathcal{N}_c$ associated with $\mathcal{R}$.

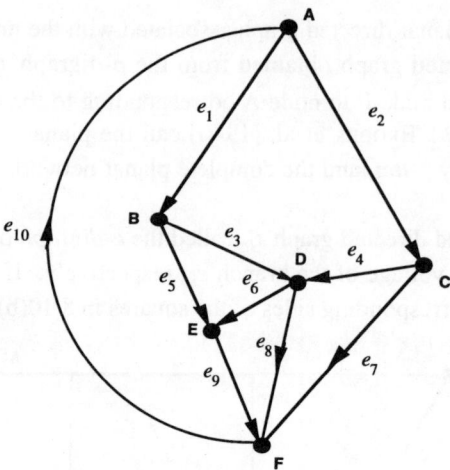

Fig. 8.12. The associated directed graph G_c of the dissected rectangle $\mathcal{R}$ of fig. 8.2.

$i_1 = 30$, $i_2 = 36$, $i_3 = 14$, $i_4 = 8$, $i_5 = 16$, $i_6 = 2$, $i_7 = 28$, $i_8 = 20$, $i_9 = 18$

and if the input current i_{10} is set to 66 (= 30 + 36), we see that KCL is satisfied for each node of G_c, a direct consequence of how the rectangle $\mathcal{R}$ was dissected. Our objective is to demonstrate that the branch currents i_k are analogous to the horizontal sides of the elements of $\mathcal{R}$, and that the branch voltages v_k to their vertical sides. To make the elements of $\mathcal{R}$ to be squares, we consider the special situation by setting all resistances r_k in G_c to 1 Ω. In this way, the branch voltages v_k will be numerically equal to the branch currents i_k and each element of $\mathcal{R}$ will be a square. Conversely, if such a resistive network solution exists, a perfect rectangle can be constructed. Therefore, there is a one-to-one correspondence between squared rectangles and certain equivalence classes of planar graphs G_c.

In the c-digraph G_c of fig. 8.12, let node F be the potential reference point for the nodal voltages v_A, v_B, v_C, v_D and v_E at the nodes A, B, C, D and E, respectively. Then the nodal equation of the c-net is found to be

$$\begin{bmatrix} 2 & -1 & -1 & 0 & 0 \\ -1 & 3 & 0 & -1 & -1 \\ -1 & 0 & 3 & -1 & 0 \\ 0 & -1 & -1 & 4 & -1 \\ 0 & -1 & 0 & -1 & 3 \end{bmatrix} \begin{bmatrix} v_A \\ v_B \\ v_C \\ v_D \\ v_E \end{bmatrix} = \begin{bmatrix} i_{10} \\ 0 \\ 0 \\ 0 \\ 0 \end{bmatrix}. \qquad (8.1)$$

Since $v_A = E$ is known and i_{10} is unknown, (8.1) can be rearranged to give

§3   The electrical network associated with a dissected rectangle

$$i_{10} = 2E - v_B - v_C \tag{8.2a}$$

$$\begin{bmatrix} 3 & 0 & -1 & -1 \\ 0 & 3 & -1 & 0 \\ -1 & -1 & 4 & -1 \\ -1 & 0 & -1 & 3 \end{bmatrix} \begin{bmatrix} v_B \\ v_C \\ v_D \\ v_E \end{bmatrix} = \begin{bmatrix} E \\ E \\ 0 \\ 0 \end{bmatrix} \tag{8.2b}$$

which can be solved to yield

$$\begin{bmatrix} v_B \\ v_C \\ v_D \\ v_E \end{bmatrix} = \begin{bmatrix} 0.46875 & 0.06250 & 0.18750 & 0.21875 \\ 0.06250 & 0.37500 & 0.12500 & 0.06250 \\ 0.18750 & 0.12500 & 0.37500 & 0.18750 \\ 0.21875 & 0.06250 & 0.18750 & 0.46875 \end{bmatrix} \begin{bmatrix} E \\ E \\ 0 \\ 0 \end{bmatrix} = \begin{bmatrix} 0.53125 \\ 0.43750 \\ 0.31250 \\ 0.28125 \end{bmatrix} E \tag{8.3}$$

where E is the source voltage. To make all the branch voltages and currents integers, set E to 64 and obtain

$$\begin{bmatrix} v_B \\ v_C \\ v_D \\ v_E \end{bmatrix} = \begin{bmatrix} 34 \\ 28 \\ 20 \\ 18 \end{bmatrix} \tag{8.4a}$$

$$i_{10} = 2E - v_B - v_C = 128 - 34 - 28 = 66. \tag{8.4b}$$

The branch-voltage vector V can be computed from

$$V = \begin{bmatrix} v_1 \\ v_2 \\ v_3 \\ v_4 \\ v_5 \\ v_6 \\ v_7 \\ v_8 \\ v_9 \\ v_{10} \end{bmatrix} = A'V_n = \begin{bmatrix} 1 & -1 & 0 & 0 & 0 \\ 1 & 0 & -1 & 0 & 0 \\ 0 & 1 & 0 & -1 & 0 \\ 0 & 0 & 1 & -1 & 0 \\ 0 & 1 & 0 & 0 & -1 \\ 0 & 0 & 0 & 1 & -1 \\ 0 & 0 & 1 & 0 & 0 \\ 0 & 0 & 0 & 1 & 0 \\ 0 & 0 & 0 & 0 & 1 \\ -1 & 0 & 0 & 0 & 0 \end{bmatrix} \begin{bmatrix} 64 \\ 34 \\ 28 \\ 20 \\ 18 \end{bmatrix} = \begin{bmatrix} 30 \\ 36 \\ 14 \\ 8 \\ 16 \\ 2 \\ 28 \\ 20 \\ 18 \\ -64 \end{bmatrix} \tag{8.5}$$

where A is a basis incidence matrix of $\mathcal{G}_c$ and V_n is the nodal voltage vector. The resistive branch-current vector I_r is numerically equal to the resistive branch-voltage vector V_r because the branch-impedance matrix Z is the identity matrix:

$$V_r = ZI_r = I_r. \tag{8.6}$$

The elements of I_r give the sides of the required squares:

$$30, 36, 14, 8, 16, 2, 28, 20, 18$$

Observe that the area of the rectangle $\mathcal{R}$, being the sum of the areas of the squares, is equal to the total power consumed by the resistive one-port network $\mathcal{N}_p$:

$$\begin{aligned} 64 \times 66 = v_A i_{10} &= 30^2 + 36^2 + 14^2 + 8^2 + 16^2 + 2^2 + 28^2 + 20^2 + 18^2 \\ &= i_1 v_1 + i_2 v_2 + i_3 v_3 + i_4 v_4 + i_5 v_5 + i_6 v_6 + i_7 v_7 + i_8 v_8 + i_9 v_9 = 4224. \end{aligned} \tag{8.7}$$

In general, if the resistances of the p-net $\mathcal{N}_p$ are not set to unity, the resulting rectangle is dissected into a number of subrectangles. As shown by DEHN [1903], a rectangle can be squared if, and only if, its sides are commensurable. Not only can it be squared, SPRAGUE [1940] and BROOKS et al. [1940] proved that it can be squared perfectly in infinitely many totally distinct ways. This will be discussed in §5. However, the basic problem of how to divide a given rectangular region to yield a perfect rectangle remains unanswered. If such a solution exists, then how to determine the perfect rectangle of least order or a reasonable estimate of that minimal order. YAGLOM [1968] demonstrated that, given an a by b rectangle with rational a/b, it can always be subdivided to yield a perfect rectangle of order at most $13a^2b^2 - 11ab - 1$. For the perfect rectangle $\mathcal{R}$ of fig. 8.10(b), Yaglom's upper bound will be

$$13a^2b^2 - 11ab - 1 = 13 \times 66^2 \times 64^2 - 11 \times 66 \times 64 - 1 = 231,901,823 \tag{8.8}$$

but the minimum order of perfect squaring such a rectangle is only 9.

In constructing the associated directed graph $\mathcal{G}_p$ of the dissected rectangle $\mathcal{R}$ of fig. 8.10(b), each horizontal line segment of $\mathcal{R}$ is represented by a node. Suppose that, instead of considering the horizontal line segments, the vertical line segments (thick lines) of $\mathcal{R}$ are represented by nodes as shown in fig. 8.13, yielding another associated directed graph $\mathcal{G}'_p$ called the *dual p-digraph* of $\mathcal{R}$, which is redrawn in fig. 8.14. Its corresponding one-port network $\mathcal{N}'_p$ is called the *dual one-port*

§ 3   The electrical network associated with a dissected rectangle   529

network or simply the *dual p-net* of $\mathcal{R}$. As before, if an excitation source is inserted at the input terminals of $\mathcal{N}'_p$, the resulting complete network $\mathcal{N}'_c$ is referred to as the *dual c-net*. The directed graph $\mathcal{G}'_c$ associated with the dual c-net $\mathcal{N}'_c$ is called the *dual c-digraph* and is presented in fig. 8.15.

The dual p-digraph $\mathcal{G}'_c$ and p-digraph $\mathcal{G}_c$ of $\mathcal{R}$ are geometrical duals of themselves. By superimposing the dual p-digraph $\mathcal{G}'_c$ of fig. 8.15 on the p-digraph $\mathcal{G}_c$ of fig. 8.12, we obtain fig. 8.16. Observe that one is the geometrical dual of the other, where the branch currents in one are the branch voltages in the other. Furthermore, the nodal equations in one are the same as the mesh equations in the other. In terms of the p-digraph and the dual p-digraph, we introduce the notion of one terminal-pair planar graph.

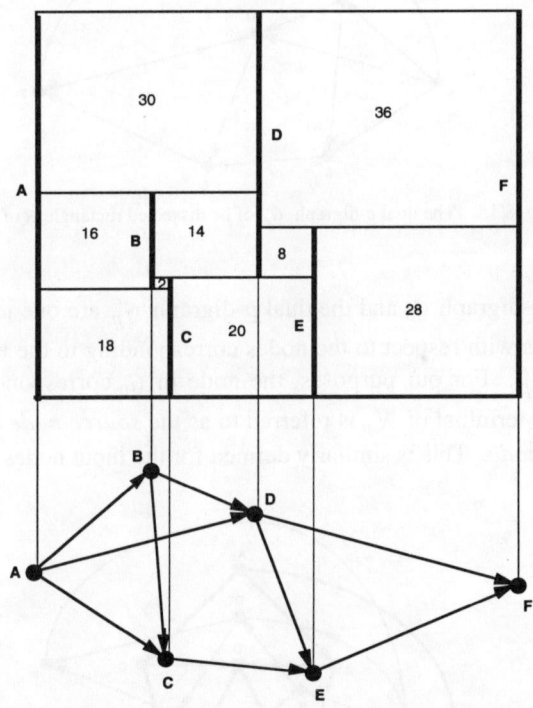

Fig. 8.13. The construction of a dual p-digraph $\mathcal{G}'_p$ associated with the dissected rectangle $\mathcal{R}$ of fig. 8.2.

DEFINITION 8.4: *One terminal-pair planar graph* or *digraph*. A graph or digraph is said to be *planar* with respect to two designated nodes if it remains planar after the insertion of an edge between the designated nodes.

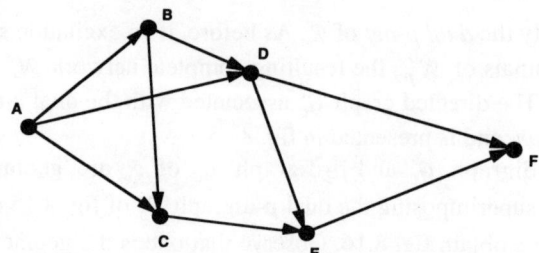

Fig. 8.14. The dual p-digraph G'_p of the dissected rectangle $\mathcal{R}$ of fig. 8.2.

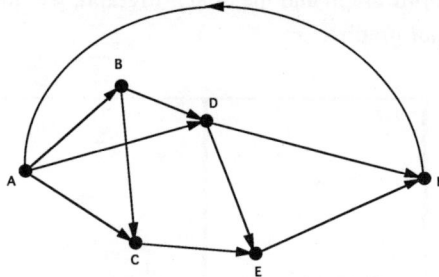

Fig. 8.15. The dual c-digraph G'_c of he dissected rectangle $\mathcal{R}$ of fig. 8.2.

Thus, the p-digraph G_p and the dual p-digraph G'_p are one terminal-pair planar directed graphs with respect to the nodes corresponding to the two input terminals of $\mathcal{N}_p$ and $\mathcal{N}'_p$. For our purposes, the node in G_p corresponding to the current injecting input terminal of $\mathcal{N}_p$ is referred to as the *source node* and the other input node the *sink node*. This is similarly defined for the input nodes of G'_p.

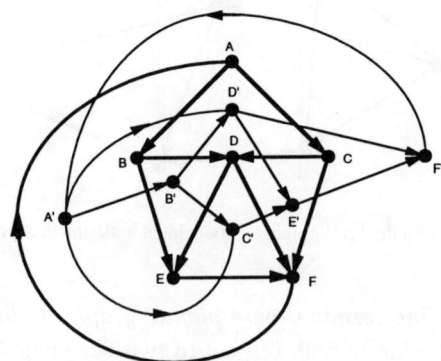

Fig. 8.16. The duality relation between G_c and G'_c.

§3 The electrical network associated with a dissected rectangle

Since the lowest order for a perfect square is 21, it is not possible to dissect any square into fewer than 21 unequal squares. However, if we allow one rectangle in the division, it is possible to dissect a square into fewer than 21 unequal squares and one rectangle. As demonstrated by DUIJVESTIJN, FEDERICO and LEEUW [1982], the lowest possible order for such a dissection is 6.

Consider the c-net $\mathcal{N}_c$ associated with the perfect rectangle of fig. 8.10(b). Set all the resistances r_i to 1 except r_1, which is rewritten as $r_1 = 1/g_1$. The indefinite-admittance matrix Y_{ind} of the modified c-net $\mathcal{N}_c$ of fig. 8.17 is found to be

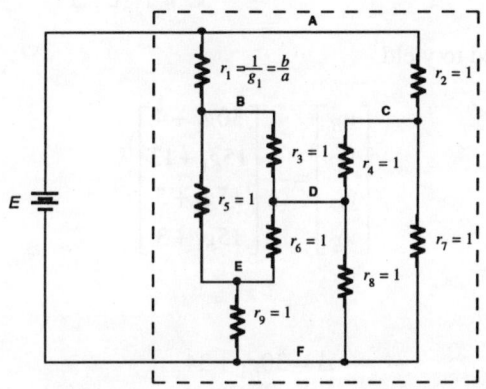

Fig. 8.17. A modified c-net $\mathcal{N}_c$ of the dissected rectangle of fig. 8.10(b).

$$Y_{\text{ind}} = \begin{bmatrix} 1+g_1 & -g_1 & -1 & 0 & 0 & 0 \\ -g_1 & 2+g_1 & 0 & -1 & -1 & 0 \\ -1 & 0 & 3 & -1 & 0 & -1 \\ 0 & -1 & -1 & 4 & -1 & -1 \\ 0 & -1 & 0 & -1 & 3 & -1 \\ 0 & 0 & -1 & -1 & -1 & 3 \end{bmatrix}. \tag{8.9}$$

Choosing node F as the potential-reference point, the nodal equation becomes

$$\begin{bmatrix} g_1+1 & -g_1 & -1 & 0 & 0 \\ -g_1 & g_1+2 & 0 & -1 & -1 \\ -1 & 0 & 3 & -1 & 0 \\ 0 & -1 & -1 & 4 & -1 \\ 0 & -1 & 0 & -1 & 3 \end{bmatrix} \begin{bmatrix} v_A \\ v_B \\ v_C \\ v_D \\ v_E \end{bmatrix} = \begin{bmatrix} i_{10} \\ 0 \\ 0 \\ 0 \\ 0 \end{bmatrix} \tag{8.10}$$

where v_A, v_B, v_C, v_D and v_E are nodal voltages from nodes A, B, C, D and E to node F, respectively. Since the nodal voltage v_A is known, (8.10) can be rewritten as

$$i_{10} = (g_1 + 1)v_A - g_1 v_B - v_C \tag{8.11}$$

$$\begin{bmatrix} g_1+2 & 0 & -1 & -1 \\ 0 & 3 & -1 & 0 \\ -1 & -1 & 4 & -1 \\ -1 & 0 & -1 & 3 \end{bmatrix} \begin{bmatrix} v_B \\ v_C \\ v_D \\ v_E \end{bmatrix} = \begin{bmatrix} g_1 \\ 1 \\ 0 \\ 0 \end{bmatrix} E \tag{8.12}$$

which can be solved to yield

$$\begin{bmatrix} v_B \\ v_C \\ v_D \\ v_E \end{bmatrix} = \frac{1}{\Delta} \begin{bmatrix} 30g_1 + 4 \\ 15g_1 + 13 \\ 15g_1 + 5 \\ 15g_1 + 3 \end{bmatrix} \tag{8.13}$$

where

$$\Delta = 30g_1 + 34 \tag{8.14}$$

is the determinant of the coefficient matrix of (8.12). From (8.11) the input current is found to be

$$i_{10} = (g_1+1)v_A - g_1 v_B - v_C = (g_1+1)E - \frac{g_1(30g_1+4) + 15g_1 + 13}{\Delta} E. \tag{8.15}$$

To simplify the solution and to clear the fractions, let $g_1 = a/b$ and $E = b\Delta$. The input current and nodal voltages are determined as

$$\begin{bmatrix} i_{10} \\ v_B \\ v_C \\ v_D \\ v_E \end{bmatrix} = \begin{bmatrix} 45a + 21b \\ 30a + 4b \\ 15a + 13b \\ 15a + 5b \\ 15a + 3b \end{bmatrix}. \tag{8.16}$$

The resistive branch-voltage vector V_r and the branch-current vector I_r are obtained as follows:

$$V_r = \begin{bmatrix} v_1 \\ v_2 \\ v_3 \\ v_4 \\ v_5 \\ v_6 \\ v_7 \\ v_8 \\ v_9 \end{bmatrix} = A'V_n = \begin{bmatrix} 1 & -1 & 0 & 0 & 0 \\ 1 & 0 & -1 & 0 & 0 \\ 0 & 1 & 0 & -1 & 0 \\ 0 & 0 & 1 & -1 & 0 \\ 0 & 1 & 0 & 0 & -1 \\ 0 & 0 & 0 & 1 & -1 \\ 0 & 0 & 1 & 0 & 0 \\ 0 & 0 & 0 & 1 & 0 \\ 0 & 0 & 0 & 0 & 1 \end{bmatrix} \begin{bmatrix} b\Delta \\ v_B \\ v_C \\ v_D \\ v_E \end{bmatrix} = \begin{bmatrix} 30b \\ 15a+21b \\ 15a-b \\ 8b \\ 15a+b \\ 2b \\ 15a+13b \\ 15a+5b \\ 15a+3b \end{bmatrix} \quad (8.17)$$

$$I_r = \begin{bmatrix} i_1 \\ i_2 \\ i_3 \\ i_4 \\ i_5 \\ i_6 \\ i_7 \\ i_8 \\ i_9 \end{bmatrix} = YV_r = \begin{bmatrix} a/b & 0 & 0 & 0 & 0 & 0 & 0 & 0 & 0 \\ 0 & 1 & 0 & 0 & 0 & 0 & 0 & 0 & 0 \\ 0 & 0 & 1 & 0 & 0 & 0 & 0 & 0 & 0 \\ 0 & 0 & 0 & 1 & 0 & 0 & 0 & 0 & 0 \\ 0 & 0 & 0 & 0 & 1 & 0 & 0 & 0 & 0 \\ 0 & 0 & 0 & 0 & 0 & 1 & 0 & 0 & 0 \\ 0 & 0 & 0 & 0 & 0 & 0 & 1 & 0 & 0 \\ 0 & 0 & 0 & 0 & 0 & 0 & 0 & 1 & 0 \\ 0 & 0 & 0 & 0 & 0 & 0 & 0 & 0 & 1 \end{bmatrix} \begin{bmatrix} 30b \\ 15a+21b \\ 15a-b \\ 8b \\ 15a+b \\ 2b \\ 15a+13b \\ 15a+5b \\ 15a+3b \end{bmatrix} = \begin{bmatrix} 30a \\ 15a+21b \\ 15a-b \\ 8b \\ 15a+b \\ 2b \\ 15a+13b \\ 15a+5b \\ 15a+3b \end{bmatrix}$$

(8.18)

where A is a basis incidence matrix, Y the branch-admittance matrix, and V_n the nodal voltage vector of the c-net $\mathcal{N}_c$.

The dissected rectangle $\mathcal{R}$ corresponding to the current distribution of (8.18) is shown in fig. 8.18. If the width of the rectangle is set equal to its height, the rectangle becomes a square, and the result is a square divided into 8 squares and one rectangle with a ratio of the horizontal to vertical side equal to g_1. To this end, let

$$30a + 15a + 21b = 30b + 15a + b + 15a + 3b \quad (8.19)$$

or

$$15a = 13b. \quad (8.20)$$

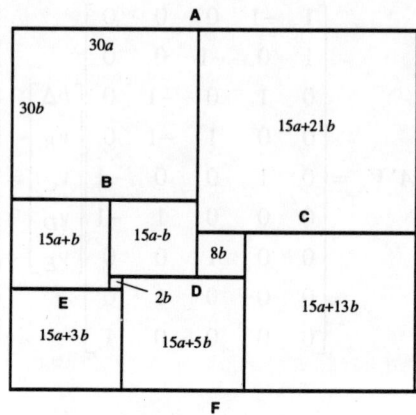

Fig. 8.18. The dissected rectangle corresponding to the current distribution of (8.18).

Thus, by letting $a = 13/15$ and $b = 1$ we obtain a dissection of a square of sides 60 into 8 unequal squares and one rectangle of sides 26 by 30. The result is shown in fig. 8.19.

Even though the lowest-order perfect square is 21, the order of the dissection of a square into unequal squares and one rectangle can be less than 21, as demonstrated in the above example. In fact, DUIJVESTIJN, FEDERICO and LEEUW [1982] showed that the lowest such order is 7, as given in fig. 8.20.

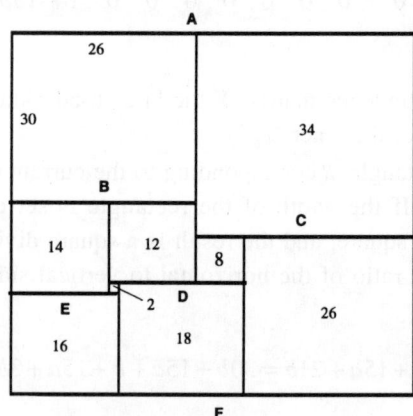

Fig. 8.19. A dissection of a square into eight squares and one rectangle.

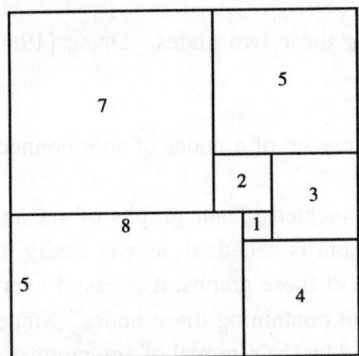

Fig. 8.20. The dissection of a square into unequal squares and one rectangle of lowest order.

§ 4. Characterization of the c-nets and c-digraphs

In the preceeding section, we demonstrated that, for each simple perfect rectangle, there is a resistive one-port network called the p-net $\mathcal{N}_p$, and its associated digraph is called the p-digraph G_p. In the present section, we shall characterize these digraphs by showing that there exists a one-to-one correspondence between squared rectangles and certain classes of planar graphs, and that each simple perfect rectangle corresponds to a 3-connected planar c-digraph G_c derived from the p-digraph G_p. These are central results of BROOKS, SMITH, STONE, and TUTTE [1940]. They further indicate that for each p-net $\mathcal{N}_p$, there is a corresponding squared rectangle. This correspondence is many to one. By deleting branches of zero currents and nodes at which all currents are zero, the *normal form* of $\mathcal{N}_p$ can be defined. Then they show that there is a one-to-one correspondence between the class of p-nets having the same normal form and the class of oriented squared rectangles.

Following the proposition of KAZARINOFF and WEITZENKAMP [1973a, b], we would like simple squared rectangles to correspond to 3-connected planar digraphs, and compound squared rectangles to 2-connected planar digraphs.

DEFINITION 8.5: *n-connectedness*. A graph or digraph is said to be *n-connected* if n is the minimum number of nodes whose removal results in a disconnected or an isolated node.

A disconnected graph is a 0-connected graph, while a connected graph with a cutpoint is 1-connected. Therefore, a nontrivial graph is connected if and only if it is 1-connected, and is 2-connected if and only if it is a block having more than one edge. A single edge, being the complete graph of order 2, is the only block not 2-connected. A graph is 2-connected if and only if, for every pair of the nodes, there is a circuit containing these two nodes. DIRAC [1960] extended this result to n-connected graphs.

THEOREM 8.1: Every set of n nodes of an n-connected graph lies on a circuit of the graph.

Examples of 3-connected planar graphs of six and eight edges are shown in fig. 8.21. Each of them is self-dual, as can easily be verified by a geometrical construction. In each of these graphs, it is easy to verify that for each set of three nodes there is a circuit containing these nodes. Since an n-node complete graph cannot be disconnected by the removal of any number of nodes less than $n - 1$, and the removal of $n - 1$ nodes results in an isolated node, the n-node complete graph is $(n - 1)$-connected.

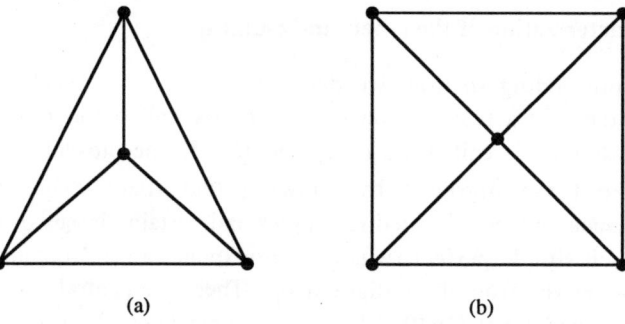

Fig. 8.21. The 3-connected planar graphs of six and eight edges.

In the preceeding section, we have demonstrated that, for each perfect rectangle $\mathcal{R}$, there corresponds a c-net $\mathcal{N}_c$ with unit resistances such that the associated c-digraph $\mathcal{G}_c$ is a planar 2-connected or 3-connected digraph. If $\mathcal{R}$ is a compound perfect rectangle, then by definition there is a perfect subrectangle $\mathcal{R}_1$ in $\mathcal{R}$. Let $\mathcal{N}_{p1}$ and $\mathcal{N}_{c1}$ be the associated p-net and c-net of $\mathcal{R}_1$, respectively. Clearly, $\mathcal{N}_{p1}$ is a subnet of the p-net $\mathcal{N}_p$ of $\mathcal{R}$, and is referred to as a *polar subnet* of $\mathcal{R}$. If $\mathcal{G}_{p1}$ is the associated digraph of $\mathcal{N}_{p1}$ or the p-digraph of $\mathcal{R}_1$, then $\mathcal{G}_{p1}$ is a subgraph of $\mathcal{G}_p$ such that $\mathcal{G}_{p1}$ and $\mathcal{G}_p - \mathcal{G}_{p1}$ meet only at the nodes corresponding to the two input terminals of $\mathcal{N}_{p1}$. Likewise, if $\mathcal{G}_{c1}$ is the associated digraph of $\mathcal{N}_{c1}$ or the c-digraph of $\mathcal{R}_1$, then $\mathcal{G}_{c1}$ is a subgraph of $\mathcal{G}_c$ such that $\mathcal{G}_{c1}$ and $\mathcal{G}_c - \mathcal{G}_{c1}$ meet only at the nodes corresponding to the two input terminals of $\mathcal{N}_{p1}$. In other words, $\mathcal{G}_c$ can be disconnected by the removal of these two nodes or $\mathcal{G}_c$ is 2-connected. For example, in fig. 8.20 if we knew a squaring of the 5 by 8 rectangle, we would obtain a compound squaring of a square. For the compound perfect square of order

25 shown in fig. 8.7, the digraph associated with the polar subnet of the unshaded subrectangle is a subgraph of the p-digraph or c-digraph of the compound perfect square such that this subgraph and its complement have only two nodes in common, and the resulting c-digraph of the square is 2-connected. On the other hand, if a perfect rectangle is simple, its c-digraph must be 3-connected. We summarize the above by stating an important result of BROOKS, SMITH, STONE, and TUTTE (1940) as:

THEOREM 8.2: Each simple perfect rectangle corresponds to a c-net with unit resistances, whose associated c-digraph is a 3-connected planar graph. Each compound perfect rectangle corresponds to a c-net with unit resistances, whose associated c-digraph is a 2-connected planar graph.

Thus, to generate simple perfect rectangles, we need only consider 3-connected planar graphs from which simple perfect rectangles are enumerated. Similarly, we use 2-connected planar graphs to generate compound perfect rectangles. We next consider the problem of finding conditions under which a rectangle can be squared. To this end, let $\mathcal{N}_c$ be the c-net with input terminals r and s associated with a squared rectangle $\mathcal{R}$. Let p and q be any two nodes in $\mathcal{N}_c$, as shown in fig. 8.22. If Y_{ind} is the indefinite-admittance matrix of $\mathcal{N}_c$, let Y_{uv} and $Y_{rp,sq}$ be the first-order and the second-order cofactors of the elements of Y_{ind}, respectively. Then the voltage across any resistor connecting the nodes p and q in $\mathcal{N}_c$ is found from (4.155) as

$$V_{pq} = I_0 \frac{Y_{rp,sq}}{Y_{uv}}. \tag{8.21}$$

Choose $I_0 = Y_{uv}$. For a squared rectangle, all resistances of the p-net are 1, so that all the elements of Y_{ind} are integers. Thus, all the branch currents I_{pq} and branch voltages V_{pq} are integers, because

$$I_{pq} = 1 \times V_{pq} = Y_{rp,sq} \tag{8.22}$$

is an integer. Thus, in general all the branch currents and voltages are rational numbers obtaining a result of DEHN [1903], which states that a rectangle can be squared if and only if its sides are commensurable. Conversely, it was first shown by SPRAGUE [1940] and, independently, by BROOKS et al. [1940] that every rectangle with commensurable sides is perfectible in an infinity of essentially different ways. Its construction will be presented in §5.

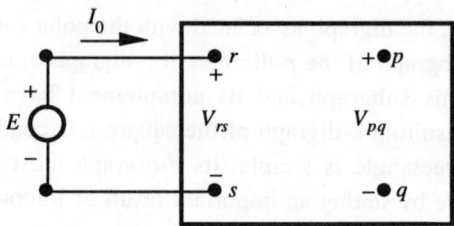

Fig. 8.22. The c-net $\mathcal{N}_c$ associated with a squared rectangle $\mathcal{R}$.

THEOREM 8.3: Every squared rectangle has commensurable sides and elements. Every rectangle with commensurable sides is perfectible in an infinity of essentially different ways.

The general problem of how to dissect a given rectangle to yield a perfect rectangle remains unsolved. YAGLOM [1968] gave an upper bound by showing that an a by b rectangle (a/b rational) can always be dissected to yield a perfect rectangle of order at most $13a^2b^2 - 11ab - 1$, which is an extremely large number. In the case of the 66 by 64 rectangle in fig. 8.10(b), the upper bound order is 231, 901, 823, but in reality this rectangle can be dissected into only 9 elements. No one has yet found any algorithm for determining or estimating the perfect squaring a rectangle of the least order. We now proceed to characterize the class of planar graphs corresponding to perfect rectangles.

DEFINITION 8.6: C_n. Let C_n ($n \geq 5$) denote the class of all n-edge one-terminal-pair planar graphs G having no parallel edges and self-loops such that each node of G that is not a terminal node is of degree at least 3 and that the insertion of an edge between the input terminals of G results in a 2-connected or 3-connected graph.

We remark that it is sufficient to consider the above class of one-terminal pair planar graphs. There is no need to include the one-terminal-pair planar digraphs, because they differ only by the edge orientations, which have no effect on the dissection of a rectangle. For $n \geq 5$, C_n is not empty, because the 4-node complete graph with two nodes designated as the input terminals is an element of C_6.

To show that if G_p is the p-digraph of a perfect rectangle $\mathcal{R}$ of order n, the associated undirected graph of G_p is in C_n. We first verify that for each node k in the associated c-digraph G_c of $\mathcal{R}$ there is a circuit containing k and the two input nodes of G_p. Such a circuit can always be found by tracing a path from k to each of the terminal nodes in G_p and including the edge in $G_c - G_p$. To confirm that in G_p each non-terminal node is of degree of at least 3 and that no parallel edges exist, we observe that if these conditions are violated then there are two edges connected either in series or in parallel. Their corresponding branches in the associated p-net

§ 4   Characterization of the c-nets and c-digraphs   539

of $\mathcal{R}$ would carry the same nonzero current, showing that two elements of $\mathcal{R}$ are equal. This is not possible for a perfect $\mathcal{R}$. Therefore, the associated undirected graph of the p-digraph of every perfect rectangle is in the class C_n. This result provides a basis of an algorithm for generating all perfect rectangles.

THEOREM 8.4: The associated undirected graph of the p-digraph of a perfect rectangle $\mathcal{R}$ of order n is in C_n. If $\mathcal{R}$ is simple, its c-digraph is 3-connected, and if it is compound its c-digraph is 2-connected.

We next show that C_n can be generated recurrently from the lower orders. However, before we do this, we first characterize the class of 3-connected graphs due to TUTTE [1961] used extensively in generating squared rectangles. His characterization is based on the special graph called the "wheel". For $n \geq 4$, the *wheel* W_n of *order n* is presented in fig. 8.23. It is the union of a circuit and a star graph. It is interesting to note that the wheel is self dual.

THEOREM 8.5: A graph having no self-loops and parallel edges is 3-connected if and only if it is either a wheel or can be obtained from a wheel by a sequence of following operations:
(a) The insertion of a new edge.
(b) The splitting of a node k of degree at least 4 into two adjacent nodes k' and k'' such that each edge originally incident at k is now incident at k' or k'' so that the degrees of k' and k'' in the resulting graph are at least 3.

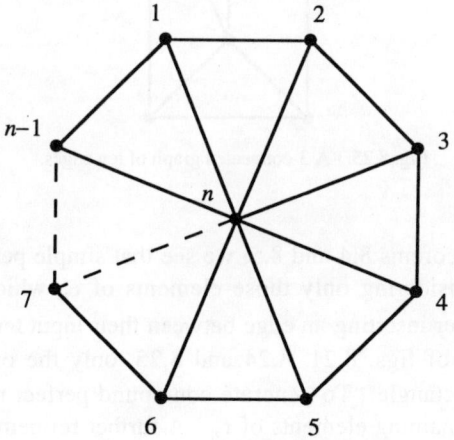

Fig. 8.23. The wheel W_n of order n.

Examples of such 3-connected graphs of six and eight edges were given earlier in fig. 8.21. Examples of 3-connected graphs of nine and ten edges are presented in figs. 8.24 and 8.25. The two graphs in fig. 8.24 are dual to each other and can be obtained from the graph of fig. 8.21(b) by the operations of inserting a new edge and/or splitting a node as described in Theorem 8.5. Likewise, the graph of fig. 8.25 can be obtained from that of fig. 8.24(a) by the insertion of a new edge.

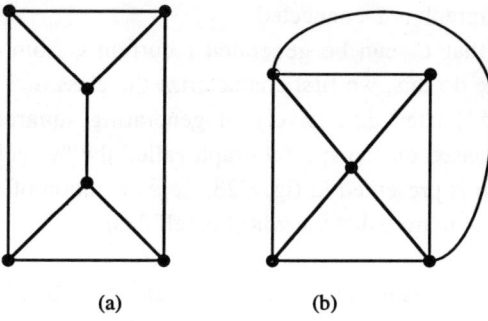

Fig. 8.24. The 3-connected graphs of nine edges.

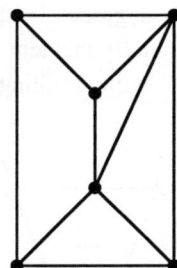

Fig. 8.25. A 3-connected graph of ten edges.

As a result of Theorems 8.4 and 8.5, we see that simple perfect rectangles can be generated by considering only those elements of C_n which will result in 3-connected graphs after inserting an edge between their input terminals. Among the 3-connected graphs of figs. 8.21, 8.24 and 8.25, only the one in fig. 8.25 can produce a perfect rectangle. To generate compound perfect rectangles, we need only consider the remaining elements of C_n. A further refinement of the elements of C_n is possible. According to KAZARINOFF and WEITZENKAMP [1973b], the elements of C_n can be broken into four parts by the following theorem.

THEOREM 8.6: Each element G of C_n satisfies exactly one of the following:

(1) The insertion of an edge between the input terminals of G results in a 3-connected graph.
(2) $G = G_1 \cup e_1$, where $G_1 \in C_{n-1}$ and e_1 is a new inserted edge incident at a terminal node p of G_1 in such a way that e_1 connects p to a terminal node of G and $G_1 \cap e_1 = p$. The other terminal node of G_1 is the second terminal of G.
(3) $G = G_2 \cup e_2$, where $G_2 \in C_n$, G and G_2 have the same input terminals, and e_2 is a new inserted edge connecting the input terminals of G_2.
(4) There exist integers m and k with $m, k \geq 5$ and elements $G_m \in C_m$ and $G_k \in C_k$ such that by inserting an edge between the input terminals of G_m the resulting graph is 3-connected, and G is the graph obtained by joining G_m and G_k at their input terminals.

We remark that, to produce all simple perfect rectangles, we consider only the one-terminal-pair planar graphs of C_n satisfying condition (1) of the theorem. To discover all compound perfect rectangles, we need to consider elements of C_n satisfying conditions (2)–(4). Since compound perfect rectangles corresponding to graphs meeting conditions (2) and (3) of the theorem consist of a square adjoined to one side of a smaller perfect rectangle, they can be found inductively and, therefore, are relatively easy. We are left with elements satisfying condition (4). KAZARINOFF and WEITZENKAMP [1973a, b] have shown that the element of C_n corresponding to a compound perfect square of order n satisfies condition (4) of the theorem. Therefore, Theorem 8.6 provides a means of generating simple and compound perfect rectangles, and have been implemented on a digital computer by KAZARINOFF and WEITZENKAMP [1973a, b].

§ 5. Perfect Subdivision of the General Rectangle

In this section, we demonstrate that there exist infinitely many totally different perfect rectangles. This leads to the conclusion that every rectangle whose sides are commensurable is perfectible. The proof of this result was given by BROOKS et al. [1940] and SPRAGUE [1940]. We shall follow the work of BROOKS et al. [1940]. We first show how to select a sequence of suitable p-nets, from which we construct a sequence of perfect squares. A subsequence of this sequence of perfect squares is a sequence of perfect squares, every two of which are totally different. This leads to the final conclusion that any rectangle whose sides are commensurable can be squared perfectly in an infinity of totally different ways.

5.1. The perfect rectangles $\mathcal{R}_n$

Consider the one-terminal pair digraph G_p of fig. 8.26, each edge of which denotes a 1-Ω resistor. The input terminals are u_0 and v_0. Let

$$\varphi_r = \frac{(2+\sqrt{3})^r - (2-\sqrt{3})^r}{2\sqrt{3}}. \tag{8.23}$$

Since $\varphi_0 = 0$, $\varphi_1 = 1$ and

$$\varphi_{r+1} - 4\varphi_r + \varphi_{r-1} = 0, \quad r \geq 1 \tag{8.24}$$

we see that φ_r is an integer.

Let I_{uv} and V_{uv} be the branch current and branch voltage of the edge (u,v). The branch currents are found to be

$$I_{u_0 u_r} = \frac{5\varphi_n + \varphi_{n-1} + 3\varphi_r - 3\varphi_{r-1}}{2}, \quad 0 < r < n$$
$$I_{u_0 u_{n+1}} = 3\varphi_n \tag{8.25a}$$

$$I_{u_r v_0} = \frac{5\varphi_n + \varphi_{n-1} - 3\varphi_r + 3\varphi_{r-1}}{2}, \quad 0 < r < n+1$$
$$I_{u_{n+1} v_0} = 2\varphi_n + \varphi_{n-1} \tag{8.25b}$$

$$I_{u_r u_{r+1}} = 3\varphi_r, \quad 0 < r < n$$
$$I_{u_n u_{n+1}} = \varphi_{n-1} - \varphi_n \tag{8.25c}$$

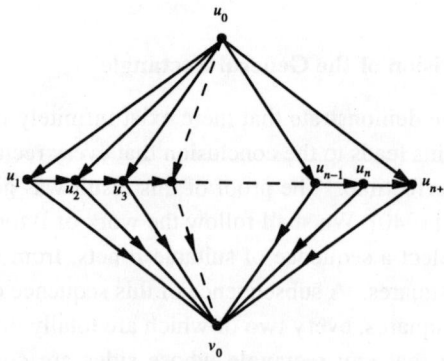

Fig. 8.26. A one-terminal pair digraph G_p.

§ 5 Perfect subdivision of the general rectangle

To verify, we consider the KCL equations at the nodes $u_1, u_2, ..., u_{n+1}$, as follows:

$$I_{u_0 u_1} = \frac{5\varphi_n + \varphi_{n-1} + 3}{2} = I_{u_1 u_2} + I_{u_1 v_0} \tag{8.26a}$$

$$I_{u_0 u_r} + I_{u_{r-1} u_r} = \frac{5\varphi_n + \varphi_{n-1} + 3\varphi_r + 3\varphi_{r-1}}{2} = I_{u_r u_{r+1}} + I_{u_r v_0}, \quad 1 < r < n \tag{8.26b}$$

$$I_{u_{n-1} u_n} = 3\varphi_{n-1} = I_{u_n u_{n+1}} + I_{u_n v_0} \tag{8.26c}$$

$$I_{u_0 u_{n+1}} + I_{u_n u_{n+1}} = 2\varphi_n + \varphi_{n-1} = I_{u_{n+1} v_0}. \tag{8.26d}$$

Likewise, since each branch is a 1-Ω resistor, $I_{uv} = V_{uv}$ and the KVL equations are satisfied, as follows:

$$V_{u_0 u_r} + V_{u_r v_0} = 5\varphi_n + \varphi_{n-1}, \quad r \neq n \tag{8.27a}$$

$$V_{u_0 u_{n+1}} - V_{u_n u_{n+1}} + V_{u_n v_0} = 5\varphi_n + \varphi_{n-1} \tag{8.27b}$$

$$V_{u_r u_{r+1}} = V_{u_r v_0} - V_{u_{r+1} v_0} = I_{u_r v_0} - I_{u_{r+1} v_0} = \frac{3\varphi_{r+1} - 6\varphi_r + 3\varphi_{r-1}}{2}$$
$$= 3\varphi_r, \quad 0 < r < n+1. \tag{8.27c}$$

The second line of (8.27c) follows from (8.24). The current I_0 at the input terminal is computed as

$$I_0 = I_{u_0 u_{n+1}} + \sum_{r=1}^{n-1} I_{u_0 u_r} = \frac{(5n+1)\varphi_n + (n+2)\varphi_{n-1}}{2}. \tag{8.28}$$

The input voltage $V_{u_0 v_0}$ from node u_0 to node v_0 is given in (8.27a) as

$$V_{u_0 v_0} = 5\varphi_n + \varphi_{n-1}. \tag{8.29}$$

For $n > 2$, it is straightforward to verify that

$$0 < I_{u_1 u_2} < I_{u_2 u_3} < \cdots < I_{u_{n-2} u_{n-1}} < -I_{u_n u_{n+1}} < I_{u_{n-1} u_n} < I_{u_n v_0} < I_{u_{n+1} v_0} < I_{u_{n-1} v_0},$$
$$< I_{u_{n-2} v_0} < \cdots < I_{u_1 v_0} < I_{u_0 u_1} < I_{u_0 u_2} < \cdots < I_{u_0 u_{n-1}} < I_{u_0 u_{n+1}} \tag{8.30}$$

Therefore, all the branch currents are different. The rectangle $\mathcal{R}_n$ associated with the p-digraph $\mathcal{G}_p$ of fig. 8.26 has the horizontal side p_n equal to I_0 and the vertical side q_n equal to $V_{u_0 v_0}$:

$$p_n = \frac{(5n+1)\varphi_n + (n+2)\varphi_{n-1}}{2} \tag{8.31}$$

$$q_n = 5\varphi_n + \varphi_{n-1} \tag{8.32}$$

showing that

$$q_n \to \infty \text{ and } \frac{p_n}{q_n} \to \infty \tag{8.33}$$

as $n \to \infty$. Therefore, this rectangle $\mathcal{R}_n$ can be dissected perfectly with the elements equal to the currents shown in (8.30). In addition, since from (8.31) and (8.32) we have

$$(n+2)q_n - 2p_n = 9\varphi_n \tag{8.34a}$$

$$(5n+1)q_n - 10p_n = -9\varphi_{n-1} \tag{8.34b}$$

and since we can prove by induction, using (8.24), that φ_n and φ_{n-1} do not contain any common factor other than 1, it follows that the common factors of p_n and q_n are divisible by 9. This result will be used later.

In the one-terminal pair digraph $\mathcal{G}_p$ of fig. 8.26, if instead we insert the edge (u_0, u_n) and remove the edge (u_0, u_r), where $1 < r \le n/2$, and go through the above analysis, the resulting rectangle $\mathcal{R}_{nr}$ is perfect. The perfect rectangle $\mathcal{R}_{n2}$ is essentially the same as $\mathcal{R}_n$.

Our conclusion is that for $n > 2$, the dissected rectangle $\mathcal{R}_n$ associated with the one-terminal pair digraph $\mathcal{G}_p$ of fig. 8.26 is perfect, and its sides are divisible by 9.

5.2. The perfect square S_n

In this section, we shall select a sequence of suitable one-terminal pair digraphs $\mathcal{G}_p$ and their associated perfect rectangles $\mathcal{R}_n$. For each $\mathcal{R}_n$ we show how to form a perfect square S_n.

Consider the two one-terminal pair digraphs $\mathcal{G}_{pa}$ and $\mathcal{G}_{pb}$ of fig. 8.27, where nodes 1 and 2 are the one-terminal pair. All the edges have conductance 1 except three which have g, as shown in the figure. The associated c-nets $\mathcal{N}_{ca}$ and $\mathcal{N}_{cb}$ of $\mathcal{G}_{pa}$ and $\mathcal{G}_{pb}$ are presented in fig. 8.28.

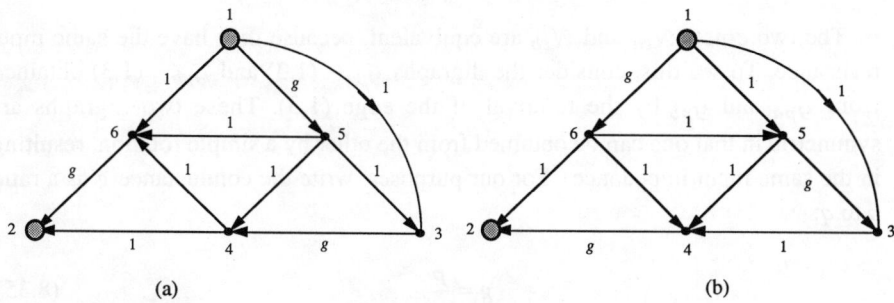

Fig. 8.27. Two equivalent one-terminal pair digraphs $\mathcal{G}_{pa}$ and $\mathcal{G}_{pb}$.

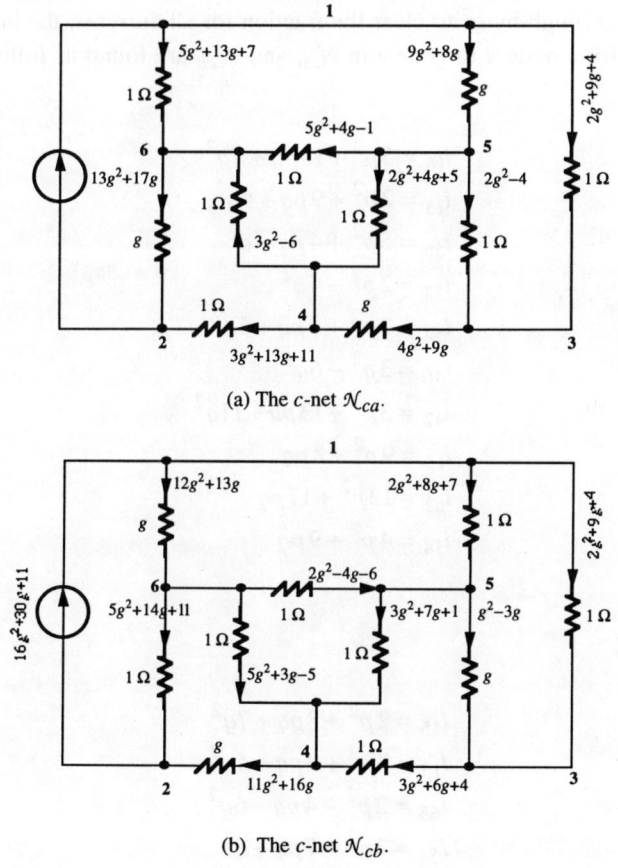

(a) The c-net $\mathcal{N}_{ca}$.

(b) The c-net $\mathcal{N}_{cb}$.

Fig. 8.28. The c-nets $\mathcal{N}_{ca}$ and $\mathcal{N}_{cb}$ associated with $\mathcal{G}_{pa}$ and $\mathcal{G}_{pb}$.

The two c-nets $\mathcal{N}_{ca}$ and $\mathcal{N}_{cb}$ are equivalent, because they have the same input resistance. To see this, consider the digraphs $\mathcal{G}_{pa} - (1,3)$ and $\mathcal{G}_{pb} - (1,3)$ obtained from $\mathcal{G}_{pa}$ and $\mathcal{G}_{pb}$ by the removal of the edge $(1,3)$. These two digraphs are symmetric in that one can be obtained from the other by a simple rotation, resulting in the same input impedance. For our purposes, write the conductance g as a ratio p to q:

$$g = \frac{p}{q}. \tag{8.35}$$

The branch currents can be determined and they are shown in fig. 8.28. It is straightforward to verify that they satisfy KCL, KVL and Ohm's law equations. By multiplying through by q^2 to clear the fraction for all currents, the branch currents i_{uv} and I_{uv} from node u to node v in $\mathcal{N}_{ca}$ and $\mathcal{N}_{cb}$ are found as follows: For $\mathcal{N}_{ca}$ we have

$$i_{16} = 5p^2 + 13pq + 7q^2 \tag{8.36a}$$
$$i_{13} = 2p^2 + 9pq + 4q^2 \tag{8.36b}$$
$$i_{56} = 5p^2 + 4pq - q^2 \tag{8.36c}$$
$$i_{53} = 2p^2 - 4q^2 \tag{8.36d}$$
$$i_{54} = 2p^2 + 4pq + 5q^2 \tag{8.36e}$$
$$i_{46} = 3p^2 - 6q^2 \tag{8.36f}$$
$$i_{42} = 3p^2 + 13pq + 11q^2 \tag{8.36g}$$
$$i_{15} = 9p^2 + 8pq \tag{8.37a}$$
$$i_{62} = 13p^2 + 17pq \tag{8.37b}$$
$$i_{34} = 4p^2 + 9pq \tag{8.37c}$$

and for $\mathcal{N}_{cb}$

$$I_{15} = 2p^2 + 8pq + 7q^2 \tag{8.38a}$$
$$I_{13} = 2p^2 + 9pq + 4q^2 \tag{8.38b}$$
$$I_{65} = 2p^2 - 4pq - 6q^2 \tag{8.38c}$$
$$I_{54} = 3p^2 + 7pq + q^2 \tag{8.38d}$$
$$I_{64} = 5p^2 + 3pq - 5q^2 \tag{8.38e}$$

$$I_{62} = 5p^2 + 14pq + 11q^2 \tag{8.38f}$$

$$I_{34} = 3p^2 + 6pq + 4q^2 \tag{8.38g}$$

$$I_{16} = 12p^2 + 13pq \tag{8.39a}$$

$$I_{53} = p^2 - 3pq \tag{8.39b}$$

$$I_{42} = 11p^2 + 16pq \tag{8.39c}$$

The input current for both c-nets is given by

$$i_{21} = I_{21} = 16p^2 + 30pq + 11q^2. \tag{8.40}$$

The terminal voltages v_{12} and V_{12} from node 1 to node 2 are obtained from $\mathcal{N}_{ca}$ and $\mathcal{N}_{cb}$ as

$$v_{12} = i_{16} \times 1 + i_{62} \frac{q}{p} = 5p^2 + 26pq + 24q^2 \tag{8.41a}$$

$$V_{12} = I_{16} \frac{q}{p} + I_{62} \times 1 = 5p^2 + 26pq + 24q^2 \tag{8.41b}$$

showing that

$$v_{12} = V_{12} = I_{16} \frac{q}{p} + I_{62} \times 1 = 5p^2 + 26pq + 24q^2 \tag{8.42}$$

as expected. The dissected rectangles $\mathcal{R}_a$ and $\mathcal{R}_b$ corresponding to the c-nets $\mathcal{N}_{ca}$ and $\mathcal{N}_{cb}$ are shown in fig. 8.29. Observe that there are three subrectangles in each of $\mathcal{R}_a$ and $\mathcal{R}_b$. The sides of these subrectangles are multiples of those of the perfect rectangle $\mathcal{R}_n$ obtained in the preceding section by setting $p = p_n$ and $q = q_n$, as follows: For $\mathcal{R}_a$ we have

$$\left(9p^2 + 8pq\right) \text{ by } \left(9pq + 8q^2\right) \Rightarrow (9p + 8q) \text{ multiplying } p \text{ by } q$$

$$\left(13p^2 + 17pq\right) \text{ by } \left(13pq + 17q^2\right) \Rightarrow (13p + 17q) \text{ multiplying } p \text{ by } q$$

$$\left(4p^2 + 9pq\right) \text{ by } \left(4pq + 9q^2\right) \Rightarrow (4p + 9q) \text{ multiplying } p \text{ by } q$$

and for $\mathcal{N}_b$ we have

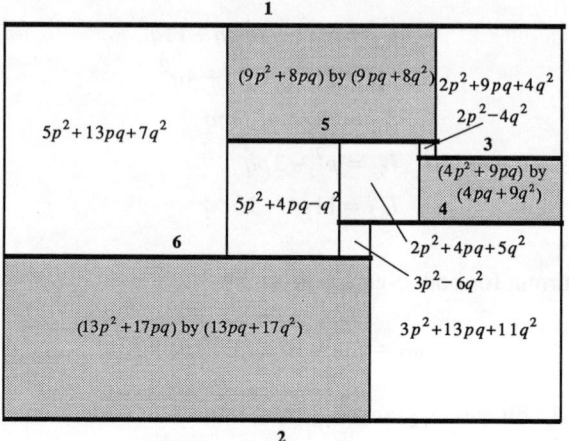

(a) The dissected rectangle $\mathcal{R}_a$.

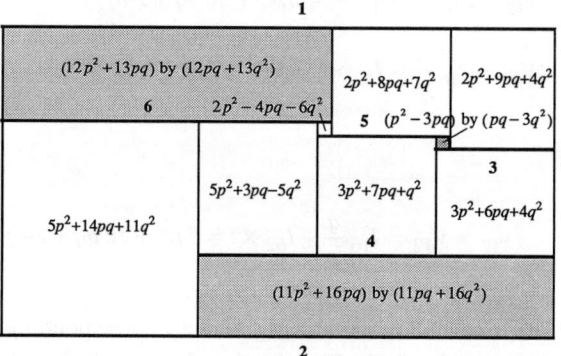

(b) The dissected rectangle $\mathcal{R}_b$.

Fig. 8.29. The dissected rectangles $\mathcal{R}_a$ and $\mathcal{R}_b$ associated with the c-nets $\mathcal{N}_{ca}$ and $\mathcal{N}_{cb}$.

$$\left(12p^2 + 13pq\right) \text{ by } \left(12pq + 13q^2\right) \Rightarrow (12p + 13q) \text{ multiplying } p \text{ by } q$$
$$\left(p^2 - 3pq\right) \text{ by } \left(pq - 3q^2\right) \Rightarrow (p - 3q) \text{ multiplying } p \text{ by } q$$
$$\left(11p^2 + 16pq\right) \text{ by } \left(11pq + 16q^2\right) \Rightarrow (11p + 16q) \text{ multiplying } p \text{ by } q.$$

By multiplying the elements of $\mathcal{R}_n$ by the above multiples, the six subrectangles can be perfectly dissected. Therefore, $\mathcal{R}_a$ and $\mathcal{R}_b$ are compound perfect rectangles.

Our next step is to construct a perfect square. To this end, let us place the two perfect rectangles $\mathcal{R}_a$ and $\mathcal{R}_b$ (shaded area) as shown in fig. 8.30 with the common

§ 5   Perfect subdivision of the general rectangle   549

area corresponding to their corner squares of side $2p^2 + 9pq + q^2$, and then add two corner squares to complete the desired square S_n, the side of which is given by

$$19p^2 + 47pq + 31q^2. \qquad (8.43)$$

By virtue of (8.33), we can choose a sufficiently large n so that

$$p_n > 180 q_n \qquad (8.44)$$

In the following, we show that, under this condition, S_n is a perfect square.

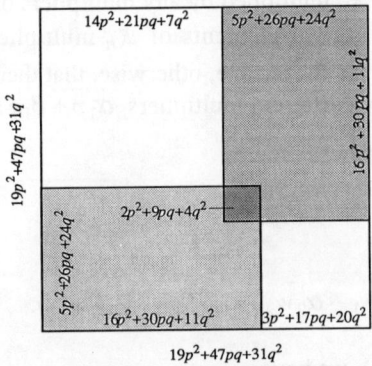

Fig. 8.30. A compound perfect square S_n constructed from $\mathcal{R}_a$ and $\mathcal{R}_b$.

With the exception of the elements of the six subrectangles in $\mathcal{R}_a$ and $\mathcal{R}_b$, all other elements in S_n can be arranged in strictly decreasing order, so no two of them are equal, as follows:

$$\begin{aligned}
14p^2 + 21pq + 7q^2 &> 5p^2 + 14pq + 11q^2 > 5p^2 + 13pq + 7q^2 > 5p^2 + 4pq - q^2 \\
&> 5p^2 + 3pq - 5q^2 > 3p^2 + 17pq + 20q^2 > 3p^2 + 13pq + 11q^2 \\
&> 3p^2 + 7pq + q^2 > 3p^2 + 6pq + 4q^2 > 3p^2 - 6q^2 \\
&> 2p^2 + 9pq + 4q^2 > 2p^2 + 8pq + 7q^2 \\
&> 2p^2 + 4pq + 5q^2 > 2p^2 - 4q^2 > 2p^2 - 4pq - 6q^2. \qquad (8.45)
\end{aligned}$$

Observe that the least element of (8.45) is greater than $(13p + 17q)q$ or

$$2p^2 - 4pq - 6q^2 > (13p + 17q)q. \qquad (8.46)$$

This shows that every element in (8.45) is greater than any element in any of the six subrectangles in $\mathcal{R}_a$ and $\mathcal{R}_b$. Recall that the elements of $\mathcal{R}_a$ and $\mathcal{R}_b$ are multiples of the elements of $\mathcal{R}_n$, the multipliers of which, arranged in strictly decreasing order and under the condition (8.44), are given by

$$13p+17q > 12p+13q > 11p+16q > 9p+8q > 4p+9q > p-3q. \tag{8.47}$$

Our conclusion is that the elements of S_n in (8.45) are all different, and none of these equals an element of any of the six perfect subrectangles in $\mathcal{R}_a$ and $\mathcal{R}_b$. The latter is equivalent to saying that no element in (8.45) equals an element of the simple perfect rectangle $\mathcal{R}_n$ multiplied by any multipliers of (8.47).

We next prove that no two elements of $\mathcal{R}_n$ multiplied by any two different multipliers of (8.47) are equal. Assume, otherwise, that there are two such elements w_1 and w_2 of $\mathcal{R}_n$ and two different multipliers $\alpha_1 p + \beta_1 q$ and $\alpha_2 p + \beta_2 q$ in (8.47) such that

$$(\alpha_1 p + \beta_1 q) w_1 = (\alpha_2 p + \beta_2 q) w_2 \tag{8.48}$$

or

$$(\alpha_1 w_1 - \alpha_2 w_2) p = (\beta_2 w_2 - \beta_1 w_1) q. \tag{8.49}$$

Now, if $\beta_2 w_2 - \beta_1 w_1 = 0$, we have

$$\alpha_1 w_1 - \alpha_2 w_2 = 0 \tag{8.50}$$

since $p \neq 0$. Substituting this in (8.48) yields

$$\alpha_1 \beta_2 - \alpha_2 \beta_1 = 0 \tag{8.51}$$

By inspection of the coefficients in (8.45), we see that

$$\alpha_1 \beta_2 - \alpha_2 \beta_1 \neq 0. \tag{8.52}$$

Thus, $\beta_2 w_2 - \beta_1 w_1 \neq 0$. Since $w_1 > 0$ and $q > w_2$ it follows from (8.47) that we have the inequality

$$0 < |\beta_2 w_2 - \beta_1 w_1| < 20q. \tag{8.53}$$

Since from (8.34), p and q are divisible by 9, and we can write $p = 9\hat{p}$ and $q = 9\hat{q}$, (8.49) becomes

$$9\hat{p}\frac{\alpha_1 w_1 - \alpha_2 w_2}{\hat{q}} = 9(\beta_2 w_2 - \beta_1 w_1) < 180q \qquad (8.54)$$

showing that

$$p < 180q. \qquad (8.55)$$

This contradicts to our assumption that we have chosen a sufficiently large n so that (8.44) holds. Thus, no two elements in the six subrectangles of $\mathcal{R}_a$ and $\mathcal{R}_b$ are equal. We now summarize the above result by stating the following theorem.

THEOREM 8.7: For sufficiently large n, the squared square S_n is perfect.

The perfect square S_n is of course not simple. It contains six perfect subrectangles corresponding to the six resistors of resistance $1/g$ in the two c-nets $\mathcal{N}_{ca}$ and $\mathcal{N}_{cb}$ of fig. 8.28.

5.3. Sequence of perfect squares

In this section, we demonstrate how to generate a sequence of perfect squares by showing that, for any large enough n, the perfect squares S_n and S_N constructed in the preceding section are totally different for all large N.

For each choice of n, let $\mathcal{R}_n$ be the simple perfect rectangle constructed in 5.1 with sides p by q, and let S_n be the corresponding perfect square constructed in 5.2 by assigning resistance q/p to six resistors. As usual, all other resistors have resistance 1. Likewise, for each choice N, let $\mathcal{R}_N$, P, Q, S_N and Q/P be the corresponding quantities. Now we can bring S_n and S_N to the same size by multiplying the elements of S_n, as given by (8.45) and (8.47), by $19P^2 + 47PQ + 31Q^2$, the side of S_N; and those of S_N by $19p^2 + 47pq + 31q^2$, the side of S_n, as given by (8.43). Let the resulting perfect squares be denoted by S'_n and S'_N. If w is an element of $\mathcal{R}_n$, then a typical element E in any of the six subrectangles of S'_N is given by

$$E = (\alpha P + \beta Q)(19p^2 + 47pq + 31q^2)w \qquad (8.56)$$

where α and β are coefficients in (8.47). For large n and N, p and P become dominant and (8.56) is bounded by

$$E < 360Pp^2 w \qquad (8.57)$$

because $|\alpha|, |\beta| \leq 17$. On the other hand, from (8.45) and (8.47), each element e of S'_n is at least as large as $P^2 px$ or

$$e > P^2 px \tag{8.58}$$

where $x = e$ if e is an element of any of the subrectangles of S'_n and $x = 1$ otherwise. Thus, by choosing large P in comparison with p, each element e of S'_n is larger than each element E in any of the subrectangles of S'_N.

We next show that each element of S'_N, other than those contained in any of the subrectangles, is greater than every element of S'_n contained in any of its subrectangles. To see this, we refer to (8.45) and notice that every such element of S'_N is as large as $P^2 p^2$ times some nonzero constant, whereas an element of any of the subrectangles of S'_n is less than $360 P^2 p$ times some nonzero constant. Thus, for sufficiently large n and N, the result follows.

Finally, we demonstrate that no element of S'_N, other than those contained in any of the subrectangles, can equal any element of S'_n not contained in any of its subrectangles. Suppose otherwise we have two such elements

$$\left(AP^2 + BPQ + CQ^2\right)\left(19p^2 + 47pq + 31q^2\right)$$
$$= \left(ap^2 + bpq + cq^2\right)\left(19P^2 + 47PQ + 31Q^2\right) \tag{8.59}$$

where A, B, C, a, b and c are integers numerically less than 22. Equation (8.59) can be rewritten as

$$P^2\left[(19A - 19a)p^2 + (47A - 19b)pq + (31A - 19c)q^2\right]$$
$$= PQ\left[(47a - 19B)p^2 + (47b - 47B)pq + (47c - 31B)q^2\right]$$
$$+ Q^2\left[(31a - 19C)p^2 + (31b - 47C)pq + (31c - 31C)q^2\right]. \tag{8.60}$$

Now, observe that $47A - 19b \neq 0$; for otherwise 19 would be divisible by A, where from (8.45) we have $0 < A < 19$. Hence, the left-hand side of (8.60) is at least as large as $P^2 pq$ times some nonzero constant. In fact, if $A \neq a$, it is as large as $P^2 p^2$. The right-hand side is at most PQp^2 times a constant. Therefore, by virtue of (8.33) it is always possible to choose a sufficiently large N so that P dominates both p and Q, violating (8.60). As a result, none of the elements of S'_N, which are not contained in any of the subrectangles, can equal any element of S'_n. This leads to the following main result of this section.

THEOREM 8.8: There is a sequence of perfect squares, every two of which are totally different.

This result can now be used to prove that any rectangle whose sides are commensurable can be squared perfectly in an infinity of totally different ways.

Without loss of generality, we may assume that the sides of a given rectangle $\mathcal{R}$ are integers h and k. By drawing lines parallel to its sides, the rectangle $\mathcal{R}$ can be divided into hk squares of side 1. Appealing to Theorem 8.8, let $\{\mathcal{T}_m\}$ be a sequence of perfect squares, every two of which are totally different. These perfect squares $\mathcal{T}_m$ can all be scaled so that they are all of side 1. For any positive integer n, replace the ith of the hk unit squares of $\mathcal{R}$ by $\mathcal{T}_{(hkn+i)}$. This results in a perfect subdivision of $\mathcal{R}$ for each n. For any two values of n, these perfect subdivisions of $\mathcal{R}$ are totally different. Thus, a rectangle can be squared perfectly if it can be squared at all. We summarize the above result by stating the following main theorem of this section.

THEOREM 8.9: Any rectangle whose sides are commensurable can be squared perfectly in an infinity of totally different ways.

§ 6. Extension to Perfect Rectangular Parallelepiped

An immediate and natural generalization of the problem of dissecting a rectangle into a finite number of different squares is to dissect a rectangular parallelepiped into a finite number of different cubes called a *perfect cubed parallelepiped*. BROOKS et al. [1940] demonstrated that there is no perfect cubed parallelepiped. Their reasoning is as follows:

First observe that in any perfect rectangle, the smallest element is not on the boundary of the rectangle. Assume otherwise that the smallest square s_1 is either at the corner or on the boundary. If s_1 is at the corner, one of the two adjacent squares of s_1 must be smaller than s_1, a contradiction. Likewise, if s_1 is on the boundary, but not at the corner, then at least one of the three adjacent squares of s_1 is smaller than s_1, again a contradiction.

Suppose that we have a perfect cubed parallelepiped $\mathcal{P}$. Let $\mathcal{R}_1$ be its base. The elements of $\mathcal{P}$ resting on $\mathcal{R}_1$ induce a dissection of $\mathcal{R}_1$ into a perfect rectangle. Without loss of generality, we can assume that more than one cube rests on $\mathcal{R}_1$. For, otherwise, we can consider the upper face of $\mathcal{P}$. Let s_1 be the smallest element of $\mathcal{R}_1$. Let c_1 be the corresponding cube of $\mathcal{P}$. Since s_1 is surrounded by larger squares on all four sides, c_1 is surrounded by larger, and therefore higher, cubes on all four faces. Hence, the upper face of c_1 is divided into a perfect rectangle $\mathcal{R}_2$ by the elements of $\mathcal{P}$ resting on it. Let s_2 be the smallest element of $\mathcal{R}_2$, which cannot be on the boundary of $\mathcal{R}_2$. A similar argument would lead to the perfect dissection of the upper face of c_2 by the elements of $\mathcal{P}$ resting on it; and so on. In this way, we generate an infinite sequence of elements c_n of $\mathcal{P}$ such that $c_{n+1} < c_n$. This is impossible because $\mathcal{P}$ is supposed to be divided into a *finite* number of cubes. Therefore, there is no perfect cubed parallelepiped.

§ 7. VLSI Layout

In Very-Large-Scale-Integrated (VLSI) circuit design, the first step is to obtain a partition of a rectangular floor into smaller ones known as the *floorplan*. In designing such a floorplan, we take into consideration the required or admissible mutual positions of the building blocks. In many cases, not all the geometrical dimensions of the building blocks are predetermined, and some can be assigned to other values to yield the so-called optimum layout. The problem is equivalent to dissecting a rectangle into subrectangles of fixed areas, so that the unutilized area considered to be *wasted* is minimized.

Consider the floorplanning problem of a microprocessor compatible 8-bit A/D converter with an 8-channel multiplexer. The circuit block diagram of this device is shown in fig. 8.31, and its corresponding floorplan is presented in fig. 8.32, where MUL stands for multiplexer module, ALD for address latch and decoder module, COM for comparator module, ST for switch tree module, CR for capacitor/resistor array module, CT for control and timing module, SAR for successive approximation registers module, TSO for three-state output module, CHA for channel and BUS for common bus module.

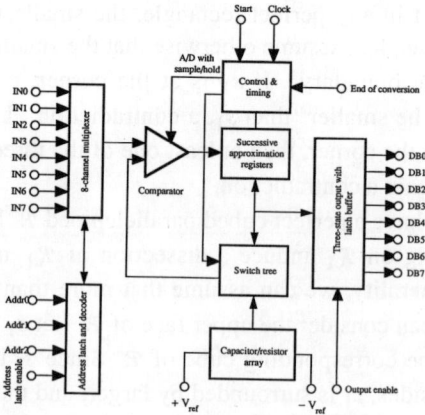

Fig. 8.31. The circuit block diagram of a microprocessor compatible 8-bit A/D converter.

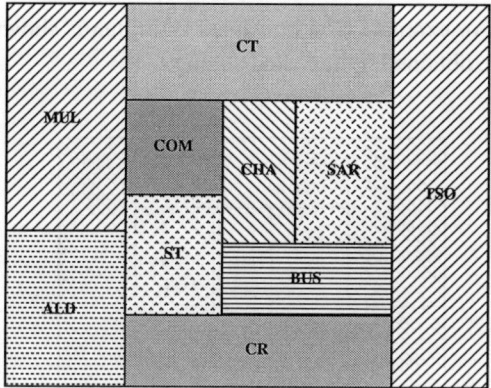

Fig. 8.32. The floorplan for the 8-bit A/D converter of fig. 8.31.

In general, there are three major steps involved in the physical layout of a VLSI circuit: Floorplan, placement, and routing. A general floorplan can be stated as follows: Given a set of modules with arbitrary geometries, estimated area, delay and power consumption, find geometries, locations, aspect ratios for all modules to meet the following constraints:

1. total estimated area and aspect ratio of the chip
2. aspect ratios of the modules
3. total estimated power dissipation and possible delay

where the *aspect ratio* of a rectangular module is defined as the ratio of its height to width. Once an appropriate floorplan is found, the next step is to place all the modules in proper locations, so that a weighted sum of total area, net length, delay, and power consumption is minimized. This is known as the *placement*. The floorplanning task is different from that of the placement in that the former can select the hierarchical partitioning of the design, the module dimension and input/out (I/O) locations, whereas the latter cannot, even though many of the floorplanning algorithms are derived from the placement algorithms. Finally, physical connections among the modules and I/O devices are accomplished through routing. It is known that the general floorplanning and placement problems are both NP-complete, meaning that no polynomial algorithm can be found to solve them.

A floorplan is said to be *slicing* if it can be obtained recursively by slicing or cutting a rectangle or subrectangle into two parts by either a horizontal or vertical line until the desired floorplan is achieved. Otherwise, it is called a *nonsliced* floorplan. Figure 8.32 is an example of a sliced floorplan, because it can be obtained by repeated slicing a rectangle or subrectangle, as indicated in fig. 8.33 by different shades. An example of a nonsliced floorplan is shown in fig. 8.34. This floorplan cannot be obtained by repeated slicing.

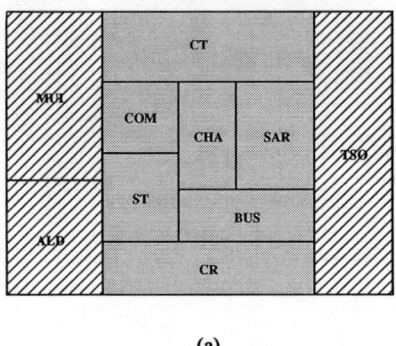

(a)

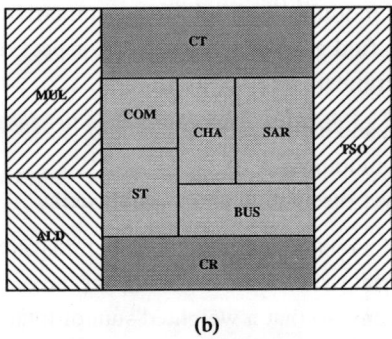

(b)

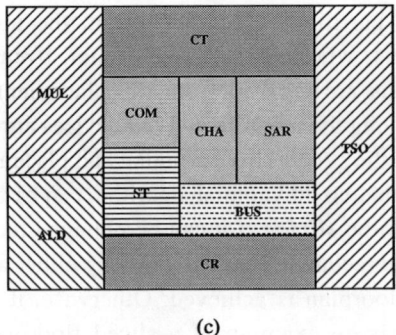

(c)

Fig. 8.33. A floorplan that can be obtained by repeated slicing.

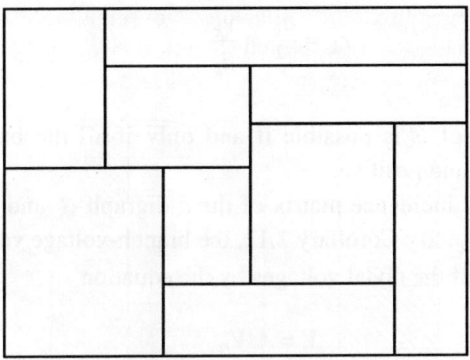

Fig. 8.34. An example of a nonsliced floorplan.

In this section, we study general floorplans and present techniques to minimize the area of a general layout obtained from an arbitrary floorplan, because area minimization is typical in VLSI design, where the cost of a chip is dominated by its area.

7.1. *The zero wasted area floorplan with continuous aspect ratios*

A floorplan is a dissection of a rectangle $\mathcal{R}$ into b subrectangles $\mathcal{R}_k$ of different sizes. Each subrectangle $\mathcal{R}_k$ is occupied in the actual layout by some physical blocks, which has a width w_k and a height h_k. Assume that the area a_k of $\mathcal{R}_k$ is fixed and its width w_k and height h_k are allowed to change continuously. Then we have

$$a_k = w_k h_k, \quad k = 1, 2, \ldots, b. \tag{8.61}$$

Our problem is to determine a width w and a height h of the dissected rectangle $\mathcal{R}$ meeting the area requirements of its elements (subrectangles).

As before, let $\mathcal{G}_p, \mathcal{G}_c, \mathcal{N}_p$ and $\mathcal{N}_c$ be the p-digraph, c-digraph, p-net and c-net of $\mathcal{R}$, respectively. In the p-net, let r_k, i_k and v_k be the kth branch resistance, current and voltage. Since $i_k = w_k$ and $v_k = h_k$ the power p_k consumed in resistor r_k corresponds to the area a_k of the subrectangle $\mathcal{R}_k$ or

$$p_k = a_k = w_k h_k = v_k i_k. \tag{8.62}$$

This is equivalent to solving the nonlinear network problem in $\mathcal{N}_c$, the branch resistances of which are either current-controlled or voltage-controlled:

$$r_k = \frac{a_k}{i_k^2} = \frac{v_k^2}{a_k}.\qquad(8.63)$$

Thus, a dissection of $\mathcal{R}$ is possible if and only if all the branch currents i_k or voltages v_k are real and positive.

Let A be a basis incidence matrix of the c-digraph $\mathcal{G}_c$ and let V_n be the nodal voltage vector of $\mathcal{N}_c$. By Corollary 2.12, the branch-voltage vector V of $\mathcal{N}_c$ can be expressed in terms of the nodal voltages by the equation

$$V = A'V_n. \qquad(8.64)$$

Write

$$I = I_p + J \qquad(8.65a)$$
$$V = V_p + E \qquad(8.65b)$$

where I_p and V_p denote the resistive branch currents i_k and branch voltages v_k, and J and E the independent current sources and voltage sources in $\mathcal{N}_c$. In the present situation, we have only one source, which can either be a current source or a voltage source. Therefore, either J or E will be zero, depending on the choice of the excitation source for $\mathcal{N}_c$. For nodal analysis, it is more convenient to apply a current source. Substituting (8.65a) for I in the Kirchhoff current law equation

$$AI = 0 \qquad(8.66)$$

we obtain

$$AI_p = -AJ. \qquad(8.67)$$

Since the branches of $\mathcal{N}_p$ are either voltage-controlled or current-controlled nonlinear resistors, the branch-current vector I_p and the branch-voltage vector V_p are related by the vector equation

$$I_p = g(V_p). \qquad(8.68)$$

Finally, combining (8.64), (8.65b), (8.67) and (8.68) yields

$$Ag(e) = -AJ \qquad(8.69)$$

where

§ 7   VLSI layout   559

$$e = A'V_n - E.$$ (8.70)

For an n-node c-net $\mathcal{N}_c$, (8.69) represents a system of $n - 1$ nonlinear nodal equations in terms of the $n - 1$ node-to-datum nodal voltages v_{nk}. If we denote the voltage vector e by x, this equation can be written in the general form

$$f(x) = 0$$ (8.71)

where

$$f(x) = Ag(x) + AJ$$ (8.72)

which can be solved using any of the standard numerical methods such as the Newton–Raphson algorithm. We refer the reader to CHUA and LIN [1975] for an excellent exposition on this subject.

We illustrate the above by the following examples from WANG and CHEN [1993].

Example 8.1: Consider the simple nonslicing floorplan $\mathcal{R}$ of fig. 8.35. Assume that the areas a_k of its subrectangles $\mathcal{R}_k$ are specified as follows:

$$a_1 = 40, \quad a_2 = 14, \quad a_3 = 8, \quad a_4 = 20, \quad a_5 = 18$$ (8.73)

We wish to find a zero wasted area floorplan.

From the floorplan $\mathcal{R}$, we first construct its p-digraph $\mathcal{G}_p$, c-digraph $\mathcal{G}_c$, p-net $\mathcal{N}_p$ and c-net $\mathcal{N}_c$. They are presented in figs. 8.36, 8.37, 8.38 and 8.39.

In the c-digraph $\mathcal{G}_c$ of fig. 8.37, choose n_4 as the reference node. Its basis incidence matrix A and the KCL equation are found to be

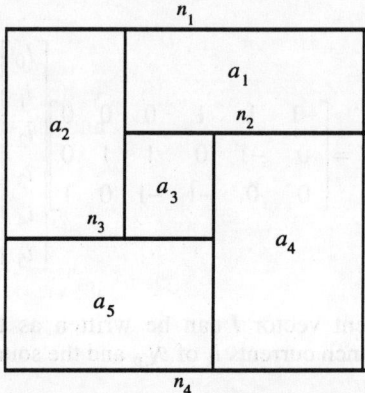

Fig. 8.35. A simple nonslicing floorplan $\mathcal{R}$.

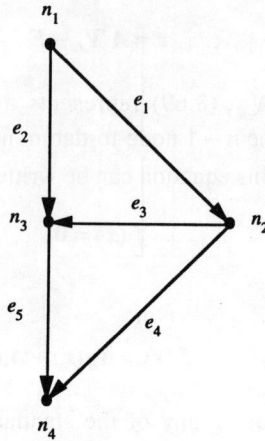

Fig. 8.36. The *p*-digraph of the rectangular floorplan $\mathcal{R}$ of fig. 8.35.

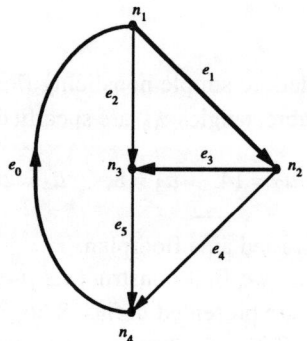

Fig. 8.37. The *c*-digraph of the rectangular floorplan $\mathcal{R}$ of fig. 8.35.

$$AI = \begin{bmatrix} -1 & 1 & 1 & 0 & 0 & 0 \\ 0 & -1 & 0 & 1 & 1 & 0 \\ 0 & 0 & -1 & -1 & 0 & 1 \end{bmatrix} \begin{bmatrix} I_0 \\ i_1 \\ i_2 \\ i_3 \\ i_4 \\ i_5 \end{bmatrix} = \mathbf{0} \qquad (8.74)$$

where the branch-current vector I can be written as the sum of two vectors corresponding to the branch currents i_k of $\mathcal{N}_p$ and the source current I_0 in $\mathcal{N}_c$:

$$I = \begin{bmatrix} I_0 \\ i_1 \\ i_2 \\ i_3 \\ i_4 \\ i_5 \end{bmatrix} = \begin{bmatrix} 0 \\ i_1 \\ i_2 \\ i_3 \\ i_4 \\ i_5 \end{bmatrix} + \begin{bmatrix} I_0 \\ 0 \\ 0 \\ 0 \\ 0 \\ 0 \end{bmatrix} = I_p + J. \tag{8.75}$$

Likewise, the branch-voltage vector V can be decomposed as

$$V = \begin{bmatrix} v_0 \\ v_1 \\ v_2 \\ v_3 \\ v_4 \\ v_5 \end{bmatrix} = \begin{bmatrix} v_0 \\ v_1 \\ v_2 \\ v_3 \\ v_4 \\ v_5 \end{bmatrix} + \begin{bmatrix} 0 \\ 0 \\ 0 \\ 0 \\ 0 \\ 0 \end{bmatrix} = V_p + E \tag{8.76}$$

where $E = 0$.

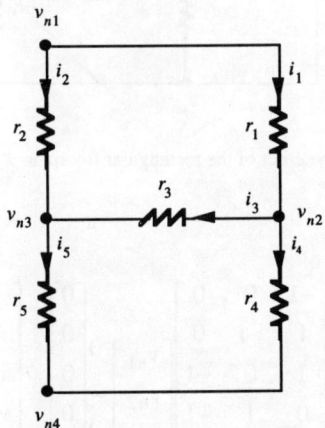

Fig. 8.38. The p-net of the rectangular floorplan $\mathcal{R}$ of fig. 8.35.

According to (8.62), the resistive branch currents i_k and voltages v_k of $\mathcal{N}_c$ are related by

$$\mathbf{I}_p = \begin{bmatrix} 0 \\ a_1 \\ \hline v_1 \\ a_2 \\ \hline v_2 \\ a_3 \\ \hline v_3 \\ a_4 \\ \hline v_4 \\ a_5 \\ \hline v_5 \end{bmatrix} = \begin{bmatrix} 0 \\ 40 \\ \hline v_1 \\ 14 \\ \hline v_2 \\ 8 \\ \hline v_3 \\ 20 \\ \hline v_4 \\ 18 \\ \hline v_5 \end{bmatrix} = g(\mathbf{V}_p). \tag{8.77}$$

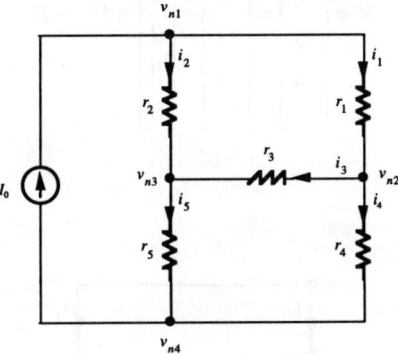

Fig. 8.39. The c-net of the rectangular floorplan $\mathcal{R}$ of fig. 8.35.

From (8.70) we have

$$e = A'V_n - E = \begin{bmatrix} -1 & 0 & 0 \\ 1 & -1 & 0 \\ 1 & 0 & -1 \\ 0 & 1 & -1 \\ 0 & 1 & 0 \\ 0 & 0 & 1 \end{bmatrix} \begin{bmatrix} v_{n1} \\ v_{n2} \\ v_{n3} \end{bmatrix} - \begin{bmatrix} 0 \\ 0 \\ 0 \\ 0 \\ 0 \\ 0 \end{bmatrix} = \begin{bmatrix} -v_{n1} \\ v_{n1} - v_{n2} \\ v_{n1} - v_{n3} \\ v_{n2} - v_{n3} \\ v_{n2} \\ v_{n3} \end{bmatrix} \tag{8.78}$$

where v_{n1}, v_{n2} and v_{n3} are the nodal voltages from nodes n_1, n_2 and n_3 to the reference node n_4 in $\mathcal{N}_c$ of fig. 8.39, and

$$AJ = \begin{bmatrix} -1 & 1 & 1 & 0 & 0 & 0 \\ 0 & -1 & 0 & 1 & 1 & 0 \\ 0 & 0 & -1 & -1 & 0 & 1 \end{bmatrix} \begin{bmatrix} I_0 \\ 0 \\ 0 \\ 0 \\ 0 \\ 0 \end{bmatrix} = \begin{bmatrix} -I_0 \\ 0 \\ 0 \end{bmatrix}. \qquad (8.79)$$

Thus, from (8.64), (8.65b), (8.77) and (8.78) we obtain

$$g(e) = g(V_p) = \begin{bmatrix} \dfrac{0}{40} \\ \dfrac{v_1}{14} \\ \dfrac{v_2}{8} \\ \dfrac{v_3}{20} \\ \dfrac{v_4}{18} \\ v_5 \end{bmatrix} = \begin{bmatrix} \dfrac{0}{40} \\ \dfrac{v_{n1} - v_{n2}}{14} \\ \dfrac{v_{n1} - v_{n3}}{8} \\ \dfrac{v_{n2} - v_{n3}}{20} \\ \dfrac{v_{n2}}{18} \\ v_{n3} \end{bmatrix}. \qquad (8.80)$$

Substituting (8.79) and (8.80) in (8.69) gives

$$Ag(e) = \begin{bmatrix} -1 & 1 & 1 & 0 & 0 & 0 \\ 0 & -1 & 0 & 1 & 1 & 0 \\ 0 & 0 & -1 & -1 & 0 & 1 \end{bmatrix} \begin{bmatrix} \dfrac{0}{40} \\ \dfrac{v_{n1} - v_{n2}}{14} \\ \dfrac{v_{n1} - v_{n3}}{8} \\ \dfrac{v_{n2} - v_{n3}}{20} \\ \dfrac{v_{n2}}{18} \\ v_{n3} \end{bmatrix}$$

$$= \begin{bmatrix} \dfrac{40}{v_{n1}-v_{n2}} + \dfrac{14}{v_{n1}-v_{n3}} \\ -\dfrac{40}{v_{n1}-v_{n2}} + \dfrac{8}{v_{n2}-v_{n3}} + \dfrac{20}{v_{n2}} \\ -\dfrac{14}{v_{n1}-v_{n3}} - \dfrac{8}{v_{n2}-v_{n3}} + \dfrac{18}{v_{n3}} \end{bmatrix}$$

$$= -\mathbf{AJ} = \begin{bmatrix} I_0 \\ 0 \\ 0 \end{bmatrix} \tag{8.81}$$

Simplifying and rearranging the terms yield

$$\begin{bmatrix} \dfrac{40}{v_{n1}-v_{n2}} + \dfrac{14}{v_{n1}-v_{n3}} \\ -\dfrac{40}{v_{n1}-v_{n2}} + \dfrac{8}{v_{n2}-v_{n3}} + \dfrac{20}{v_{n2}} \\ -\dfrac{14}{v_{n1}-v_{n3}} - \dfrac{8}{v_{n2}-v_{n3}} + \dfrac{18}{v_{n3}} \end{bmatrix} = \begin{bmatrix} I_0 \\ 0 \\ 0 \end{bmatrix}. \tag{8.82}$$

Equation (8.82) constitutes the three nodal equations in the three nodal voltages v_{n1}, v_{n2} and v_{n3}. It is clearly assumes the same general form of (8.71), being a *system of nonlinear algebraic equations*. Except in very special cases, the solutions of such a system cannot be obtained in *closed analytic form*. Thus, they must in general be solved by appropriate *numerical iteration techniques*. The nodal voltages are found to be

$$\begin{bmatrix} v_{n1} \\ v_{n2} \\ v_{n3} \end{bmatrix} = \begin{bmatrix} \dfrac{100}{I_0} \\ \dfrac{50}{I_0} \\ \dfrac{30}{I_0} \end{bmatrix}. \tag{8.83}$$

The branch voltages and the branch currents are obtained from (8.64) and (8.68), as follows:

$$V = A'V_n = \begin{bmatrix} -1 & 0 & 0 \\ 1 & -1 & 0 \\ 1 & 0 & -1 \\ 0 & 1 & -1 \\ 0 & 1 & 0 \\ 0 & 0 & 1 \end{bmatrix} \begin{bmatrix} v_{n1} \\ v_{n2} \\ v_{n3} \end{bmatrix} = \begin{bmatrix} -v_{n1} \\ v_{n1} - v_{n2} \\ v_{n1} - v_{n3} \\ v_{n2} - v_{n3} \\ v_{n2} \\ v_{n3} \end{bmatrix} = \frac{1}{I_0} \begin{bmatrix} -100 \\ 50 \\ 70 \\ 20 \\ 50 \\ 30 \end{bmatrix} \quad (8.84)$$

$$I = I_p + J = g(V_p) + J = \begin{bmatrix} 0 \\ \dfrac{40}{v_1} \\ \dfrac{14}{v_2} \\ \dfrac{8}{v_3} \\ \dfrac{20}{v_4} \\ \dfrac{18}{v_5} \end{bmatrix} + \begin{bmatrix} I_0 \\ 0 \\ 0 \\ 0 \\ 0 \\ 0 \end{bmatrix} = I_0 \begin{bmatrix} 1 \\ 0.8 \\ 0.2 \\ 0.4 \\ 0.4 \\ 0.6 \end{bmatrix}. \quad (8.85)$$

Since the width w_k and height h_k of the subrectangle $\mathcal{R}_k$ of $\mathcal{R}$ correspond to the current i_k and voltage v_k of the kth branch, we obtain

$$w_1 = 0.8 I_0, \quad h_1 = \frac{50}{I_0} \quad (8.86a)$$

$$w_2 = 0.2 I_0, \quad h_2 = \frac{70}{I_0} \quad (8.86b)$$

$$w_3 = 0.4 I_0, \quad h_3 = \frac{20}{I_0} \quad (8.86c)$$

$$w_4 = 0.4 I_0, \quad h_4 = \frac{50}{I_0} \quad (8.86d)$$

$$w_5 = 0.6 I_0, \quad h_5 = \frac{30}{I_0}. \quad (8.86e)$$

The corresponding dissection is shown in fig. 8.40. Observe that for each choice of the source current I_0, there corresponds a dissection of $\mathcal{R}$. Although there are infinitely many possibilities, all of them are within a multiplicative constant. In this sense, the dissection is unique within a scaling factor.

The total area a of $\mathcal{R}$, being equal to the total power consumed by all the resistors, is equal to the sum of the products $w_k h_k$, and is given by

$$a = \sum_{k=1}^{5} w_k h_k = 80 \times 0.5 + 20 \times 0.7 + 40 \times 0.2 + 40 \times 0.5 + 60 \times 0.3 = 100 \quad (8.87)$$

giving the width w and height h of $\mathcal{R}$ as

$$w = I_0, \quad h = \frac{100}{I_0}. \quad (8.88)$$

For $I_0 = 10$, the corresponding zero wasted area floorplan $\mathcal{R}$ is shown in fig. 8.41.

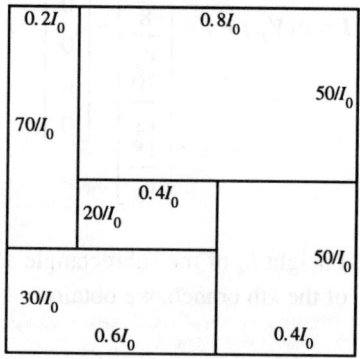

Fig. 8.40. A zero wasted area floorplan $\mathcal{R}$.

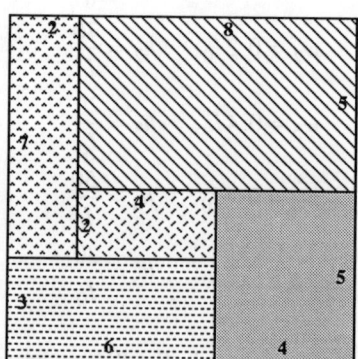

Fig. 8.41. A zero wasted area floorplan $\mathcal{R}$ of fig. 8.40 for $I_0 = 10$.

Example 8.2: Consider the floorplan $\mathcal{R}$ of fig. 8.32, which is redrawn in fig. 8.42, for the microprocessor compatible 8-bit A/D converter with 8-channel multiplexer shown in fig. 8.31. Let the areas a_k of the building blocks be specified as follows:

$$a_1 = 15, \quad a_2 = 5, \quad a_3 = 30, \quad a_4 = 1.5, \quad a_5 = 4.5, \quad (8.89)$$
$$a_6 = 7.5, \quad a_7 = 5.5, \quad a_8 = 16, \quad a_9 = 5, \quad a_{10} = 10.$$

We wish to find a zero wasted area floorplan.

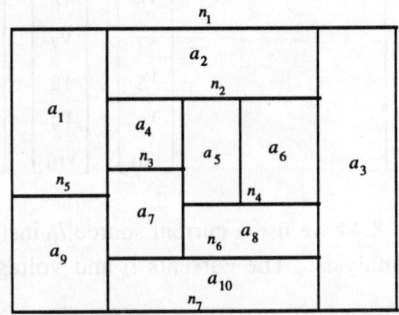

Fig. 8.42. A floorplan for a 8-bit A/D converter of fig. 8.31.

The first step is to construct a c-digraph $\mathcal{G}_c$ of $\mathcal{R}$ as shown in fig. 8.43. The corresponding c-net $\mathcal{N}_c$ is presented in fig. 8.44. As before, let v_k and i_k denote the kth branch voltage and current. From figs. 8.43 and 8.44, the KCL equation becomes

$$AI = \begin{bmatrix} -1 & 1 & 1 & 1 & 0 & 0 & 0 & 0 & 0 & 0 \\ 0 & 0 & -1 & 0 & 1 & 1 & 1 & 0 & 0 & 0 \\ 0 & 0 & 0 & 0 & -1 & 0 & 0 & 1 & 0 & 0 \\ 0 & 0 & 0 & 0 & 0 & -1 & -1 & 0 & 1 & 0 \\ 0 & -1 & 0 & 0 & 0 & 0 & 0 & 0 & 1 & 0 \\ 0 & 0 & 0 & 0 & 0 & 0 & 0 & -1 & -1 & 0 & 1 \end{bmatrix} \begin{bmatrix} I_0 \\ i_1 \\ i_2 \\ i_3 \\ i_4 \\ i_5 \\ i_6 \\ i_7 \\ i_8 \\ i_9 \\ i_{10} \end{bmatrix} = 0 \quad (8.90)$$

where from (8.65) the branch-current vector I and branch-voltage vector V of $\mathcal{N}_c$ can be decomposed as

$$I = \begin{bmatrix} I_0 \\ i_1 \\ i_2 \\ i_3 \\ i_4 \\ i_5 \\ i_6 \\ i_7 \\ i_8 \\ i_9 \\ i_{10} \end{bmatrix} = \begin{bmatrix} 0 \\ i_1 \\ i_2 \\ i_3 \\ i_4 \\ i_5 \\ i_6 \\ i_7 \\ i_8 \\ i_9 \\ i_{10} \end{bmatrix} + \begin{bmatrix} I_0 \\ 0 \\ 0 \\ 0 \\ 0 \\ 0 \\ 0 \\ 0 \\ 0 \\ 0 \\ 0 \end{bmatrix} = I_p + J, \quad V = \begin{bmatrix} v_0 \\ v_1 \\ v_2 \\ v_3 \\ v_4 \\ v_5 \\ v_6 \\ v_7 \\ v_8 \\ v_9 \\ v_{10} \end{bmatrix} = \begin{bmatrix} v_0 \\ v_1 \\ v_2 \\ v_3 \\ v_4 \\ v_5 \\ v_6 \\ v_7 \\ v_8 \\ v_9 \\ v_{10} \end{bmatrix} + \begin{bmatrix} 0 \\ 0 \\ 0 \\ 0 \\ 0 \\ 0 \\ 0 \\ 0 \\ 0 \\ 0 \\ 0 \end{bmatrix} = V_p + E. \tag{8.91}$$

Observe that in fig. 8.44 we use a current source I_0 instead of a voltage source E, because of nodal analysis. The currents i_k and voltages v_k of the nonlinear resistors are related by

$$I_p = \begin{bmatrix} 0 & \dfrac{15}{v_1} & \dfrac{5}{v_2} & \dfrac{30}{v_3} & \dfrac{1.5}{v_4} & \dfrac{4.5}{v_5} & \dfrac{7.5}{v_6} & \dfrac{5.5}{v_7} & \dfrac{16}{v_8} & \dfrac{5}{v_9} & \dfrac{10}{v_{10}} \end{bmatrix}' = g(V_p) \tag{8.92}$$

where the prime denotes the matrix transpose. From (8.69) the nodal equation becomes

$$Ag(e) = Ag(A'V_n) = -AJ \tag{8.93}$$

where

$$V_n = \begin{bmatrix} v_{n1} & v_{n2} & v_{n3} & v_{n4} & v_{n5} & v_{n6} \end{bmatrix}' \tag{8.94}$$

$$J = \begin{bmatrix} I_0 & 0 & 0 & 0 & 0 & 0 & 0 & 0 & 0 & 0 \end{bmatrix}' \tag{8.95}$$

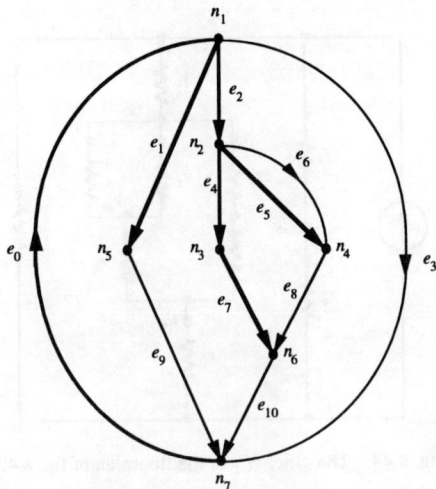

Fig. 8.43. The c-digraph $\mathcal{G}_c$ of the floorplan of fig. 8.42.

obtaining from (8.70)

$$e = A'V_n - E = \begin{bmatrix} -1 & 0 & 0 & 0 & 0 & 0 \\ 1 & 0 & 0 & 0 & -1 & 0 \\ 1 & -1 & 0 & 0 & 0 & 0 \\ 1 & 0 & 0 & 0 & 0 & 0 \\ 0 & 1 & -1 & 0 & 0 & 0 \\ 0 & 1 & 0 & -1 & 0 & 0 \\ 0 & 1 & 0 & -1 & 0 & 0 \\ 0 & 0 & 1 & 0 & 0 & -1 \\ 0 & 0 & 0 & 1 & 0 & -1 \\ 0 & 0 & 0 & 0 & 1 & 0 \\ 0 & 0 & 0 & 0 & 0 & 1 \end{bmatrix} \begin{bmatrix} v_{n1} \\ v_{n2} \\ v_{n3} \\ v_{n4} \\ v_{n5} \\ v_{n6} \end{bmatrix} - \begin{bmatrix} 0 \\ 0 \\ 0 \\ 0 \\ 0 \\ 0 \\ 0 \\ 0 \\ 0 \\ 0 \\ 0 \end{bmatrix} = \begin{bmatrix} v_{n1} \\ v_{n1} - v_{n5} \\ v_{n1} - v_{n2} \\ v_{n1} \\ v_{n2} - v_{n3} \\ v_{n2} - v_{n4} \\ v_{n2} - v_{n4} \\ v_{n3} - v_{n6} \\ v_{n4} - v_{n6} \\ v_{n5} \\ v_{n6} \end{bmatrix} = \begin{bmatrix} v_0 \\ v_1 \\ v_2 \\ v_3 \\ v_4 \\ v_5 \\ v_6 \\ v_7 \\ v_8 \\ v_9 \\ v_{10} \end{bmatrix} = V_p$$

(8.96)

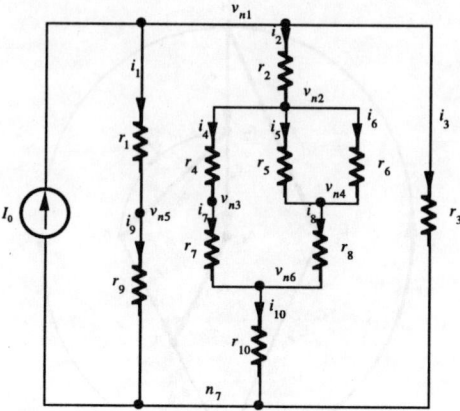

Fig. 8.44. The c-net $\mathcal{N}_c$ of the floorplan of fig. 8.42.

Substituting these in (8.93) yields

$$Ag(e) = \begin{bmatrix} -1 & 1 & 1 & 1 & 0 & 0 & 0 & 0 & 0 & 0 & 0 \\ 0 & 0 & -1 & 0 & 1 & 1 & 1 & 0 & 0 & 0 & 0 \\ 0 & 0 & 0 & 0 & -1 & 0 & 0 & 1 & 0 & 0 & 0 \\ 0 & 0 & 0 & 0 & 0 & -1 & -1 & 0 & 1 & 0 & 0 \\ 0 & -1 & 0 & 0 & 0 & 0 & 0 & 0 & 0 & 1 & 0 \\ 0 & 0 & 0 & 0 & 0 & 0 & 0 & -1 & -1 & 0 & 1 \end{bmatrix} \begin{bmatrix} \dfrac{0}{15} \\ \dfrac{v_{n1} - v_{n5}}{5} \\ \dfrac{v_{n1} - v_{n2}}{30} \\ \dfrac{v_{n1}}{1.5} \\ \dfrac{v_{n2} - v_{n3}}{4.5} \\ \dfrac{v_{n2} - v_{n4}}{7.5} \\ \dfrac{v_{n2} - v_{n4}}{5.5} \\ \dfrac{v_{n3} - v_{n6}}{16} \\ \dfrac{v_{n4} - v_{n6}}{5} \\ \dfrac{v_{n5}}{10} \\ v_{n6} \end{bmatrix}$$

$$= \begin{bmatrix} \dfrac{15}{v_{n1}-v_{n5}} + \dfrac{5}{v_{n1}-v_{n2}} + \dfrac{30}{v_{n1}} \\ -\dfrac{5}{v_{n1}-v_{n2}} + \dfrac{1.5}{v_{n2}-v_{n3}} + \dfrac{4.5}{v_{n2}-v_{n4}} + \dfrac{7.5}{v_{n2}-v_{n4}} \\ -\dfrac{1.5}{v_{n2}-v_{n3}} + \dfrac{5.5}{v_{n3}-v_{n6}} \\ -\dfrac{4.5}{v_{n2}-v_{n4}} - \dfrac{7.5}{v_{n2}-v_{n4}} + \dfrac{16}{v_{n4}-v_{n6}} \\ -\dfrac{15}{v_{n1}-v_{n5}} + \dfrac{5}{v_{n5}} \\ -\dfrac{5.5}{v_{n3}-v_{n6}} - \dfrac{16}{v_{n4}-v_{n6}} + \dfrac{10}{v_{n6}} \end{bmatrix} = \begin{bmatrix} I_0 \\ 0 \\ 0 \\ 0 \\ 0 \\ 0 \end{bmatrix} = -AJ. \quad (8.97)$$

This is a system of six nonlinear nodal equations in six nodal voltages v_{nk} of the form of (8.71):

$$\begin{bmatrix} \dfrac{15}{v_{n1}-v_{n5}} + \dfrac{5}{v_{n1}-v_{n2}} + \dfrac{30}{v_{n1}} \\ -\dfrac{5}{v_{n1}-v_{n2}} + \dfrac{1.5}{v_{n2}-v_{n3}} + \dfrac{4.5}{v_{n2}-v_{n4}} + \dfrac{7.5}{v_{n2}-v_{n4}} \\ -\dfrac{1.5}{v_{n2}-v_{n3}} + \dfrac{5.5}{v_{n3}-v_{n6}} \\ -\dfrac{4.5}{v_{n2}-v_{n4}} - \dfrac{7.5}{v_{n2}-v_{n4}} + \dfrac{16}{v_{n4}-v_{n6}} \\ -\dfrac{15}{v_{n1}-v_{n5}} + \dfrac{5}{v_{n5}} \\ -\dfrac{5.5}{v_{n3}-v_{n6}} - \dfrac{16}{v_{n4}-v_{n6}} + \dfrac{10}{v_{n6}} \end{bmatrix} = \begin{bmatrix} I_0 \\ 0 \\ 0 \\ 0 \\ 0 \\ 0 \end{bmatrix} \quad (8.98)$$

and can be solved by the classical Newton–Raphson algorithm (see, for example, CHUA and LIN [1975]) to give

$$V_n = \begin{bmatrix} v_{n1} \\ v_{n2} \\ v_{n3} \\ v_{n4} \\ v_{n5} \\ v_{n6} \end{bmatrix} = \begin{bmatrix} \dfrac{100}{I_0} \\ \dfrac{90}{I_0} \\ \dfrac{75}{I_0} \\ \dfrac{60}{I_0} \\ \dfrac{25}{I_0} \\ \dfrac{20}{I_0} \end{bmatrix}. \qquad (8.99)$$

From (8.96) the branch voltages v_k are obtained as

$$V = \begin{bmatrix} v_0 \\ v_1 \\ v_2 \\ v_3 \\ v_4 \\ v_5 \\ v_6 \\ v_7 \\ v_8 \\ v_9 \\ v_{10} \end{bmatrix} = \begin{bmatrix} v_{n1} \\ v_{n1} - v_{n5} \\ v_{n1} - v_{n2} \\ v_{n1} \\ v_{n2} - v_{n3} \\ v_{n2} - v_{n4} \\ v_{n2} - v_{n4} \\ v_{n3} - v_{n6} \\ v_{n4} - v_{n6} \\ v_{n5} \\ v_{n6} \end{bmatrix} = \frac{1}{I_0} \begin{bmatrix} 100 \\ 75 \\ 10 \\ 100 \\ 15 \\ 30 \\ 30 \\ 55 \\ 40 \\ 25 \\ 20 \end{bmatrix}. \qquad (8.100)$$

The branch currents i_k are found from (8.65a) and (8.92) as

§ 7   VLSI layout

$$I = I_p + J = g(V_p) + J = \begin{bmatrix} \dfrac{I_0}{15} \\ v_1 \\ \dfrac{v_1}{5} \\ \dfrac{v_2}{30} \\ \dfrac{v_3}{1.5} \\ \dfrac{v_4}{4.5} \\ \dfrac{v_5}{7.5} \\ \dfrac{v_6}{5.5} \\ \dfrac{v_7}{16} \\ \dfrac{v_8}{5} \\ \dfrac{v_9}{10} \\ v_{10} \end{bmatrix} = \begin{bmatrix} I_0 \\ 0.2 I_0 \\ 0.5 I_0 \\ 0.3 I_0 \\ 0.1 I_0 \\ 0.15 I_0 \\ 0.25 I_0 \\ 0.1 I_0 \\ 0.4 I_0 \\ 0.2 I_0 \\ 0.5 I_0 \end{bmatrix}. \qquad (8.101)$$

Thus, the width w_k and height h_k of the desired subrectangles $\mathcal{R}_k$, which correspond to branch currents i_k and branch voltage v_k, are given by

$$w_1 = 0.2 I_0, \quad h_1 = \frac{75}{I_0}, \quad w_2 = 0.5 I_0, \quad h_2 = \frac{10}{I_0}, \qquad (8.102a)$$

$$w_3 = 0.3 I_0, \quad h_3 = \frac{100}{I_0}, \quad w_4 = 0.1 I_0, \quad h_4 = \frac{15}{I_0}, \qquad (8.102b)$$

$$w_5 = 0.15 I_0, \quad h_5 = \frac{30}{I_0}, \quad w_6 = 0.25 I_0, \quad h_6 = \frac{30}{I_0}, \qquad (8.102c)$$

$$w_7 = 0.1 I_0, \quad h_7 = \frac{55}{I_0}, \quad w_8 = 0.4 I_0, \quad h_8 = \frac{40}{I_0}, \qquad (8.102d)$$

$$w_9 = 0.2 I_0, \quad h_9 = \frac{25}{I_0}, \quad w_{10} = 0.5 I_0, \quad h_{10} = \frac{20}{I_0}. \qquad (8.102e)$$

The corresponding general floorplan $\mathcal{R}$ is presented in fig. 8.45. For $I_0 = 10$, it is shown in fig. 8.46. Observe that for each choice of I_0, there corresponds a dissection, and they are all within a multiplicative constant.

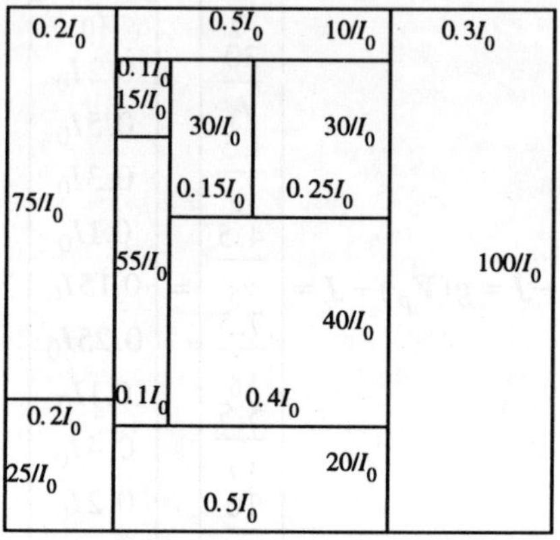

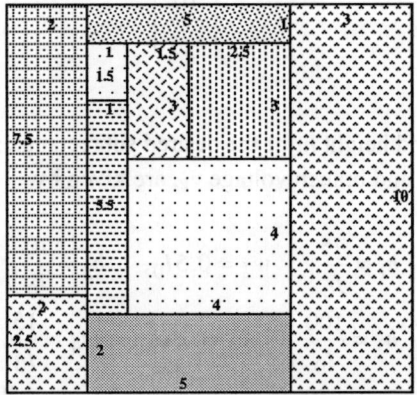

Fig. 8.45. A zero wasted area floorplan for the 8-bit A/D converter of fig. 8.31.

Fig. 8.46. A zero wasted area floorplan of fig. 8.45 for $I_0 = 10$.

7.2. *Solving the nonlinear nodal equations*

In the preceding section, we showed that solving the zero wasted area floorplan problem is equivalent to that of solving a system of $n - 1$ nonlinear nodal equations of the form

$$Ag(e) = -AJ \tag{8.103}$$

where

$$e = A'V_n - E \tag{8.104}$$

contains the nodal voltage vector V_n in $n - 1$ nodal voltages v_{nk}, $k = 1, 2, \ldots, n - 1$. This equation is of the general form

$$f(V_n) = Ag(e) + AJ = 0 \tag{8.105}$$

and can be solved by the $(n - 1)$-dimensional Newton–Raphson algorithm.

In the following, we derive the Newton–Raphson algorithm for solving the system of nonlinear equations (8.103). First, let us assume that

$$V_n = \begin{bmatrix} v_{n1}^{(j)} & v_{n2}^{(j)} & \cdots & v_{n(n-1)}^{(j)} \end{bmatrix}' \equiv V_n^{(j)} \tag{8.106}$$

is the value of V_n obtained at the jth iteration. Now expand each equation of (8.105), written as

$$f(V_n) = \begin{bmatrix} f_1(v_{n1}, v_{n2}, \ldots, v_{n(n-1)}) \\ f_2(v_{n1}, v_{n2}, \ldots, v_{n(n-1)}) \\ \vdots \\ f_{n-1}(v_{n1}, v_{n2}, \ldots, v_{n(n-1)}) \end{bmatrix} \equiv \begin{bmatrix} f_1(V_n) \\ f_2(V_n) \\ \vdots \\ f_{n-1}(V_n) \end{bmatrix} \tag{8.107}$$

by the Taylor series for functions of several variables about the point $V_n^{(j)}$, and obtain

$$f_i(V_n) = f_i(V_n^{(j)}) + \left.\frac{\partial f_i(V_n)}{\partial v_{n1}}\right|_{V_n = V_n^{(j)}} \left[v_{n1} - v_{n1}^{(j)}\right] + \left.\frac{\partial f_i(V_n)}{\partial v_{n2}}\right|_{V_n = V_n^{(j)}} \left[v_{n2} - v_{n2}^{(j)}\right]$$

$$+ \cdots + \left.\frac{\partial f_i(V_n)}{\partial v_{n(n-1)}}\right|_{V_n = V_n^{(j)}} \left[v_{n(n-1)} - v_{n(n-1)}^{(j)}\right] \tag{8.108}$$

$$+ \text{ higher-order terms involving } \left[v_{nk} - v_{nk}^{(j)}\right]^m,$$

$$m > 1, \quad i, k = 1, 2, \ldots, n - 1.$$

If $V_n = V_n^{(j+1)}$ is the value of V_n at the $(j+1)$th iteration, from (8.108) we have

$$f_i(V_n^{(j+1)}) = f_i(V_n^{(j)}) + \frac{\partial f_i(V_n)}{\partial v_{n1}}\bigg|_{V_n=V_n^{(j)}} \left[v_{n1}^{(j+1)} - v_{n1}^{(j)}\right]$$

$$+ \frac{\partial f_i(V_n)}{\partial v_{n2}}\bigg|_{V_n=V_n^{(j)}} \left[v_{n2}^{(j+1)} - v_{n2}^{(j)}\right] + \cdots$$

$$+ \frac{\partial f_i(V_n)}{\partial v_{n(n-1)}}\bigg|_{V_n=V_n^{(j)}} \left[v_{n(n-1)}^{(j+1)} - v_{n(n-1)}^{(j)}\right] \tag{8.109}$$

$$+ \text{ higher-order terms involving } \left[v_{nk}^{(j+1)} - v_{nk}^{(j)}\right]^m,$$

$$m > 1, \quad i, k = 1, 2, \ldots, n-1.$$

Assume that $V_n^{(j)}$ is a good guess to the correct solution so that each element of the vector $V_n^{(j+1)} - V_n^{(j)}$ is a small number. Therefore, we can ignore all higher-order terms in (8.109), resulting in the following approximation:

$$f_i(V_n^{(j+1)}) \cong f_i(V_n^{(j)}) + \frac{\partial f_i(V_n)}{\partial v_{n1}}\bigg|_{V_n=V_n^{(j)}} \left[v_{n1}^{(j+1)} - v_{n1}^{(j)}\right]$$

$$+ \frac{\partial f_i(V_n)}{\partial v_{n2}}\bigg|_{V_n=V_n^{(j)}} \left[v_{n2}^{(j+1)} - v_{n2}^{(j)}\right]$$

$$+ \cdots + \frac{\partial f_i(V_n)}{\partial v_{n(n-1)}}\bigg|_{V_n=V_n^{(j)}} \left[v_{n(n-1)}^{(j+1)} - v_{n(n-1)}^{(j)}\right], \quad i = 1, 2, \ldots, n-1. \tag{8.110}$$

Since we wish $V_n^{(j+1)}$ to be a solution of (8.105), we set

$$f_i(V_n^{(j+1)}) = 0, \quad i = 1, 2, \ldots, n-1 \tag{8.111}$$

and obtain the following system of equations:

$$f_i(V_n^{(j)}) + \frac{\partial f_i(V_n)}{\partial v_{n1}}\bigg|_{V_n=V_n^{(j)}} \left[v_{n1}^{(j+1)} - v_{n1}^{(j)}\right] + \frac{\partial f_i(V_n)}{\partial v_{n2}}\bigg|_{V_n=V_n^{(j)}} \left[v_{n2}^{(j+1)} - v_{n2}^{(j)}\right]$$

$$+ \cdots + \left.\frac{\partial f_i(V_n)}{\partial v_{n(n-1)}}\right|_{V_n=V_n^{(j)}} \left[v_{n(n-1)}^{(j+1)} - v_{n(n-1)}^{(j)} \right] = 0, \quad i=1,2,\ldots,n-1. \qquad (8.112)$$

This represents a system of $n-1$ nodal equations and can be put in matrix form as follows:

$$\mathcal{J}(V_n^{(j)})\left[V_n^{(j+1)} - V_n^{(j)} \right] = -f(V_n^{(j)}) \qquad (8.113)$$

where

$$\mathcal{J}(V_n^{(j)}) \equiv \begin{bmatrix} \dfrac{\partial f_1(V_n)}{\partial v_{n1}} & \dfrac{\partial f_1(V_n)}{\partial v_{n2}} & \cdots & \dfrac{\partial f_1(V_n)}{\partial v_{n(n-1)}} \\ \dfrac{\partial f_2(V_n)}{\partial v_{n1}} & \dfrac{\partial f_2(V_n)}{\partial v_{n2}} & \cdots & \dfrac{\partial f_2(V_n)}{\partial v_{n(n-1)}} \\ \vdots & \vdots & \cdots & \vdots \\ \dfrac{\partial f_{n-1}(V_n)}{\partial v_{n1}} & \dfrac{\partial f_{n-1}(V_n)}{\partial v_{n2}} & \cdots & \dfrac{\partial f_{n-1}(V_n)}{\partial v_{n(n-1)}} \end{bmatrix} \qquad (8.114)$$

is the *Jacobian matrix* of $f(V_n)$ evaluated at $V_n = V_n^{(j)}$. From (8.113) we can solve $V_n^{(j+1)}$ to yield

$$V_n^{(j+1)} = V_n^{(j)} - \left[J(V_n^{(j)}) \right]^{-1} f(V_n^{(j)}). \qquad (8.115)$$

Equation (8.115) is the well-known *Newton–Raphson equation*, which is a good general-purpose algorithm for solving a system of nonlinear nodal equations. The algorithm requires the evaluation of $(n-1)^2$ partial derivatives $\partial f_i(V_n)/\partial v_{nk}$ at each iteration in order to obtain the Jacobian matrix $\mathcal{J}(V_n^{(j)})$. This is a cumbersome process even with the aid of a computer. To circumvent this difficulty, many modifications of the Newton–Raphson algorithm have been proposed to avoid the need to evaluate the Jacobian matrix at each iteration known as the *quasi-Newton methods*. However, they are beyond the scope of this book. We refer the interested reader to ORTEGA and RHEINBOLDT [1970] and POLAK [1971] for details.

Substituting $f(V_n)$ from (8.105) in (8.115) gives

$$V_n^{(j+1)} = V_n^{(j)} - \left[A\left.\frac{\partial g(e)}{\partial e}\right|_{e=A'V_n^{(j)}-E} A' \right]^{-1} \left[Ag(A'V_n^{(j)} - E) + AJ \right] \qquad (8.116)$$

where the Jacobian of $f(V_n)$ is given by

$$\mathcal{J}(V_n) = A \left. \frac{\partial g(e)}{\partial e} \right|_{e=A'V_n - E} A' \tag{8.117}$$

$$\frac{\partial g(e)}{\partial e} \equiv \begin{bmatrix} \frac{\partial g_1}{\partial e_1} & \frac{\partial g_1}{\partial e_2} & \cdots & \frac{\partial g_1}{\partial e_{n-1}} \\ \frac{\partial g_2}{\partial e_1} & \frac{\partial g_2}{\partial e_2} & \cdots & \frac{\partial g_2}{\partial e_{n-1}} \\ \vdots & \vdots & \cdots & \vdots \\ \frac{\partial g_{n-1}}{\partial e_1} & \frac{\partial g_{n-1}}{\partial e_2} & \cdots & \frac{\partial g_{n-1}}{\partial e_{n-1}} \end{bmatrix} \tag{8.118a}$$

$$g = \begin{bmatrix} g_1 & g_2 & \cdots & g_{n-1} \end{bmatrix}' \tag{8.118b}$$

$$e = \begin{bmatrix} e_1 & e_2 & \cdots & e_{n-1} \end{bmatrix}'. \tag{8.118c}$$

Observe that in writing (8.115) it is advantageous to recast it into the form

$$\mathcal{J}(V_n^{(j)})V_n^{(j+1)} = \mathcal{J}(V_n^{(j)})V_n^{(j)} - f(V_n^{(j)}) \tag{8.119}$$

so that we can solve for $V_n^{(j+1)}$ by Gaussian elimination, because inverting a matrix by Gaussian elimination requires approximately three times as many multiplications as solving a corresponding system of linear equations.

A critical problem in applying the New–Raphson algorithm is the choice of the initial value of V_n. For some initial guesswork, it is possible for the iteration to diverge. However, it can be rigorously proved that, if the initial guess is close to a correct solution, the Newton–Raphson algorithm will always converge and its convergence rate is quadratic. The following theorem will show how close the initial guess must be in order to assure convergence. Its proof can be found in ORTEGA and RHEINBOLDT [1970].

THEOREM 8.10: (*Newton–Raphson–Kantorovich Theorem*) Let

$$V_n^{(j)} = \begin{bmatrix} v_{n1}^{(j)} & v_{n2}^{(j)} & \cdots & v_{n(n-1)}^{(j)} \end{bmatrix}' \tag{8.120}$$

be the value obtained from (8.115) after the jth iteration. The initial guess corresponds to $j = 0$. Let the inverse of the Jacobian matrix at $V_n^{(0)}$ be

$$\mathcal{J}^{-1}(V_n^{(0)}) = \left[\frac{A_{ij}(V_n^{(0)})}{D(V_n^{(0)})}\right]' \tag{8.121}$$

where $A_{ij}(V_n^{(0)})$ is the cofactor of the ith row and the jth column element of $\mathcal{J}(V_n^{(0)})$ and $D(V_n^{(0)})$ is the determinant of $\mathcal{J}(V_n^{(0)})$. Assume that the following conditions are satisfied:

$$\sum_{i=1}^{n-1}\left[v_{ni}^{(1)} - v_{ni}^{(0)}\right]^2 \leq A^2 \tag{8.122}$$

$$\frac{1}{|D(V_n^{(0)})|}\left[\sum_{i,j=1}^{n-1} A_{ij}^2(V_n^{(0)})\right]^{1/2} \leq B \tag{8.123}$$

$$\sum_{i,j,k=1}^{n-1}\left[\frac{\partial^2 f_i(V_n)}{\partial v_{nj}\partial v_{nk}}\right]^2_{V_n=V_n^{(0)}} \leq K^2 \tag{8.124}$$

where A, B and K are nonnegative constants such that

$$ABK \equiv C_0 \leq \frac{1}{2}. \tag{8.125}$$

Then the sequence defined by (8.115) converges to a solution

$$\tilde{V}_n = \begin{bmatrix} \tilde{v}_{n1} & \tilde{v}_{n2} & \cdots & \tilde{v}_{n(n-1)} \end{bmatrix}' \tag{8.126}$$

such that the vector norm

$$\|\tilde{V}_n - V_n^{(0)}\| \equiv \sqrt{\sum_{i=1}^{n-1}\left[\tilde{v}_{ni} - v_{ni}^{(0)}\right]^2} \leq \frac{1-\sqrt{1-2C_0}}{C_0} A \tag{8.127}$$

We remark that in the theorem, we used the term "norm" of a matrix or vector. There are many ways of defining the norms of a matrix. A *norm* of a square matrix $M = [m_{ij}]$ of order n, written as $\|M\|$, is defined as one of the following nonnegative numbers:

$$\|M\| = n_1,\ n_2,\ n_3,\ \text{or}\ n_4 \tag{8.128}$$

where

$$n_1 = \max_i \sum_{j=1}^{n} |m_{ij}| \tag{8.129a}$$

$$n_2 = \max_j \sum_{i=1}^{n} |m_{ij}| \tag{8.129b}$$

$$n_3 = \sum_{i=1}^{n} \sum_{j=1}^{n} |m_{ij}| \tag{8.129c}$$

$$n_4 = \sqrt{\sum_{i=1}^{n} \sum_{j=1}^{n} |m_{ij}|^2}. \tag{8.129d}$$

Another norm is the *Hilbert norm* defined as

$$\|M\| \equiv n_5 \equiv \max_{1 \leq i \leq n} \sqrt{\lambda_i} \tag{8.130}$$

where λ_i is the ith eigenvalue of $M'M$. It can be shown that

$$n_5 \leq \min\ (n_1, n_2). \tag{8.131}$$

If both n_1 and n_2 are less than some specified criterion, say ε, then n_5 is also less than ε. To avoid the eigenvalue evaluation for n_5 and yet to be able to guarantee that n_5 is less than ε, it is convenient to use the real nonnegative number min (n_1, n_2) as a "norm" of M in testing for convergence.

The concept of norms of a square matrix can be extended to vectors. A *norm* of a vector

$$x = \begin{bmatrix} x_1 & x_2 & \cdots & x_n \end{bmatrix}' \tag{8.132}$$

written as $\|x\|$, is defined as one of the following nonnegative numbers:

$$\|x\| = \tilde{n}_1, \tilde{n}_2, \text{ or } \tilde{n}_3 \tag{8.133}$$

where

$$\tilde{n}_1 = \sqrt{|x_1|^2 + |x_2|^2 + \cdots + |x_n|^2} \tag{8.134a}$$

$$\tilde{n}_2 = \sum_{i=1}^{n} |x_i| \tag{8.134b}$$

$$\tilde{n}_3 = \max_i |x_i|. \tag{8.134c}$$

Among these norms, the norm $\tilde{n}_1$ is most commonly used in the explicit calculations. It can be shown that the various norms just defined are equivalent. The concept of norms is most useful to prove convergence, as stated in Theorem 8.10.

7.3. *Floorplan area optimization with constrained aspect ratio*

In 7.1 we have shown that if there is no constraint on the aspect ratios of individual subrectangles, a zero wasted area floorplan can be achieved if the following corresponding nonlinear nodal equations have real and positive solution:

$$Ag(A'V_n - E) = -AJ \tag{8.135}$$

using Newton–Raphson algorithm (8.115) if necessary. Once the nodal voltage vector V_n is known, the branch-voltage and branch-current vectors V and I can be calculated from (8.64), (8.65) and (8.68) as

$$V = A'V_n \tag{8.136}$$
$$I = g(A'V_n - E) + J. \tag{8.137}$$

In practice, however, a module may have physical constraints on its dimensions such as height and width. In most of these cases, a zero wasted area floorplan may not be achievable. In this section, we present an optimization procedure to find a minimum wasted area floorplan.

Given a b-module floorplan with a preassigned topology, let the height and width of the ith module M_i be denoted by y_i and x_i, respectively. Our objective is to find the optimum dimensions y_i and x_i of all modules, so that they satisfy the following constraints:

$$x_i y_i \geq a_i \qquad (8.138a)$$
$$x_i \geq w_i \qquad (8.138b)$$
$$y_i \geq h_i \qquad (8.138c)$$

for $i = 1, 2, \ldots, b$, where a_i, w_i and h_i are the minimum required area, width and height of module M_i, respectively. This is equivalent to finding the branch voltages and branch currents corresponding to y_i and x_i of an associated nonlinear network so that its total power consumption is minimized subject to the constraints (8.138).

To this end, let $\mathcal{N}_c$ and $\mathcal{G}_c$ be the c-net and c-digraph associated with a floorplan $\mathcal{R}$. Since the area of each subrectangle of $\mathcal{R}$ is required only to be not less than, not equal to, a_i, and since the formulation of the nodal equations requires a preassigned functional relationship between the branch voltages and branch currents, for our purposes the nodal equations are inappropriate. We have to start from the KCL and KVL equations. As in (8.65), write

$$I = I_p + J \qquad (8.139a)$$
$$V = V_p + E \qquad (8.139b)$$

where I_p and V_p denote the resistive branch currents i_k and branch voltages v_k of $\mathcal{N}_c$, and J and E the independent current sources and the voltage sources in $\mathcal{N}_c$. In the present situation, we have only one source, which can either be a current source or a voltage source. Therefore, either J or E will be zero, depending on the choice of the excitation source for $\mathcal{N}_c$.

Let Q_f be the fundamental cutset matrix with respect to a tree t of $\mathcal{G}_c$. Substituting (8.139a) for I in the KCL equation

$$Q_f I = 0 \qquad (8.140)$$

we obtain

$$Q_f I_p = - Q_f J \qquad (8.141)$$

Partition Q_f and I_p in accordance with the branches of t and links of its cotree $\bar{t}$.

$$\begin{bmatrix} 1_t & Q_{f\bar{t}} \end{bmatrix} \begin{bmatrix} I_{pt} \\ I_{p\bar{t}} \end{bmatrix} = -Q_f J \qquad (8.142)$$

§ 7 VLSI layout

giving

$$I_{pt} = -Q_{f\bar{t}}I_{p\bar{t}} - Q_f J. \tag{8.143}$$

Likewise, let B_f be the fundamental circuit matrix with respect to t. Substituting (8.139b) for V in the KVL equation

$$B_f V = 0 \tag{8.144}$$

we obtain

$$B_f V_p = -B_f E. \tag{8.145}$$

As in (8.142), partition B_f and V_p in accordance with the branches of t and the links of its cotree $\bar{t}$.

$$\begin{bmatrix} B_{ft} & 1_{\bar{t}} \end{bmatrix} \begin{bmatrix} V_{pt} \\ V_{p\bar{t}} \end{bmatrix} = -B_f E \tag{8.146}$$

yielding

$$V_{p\bar{t}} = -B_{ft}V_{pt} - B_f E. \tag{8.147}$$

If I_0 and V_0 are the input current and voltage of the c-net $\mathcal{N}_c$, our objective is to minimize

$$f(V_{pt}, I_{p\bar{t}}) = V_0 I_0 \tag{8.148}$$

subject to

$$\begin{bmatrix} V_{p\bar{t}} + B_{ft}V_{pt} + B_f E \\ I_{pt} + Q_{f\bar{t}}I_{p\bar{t}} + Q_f J \end{bmatrix} = 0 \tag{8.149}$$

$$\tilde{V}_p \geq H_m \tag{8.150}$$

$$\tilde{I}_p \geq W_m \tag{8.151}$$

$$p \geq a \tag{8.152}$$

where

$$V_p \equiv \begin{bmatrix} V_{pt} \\ V_{p\bar{t}} \end{bmatrix} = \begin{bmatrix} 0 \\ v_1 \\ v_2 \\ \vdots \\ v_b \end{bmatrix} \equiv \begin{bmatrix} 0 \\ \tilde{V}_p \end{bmatrix} \qquad (8.153)$$

$$I_p \equiv \begin{bmatrix} I_{pt} \\ I_{p\bar{t}} \end{bmatrix} = \begin{bmatrix} 0 \\ i_1 \\ i_2 \\ \vdots \\ i_b \end{bmatrix} \equiv \begin{bmatrix} 0 \\ \tilde{I}_p \end{bmatrix} \qquad (8.154)$$

are the resistive branch-voltage and branch-current vectors of $\mathcal{G}_c$,

$$p = \begin{bmatrix} v_1 i_1 & v_2 i_2 & \cdots & v_b i_b \end{bmatrix}' \qquad (8.155)$$

$$a = \begin{bmatrix} a_1 & a_2 & \cdots & a_b \end{bmatrix}' \qquad (8.156)$$

are the power and area vectors of the modules M_i and

$$H_m = \begin{bmatrix} h_1 & h_2 & \cdots & h_b \end{bmatrix}' \qquad (8.157)$$

$$W_m = \begin{bmatrix} w_1 & w_2 & \cdots & w_b \end{bmatrix}' \qquad (8.158)$$

denote their minimum-height and minimum-width vectors. In (8.150), we define $\tilde{V}_p \geq H_m$ if and only if $v_k \geq h_k$ for all k. This is similarly valid for $\tilde{I}_p \geq W_m$ and $p \geq a$. The formulation described above is a standard optimization problem.

An optimization problem begins with a set of independent variables or parameters, and often includes constraints that define the acceptable values of the variables. Another essential component is a measure of "goodness" called the *objective function*, depending in some way on the variables. The solution of an optimization problem is a set of allowed values for which the objective function assumes an "optimal" value. An optimization problem can generally be formulated as follows:

Minimize $f(x)$
subject to $g_i(x) = 0, \quad i = 1, 2, \ldots, p$
$g_i(x) \leq 0, \quad i = p+1, p+2, \ldots, b$

where x is the unknown vector, $f(x)$ is the objective function, and $g_i(x)$ are the constraint functions.

The problem can be solved by either a direct method or an indirect method. In the direct method, the constraints are handled explicitly, whereas in most indirect methods, the problem is solved by a sequence of unconstrained minimization problems. For problems involving explicit nonlinear equations such as given in the above, the penalty function method is convenient and is expected to work most efficiently. In the present situation, we apply suitable penalty functions to transform the constrained floorplan area minimization problem to an unconstrained minimization problem.

The first step is to find a suitable penalty function that will incur a positive penalty for infeasible points and none otherwise. This function is then placed into an objective function, so that any violation of the constraints will result in penalty. A suitable penalty function $\Theta(x)$ is found to be

$$\Theta(x) = \sum_{i=1}^{p} \Psi[f_i(x)] + \sum_{i=p+1}^{b} \Phi[f_i(x)] \tag{8.159}$$

where

$$\begin{cases} \Psi(z) = 0, & z = 0 \\ \Psi(z) > 0, & z \neq 0 \end{cases} \tag{8.160a}$$

$$\begin{cases} \Phi(z) = 0, & z \leq 0 \\ \Phi(z) > 0, & z > 0 \end{cases} \tag{8.160b}$$

These functions typically assume the forms

$$\Psi(z) = |z|^k \tag{8.161a}$$

$$\Phi(z) = [\max(0, z)]^k \tag{8.161b}$$

where k is a positive integer. Thus, the above constrained problem can be converted into an unconstrained one, as follows: Choose $k = 2$.

Minimize $\quad f(x) + \mu \Theta(x)$
subject to $\quad x \in \Re^n$, the n-dimensional real space
where

$$\Theta(x) = \sum_{i=1}^{p} |f_i(x)|^2 + \sum_{i=p+1}^{b} \{\max[0, f_i(x)]\}^2. \qquad (8.162)$$

Let $x^{(j)}$ be the optimal solution at the jth iteration for the penalty parameter $\mu^{(j)}$. If $\mu^{(j)} \Theta(x^{(j)})$ is less than the preassigned tolerance, change the penalty parameter to $\mu^{(j+1)} = \beta \mu^{(j)}$ for some scalar β, and continue the process.

In terms of our floorplan area minimization problem, the objective function is chosen to be

$$f(V_{pt}, I_{p\tilde{\imath}}) = \left(V_0 I_0 - \sum_{i=1}^{b} a_i \right)^2. \qquad (8.163)$$

This function is to be minimized. The equality constraint is obtained from (8.149) as

$$g(V_{pt}, I_{p\tilde{\imath}}) \equiv [g_1 \quad g_2 \quad \cdots \quad g_{b+1}]' \equiv \begin{bmatrix} V_{p\tilde{\imath}} + B_{ft} V_{pt} + B_f E \\ I_{pt} + Q_{ft} I_{p\tilde{\imath}} + Q_f J \end{bmatrix} = 0. \qquad (8.164)$$

From (8.150) and (8.151), the minimum dimensional constraint function is given by

$$q_1(V_{pt}, I_{p\tilde{\imath}}) \equiv [q_{11} \quad q_{12} \quad \cdots \quad q_{1,2b}]' \equiv \begin{bmatrix} H_m - \tilde{V}_p \\ W_m - \tilde{I}_p \end{bmatrix} \le 0 \qquad (8.165)$$

and from (8.152) the area constraint becomes

$$q_2(V_{pt}, I_{p\tilde{\imath}}) \equiv [q_{21} \quad q_{22} \quad \cdots \quad q_{2b}]' = a - p \le 0. \qquad (8.166)$$

The desired penalty function can be written as

$$\Theta(V_{pt}, I_{p\bar{\imath}}) = \sum_{i=1}^{b+1} g_i^2(V_{pt}, I_{p\bar{\imath}}) + \sum_{i=1}^{2b} \left\{ \max\left[0, q_{1i}(V_{pt}, I_{p\bar{\imath}})\right]\right\}^2$$
$$+ \sum_{j=1}^{b} \left\{ \max\left[0, q_{2j}(V_{pt}, I_{p\bar{\imath}})\right]\right\}^2. \quad (8.167)$$

Since for any solution $(V_{pt}, I_{p\bar{\imath}})$, (8.164) can always be satisfied by choosing

$$\begin{bmatrix} V_{p\bar{\imath}} \\ I_{pt} \end{bmatrix} = -\begin{bmatrix} B_{ft}V_{pt} + B_f E \\ Q_{fi}I_{p\bar{\imath}} + Q_f J \end{bmatrix} \quad (8.168)$$

the first summation on the right-hand side of (8.167) can be set to zero, and the penalty function is simplified to

$$\Theta(V_{pt}, I_{p\bar{\imath}}) = \sum_{i=1}^{2b} \left\{\max\left[0, q_{1i}(V_{pt}, I_{p\bar{\imath}})\right]\right\}^2 + \sum_{j=1}^{b} \left\{\max\left[0, q_{2j}(V_{pt}, I_{p\bar{\imath}})\right]\right\}^2. \quad (8.169)$$

Thus, the unconstrained floorplan area minimization problem can be stated as follows:

Minimize
$$f(V_{pt}, I_{p\bar{\imath}}) + \mu \Theta(V_{pt}, I_{p\bar{\imath}})$$
subject to
$$\begin{bmatrix} V_{pt} \\ I_{p\bar{\imath}} \end{bmatrix} \in \Re^{b+1}$$

where μ is a penalty parameter, and $f(V_{pt}, I_{p\bar{\imath}})$ and $\Theta(V_{pt}, I_{p\bar{\imath}})$ are defined in (8.163) and (8.169), respectively.

We illustrate the above result by an example of WANG and CHEN [1993].

Example 8.3: Consider the floorplan $\mathcal{R}$ of fig. 8.32, which is redrawn in fig. 8.47, for the microprocessor compatible 8-bit A/D converter with 8-channel multiplexer shown in fig. 8.31. The areas a_k of the building blocks are required to satisfy the following constraints:

$$a_1 \geq 15, \quad a_2 \geq 5, \quad a_3 \geq 30, \quad a_4 \geq 1.5, \quad a_5 \geq 4.5,$$
$$a_6 \geq 7.5, \quad a_7 \geq 5.5, \quad a_8 \geq 16, \quad a_9 \geq 5, \quad a_{10} \geq 10. \quad (8.170)$$

In addition, all the height and width are required to be greater than or equal to 2. We wish to find a minimum wasted area floorplan.

The objective function is given by

$$f(V_{pt}, I_{p\tilde{i}}) = \left(V_0 I_0 - \sum_{i=1}^{10} a_i\right)^2 = (V_0 I_0 - 100)^2. \tag{8.171}$$

The minimum dimensional constraint function is obtained from (8.165) as

$$q_1(V_{pt}, I_{p\tilde{i}}) \equiv \begin{bmatrix} q_{11} & q_{12} & \cdots & q_{1,20} \end{bmatrix}' \equiv \begin{bmatrix} H_m - \tilde{V}_p \\ W_m - \tilde{I}_p \end{bmatrix} \leq 0 \tag{8.172}$$

where

$$\tilde{V}_p = \begin{bmatrix} v_1 & v_2 & v_4 & v_5 & v_7 & v_3 & v_6 & v_8 & v_9 & v_{10} \end{bmatrix}' \tag{8.173a}$$

$$\tilde{I}_p = \begin{bmatrix} i_1 & i_2 & i_4 & i_5 & i_7 & i_3 & i_6 & i_8 & i_9 & i_{10} \end{bmatrix}' \tag{8.173b}$$

$$H_m = \begin{bmatrix} 2 & 2 & 2 & 2 & 2 & 2 & 2 & 2 & 2 \end{bmatrix}' \tag{8.174a}$$

$$W_m = \begin{bmatrix} 2 & 2 & 2 & 2 & 2 & 2 & 2 & 2 & 2 \end{bmatrix}'. \tag{8.174b}$$

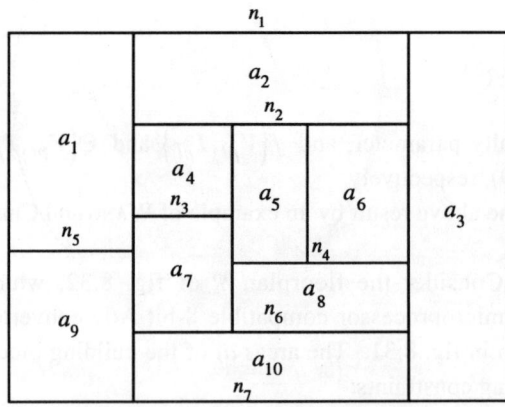

Fig. 8.47. A floorplan for a 8-bit A/D converter of fig. 8.31.

From Example 8.2, the c-digraph G_c of the floorplan of fig. 8.47 is redrawn in fig. 8.48. In G_c choose e_0, e_1, e_2, e_4, e_5, and e_7 to be the branches of a tree t. Then the links of its cotree $\bar{t}$ are given by e_3, e_6, e_8, e_9, and e_{10}. With respect to this tree t, the f-cutset matrix Q_f and the f-circuit matrix B_f are found to be

$$Q_f = \begin{bmatrix} \mathbf{1}_6 & Q_{f\bar{t}} \end{bmatrix} = \begin{bmatrix} 1 & 0 & 0 & 0 & 0 & 0 & -1 & 0 & 0 & -1 & -1 \\ 0 & 1 & 0 & 0 & 0 & 0 & 0 & 0 & 0 & -1 & 0 \\ 0 & 0 & 1 & 0 & 0 & 0 & 0 & 0 & 0 & 0 & -1 \\ 0 & 0 & 0 & 1 & 0 & 0 & 0 & 0 & 1 & 0 & -1 \\ 0 & 0 & 0 & 0 & 1 & 0 & 0 & 1 & -1 & 0 & 0 \\ 0 & 0 & 0 & 0 & 0 & 1 & 0 & 0 & 1 & 0 & -1 \end{bmatrix} \qquad (8.175)$$

$$B_f = \begin{bmatrix} B_{f t} & \mathbf{1}_5 \end{bmatrix} = \begin{bmatrix} 1 & 0 & 0 & 0 & 0 & 0 & 1 & 0 & 0 & 0 & 0 \\ 0 & 0 & 0 & 0 & -1 & 0 & 0 & 1 & 0 & 0 & 0 \\ 0 & 0 & 0 & -1 & 1 & -1 & 0 & 0 & 1 & 0 & 0 \\ 1 & 1 & 0 & 0 & 0 & 0 & 0 & 0 & 0 & 1 & 0 \\ 1 & 0 & 1 & 1 & 0 & 1 & 0 & 0 & 0 & 0 & 1 \end{bmatrix} \qquad (8.176)$$

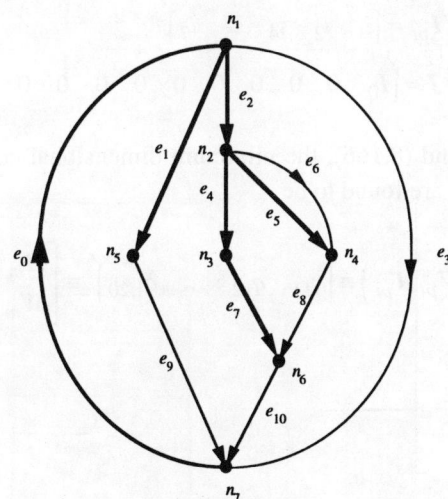

Fig. 8.48. The c-digraph G_c of the floorplan of fig. 8.42.

From (8.164), the equality constraint becomes

$$g(V_{pt}, I_{p\bar{t}}) \equiv \begin{bmatrix} g_1 \\ g_2 \\ g_3 \\ g_4 \\ g_5 \\ g_6 \\ g_7 \\ g_8 \\ g_9 \\ g_{10} \\ g_{11} \end{bmatrix} \equiv \begin{bmatrix} V_{p\bar{t}} + B_{ft}V_{pt} + B_f E \\ I_{pt} + Q_{ft}I_{p\bar{t}} + Q_f J \end{bmatrix} = \begin{bmatrix} v_3 + V_0 \\ v_6 - v_5 \\ v_8 - v_4 + v_5 - v_7 \\ v_1 + v_9 + V_0 \\ v_2 + v_4 + v_7 + v_{10} + V_0 \\ -i_3 - i_9 - i_{10} + I_0 \\ i_1 - i_9 \\ i_2 - i_{10} \\ i_4 + i_8 - i_{10} \\ i_5 + i_6 - i_8 \\ i_7 + i_8 - i_{10} \end{bmatrix} = \mathbf{0}$$

(8.177)

where $E = 0$ and

$$V_{p\bar{t}} = \begin{bmatrix} v_3 & v_6 & v_8 & v_9 & v_{10} \end{bmatrix}'$$ (8.178a)

$$I_{pt} = \begin{bmatrix} i_1 & i_2 & i_4 & i_5 & i_7 \end{bmatrix}'$$ (8.178b)

$$J = \begin{bmatrix} I_0 & 0 & 0 & 0 & 0 & 0 & 0 & 0 & 0 & 0 \end{bmatrix}'.$$ (8.178c)

From (8.164) and (8.166), the minimum dimensional constraint function and the area constraints are found to be

$$q_1(V_{pt}, I_{p\bar{t}}) \equiv \begin{bmatrix} q_{11} & q_{12} & \cdots & q_{1,20} \end{bmatrix}' \equiv \begin{bmatrix} H_m - \tilde{V}_p \\ W_m - \tilde{I}_p \end{bmatrix}$$ (8.179a)

where

$$H_m - \tilde{V}_p = \begin{bmatrix} 2-v_1 \\ 2-v_2 \\ 2-v_4 \\ 2-v_5 \\ 2-v_7 \\ 2-v_3 \\ 2-v_6 \\ 2-v_8 \\ 2-v_9 \\ 2-v_{10} \end{bmatrix}, \quad W_m - \tilde{I}_p = \begin{bmatrix} 2-i_1 \\ 2-i_2 \\ 2-i_4 \\ 2-i_5 \\ 2-i_7 \\ 2-i_3 \\ 2-i_6 \\ 2-i_8 \\ 2-i_9 \\ 2-i_{10} \end{bmatrix} \quad (8.179b)$$

and

$$q_2(V_{pt}, I_{p\tilde{i}}) \equiv \begin{bmatrix} q_{21} \\ q_{22} \\ q_{23} \\ q_{24} \\ q_{25} \\ q_{26} \\ q_{27} \\ q_{28} \\ q_{29} \\ q_{2,10} \end{bmatrix} \equiv a - p = \begin{bmatrix} 15 - v_1 i_1 \\ 5 - v_2 i_2 \\ 1.5 - v_4 i_4 \\ 4.5 - v_5 i_5 \\ 5.5 - v_7 i_7 \\ 30 - v_3 i_3 \\ 7.5 - v_6 i_6 \\ 16 - v_8 i_8 \\ 5 - v_9 i_9 \\ 10 - v_{10} i_{10} \end{bmatrix} \leq 0. \quad (8.179c)$$

Finally, from (8.169), the penalty function $\Theta(V_{pt}, I_{p\tilde{i}})$ is found to be

$$\Theta(V_{pt}, I_{p\tilde{i}}) = \sum_{i=1}^{20} \left\{ \max\left[0, q_{1i}(V_{pt}, I_{p\tilde{i}})\right] \right\}^2 + \sum_{j=1}^{10} \left\{ \max\left[0, q_{2j}(V_{pt}, I_{p\tilde{i}})\right] \right\}^2. \quad (8.180)$$

The corresponding floorplan area minimization problem can now be stated as follows:

Minimize

$$f(V_{pt}, I_{\bar{t}}) + \mu \Theta(V_{pt}, I_{\bar{t}})$$

subject to

$$\begin{bmatrix} V_{pt} \\ I_{p\bar{t}} \end{bmatrix} \in \Re^{b+1}$$

In Example 8.2, when there are no dimensional constraints, a zero wasted area floorplan is possible. The corresponding tree-branch voltages and cotree-link currents were found to be

$$\begin{bmatrix} V^{(0)}_{pt} \\ I^{(0)}_{p\bar{t}} \end{bmatrix} = \begin{bmatrix} v^{(0)}_0 \\ v^{(0)}_1 \\ v^{(0)}_2 \\ v^{(0)}_4 \\ v^{(0)}_5 \\ v^{(0)}_7 \\ i^{(0)}_3 \\ i^{(0)}_6 \\ i^{(0)}_8 \\ i^{(0)}_9 \\ i^{(0)}_{10} \end{bmatrix} = \begin{bmatrix} 10 \\ 7.5 \\ 1 \\ 1.5 \\ 3 \\ 5.5 \\ 3 \\ 2.5 \\ 4 \\ 2 \\ 5 \end{bmatrix}. \quad (8.181)$$

Using this as our initial solution and choosing the initial penalty parameter $\mu^{(0)} = 0.1$ and $\beta = 10$ with kth penalty parameter $\mu^{(k)} = \beta \mu^{(k-1)}$, an optimum solution was found after 8 iterations, using 5.42-second cpu time in a Sun workstation. The results are shown in Table 8.1. The corresponding minimum wasted area floorplan is presented in fig. 8.49. The total floorplan area is 119.14, of which 19.14 is wasted.

Table 8.1. Optimization sequences for Example 8.3.

Iteration	β	OF	Penalty
1	0.1	0.18569134175777E-01	0.20880711078644E+00
2	1.0	0.15997153520584E+00	0.20362472534180E+01
3	10.0	0.12812318801880E+01	0.16817169189453E+02
4	100.0	0.66730947494507E+01	0.66533477783203E+02
5	1000.0	0.14676589012146E+02	0.67054031372070E+02
6	10000.0	0.18562286376953E+02	0.94149866104126E+01
7	100000.0	0.19051223754883E+02	0.12082252502441E+01
8	1000000.0	0.19140869140625E+02	0.00000000000000E+00

The final dimensions are:

V(1)= 0.7544352E+01   I(1)= 0.2000280E+01   A= 0.1509082E+02
V(2)= 0.2001245E+01   I(2)= 0.6600647E+01   A= 0.1320951E+02
V(3)= 0.1036409E+02   I(3)= 0.2894621E+01   A= 0.3000011E+02
V(4)= 0.2061910E+01   I(4)= 0.2000202E+01   A= 0.4124236E+01
V(5)= 0.2884552E+01   I(5)= 0.2000106E+01   A= 0.5769409E+01
V(6)= 0.2884552E+01   I(6)= 0.2600339E+01   A= 0.7500813E+01
V(7)= 0.4300841E+01   I(7)= 0.2000202E+01   A= 0.8602551E+01
V(8)= 0.3478199E+01   I(8)= 0.4600445E+01   A= 0.1600125E+02
V(9)= 0.2819737E+01   I(9)= 0.2000280E+01   A= 0.5640265E+01
V(10)= 0.2000093E+01   I(10)= 0.6600647E+01   A= 0.1320190E+02

THE TOTAL AREA FOR THE FLOORPLAN = 0.1191408691E+03

THE TOTAL WASTED AREA = 0.1914082336E+02

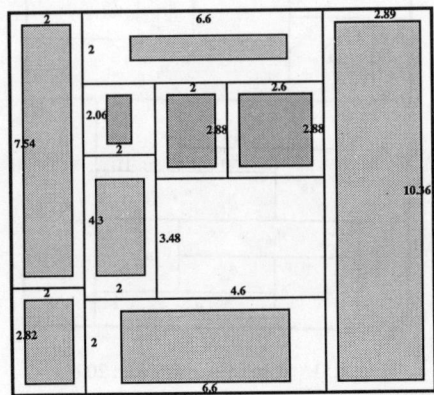

Fig. 8.49. A minimum wasted area floorplan of fig. 8.42.

The following example illustrates the ability of the algorithm to handle a large floorplan, and is taken from WANG [1993].

Example 8.4: Consider a 100-block nonslicing floorplan which is divided into five sub-floorplans FPi (i = 1, 2, 3, 4, 5) as shown in fig. 8.50. Each sub-floorplan FPi is composed of 20 blocks, and all of their topology is the same as shown in fig. 8.51 with the area constraints a_k of the blocks given by

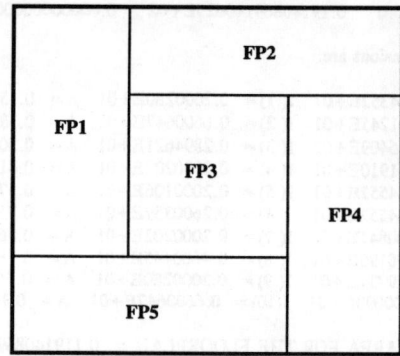

Fig. 8.50. A 100-block floorplan divided into five sub-floorplans.

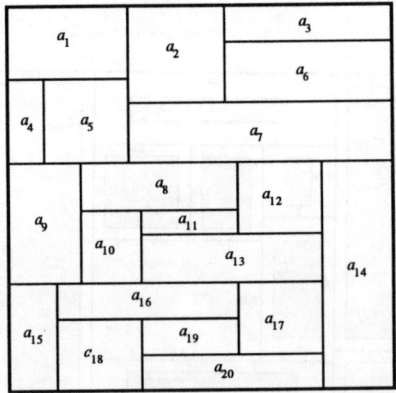

Fig. 8.51. A sub-floorplan with 20 blocks.

$a_1 = 8$, $a_2 = 6$, $a_3 = 8$, $a_4 = 2$, $a_5 = 6$, $a_6 = 4$, $a_7 = 6$, (8.182a)

$a_8 = 1.5$, $a_9 = 4$, $a_{10} = 1.5$, $a_{11} = 1$, $a_{12} = 3$, $a_{13} = 5$, $a_{14} = 12$, (8.182b)

$a_{15} = 4$, $a_{16} = 5$, $a_{17} = 6$, $a_{18} = 6$, $a_{19} = 6$, $a_{20} = 5$. (8.182c)

It is further stipulated that all height and width be no less than 2 and 0.5, respectively, for sub-floorplans FP1 and FP4, no less than 0.5 and 2 for FP2 and FP5, respectively; and no less than 1 for FP3.

We first use the algorithm developed in the preceding sections to obtain a zero wasted area floorplan without any dimensional constraints. It took only 0.08-second cpu time in a Sun workstation to accomplish this task and the result is presented in fig. 8.52. Using these as the initial values and constructing a corresponding unconstrained floorplan area minimization problem with an appropriate penalty function, the optimization process generates a layout in 110.33 seconds. The corresponding minimum wasted area floorplan is shown in fig. 8.53.

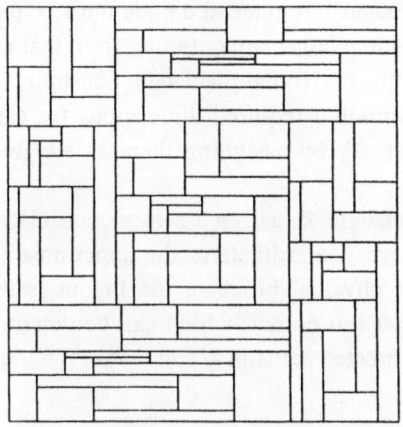

Fig. 8.52. A 100-block zero wasted area floorplan.

Fig. 8.53. A minimum wasted area floorplan for Example 8.4.

7.4. *From digraphs to layout*

In the preceding section, we formulated the problem of floorplan area minimization with constrained aspect ratios as an optimization problem with a set of equality and inequality constraints. In the present section, we demonstrate that the simplest and least complex method in obtaining a feasible, but not necessarily minimum area, solution is to compute the length of the longest paths in the p-digraph and the dual p-digraph.

Recall that for a dissected rectangle $\mathcal{R}$ there corresponds two digraphs called the p-digraph $\mathcal{G}_p$ and dual p-digraph $\mathcal{G}'_p$. The p-digraph $\mathcal{G}_p$ defines the vertical placement constraints by specifying the above/below relationships. They are reflected by the incidence of edges at each node of $\mathcal{G}_p$. For example, subrectangles whose corresponding edges terminate at a node must be placed to locations above those whose corresponding edges are outgoing from that node. Likewise, the dual p-digraph $\mathcal{G}'_p$ defines the horizontal placement constraints by specifying left/right relationships. The minimum required dimensions for each subrectangle can be introduced into $\mathcal{G}_p$ and $\mathcal{G}'_p$ by assigning them as weights to the corresponding edges.

In the dissected rectangle $\mathcal{R}$, assign a zero x-coordinate to the leftmost vertical line segment and a zero y-coordinate to the uppermost horizontal line segment. The position of each physical block in the layout is then determined by the coordinates of its upper left corner, which can be determined by calculating the length of the longest directed paths in $\mathcal{G}_p$ and $\mathcal{G}'_p$. We illustrate this procedure by the following example.

Example 8.5: Consider again the dissected rectangle $\mathcal{R}$ of fig. 8.47, which is redrawn in fig. 8.54. Assume that the dimensions of each physical block B_k, which has a width w_{mk} and a height h_{mk} and is to be placed inside the subrectangle $\mathcal{R}_k$ with area a_k of the floorplan of fig. 8.54, are specified as follows:

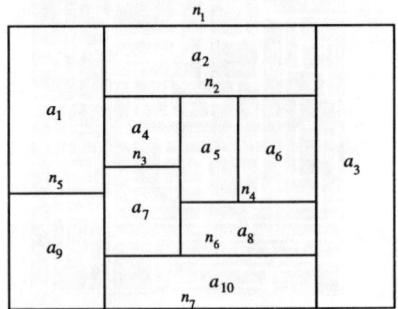

Fig. 8.54. A rectangle $\mathcal{R}$ with minimum required dimensions for its subrectangles.

$$\begin{aligned}
&w_{m1}=5, \quad h_{m1}=2, \quad w_{m2}=1, \quad h_{m2}=4, \quad w_{m3}=6,\\
&h_{m3}=2, \quad w_{m4}=1, \quad h_{m4}=1, \quad w_{m5}=2, \quad h_{m5}=1,\\
&w_{m6}=2, \quad h_{m6}=2, \quad w_{m7}=4, \quad h_{m7}=2, \quad w_{m8}=4,\\
&h_{m8}=3, \quad w_{m9}=2, \quad h_{m9}=2, \quad w_{m10}=2, \quad h_{m10}=4.
\end{aligned} \qquad (8.183)$$

From fig. 8.54, we next construct its p-digraph G_p, which was done in Example 8.3 and was given in fig. 8.48. Assigning the minimum required height h_{mk} to the edges of G_p results in the weighted p-digraph of fig. 8.55. Likewise, the weighted dual p-digraph G'_p of the dissected rectangle $\mathcal{R}$ of fig. 8.54 is constructed in fig. 8.56, where the weight corresponds to the minimum required width w_{mk}.

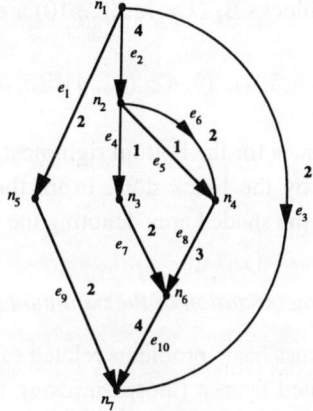

Fig. 8.55. A weighted p-digraph G_p of fig. 8.54.

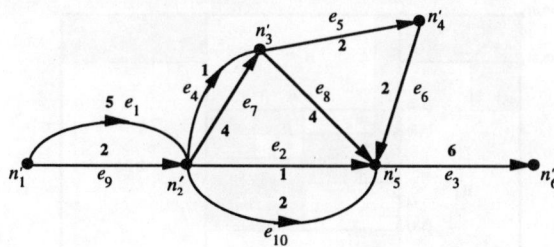

Fig. 8.56. A weighted dual p-digraph G'_p of fig. 8.54.

The longest directed paths from node n_1 to nodes n_k ($k = 2, 3, 4, 5, 6, 7$) in G_p are given by

$$e_2, \ e_2e_4, \ e_2e_6, \ e_1, \ e_2e_6e_8, \ e_2e_6e_8e_{10} \qquad (8.184)$$

having lengths 4, 5, 6, 2, 9 and 13, respectively. Similarly, the longest directed paths from node n'_1 to nodes n'_k ($k = 2, 3, 4, 5, 6$) are found to be

$$e_1, \ e_1e_7, \ e_1e_7e_5, \ e_1e_7e_8, \ e_1e_7e_8e_3 \qquad (8.185)$$

having lengths 5, 9, 11, 13, and 19, respectively. Thus, the coordinates of the upper left corner of each of the blocks B_k ($k = 1, 2, ..., 10$) are obtained as

$$(0, 0), \ (5, 0), \ (13, 0), \ (5, 4), \ (9, 4), \ (11, 4), \ (5, 5), \ (9, 6), \ (0, 2), \ (5, 9)$$

respectively. The coordinates for the bottom rightmost corner are (19,13). They are represented in fig. 8.57 by the black dots. From these coordinates, a layout is obtained in fig. 8.57 with the shaded area denoting the physical blocks.

7.5. Graph-theoretic characterization of the minimum area layout

In this section, we discuss basic problems related to the minimization of the area of a general layout obtained from a floorplan, using the graph model as our tool. We present a theorem that separates the product of the longest directed paths from the product of the shortest directed paths in a family of dual p-digraph pairs, and characterize a zero wasted area layout due to WIMER, KOREN and CEDERBAUM [1988].

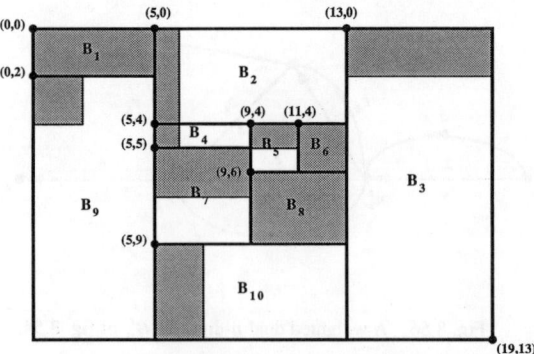

Fig. 8.57. A layout derived from the floorplan of fig. 8.54.

Recall that for every floorplan $\mathcal{R}$, there corresponds a p-digraph G_p and a dual p-digraph G'_p. The physical dimensions of the blocks are introduced into G_p and G'_p by assigning them as weights to the corresponding edges. The position of every block in the layout is then determined by the coordinates of its upper left corner corresponding to the length of the longest directed paths from the source node to the corresponding nodes of the blocks in G_p and G'_p.

Consider a family $\{(G_p, G'_p)\}$ of dual pairs (G_p, G'_p) of p-digraphs G_p and their dual p-digraphs G'_p. Let

$$h = \begin{bmatrix} h_1 & h_2 & \ldots & h_b \end{bmatrix} \tag{8.186}$$

$$w = \begin{bmatrix} w_1 & w_2 & \ldots & w_b \end{bmatrix} \tag{8.187}$$

be an arbitrary assignment of weights to the b-edge digraphs G_p and G'_p, respectively, such that for any two dual edges, the product of their weights equals a preassigned constant a_i:

$$h_j w_j = a_j, \quad 1 \leq j \leq b. \tag{8.188}$$

Let $W_\ell(h)$ and $H_\ell(w)$ be the length of the longest directed paths in G_p and G'_p from the source node to the sink node, and let $W_s(h)$ and $H_s(w)$ be the length of the shortest directed paths in G_p and G'_p from the source node to the sink node, respectively. If w_i is interpreted as the width of a building block B_i, h_i as its height, and a_i as its area, using the procedure outlined in the preceding section, a layout can be derived from G_p and G'_p. Since this is a valid layout, no two blocks overlap. We conclude that for every edge length assignment satisfying (8.188) we have

$$W_\ell(h) H_\ell(w) \geq \sum_{j=1}^{b} h_j w_j = \sum_{j=1}^{b} a_j. \tag{8.189}$$

Since this is true for every element in $\{(G_p, G'_p)\}$, it must be true for the pair with the minimum product of the length of their longest directed paths, or

$$\min_{(G_p, G'_p) \in \{(G_p, G'_p)\}} W_\ell(h) H_\ell(w) \geq \sum_{j=1}^{b} h_j w_j = \sum_{j=1}^{b} a_j. \tag{8.190}$$

Likewise, if we use the length of the shortest directed paths in G_p and G'_p to determine the coordinates of the upper left corner of a block instead of using the length of the longest directed paths, we generate an invalid layout in which there is no vacant area, some blocks may overlap, and some may extend beyond the right and bottom edges of the layout $\mathcal{R}$, and obtain

$$W_s(h)H_s(w) \leq \sum_{j=1}^{b} h_j w_j = \sum_{j=1}^{b} a_j. \tag{8.191}$$

Since this is true for every element in $\{(\mathcal{G}_p, \mathcal{G}'_p)\}$, it must also be true for the maximum product of the length of the shortest directed paths, or

$$\max_{(\mathcal{G}_p, \mathcal{G}'_p) \in \{(\mathcal{G}_p, \mathcal{G}'_p)\}} W_s(h)H_s(w) \leq \sum_{j=1}^{b} h_j w_j = \sum_{j=1}^{b} a_j. \tag{8.192}$$

Combining this with (8.190) yields

$$\min_{(\mathcal{G}_p, \mathcal{G}'_p) \in \{(\mathcal{G}_p, \mathcal{G}'_p)\}} W_\ell(h)H_\ell(w) \geq \sum_{j=1}^{b} a_j \geq \max_{(\mathcal{G}_p, \mathcal{G}'_p) \in \{(\mathcal{G}_p, \mathcal{G}'_p)\}} W_s(h)H_s(w). \tag{8.193}$$

In words, (8.193) can be stated as follows:

THEOREM 8.11: *In a family of dual pairs for which the product of their dual edges' length is invariant, there exists a constant which separates the product of the length of their shortest directed paths from the product of the length of their longest directed paths.*

For each dual pair $(\mathcal{G}_p, \mathcal{G}'_p)$ in the family $\{(\mathcal{G}_p, \mathcal{G}'_p)\}$, there is a layout implied by this pair. We next characterize such a pair so that a zero wasted area layout is possible. The necessary and sufficient conditions for $(\mathcal{G}_p, \mathcal{G}'_p)$ to represent such an optimal layout is given by the following theorem due to WIMER, KOREN and CEDERBAUM [1988].

THEOREM 8.12: *The necessary and sufficient condition for a layout generated by a dual pair of digraphs to be a zero wasted area layout is that every directed path from the source node to the sink node is of maximal length.*

Proof. Necessity. Let $(\mathcal{G}_p, \mathcal{G}'_p)$ be the dual pair of some zero wasted area layout, the area of which is A. Let a_j be the area of the jth rectangular module (block), whose width is w_j and height h_j, $1 \leq j \leq b$. Choose a longest directed path P_1 of length ℓ_1 from the source node to the sink node in $\mathcal{G}_p$. Assume to the contrary that $\mathcal{G}_p$ contains a directed path

$$P_2 = (j_1, j_2)(j_2, j_3) \cdots (j_{i-1}, j_i)(j_i, j_{i+1}) \cdots (j_{m-1}, j_m) \tag{8.194}$$

from the source node to the sink node of length $\ell_2 = m - 1 < \ell_1$, $m \geq 2$. Let j_i be the last node on P_2 such that the subpath $(j_1, j_2)(j_2, j_3) \cdots (j_{i-1}, j_i)$ is a longest directed path from j_1 to j_i. Clearly, $j_i \neq j_m$, because P_2 is not a longest directed path. Then the edge (j_i, j_{i+1}) cannot be a longest directed path from j_i to j_{i+1}, and a longer directed path from j_i to j_{i+1} exists. Therefore, (j_i, j_{i+1}) does not belong to any longest directed path from the source node to the sink node in $\mathcal{G}_p$.

Let P_3 be a directed path of maximum length ℓ_3 from the source node to the sink node in $\mathcal{G}_p$ containing the edge (j_i, j_{i+1}). Since $\ell_3 < \ell_1$, we have

$$\ell_1 - \ell_3 = \varepsilon > 0. \tag{8.195}$$

We now transform the dual pair $(\mathcal{G}_p, \mathcal{G}'_p)$ to another one $(\mathcal{H}_p, \mathcal{H}'_p)$, as follows: $\mathcal{H}_p$ is obtained from $\mathcal{G}_p$ by replacing the edge $e_G = (j_i, j_{i+1})$ of length $\ell(e_G)$ by two series edges (heavy lines) e_{G1} $e_{G2} = (j_i, k)(k, j_{i+1})$ of lengths ε and $\ell(e_G)$, respectively, as depicted in fig. 8.58, whereas $\mathcal{H}'_p$ is obtained from $\mathcal{G}'_p$ by replacing the dual edge e_H of $e_G = (j_i, j_{i+1})$ by two parallel edges e_{H1} and e_{H2} of the same length $\ell(e_H)$ as that of e_H. These two parallel edges e_{H1} and e_{H2} of $\mathcal{H}'_p$ become the dual edges of $e_{G1} = (j_i, k)$ and $e_{G2} = (k, j_{i+1})$ in $\mathcal{H}_p$, respectively, ensuring that $\mathcal{H}_p$ and $\mathcal{H}'_p$ are the dual pair corresponding to a floorplan.

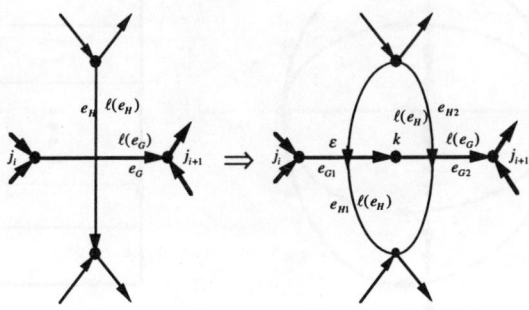

Fig. 8.58. Dual digraph pair transformation used in the proof of Theorem 8.12.

Observe that the length of the longest directed paths from the source node to the sink node in G_p and G'_p are the same as that of $\mathcal{H}_p$ and $\mathcal{H}'_p$. Hence, the transformed dual pair $(\mathcal{H}_p, \mathcal{H}'_p)$ imposes a layout with $b + 1$ blocks and having the same area A where e_{G1} and e_{H1} are the dual edges corresponding to the additional block, obtaining the inequality

$$A \geq \sum_{i=1}^{b+1} w_i h_i = \varepsilon \ell(e_H) + \sum_{i=1}^{b} w_i h_i > \sum_{i=1}^{b} a_i = A \qquad (8.196)$$

which is impossible. Therefore, all the directed paths from the source node to the sink node in G_p and G'_p are of maximum length, and the condition stated in the theorem is necessary.

Sufficiency. We prove sufficiency by induction over the number k of edges in a longest directed path from the source node to the sink node in G_p. For $k = 1$, all the longest directed paths contain only one edge, being parallel edges of equal length W, as shown in fig. 8.59. Their dual edges in $\mathcal{H}'_p$ form a single longest directed path of length H from the source node to the sink node. Thus, we obtain

$$A = WH = W \sum_{i=1}^{b} h_i = \sum_{i=1}^{b} w_i h_i = \sum_{i=1}^{b} a_i. \qquad (8.197)$$

This shows that there is no wasted area.

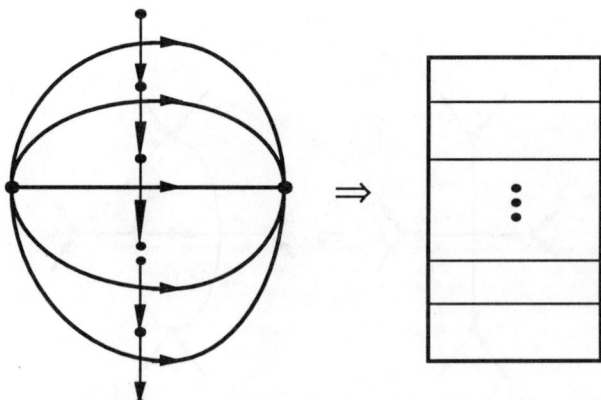

Fig. 8.59. The number of edges in a longest directed path is one.

To complete our proof, we assume that the theorem is sufficient if the number of edges contained in a longest directed path from the source node to the sink node in G_p is not greater than $m - 1$, $m \geq 2$. We show that it is also true if G_p contains an m-edge longest directed path. This is accomplished by reducing the m-edge directed path to a $(m-1)$-edge directed path by applying a finite series of transformations to G_p and G'_p. To this end, let e_j be the m edges of length h_j incident at the sink node, as depicted in fig. 8.60(a). Let

$$h_\ell = \min_{1 \leq j \leq m} h_j. \tag{8.198}$$

Now we replace every edge e_j for which $h_j > h_\ell$ by two series edges e_{j1} and e_{j2}, the length h_{j1} and h_{j2} of which are $h_{j1} = h_\ell$ and $h_{j2} = h_j - h_\ell$, respectively. The corresponding dual edges in G'_p are replaced by two parallel edges having the same length as shown in fig. 8.60(b). Observe that the layout imposed by the new dual pairs has the property that the length of the longest directed paths in the new dual pair of fig. 8.60(b) is the same as that in the original pair of fig. 8.60(a), and that the sum of block areas in the new dual pair equals that of the original pair. Now the nodes of the parallel edges of G'_p in fig. 8.60(b) are each split into two to produce a new dual p-digraph $\tilde{G}'_p$, so that the parallel edges can be equally divided between the two sets of nodes, as shown in fig. 8.60(c), resulting in two disjoint but equally long directed paths in $\tilde{G}'_p$. Finally, the new nodes just introduced in G_p and the initial node of the edge e_ℓ are coalesced to a single node, as indicated in fig. 8.60(d), yielding a separable p-digraph $\tilde{G}_p$ with components $\tilde{G}_{p1}$ and $\tilde{G}_{p2}$. Let $\tilde{G}'_{p1}$ and $\tilde{G}'_{p2}$ denote the corresponding digraphs in $\tilde{G}'_p$. The layout imposed by the component dual pair $(\tilde{G}_{p2}, \tilde{G}'_{p2})$ has zero wasted area as previously shown in fig. 8.59. Since the longest directed paths from the source node to the sink node in $\tilde{G}'_{p1}$ and $\tilde{G}'_{p2}$ have the same length, the two layouts induced by the dual pairs $(\tilde{G}_{p1}, \tilde{G}'_{p1})$ and $(\tilde{G}_{p2}, \tilde{G}'_{p2})$ have the same height. Therefore, they can be matched without any wasting area. If the layout induced by the dual pair $(\tilde{G}_{p1}, \tilde{G}'_{p1})$ has zero wasted area, we are finished. Otherwise, the number of edges in $\tilde{G}_{p1}$ and $\tilde{G}'_{p1}$ is at least one less than that of G_p and G'_p, and the number of edges in one directed path is at least one less than that of the corresponding directed path in G_p. Now the process can be repeated by applying the transformation to $(\tilde{G}_{p1}, \tilde{G}'_{p1})$. Then after a finite number of such steps, the maximum number of edges of any directed path in the resulting dual pair must not exceed $m - 1$. By induction hypothesis, such a dual pair imposes a zero wasted area layout, and sufficiency follows. This completes the proof of the theorem.

In fact, if a zero wasted area layout exists, then a whole family of zero wasted area layouts can be obtained by applying a uniform stretch in one dimension while the other dimension is contracted by a reciprocal magnitude. If W, H and A denote the width, height, and area of the layout induced by the dual pair $(\mathcal{G}_p, \mathcal{G}'_p)$, respectively, and $\mu = H/W$ is its aspect ratio, then for every $\lambda > 0$ there exists a zero wasted area layout of area A and aspect ratio λ. This can easily be accomplished by multiplying the width of each original block by $\sqrt{\mu/\lambda}$ and the height by $\sqrt{\lambda/\mu}$, yielding a layout having the same block areas as the original one. The aspect ratio of the resulting layout is found to be

$$\frac{H\sqrt{\lambda/\mu}}{W\sqrt{\mu/\lambda}} = \frac{\lambda}{\mu}\frac{H}{W} = \frac{\lambda}{\mu}\mu = \lambda \qquad (8.199)$$

as expected.

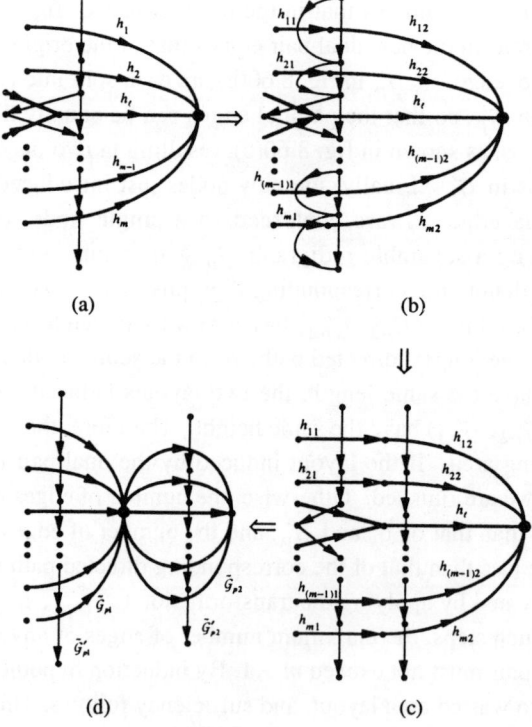

Fig. 8.60. A process of reducing the number of edges in a longest directed path.

§ 8. Conclusions

In this chapter, we discussed the problem of dividing a rectangle into a finite number of non-overlapping squares, no two of which are equal. If all the elements are unequal, the resulting dissection is perfect. It was shown that there are no perfect rectangles of order less than 9, and exactly two of order 9. A simple squared rectangle is a squared rectangle with no subset of its elements arranged in the form of a rectangle. It can be shown that there is no simple perfect square of order less than 21, and no compound perfect square of order less than 22. Therefore, there is no perfect square of order less than 21. We presented the simple perfect square of lowest order 21. Also, we indicated that there are no compound perfect squares of order less than 24 and there is one and only one compound perfect square of order 24.

There are 2 simple perfect rectangles of order 9, 6 of order 10, 22 of order 11, 67 of order 12 and 214 of order 13. In addition, there are 1 simple imperfect rectangle of order 9, 0 of order 10, 0 of order 11, 9 of order 12 and 33 of order 13. Even with these low orders, it is impractical to draw out all of them. To overcome this difficulty, we presented a coding system known as the Bouwkamp code.

The problem of dissecting a rectangle or square into squares is closely related to the solution of a resistive network called the p-net and dual p-net. Their associated digraphs are known as the p-digraph and dual p-digraph. If an excitation source is applied to a p-net or dual p-net, the resulting net becomes the c-net or dual c-net, and their corresponding digraphs are referred to as the c-digraph or dual c-digraph. We demonstrated that the branch currents and voltages in the c-net correspond to the width and height of the subrectangles of a dissection. Furthermore, we characterized these digraphs by showing that there exists a one-to-one correspondence between squared rectangles and certain classes of planar graphs, and that each simple perfect rectangle corresponds to a 3-connected planar c-digraph derived from the p-digraph. For each p-net, there is a corresponding squared rectangle. This correspondence is many to one. By deleting branches of zero currents and nodes at which all currents are zero, the normal form of the p-net can be defined. Then there is a one-to-one correspondence between the class of p-nets having the same normal form and the class of oriented squared rectangles. We correlate the simple squared rectangles to 3-connected planar digraphs, and compound squared rectangles to 2-connected planar digraphs. The main result is that every squared rectangle has commensurable sides and elements, and every rectangle with commensurable sides is perfectible in an infinity of essentially different ways.

In VLSI design, the first step is to obtain a partition of a rectangular floor called the floorplan. This problem is equivalent to dissecting a rectangle into subrectangles of fixed areas. If the aspect ratios of the subrectangles are allowed to change continuously, it is possible to find a zero wasted area floorplan. In fact, if a zero wasted area layout exists, then a whole family of zero wasted area layouts can be obtained by applying a uniform stretch in one dimension while the other dimension is contracted by a reciprocal magnitude. The solution of the zero wasted area floorplan problem is equivalent to that of solving a system of nonlinear nodal equations associated with a resistive network, the solution of which can be obtained numerically by using the multidimensional Newton–Raphson algorithm, if necessary. In practice, however, a module may have physical constraints on its dimensions such as height and width. In most of these cases, a zero wasted area floorplan may not be achievable. For this we introduced an optimization procedure to find a minimum wasted area floorplan.

Finally, we discussed the basic problems related to the minimization of the area of a general layout obtained from a floorplan by using the graph model as our tool. We showed that in a family of dual pairs of p-digraphs and their dual p-digraphs for which the product of their dual edges' length is invariant, there exists a constant which separates the product of the length of their shortest directed paths from the product of the length of their longest directed paths, and that the necessary and sufficient condition for a layout generated by a dual pair of digraphs to be a zero wasted area layout is that every directed path from source node to the sink node is of maximal length.

Problems

8.1. Write the Bouwkamp code for the following simple perfect 2 × 1 rectangle of order 23 (FEDERICO [1970]).

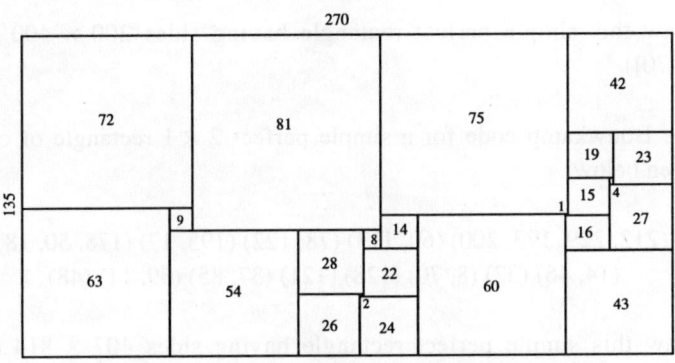

Fig. 8.61. A simple perfect 2 × 1 rectangle of order 23.

8.2. The Bouwkamp code for a simple perfect 2 × 1 rectangle of order 24 is given below:

(92, 71, 93, 53, 43) (10, 33) (40, 23) (21, 50) (17, 39)
(84, 29) (45, 83, 22) (61) (55, 24) (7, 38) (31)

Draw this simple perfect rectangle having sides 176 × 352 (FEDERICO [1970]).

8.3. The Bouwkamp code for a simple perfect 2 × 1 rectangle of order 24 is given below:

(101, 88, 97, 74) (23, 51) (35, 53) (47, 45, 28) (79, 22)
(57) (24, 55) (39, 14) (38, 7) (11, 36) (31) (25)

Draw this simple perfect rectangle having sides 180 × 360 (FEDERICO [1970]).

8.4. The Bouwkamp code for a 2 × 1 simple perfect rectangle of order 24 is given below:

(104, 79, 82, 73, 62) (11, 51) (44, 40) (25, 28, 26)
(47, 35) (96, 33) (2, 24) (30) (4, 87) (83) (71) (63)

Draw this simple perfect rectangle having sides 200 × 400 (FEDERICO [1970]).

8.5. The Bouwkamp code for a simple perfect 2 × 1 rectangle of order 25 is given below:

(212, 229, 173, 200) (60, 113) (78, 122) (195, 17) (178, 50, 18)
(14, 46) (32) (8, 70) (128) (121) (37, 85) (59, 11) (48)

Draw this simple perfect rectangle having sides 407 × 814 (FEDERICO [1970]).

8.6. The Bouwkamp code for a simple perfect 2 × 1 rectangle of lowest order 22 is given below:

(83, 49, 71, 69) (27, 22) (2, 67) (30, 65) (7, 20)
(53, 24, 13) (11, 17, 5) (35) (29, 6) (23)

Draw this lowest order simple perfect rectangle having sides 136 × 272 (DUIJVESTIJN [1979]).

8.7. In all perfect squared rectangles of small order 9, 10 or 11, the largest subsquare appears in a corner. This phenomenon tends to persist, although occasionally the largest square appears at the "middle" of one of the sides. The Bouwkamp code for a simple perfect squared rectangle of order 22 in which the largest subsquare appears in the "center" of the dissection is given below:

(419, 366, 174, 156, 255) (18, 138) (192) (39, 216) (177) (53, 505)
(472) (393) (7, 386) (133, 379) (359, 113) (246)

Draw this "centered" simple perfect rectangle having sides 1370 × 1250 (KAZARINOFF and WEITZENKAMP [1973b]).

8.8. Figure 8.62 shows squared rectangles containing one rectangle with nonnegative integers a and b. Can these squared rectangles be made squared squares containing one rectangle? (DUIJVESTIJN, FEDERICO, and LEEUW [1982]).

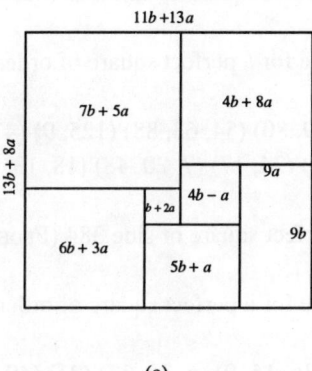

(a)

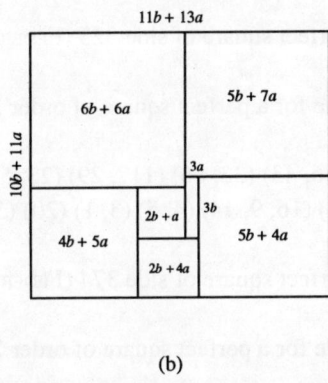

(b)

Fig. 8.62. Two squared rectangles each of which contains one rectangle.

8.9. The Bouwkamp code for a simple perfect rectangle of order 18 is given below:

(129, 61, 70) (23, 29, 9) (20, 59) (17, 6) (11, 44) (28) (157) (5, 54) (49) (103)

Draw this simple perfect rectangle.

8.10. The Bouwkamp code for a perfect square of order 25 is given below:

(124, 111) (43, 35, 33) (55, 39, 30) (2, 31) (8, 29) (81)
(16, 9, 14) (4, 5) (60) (3, 1) (20) (56, 18) (38)

Draw this simple perfect square of side 235 (FEDERICO [1963]).

8.11. The Bouwkamp code for a perfect square of order 26 is given below:

(179, 205) (99, 80) (54, 63, 88) (125, 9) (47, 25) (41, 58)
(22, 91) (69) (24, 17) (7, 20, 48) (18, 13) (5, 28) (23)

Draw this simple perfect square of side 384 (FEDERICO [1963]).

8.12. The Bouwkamp code for a perfect square of order 27 is given below:

(196, 60, 69) (36, 15, 9) (6, 35, 37) (21) (49, 8) (41, 2) (39)
(71, 58) (129, 67) (13, 45) (52, 32) (77) (62, 5) (57)

Draw this simple perfect square of side 325 (FEDERICO [1963]).

8.13. The Bouwkamp code for a perfect square of order 28 is given below:

(146, 109, 119) (66, 43) (33, 86) (117, 29) (23, 53) (88, 30) (169)
(55, 39, 111) (16, 9, 14) (4, 5) (3, 1) (20) (56, 18) (38)

Draw this simple perfect square of side 374 (FEDERICO [1963]).

8.14. The Bouwkamp code for a perfect square of order 29 is given below:

(146, 109, 119, 55, 39) (16, 9, 14) (4, 5) (3, 1) (20) (56, 18) (38) (66, 43)
(94) (33, 86) (117, 29) (23, 53) (88, 30) (263) (205)

Draw this simple perfect square of side 468 (FEDERICO [1963]).

8.15. The Bouwkamp code for a perfect square of order 30 is given below:

(128, 114, 149, 81, 94) (68, 13) (107) (14, 24, 22, 54) (106, 36) (2, 20)
(26) (85, 80, 52) (8, 12) (70) (66) (28, 131) (5, 103) (332) (234)

Draw this simple perfect square of side 566 (FEDERICO [1963]).

8.16. The Bouwkamp code for a perfect square of order 37 is given below:

(728, 378, 406, 435) (350, 28) (405, 29) (464) (468, 347, 83) (184, 206, 98)
(10, 454) (108) (162, 22) (336) (245, 102) (20, 142) (122)
(210, 54) (56, 189) (250, 594) (571, 133) (438, 94) (344)

Draw this simple perfect square of side 1947 (WILLCOCKS [1951]).

8.17. Consider the simple nonslicing floorplan $\mathcal{R}$ of fig. 8.35 with the following areas for its subrectangles:

$$a_1 = 50, \quad a_2 = 24, \quad a_3 = 18, \quad a_4 = 30, \quad a_5 = 28 \qquad (8.200)$$

Determine a zero wasted area floorplan.

8.18. Repeat Example 8.2 for the following building block areas:

$$\begin{aligned} a_1 &= 16, & a_2 &= 25, & a_3 &= 30, & a_4 &= 2, & a_5 &= 5 \\ a_6 &= 8, & a_7 &= 6, & a_8 &= 18, & a_9 &= 6, & a_{10} &= 11 \end{aligned} \qquad (8.201)$$

8.19. Repeat Example 8.3 if all the heights and widths are required to be greater than or equal to 3, everything else being the same.

8.20. Repeat Example 8.3 for the following constraints imposed on the areas of the building blocks, everything else being the same:

$$\begin{aligned} a_1 &\geq 16, & a_2 &\geq 25, & a_3 &\geq 30, & a_4 &\geq 2, & a_5 &\geq 5 \\ a_6 &\geq 8, & a_7 &\geq 6, & a_8 &\geq 18, & a_9 &\geq 6, & a_{10} &\geq 11 \end{aligned} \qquad (8.202)$$

CHAPTER 9

GRAPHICAL MATRICES

In ch. 2, we presented the incidence matrix, the circuit matrix, and the cut matrix associated with a graph or a directed graph. If an incidence matrix is given, the topological graph can easily be drawn directly from the graph or directed graph. A fundamental problem came to focus immediately is that whether a matrix of integers mod 2 is the basis circuit or cut matrix of a graph, or a real matrix of 1, −1 and 0 is the basis circuit or cut matrix of a directed graph. WHITNEY [1935] showed that there are matrices of integers mod 2 that are not circuit or cut matrix of any graph. To see this, let us consider the matrix

$$F = \begin{bmatrix} 1 & 0 & 0 & 0 & 1 & 1 & 1 \\ 0 & 1 & 0 & 1 & 0 & 1 & 1 \\ 0 & 0 & 1 & 1 & 1 & 0 & 1 \end{bmatrix}. \tag{9.1}$$

If this matrix is the cut matrix of some graph, it should be reducible to an incidence matrix with at most two 1's in each column by elementary row operations. This matrix cannot be reduced to such a matrix. In this chapter, we study this fundamental problem and present necessary and sufficient conditions under which a matrix is the circuit or cut matrix of a graph or directed graph due to TUTTE [1958, 1959].

§ 1. Totally Unimodular Matrix

Consider the directed graph G_d of fig. 2.20, which is redrawn in fig. 9.1. A basis cutset matrix Q_{d1} with respect to the four chosen cutsets as indicated is given by

$$Q_{d1} = \begin{bmatrix} 1 & 0 & 0 & 0 & 1 & -1 & 0 & 0 \\ 0 & 1 & 0 & 0 & 0 & 1 & -1 & 0 \\ 0 & 0 & 1 & 0 & 0 & 0 & 1 & -1 \\ 0 & 0 & 0 & 1 & -1 & 0 & 0 & 1 \end{bmatrix}. \tag{9.2}$$

Now suppose that we remove all the directions in G_d. This results in the associated undirected graph G of fig. 9.2. If the same set of cutsets as in G_d is chosen to construct a basis cutset matrix Q_1, retaining the same row and column ordering, we obtain

$$Q_1 = \begin{bmatrix} 1 & 0 & 0 & 0 & 1 & 1 & 0 & 0 \\ 0 & 1 & 0 & 0 & 0 & 1 & 1 & 0 \\ 0 & 0 & 1 & 0 & 0 & 0 & 1 & 1 \\ 0 & 0 & 0 & 1 & 1 & 0 & 0 & 1 \end{bmatrix} \quad (9.3)$$

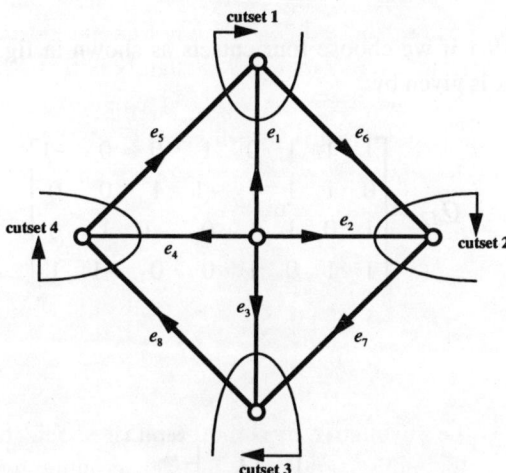

Fig. 9.1. A directed graph G_d with four chosen cutsets.

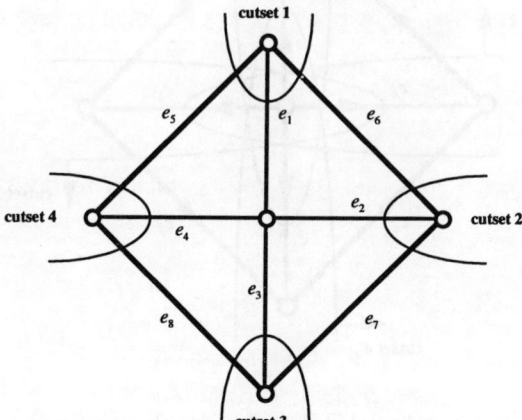

Fig. 9.2. The associated undirected graph G of the directed graph G_d of fig. 9.1.

Observe that both matrices have many properties in common. First of all, they have the nonzero elements in the same positions, and have the same rank. Since the major submatrices of Q_{d1} and Q_1 correspond to trees of G_d and G, respectively, the major submatrices of Q_{d1} and Q_1 correspond. In addition, the ranks of corresponding submatrices are unaltered. For example, the submatrix of Q_{d1}, formed by the first three rows and the last three columns is nonsingular, so is the corresponding submatrix in Q_1. In other words, the matrix Q_1 is very special in that some of its 1's may be appropriately replaced by -1's, so that the ranks of its submatrices remain unaltered. This is not always true for other cutset or circuit matrices.

Again, in fig. 9.1 if we choose four cutsets as shown in fig. 9.3, the resulting basis cutset matrix is given by

$$Q_{d2} = \begin{bmatrix} 1 & 1 & 1 & 0 & 1 & 0 & 0 & -1 \\ 0 & 1 & 1 & 1 & -1 & 1 & 0 & 0 \\ 1 & 0 & 1 & 1 & 0 & -1 & 1 & 0 \\ 1 & 1 & 0 & 1 & 0 & 0 & -1 & 1 \end{bmatrix} \tag{9.4}$$

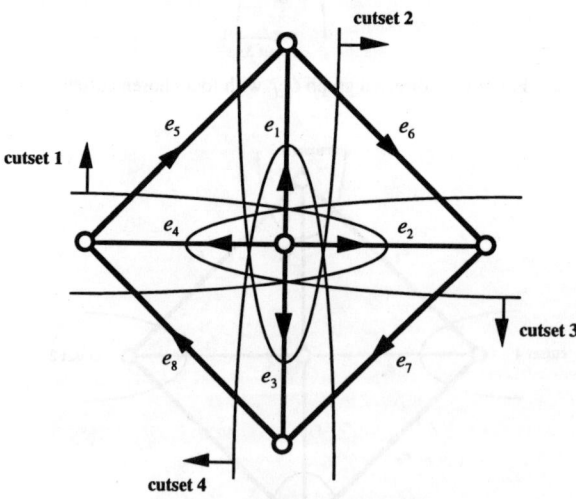

Fig. 9.3. A directed graph G_d with four chosen cutsets.

The associated undirected graph G is presented in fig. 9.4, and the corresponding basis cutset matrix is found to be

$$Q_2 = \begin{bmatrix} 1 & 1 & 1 & 0 & 1 & 0 & 0 & 1 \\ 0 & 1 & 1 & 1 & 1 & 1 & 0 & 0 \\ 1 & 0 & 1 & 1 & 0 & 1 & 1 & 0 \\ 1 & 1 & 0 & 1 & 0 & 0 & 1 & 1 \end{bmatrix}. \tag{9.5}$$

In this case, the matrix Q_2 cannot be replaced by a matrix of 1's, -1's, and 0's such that ranks of submatrices remain invariant. For example, the mod 2 rank of the submatrix of Q_2 formed by the first three rows and columns 3, 4 and 6 is 2, whereas the rank of the corresponding submatrix in Q_{d2} is 3.

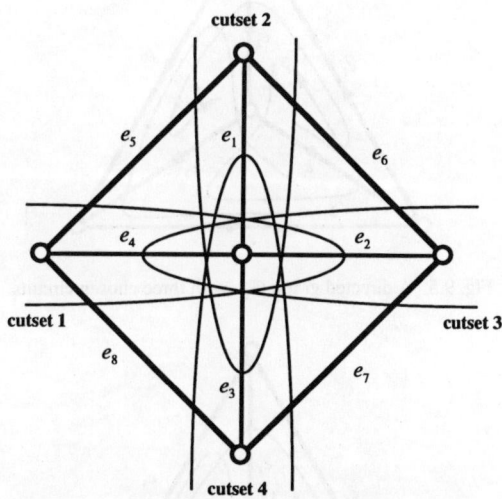

Fig. 9.4. The associated undirected graph G of the directed graph G_d of fig. 9.3.

As an example for the circuit matrix, consider the directed graph G_d of fig. 9.5. Choosing the three circuits as indicated, the basis circuit matrix is given by

$$B_d = \begin{bmatrix} 1 & 0 & 1 & 0 & -1 & 1 \\ 0 & 1 & 1 & 1 & 0 & -1 \\ 1 & -1 & 0 & 1 & 1 & 0 \end{bmatrix}. \tag{9.6}$$

Since the determinant of the major submatrix composed of the columns 1, 2 and 4 has a value 2, the submatrix is nonsingular. However, if we remove all the directions of the edges and choose the same set of circuits, the resulting undirected graph G is shown in fig. 9.6. The corresponding circuit matrix B is obtained from B_d by removing all the negative signs, and is given by

$$B = \begin{bmatrix} 1 & 0 & 1 & 0 & 1 & 1 \\ 0 & 1 & 1 & 1 & 0 & 1 \\ 1 & 1 & 0 & 1 & 1 & 0 \end{bmatrix}. \tag{9.7}$$

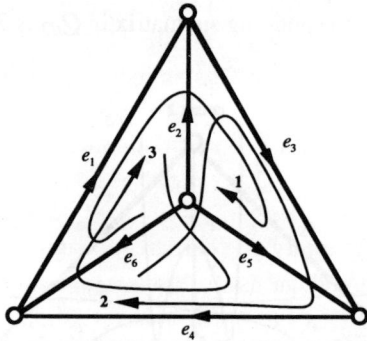

Fig. 9.5. A directed graph G_d with three chosen circuits.

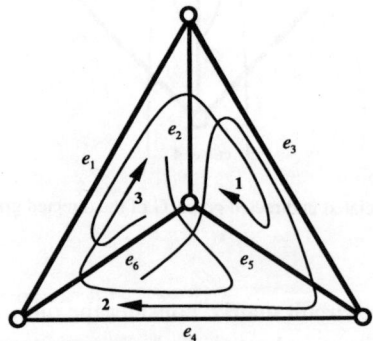

Fig. 9.6. The associated undirected graph G of the directed graph G_d of fig. 9.5.

Observe that the mod 2 rank of B is only 2, because the sum of its rows over the field mod 2 is zero, whereas the rank of B_d is 3. Therefore, it is not possible to replace an appropriate set of 1's in B by -1, so that the ranks of the corresponding submatrices are unaltered. Notice that B is not a basis circuit matrix, whereas B_d is. We now study conditions under which such a replacement is possible. Before we do this, we review the notion of a *totally unimodular matrix* as given in Definition 2.22. A totally unimodular matrix is also called an *E-matrix* by CEDERBAUM [1958], and plays a vital role in our discussion.

The transpose, the matrix obtained by any permutation of rows and columns, and all submatrices of a totally unimodular matrix are totally unimodular, so do matrices obtained by multiplying rows or columns by -1. The reason is that, apart from the sign, any element or subdeterminant of the new matrix is numerically equal to the corresponding element or subdeterminant of the original matrix. In addition, as established in Corollary 2.14, if a square nonsingular matrix is totally unimodular, so is its inverse. Also, the direct sum of two totally unimodular matrices is totally unimodular, and, as stated in Theorem 2.19, the incidence matrix, the f-circuit matrix, and the f-cut matrix are totally unimodular. The following theorem characterizes a totally unimodular matrix due to CEDERBAUM [1958].

THEOREM 9.1: The necessary and sufficient conditions for a real matrix F of order n to be a totally unimodular matrix are that on assuming any $n-1$ of the $2n$ variables to be zero, there exists a pair of n-vectors X and Y with $X \neq 0$ satisfying the equation

$$FX = Y \qquad (9.8)$$

in which the remaining $n+1$ unspecified variables are $1, -1, 0$.

Proof. Necessity. Let $F = [f_{ij}]$ be a totally unimodular matrix. Set any $n-1$ variables of the $2n$ variables x_i and y_i in the transposes of the n-vectors

$$X' = \begin{bmatrix} x_1 & x_2 & \cdots & x_n \end{bmatrix} \qquad (9.9a)$$

$$Y' = \begin{bmatrix} y_1 & y_2 & \cdots & y_n \end{bmatrix} \qquad (9.9b)$$

to zero, where r of these zeros are y's ($0 \leq r \leq n-1$) and the other $n-r-1$ are x's. We now permute the rows and columns of F so that the vanishing y's occupy the first r positions with $y_1 = y_2 = \cdots = y_r = 0$, and the vanishing x's occupy the last

$n-r-1$ positions with $x_{r+2} = x_{r+3} = \cdots = x_n = 0$ without changing the unimodular character of F. Therefore, the system (9.8) can be partitioned as

$$\begin{bmatrix} 0_r \\ Y_2 \end{bmatrix} = \begin{bmatrix} F_{11} & F_{12} \\ F_{21} & F_{22} \end{bmatrix} \begin{bmatrix} X_1 \\ 0_{n-r-1} \end{bmatrix} \tag{9.10}$$

where

$$X_1' = \begin{bmatrix} x_1 & x_2 & \cdots & x_{r+1} \end{bmatrix} \tag{9.11a}$$

$$Y_2' = \begin{bmatrix} y_{r+1} & y_{r+2} & \cdots & y_n \end{bmatrix} \tag{9.11b}$$

In explicit form, the first r equations of the system become

$$F_{11} X_1 = 0. \tag{9.12}$$

If $F_{11} = 0$, let

$$X_1 = \begin{bmatrix} 1 & 0 & \cdots & 0 \end{bmatrix}. \tag{9.13}$$

Then we have

$$Y_2 = F_{21} X_1 \tag{9.14}$$

proving that the elements of Y_2 are 1, −1, or 0. If $F_{11} \neq 0$, let its rank be ρ, $0 < \rho \leq r$. Now we can permute the rows and columns of F_{11}, so that the top left submatrix of order ρ is nonsingular. In partitioned form, (9.12) can be rewritten as

$$\begin{bmatrix} F_{11,11} & F_{11,12} \\ F_{11,21} & F_{11,22} \end{bmatrix} \begin{bmatrix} X_{1,1} \\ X_{1,2} \end{bmatrix} = 0 \tag{9.15}$$

where $F_{11,11}$ is nonsingular and of order ρ. Set

$$X_{1,2} = \begin{bmatrix} 1 & 0 & \cdots & 0 \end{bmatrix} \tag{9.16}$$

and we obtain

$$F_{11,11}X_{1,1} = -W_1 \qquad (9.17)$$

where W_1 is the first column of $F_{11,12}$. Appealing to Cramer's rule, the elements x_i of $X_{1,1}$ are found to be

$$x_i = -\frac{\det \tilde{F}_{11,11}}{\det F_{11,11}}, \qquad i = 1, 2, \ldots, \rho \qquad (9.18)$$

where $\tilde{F}_{11,11}$ is the matrix obtained from $F_{11,11}$ by replacing column i by W_1. Apart from a permutation of the columns, $\tilde{F}_{11,11}$ is a submatrix of $[F_{11,11} \; F_{11,12}]$. Since F is totally unimodular, the submatrix $\tilde{F}_{11,11}$ is also totally unimodular, so is $\tilde{F}_{11,11}$. Therefore, $X_{1,1}$ is a matrix of 1, −1, or 0, so is $X \neq 0$. We next partition the other equations in (9.10) accordingly, and obtain

$$Y_2 = [F_{21,11} \; F_{21,12}] \begin{bmatrix} X_{1,1} \\ X_{1,2} \end{bmatrix} = F_{21,11}X_{1,1} + W_2[x_{\rho+1}] \qquad (9.19)$$

where $x_{\rho+1} = 1$, and W_2 denotes the first column of $F_{21,12}$. To find y_p, $p = r+1$, $r+2, \ldots, n$ we augment the first ρ equations of (9.14) by the pth equation of (9.19) to yield

$$\begin{bmatrix} F_{11,11} & W_1 \\ F_{21,11p} & f_{p,\rho+1} \end{bmatrix} \begin{bmatrix} X_{1,1} \\ x_{\rho+1} \end{bmatrix} = \begin{bmatrix} 0 \\ y_p \end{bmatrix} \qquad (9.20)$$

where $F_{21,11p}$ is the pth row of $F_{21,11}$ and $f_{p,\rho+1}$ is the pth row and $(\rho+1)$th column element of F. Let Δ_p be the determinant of the coefficient matrix of (9.20). If $\Delta_p = 0$, the rows of the coefficient matrix are dependent and we can choose $y_p = 0$. Otherwise, solve (9.20) by Cramer's rule for $x_{\rho+1}$ to give

$$x_{\rho+1} = \frac{\det F_{11,11}}{\Delta_p} y_p \qquad (9.21)$$

or, since $x_{\rho+1} = 1$, we obtain

$$y_p = \frac{\Delta_p}{\det F_{11,11}} \qquad (9.22)$$

Because F is totally unimodular, $y_p = \pm 1$. Therefore, the required X and Y have been found and the proof of necessity is complete.

Sufficiency. The sufficiency of the conditions is established by induction over the order of the submatrices of F. To this end, let us assume that by setting any $n-1$ variables x_i and y_i to zero, there exists a pair of n-vectors X and Y with $X \neq 0$ satisfying (9.8) such that the remaining $n+1$ unspecified variables are $1, -1$, or 0.

To show that every element of F, being subdeterminant of order 1, is $1, -1$, or 0, choose X to be an n-vector, whose elements are all zero except the ith row element x_i which is nonzero. By hypothesis, there exists a pair of X and Y with $X \neq 0$ satisfying (9.8) such that x_i and all the elements y_k of Y are $1, -1$, or 0. By multiplying all equations of (9.8) by -1 if necessary, without loss of generality we may assume that $x_i = 1$, obtaining

$$y_k = f_{ki}, \quad k = 1, 2, \ldots, n \tag{9.23}$$

showing that all the elements f_{ki} of F are $1, -1$, or 0. Thus, the conditions are sufficient for all submatrices of order 1. To complete the proof, we assume that all the square submatrices of order $\rho \geq 2$ or less have determinants $1, -1$, or 0. We show that all square submatrices of order $\rho + 1$ have determinants $1, -1$, or 0.

Consider a nonsingular submatrix of order $\rho + 1$. Without loss of generality, we may assume that this submatrix occupies the top left position of F, whose determinant is $\Delta^{(\rho+1)}$. Now in (9.8) set the first ρ elements of Y and the last $n - \rho - 1$ elements of X to zero, and partition (9.8) accordingly, yielding

$$Y = \begin{bmatrix} 0_\rho \\ Y_2 \end{bmatrix} = \begin{bmatrix} F_{11} & F_{12} \\ F_{21} & F_{22} \end{bmatrix} \begin{bmatrix} X_1 \\ 0_{n-\rho-1} \end{bmatrix} = FX \tag{9.24}$$

where Y_2 is an $(n - \rho)$-vector and X_1 an $(\rho + 1)$-vector, the first $\rho + 1$ equations of which can be written as

$$\begin{bmatrix} 0_\rho \\ y_{\rho+1} \end{bmatrix} = \begin{bmatrix} F_{11} \\ F_{21,1} \end{bmatrix} X_1 \tag{9.25}$$

where $F_{21,1}$ denotes the first row of F_{21}. Since by assumption there exists a pair of n-vectors X and Y of $1, -1$, and 0 with $X \neq 0$ satisfying (9.24), and since the coefficient matrix on the right-hand side of (9.25) is the nonsingular matrix of order $\rho + 1$ under consideration, $y_{\rho+1} \neq 0$ and so $y_{\rho+1} = \pm 1$. For, otherwise, (9.25) can only have the trivial solution $X_1 = 0$, implying that $X = 0$, a contradiction. Solving (9.25) for X_1 gives

$$X_1 = \begin{bmatrix} F_{11} \\ F_{21,1} \end{bmatrix}^{-1} \begin{bmatrix} 0_\rho \\ y_{\rho+1} \end{bmatrix} = \frac{1}{\tilde{\Delta}} \begin{bmatrix} \tilde{\Delta}_{(\rho+1)1} \\ \tilde{\Delta}_{(\rho+1)2} \\ \vdots \\ \tilde{\Delta}_{(\rho+1)(\rho+1)} \end{bmatrix} y_{\rho+1} = \pm \frac{1}{\tilde{\Delta}} \begin{bmatrix} \tilde{\Delta}_{(\rho+1)1} \\ \tilde{\Delta}_{(\rho+1)2} \\ \vdots \\ \tilde{\Delta}_{(\rho+1)(\rho+1)} \end{bmatrix} \quad (9.26)$$

where

$$\tilde{\Delta} = \det \begin{bmatrix} F_{11} \\ F_{21,1} \end{bmatrix} \quad (9.27)$$

and $\tilde{\Delta}_{ij}$ is the cofactor of the ith row and jth column element of the coefficient matrix in the determinant $\tilde{\Delta}$. Since, by induction hypothesis, all the cofactors $\tilde{\Delta}_{(\rho+1)m}$ ($m = 1, 2, ..., \rho + 1$) in (9.27), apart from the sign, are determinants of the submatrices of F of order ρ, they have values 1, -1, or 0. Since not all the elements of X_1 can be zero, we have for the nonvanishing x_i, $1 \le i \le \rho + 1$,

$$x_i = \pm \frac{\tilde{\Delta}_{(\rho+1)i}}{\tilde{\Delta}} \quad (9.28)$$

By induction hypothesis, the nonvanishing x_i has value 1 or -1, so is $\tilde{\Delta}_{(\rho+1)i}$. We conclude from (9.28) that

$$\tilde{\Delta} = \pm 1 \quad (9.29)$$

Thus, by induction, the matrix F is totally unimodular. This completes the proof of the theorem.

THEOREM 9.2: A matrix containing the matrix

$$N = [n_{ij}]_{3 \times 3} = \begin{bmatrix} \pm 1 & 0 & \pm 1 & \pm 1 \\ \pm 1 & \pm 1 & 0 & \pm 1 \\ \pm 1 & \pm 1 & \pm 1 & 0 \end{bmatrix} \quad (9.30)$$

as a submatrix cannot be a totally unimodular matrix, where ± 1 means that the entries may either be 1 or -1.

Proof. It is sufficient to show that N is not totally unimodular. Since by multiplying the rows and columns by -1 will not change the unimodular character of a matrix, without loss of generality, we may assume that all the nonzero elements of the first row to be 1 by multiplying the columns by -1 if necessary. Likewise, by multiplying the 2nd or 3rd rows or both by -1 if necessary, we can make all the entries of the first column to be 1. Now we claim that if N is totally unimodular, n_{24} and n_{33} cannot be -1 and n_{22} and n_{32} must have the same sign. For instance, if $n_{22} = -1$, then

$$\det \begin{bmatrix} n_{21} & n_{22} \\ n_{31} & n_{32} \end{bmatrix} = \det \begin{bmatrix} 1 & -1 \\ 1 & \pm 1 \end{bmatrix} \quad (9.31)$$

whose value is 2 if $n_{32} = 1$. Thus, n_{22} and n_{32} must have the same sign. By multiplying the 2nd column by -1 if necessary, we can assume that $n_{22} = n_{32} = 1$. If n_{24} or $n_{33} = -1$, then

$$\det \begin{bmatrix} n_{11} & n_{14} \\ n_{21} & n_{24} \end{bmatrix} \text{ or } \det \begin{bmatrix} n_{11} & n_{13} \\ n_{31} & n_{33} \end{bmatrix} = \det \begin{bmatrix} 1 & 1 \\ 1 & -1 \end{bmatrix} = -2 \quad (9.32)$$

a contradiction. Therefore, it is sufficient to consider the matrix

$$N = \begin{bmatrix} 1 & 0 & 1 & 1 \\ 1 & 1 & 0 & 1 \\ 1 & 1 & 1 & 0 \end{bmatrix}. \quad (9.33)$$

The determinant of the major submatrix formed by the last three columns is 2, so N is not a totally unimodular matrix. This completes the proof of the theorem.

In an attempt to characterize a real matrix of $1, -1$, or 0 to be the circuit or cutset matrix of a directed graph, one frequently encounters this particular matrix N. See, for example, WHITNEY [1935], GOULD [1957], CEDERBAUM [1958], and TUTTE [1958].

THEOREM 9.3: If a symmetric matrix K can be decomposed as the product of three matrices

$$K = FDF' \quad (9.34)$$

where F is a totally unimodular matrix and D a diagonal matrix with positive constant, then each principal minor of K is not less than the modulus of any other minor formed from the same rows or columns.

Proof. Using the symbol of §4.2, let $K(I_u, J_v)$ be the submatrix of K consisting of the rows and columns corresponding to the integers in the sets I_u and J_v. For a square submatrix of order n of K, we have

$$K(I_u, J_u) = F(I_u, K_m) D F'(K_m, J_u) \qquad (9.35)$$

where m is the order of D, and $F(I_u, K_m)$ is the submatrix of F composed of the rows and columns corresponding to the integers in I_u and K_m. Applying the Binet–Cauchy theorem of Theorem 2.17, its determinant can be expressed in terms of the minors of the component matrices as

$$\det K(I_u, J_u) = \sum_{L_u} \sum_{M_u} \det F(I_u, L_u) \det D(L_u, M_u) \det F'(M_u, J_u)$$

$$= \sum_{L_u} \det F(I_u, L_u) \det D(L_u, L_u) \det F'(L_u, J_u)$$

$$= \sum_{L_u} \left[\det F(I_u, L_u) \det F(J_u, L_u) \right] \det D(L_u, L_u) \qquad (9.36)$$

For the principal minors of order n, (9.36) becomes

$$\det K(I_u, I_u) = \sum_{L_u} \det F(I_u, L_u) \det D(L_u, L_u) \det F'(L_u, I_u)$$

$$= \sum_{L_u} \left[\det F(I_u, L_u) \right]^2 \det D(L_u, L_u) \qquad (9.37)$$

By comparing (9.36) and (9.37) and taking into account the fact that $\det D(L_u, L_u)$ is positive and $\det F(I_u, L_u)$ and $\det F(J_u, L_u)$ are $1, -1$ or 0, we see that all terms of (9.37) are not less than the moduli of the corresponding terms of (9.36). The theorem follows. This completes the proof of the theorem.

The condition given in Theorem 9.3 is only necessary. However, it is not sufficient. For example, each of the principal minor of the symmetric matrix

$$\begin{bmatrix} 20 & -10 & 0 \\ -10 & 35 & -30 \\ 0 & -30 & 36 \end{bmatrix} \tag{9.38}$$

is not less than the modulus of any other minor formed from the same rows or columns, but it can be shown that this matrix cannot be decomposed in the product of (9.34). Nevertheless, the above congruent transformation of a positive definite matrix K by a square nonsingular totally unimodular matrix into a diagonal form, if at all possible, is essentially unique, as can be seen by the following theorem.

THEOREM 9.4: If a symmetric matrix K has the decompositions

$$K = F_1 D_1 F'_1 = F_2 D_2 F'_2 \tag{9.39}$$

where F_1 and F_2 are square nonsingular totally unimodular matrices, and D_1 and D_2 are diagonal matrices with positive constant, then they may differ from each other only in the order and the sign of the columns of F_1 or F_2 and the order of the elements of D_1 or D_2.

Proof. From (9.39) we obtain

$$D_1 = F_1^{-1} F_2 D_2 F'_2 (F'_1)^{-1} = H_1 D_2 H'_1 \tag{9.40}$$

where

$$H_1 = F_1^{-1} F_2 = \left[h_{ij}^{(1)} \right]_{n \times n} \tag{9.41}$$

and n is the order of F_1 and F_2. The ith diagonal element $d_{ii}^{(1)}$ of D_1 is given by

$$d_{ii}^{(1)} = \sum_{j=1}^{n} \left[h_{ij}^{(1)} \right]^2 d_{jj}^{(2)} \tag{9.42}$$

where $d_{jj}^{(2)}$ is the jth diagonal element of D_2. Since H_1 is nonsingular with integer elements, for each j there is at least one nonvanishing $h_{ij}^{(1)}$ or $\left[h_{ij}^{(1)}\right]^2 \geq 1$. Thus, the trace of D_1, being the sum of the diagonal elements, is not less than that of D_2:

$$\text{trace } D_1 = \sum_{i=1}^{n} d_{ii}^{(1)} = \sum_{i=1}^{n}\sum_{j=1}^{n}\left[h_{ij}^{(1)}\right]^2 d_{jj}^{(2)} = \sum_{j=1}^{n}\sum_{i=1}^{n}\left[h_{ij}^{(1)}\right]^2 d_{jj}^{(2)}$$

$$= \sum_{j=1}^{n} d_{jj}^{(2)} \sum_{i=1}^{n}\left[h_{ij}^{(1)}\right]^2 \geq \sum_{j=1}^{n} d_{jj}^{(2)} = \text{trace } D_2. \tag{9.43}$$

Likewise, from (9.39) we can express D_2 in terms of D_1 and obtain

$$D_2 = F_2^{-1} F_1 D_1 F'_1 (F'_2)^{-1} = H_2 D_1 H'_2 \tag{9.44}$$

where

$$H_2 = F_2^{-1} F_1 = \left[h_{ij}^{(2)}\right]_{n \times n}. \tag{9.45}$$

Following (9.43) yields

$$\text{trace } D_2 = \sum_{i=1}^{n} d_{ii}^{(2)} = \sum_{i=1}^{n}\sum_{j=1}^{n}\left[h_{ij}^{(2)}\right]^2 d_{jj}^{(1)}$$

$$= \sum_{j=1}^{n} d_{jj}^{(1)} \sum_{i=1}^{n}\left[h_{ij}^{(2)}\right]^2 \geq \sum_{j=1}^{n} d_{jj}^{(1)} = \text{trace } D_1. \tag{9.46}$$

Combining (9.43) and (9.46) gives

$$\text{trace } D_1 = \text{trace } D_2 \tag{9.47}$$

or from (9.43) and (9.46)

$$\text{trace } D_1 = \sum_{j=1}^{n} d_{jj}^{(2)} \sum_{i=1}^{n}\left[h_{ij}^{(1)}\right]^2 = \text{trace } D_2 = \sum_{j=1}^{n} d_{jj}^{(1)} \sum_{i=1}^{n}\left[h_{ij}^{(2)}\right]^2 \tag{9.48}$$

This is possible only if

$$\sum_{i=1}^{n}\left[h_{ij}^{(1)}\right]^{2} = \sum_{i=1}^{n}\left[h_{ij}^{(2)}\right]^{2} = 1 \qquad (9.49)$$

showing that for each j only one $\left[h_{ij}^{(1)}\right]^{2}$ or $\left[h_{ij}^{(2)}\right]^{2}$ can be nonzero, the value of which must necessarily be 1. For, otherwise, the equality sign in (9.47) cannot be maintained. In other words, in any column and row of H_1 and H_2 there is exactly one nonzero element equal to 1 or -1. They differ from the identity matrix only in the order and sign of its rows or columns. Premultiplication of a matrix by such a matrix results in a permutation and/or a change of sign of some of its rows, whereas postmultiplication of a matrix by such a matrix leads to the same operations on its columns. Therefore, from (9.39), D_1 and D_2 differ only in the order of the diagonal elements.

Finally, from (9.41) and (9.45) we have

$$F_2 = F_1 H_1, \qquad F_1 = F_2 H_2. \qquad (9.50)$$

This shows that F_1 and F_2 may differ only in the order and sign of their columns. This completes the proof of the theorem.

For a matrix of 1, -1 and 0, with two or less nonzero entries in each column, necessary and sufficient conditions for the matrix to be totally unimodular were first described by HOFFMAN and KRUSKAL [1956] and HELLER and TOMPKINS [1956]. Since then there has been considerable interest in characterizing the class of totally unimodular matrices, some of the early significant developments may be found in HOFFMAN and HELLER [1962], GHOUILA-HOURI [1962], HELLER [1957, 1963, 1964], CAMION [1965] and TUTTE [1966]. In the following, we present another characterization of these matrices due to CHANDRASEKARAN [1969].

THEOREM 9.5: A matrix F of 1, -1 and 0 is totally unimodular if and only if for every nonsingular submatrix $H = \left[h_{ij}\right]$ of order m of F, the greatest common divisor of

$$\sum_{j \in J} \lambda_j h_{1j}, \quad \sum_{j \in J} \lambda_j h_{2j}, \quad ..., \quad \sum_{j \in J} \lambda_j h_{mj} \qquad (9.51)$$

is 1 for any $\lambda_j = \pm 1$ or 0, not all of which are zero, where J is the index set of the columns of H.

Proof. Necessity. Let λ_u be one of the coefficients λ_j for which $\lambda_u = 1$ or -1. If there exists a $\lambda_u = 1$, add all other columns of H corresponding to $\lambda_j = 1$ to the column corresponding to λ_u, and also add all columns corresponding to $\lambda_j = -1$ multiplied by -1 to the column corresponding to λ_u. Denote the resulting matrix by $\hat{H}$. Clearly, the elements of the column corresponding to λ_u in $\hat{H}$ are those given in (9.51). Let the greatest common divisor of the elements of (9.51) be a positive integer k. Since the addition of a column multiplied by 1 or -1 to another column in a matrix will not change the determinant of the matrix, we have

$$\det H = \det \hat{H} = Mk \qquad (9.52)$$

where the integer M denotes the determinant of the matrix obtained from $\hat{H}$ after the common factor k has been factored out from the column corresponding to λ_u, yielding

$$Mk = \pm 1 \qquad (9.53)$$

because F is totally unimodular. This is possible only if $M = \pm 1$ and $k = 1$. If there exists no $\lambda_u = 1$, choose any column corresponding to a $\lambda_u = -1$ and add all columns corresponding to other $\lambda_j = -1$ to this column, resulting in

$$\det H = \det \hat{H} = Mk \qquad (9.54)$$

where, as before, the integer M denotes the determinant of the matrix obtained from $\hat{H}$ after the common factor k has been factored out from the special column corresponding to $\lambda_u = -1$, or $M = \pm 1$ and $k = 1$. Therefore, if F is totally unimodular, then for every nonsingular submatrix H the greatest common divisor of (9.51) is 1 for any $\lambda_j = \pm 1$ or 0, not all of which is zero.

Sufficiency. We prove the sufficiency by induction over the order m of the submatrix H. Clearly, it is true for $m = 1$. Assume that the theorem is true for any submatrix of order m. We show that it is also true for any submatrix H of order $m + 1$. If $\det H = 0$ or ± 1, we are finished. Thus, let $\det H \neq 0$ or ± 1. Consider the matrix product of H and its adjoint matrix adj H,

$$H \text{ adj } H = (\det H)U \qquad (9.55)$$

Since the elements of adj H are cofactors of the elements of H, apart from the sign, they are determinants of the submatrices of order m of H. By induction hypothesis, they all have value 0 or ±1. Thus, we can pick any column X of adj H, the elements of which would be a set of $\lambda_j = \pm 1$ or 0, not all zero, such that the greatest common divisors of the elements of the vector HX is det $H \neq 1$. But these are the same elements as given in (9.51). By hypothesis, their greatest common divisor is 1 or det $H = 1$, a contradiction. Thus, det $H = 0$ or ±1. This completes the proof of the theorem.

§ 2. Regular and Binary Matrices

In this section, we consider the problem of replacing an appropriate set of -1's by 1's in a matrix of 1, -1, and 0, leaving the ranks of submatrices unaltered, where the rank of the derived matrix is with respect to modulo 2 algebra. Let

$$F_d = \left[f_{ij}^{(d)} \right]_{m \times n} \tag{9.56}$$

be a real matrix of integers (positive, negative, and zero) of rank m ($m \leq n$), and let

$$F = \left[f_{ij} \right]_{m \times n} \tag{9.57}$$

be a matrix of integers mod 2 of rank m ($m \leq n$). Then the linear vector space $\mathcal{V}(F_d)$ generated by the rows of F_d is the set of all linear combinations of the rows of F_d with real integral coefficients, whereas the linear vector space $\mathcal{V}(F)$ generated by the rows of F is the set of all linear combinations of the rows of F with coefficient 1 or 0 of the mod 2 algebra. We admit no column operations other than permutations.

DEFINITION 9.1: *Elementary vector*. A nonzero vector F_1 of either of the linear vector spaces $\mathcal{V}(F_d)$ and $\mathcal{V}(F)$ is said to be *elementary* if there is no other vector in the space which has nonzero entries only at a proper subset of the positions in which F_1 has nonzero entries.

Consider, for example, the matrix mod 2

$$N = \begin{bmatrix} 1 & 0 & 1 & 1 \\ 1 & 1 & 0 & 1 \\ 1 & 1 & 1 & 0 \end{bmatrix}. \tag{9.58}$$

§ 2 Graph realization

The linear vector space $\mathcal{V}(N)$ generated by all linear combinations of the rows of N with coefficients 1 or 0 is represented by the rows of the matrix

$$\begin{bmatrix} 0 & 0 & 0 & 0 \\ 1 & 0 & 1 & 1 \\ 1 & 1 & 0 & 1 \\ 1 & 1 & 1 & 0 \\ 0 & 1 & 1 & 0 \\ 0 & 1 & 0 & 1 \\ 0 & 0 & 1 & 1 \\ 1 & 0 & 0 & 0 \end{bmatrix} \tag{9.59}$$

The vectors

$$\begin{bmatrix} 0 & 1 & 0 & 1 \end{bmatrix}, \quad \begin{bmatrix} 1 & 0 & 0 & 0 \end{bmatrix} \tag{9.60}$$

are elementary, because there are no other nonzero vectors in the space $\mathcal{V}(N)$, the nonzero entries of which occur only at a proper subset of the positions where these vectors have 1's; whereas the vectors

$$\begin{bmatrix} 1 & 1 & 0 & 1 \end{bmatrix}, \quad \begin{bmatrix} 1 & 1 & 1 & 0 \end{bmatrix} \tag{9.61}$$

are not, because there exist vectors

$$\begin{bmatrix} 0 & 1 & 0 & 1 \end{bmatrix}, \quad \begin{bmatrix} 0 & 1 & 1 & 0 \end{bmatrix} \tag{9.62}$$

in the space $\mathcal{V}(N)$, the nonzero entries of which occur at a proper subset of the positions in which the vectors in (9.61) have 1's.

DEFINITION 9.2: *Primitive vector.* A vector of the linear vector space $\mathcal{V}(F_d)$ is *primitive* if it is elementary and all of its entries are 1, −1, or 0.

DEFINITION 9.3: *Real regular matrix.* A real matrix F_d of integral elements is said to be *regular* if, for every elementary vector in the linear vector space $\mathcal{V}(F_d)$, there exists a primitive vector in $\mathcal{V}(F_d)$, the nonzero elements of which appear in the same positions as in the elementary vector.

Consider the directed planar graph of Fig. 9.7. The basis circuit matrix corresponding to the four meshes is found to be

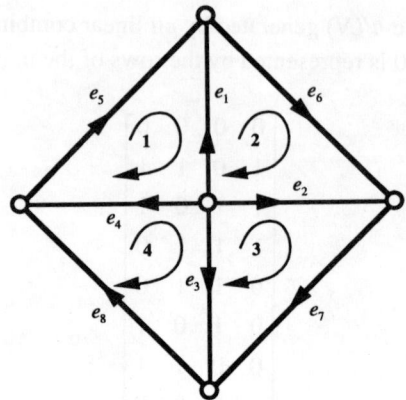

Fig. 9.7. A directed graph with four chosen circuits.

$$B_p = \begin{bmatrix} -1 & 0 & 0 & 1 & 1 & 0 & 0 & 0 \\ 1 & -1 & 0 & 0 & 0 & 1 & 0 & 0 \\ 0 & 1 & -1 & 0 & 0 & 0 & 1 & 0 \\ 0 & 0 & 1 & -1 & 0 & 0 & 0 & 1 \end{bmatrix} \quad (9.63)$$

Each row of this matrix is a primitive vector in the linear vector space $V(B_p)$ generated by all linear combinations of its rows with real integral coefficients. This matrix is also regular, because to every elementary vector in $V(B_p)$ there corresponds a primitive vector in $V(B_p)$. For example, for any real integer a, the row vector

$$[a \quad 0 \quad 0 \quad -a \quad -a \quad 0 \quad 0 \quad 0] \quad (9.64)$$

is an elementary vector in $V(B_p)$. The corresponding primitive vector in $V(B_p)$ is given by

$$[-1 \quad 0 \quad 0 \quad 1 \quad 1 \quad 0 \quad 0 \quad 0] \quad (9.65)$$

Recall that in Definition 5.8 a matrix of order $m \times n$ ($m \leq n$) containing the identity matrix of order m is referred to as a matrix in *normal form*. The basis circuit matrix B_p of (9.63) is a matrix in normal form, because it contains the identity matrix of order 4 composed of the last four columns. Thus, a real matrix F_d of integral elements in normal form is a regular matrix if for every linear combination F_1 of the rows of F_d with 1, −1, and 0, either the elements of F_1 are 1, −1, and 0; or there exists another such linear combination F_2 with coefficients 1, −1, and 0 having 1 and −1 appearing at a (not necessarily proper) subset of the position in which F_1 has nonzero elements. The importance of the regular matrices is contained in the following.

THEOREM 9.6: The basis incidence matrix, the f-circuit matrix and the f-cut matrix of a directed graph are all regular matrices.

Proof. Let A_d, B_{df} and Q_{df} be the basis incidence matrix, the f-circuit matrix and the f-cut matrix of a directed graph G_d. Since the linear vector spaces generated by A_d and Q_{df} are the same, it suffices to consider B_{df} and Q_{df}. Write

$$Q_{df} = [Q_{d11} \quad U_r], \quad B_{df} = [U_m \quad B_{d12}] \tag{9.66}$$

where r and m denote the rank and nullity of G_d, respectively.

For any linear combination Q_{d1} of rows of Q_{df} with coefficients 1, −1, and 0; let $i_1, i_2, ..., i_k$ be the cuts with coefficient 1 and $j_1, j_2, ..., j_p$ be the cuts with coefficient −1. If Q_{di_u} denotes the i_uth row of Q_{df}, we can write

$$Q_{d1} = \sum_{u=1}^{k} Q_{di_u} - \sum_{u=1}^{p} Q_{dj_u}. \tag{9.67}$$

Now consider the associated undirected graph G of G_d. The f-cut matrix Q_f of G with respect to the same set of cuts and with rows and columns arranged in the same order as in Q_{df} can be similarly partitioned. The corresponding linear combination Q_1 of rows of Q_f over the field of integers mode 2 can be written as

$$Q_1 = \sum_{u=1}^{k} Q_{i_u} + \sum_{u=1}^{p} Q_{j_u}. \tag{9.68}$$

Recall that Q_1 over the field of integers mod 2 denotes a cutset or an edge-disjoint union of cutsets in G. Since each vector on the right-hand side of (9.67) has entries 1, −1, and 0, it can produce a zero entry in Q_{d1} only if either all the component entries are zero or an even number of nonzero entries has been added. Under this situation, the corresponding entry in Q_1 is also zero, because the sum of an even number of 1's is zero in mod 2 algebra. Therefore, wherever Q_{d1} has a zero entry, the corresponding entry in Q_1 is also zero, but Q_1 may have more zeros than Q_{d1}.

Since the vectors on the right-hand side of Q_1 are linearly independent, $Q_1 \neq 0$ if $Q_{d1} \neq 0$. If Q_1 denotes a cutset, then the row vector Q_{d1} of the oriented cutset in G_d corresponding to Q_1 is in $V(Q_{df})$ and contains entries 1, −1, and 0. This row vector Q_{d1} has zeros wherever Q_1 has zeros. If Q_1 denotes an edge-disjoint union of cutsets, we can always find a cutset in the union and let Q_2 be its corresponding row vector in $V(Q_f)$. The row vector Q_{d2} of the oriented cutset in G_d corresponding to Q_2 is in $V(Q_{df})$ and contains entries 1, −1, and 0. Thus, to every elementary vector in $\mathcal{U}(Q_{df})$ there corresponds a primitive vector in $V(Q_{df})$ with nonzero entries in the same positions, showing that Q_{df} is regular. Similarly, we can show that B_{df} is regular. This completes the proof of the theorem.

We remark that the theorem is true for most basis cut matrices and basis circuit matrices, and is not only restricted to the f-cut matrix and f-circuit matrix. The restriction is imposed to ensure that the corresponding mod 2 matrices have linearly independent rows, so that Q_1 will not be a trivial cut, such as occurring in (9.7). In other words, the cut and circuit matrices in normal form are regular matrices, but a general regular matrix in normal form may not be the cut or circuit matrix of a directed graph. We next show that every regular matrix in normal form is totally unimodular. The assumption that we only consider regular matrices in normal form is not really a serious restriction, as can be seen in the following theorem.

THEOREM 9.7: To every regular matrix there corresponds a regular matrix in normal form which generates the same linear vector space with at most a permutation of the coordinates.

Proof. Let F_d be a regular matrix of order $m \times n$ and of rank m. By permuting the columns if necessary, without loss of generality we may assume that the submatrix formed by the first m columns is a major submatrix. This matrix can be partitioned as

$$F_d = \begin{bmatrix} F_{d11} & F_{d12} \end{bmatrix} \tag{9.69}$$

where F_{d11} is nonsingular and of order m. By performing the elementary row operations of (i) multiplying a row by a nonzero integer, (ii) adding to one row of an integral multiple of another row, and (iii) permuting rows, F_d can be brought to the form

$$\tilde{F}_d = \begin{bmatrix} D_m & \tilde{F}_{d12} \end{bmatrix} \tag{9.70}$$

where D_m is a diagonal matrix with nonzero integral diagonal entries. Since each row of $\tilde{F}_d$ is a linear combination of the rows of F_d with integral coefficients, it belongs to the linear vector space $\mathcal{V}(F_d)$ generated by the rows of F_d. If the ith row vector of $\tilde{F}_d$ is not elementary, there corresponds an elementary vector in $\mathcal{V}(F_d)$. Hence, the rows of $\tilde{F}_d$ that are not elementary can all be replaced by elementary vectors. Without loss of generality, we may, therefore, assume that every row of $\tilde{F}_d$ is elementary. But, by definition of a regular matrix, to each row of $\tilde{F}_d$ there corresponds a primitive vector in $\mathcal{V}(F_d)$ which preserves all the zeros in $\tilde{F}_d$. Since F_{d11} is nonsingular, none of these primitive vectors can have zeros in all the first m columns. As a result, we can replace rows of $\tilde{F}_d$ by their corresponding primitive vectors. The resulting matrix contains a diagonal matrix with 1 or -1 along its main diagonal. Since by multiplying a primitive vector by -1 does not change its primitive character, we may further assume that the resulting matrix contains the identity matrix of order m, and can be partitioned as

$$F_{dn} = \begin{bmatrix} U_m & F_{dn12} \end{bmatrix}. \qquad (9.71)$$

Since the rows of F_{dn} generates the same linear vector space $\mathcal{V}(F_{dn}) = \mathcal{V}(F_d)$, it follows that F_{dn} is a regular matrix in normal form. This completes the proof of the theorem.

THEOREM 9.8: A submatrix formed by the rows of a regular matrix in normal form constitutes a regular matrix. The rows of a regular matrix in normal form are primitive vectors.

Proof. Let $F_{dr_1}, F_{dr_2}, \ldots, F_{dr_k}$ ($1 \leq k \leq m$) be any k rows of a regular matrix F_d in normal form. Consider any linear combination W of these row vectors $F_{dr_1}, F_{dr_2}, \ldots, F_{dr_k}$ with coefficients 1, -1, and 0. There is a corresponding primitive vector in the linear vector space $\mathcal{V}(F_d)$ generated by the rows of F_d. This primitive vector cannot depend on any other row vector F_{dp}. For, otherwise, it would have a ± 1 in column p, where the original vector W has only 0. Thus, a submatrix formed by the rows of a regular matrix in normal form is regular.

Next, consider the special situation where a submatrix is composed of only one row. This row matrix F_{dh} is a regular matrix in normal form. Hence, there is a corresponding primitive vector in the linear vector space generated by $\mathcal{V}(F_{dh}) = aF_{dh}$ where a is any real integer. In other words, the primitive vector can differ from F_{dh} only by a multiplicative real integer b. Since the primitive vector has entries 1, -1, and 0 and since F_{dh} has a nonzero entry 1, being in normal form, we have $b = \pm 1$, showing that F_{dh} itself is a primitive vector. Thus, the rows of a regular matrix in normal form are primitive vectors. This completes the proof of the theorem.

THEOREM 9.9: A regular matrix in normal form is totally unimodular. A totally unimodular matrix with linearly independent rows is regular.

Proof. Necessity. We prove the theorem by induction over the orders of the square submatrices of a given regular matrix F_d in normal form. By Theorem 9.8, all the elements of F_d are $1, -1,$ and 0. Therefore, the submatrices of order 1 have determinants $1, -1,$ and 0. Now suppose that all square submatrices of orders up to k, $k \geq 2$, have determinants $1, -1,$ and 0. We show that it is also true for any nonsingular submatrix W of order $k+1$. Write

$$W = \begin{bmatrix} w_{11} & w_{12} & \cdots & w_{1k} & w_{1(k+1)} \\ w_{21} & w_{22} & \cdots & w_{2k} & w_{2(k+1)} \\ \vdots & \vdots & \cdots & \vdots & \vdots \\ w_{k1} & w_{k2} & \cdots & w_{kk} & w_{k(k+1)} \\ w_{(k+1)1} & w_{(k+1)2} & \cdots & w_{(k+1)k} & w_{(k+1)(k+1)} \end{bmatrix}. \quad (9.72)$$

By permuting the rows and columns if necessary, we may assume that the submatrix of W formed by the first k rows and columns is nonsingular. Consider the row vector

$$R = \begin{bmatrix} \Delta_{1(k+1)} & \Delta_{2(k+1)} & \cdots & \Delta_{k(k+1)} & \Delta_{(k+1)(k+1)} \end{bmatrix} \quad (9.73)$$

formed by the cofactors of elements of the last column of W, where Δ_{ij} is the cofactor of the element w_{ij} in $\Delta = \det W$. Thus, it is the last row in the inverse W^{-1} of W multiplied by Δ. By induction hypothesis, we have

$$\Delta_{i(k+1)} = 1, \ -1, \text{ or } 0, \quad i = 1, 2, \ldots, k+1 \quad \text{and} \quad \Delta_{(k+1)(k+1)} \neq 0. \quad (9.74)$$

Recall that the sum of products of elements from a row or column times the corresponding cofactors of the elements of another row or column of a matrix is zero. We obtain the row vector formed by the matrix product

$$RW = [0 \ 0 \ \cdots \ 0 \ \Delta]. \quad (9.75)$$

§ 2 Regular and binary matrices

In other words, this row vector RW is obtained as a linear combination of the rows of W with coefficients $1, -1$, and 0. If we replace the last row of W by RW, the new matrix becomes

$$W_1 = \begin{bmatrix} w_{11} & w_{12} & \cdots & w_{1k} & w_{1(k+1)} \\ w_{21} & w_{22} & \cdots & w_{2k} & w_{2(k+1)} \\ \vdots & \vdots & \cdots & \vdots & \vdots \\ w_{k1} & w_{k2} & \cdots & w_{kk} & w_{k(k+1)} \\ 0 & 0 & \cdots & 0 & \Delta \end{bmatrix} \tag{9.76}$$

the determinant of which is given by

$$\det W_1 = \Delta \Delta_{(k+1)(k+1)}. \tag{9.77}$$

Thus, the last row of W_1 is a linear combination of the rows of W with coefficients $1, -1$, and 0. Let F_{d1} be the row vector obtained by the same linear combination of the rows of F_d corresponding to those rows of W. If $\Delta \neq \pm 1$, there exists a primitive vector with no additional nonzero entries since F_d is a regular matrix in normal form. According to Theorem 9.8, this primitive vector can be expressed as a linear combination of the same rows with coefficients $1, -1$, and 0, because these rows themselves constitute a regular matrix. This primitive vector cannot have a 0 in the position occupied by Δ since W is nonsingular. Using the coefficients of this linear combination that yields the primitive vector on W is equivalent to replacing the last row of W by the linear combination of rows of W as defined by the primitive vector. Since all the coefficients are $1, -1$, and 0, the determinant changes at most in sign. The last row now becomes $\begin{bmatrix} 0 & 0 & \cdots & 0 & \pm 1 \end{bmatrix}$. Expanding the new determinant of the new matrix by the last row gives

$$\det W = \pm \Delta_{(k+1)(k+1)} = \pm 1. \tag{9.78}$$

The theorem follows. The completes the proof of necessity.

Sufficiency. Let F_d be a totally unimodular matrix of order $m \times n, m \leq n$. Since the rows of F_d are linearly independent, F_d does not contain any zero row. Consider any linear combination Y of the rows of F_d, which can be represented as the product of a row vector X of $1, -1$, and 0 and F_d, or

$$XF_d = Y \tag{9.79}$$

Upon taking transposes on both side, we obtain

$$F'_d X' = Y' \tag{9.80}$$

Write

$$X = \begin{bmatrix} x_1 & x_2 & \cdots & x_m \end{bmatrix} \tag{9.81a}$$

$$Y = \begin{bmatrix} y_1 & y_2 & \cdots & y_m \end{bmatrix}. \tag{9.81b}$$

If Y is not a row vector of 1, -1, and 0, we must show that there exists another row vector $\tilde{X} \neq \mathbf{0}$ with elements 1, -1, and 0, such that

$$F'_d \tilde{X}' = \tilde{Y}' \neq \mathbf{0} \tag{9.82}$$

is a column vector of 1, -1, and 0, and has zeros wherever Y' has zeros. Write

$$F_d = \begin{bmatrix} f_{ij} \end{bmatrix}_{m \times n} \tag{9.83a}$$

$$\tilde{X} = \begin{bmatrix} \tilde{x}_1 & \tilde{x}_2 & \cdots & \tilde{x}_m \end{bmatrix} \tag{9.83b}$$

$$\tilde{Y} = \begin{bmatrix} \tilde{y}_1 & \tilde{y}_2 & \cdots & \tilde{y}_m \end{bmatrix} \tag{9.83c}$$

If Y contains no zeros at all, let

$$X = \begin{bmatrix} 1 & 0 & \cdots & 0 \end{bmatrix}. \tag{9.84}$$

Then Y' becomes the first column of F'_d or Y is the first row of F_d, which contains only the elements 1, -1, and 0. In general, we may assume that Y contains ρ zeros. By permuting rows of F_d if necessary, we may assume that the first ρ elements in Y are zeros. The resulting first ρ equations of (9.80) become

$$f_{11} x_1 + f_{21} x_2 + \cdots + f_{m1} x_m = 0$$
$$f_{12} x_1 + f_{22} x_2 + \cdots + f_{m2} x_m = 0 \tag{9.85}$$

$$f_{1\rho}x_1 + f_{2\rho}x_2 + \cdots + f_{m\rho}x_m = 0.$$

Since these equations are satisfied by the given linear combination X, the columns of the coefficient matrix of (9.85) are linearly dependent, meaning that its rank r is less than m or $r < m$. Again, by permuting the rows and columns if necessary, we may assume that the leading square submatrix of order r is nonsingular. In partitioned form, (9.85) can be written in matrix form as

$$\begin{bmatrix} F_{11} & F_{12} \\ F_{21} & F_{22} \end{bmatrix} \begin{bmatrix} X_1' \\ X_2' \end{bmatrix} = \mathbf{0} \qquad (9.86)$$

where F_{11} is a nonsingular submatrix of order r, $r < m$, and the coefficient matrix of (9.86) is of order $\rho \times m$. Now, consider the partitioned equation

$$\begin{bmatrix} F_{11} & F_{12} \\ F_{21} & F_{22} \end{bmatrix} \begin{bmatrix} \tilde{X}_1' \\ \tilde{X}_2' \end{bmatrix} = \mathbf{0}. \qquad (9.87)$$

Let

$$\tilde{X}_2 = \begin{bmatrix} 1 & 0 & \cdots & 0 \end{bmatrix} \neq \mathbf{0}. \qquad (9.88)$$

Substituting this in (9.87) gives

$$F_{11}\tilde{X}_1' = -f_{r+1} \qquad (9.89)$$

where f_{r+1} denotes the first column of F_{12}. This equation can be solved for the elements $\tilde{x}_i$ ($i = 1, 2, ..., r$) of $\tilde{X}_1'$ by Cramer's rule to yield

$$\tilde{x}_i = -\frac{\tilde{\Delta}_i}{\tilde{\Delta}}, \qquad i = 1, 2, ..., r. \qquad (9.90)$$

where $\tilde{\Delta} = \det F_{11}$ and $\tilde{\Delta}_i$ is the determinant of the matrix obtained from F_{11} by replacing column i of F_{11} by f_{r+1}. Since F_d is totally unimodular, $\tilde{\Delta} = \det F_{11} = 1$ or -1 and $\tilde{\Delta}_i = 1, -1,$ or 0. Thus, $\tilde{x}_i = 1, -1,$ or 0 for $i = 1, 2, ..., r$.

To compute $\tilde{y}_j$ ($j = \rho + 1, \rho + 2, ..., n$), we augment the first r equations of (9.87) by the jth equation of the original system (9.85) to give

$$\begin{bmatrix} F_{11} & f_{r+1} \\ f'_j & f_{j(r+1)} \end{bmatrix} \begin{bmatrix} \tilde{X}'_1 \\ \tilde{x}_{r+1} \end{bmatrix} = \begin{bmatrix} 0 \\ \tilde{y}_j \end{bmatrix} \tag{9.91}$$

since all $\tilde{x}_i$ other than those in $\tilde{X}'_1$ and $\tilde{x}_{r+1}$ are zeros, where

$$f'_j = \begin{bmatrix} f_{1j} & f_{2j} & \cdots & f_{rj} \end{bmatrix}. \tag{9.92}$$

If the coefficient matrix of (9.91) is singular, take $\tilde{y}_j = 0$. Otherwise, (9.91) can be solved by Cramer's rule to yield

$$\tilde{x}_{r+1} = \tilde{y}_j \frac{\tilde{\Delta}}{\Delta} \tag{9.93}$$

where Δ is the determinant of the coefficient matrix of (9.91), or

$$\tilde{y}_j = \frac{\Delta}{\tilde{\Delta}} \tag{9.94}$$

where from (9.88) $x_{r+1} = 1$. Since F_d is totally unimodular and F_{11} is nonsingular, $\tilde{\Delta} = \pm 1$ and $\Delta = 1, -1$, or 0. Since by hypothesis the rows of F_d are linearly independent, so do the columns of F'_d, or $\tilde{Y} \neq 0$. Therefore, there exist row vectors $\tilde{X} \neq 0$ and $\tilde{Y} \neq 0$ with elements 1, -1, and 0 satisfying (9.82) such that $\tilde{Y}$ has zero entries whenever Y has. In other words, F_d is a regular matrix. This completes the proof of the theorem.

We remark that every submatrix of a regular matrix in normal form is itself regular but it may not be in normal form, and is, therefore, totally unimodular, a proof of which is similar to that of Theorem 9.9 and is omitted. Conversely, every submatrix of a totally unimodular matrix with linearly independent rows is regular.

DEFINITION 9.4: *Binary matrix.* A matrix F_d of integers 1, -1, and 0 is said to be *binary* if the replacement of -1's by 1's results in a matrix F with elements 1 and 0 over the field of integers modulo 2, such that the ranks of the corresponding submatrices in F_d and F remains unaltered.

THEOREM 9.10: A totally unimodular matrix is binary, so is a regular matrix in normal form.

Proof. Since by Theorem 9.9, every regular matrix in normal form is totally unimodular, it is sufficient to show that every totally unimodular matrix is binary.

Let M be a submatrix of order n of a totally unimodular matrix F_d over the real field of integers. Let $\hat{M}$ be the corresponding submatrix in F obtained from F_d by replacing -1's by 1's over the field of integers modulo 2. Write

$$M = [m_{ij}], \quad \hat{M} = [\hat{m}_{ij}]. \tag{9.95}$$

Appealing to the general expansion formula for a determinant, the determinants of these submatrices are found to be

$$\det M = \sum_{(j_1 j_2 \cdots j_n)} \varepsilon_{j_1 j_2 \cdots j_n} m_{1 j_1} m_{2 j_2} \cdots m_{n j_n} \tag{9.96}$$

$$\det \hat{M} = \sum_{(j_1 j_2 \cdots j_n)} \hat{m}_{1 j_1} \hat{m}_{2 j_2} \cdots \hat{m}_{n j_n} \tag{9.97}$$

where the sum is over all permutations $(j_1 j_2 \cdots j_n)$ of $(1, 2, \ldots, n)$, and $\varepsilon_{j_1 j_2 \cdots j_n}$ is $+1$ or -1 depending on whether the permutation is even or odd, respectively. Since F_d is totally unimodular, $m_{ij} = 1, -1$, or 0, so is $\det M$. Thus, a term in the summation of (9.96) is ± 1 if and only if the corresponding term in the summation of (9.97) is 1. For M to be nonsingular, the number of $+1$ and -1 terms in the summation of (9.96) must differ exactly by one, because $\det M = \pm 1$. This shows that there is an odd number of 1's in the summation of (9.97), and, therefore, $\hat{M}$ is also nonsingular over the field of integers modulo 2. Likewise, if M is singular, the number of $+1$ terms in the summation of (9.96) must equal that of the -1 terms, resulting in an even number of 1's in the summation of (9.97). Conversely, this is also true. Consequently, a totally unimodular matrix is binary, so is a regular matrix in normal form. This completes the proof of the theorem.

§ 3. Circuit and Cutset Matrices

In the foregoing, we discussed the conditions under which a real matrix of integers is a regular or binary matrix. In the present section, we characterize a matrix of integers mod 2 to be the circuit or cutset matrix of a graph. To this end, we define the notion of regular matrix mod 2.

DEFINITION 9.5: *Regular matrix mod 2.* A matrix over the field of integers modulo 2 is said to be *regular* if the replacement of a suitable set of 1's by -1's makes it regular over the real field.

As indicated in Theorem 9.2, a real matrix of integers is not regular if it contains the following matrix

$$N_d = \begin{bmatrix} \pm 1 & 0 & \pm 1 & \pm 1 \\ \pm 1 & \pm 1 & 0 & \pm 1 \\ \pm 1 & \pm 1 & \pm 1 & 0 \end{bmatrix} \tag{9.98}$$

or its transpose N_d' as a submatrix. Thus, if a matrix of integers mod 2 contains the matrix

$$N = \begin{bmatrix} 1 & 0 & 1 & 1 \\ 1 & 1 & 0 & 1 \\ 1 & 1 & 1 & 0 \end{bmatrix} \tag{9.99}$$

obtained from N_d by replacing entries ± 1 by 1's, or its transpose N' as a submatrix, it cannot be regular. On the other hand, if it does not contain such a submatrix, we cannot conclude that it is regular. It is possible that some linear combinations of its rows may result in such a submatrix. It appears that we must check all linear combinations in order to ascertain the nonexistence of such a submatrix. This is not all necessary, if we put the given matrix in normal form.

To normalize a matrix F, it is sufficient to premultiply it by the inverse of a major submatrix. The product of this inverse and the major submatrix results in the identity matrix as a submatrix of F. This new matrix is referred to as a *normal form* of F. As an illustration, consider the matrix of integers mod 2

$$F = \begin{bmatrix} 1 & 1 & 1 & 0 & 1 & 0 & 0 & 1 \\ 0 & 1 & 1 & 1 & 1 & 1 & 0 & 0 \\ 1 & 0 & 1 & 1 & 0 & 1 & 1 & 0 \\ 1 & 1 & 0 & 1 & 0 & 0 & 1 & 1 \end{bmatrix}. \tag{9.100}$$

The submatrix composed of the first four columns

$$F_m = \begin{bmatrix} 1 & 1 & 1 & 0 \\ 0 & 1 & 1 & 1 \\ 1 & 0 & 1 & 1 \\ 1 & 1 & 0 & 1 \end{bmatrix} \qquad (9.101)$$

is a major submatrix, the inverse of which is found to be

$$F_m^{-1} = \begin{bmatrix} 1 & 0 & 1 & 1 \\ 1 & 1 & 0 & 1 \\ 1 & 1 & 1 & 0 \\ 0 & 1 & 1 & 1 \end{bmatrix} \qquad (9.102)$$

Premultiplying F by F_m^{-1} yields a matrix in normal form

$$F_m^{-1}F = \begin{bmatrix} 1 & 0 & 0 & 0 & 1 & 1 & 0 & 0 \\ 0 & 1 & 0 & 0 & 0 & 1 & 1 & 0 \\ 0 & 0 & 1 & 0 & 0 & 0 & 1 & 1 \\ 0 & 0 & 0 & 1 & 1 & 0 & 0 & 1 \end{bmatrix} \qquad (9.103)$$

THEOREM 9.11: A matrix over the integers modulo 2 is regular if and only if no normal forms of the matrix contains either of the matrices

$$N = \begin{bmatrix} 1 & 0 & 1 & 1 \\ 1 & 1 & 0 & 1 \\ 1 & 1 & 1 & 0 \end{bmatrix} \quad \text{or} \quad N' = \begin{bmatrix} 1 & 1 & 1 \\ 0 & 1 & 1 \\ 1 & 0 & 1 \\ 1 & 1 & 0 \end{bmatrix} \qquad (9.104)$$

as a submatrix.

Proof. Let F be a regular matrix mod 2 in normal form. If F contains either of the two matrices of (9.104) as a submatrix, then there exists a suitable set of 1's, the replacement of which by -1 yields a real regular matrix F_d in normal form. It follows that F_d contains the matrix of (9.30) or its transpose as a submatrix. Since according to Theorem 9.9 every real regular matrix in normal form is totally unimodular, F_d is totally unimodular. By Theorem 9.2, this is not possible. Therefore, F cannot contain either of the two matrices of (9.104) as a submatrix, and the theorem is necessary. The proof of sufficiency is too long to be presented here, and the reader is referred to TUTTE [1958, 1959] for details.

A complete compact and elegant description of a matrix of integers mod 2 to be the cutset or circuit matrix of a graph was first given by TUTTE [1958], and its characterization is stated in terms of the nonexistence of certain submatrices associated with the two basic nonplanar graphs of KURATOWSKI [1930]. To this end, we first discuss the two basic nonplanar graphs.

An edge of a graph is said to be *subdivided* if it is replaced by a path of length of at least 2 as shown in fig. 9.8. The forbidden subgraphs used to characterize a planar graph are the 5-node complete graph K_5 known as the *complete pentagon* and the 6-node complete bipartite graph $K_{3,3}$ known as the *Thomsen graph* of fig. 9.9 and their variants. The variants of K_5 and $K_{3,3}$ are defined below.

DEFINITION 9.6: *Homeomorphism*. Two graphs are said to be *homeomorphic* if both can be obtained from the same graph by a sequence of subdivisions of the edges.

The two graphs G_1 and G_2 of fig. 9.10 are homeomorphic because both can be obtained from graph K_5 of fig. 9.9(a) by a sequence of subdivisions of edges. For example, if we replace the paths (1, 6)(6, 7)(7, 5) and (1, 8)(8, 9)(9, 10)(10, 3) in G_1 by the edges (1, 5) and (1, 3), respectively, we obtain the graph K_5. Likewise, if we replace the paths (1, 6)(6, 7)(7, 4) and (2, 8)(8, 9)(9, 10)(10, 3) by the edges (1, 4) and (2, 3) G_2, we obtain K_5. Two circuits of different lengths are homeomorphic because the longer circuit can be obtained from the shorter one by a sequence of subdivisions.

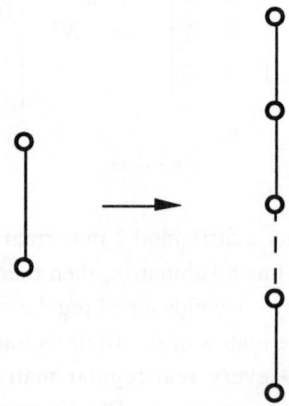

Fig. 9.8. The replacement of an edge by a path of length at least 2.

§ 3 Circuit and cutset matrices 643

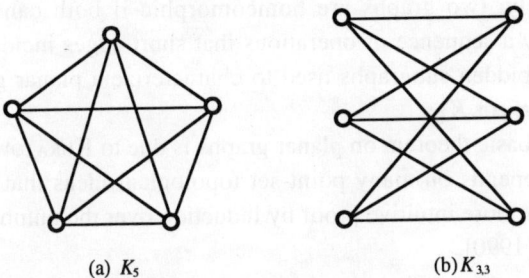

(a) K_5 (b) $K_{3,3}$

Fig. 9.9. The forbidden subgraphs K_5 and $K_{3,3}$ used to characterize a planar graph.

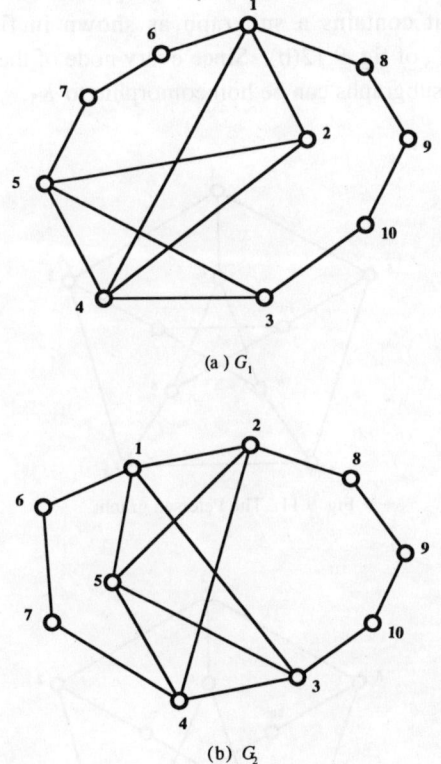

(a) G_1

(b) G_2

Fig. 9.10. Two homeomorphic graphs G_1 and G_2.

More intuitively, two graphs are homeomorphic if both can be reduced to the same graph by a sequence of operations that short edges incident to nodes of degree 2. The forbidden subgraphs used to characterize a planar graph are those homeomorphic to K_5 or $K_{3,3}$.

The following basic theorem on planar graphs is due to KURATOWSKI [1930], the proof of which depends on many point-set topological ideas that have not been developed here. A more intuitive proof by induction over the number of edges can be found in CHEN [1990].

THEOREM 9.12: A graph is planar if and only if it has no subgraph homeomorphic to K_5 or $K_{3,3}$.

As an illustration, consider the Petersen graph of fig. 9.11. This graph is nonplanar because it contains a subgraph as shown in fig. 9.12(a) which is homeomorphic to $K_{3,3}$ of fig. 9.12(b). Since every node of the Petersen graph is of degree 3, none of its subgraphs can be homeomorphic to K_5.

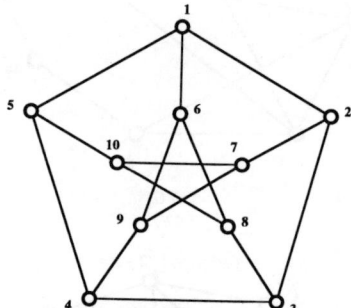

Fig. 9.11. The Petersen graph.

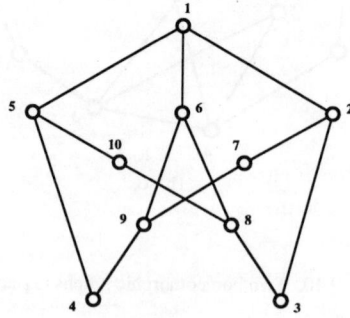

(a)

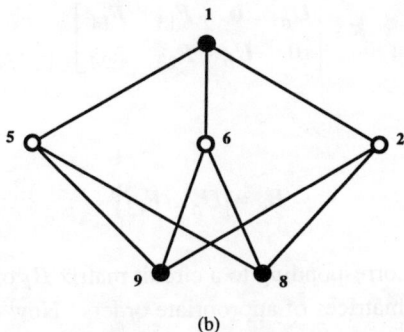

Fig. 9.12. A subgraph (a) of the Petersen graph which is homeomorphic to $K_{3,3}$ (b).

With these preliminaries, we are now in a position to state a fundamental result of this chapter.

THEOREM 9.13: A matrix over the field of integers modulo 2 is the cutset matrix (circuit matrix) of a graph if and only if it is regular and no normal form of the matrix contains as a submatrix the circuit matrix (cutset matrix) of either of the two basic nonplanar graphs K_5 and $K_{3,3}$.

Proof. Necessity. Let F be the mod 2 cutset matrix of graph G. By assigning directions to the edges of G, we obtain an associated directed graph G_d of G. Let F_d be the real cutset matrix of G_d with respect to the corresponding set of cutsets of G. Clearly, F_d is obtainable from F by the replacement of a suitable set of 1's by -1's. According to Theorem 9.7, to every real regular matrix there corresponds a real regular matrix in normal form which generates the same linear vector space with at most a permutation of coordinates. Without loss of generality, we may assume that F_d is a real cutset matrix in normal form. Such a cutset matrix F_d contains the identity matrix, the columns of which corresponds to the branches of a tree. Therefore, F_d can be considered as the f-cutset matrix with respect to this tree. Appealing to Theorem 9.6 shows that F_d is a real regular matrix. Thus, F is a regular matrix mod 2.

To show that no normal form of F can contain as a submatrix the circuit matrix of either of the two basic nonplanar graphs K_5 and $K_{3,3}$ of fig. 9.9, we first observe that a circuit matrix of either K_5 or $K_{3,3}$ cannot be realized as a cutset matrix of any graph. Suppose that a normal form of F contains such a circuit matrix as a submatrix. By permuting the rows and columns if necessary, F can be partitioned as

$$F = \begin{bmatrix} U_a & 0 & F_{13} & F_{14} \\ 0 & U_b & F_{23} & F_{24} \end{bmatrix} \tag{9.105}$$

where

$$B_k = \begin{bmatrix} U_a & F_{13} \end{bmatrix} \tag{9.106}$$

denotes the submatrix corresponding to a circuit matrix B_k of either K_5 or $K_{3,3}$, and U_a and U_b are identity matrices of appropriate orders. Now consider the operation of deleting an edge corresponding to a column of U_b and then let the two endpoints of the edge coalesce. If we apply this operation to all the columns of U_b, the cutset matrix F_1 of the resulting graph G_a is the matrix obtained from F by deleting the rows and columns corresponding to those of U_b, or

$$F_1 = \begin{bmatrix} U_a & F_{13} & F_{14} \end{bmatrix}. \tag{9.107}$$

By removing the edges corresponding to the columns of F_{14}, yields graph G_k, the cutset matrix of which is B_k. This is impossible because a circuit matrix B_k of K_5 or $K_{3,3}$ cannot be realized as a cutset matrix of any graph. For, otherwise, a nonplanar graph would have a dual. Therefore, no normal form of F can contain B_k.

Similarly, we can show that if a matrix of integers mod 2 is the circuit matrix of a graph, then it must be regular and no normal form of the matrix can contain as a submatrix a cutset matrix of either of the two basic nonplanar graphs K_5 or $K_{3,3}$. This completes the proof of necessity.

Sufficiency. The original proof by TUTTE [1958, 1959] is very complicated. BELEVITCHE [1963] suggests the use of the direct constructive procedure of GOULD [1958]. However, the process requires the consideration of every nonrealizable possibilities arising from a combination of all fictitious path sets. MAYEDA [1970] proposes to prove sufficiency by contradiction. His proof is outlined below.

A matrix mod 2 in normal form is said to be a *minimum nonrealizable matrix* if it satisfies the following three conditions: (1) it is not a cutset (circuit) matrix of any graph, (2) it does not satisfy the conditions of the theorem, and (3) by deleting any column other than those corresponding to the identity matrix, or by deleting any row of any of its normal form, the resulting matrix is realizable as a cutset (circuit) matrix of a graph. The sufficiency of the theorem is proved by showing that a minimum nonrealizable matrix does not exist. This approach is logical. However, there are some flaws in the original proof of MAYEDA [1970]. It was recently pointed out by KAJITANI [1994] that Mayeda's proof does not exhaust all the types of graphs that he needs to consider in his proof, and a counterexample was found by Ueno. Therefore, additional work is required to make it complete.

For a vigorous complete proof of sufficiency, we refer to Tutte's original paper, which requires many algebraic topological concepts that have not been developed here, aside from being extremely long. Therefore, we are satisfied with a statement of the result.

DEFINITION 9.7: *Graphic matrix.* A matrix of integers mod 2 is said to be *graphic* if it can be realized as the circuit or cut matrix of a graph.

THEOREM 9.13 states that a matrix of integers mod 2 is graphic if and only it is regular and no normal form of the matrix contains a circuit or cutset matrix of either of the complete pentagon and Thomsen graph. Invoking Theorem 9.11, we see that a matrix of integers mod 2 is graphic if and only if none of its normal forms contains the matrix

$$N = \begin{bmatrix} 1 & 0 & 1 & 1 \\ 1 & 1 & 0 & 1 \\ 1 & 1 & 1 & 0 \end{bmatrix} \quad \text{or} \quad N' = \begin{bmatrix} 1 & 1 & 1 \\ 0 & 1 & 1 \\ 1 & 0 & 1 \\ 1 & 1 & 0 \end{bmatrix} \tag{9.108}$$

or a circuit or cutset matrix of either of the complete pentagon and Thomsen graph.

A partition of the edges of the complete pentagon into tree t_1 (solid lines) and cotree $\bar{t}_1$ (broken lines) is shown in fig. 9.13. The f-circuit matrix with respect to this tree is obtained as

$$B_{1f} = \begin{bmatrix} 1 & 0 & 0 & 0 & 0 & 0 & 1 & 1 & 1 & 1 \\ 0 & 1 & 0 & 0 & 0 & 0 & 1 & 1 & 0 & 0 \\ 0 & 0 & 1 & 0 & 0 & 0 & 0 & 0 & 1 & 1 \\ 0 & 0 & 0 & 1 & 0 & 0 & 1 & 1 & 1 & 0 \\ 0 & 0 & 0 & 0 & 1 & 0 & 0 & 1 & 1 & 1 \\ 0 & 0 & 0 & 0 & 0 & 1 & 0 & 1 & 1 & 0 \end{bmatrix} = [U_6 \quad B_{1t}] \tag{9.109}$$

and from (5.3b) the f-cutset is found from (9.109) to be

$$Q_{1f} = [B'_{1t} \quad U_4] = \begin{bmatrix} 1 & 1 & 0 & 1 & 0 & 0 & 1 & 0 & 0 & 0 \\ 1 & 1 & 0 & 1 & 1 & 1 & 0 & 1 & 0 & 0 \\ 1 & 0 & 1 & 1 & 1 & 1 & 0 & 0 & 1 & 0 \\ 1 & 0 & 1 & 0 & 1 & 0 & 0 & 0 & 0 & 1 \end{bmatrix} = [Q_{1\bar{\imath}} \quad U_4]. \quad (9.110)$$

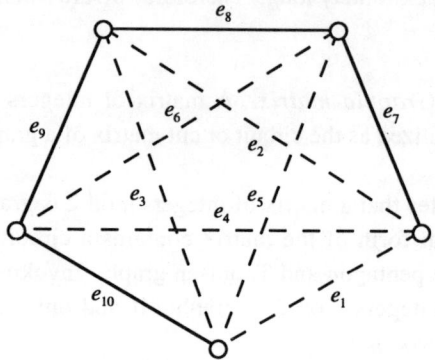

Fig. 9.13. The complete pentagon with a chosen tree t_1.

There are many other linear trees which are isomorphic to the tree t_1 in fig. 9.13, each of which will result in the same circuit and cutset matrices in normal form except for a permutation of rows and columns. A tree t_2 not isomorphic to t_1 is shown in fig. 9.14. The circuit and cutset matrices with respect to t_2 are found to be

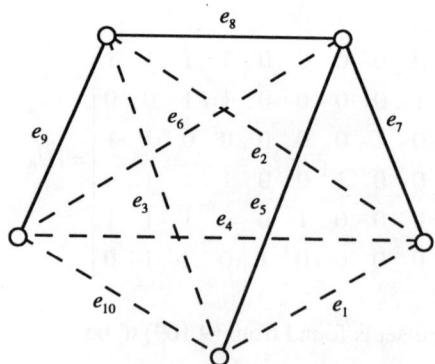

Fig. 9.14. The complete pentagon with a chosen tree t_2.

$$B_2 = \begin{bmatrix} 1 & 0 & 0 & 0 & 0 & 0 & 1 & 0 & 1 & 0 \\ 0 & 1 & 0 & 0 & 0 & 0 & 1 & 1 & 0 & 1 \\ 0 & 0 & 1 & 0 & 0 & 0 & 1 & 1 & 0 & 0 \\ 0 & 0 & 0 & 1 & 0 & 0 & 0 & 1 & 0 & 1 \\ 0 & 0 & 0 & 0 & 1 & 0 & 0 & 1 & 1 & 0 \\ 0 & 0 & 0 & 0 & 0 & 1 & 0 & 1 & 1 & 1 \end{bmatrix} = \begin{bmatrix} U_6 & B_{2t} \end{bmatrix} \qquad (9.111)$$

$$Q_2 = \begin{bmatrix} B'_{2t} & U_4 \end{bmatrix} = \begin{bmatrix} 1 & 1 & 1 & 0 & 0 & 0 & 1 & 0 & 0 & 0 \\ 0 & 1 & 1 & 1 & 1 & 1 & 0 & 1 & 0 & 0 \\ 1 & 0 & 0 & 0 & 1 & 1 & 0 & 0 & 1 & 0 \\ 0 & 1 & 0 & 1 & 0 & 1 & 0 & 0 & 0 & 1 \end{bmatrix} = \begin{bmatrix} Q_{2\bar{t}} & U_4 \end{bmatrix}. \qquad (9.112)$$

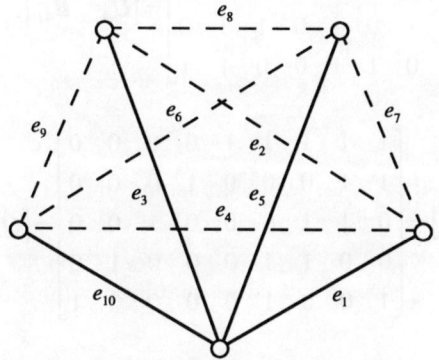

Fig. 9.15. The complete pentagon with a chosen tree t_3.

Finally, for the star tree t_3 of fig. 9.15, which is neither isomorphic to t_1 nor to t_2, the corresponding f-circuit and f-cutset matrices are obtained as

$$B_3 = \begin{bmatrix} 1 & 0 & 0 & 0 & 0 & 0 & 0 & 1 & 1 & 0 \\ 0 & 1 & 0 & 0 & 0 & 0 & 1 & 0 & 1 & 0 \\ 0 & 0 & 1 & 0 & 0 & 0 & 1 & 1 & 0 & 0 \\ 0 & 0 & 0 & 1 & 0 & 0 & 1 & 0 & 0 & 1 \\ 0 & 0 & 0 & 0 & 1 & 0 & 0 & 0 & 1 & 1 \\ 0 & 0 & 0 & 0 & 0 & 1 & 0 & 1 & 0 & 1 \end{bmatrix} = \begin{bmatrix} U_6 & B_{3t} \end{bmatrix} \qquad (9.113)$$

$$Q_3 = \begin{bmatrix} B'_{3t} & U_4 \end{bmatrix} = \begin{bmatrix} 0 & 1 & 1 & 1 & 0 & 0 & 1 & 0 & 0 & 0 \\ 1 & 0 & 1 & 0 & 0 & 1 & 0 & 1 & 0 & 0 \\ 1 & 1 & 0 & 0 & 1 & 0 & 0 & 0 & 1 & 0 \\ 0 & 0 & 0 & 1 & 1 & 1 & 0 & 0 & 0 & 1 \end{bmatrix} = \begin{bmatrix} Q_{3\bar{t}} & U_4 \end{bmatrix}. \quad (9.114)$$

All these matrices are matrices of integers mod 2 in normal form, and their columns do not correspond to the same edge order in the graph. However, they are all arranged in the order of cotree links and tree branches, depending on the choice of the trees.

The Thomsen graph admits three non-isomorphic trees t_i ($i = 4, 5, 6$). For the tree t_4 of fig. 9.16, the f-circuit and f-cutset matrices are given by

$$B_4 = \begin{bmatrix} 1 & 0 & 0 & 0 & 1 & 1 & 0 & 0 & 1 \\ 0 & 1 & 0 & 0 & 1 & 1 & 1 & 0 & 0 \\ 0 & 0 & 1 & 0 & 1 & 0 & 1 & 1 & 0 \\ 0 & 0 & 0 & 1 & 1 & 0 & 0 & 1 & 1 \end{bmatrix} = \begin{bmatrix} U_4 & B_{4t} \end{bmatrix} \quad (9.115)$$

$$Q_4 = \begin{bmatrix} B'_{4t} & U_5 \end{bmatrix} = \begin{bmatrix} 1 & 1 & 1 & 1 & 1 & 0 & 0 & 0 & 0 \\ 1 & 1 & 0 & 0 & 0 & 1 & 0 & 0 & 0 \\ 0 & 1 & 1 & 0 & 0 & 0 & 1 & 0 & 0 \\ 0 & 0 & 1 & 1 & 0 & 0 & 0 & 1 & 0 \\ 1 & 0 & 0 & 1 & 0 & 0 & 0 & 0 & 1 \end{bmatrix} = \begin{bmatrix} Q_{4\bar{t}} & U_5 \end{bmatrix}. \quad (9.116)$$

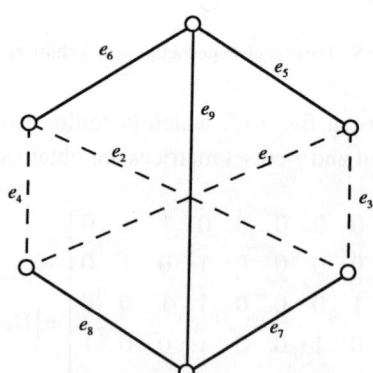

Fig. 9.16. The Thomsen graph with a chosen tree t_4.

For the tree t_5 of fig. 9.17, the f-circuit and f-cutset matrices are found to be

$$B_5 = \begin{bmatrix} 1 & 0 & 0 & 0 & 1 & 1 & 1 & 1 & 1 \\ 0 & 1 & 0 & 0 & 1 & 1 & 1 & 0 & 0 \\ 0 & 0 & 1 & 0 & 0 & 1 & 1 & 1 & 0 \\ 0 & 0 & 0 & 1 & 0 & 0 & 1 & 1 & 1 \end{bmatrix} = [U_4 \quad B_{5t}] \quad (9.117)$$

$$Q_5 = [B'_{5t} \quad U_5] = \begin{bmatrix} 1 & 1 & 0 & 0 & 1 & 0 & 0 & 0 & 0 \\ 1 & 1 & 1 & 0 & 0 & 1 & 0 & 0 & 0 \\ 1 & 1 & 1 & 1 & 0 & 0 & 1 & 0 & 0 \\ 1 & 0 & 1 & 1 & 0 & 0 & 0 & 1 & 0 \\ 1 & 0 & 0 & 1 & 0 & 0 & 0 & 0 & 1 \end{bmatrix} = [Q_{5\bar{t}} \quad U_5]. \quad (9.118)$$

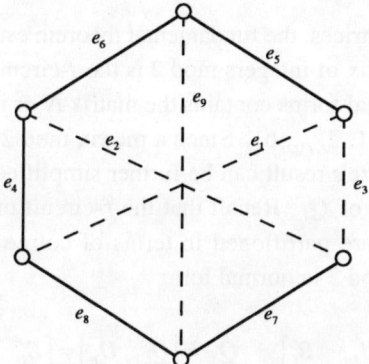

Fig. 9.17. The Thomsen graph with a chosen tree t_5.

Finally, for the tree t_6 of fig. 9.18, the f-circuit and f-cutset matrices are given by

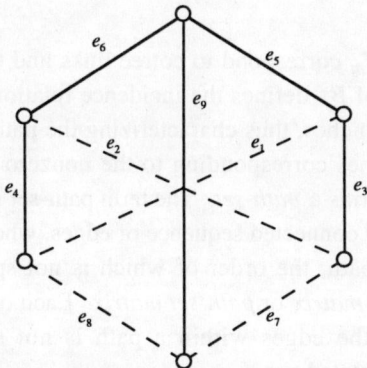

Fig. 9.18. The Thomsen graph with a chosen tree t_6.

$$B_6 = \begin{bmatrix} 1 & 0 & 0 & 0 & 1 & 1 & 1 & 0 & 0 \\ 0 & 1 & 0 & 0 & 0 & 1 & 1 & 1 & 0 \\ 0 & 0 & 1 & 0 & 0 & 0 & 1 & 1 & 1 \\ 0 & 0 & 0 & 1 & 1 & 1 & 0 & 0 & 1 \end{bmatrix} = [U_4 \quad B_{6t}] \qquad (9.119)$$

$$Q_6 = [B'_{6t} \quad U_5] = \begin{bmatrix} 1 & 0 & 0 & 1 & 1 & 0 & 0 & 0 & 0 \\ 1 & 1 & 0 & 1 & 0 & 1 & 0 & 0 & 0 \\ 1 & 1 & 1 & 0 & 0 & 0 & 1 & 0 & 0 \\ 0 & 1 & 1 & 0 & 0 & 0 & 0 & 1 & 0 \\ 0 & 0 & 1 & 1 & 0 & 0 & 0 & 0 & 1 \end{bmatrix} = [Q_{6\bar{t}} \quad U_5] . \qquad (9.120)$$

In terms of these matrices, the fundamental theorem established by TUTTE [1958, 1959] states that a matrix of integers mod 2 is the f-circuit matrix of a graph if and only if none of its normal forms contains the matrix N or its transpose N', or the f-cutset matrices Q_i ($i = 1, 2, ..., 6$). Since a matrix mod 2 in normal form contains the identity matrix, Tutte's result can be further simplified by considering only the forbidden submatrices of Q_i. Recall that the f-circuit matrix B_f and the f-cutset matrix Q_f of a graph are partitioned in terms of cotree links and tree branches, resulting in matrices mod 2 in normal form

$$B_f = [U_m \quad B_t], \qquad Q_f = [Q_{\bar{t}} \quad U_r] = [B'_t \quad U_r] \qquad (9.121)$$

where

$$Q_{\bar{t}} = B'_t \qquad (9.122)$$

Since the columns of U_m correspond to cotree links and the columns of B_t to tree branches, the kth row of B_f defines the incidence relation of the kth fundamental circuit with the tree branches, thus characterizing the path of the kth circuit in the tree. The set of branches corresponding to the nonzero elements in a row of B_t belongs to a path and forms a *path-set*. The term path-set is different from a path in that a path is an ordered connected sequence of edges, whereas a path-set is a set of edges belonging to a path, the order of which is not specified. Therefore, the matrix B_t is called *tree-matrix* or *path-set matrix*. Each of its rows defines a path-set, but the order of the edges within a path is not specified. Using these, Theorem 9.13 can be restated as

§ 4. 2-Isomorphism

THEOREM 9.14: A matrix over the field of integers modulo 2 is the cutset matrix of a graph if and only if none of its normal forms contains as a submatrix any of the matrices N, N' and B_{it} ($i = 1, 2, ..., 6$), and is the circuit matrix of a graph if and only if none of its normal forms contains as a submatrix any of the matrices N, N' and $Q_{i\bar{t}}$, as defined in (9.108)–(9.120).

§ 4. 2-Isomorphism

Before we discuss the problem of realizing a matrix of integers mod 2 to be the cutset or circuit matrix of a graph, we first characterize the class of graphs realized by a graphic matrix. In §2.2 of ch. 1, we introduced the notion of isomorphism to mean that two graphs are identical or the same. In this section, we present two types of equivalence operations that transform a graph into another one without altering its circuits. This whole class of equivalencies was first investigated by WHITNEY [1933b], who named the equivalence 2-isomorphism.

DEFINITION 9.8: *2-Isomorphism*. Two graphs G_1 and G_2 are said to be *2-isomorphic* if one can be made isomorphic to the other by a series of the following operations:

(a) Decompose a separable graph into blocks and then reconnect them at the single nodes to form a cutpoint.
(b) If g is a proper nonnull subgraph of G_1 or G_2 such that it has only two nodes i and j in common with its complement $\bar{g}$, decompose G_1 or G_2 into g and $\bar{g}$ by splitting off the nodes i and j and then reconnect at the same nodes after turning around at these nodes for either g or $\bar{g}$.

To illustrate the concept of 2-isomorphism, consider the two graphs G_1 and G_2 of fig. 9.19. To show that these two graphs are 2-isomorphic, we choose the subgraph $g = abcdefghij$ and its complement $\bar{g} = kmnpqrt$ in G_1. We next decompose G_1 into g and $\bar{g}$ by splitting off the nodes i and j and then reconnect at the same nodes after turning around at these nodes for g. The resulting graph G_3 is shown in Fig. 9.20. In G_3, let $g = abcdefghijkmn$ and $\bar{g} = pqrt$. Turning around and reconnecting at the nodes u and j for $\bar{g}$ yields the graph G_4 of fig. 9.21. Finally, let $g = kmn$ and $\bar{g} = abcdefghijpqrt$. Turning around and reconnecting at the nodes i and u for g gives the desired graph G_2 of fig. 9.19(b). Thus, G_1 and G_2 are 2-isomorphic.

The most important result in characterizing 2-isomorphic graphs is given by WHITNEY [1933b], and is stated as a theorem.

THEOREM 9.15: Two graphs are 2-isomorphic if and only if there is a one-to-one correspondence between their edges such that circuits in one correspond to circuits formed by the corresponding edges in the other.

The necessary part of the theorem is fairly evident. The proof of the sufficient part is too long to be given here. For the interested reader, we refer to the original paper by WHITNEY [1933b]. In the two 2-isomorphic graphs G_1 and G_2 of fig. 9.19, the corresponding edges are designated by the same edge symbols. If we pick, for instance, the circuit $L = ieakmptr$ in G_1, the subgraph formed by the corresponding edges in G_2 is also a circuit. This holds for all such circuits and we say that *circuits correspond to circuits*.

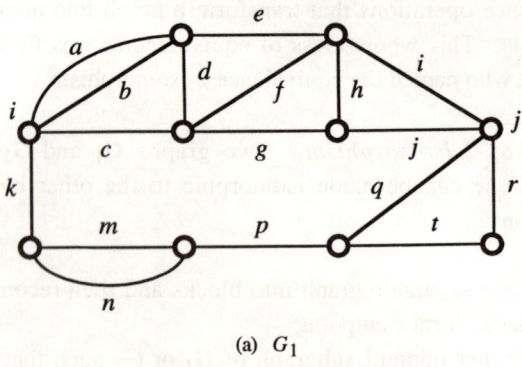

(a) G_1

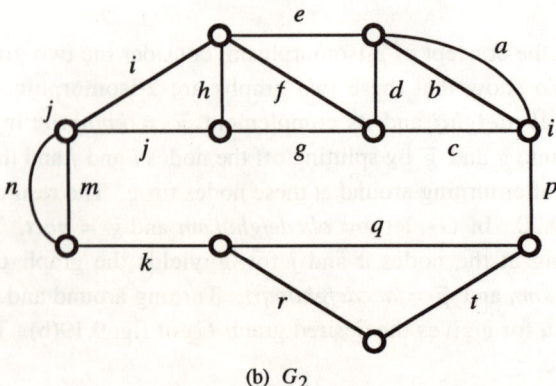

(b) G_2

Fig. 9.19. Two 2-isomorphic graphs G_1 and G_2.

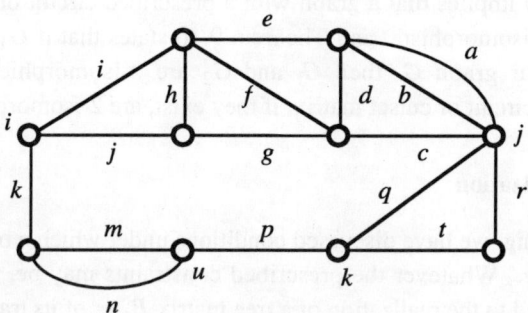

Fig. 9.20. Graph G_3.

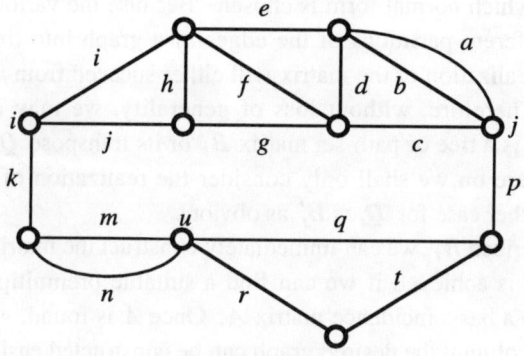

Fig. 9.21. Graph G_4.

The 2-isomorphism also defines the class of graphs that are duals of a planar graph, and this result is also due to WHITNEY [1933a].

THEOREM 9.16: *The duals of a planar graph are unique within 2-isomorphism.*

Proof. Let G_1 and G_2 be two duals of a planar graph G. Take any circuit L_1 of G_1. Then we can show that the subgraph C of G corresponding to the edges of L_1 is a cutset of G. Likewise, we can show that the subgraph L_2 of G_2 corresponding to the edges C of G is a circuit. Thus, circuits of G_1 correspond to circuits of G_2, and by Theorem 9.15, graphs G_1 and G_2 are 2-isomorphic. This completes the proof of the theorem.

Theorem 9.15 implies that a graph with a prescribed circuit or cutset matrix is unique within 2-isomorphism, and Theorem 9.16 states that if G_1 and G_2 are two duals of a planar graph G, then G_1 and G_2 are 2-isomorphic. Thus, all the realizations of a circuit or cutset matrix, if they exist, are 2-isomorphic.

§5. Graph Realization

In the foregoing, we have discussed conditions under which a matrix of integers mod 2 is graphic. Whatever the prescribed constraints may be, the problem can always be reduced to the realization of a tree matrix B_t or of its transpose, as stated in (9.121) and (9.122), by the reduction to a normal form by premultiplying the matrix by the inverse of a major submatrix. In fact, a matrix mod 2 in normal form is graphic if and only if its tree matrix or transpose is realizable as a tree. It is immaterial as to which normal form is chosen. Because the various normal forms correspond to different partitions of the edges of a graph into tree branches and cotree links, the realization of the matrix will either succeed from all normal forms or fail for all. Therefore, without loss of generality, we may assume that the prescribed matrix is a tree or path-set matrix B_t or its transpose $Q_t = B_t'$. For our purposes, from here on we shall only consider the realization of B_t, leaving the extension to the other case for $Q_t = B_t'$ as obvious.

From the prescribed B_t, we can immediately construct the matrix B_f by (9.121), and the synthesis is achieved if we can find a suitable premultiplier, which will transform B_f into a basis incidence matrix A. Once A is found, which contains at most two 1's per column, the desired graph can be constructed easily. If the matrix B_t contains at most two 1's per column, we are finished. Otherwise, a suitable premultiplier can be searched systematically. This is equivalent to mod 2 addition of one row to another. Since B_f contains the unit matrix U_m, the row operations would be restricted in that each column of U_m cannot contain more than two 1's. As a result, the allowed transformations are those replacing certain rows of B_f by the sum of two rows. However, such a procedure as studied by BAYARD [1954] and OKADA [1954] becomes rapidly prohibitive for large matrices.

Our starting point is a path-set matrix B_t. We first make the following reductions, which are not indispensable, to streamline the procedure:

(a) Suppression of trivial rows or columns. If a row (column) of B_t contains a single nonzero element, the corresponding tree branch is connected in parallel (series) with the defining link of the f-circuit.

(b) Suppression of duplicate rows or columns. If two rows are identical, the defining links of the two corresponding f-circuits are connected in parallel. If two columns are identical, the two corresponding tree branches are connected in series.

We consider the rows of the f-circuit matrix $\boldsymbol{B}_f$. Start from the first f-circuit realized by connecting its branches in some arbitrary order. The realization is continued by adding an f-circuit to the previously constructed subgraph. At the kth stage, the kth f-circuit is added to the subgraph already realizing the first $k-1$ f-circuits. The addition of the kth f-circuit is possible only if its intersection with the subgraph is a path-set. Since by Theorem 9.15 the realized subgraph is unique within 2-isomorphism, the realization of the kth f-circuit amounts to a rearrangement of a set of branches into a path by the two equivalence operations of Definition 9.8. This procedure due to LÖFGREN [1951] terminates in m steps or fails at some stage. Observe that the order in which the f-circuits are taken is immaterial, and so is the possible choice between alternative rearrangements. Instead of using the f-circuit matrix, a dual procedure to Löfgren's operation on the f-cutset matrix was proposed by MAYEDA [1960a, 1960b], and the resulting graph can then be constructed in a dual manner.

Instead of considering the f-circuit matrix $\boldsymbol{B}_f$, we can grow a tree from the path-sets defined by the rows of the tree matrix $\boldsymbol{B}_t$. The order in which the branches of a path-set are connected in the path is initially unknown but is progressively determined by comparison with other path-sets. We can base such a comparison exclusively on the following three intuitive properties first proposed by AUSLANDER and TRENT [1959].

(a) All branches of a path-set must appear in the corresponding path, head to toe in some order.
(b) The collection of branches appearing in two path-sets must appear head to toe in exactly the same order in both paths.
(c) No set of branches in the tree can form a circuit.

Property (c) is difficult to implement. GOULD [1958] provides the following three rather intuitive consequences of (c):

(d) The intersection of two path-sets is a path-set.
(e) If $\{e_1, e_2, b_1\}, \{e_2, e_3, b_2\}, ..., \{e_{u-1}, e_u, b_{u-1}\}, \{e_u, e_1, b_u\}$ $(u \geq 3)$ are path-sets, and e_i are non-empty, then $\{e_i, e_k\}$ is a path-set for all i, k $(i \neq k)$.

(f) If $\{e_1, e_2\}$ and $\{e_2, e_3\}$ are path-sets with e_1, e_2 and e_3 non-empty, and if a path-set exists containing e_1, e_2 and e_3, then e_1, e_3 and $\{e_1, e_2, e_3\}$ are path-sets, where e_2 is between e_1 and e_3 and adjacent to both, and e_2 is already a path-set as the intersection of $\{e_1, e_2\}$ and $\{e_2, e_3\}$.

Property (e) follows from the fact that the cyclically overlapping sequence $e_1 e_2 \dots e_u e_1$ is recognized as forming a star, as shown in fig. 9.22. As a result, any pair of subpaths incident to the common node forms a path. Also, by Property (d) the e_i are themselves path-sets. This result is not necessarily true if $b_1, b_2, \dots, b_u$ are not disjoint, but b_i may be empty. GOULD [1958] has shown that a tree, if it exists at all, can be constructed by applying the Properties (a), (b) and (d). His procedure is conveniently subdivided into the following three stages:

(i) Consider only the maximal path-sets. A *maximal path-set* is a path-set not included in other path-sets. Apply Property (e) to all pertinent combinations of maximal path-sets. Property (d) is then applied to all combinations of path-sets.

(ii) Each maximal path-set is considered separately and ordered by the application of Properties (a), (b) and (f).

(iii) The various ordered maximal path-sets are grafted on each other to form the tree.

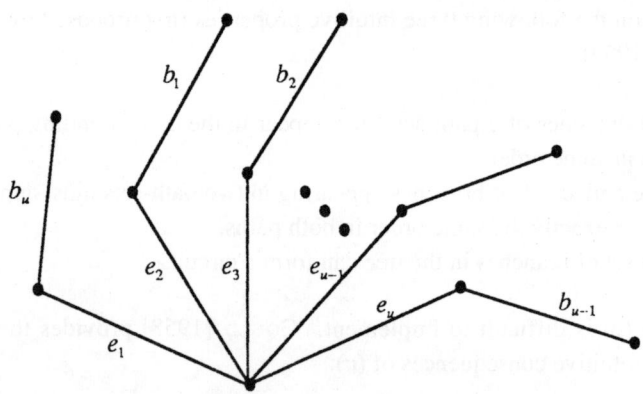

Fig. 9.22. A cyclically overlapping sequence forms a star.

In Stage (ii), once the stars resulting from the mutual intersections of all maximal path-sets are obtained, they are complete in that all intersections of subpaths belonging to different maximal path-sets are included in the intersections of the maximal path-sets. Therefore, further application of Property (e) to subpaths can generate no new path-sets. If Gould's procedure fails, the tree is not realizable. If it succeeds, the result is a tree satisfying Properties (a)–(c).

To simplify the notation, we use Gould's representation of path-sets. An unordered set of edges of a path is represented by enclosing the edge symbols, separated by commas, within braces. An ordered set of edges of a path is represented by enclosing the edge symbols, appropriately ordered, within brackets. For example, $\{a, b, c, d, e\}$ is an unordered set of edges of a path containing the edges a, b, c, d and e, whereas $[abcde]$ is an ordered set of edges of a path $abcde$. The expressions $[abcde]$ and $[edcba]$ are, of course, equivalent. Clearly, $[abcde]$ implies $\{a, b, c, d, e\}$ but the converse is generally true if and only if it contains not more than two elements. For convenience, write $\{a, b\}$ as $[ab]$. Thus, $\{a, b, \{c, d, e\}\}$ denotes a path-set containing the edges a, b, c, d and e, in which the subset $\{c, d, e\}$ is also a path-set; and $[ab[cde]]$ is a path-set, in which the edges a, b, c, d and e appear either in the order $[abcde]$ or $[abedc]$. Likewise, $\{a, b, [cde]\}$ is a path-set containing the edges a, b, c, d and e, in which the edges $c, d,$ and e appear consecutively in that order; whereas $\{a\{b, c, d\}e\}$ is a path-set containing the edges a, b, c, d and e, the terminal edges of which are a and e. Finally, we adopt a special convention for the use of parentheses within brackets. The edge symbols outside the parentheses indicate an ordered set of edges in the usual way, while the symbols within the parentheses denote another ordered set which joins the first. For example, $[ab(cd)e]$ is a subgraph in which the edges a, b and e in that order form a path and edges c and d form another path, and in which edges b, c and e are incident with a common node. Thus, $[ab(cd)e]$ denotes the path-sets $[abe]$, $[abcd]$ and $[dce]$. Observe that while a sequence of edge symbols occurring within brackets but not within parentheses may be reversed without altering its meaning, a sequence occurring within parentheses may not be reversed. The expression $[ab(cd)e]$ is equivalent $[e(cd)ba]$ but not to $[ab(dc)e]$, which denotes a subtree with edges b, d and e incident with a common node. As a result, it is possible to indicate the entire structure of any tree or multitree by means of a single expression. As an example, consider the expression $[a(b)cd(ef)g(h)(i)j]$, which represents the tree structure of fig. 9.23. This tree can also be represented by a number of different but equivalent expressions such as $[a(cd(g(h)(i)j)ef)b]$ and $[h(i)(j)g(dc(a)b)ef]$.

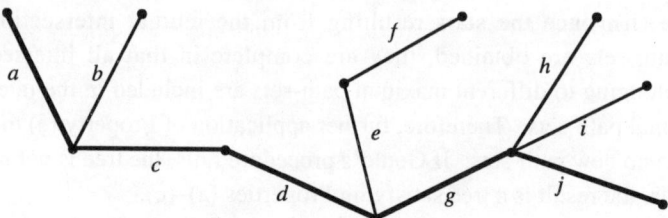

Fig. 9.23. A tree represented by [a(b)cd(ef) g(h)(i)j].

We illustrate the above result by the following example of GOULD [1958].

Example 9.1: We wish to realize a graph, the basis circuit matrix of which is specified as

$$B_f = [U_{16} \quad B_t] \tag{9.123}$$

where the tree or path-set matrix B_t is given by

$$B_t = \begin{bmatrix}
 & q & r & s & t & u & v & w & x & y & z & \alpha & \beta & \chi & \delta & \varepsilon \\
1 & 0 & 0 & 0 & 0 & 1 & 0 & 0 & 0 & 0 & 0 & 0 & 0 & 0 & 0 \\
1 & 0 & 1 & 0 & 0 & 0 & 1 & 0 & 0 & 0 & 1 & 0 & 1 & 0 & 0 \\
0 & 0 & 0 & 0 & 1 & 0 & 0 & 1 & 0 & 0 & 0 & 0 & 0 & 0 & 0 \\
1 & 0 & 0 & 0 & 0 & 1 & 0 & 0 & 0 & 0 & 0 & 0 & 0 & 0 & 0 \\
0 & 0 & 1 & 0 & 0 & 0 & 0 & 0 & 0 & 1 & 0 & 0 & 0 & 0 \\
0 & 0 & 0 & 1 & 0 & 0 & 0 & 0 & 0 & 1 & 0 & 0 & 0 & 0 & 0 \\
1 & 0 & 0 & 0 & 0 & 1 & 0 & 1 & 0 & 0 & 0 & 0 & 0 & 0 & 1 \\
0 & 0 & 0 & 0 & 0 & 0 & 0 & 1 & 0 & 0 & 0 & 0 & 0 & 1 & 0 \\
1 & 0 & 1 & 0 & 0 & 0 & 1 & 0 & 0 & 0 & 1 & 1 & 1 & 0 & 0 \\
0 & 0 & 0 & 0 & 0 & 0 & 0 & 1 & 0 & 0 & 0 & 0 & 0 & 0 & 1 \\
0 & 1 & 0 & 0 & 0 & 0 & 0 & 0 & 1 & 0 & 0 & 0 & 0 & 0 & 1 \\
0 & 0 & 0 & 1 & 0 & 0 & 0 & 0 & 0 & 0 & 0 & 0 & 0 & 0 & 0 \\
0 & 0 & 0 & 0 & 1 & 0 & 0 & 1 & 0 & 0 & 0 & 0 & 0 & 1 & 0 \\
0 & 0 & 1 & 0 & 0 & 0 & 0 & 0 & 0 & 0 & 0 & 0 & 1 & 0 & 0 \\
0 & 1 & 0 & 0 & 0 & 0 & 0 & 0 & 0 & 0 & 0 & 1 & 0 & 0 & 1 \\
1 & 0 & 1 & 0 & 0 & 0 & 0 & 0 & 0 & 0 & 1 & 0 & 1 & 0 & 0
\end{bmatrix} \tag{9.124}$$

From the tree matrix B_t, the maximal path-sets are found to be

$$\{t, z\}, \quad \{q, v, x, \varepsilon\}, \quad \{q, s, w, \alpha, \beta, \chi\}, \qquad (9.125a)$$

$$\{r, y, \varepsilon\}, \quad \{u, x, \delta\}, \quad \{r, \beta, \varepsilon\} \qquad (9.125b)$$

which correspond to the rows 6, 7, 9, 11, 13 and 15. Since the maximal path-sets

$$\{q, v, x, \varepsilon\}, \quad \{q, s, w, \alpha, \beta, \chi\}, \quad \{r, \beta, \varepsilon\} \qquad (9.126)$$

meet the conditions of Property (e), $[q\varepsilon]$, $[\beta\varepsilon] = [\varepsilon\beta]$ and $[q\beta] = [\beta q]$ are path-sets. They generate a subtree $[q(\varepsilon)\beta]$ as shown in fig. 9.24.

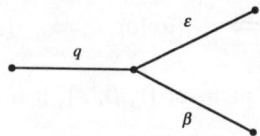

Fig. 9.24. A subtree $[q(\varepsilon)\beta]$.

To generate other path-sets, we apply Property (d) to the path-sets corresponding to the rows of the path-set matrix B_t. Only the intersection of the path-sets corresponding to rows 11 and 15 is non-trivial, yielding the path-set $[r\varepsilon]$. Therefore, we establish the existence of the following path-sets:

$$[qv], \quad \{q, s, w, \alpha, \chi\}, \quad [ux], \quad [qv], \quad [s\alpha], \quad [tz], \quad \{q, v, x, \varepsilon\} \qquad (9.127a)$$

$$[x\delta], \quad \{q, s, w, \alpha, \beta, \chi\}, \quad [x\varepsilon], \quad \{r, y, \varepsilon\}, \quad [t], \quad \{u, x, \delta\} \qquad (9.127b)$$

$$[s\chi], \quad \{r, \beta, \varepsilon\}, \quad \{q, s, \alpha, \chi\}, \quad [q\varepsilon], \quad [q\beta], \quad [\beta\varepsilon], \quad [r\varepsilon] \qquad (9.127c)$$

the maximal path-sets of which are given in (9.125). Each maximal path-set is now ordered by combining into its expression the expression of every path-set included in it. For the maximal path-set $\{q, v, x, \varepsilon\}$, it includes the path-sets $[qv]$, $[x\varepsilon]$ and $[q\varepsilon]$, and can be ordered as follows:

$$\{q, v, x, \varepsilon\} \Rightarrow \{[qv], x, \varepsilon\} \Rightarrow \{[qv][x\varepsilon]\} \Rightarrow [vq\varepsilon x]. \qquad (9.128)$$

For the maximal path-set $\{q, s, w, \alpha, \beta, \chi\}$, it includes the path-sets $[s\alpha]$, $\{q, s, w, \alpha, \chi\}$, $[s\chi]$, $\{q, s, \alpha, \chi\}$ and $[q\beta]$, and can be ordered as

$$\{q, s, w, \alpha, \beta, \chi\} \Rightarrow \{[s\alpha], q, w, \beta, \chi\} \Rightarrow [\{[s\alpha], q, w, \chi\}\beta]$$

$$\Rightarrow [\{[\chi s\alpha], q, w\}\beta] \Rightarrow [[[[\chi s\alpha]q]w]\beta] \Rightarrow [w[\chi s\alpha]q\beta]. \quad (9.129)$$

For the maximal path-set $\{u, x, \delta\}$, it includes the path-sets $[ux]$ and $[x\delta]$ and can be ordered as

$$\{r, y, \varepsilon\} \Rightarrow [[r\varepsilon]y]. \quad (9.130)$$

For the maximal path-set $\{u, x, \delta\}$, it includes the path-sets $[ux]$ and $[x\delta]$ and can be ordered as

$$\{u, x, \delta\} \Rightarrow [[ux]\delta] \Rightarrow [ux\delta] \quad (9.131)$$

and, finally, for the maximal path-set $\{r, \beta, \varepsilon\}$, it includes the path-sets $[\beta\varepsilon]$ and $[r\varepsilon]$, and can be ordered as

$$\{r, \beta, \varepsilon\} \Rightarrow [r[\beta\varepsilon]] \Rightarrow [r\varepsilon\beta]. \quad (9.132)$$

The maximal path-set $[tz]$ is, of course, already ordered, which includes the path-set $[t]$.

Observe that two of the maximal path-sets $[w[\chi s\alpha]q\beta]$ and $[[r\varepsilon]y]$ of (9.129) and (9.130) are not yet ordered. The edges χ, s and α are connected in series, so are the edges r and ε. They can be ordered arbitrarily as $[w\chi s\alpha q\beta]$ and $[\varepsilon ry]$.

Our next step is to combine these ordered maximal path-sets by taking into consideration that $[q(\varepsilon)\beta]$ of fig. 9.24 is a subtree. Then the other known subtrees are found to be

$$[tz] \quad [vq(\beta)\varepsilon x] \quad [w\chi s\alpha q(\varepsilon)\beta] \quad [\varepsilon ry] \quad [ux\delta] \quad [r\varepsilon(\beta)q] \quad (9.133)$$

Since $[w\chi s\alpha q(\varepsilon)\beta] = [w\chi s\alpha q(\beta)\varepsilon]$, the subtrees $[vq(\beta)\varepsilon x]$ and $[w\chi s\alpha q(\beta)\varepsilon]$ can be combined to yield

$$[w\chi s\alpha(v)q(\beta)\varepsilon x] \quad (9.134)$$

which has more than two edges in common with the subtree $[r\varepsilon(\beta)q] = [q(\beta)\varepsilon r]$. These two subtrees can be combined to give

$$[w\chi s\alpha(v)q(\beta)\varepsilon(r)x]. \quad (9.135)$$

This in turn has two edges in common with $[\varepsilon ry]$ to yield

$$[w\chi s\alpha(v)q(\beta)\varepsilon(ry)x] \qquad (9.136)$$

which has one edge in common with the subtree $[ux\delta]$. Since $[ux\delta]$ is included in a series connection, they may be combined arbitrarily to produce

$$[w\chi s\alpha(v)q(\beta)\varepsilon(ry)(u)x\delta]. \qquad (9.137)$$

This subtree has no edge in common with the remaining subtree $[tz]$, the graph representing the circuit matrix B_f is, therefore, a two-tree as shown in fig. 9.25.

Example 9.2: We wish to realize a graph, the basis circuit matrix of which is specified as

$$B = \begin{bmatrix} 1 & 0 & 0 & 0 & 0 & 0 & 1 & 1 & 0 & 1 & 1 \\ 0 & 1 & 0 & 0 & 0 & 1 & 1 & 0 & 0 & 1 & 1 \\ 0 & 0 & 1 & 0 & 0 & 0 & 0 & 0 & 1 & 0 & 1 \\ 0 & 0 & 0 & 1 & 0 & 1 & 1 & 1 & 1 & 0 & 0 \\ 0 & 0 & 0 & 0 & 1 & 0 & 1 & 1 & 0 & 1 & 1 \end{bmatrix}. \qquad (9.138)$$

This matrix is not realizable because it contains a submatrix

$$\begin{bmatrix} 1 & 1 & 0 & 1 \\ 1 & 1 & 1 & 0 \\ 0 & 1 & 1 & 1 \end{bmatrix} \qquad (9.139)$$

formed by the rows 2, 4 and 5 and columns 6, 7, 8 and 11. This submatrix is the matrix N of (9.108) within a permutation of the columns. Thus, by Theorem 9.14, the matrix B is not graphic.

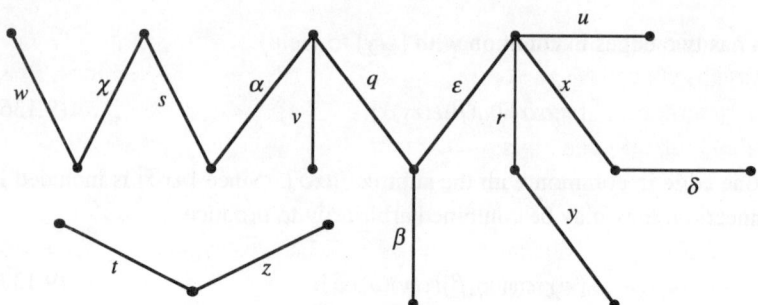

Fig. 9.25. A two-tree realization of the tree matrix of (9.124).

§ 6. Conclusions

In this chapter, we have studied the fundamental problem of characterizing a matrix of integers modulo 2 to be the basis circuit or cut matrix of a graph, or a real matrix of 1, − 1 and 0 to be the basis circuit or cut matrix of a directed graph.

We first reviewed the notion of a totally unimodular matrix and presented the necessary and sufficient conditions for a real matrix to be totally unimodular. As a result, a totally unimodular matrix cannot contain certain matrices as its submatrix. We then defined the regular matrix, and demonstrated that the basis incidence matrix, the f-circuit matrix and the f-cut matrix of a directed graph are all regular matrices, and to every regular matrix there corresponds a regular matrix in normal form which generates the same linear vector space with at most a permutation of the coordinates. In addition, we showed that a regular matrix in normal form is totally unimodular, and a totally unimodular matrix with linearly independent rows is regular. A binary matrix F_d is a matrix of integers 1, − 1, and 0, such that the replacement of − 1's by 1's in F_d results in a matrix F with elements 1 and 0 over the field of integers modulo 2, which preserves the ranks of the corresponding submatrices in F_d and F. Therefore, a totally unimodular matrix is binary, so does a regular matrix in normal form. A regular matrix mod 2 is a matrix such that the replacement of a suitable set of 1's by − 1's results in a regular matrix over the real field. It follows that a matrix over the integers modulo 2 is regular if and only if no normal forms of the matrix contains certain matrices as its submatrix.

The characterization of a matrix of integers mod 2 to be the cutset or circuit matrix of a graph is given in terms of the nonexistence of certain submatrices associated with the two basic nonplanar graphs of Kuratowski. Specifically, it states that a matrix over the field of integers modulo 2 is the cutset matrix (circuit matrix) of a graph if and only if it is regular and no normal form of the matrix contains as a submatrix the circuit matrix (cutset matrix) of either of the two basic nonplanar graphs K_5 and $K_{3,3}$. A graphic matrix is one that can be realized as the circuit or cut matrix of a graph. Thus, a matrix over the field of integers modulo 2 is graphic if and only if none of its normal forms contains as a submatrix any of the eight special matrices. However, all the realizations of a graphic matrix are within 2-isomorphic.

By premultiplying a given matrix by the inverse of a major submatrix, the realization of a graphic matrix can always be reduced to that of a tree matrix or its transpose. In fact, a matrix mod 2 in normal form is graphic if and only if its tree matrix or transpose is realizable as a tree. It is immaterial as to which normal form is chosen. Because the various normal forms correspond to different partitions of the edges of a graph into tree branches and cotree links, the realization of the matrix will either succeed from all normal forms or fail for all. Therefore, without loss of generality, we may assume that the prescribed matrix is a tree or path-set matrix or its transpose. To complete the realization, we presented a systematic procedure for the realization of a tree matrix, resulting in the final realization of the circuit or cut matrix.

Problems

9.1. Determine whether the following matrix over the field of integers modulo 2 can be realized as the cutset matrix of a graph:

$$\begin{bmatrix} 1 & 1 & 0 & 0 & 0 \\ 0 & 1 & 1 & 0 & 1 \\ 0 & 0 & 0 & 1 & 1 \end{bmatrix} \tag{9.140}$$

9.2. Determine whether the following matrix of 1, −1, and 0 in normal form can be realized as the circuit matrix of a directed graph:

$$\begin{bmatrix} 1 & 0 & 0 & 0 & -1 & 1 & 0 & 0 \\ 0 & 1 & 0 & 0 & 0 & -1 & 0 & 0 \\ 0 & 0 & 1 & 0 & 0 & -1 & 1 & 0 \\ 0 & 0 & 0 & 1 & 0 & -1 & 1 & 1 \end{bmatrix} \tag{9.141}$$

9.3. Determine whether the following matrix of 1, −1, and 0 in normal form can be realized as the cutset matrix of a directed graph:

$$\begin{bmatrix} 1 & 0 & 0 & 0 & 1 & 0 & 0 & 0 \\ -1 & 1 & 1 & 1 & 0 & 1 & 0 & 0 \\ 0 & 0 & -1 & -1 & 0 & 0 & 1 & 0 \\ 0 & 0 & 0 & -1 & 0 & 0 & 0 & 1 \end{bmatrix} \quad (9.142)$$

9.4. Find the directed graphs of the following realizable circuit matrices:

$$\begin{bmatrix} 1 & -1 & -1 & 0 & -1 & 0 & 0 & 0 & 0 \\ 0 & 1 & 0 & 1 & 0 & 0 & 0 & 0 & -1 \\ 0 & 0 & 1 & -1 & 0 & 0 & -1 & 1 & 0 \\ 0 & 0 & 0 & 0 & 1 & 1 & 0 & 0 & 1 \end{bmatrix} \quad (9.143)$$

$$\begin{bmatrix} 0 & 1 & 0 & 0 & -1 & 0 & 0 & 0 & 1 \\ 0 & -1 & 1 & 1 & 0 & 0 & 0 & 0 & 0 \\ -1 & 0 & -1 & 0 & 0 & 0 & 0 & 1 & 0 \\ 0 & 0 & 0 & 0 & 0 & 0 & -1 & -1 & 0 \\ 0 & 0 & 0 & -1 & 0 & 1 & 1 & 0 & -1 \end{bmatrix} \quad (9.144)$$

9.5. Are the matrices in Problem 9.4 realizable as the cut matrices of any directed graphs?

9.6. A cutset matrix of a connected directed graph is given below:

$$\begin{bmatrix} 1 & 0 & 0 & 0 & -1 & 0 & 0 & 0 \\ 0 & 1 & 0 & 0 & -1 & 1 & 0 & 0 \\ 0 & 0 & 1 & 0 & 0 & 1 & 0 & 1 \\ 1 & 0 & 1 & 1 & -1 & 1 & -1 & 0 \end{bmatrix} \quad (9.145)$$

Find a directed graph realizing the cutset matrix.

9.7. Determine whether the following two graphs are isomorphic:

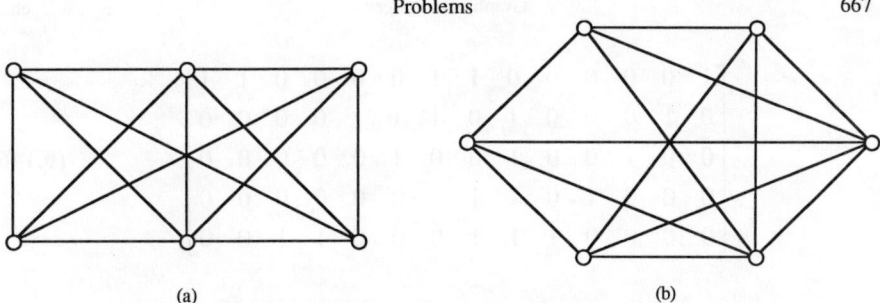

Fig. 9.26. Two graphs used to test for isomorphism.

9.8. Determine whether the following two graphs are 2-isomorphic:

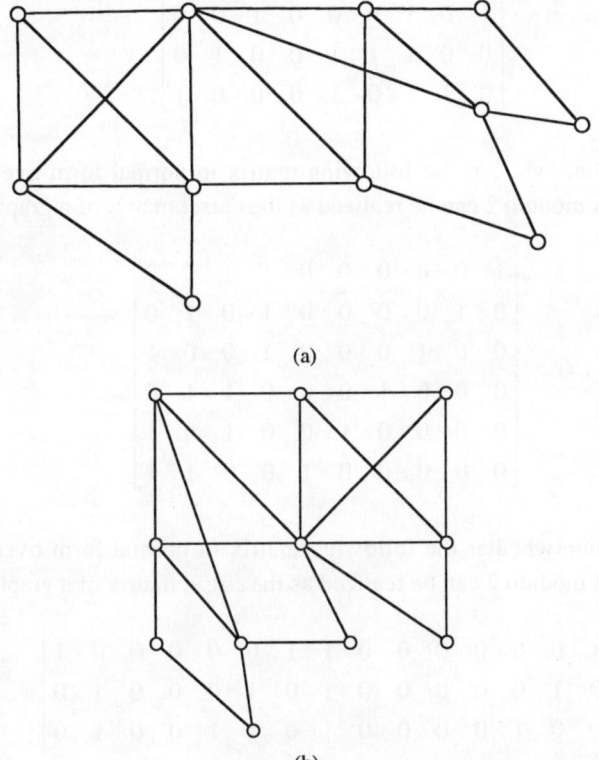

Fig. 9.27 Two graphs used to test for 2-isomorphism.

9.9. Determine whether the following matrix in normal form over the field of integers modulo 2 can be realized as the circuit matrix of a graph:

$$\begin{bmatrix} 1 & 0 & 0 & 0 & 0 & 0 & 1 & 1 & 0 & 0 & 0 & 0 & 1 & 1 \\ 0 & 1 & 0 & 0 & 0 & 1 & 0 & 1 & 1 & 1 & 0 & 0 & 0 & 0 \\ 0 & 0 & 1 & 0 & 0 & 1 & 0 & 0 & 1 & 0 & 0 & 0 & 0 & 0 \\ 0 & 0 & 0 & 1 & 0 & 1 & 1 & 0 & 0 & 0 & 0 & 0 & 0 & 0 \\ 0 & 0 & 0 & 0 & 1 & 1 & 1 & 0 & 0 & 0 & 1 & 1 & 0 & 0 \end{bmatrix} \quad (9.146)$$

9.10. Determine whether the following matrix in normal form over the field of integers modulo 2 can be realized as the cutset matrix of a graph:

$$\begin{bmatrix} 1 & 1 & 1 & 0 & 0 & 1 & 0 & 0 & 0 \\ 1 & 0 & 1 & 0 & 0 & 0 & 1 & 0 & 0 \\ 0 & 0 & 1 & 1 & 1 & 0 & 0 & 1 & 0 \\ 0 & 1 & 1 & 0 & 1 & 0 & 0 & 0 & 1 \end{bmatrix} \quad (9.147)$$

9.11. Determine whether the following matrix in normal form over the field of integers modulo 2 can be realized as the cutset matrix of a graph:

$$\begin{bmatrix} 1 & 0 & 0 & 0 & 0 & 0 & 1 & 1 & 0 & 0 \\ 0 & 1 & 0 & 0 & 0 & 0 & 1 & 0 & 1 & 0 \\ 0 & 0 & 1 & 0 & 0 & 0 & 1 & 0 & 0 & 1 \\ 0 & 0 & 0 & 1 & 0 & 0 & 0 & 1 & 1 & 0 \\ 0 & 0 & 0 & 0 & 1 & 0 & 0 & 1 & 0 & 1 \\ 0 & 0 & 0 & 0 & 0 & 1 & 0 & 0 & 1 & 1 \end{bmatrix} \quad (9.148)$$

9.12. Determine whether the following matrix in normal form over the field of integers modulo 2 can be realized as the cutset matrix of a graph:

$$\begin{bmatrix} 1 & 0 & 0 & 0 & 0 & 0 & 0 & 1 & 1 & 0 & 0 & 0 & 0 & 1 \\ 0 & 1 & 0 & 0 & 0 & 0 & 0 & 1 & 0 & 1 & 0 & 0 & 0 & 1 & 0 \\ 0 & 0 & 1 & 0 & 0 & 0 & 0 & 1 & 0 & 1 & 1 & 0 & 0 & 1 & 0 \\ 0 & 0 & 0 & 1 & 0 & 0 & 0 & 1 & 1 & 1 & 1 & 1 & 0 & 1 & 0 \\ 0 & 0 & 0 & 0 & 1 & 0 & 0 & 1 & 1 & 1 & 1 & 1 & 1 & 0 & 0 \\ 0 & 0 & 0 & 0 & 0 & 1 & 0 & 1 & 1 & 1 & 0 & 1 & 1 & 0 & 0 \\ 0 & 0 & 0 & 0 & 0 & 0 & 1 & 1 & 1 & 0 & 0 & 1 & 1 & 0 & 1 \end{bmatrix} \quad (9.149)$$

9.13. Let F be a square nonsingular unimodular matrix and D a diagonal matrix with positive real elements. Show that if k_{pq} is the nonvanishing element with the smallest absolute value or one of such smallest elements in the matrix triple product

$$K = FDF' = \left[k_{ij} \right] \qquad (9.150)$$

then $|k_{pq}|$ is equal to some element of D.

9.14. A real symmetric matrix is said to be *paramount* if each principal minor of the matrix is not less in magnitude than any other minor built from the same rows. Show that the matrix K of (9.150) is paramount.

9.15. Decompose the matrix

$$K = \begin{bmatrix} 6 & 4 & 3 \\ 4 & 6 & 4 \\ 3 & 4 & 4 \end{bmatrix} \qquad (9.151)$$

into the product $K = FDF'$, where F is a square nonsingular unimodular matrix and D a diagonal matrix with positive real elements.

9.16. Decompose the matrix

$$K = \begin{bmatrix} 7 & 1 & 2 & 3 \\ 1 & 12 & 4 & 5 \\ 2 & 4 & 15 & 6 \\ 3 & 5 & 6 & 18 \end{bmatrix} \qquad (9.152)$$

into the product $K = FDF'$, where F is a square nonsingular unimodular matrix and D a diagonal matrix with positive real elements.

9.17. Find a directed graph of the following realizable circuit matrix in normal form:

$$\begin{bmatrix} 1 & 0 & 0 & 0 & 0 & 1 & 1 & 0 & -1 & 0 & 0 & 0 \\ 0 & 1 & 0 & 0 & 0 & 1 & 0 & 0 & 0 & -1 & 0 & -1 \\ 0 & 0 & 1 & 0 & 0 & 0 & 1 & -1 & 0 & 1 & -1 & 1 \\ 0 & 0 & 0 & 1 & 0 & 0 & 0 & -1 & 1 & 0 & -1 & 0 \\ 0 & 0 & 0 & 0 & 1 & 0 & 1 & -1 & 0 & 0 & 0 & 1 \end{bmatrix} \quad (9.153)$$

9.18. Find a directed graph of the following realizable cutset matrix in normal form:

$$\begin{bmatrix} -1 & -1 & 0 & 0 & 0 & 1 & 0 & 0 & 0 & 0 & 0 & 0 \\ -1 & 0 & -1 & 0 & -1 & 0 & 1 & 0 & 0 & 0 & 0 & 0 \\ 0 & 0 & 1 & 1 & 1 & 0 & 0 & 1 & 0 & 0 & 0 & 0 \\ 1 & 0 & 0 & -1 & 0 & 0 & 0 & 0 & 1 & 0 & 0 & 0 \\ 0 & 1 & -1 & 0 & 0 & 0 & 0 & 0 & 0 & 1 & 0 & 0 \\ 0 & 0 & 1 & 1 & 0 & 0 & 0 & 0 & 0 & 0 & 1 & 0 \\ -1 & 1 & -1 & 0 & -1 & 0 & 0 & 0 & 0 & 0 & 0 & 1 \end{bmatrix} \quad (9.154)$$

9.19. Find a directed graph of the following realizable circuit matrix:

$$\begin{bmatrix} -1 & -1 & 0 & 0 & 0 & 0 & 0 & 0 & 0 & 1 \\ -1 & 0 & -1 & 1 & 0 & 0 & 0 & 0 & 0 & 0 \\ 0 & -1 & 1 & 0 & 0 & 0 & 0 & 0 & 1 & 0 \\ 0 & -1 & 1 & 0 & 1 & 0 & 0 & 0 & 0 & 0 \\ 0 & -1 & 1 & 0 & 0 & 0 & 1 & 0 & 0 & 0 \\ 0 & 0 & 0 & 0 & 0 & -1 & 0 & 1 & 0 & 0 \end{bmatrix} \quad (9.155)$$

9.20. Find a directed graph of the following realizable cutset matrix:

$$\begin{bmatrix} 1 & 0 & 0 & 1 & 0 & 0 & 0 & 0 & 0 & 1 \\ -1 & 1 & 1 & 0 & 0 & 0 & 0 & 0 & 0 & 0 \\ 0 & 0 & -1 & -1 & 1 & 0 & 1 & 0 & 1 & 0 \\ 0 & 0 & 0 & 0 & 0 & 1 & 0 & 1 & 0 & 0 \end{bmatrix} \quad (9.156)$$

BIBLIOGRAPHY

VAN AARDENNE-EHRENFEST, T. and N. G. DE BRUIJN, Circuits and trees in oriented linear graphs, Simon Stevin **28** (1951) 203-217.
AITKEN, A. C., Determinants and Matrices (Interscience, New York) 9th ed., 1962.
ASH, R. B., Topology and the solution of linear systems, J. Franklin Inst. **268** (1959) 453-463.
ASH, R. B. and W. H. KIM, On realizability of a circuit matrix, IRE Trans. Circuit Theory **CT-6** (1959) 219-223.
AUSLANDER, L. and M. TRENT, Incidence matrices and linear graphs, J. Math. Mech. **8** (1959) 827-835.
AVONDO-BODINO, G., Economic Applications of the Theory of Graphs (Gordon and Breach, New York), 1962.
AYOUB, J. N. and I. T. FRISCH, Degree realization of undirected graphs in reduced form, J. Franklin Inst. **289** (1970) 303-312.
BÄBLER, F., Über eine spezielle Klasse Euler'scher Graphen, Comment. Math. Helv. **27** (1953) 81-100.
BALABANIAN, N. and T. A. BICKART, Electrical Network Theory (Wiley, N.Y.), 1969.
BARABASCHI, S. and E. GATTI, Modern methods of analysis for active electrical networks with particular regard to feedback systems, Part I, Energia Nucleare **2** (1954) 105-119.
BARON, G. and W. IMRICH, On the maximal distance of spanning trees, J. Combinatorial Theory **5** (1968) 378-385.
BASHKOW, T. R., The A matrix, new network description, IRE Trans. Circuit Theory **CT-4** (1957) 117-119.
BAYARD, M., Théorie des réseaux de Kirchhoff (Ed. Rev. d'Optique, Paris), 1954.
BEHZAD, M. and G. CHARTRAND, No graph is perfect, Amer. Math. Monthly **74** (1967) 962-963.
BEINEKE, L. W. and F. HARARY, Local restrictions for various classes of directed graphs, J. London Math. Soc. **40** (1965) 87-95.
BELEVITCH, V., On the realizability of graphs with prescribed circuit matrices, in Switching Theory in Space Technology (Stanford University Press, Stanford, CA), 1963, pp. 126-144.
BERGE, C., Theory of Graphs and Its Applications (Methuen, London), 1962.
BERGER, I. and A. NATHAN, The algebra of sets of trees, k-trees, and other configurations, IEEE Trans. Circuit Theory **CT-15** (1968) 221-228.
BERS, A., The degrees of freedom in RLC networks, IRE Trans. Circuit Theory **CT-6** (1959) 91-95.
BOISVERT, M. and L. P. A. ROBICHAUD, Direct analysis of electrical networks, Rapport de Recherches No. 9, Dept. de Genie Electrique, Université Laval, Quebec, Canada, October 1956.
BOTT, R. and J. P. MAYBERRY, Matrices and trees, in: O. Morgenstern (ed.), Economic Activity Analysis (Wiley, New York), 1954, 391-400.
BÖTTGER, G. and H. HARDERS, Note on a problem by S. L. Hakimi concerning planar graphs without parallel elements, J. Soc. Indust. Appl. Math. **12** (1964) 838-839.

BOUWKAMP, C. J., On the dissection of rectangles into squares I, Koninkl. Nederl. Akad. Wetensch. Proc. Ser. A. **49** (1946) 1176–1188; II, III, **50** (1947) 58–71, 72–78.
BRANIN Jr., F. H., The relation between Kron's method and the classical methods of network analysis, Matrix Tensor Quart. **13** (1962) 69–105.
BROOKS, R. L., C. A. B. SMITH, A. H. STONE and W. T. TUTTE, The dissection of rectangles into squares, Duke Math. J. **7** (1940) 312–340.
BROWN, D. P., Derivative-explicit differential equations for RLC graphs, J. Franklin Inst. **275** (1963) 503–514.
BROWN, D. P., Topological properties of resistance matrices, SIAM J. Appl. Math. **16** (1968) 387–394.
BROWN, D. P. and A. BUDNER, A note on planar graphs, J. Franklin Inst. **280** (1965) 222–230.
BROWNELL, R. A., Growing the trees of a graph, Proc. IEEE **56** (1968) 1121–1123.
BROWNLEE, A., Directed graph realization of degree pairs, Amer. Math. Monthly **75** (1968) 36–38.
BRYANT, P. R., A topological investigation of network determinants, Proc. IEE (London) **106C** (1959a) 16–22.
BRYANT, P. R., The order of complexity of electrical networks, Proc. IEE (London) **106C** (1959b) 174–188.
BRYANT, P. R., Problems in electrical network theory, Ph. D. dissertation, Cambridge University, Cambridge (1959c).
BRYANT, P. R., The degrees of freedom in RLC networks, IRE Trans. Circuit Theory **CT-7** (1960) 173–174 and 357.
BRYANT, P. R., The algebra and topology of electrical networks, Proc. IEE (London) **108C** (1961) 215–229.
BRYANT, P. R., The explicit form of Bashkow's A matrix, IRE Trans. Circuit Theory **CT-9** (1962) 303–306.
BRYANT, P. R. and J. Tow, The A-matrix of linear passive reciprocal networks, J. Franklin Inst. **293** (1972) 401–419.
BUSACKER, R. G. and T. L. SAATY, Finite Graphs and Networks: An Introduction with Applications (McGraw-Hill, New York), 1965.
CAMION, P., Characterization of totally unimodular matrices, Proc. Amer. Math. Soc. **16** (1965) 1068–1073.
CARLIN, H. J. and A. B. GIORDANO, Network Theory (Prentice-Hall, N.J.), 1964.
CAYLEY, A., A theorem on trees, Quart. J. Math. **23** (1889) 376–378.
CAYLEY, A., The Collected Papers of A. Cayley, Cambridge **13** (1897) 26–28.
CEDERBAUM, I., Invariance and mutual relations of electrical network determinants, J. Math. Phys. **35** (1956) 236–244.
CEDERBAUM, I., Matrices all of whose elements and subdeterminants are 1, − 1 or 0, Math. and Phys. **36** (1958) 351–361.
CHANDRASEKARAN, R., Total unimodularity of matrices, SIAM J. Appl. Math. **17** (1969) 1032–1034.
CHEN, W. K., The inversion of matrices by flow graphs, J. Soc. Indust. Appl. Math. **12** (1964a) 676–685.
CHEN, W. K., On signal-flow graphs, Proc. IEEE **52** (1964b) 967.
CHEN, W. K., Flow graphs: some properties and methods of simplification, IEEE Trans. Circuit Theory **CT-12** (1965a) 128–130.
CHEN, W. K., On the modifications of flow graphs, J. Soc. Indust. Appl. Math. **13** (1965b) 493–505.
CHEN, W. K., Matrix graphs and bipartite graphs, IEEE Trans. Circuit Theory **CT-12** (1965c) 268–269.
CHEN, W. K., Topological analysis for active networks, IEEE Trans. Circuit Theory **CT-12** (1965d) 85–91.

CHEN, W. K., On directed trees and directed k-trees of a digraph and their generation, SIAM J. Appl. Math. **14** (1966a) 550–560.
CHEN, W. K., On the realization of directed trees and directed 2-trees, IEEE Trans. Circuit Theory **CT-13** (1966b) 230–232.
CHEN, W. K., Note on topological analysis for active networks, IEEE Trans. Circuit Theory **CT-13** (1966c) 438–439.
CHEN, W. K., A generalization of the equicofactor matrix, IEEE Trans. Circuit Theory **CT-13** (1966d) 440–442.
CHEN, W. K., On the realization of a (p, s)-digraph with prescribed degrees, J. Franklin Inst. **281** (1966e) 406–422.
CHEN, W. K., On directed graph solutions of linear algebraic equations, SIAM Rev. **9** (1967a) 692–707.
CHEN, W. K., Unified theory on topological analysis of linear systems, Proc. IEE (London) **114** (1967b) 1630–1636.
CHEN, W. K., Hamilton circuits in directed-tree graphs, IEEE Trans. Circuit Theory **CT-14** (1967c) 231–233.
CHEN, W. K., Iterative procedure for generating trees and directed trees, Electronics Letters **4** (1968a) 516–518.
CHEN, W. K., On unisignants and their evaluation, SIAM J. Appl. Math. **16** (1968b) 603–619.
CHEN, W. K., A physical interpretation of the multiple-node removal algorithm of a flow graph, Proc. IEEE **57** (1969a) 1691–1692.
CHEN, W. K., Conversion of an arbitrary node into a source node in a flow graph, Electronics Letters **5** (1969b) 338–339.
CHEN, W. K., Modification of topological formulas for active two-port networks, Proc. IEEE **57** (1969c) 2166–2167.
CHEN, W. K., On generation of trees without duplications, Proc. IEEE **57** (1969d) 1292–1293.
CHEN, W. K., Generation of trees and cotrees of a graph by decomposition, Proc. IEE (London) **116** (1969e) 1639–1643.
CHEN, W. K., Computer generation of trees and cotrees in a cascade of multiterminal networks, IEEE Trans. Circuit Theory **CT-16** (1969f) 518–526.
CHEN, W. K., Unified theory on the generation of trees of a graph Part I. The Wang algebra formulation, Int. J. Electronics **27** (1969g) 101–117.
CHEN, W. K., Unified theory on the generation of trees of a graph Part II. The matrix formulation, Int. J. Electronics **27** (1969h) 319–336.
CHEN, W. K., Graph-theoretic considerations on the invariance and mutual relations of the determinants of the generalized network matrices and their generalized cofactors, Quart. J. Math. Oxford (2), **21** (1970a) 459–479.
CHEN, W. K., On vector spaces associated with a graph, Proc. 13th Midwest Symposium on Circuit Theory, University of Minnesota, Minneapolis, Minnesota, 1970b, XI.2.1–XI.2.10.
CHEN, W. K., On the nonsingular submatrices of the incidence matrix of a graph over the real field, J. Franklin Inst. **289** (1970c) 155–166.
CHEN, W. K., On vector spaces associated with a graph, SIAM J. Appl. Math. **20** (1971a) 526–529.
CHEN, W. K., On d-invariant transformations of (p, s)-digraphs, J. Franklin Inst. **291** (1971b) 89–100.
CHEN, W. K., Unified theory on the generation of trees of a graph Part III. Decomposition and elementary transformations, Int. J. Electronics **31** (1971c) 301–319.
CHEN, W. K., Topological formulations and the order of complexity of active networks: A unified survey, Networks **2** (1972a) 237–260.
CHEN, W. K., Characterizations of complete directed trees and two-trees, IEEE Trans. Circuit Theory **CT-19** (1972b) 241–247.

CHEN, W. K., Codifying trees and multitrees of a complete graph, J. Franklin Inst. **293** (1972c) 11–23.

CHEN, W. K., On matrices and composite graphs of active networks, IEEE Trans. Circuit Theory **CT-19** (1972d) 510–511.

CHEN, W. K., On equivalence of realizability conditions of a degree sequence, IEEE Trans. Circuit Theory **CT-20** (1973) 260–262.

CHEN, W. K., Theory of Nets: Flows in Networks (John Wiley, New York), 1990.

CHEN, W. K. and F. N. T. CHAN, On the unique solvability of linear active networks, IEEE Trans. Circuits Systems **CAS-21** (1974) 26–35.

CHEN, W. K. and I. C. GOYAL, Tables of essential complementary partitions, IEEE Trans. Circuit Theory **CT-18** (1971) 562–563.

CHEN, W. K. and H. C. LI, Computer generation of directed trees and complete trees, Int. J. Electronics **34** (1973) 1–21.

CHEN, W. K. and S. K. MARK, On the algebraic relationships of trees, co-trees, circuits, and cutsets of a graph, IEEE Trans. Circuit Theory **CT-16** (1969) 176–184.

CHOW, Y. and E. CASSIGNOL, Linear Signal-Flow Graphs and Applications (Wiley, New York), 1962.

CHUA, L. O. and P. M. LIN, Computer-Aided Analysis of Electronic Circuits (Prentice-Hall, Englewood Cliffs, N.J.), 1975.

CIESIELSKI, M. J. and E. KINNEN, Digraph relaxation for 2-dimensional placement of IC blocks, IEEE Trans. Comput.-Aided Des. Integr. Circuits Syst. **CAD-6** (1987) 55–66.

CLARKE, L. E., On Cayley's formula for counting trees, J. London Math. Soc. **33** (1958) 471–474.

COATES, C. L., General topological formulas for linear network functions, IRE Trans. Circuit Theory **CT-5** (1958) 30–42.

COATES, C. L., Flow-graph solutions of linear algebraic equations, IRE Trans. Circuit Theory **CT-6** (1959) 170–187.

CODDINGTON, E. A. and N. LEVINSON, Theory of Ordinary Differential Equations (McGraw-Hill), New York), 1955.

COPI, I. M., Matrix development of the calculus of relations, J. Symbolic Logic **31** (1958) 193–203.

CUMMINS, R. L., Hamilton circuits in tree graphs, IEEE Trans. Circuit Theory **CT-13** (1966) 82–90.

DEHN, M., Über die Zerlegung von Rechtecken in Rechtecke, Math. Ann. **57** (1903) 69–70.

DERVISOGLU, A., Bashkow's, "A matrix for active RLC networks" IEEE Trans. Circuit Theory **CT-11** (1964) 404–406.

DESOER, C. A., The optimum formula for the gain of a flow graph or a simple derivation of Coates' formula, Proc. IRE **48** (1960) 883–889.

DIJKSTRA, E. W., Some theorems on spanning subtrees of a graph, Indag. Math. **22** (1960) 196–199.

DIRAC, G. A., 4-chrome Graphen und vollstandige 4-Graphen, Math. Nachr. **22** (1960) 51–60.

DOYLE, T. C., Topological and dynamical invariant theory of an electrical network, J. Math. Phys. **34** (1955) 81–94.

DUFFIN, R. J., An analysis of the Wang algebra of networks, Trans. Amer. Math. Soc. **93** (1959) 114–131.

DUIJVESTIJN, A. J. W., Electronic computation of squared rectangles, dissertation, Technische Hogeschool, Eindhoven, Netherlands; also in Phillips Res. Rept. **17** (1962) 523–612.

DUIJVESTIJN, A. J. W., Simple perfect squared square of lowest order, J. Comb. Theory, Series B **25** (1978) 240–243.

DUIJVESTIJN, A. J. W., A lowest order simple perfect 2×1 squared rectangle, J. Combinatorial Theory B **26** (1979) 371–374.

DUIJVESTIJN, A. J. W., P. J. FEDERICO and P. LEEUW, Compound perfect squares, Math. Monthly **89** (1982) 15–32.
ECKMANN, B., Harmonische Funktionen und Randwert Aufgaben in einem Komplex, Comm. Math. Helvetici **17** (1944–1945) 240–255.
EDMONDS, J., Existence of k-edge connected ordinary graphs with prescribed degrees, J. Res. Natl. Bur. Stand. **68B** (1964) 73–74.
ERDÖS, P. and T. GALLAI, Gráfok elöirt fokú pontokkal, Mat. Lapok **11** (1960) 264–274.
EULER, L., Solutio problematis ad geometriam situs pertinentis, Comment. Academiae Sci. I. Petropolitanae **8** (1736) 128–140.
FEDERICO, P. J., Note on some low-order perfect squared squares, Can. J. Math. **15** (1963) 350–362.
FEDERICO, P. J., Some simple perfect 2×1 rectangles, J. Combinatorial Theory **8** (1970) 244–246.
FEDERICO, P. J., Squaring rectangles and squares: A historical review with annotated bibliography, in Graph Theory and Related Topics, G. A. Bondy and U. S. R. Murty, eds. (Academic Press, New York), 1979, pp. 173–196.
FETTWEIS, A., On the algebraic derivation of the state equations, IEEE Trans. Circuit Theory **CT-16** (1969) 171–175.
FEUSSNER, W., Über Stromverzweigung in Netzförmigen Leitern, Ann. Phys., 4th series, **9** (1902) 1304–1329.
FEUSSNER, W., Zur Berechnung der Stromstärke in Netzförmigen Leitern, Ann. Phys., 4th series, **15** (1904) 385–394.
FLAMENT, C., Applications of Graph Theory to Group Structure (Prentice-Hall, N.J.), 1963.
FORD Jr., L. R. and D. R. FULKERSON, Flows in Networks (Princeton University Press, N.J.), 1962.
FRANKLIN, P., The electric currents in a network, J. Math. Phys. **4** (1925) 97–102.
FU, Y., Realization of circuit matrices, IEEE Trans. Circuit Theory **CT-12** (1965), 604–607.
FULKERSON, D. R., Zero-one matrices with zero trace, Pacific J. Math. **10** (1960) 831–836.
FULKERSON, D. R., A. J. HOFFMAN and M. H. MCANDREW, Some properties of graphs with multiple edges, Can. J. Math. **17** (1965) 166–177.
GALE, D., A theorem on flows in networks, Pacific J. Math. **7** (1957) 1073–1082.
GARDNER, M. F. and J. L. BARNES, Transients in Linear Systems (Wiley, New York), 1945.
GHOUILA-HOURI, A., Caractérisation des matrices totalement unimodulaires, C.R. Acad. Sci. Paris **254** (1962) 1192–1194.
GOULD, R. L., Application of graph theory to the synthesis of contact networks, Proc. Int. Symposium on Switching Circuits, Harvard Univ., Cambridge, Mass., April 1957.
GOULD, R., Graphs and vector spaces, J. Math. Phys. **37** (1958) 193–214.
GUILLEMIN, E. A., Communication Networks (Wiley, New York) 1931, vol. 1.
GUILLEMIN, E. A., How to grow your own trees from cut-set or tie-set matrices, IRE Trans. Circuit Theory **CT-6** (Supplement) (1959) 110–126.
HAKIMI, S. L., On the realizability of a set of trees, IRE Trans. Circuit Theory **CT-8** (1961) 11–17.
HAKIMI, S. L., On realizability of a set of integers as degrees of the vertices of a linear graph I, J. Soc. Indust. Appl. Math. **10** (1962) 496–506.
HAKIMI, S. L., On realizability of a set of integers as degrees of the vertices of a linear graph II. Uniqueness, J. Soc. Indust. Appl. Math. **11** (1963) 135–147.
HAKIMI, S. L., On the degrees of the vertices of a directed graph, J. Franklin Inst. **279** (1965) 290–308.
HAKIMI, S. L., On the existence of graphs with prescribed degrees and connectivity, SIAM J. Appl. Math. **26** (1974) 154–164.
HAKIMI, S. L., and S. S. YAU, Distance matrix of a graph and its realizability, Quart. Appl. Math. **22** (1964) 305–317.
HARARY, F., A graph theoretic method for the complete reduction of a matrix with a view toward finding its eigenvalues, J. Math. and Phys. **38** (1959) 104–111.

HARARY, F., The determinant of the adjacency matrix of a graph, SIAM Rev. **4** (1962a) 202–210.
HARARY, F., A graph theoretic approach to matrix inversion by partitioning, Numer. Math. **4** (1962b) 128–135.
HARARY, F. (ed.), Graph Theory and Theoretical Physics (Academic Press, New York), 1967.
HARARY, F., Graph Theory (Addison-Wesley, Mass.), 1969.
HARARY, F. and R. Z. NORMAN, Graph Theory as a Mathematical Model in Social Science (Institute for Social Research, University of Michigan, Ann Arbor), 1953.
HAVEL, V., Poznámka o existenci konecnych grafù, Casopis Pest. Mat. **80** (1955) 477–480.
HEAWOOD, P. J., Map colour theorems, Quart. J. Math. **24** (1890) 332–338.
HELLER, I., On linear systems with integral valued solutions, Pacific J. Math. **7** (1957) 1351–1364.
HELLER, I., On a class of equivalent systems of linear inequalities, Pacific J. Math. **13** (1963) 1209–1227.
HELLER, I., On linear programs equivalent to the transportation problem, SIAM J. Appl. Math. **12** (1964) 31–42.
HELLER, I. and C. B. TOMPKINS, An extension of a theorem of Dantzig's, in Linear Inequalities and Related Systems (Princeton University Press, Princeton, NJ), 1956, pp. 247–254.
HOFFMAN, A. J. and I. HELLER, On unimodular matrices, Pacific J. Math. **12** (1962) 1321–1327.
HOFFMAN, A. J. and J. B. KRUSKAL, Integral boundary points of convex polyhedra, in Linear Inequalities and Related Systems (Princeton University Press, Princeton, NJ), 1956, pp. 223–246.
HOHN, F. E. and L. SCHISSLER, Boolean matrices and the design of combinational relay switching circuits, Bell System Tech. J. **34** (1955) 177–202.
HOHN, F. E., S. SESHU and D. D. AUFENKAMP, The theory of nets, IRE Trans. Electronic Computers **EC-6** (1957) 154–161.
HOLZMANN, C. A. and F. HARARY, On the tree graph of a matroid, SIAM J. Appl. Math. **22** (1972) 187–193.
HOSKINS, R. F., Signal flow-graph analysis and feedback theory, Proc. IEE (London) **108C** (1960) 12–19.
HOSKINS, R. F., Flow graphs, signal flow graphs, and graph determinants, Proc. IEE (London) **109C** (1961) 263–269.
INGRAM, W. H. and C. M. CRAMLET, On the foundations of electrical network theory, J. Math. Phys. **23** (1944) 134–155.
JEANS, J. H., The Mathematical Theory of Electricity and Magnetism (Cambridge University Press, England), 1925.
KAC, M. and J. C. WARD, A combinatorial solution of the 2-dimensional Ising model, Phys. Rev. **88** (1952) 1332–1337.
KAJITANI, Y., On the realizability of fundamental circuit matrices, J. Franklin Inst. **290** (1970) 355–363.
KAJITANI, Y., private discussion and communication, 1994.
KAMAE, T., The existence of a Hamilton circuit in a tree graph, IEEE Trans. Circuit Theory **CT-14** (1967a) 279–283.
KAMAE, T., A graph-theoretical proof of an adjacency property of major submatrices, SIAM J. Appl. Math. **15** (1967b) 1390–1399.
KAUFMANN, A., Graphs, Dynamic Programming, and Finite Games (Academic Press, N.Y.), 1967.
KAZARINOFF, N. D. and R. WEITZENKAMP, On existence of compound perfect squared squares of small order, J. Combinatorial Theory, Series B **14** (1973a) 163–179.
KAZARINOFF, N. D. and R. WEITZENKAMP, Squaring rectangles and squares, Math. Monthly **80** (1973b) 877–888.
KIRCHHOFF, G., Über die Auflösung der Gleichungen, auf welche man bei der Untersuchungen

der linearen Verteilung galvanischer Ströme geführt wird, Poggendorf Ann. Phys. **72** (1847) 497–508.
KISHI, G. and KAJITANI, Y., On Hamilton circuits in tree graphs, IEEE Trans. Circuit Theory **CT-15** (1968) 42–50.
KLEITMAN, D. J., Minimal number of multiple edges in realization of an incidence sequence without loops, SIAM J. Appl. Math. **18** (1970) 25–28.
KÖNIG, D., Über Graphen und ihre Anwendung auf Determinantentheorie und Mengenlehre, Math. Ann. **77** (1916) 453–465.
KÖNIG, D., Theorie der endlichen und unendlichen Graphen (Chelsea, New York), 1950.
KOTZIG, A., On decomposition of a tree into the minimal number of paths, Matematicky časopis **17** (1967) 76–78.
KRON, G., Tensor Analysis of Networks (Wiley, New York), 1939, 102–104.
KU, Y. H., Extension of Maxwell's rule for analyzing electrical networks, Science Report of National Tsing Hua University, Series A, **1** (1932) No. 6.
KU, Y. H., Résumé of Maxwell's and Kirchhoff's rules for network analysis, J. Franklin Inst. **253** (1952) 211–224.
KUH, E. S. and R. A. ROHRER, The state-variable approach to network analysis, Proc. IEEE **53** (1965) 672–686.
KUH, E. S., D. M. LAYTON and J. TOW, Network analysis and synthesis via state variables, in Network and Switching Theory, G. Biorci (ed.), (Academic Press, New York), pp. 135–155, 1968.
KURATOWSKI, K., Sur le problème des courbes gauches en topologie, Fund. Math. **15–16** (1930) 271–283.
LEMPEL, A., A note on orientation of graphs, Amer. Math. Monthly **75** (1968) 865–867.
LI, H. C., Computer-aided enumeration of trees, M.S. Thesis, Dept. of Elec. Engrg., Ohio Univ., Athens, Ohio, March 1971.
LÖFGREN, L., Irredundant and redundant Boolean branch-networks, IRE Trans. Circuit Theory **CT-6** (Supplement) (1959) 158–175.
LORENS, C. S., Flowgraphs (McGraw-Hill, New York), 1964.
LOVÁSZ, L., On decomposition of graphs, Studia Sci. Math. Hungar. **1** (1966) 237–238.
LUCE, R. D. and A. D. PERRY, A method of matrix analysis of group structure, Psychometrika **14** (1949) 95–116.
MACLANE, S. and G. BIRKHOFF, Algebra (Macmillan, New York), 1967.
MALIK, N. R. and H. W. HALE, Equations for active networks: existence of unique solutions, IEEE Trans. Circuit Theory **CT-14** (1967) 37–43.
MARK, S. K., On the order of complexity of active electrical networks, Ph.D. dissertation, Ohio University, Athens, June 1971.
MASON, S. J., Feedback theory – some properties of signal flow graphs, Proc. IRE **41** (1953) 1144–1156.
MASON, S. J., Feedback theory – further properties of signal flow graphs, Proc. IRE **44** (1956) 920–926.
MASON, S. J., Topological analysis of linear nonreciprocal networks, Proc. IRE **45** (1957) 829–838.
MASON, S. J. and H. J. ZIMMERMANN, Electronic Circuits, Signals, and Systems (Wiley, New York), 1960, chs. 4 and 5.
MAXWELL, J. C., Electricity and Magnetism (Clarendon Press, Oxford), 1892, ch. 6 and appendix.
MAYEDA, W., Topological formulas for nonreciprocal networks and networks with transformers, Proc. Natl. Electronics Conf. **14** (1958) 631–643.
MAYEDA, W., Necessary and sufficient conditions for realizability of cut-set matrices, IRE Trans. Circuit Theory **CT-7** (1960a) 79–81.
MAYEDA, W., Synthesis of switching functions by graph theory, IBM J. **4** (1960b) 321–328.
MAYEDA, W., Realizability of fundamental cut-set matrices of oriented graphs, IEEE Trans. Circuit Theory **CT-10** (1963) 133–134.

MAYEDA, W., A proof of Tutte's realizability condition, IEEE Trans. Circuit Theory **CT-17** (1970) 506–511.
MAYEDA, W. and S. SESHU, Topological formulas for network functions, Engineering Experiment Station Bulletin 446, University of Illinois, Urbana, 1957.
MAYEDA, W. and S. SESHU, Generation of trees without duplications, IEEE Trans. Circuit Theory **CT-12** (1965) 181–185.
MAYEDA, W., S. L. HAKIMI, W. K. CHEN and N. DEO, Generation of complete trees, IEEE Trans. Circuit Theory **CT-15** (1968) 101–105.
MCCLENAHAN, J. O. and S. P. CHAN, Computer analysis of general linear networks using digraphs, Int. J. Electronics **33** (1972) 153–191.
MCILROY, M. D., Generator of spanning trees, Comm. Assoc. Computing Machinery **12** (1969) 511, Algorithm 354.
MENON, V. V., On the existence of trees with given degrees, Sankhya, Series A, **26** (1964) 63–68.
MILIC, M. M., General passive networks – solvability, degeneracies, and order of complexity, IEEE Trans. Circuits Systems **CAS-21** (1974) 177–183.
MORON, Z., O rozkladach prostokatów na kwadraty, Przleglad. Matem. – Fizyczny, **3** (1925) 152–153.
MOWSHOWITZ, A., The characteristic polynomial of a graph, J. Combinatorial Theory **12(B)** (1972) 177–193.
NATHAN, A., A proof of the topological rules of signal-flow-graph analysis, Proc. IEE (London) **109C** (1961) 83–85.
NATHAN, A., A proof of the generalized topological Kirchhoff's rules, Proc. IEE (London) **109C** (1962) 45–50.
NATHAN, A., Topological rules for linear networks, IEEE Trans. Circuit Theory **CT-12** (1965) 344–358.
NERODE, A. and H. SHANK, An algebraic proof of Kirchhoff's network theorem, Amer. Math. Monthly **68** (1961) 244–247.
NEVILLE, E. H., The codifying of tree-structure, Proc. Cambridge Philos. Soc. **49** (1953) Part 3, 381–385.
NUMATA, J. and M. IRI, Mixed-type topological formulas for general linear networks, IEEE Trans. Circuit Theory **CT-20** (1973) 488–494.
OHM, G. S., The Galvanic Circuit Investigated Mathematically (Original German edition: Berlin, 1827); translation by W. Francis (Van Nostrand, 2nd edition, 1905).
OKADA, S., Topology applied to switching circuits, Proc. Brooklyn Symposium on Information Networks, New York, 1954, pp. 267–290.
OKADA, S., On node and mesh determinants, Proc. IRE **43** (1955) 1527.
ORE, O., Theory of Graphs (American Math. Soc., R.I.), 1962; Colloq. Publ. **38**.
ORE, O. and J. STEMPLE, Numerical calculations on the four-color problem, J. Combinatorial Theory **8** (1970) 65–78.
ORTEGA, J. M. and W. C. RHEINBOLDT, Iterative Solution of Nonlinear Equations in Several Variables (Academic Press, New York), 1970.
OWENS, A. B., On determining the minimum number of multiple edges for an incidence sequence, SIAM J. Appl. Math. **18** (1970) 238–240.
OWENS, A. B., On the planarity of regular incidence sequences, J. Combinatorial Theory **11** (1971) 201–212.
OWENS, A. B. and H. M. TRENT, On determining minimal singularities for the realizations of an incidence sequence, SIAM J. Appl. Math. **15** (1967) 406–418.
PARKER, S. R. and H. J. LOHSE, A direct procedure for the synthesis of network graphs from a given fundamental loop or cutset matrix, IEEE Trans. Circuit Theory **CT-16** (1969) 221–223.
PAUL Jr., A. J., Generation of directed trees and 2-trees without duplication, IEEE Trans. Circuit Theory **CT-14** (1967) 354–356.

PEIKARI, B., Fundamentals of Network Analysis and Synthesis (Prentice-Hall, Englewood Cliffs, N.J.), 1974.
PERCIVAL, W. S., The solution of passive electrical networks by means of mathematical trees, Proc. IEE (London) **100C** (1953) 143–150.
PERCIVAL, W. S., Improved matrix and determinant methods for solving networks, Proc. IEE (London) **101** (1954) Part IV, 258–265.
PERCIVAL, W. S., The graphs of active networks, Proc. IEE (London) **102C** (1955) 270–278.
PEREIRA, J. M. S. S., A note on the tree realizability of a distance matrix, J. Combinatorial Theory **6** (1969) 303–310.
POLAK, E., Computational Methods in Optimization: A Unified Approach (Academic Press, New York), 1971.
PÓLYA, G. and G. SZEGÖ, Aufgaben und Lehrsätze aus der Analysis, vols. 19 and 20 (Springer-Verlag, Berlin), 1945, p. 98.
PONSTEIN, J., Matrix description of networks, J. Soc. Indust. Appl. Math. **9** (1961) 233–268.
PONSTEIN, J., Self-avoiding paths and the adjacency matrix of a graph, SIAM J. Appl. Math. **14** (1966) 600–609.
PONTRYAGIN, L. S., Ordinary Differential Equations (Addison-Wesley, Reading, Mass.), 1962.
PREAS, B. T. and C. W. GWYN, Methods for hierarchical automatic layout of custom LSI-integrated circuits, Proc. 15th Design Automation Conf., 1978, pp. 206–212.
PURSLOW, E. J. and R. SPENCE, Order of complexity of active networks, Proc. IEE (London) **114** Pt. 2 (1967) 195–198.
REED, M. B., Foundation for Electric Network Theory (Prentice-Hall, N.J.), 1961.
REZA, F. M., Order of complexity and minimal structures in network analysis, Proc. Symposium on Circuit Analysis, University of Illinois, Urbana, Illinois, 1955, 7.1–7.33.
RIEGLE, D. E. and P. M. LIN, Matrix signal flow graphs and an optimum topological method for evaluating their gains, IEEE Trans. Circuit Theory **CT-19** (1972) 427–435.
ROBICHAUD, L. P. A., M. BOISVERT and J. ROBERT, Signal Flow Graphs and Applications (Prentice-Hall, N.J.), 1962.
ROHRER, R. A., Circuit Theory: An Introduction to the State Variable Approach (McGraw-Hill, New York), 1970.
ROTH, J. P., An application of algebraic topology to numerical analysis: on the existence of a solution to the network problem, Proc. Natl. Acad. Sci. U.S. **41** (1955) 518–521.
ROTH, J. P., An application of algebraic topology: Kron's method of tearing, Quart. Appl. Math. **17** (1959) 1–24.
RYSER, H. J., Combinatorial properties of matrices of zeros and ones, Can. J. Math. **9** (1957) 371–377.
SALTZER, C., The second fundamental theorem of electrical networks, Quart. Appl. Math. **11** (1953) 119–123.
SENIOR, J. K., Unimerism, J. Chem. Phys. **19** (1951a) 865–873.
SENIOR, J. K., Partitions and their representative graphs, Amer. J. Math. **73** (1951b) 663–689.
SESHU, S., The mesh counterpart of Shekel's theorem, Proc. IRE **43** (1955) 342.
SESHU, S., Note on the realizability of directed graphs, IRE Trans. Circuit Theory **CT-9** (1962) 412–413.
SESHU, S. and N. BALABANIAN, Linear Network Analysis (Wiley, New York), 1959.
SESHU, S. and M. B. REED, Singular transformations in network theory, Proc. Natl. Electronics Conf. **11** (1955) 531–543.
SESHU, S. and M. B. REED, On the cut sets of electrical networks, Proc. 2nd Midwest Symposium on Circuit Theory, Michigan State University, East Lansing, Michigan, 1956, 1.1–1.13.
SESHU, S. and M. B. REED, Linear Graphs and Electrical Networks (Addison-Wesley, Mass.), 1961.

SESHU, S. and R. WAXMAN, Fault isolation in conventional linear systems – A feasibility study, IEEE Trans. Reliability **R-15** (1966) 11–16.
SHANNON, C. E., The theory and design of linear differential equation machines, OSRD Rept. 411, Sec. D-2 (Fire Control) of the U.S. National Defense Research Committee, January 1942.
SHARPE, G. E. and B. SPAIN, On the solution of networks by means of the equicofactor matrix, IRE Trans. Circuit Theory **CT-7** (1960) 230–239.
SHEKEL, J., Two network theorems concerning change of voltage reference terminal, Proc. IRE **42** (1954) 1125.
SHEN, D. W. C., Generalised star and mesh transformations, Phil. Mag., Series 7, **38** (1947) 267–275.
SHRIVER, B., P. J. EBERLEIN and R. D. DIXON, Permanent function of a square matrix I and II, Comm. Assoc. Computing Machinery **12** (1969) 634, Algorithm 361.
SLEPIAN, P., Mathematical Foundations of Network Analysis (Springer-Verlag, New York) 1968, p. 46 and pp. 177–186.
SPRAGUE, R., Beispiel einer Zerlegung des Quadrats in lauter verschiedene Quadrate, Math. Z. **45** (1939) 607–608.
SPRAGUE, R., Über die Zerlegung von Rechtecken in lauter verschiedene Quadrate, J. Reine Angew. Math. **182** (1940) 60–64.
SYLVESTER, J. J., On the change of systems of independent variables, Quart. J. Pure Appl. Math. **1** (1855) 42–56.
TALBOT, A., Topological analysis of general linear networks, IEEE Trans. Circuit Theory **CT-12** (1965) 170–180.
TALBOT, A., Topological analysis for active networks, IEEE Trans. Circuit Theory **CT-13** (1966) 111–112.
TANG, H. and W. K. CHEN, Generation of rectangular duals of a planar triangulated graph by elementary transformations, Proc. of IEEE Int. Symp. on Circuits and Systems, New Orleans, LA, 1990, pp. 2857–2860.
TELLEGEN, B. D. H., A general network theorem, with applications, Philips Res. Rept. **7** (1952) 259–269.
TING, S. L., On the general properties of electric network determinants and the rules for finding the denominator and the numerators, Chinese J. Phys. **1** (1935) 18–40.
TOW, J., Order of complexity of linear active networks, Proc. IEE (London) **115** Pt. 2 (1968) 1259–1262.
TRAKHTENBROT, B. A., Sintez bezpovorotnykh Skhem, Doklady Akad. Nauka SSSR **103** (1955) 973–976.
TRENT, H. M., Isomorphisms between oriented linear graphs and lumped physical systems, J. Acoust. Soc. Amer. **27** (1955) 500–527.
TRUXAL, J. G., Automatic Feedback Control System Synthesis (McGraw-Hill, New York), 1955.
TSAI, C. T., Short-cut methods for expanding the determinants involved in network problems, Chinese J. Phys. **3** (1939) 148–181.
TSANG, N. F., On electrical network determinants, J. Math. Phys. **33** (1954) 185–193.
TUTTE, W. T., The dissection of equilateral triangles into equilateral triangles, Proc. Cambridge Philos. Soc. **44** (1948) 463–482.
TUTTE, W. T., The 1-factors of oriented graphs, Proc. Amer. Math. Soc. **4** (1953) 922–931.
TUTTE, W. T., A homotopy theorem for matroids, I, II, Trans. Amer. Math. Soc. **88** (1958) 144–174.
TUTTE, W. T., Matroids and graphs, Trans. Amer. Math. Soc. **90** (1959) 527–552.
TUTTE, W. T., An algorithm for determining whether a given binary matroid is graphic, Proc. Amer. Math. Soc. **11** (1960) 905–917.
TUTTE, W. T., A theory of 3-connected graphs, Indag. Math. **23** (1961) 441–455.
TUTTE, W. T., Matroid theory, RAND Rep.-R-448-PR, The RAND Corporation, Santa Monica, 1966.

VEBLEN, O., Analysis Situs (Amer. Math. Soc., R.I.) 2nd ed., 1931; Colloq. Publ. **5**.
VEBLEN, O. and P. FRANKLIN, On matrices whose elements are integers, Ann. Math. **23** (1921) 1–15.
WALTHER, H., Über die Nichtexistenz eines Knotenpunktes, durch den alle längsten Wege eines Graphen gehen, J. Combinatorial Theory **6** (1969) 1–6.
WANG, K. T., On a new method for the analysis of electric network, Natl. Res. Inst. Engrg. Academia Sinica, Memoir **2** (1934) 1–11.
WANG, K., Floorplan and placement algorithms in very large scale integrated (VLSI) circuit designs, Ph.D. dissertation, University of Illinois, Chicago, December 1993.
WANG, K. and W. K. CHEN, A class of zero wasted floorplan for VLSI design, Proc. of IEEE Int. Symp. on Circuits and Systems, Chicago, IL, 1993a, pp. 1762–1765.
WANG, K. and W. K. CHEN, Optimal area minimization for nonslicing floorplan, Proc. of 36th Midwest Symp. on Circuits and Systems, Detroit, MI, 1993b, pp. 522–525.
WATANABE, H., A computational method for network topology, IRE Trans. Circuit Theory **CT-7** (1960) 296–302.
WEINBERG, L., Number of trees in a graph, Proc. IRE **46** (1958) 1954–1955.
WEYL, H., Repartition de corriente et uno red conductora, Revista matemática, Hispano-Americana **5** (1923) 153–164.
WHITNEY, H., Planar graphs, Fund. Math. **21** (1933a) 73–84.
WHITNEY, H., 2-isomorphic graphs, Amer. J. Math. **55** (1933b) 245–254.
WHITNEY, H., On the abstract properties of linear dependence, Amer. J. Math. **57** (1935) 509–533.
WILLCOCKS, T. H., Problem 7795 and solution, Fairy Chess Rev. **7** (1948) 97, 106.
WILLCOCKS, T. H., A note on some perfect squared squares, Can. J. Math. **3** (1951) 304–308.
WIMER, S., I. KOREN, and I. CEDERBAUM, Floorplans, planar graphs, and layouts, IEEE Trans. Circuits Syst. **35** (1988) 267–278.
YAGLOM, I. M., How to cut up a square? (Nauka, Moskva), 1968.

Other Books on Graph Theory and Its Applications

ALAVI, Y., D. R. LICK and A. T. WHITE (eds.), Graph Theory and Applications (Springer, Berlin), 1972.
ANDERSON, S., Graph Theory and Finite Combinatorics (Markham, Chicago), 1970.
BARI, R. A. and F. HARARY (eds.), Graphs and Combinatorics (Springer, Berlin), 1974.
BECKENBACH, E. F. (ed.), Applied Combinatorial Mathematics (Wiley, New York), 1964.
BELLMAN, R., K. L. COOKE and J. A. LOCKETT, Algorithms, Graphs, and Computers (Academic Press, New York), 1970.
BERGE, C., Graphs and Hypergraphs (North-Holland, Amsterdam), 1973.
BIGGS, N., Algebraic Graph Theory (Cambridge University Press, London), 1974.
BOSE, R. C. and T. A. DOWLING (eds.), Combinatorial Mathematics and Its Applications (University of North Carolina Press, North Carolina), 1969.
BRUTER, C. P. (eds.), Théorie des Matroïdes (Springer, Berlin), 1971.
CAPOBIANCO, M., J. B. FRECHEN and M. KROLIK (eds.), Recent Trends in Graph Theory (Springer, Berlin), 1971.
CHAN, S. P., Introductory Topological Analysis of Electrical Networks (Holt, Rinehart and Winston, New York), 1969.
CHAN, S. P., W. K. CHEN, P. M. LIN, W. MAYEDA and B. R. MYERS, Network Topology and Its Engineering Applications (National Taiwan University Press, Taipei), 1975.
CHARTRAND, G. and M. BEHZAD, Introduction to the Theory of Graphs (Allyn and Bacon, Boston, Mass.), 1971.
CHARTRAND, G. and S. F. KAPOOR (eds.), The Many Facets of Graph Theory (Springer, Berlin), 1969.

DEO, N., Graph Theory with Applications to Engineering and Computer Science (Prentice-Hall, Englewood Cliffs, New Jersey), 1974.
DYNKIN, E. B. and W. A. USPENSKI, Mathematische Unterhaltungen: Mehrfarbenprobleme (VEB Deutscher Verlag der Wissenschaften, Berlin), 1955.
ELMAGHRABY, S. E., Some Network Models in Management Science (Springer, Berlin), 1970.
ERDÖS, P. and G. KATONA (eds.), Theory of Graphs (Academic Press, New York), 1968.
FIEDLER, M. (ed.), Theory of Graphs and Its Applications (Academic Press, New York), 1964.
FRANKLIN, P., The Four Color Problem (Yeshiva University, New York), 1941.
FULKERSON, D. R. (ed.), Studies in Graph Theory (Mathematical Association of America, Washington, D. C.) Parts I and II, 1975.
HAPP, H. H., Diakoptics and Networks (Academic Press, New York), 1971.
HARARY, F. (ed.), Proof Techniques in Graph Theory (Academic Press, New York), 1969.
HARARY, F. (ed.), New Directions in the Theory of Graphs (Academic Press, New York), 1973.
HARARY, F. and E. M. PALMER, Graphical Enumeration (Academic Press, New York), 1973.
HARARY, F., R. Z. NORMAN and D. CARTWRIGHT, Structural Models: An Introduction to the Theory of Directed Graphs (Wiley, New York), 1965.
HARRIS, B. (ed.), Graph Theory and Its Applications (Academic Press, New York), 1970.
HENLEY, E. J. and R. A. WILLIAMS, Graph Theory in Modern Engineering (Academic Press, New York), 1973.
IRI, M., Network Flow, Transportation and Scheduling: Theory and Algorithms (Academic Press, New York), 1969.
JOHNSON, D. E. and J. R. JOHNSON, Graph Theory with Engineering Applications (Ronald, New York), 1972.
KIM, W. H. and R. T. CHIEN, Topological Analysis and Synthesis of Communication Networks (Columbia University Press, New York), 1962.
LIU, C. L., Introduction to Combinatorial Mathematics (McGraw-Hill, New York), 1968.
MALKEVITCH, J. and W. MEYER, Graphs, Models and Finite Mathematics (Prentice-Hall, Englewood Cliffs, New Jersey), 1974.
MARSHALL, C. W., Applied Graph Theory (Wiley-Interscience, New York), 1971.
MAXWELL, L. M. and M. B. REED, The Theory of Graphs: A Basis for Network Theory (Pergamon Press, Oxford), 1971.
MAYEDA, W., Graph Theory (Wiley-Interscience, New York), 1972.
MCDONOUGH, T. P. and V. C. MAVRON (eds.), Combinatorics (Cambridge University Press, London), 1974.
MOON, J. W., Topics on Tournaments (Holt, Rinehart and Winston, New York), 1968.
MOON, J. W., Counting Labelled Trees (Canadian Math. Congress, Montreal) 1970; Math. Mono. No. 1.
MULLIN, R. C., K. B. REID and D. P. ROSELLE (eds.), Proceedings of the Louisiana Conference on Combinatorics, Graph Theory and Computing (Utilitas Mathematica, University of Manitoba, Canada), 1970.
ORE, O., Graphs and Their Uses (Random House, New York), 1963.
ORE, O., The Four-Color Problem (Academic Press, New York), 1967.
PICARD, C. F., Graphes et Questionnaires (Gauthier-Villars, Paris) 1972, Vols. 1 and 2.
POTTS, R. B. and R. M. OLIVER, Flows in Transportation Networks (Academic Press, New York), 1972.
READ, R. C. (ed.), Graph Theory and Computing (Academic Press, New York), 1972.
RINGEL, G., Färbungsprobleme auf Flächen und Graphen (VEB Deutscher Verlag der Wissenschaften, Berlin), 1959.
RINGEL, G., Map Color Theorem (Springer, Berlin), 1974.
ROSENSTIEHL, P. (ed.), Theory of Graphs: International Symposium (Gordon and Breach, New York), 1967.

Roy, B., Algèbre Moderne et Théorie des Graphes (Dunod, Paris), Vol. 1, 1969 and Vol. 2, 1970.
Sachs, H. (ed.), Beiträge zur Graphentheorie (Teubner, Leipzig), 1968.
Sachs, H., Einführung in die Theorie der endlich Graphen (Teubner, Leipzig), 1970.
Sainte-Laguë, A., Les Réseaux (ou Graphes), Mémorial des Sciences Mathématiques, Paris **18** (1926).
Sedlacek, J., Einführung in die Graphentheorie (Teubner, Leipzig), 1968.
Turner, J. and W. H. Kautz, A survey of progress in graph theory in the Soviet Union, SIAM Rev., supplement issue, **12** (1970).
Tutte, W. T., Connectivity in Graphs (University of Toronto Press, Toronto), 1966.
Tutte, W. T. (ed.), Recent Progress in Combinatorics (Academic Press, New York), 1969.
Tutte, W. T., Introduction to the Theory of Matroids (American Elsevier, New York), 1971.
Wagner, K., Graphentheorie (Bibliographisches Institut, Mannheim), 1970.
White, A. T., Graphs, Groups and Surfaces (North-Holland, Amsterdam), 1973.
Wilson, R. J., Introduction to Graph Theory (Academic Press, New York), 1972.
Zykov, A. A., Theory of Finite Graphs (Nauka, Novosibirsk) 1969 (in Russian).

SYMBOL INDEX

The symbols which occur most often are listed here, separated into three categories: Roman letters, Greek letters, matrices and vectors, and operations on graphs and sets.

Roman letters

am	ammeter branch, 303
b	number of edges, 37
b_{lg}	number of inductive edges in the subgraph g, 507
b_{cg}	number of capacitive edges in the subgraph g, 507
B_d	directed bipartite graph, 400
c	number of components, 37
C	cut, 47
$\mathscr{C}, \mathscr{C}^*$	codes of a tree, 326
$C[P'(H_k), P''(H_k)]$	essential complementary partitions, 354
$C_j[P'(H_k), P''(H_k)]$	complementary partitions, 354
cs	current-source branch, 303
$C[V(Z)]$	sum of cotree-impedance products, 113
$C[W_{ab,\,cd}(Y)]$	sum of 2-cotree-admittance products, 120
$C[W_{ab,\,cd}(Z)]$	sum of 2-cotree-impedance products, 120
$\mathscr{D}, \mathscr{D}^*$	codes of a tree, 329
$\det G$	graph determinant, 155, 169
$d(i)$	degree of node i, 12
$d^+(i)$	outgoing degree of node i, 29
$d^-(i)$	incoming degree of node i, 29
$D_u(R'), D_v(R'')$	set operators, 356
e, e_k, e_{ij}	edges, 6, 199
E	edge set or Euler line, 3, 385
$f(G_s)$	product of weights, 140, 233
$G, G(V, E)$	graph, 3
$\mathscr{G}$	set of subgraphs of G, 333
$\mathcal{G}_c$	associated directed graph of $\mathcal{N}_c$ or c-digraph,

Symbol index

	525, 526
$\mathcal{G}_c'$	dual c-digraph, 529
G_c'	modified Coates graph, 190
S_i^+	set of edges outgoing from node i, 201
S_i^-	set of edges terminating at node i, 201
G_m'	modified Mason graph, 197
$G_m, G_m(A)$	Mason graph, 168
$g_{rp,\,sq}$	(open-circuit) voltage-gain or transfer voltage-ratio function, 123, 289
$\mathcal{G}_p$	associate directed graph or p-digraph, 524, 535, 596
$\mathcal{G}_p'$	dual p-digraph, 528, 596
$(\mathcal{G}_p, \mathcal{G}_p')$	dual pair, 600
$\overline{G}_s$	complement of G_s in G, 5
G_u	associated undirected graph, 25
$G(g_1; g_2)$	graph derived from G, 279, 495
$G(I)$	graph derived from G by identifying the nodes in I, 340
$G^*[V_p]$	graph derived from G by identifying the nodes not in V_p, 252, 259
$G[V_s]$	sectional subgraph defined by V_s, 16, 25
$G(Y)$	associated directed graph of Y, 232
$G_C(\hat{Y})$	C-subgraph, 507
$G_C^*(\hat{Y})$	graph derived from $G_C(\hat{Y})$, 507
$G_L(\hat{Y})$	L-subgraph, 507
$G_L^*(\hat{Y})$	graph derived from $G_L(\hat{Y})$, 507
$G_u(Y)$	associated graph of Y, 235
h	1-factor, 143
H_{ij}	1-factorial connection, 146
$H_{rp,\,sq}$	transmission, 303
$H(g)$	set of edges of g, 336
I_u, J_u	sets of u integers, 78
(i, j)	edge or directed edge, 3, 24
$i(t), i(s)$	branch current, 58, 59
$k(B), k(Q)$	magnitude squared of a nonzero major determinant, 81
L	circuit or directed circuit, 9, 28
m	nullity, 11, 25, 507
$M_{uv}(B), M_{uv}(Q)$	generalized cofactors, 97
n	number of nodes, 37

Symbol	Description
$\mathcal{N}_c$	completed net or c-net, 525
$\mathcal{N}'_c$	dual c-net, 529
$\mathcal{N}_p$	polar net or p-net, 525, 535
$\mathcal{N}'_p$	dual one-port network or dual p-net, 528
$\mathcal{N}_{p1}$	polar subnet, 536
p	differential operator, 501
$1/p$	integral operator, 501
P^k	proper path, 284
P_{ij}	path or directed path, 9, 28
per A	permanent of A, 145
$P(H_k)$	partition, 353
$p(t)$	power function, 76
q, q_g	number of even components, 145
Q_i	incidence cut of node i, 47
r	rank, 11, 25, 507
$\mathcal{R}$	dissected rectangle, 524
R	semifactor, 190
$\mathcal{R}_n$	perfect rectangle, 542
$R(j_1; j_2;...; j_k)$	k-semifactor, 194
$\mathscr{S}$	set of all subsets of $\mathscr{G}$, 333
S_n	perfect square, 544
S_0	empty set, 333
S_i^+	set of edges outgoing from node i, 201
S_i^-	set of edges terminating at node i, 201
S^{uv}	$S-\{u, v\}$, 100
SDR	set of distinct representatives, 354
t	tree, 39, time variable, 464, passive tree, 512
$\bar{t}$	cotree, 42
T	set of trees, 340
$\bar{T}$	set of cotrees, 340
t_n	directed tree, 233
T_n	set of directed trees t_n, 269
$t_{ab, cd}$	2-tree or directed 2-tree, 114, 244
$\bar{t}_{ab, cd}$	2-cotree, 119, 120
$t_{ab, cd, ef}$	3-tree or directed 3-tree, 291
$t(E_q)$	a special tree, 374
$T(E_q)$	set of trees $t(E_q)$, 374
$T(G)$	tree graph, 380

Symbol index

$T_n(G_d)$	directed-tree graph, 380
$\bar{t}(I)$	multi-tree, 340
$t(I)$	multi-cotree, 340
$T(I)$	set of multi-trees $t(I)$, 340
$\bar{T}(I)$	set of multi-cotrees $\bar{t}(I)$, 340
$U, U(G)$	unisignant, 273
$U_{r,s,p}$	sum of directed 3-tree products, 295
$u(t)$	forcing function or excitation, 465
V	node set, 3
$\mathscr{V}_B$	B-space, 57
V_k	sum of directed-tree products, 295
$\mathscr{V}_Q$	Q-space, 57
$\mathscr{V}_G, \mathscr{V}(G)$	b-dimensional vector space, 57, 335
vm	voltmeter branch, 303
vs	voltage-source branch, 303
$v(t), v(s)$	branch voltage, 58, 59
$V(Y)$	sum of tree-admittance products, 112
$V(Z)$	sum of tree-impedance products, 112
W_n	wheel of order n, 539
$W_{rp,sq}$	sum of directed 2-tree products, 294
$W_{ab,cd}(Y)$	sum of 2-tree-admittance products, 115
$W_{ab,cd}(Z)$	sum of 2-tree-impedance products, 115
$x(t)$	state variable, 464
Y_{ij}	(first-order) cofactor, 225
$y_{rp,sq}$	(short-circuit) transfer-admittance function, 123, 290
$Y_{rp,sq}$	second-order cofactor, 227
$y(t)$	output variable, 465
$z_{rp,sq}$	(open-circuit) transfer-impedance function, 123, 283
$z_{rr,ss}$	driving-point impedance function, 124, 283

Greek letters

$\alpha_{rp,sq}$	(short-circuit) current gain or transfer current-ratio function, 123, 289
$\alpha(S)$	number of elements of S, 333, 353, 399
$\emptyset$	null graph, 4
ω	frequency, 127
δ_{ij}	Kronecker's delta, 177
Δ_{ij}	cofactor, 118, 121
λ	eigenvalue, 206

688   Symbol index

μ aspect ratio, 604
σ order of complexity, 502

Matrices and vectors

A^+ matrix obtained from A by replacing -1's by 0's, 370
A matrix or a basis incidence matrix, 39, 142, 323
A_a incidence matrix, 37, 323
A_{-i} deleting the ith row from A, 95, 371
A parameter matrix, 465, 491
B basis circuit matrix, 41, 323
B_a circuit matrix, 41, 323
B_f f-circuit matrix, 43, 324
B_p basis circuit matrix of a planar graph, 83
B parameter matrix, 465, 491
BZB' network matrix, 85
$C(G)$ connection matrix, 146
C parameter matrix, 465, 491
$D(G)$ directed-tree matrix, 236
$D(G_u)$ tree matrix, 239
D parameter matrix, 465, 491
E branch voltage-source vector, 60
E independent voltage-source vector, 558
E_m loop voltage-source vector, 64
$F(I_u, J_u)$ submatrix of F, 78, 365
$F(J_p)$ submatrix or major submatrix, 78, 365
$H(p)$ operator matrix, 502
I branch-current vector, 59, 466, 558, 560
I_m loop-current vector, 64
I_p resistive branch-current vector, 558
J branch current-source vector, 60
J independent current-source vector, 558
J_c cut current-source vector, 71
J_n nodal current-source vector, 71
$\mathcal{J}(V_n)$ Jacobian matrix, 577
$\|M\|$ norm of a square matrix M, 580
P node-to-datum path matrix, 54
Q basis cut matrix, 50, 323
Q_a cut matrix, 49, 323
Q_f f-cutset matrix, 51, 324

QYQ'	network matrix, 85
U_n	identity matrix of order n, 43, 51
$U(G)$	unisignant matrix of G, 273
$u(t)$	input vector, 465
V	branch-voltage vector, 59, 466, 558
V_c	cut-voltage vector, 70
V_n	nodal voltage vector, 73, 558, 575
V_p	resistive branch-voltage vector, 558
$x, {}^l(t)$	state vector, 465
$\dot{x}, \dot{x}(t)$	time derivative of $x(t)$, 465
$y, y(t)$	output vector, 465
$Y, \hat{Y}$	branch-admittance or indefinite-admittance matrix, 60, 230, 506
Y_c	cut-admittance matrix, 71
Y_n	node-admittance matrix, 71
Y_{sc}	short-circuit admittance matrix, 294
$Y(G)$	associated equicofactor matrix, 232
Z	branch-impedance matrix, 60
Z_m	loop-impedance matrix, 64
Z_{oc}	open-circuit impedance matrix, 294

Operations on graphs and sets

$G_1 = G_2$	isomorphic graphs, 7
$K_p J_v / I_u$	$(K_p - I_u) \cup J_v$, 366
$\subseteq$	subset of, 353
$G_1 \cup G_2$	sum graph, 14
$G_1 \cap G_2$	intersection graph, 14
$G_1 \oplus G_2$	ring sum of graphs, 15
$G_1 - G_2$	removal of subgraph, 15
$S_1 @ S_2$	Wang product, 203, 333
$S_1 \times S_2$	Cartesian product, 347
$S_1 \oplus S_2$	symmetric difference, 14, 333
$S_1 \cup S_2$	set union, 13
$S_1 \cap S_2$	intersection of sets, 14
$S_1 - S_2$	difference of sets, 14

SUBJECT INDEX

A
Abstract directed graph, 24
Abstract graph, 3
Accessible permutation, 217
Active edge, 506
Acyclic directed graph, 32
Adjacency matrix, 210
Adjacent nodes, 5
Admittance, 112
  distor, 301
  unistor, 302
Alternating edge train, 461
Alternating path, 419
Arc, 3
Articulation point, 19
Aspect ratio, 555
Associated directed graph, 232, 236, 524
Associated graph, 235
Associated matrix, 236
Associated symmetric directed graph, 242
Associated undirected graph, 25

B
Basic value, of a generalized cofactor, 99
  of a network determinant, 86
Basis circuit matrix, of a directed graph, 45
  of a graph, 323
Basis cut matrix, of a directed graph, 50
  of a graph, 323
Basis incidence matrix, of a directed graph, 39
  of a graph, 323
Binary matrix, 628, 638
Binet–Cauchy theorem, 78
Bipartite graph, 22
Block, 22, 27
Block graph, 35
Bouwkamp code, 521, 523
Branch, 3, 40, 58
  meter, 303
  source, 303
Branch-admittance matrix, 60

Branch-impedance matrix, 60

C
Canonical form, (1,0)-graph in, 430
Capacitive edge, 467, 507
Cartesian product, 347
0-Cell, 3
Chord, 42
Circuit, 9, 27
  C-, 507
  directed (see directed circuit)
  f-, of a directed graph, 42
    of a graph, 323
  fundamental, 42
  generalized (see generalized circuit)
  Hamilton, 380
  L-, 507
  length of, 9
  oriented, 41
Circuit matrix, 639
Circuit edge, 10, 27
Circuit-edge incidence matrix, 41
Circuit matrix (complete), of a directed graph, 41
  of a graph, 323
f-Circuit matrix, 43
Circuit rank, 11
Circuits correspond to circuits, 654
Closed analytic form, 564
Coates graph, 140, 142, 262
  decomposable, 162
  determinant of, 155
  modified, 190
Cofactor, 225
  first-order, 225
  generalized, 97, 114
    basic value of, 99
  of a directed path, 175
  of a proper transmission path, 311
  second-order, 227
Coforest, 340

Coincidence, 41, 48
Complement, 5
Complementary partitions, 354
Complementary subgraphs, 5, 25
Complete directed graph, 240
Complete graph, 242
Complete pentagon, 642
Completed net, 525
Compound imperfect square, 520
Compound perfect square, 520
Compound squared rectangle, 520
Component, 10, 27
    even, 145
    odd, 145
    strong, 31
Congruent transformation, 624
2-Connected, 53
    planar digraphs, 535
3-Connected planar c-digraph, 535
n-Connected graph, 535
n-Connectedness, 535
Connectedness, 8
    cyclic, 21
    strong, 30
Connection, 143
    1-factorial, 146
        cyclic, 266
    k-factorial, 217
    one-, 147
Connectivity, 11
Constrained aspect ratio, 581
Continuous aspect ratio, 557
Converse deficiency, 163
Coordinates of a vector, 64
Corresponding directed graph, 400
Cotree, 42, 338, 368
    admittance product, 113
    impedance product, 113
    k-, 340
    multi-, 340
    product, 113
2-Cotree, 119
    admittance product, 120
    impedance product, 120
    product, 120
Current, 58
    loop, 64
    mesh, 64
    reference of, 59
Current generator (ideal), 59
Cut, 47, 323

C-, 507
    incidence, 47
L-, 507
    orientation of, 48
    oriented, 48
Cut-admittance matrix, 71
Cut-edge incidence matrix, 49
Cut matrix (complete), of a directed
    graph, 49
    of a graph, 323
Cut transformation, 70, 83
Cut voltage, 70
Cutpoint, 19, 27
Cutset, 46
    f-, of a directed graph, 51
    of a graph, 323
    fundamental, 51
    matrix, 639
f-Cutset matrix, 51
Cycle, 9
    P-set of, 143
Cycle rank, 11
Cyclic 1-factor, 263
Cyclically connected directed graph, 26
Cyclically connected graph, 21
Cyclomatic number, 11

D

Decomposition of a separable graph, 22
Deficiency, 163
Degree, 12
    incoming, 29
    negative, 29
    outgoing, 29
    positive, 29
Degree invariant transformation, 119
Degree pair, 398
    realization of, 399
    weighted, 453
Degree of freedom, 500
Determinant, of the Coates graph, 155
    of the electrical network, 86
    of the Mason graph, 169
Difference, of sets, 14
    of subgraphs, 15
Digraph, 529
c-Digraph, 525
p-Digraph, 399, 524, 535, 596
    weighted, 453
(p, s)-Digraph, 399
    corresponding, 499

distance between, 420
symmetric, 427
symmetric subgraph of, 428
weighted, 452
n-Dimensional real space, 586
Direct method, 585
Directed bipartite graph, 219, 400
corresponding, 400
Directed circuit, 28
length of, 28
Directed edge, 24
Directed-edge sequence, 27
closed, 27
length of, 27
open, 27
Directed-edge train, 28
Directed Euler line, 386
Directed graph, 23
abstract, 24
acyclic, 32
associated, 232, 236
associated symmetric, 242
bipartite, 219, 400
complete, 240
connected, 26
corresponding, 400
cyclically connected, 26
isomorphic, 26
nonseparable, 26
planar, 26
regular, 29
semi-regular, 412
separable, 26
strongly connected, 30
symmetric, 31
Directed path, 28
expansion in, 280
length of, 28
Directed tree, 233, 262, 370
realization of, 314
normal, 512
Directed 2-tree, 244, 266, 370
Directed 3-tree, 291
Directed k-tree, 291
Directed-tree graph, 380
Directed-tree matrix, 236
Distance, 374
between (p, s)-digraphs, 420
between subgraphs, 374
between two nodes, 22, 27, 395
Distance matrix of a tree, 396

Distor, 301
admittance, 301
Dual c-digraph, 529
Dual p-digraph, 528, 596
Dual c-net, 529
Dual p-net, 529
Dual one-port network, 528
Dual set, 405
Dynamical variable, 500

E
Edge, 3, 24
active, 506
capacitive, 467, 507
circuit, 10, 27
directed, 24
directed away, 24
directed toward, 24
gain, 150
inductive, 467, 507
last exit, 386
noncircuit, 10, 27
oriented, 24
outgoing from, 24
parallel, 3, 24
passive, 506
removal of, 13
resistive, 467, 507
shorting an, 15
terminal, 328
terminating at, 24
uncoupled, 101
Edge-disjoint subgraphs, 5
Edge sequence, 8, 26
closed, 8
directed- (*see* directed-edge sequence)
length of, 8, 27
open, 8
Edge train, 8, 27
alternating, 461
directed-, 28
Eigenvalue, 206
Element, 3, 518
Elementary directed-tree transformation, 380
Elementary (p, s) d-invariant transformation, 419
Elementary row operations, 632
Elementary transformation, 315
Elementary tree transformation, 374
Elementary vector, 628
Endpoint, 3, 8, 24

Subject index

Energy function, 76
Equicofactor matrix, 225
   associated, 232
Equivalence relation, 10, 333
Essential complementary partitions, 354
Euler line, 385
   directed, 386

F

1-Factor, 143, 262
   cyclic, 263
n-Factor, 219
1-Factorial connection, 146
   cyclic, 266
k-Factorial connection, 217
Factorization, 185
Feedback loop, 175
First Betti number, 11
Floorplan, 554
   area optimization, 581
   zero-wasted area, 557
Flow graph, 140, 151
Forest, 134, 340
   normal, 472
Forward path, 175
Four-Color Conjecture, 2
Function, driving-point admittance, 124, 293
   driving-point impedance, 124, 283
   incidence, 107
   open-circuit voltage-gain, 289
   open-circuit transfer impedance, 283
   open-circuit transfer voltage-ratio, 289
   short-circuit current-gain, 289
   short-circuit transfer admittance, 290
   short-circuit transfer current-ratio, 289
   transfer admittance, 123
   transfer current-ratio, 123
   transfer impedance, 123, 283
   transfer voltage-ratio, 123, 289
   voltage-gain, 289
Fundamental circuits, 42, 323
Fundamental circuit matrix, 43
Fundamental cutset matrix, 51
Fundamental cutsets, 51, 323

G

Generalized circuit, 391
   odd, 391
Generalized cofactor, 97, 114
   basic value of, 99
Generalized directed k-tree, 393

Generalized k-tree, 393
   odd, 393
Geometric diagram, 3, 24
Graph, 3, 428
   abstract, 3
   associated, 235
   associated undirected, 25
   bipartite, 22
   block, 35
   Coates (*see* Coates graph)
   complete, 242
   connected, 10
   cyclically connected, 21
   determinant of, 155, 169
   directed-tree, 380
   finite, 4
   flow, 140, 151
   infinite, 4
   intersection, 14
   isomorphic, 7
   labeled, 8
   linear, 1
   Mason (*see* Mason graph)
   mixed, 32
   nonseparable, 19, 20, 26
   null, 4
   (p, s)-, 428
   planar, 17
   realization, 656
   regular, 12
   separable, 19, 20
   signal-flow, 140, 167
   sum, 14
   tree, 380
   weighted, 8
Graphic matrix, 647

H

Hamilton circuit, 380
Hermitian part, 127
Hilbert norm, 580
Homeomorphism, 642

I

Impedance, 112
Imperfect rectangle, 519
Imperfect square, 520
Incidence, in a directed graph, 24
   in a graph, 3
Incidence cut, 47
Incidence matrix (complete), 37, 323

Incoming degree, 29
Indefinite-admittance matrix, 230
Indirect method, 585
Inductive edge, 467, 507
Initial node, 8, 24, 27
Interchange, of the incoming edges, 164
  of the outgoing edges, 164
(p, s)-Interchange, 449
Intersection graph, 14
Intersection of sets, 14
d-Invariance, 419
Inversion, in a Coates graph, 165
  in a Mason graph, 186
  of a matrix, 210
Isolated node, 4
Isomorphism, 6
  of directed graphs, 26
  of graphs, 7
2-Isomorphism, 653

J
Jacobi's theorem, 79
Jacobian matrix, 577
Jeans' theorem, 229
Junction, 3

K
Kirchhoff's current law, 59, 466
Kirchhoff's voltage law, 59, 466
Königsberg Bridge Problem, 1

L
Last exit edge of a node, 386
Line, 3
Linear dependence, of sets, 335
  of subgraphs, 334
Linear graph, 1
Linear independence, of sets, 335
  of subgraphs, 334
Linear subgraph, 218
Linear vector space, 57, 628
Link, 42
Loop, 9
  dynamically independent, 510
Loop-impedance matrix, 64
Loop transformation, 64, 83
Loop transmission, 175

M
Major determinant, 78
Major submatrix, 40,

Mason graph, 140, 168
  decomposable, 184
  determinant of, 169
  factorization of, 185
  modified, 197
  normalized, 177
E-Matrix, 617
Matrix, adjacency, 210
  associated, 236
  associated equicofactor, 232
  basis circuit, 45, 323
  basis cut, 50, 323
  basis incidence, 39, 323
  branch-admittance, 60
  branch-impedance, 60
  circuit (complete), 41, 323
  f-circuit, 43, 323
  circuit-edge incidence, 41
  current-transfer, 494, 495
  cut (complete), 49, 323
  cut-admittance, 71
  cut-edge incidence, 46, 49
  f-cutset, 51, 323
  decomposable, 220
  directed-tree, 236
  distance, 396
  equicofactor, 225
  fundamental circuit matrix, 43
  fundamental cutset, 51
  in normal form, 366
  in proper form, 365
  incidence (complete), 37, 323
  indecomposable, 220
  indefinite-admittance, 230
  irreducible, 220
  loop-impedance, 64
  network, 85
  node-admittance, 71, 224, 230
  node-edge incidence, 37
  node-to-datum path, 54
  open-circuit impedance, 294
  p-, 448
  parameter, 465, 491
  (p, s)-, 448
  primitive connection, 142
  rank, of a circuit, 44
    of a cut, 50
    of an incidence, 38
  reducible, 220
  short-circuit admittance, 294
  totally unimodular, 80

Subject index

tree, 239, 396
  unisignant, 273
    in modified form, 273, 277
  variable adjacency, 142, 199
  voltage-transfer, 494, 495
Matrix in normal form, 630
Matrix inversion, 210
Maximal path-set, 658
Maximal set, 462
Mesh, 17, 83
Mesh discriminant, 114
Miller integrator, 226
Minimum area layout, 598
Minimum nonrealizable matrix, 646
Minimum-height vector, 584
Minimum-width vector, 584
Mixed graph, 32
Modified Coates graph, 190
Modified Mason graph, 197
Multi-cotree, 340
Multi-tree, 340

N

Natural frequency, 77, 502
Negative degree, 29
  c-Net, 525
  p-Net, 525, 535
Network, 58
  determinant, 86
  electrical, 59
  reciprocal, 290
  RLC, 91
  RLC two-port, 123
  $RLCM$, 130
  two-port, 122
Network matrix, 85
Newton–Raphson equation, 577
Newton–Raphson–Kantorovich theorem, 578
Node, 3, 24
  adjacent, 5
  degree of, 12
  distance between, 22, 27, 395
  elimination of, 155, 177
  initial, 8, 24, 27
  isolated, 4
  reference, 39
  removal of, 13
  source, 150
  terminal, 8, 24, 27
Node-admittance matrix, 71, 224, 230
Node discriminant, 114

Node-disjoint subgraphs, 5
Node-edge incidence matrix, 37
Node expansion, 318
Node-pair, expansion on, 280
Node-pair transformation, 70
Node-pair voltage, 70
Node splitting, 157
Node-to-datum path matrix, 54
Node-to-datum voltage, 70
Noncircuit edge, 10, 27
Non-orientable manifold, 82
Nonseparable graph, 19, 20, 26
Nonsliced floorplan, 555
Norm of a square matrix, 580
Norm of a vector, 580
Normal directed tree, 512
Normal forest, 472
Normal form, 535, 630, 640
Normal tree, 323, 472
Null graph, 4
Nullity, 11, 25
Numerical iteration techniques, 564

O

Objective function, 584
Ohmicness, 128
Ohm's law, 59, 466
One terminal-pair planar graph, 529
One-connection, 147
Open-circuit impedance matrix, 294
Open-circuit transfer impedance function
  (see function)
Open-circuit voltage-gain function (see
  function)
Opposite orientations, 41, 48
Optimization problem, 584
Optimum layout, 554
Order n, 539
Order of complexity, 502
Oriented circuit, 41
Oriented cut, 48
Oriented edge, 24
Outgoing degree, 29
Output equation, 465
Outside region, 17

P

Parallel edges, 3, 24
Parameter matrix, 465, 491
Parameters, open-circuit impedance, 294
  short-circuit admittance, 294

y, 294
z, 294
Paramount, 669
Partition, 353
    complementary, 354
    essential complementary, 354
Passive edge, 506
Passive tree, 512
Path, 9, 27
    alternating, 419
    directed (*see* directed path)
    forward, 175
    length of, 9
    proper, 284
    proper transmission, 311
    transmission, 175
    value of a proper transmission, 311
Path-set, 652
Path-set matrix, 652
Penalty function method, 585
Perfect cubed parallelepiped, 553
Perfect rectangle, 519
Perfect square, 521
Perfect squared rectangle, 519
Perfectly coupled inductors, 130
Permanent, 145
Placement, 555
Planar, 529
Planar graph, 17, 26
Point, 3
Polar net, 525
Polar subnet, 536
Port, 122
Positive degree, 29
Power absorbed in a branch, 76
Power function, 76
Primary variable, 76
Primitive connection matrix, 142
Primitive vector, 629
Proper path, 284
Proper transmission-path value, 311

Q
Quasi-Newton methods, 577

R
Rank, 11, 25
Real regular matrix, 629
Realization, of degrees, 427
    of degree pairs, 399
    of directed trees, 314

Reference node, 39
Reference of current or voltage, 59
Reference system, 64
    coordinates of a vector with respect to, 64
Region, 17
    outside, 17
Regular matrix, 628
Regular directed graph, 29
Regular graph, 12
Regular matrix mod 2, 640
Regular polyhedra, 13
Resistive edge, 467, 507
Ring sum, 14, 333
Rooted tree, 233
Routing, 555

S
SDR, 354
Secondary variable, 76
Sectional subgraph, of a directed graph, 25
    of a graph, 16
Seg (segregate), 47
Self-loop, 3, 24
    removal of, 186
Semifactor, 190, 262
    1-, 266
    k-, 194
Semi-regular directed graph, 412
Separable graph, 19, 20, 26
Set of distinct representatives (SDR), 354
Set union, 13
Short-circuit admittance matrix, 294
Short-circuit current-gain function (*see* function)
Short-circuit transfer-admittance function (*see* function)
Signal, 150
    nodal, 150, 168
Signal-flow graph, 140, 167
Simple imperfect square, 520
Simple perfect square, 520
Simple squared rectangle, 520
0-Simplex, 3
1-Simplex, 3
Sink node, 530
Sliced floorplan, 555
Solving the nonlinear nodal equations, 574
Source, 59
Source node, 150, 530
B-Space, 57

Subject index

Q-Space, 57
Spanning subgraph, 4
Spanning tree, 39
Squared rectangle, 518
Squared rectangle of order n, 518
Squared square, 520
Star-mesh transformation, 255, 259
State, 465
State equations, 465
    explicit form of, 479
    in normal form, 465
State space, 502
State variables, 464
    complete set of, 465
    dynamical, 500
    dynamically independent, 500
    static, 500
Strong component, 31
Strong connectedness, 30
Subdivided edge, 642
Subgraph, 4, 25
    C-, 507
    complementary, 5, 25
    edge-disjoint, 5
    L-, 507
    linear, 218
    linear dependence of, 334
    linear independence of, 334
    node-disjoint, 5
    product, 315
    proper, 4
    sectional, 16, 25
    spanning, 4
    symmetric, 428
Subrectangle, 520
Subtree, 233
Sum graph, 14
Symmetric difference, 14, 333
Symmetric (p, s)-digraph, 427
Symmetric directed graph, 31
Symmetric subgraph, 428
System of equations, branch-current, 63
    branch-voltage, 63
    cut, 71
    loop, 64
    nodal, 71
System of nonlinear algebraic equations, 564

T

Tellegen's theorem, 76
Terminal edge, 328

Terminal node, 8, 24, 27
Thomsen graph, 642
Topological formulas, 122, 224
Totally unimodular matrix, 80, 612, 617
Trace, 625
Transformation, cut, 70, 83
    degree invariant, 419
    elementary, 315
    elementary directed-tree, 380
    elementary (p, s) d-invariant, 419
    elementary tree, 374
    loop, 64, 83
    node-pair, 70
    star-mesh, 255, 259
Transmittance, 91
Transmission (graph), 303
    in a Coates graph, 154
    in a Mason graph, 174
    loop, 175
    path, 175
Transmission law, 301
Transmittance, 150
Tree, 39, 320, 338
    admittance product, 112
    directed, 233, 262, 370
    directed 2-, 244, 266, 370
    directed 3-, 291
    directed k-, 291
    generalized directed k-, 393
    generalized k-, 393
    impedance product, 112
    k-, 340
    length of, 395
    multi-, 340
    normal, 323, 472
    odd generalized k-, 393
    passive, 512
    product, 112
    rooted, 233
    spanning, 39
2-Tree, 114
    admittance product, 115
    directed, 244
    impedance product, 115
    product, 115
k-Tree, 340
Tree graph, 380
Tree matrix, 239, 652

U

U_{ij} in modified form, 227

Uncoupled edges, 101
Unique realization of degrees, 440
Unisignant (determinant), 230, 273
Unisignant matrix, 273
  in modified form, 273, 277
Unistor, 302

V
Variable adjacency matrix, 142, 199
Vector, branch-current, 59
  branch current-source, 60, 466
  branch-voltage, 59, 466
  branch voltage-source, 60
  cut current-source, 71
  cut-voltage, 70
  input, 465
  loop-current, 64
  loop voltage-source, 64
  minimum-height, 584
  minimum-width, 584
  nodal current-source, 71
  output, 465
  state, 465
Vertex, 3

Very-large-scale-integrated (VLSI) circuit, 554
Voltage, 58
  cut, 70
  node-pair, 70
  node-to-datum, 70
  reference, 59
Voltage generator (ideal), 59

W
Wang algebra, 333
Wang product, 203, 333
Wasted area, 554
Weighted degree pair, 453
Weighted graph, 8
Weighted p-digraph, 453
Weighted (p, s)-digraph, 452
Wheel, 539
Window, 17

X
||x||, 580

Z
Zero wasted area floorplan, 557